Metaheuristics: Computer Decision-Making

Applied Optimization

Volume 86

Series Editors:

Panos M. Pardalos
University of Florida, U.S.A.

Donald W. Hearn
University of Florida, U.S.A.

Metaheuristics: Computer Decision-Making

Mauricio G. C. Resende
AT&T Labs – Research,
NJ, U.S.A.

Jorge Pinho de Sousa
INESC,
Porto, Portugal

with assistance from
Ana Viana

Kluwer Academic Publishers
Boston/ Dordrecht/ London

Distributors for North, Central and South America:
Kluwer Academic Publishers
101 Philip Drive
Assinippi Park
Norwell, Massachusetts 02061 USA
Telephone (781) 871-6600
Fax (781) 871-6528
E-Mail <kluwer@wkap.com>

Distributors for all other countries:
Kluwer Academic Publishers Group
Post Office Box 322
3300 AH Dordrecht, THE NETHERLANDS
Telephone 31 78 6576 000
Fax 31 78 6576 474
E-Mail <orderdept@wkap.nl>

Electronic Services <http://www.wkap.nl>

Library of Congress Cataloging-in-Publication

Resende, Mauricio G.C./ Pinho de Sousa, Jorge
Metaheuristics: Computer Decision-Making
ISBN 978-1-4419-5403-9

Printed in the United Kingdom by Biddles/IBT Global

Contents

Preface

Combinatorial optimization is the process of finding the best, or optimal, solution for problems with a discrete set of feasible solutions. Applications arise in numerous settings involving operations management and logistics, such as routing, scheduling, packing, inventory and production management, location, logic, and assignment of resources. The economic impact of combinatorial optimization is profound, affecting sectors as diverse as transportation (airlines, trucking, rail, and shipping), forestry, manufacturing, logistics, aerospace, energy (electrical power, petroleum, and natural gas), telecommunications, biotechnology, financial services, and agriculture.

While much progress has been made in finding exact (provably optimal) solutions to some combinatorial optimization problems, using techniques such as dynamic programming, cutting planes, and branch and cut methods, many hard combinatorial problems are still not solved exactly and require good heuristic methods. Moreover, reaching "optimal solutions" is in many cases meaningless, as in practice we are often dealing with models that are rough simplifications of reality. The aim of heuristic methods for combinatorial optimization is to quickly produce good-quality solutions, without necessarily providing any guarantee of solution quality. Metaheuristics are high level procedures that coordinate simple heuristics, such as local search, to find solutions that are of better quality than those found by the simple heuristics alone. Modern metaheuristics include simulated annealing, genetic algorithms, tabu search, GRASP, scatter search, ant colony optimization, variable neighborhood search, and their hybrids. In many practical problems they have proved to be effective and efficient approaches, being flexible to accommodate variations in problem structure and in the objectives considered for the evaluation of solutions. For all these reasons, metaheuristics have probably been one of the most stimulating research topics in optimization for the last two decades.

Metaheuristics: Computer Decision-Making grew out of the *4th Metaheuristics International Conference* (MIC2001), held in Porto, Portugal, on July 16–20, 2001, and chaired by Jorge Pinho de Sousa. Kluwer Academic Publishers published three books developed from previous editions of the Metaheuristic International Conference (MIC'95 [1] held in Breckenridge, United States, in 1995; MIC'97 [2] held in Sophia-Antipolis, France, in 1997; and MIC'99 [3] held in Angra dos Reis, Brazil, in 1999). Though this book is not a conference proceedings, it does characterize the type of research presented at MIC2001.

Metaheuristics: Computer Decision-Making exemplifies how much the field has matured in the last few years. The volume has 33 papers, covering a broad range of metaheuristics and related topics, as well as applications.
Metaheuristics and related topics include:

- genetic algorithms: Chapters 6, 9, 10, 13, 14, 17, 20, 21, and 24;
- tabu search: Chapters 15, 18, 27, 28, and 31;
- greedy randomized adaptive search procedures (GRASP): Chapters 26, 28, 30, and 31;
- local search: Chapters 19, 25, 27, and 32;
- path-relinking: Chapters 1, 28, and 30;
- ant colony systems: Chapters 5, 22, and 33;
- memetic algorithms: Chapters 4 and 29;
- variable neighborhood search: Chapters 7 and 23;
- simulated annealing: Chapters 16 and 21;
- neural networks: Chapters 8 and 20;
- population reinforcement optimization based exploration (PROBE): Chapter 2;
- Lagrangian heuristics: Chapter 3;
- scatter search: Chapter 13; and
- constraint programming: Chapter 25.

Applications and problems types include:

- assignment and partitioning problems:
  - generalized assignment: Chapter 1,
  - quadratic assignment: Chapter 12,
  - linear assignment: Chapter 22,

[1] I.H. Osman and J.P. Kelly (eds.), *Meta-heuristics: Theory and Applications*, Kluwer, 1996.
[2] S. Voss, S. Martello, I.H. Osman, and C. Roucairol (eds.), *Meta-heuristics: Advances and Trends in Local Search Paradigms for Optimization*, Kluwer, 1999.
[3] C.C. Ribeiro and P. Hansen (eds.), *Essays and Surveys in Metaheuristics*, Kluwer, 2002.

  - number partitioning: Chapter 4,
  - multi-constraint knapsack: Chapter 2, and
  - linear ordering: Chapter 3;

- scheduling:
  - nurse rostering: Chapter 7,
  - school timetabling: Chapters 8 and 31,
  - discrete-continuous scheduling: Chapter 18, and
  - discrete-lot sizing and scheduling: Chapter 27;
- tree and graph problems:
  - Steiner problem in graphs: Chapter 28,
  - communication network design: Chapter 29,
  - capacitated minimum spanning tree: Chapter 30,
  - *k*-cardinality tree: Chapter 23,
  - graph drawing: Chapter 19, and
  - graph coloring: Chapter 15;
- statistics:
  - least squares estimation: Chapter 6, and
  - generalized vector-valued regression: Chapter 20;
- cutting problems:
  - strip packing with guillotine patterns: Chapter 24, and
  - cutting stock: Chapter 32;
- mathematical programming:
  - multiobjective optimization: Chapter 10,
  - integer programming: Chapter 26, and
  - minimax problem: Chapter 17;
- vehicle routing: Chapters 5 and 33;
- single source capacitated location: Chapter 9;
- industrial applications:
  - power systems: Chapter 21,
  - petroleum reservoir optimization: Chapter 13, and
  - classification from biological databases: Chapter 16; and

- heuristic search software framework: Chapter 11.

The editors would like to thank the authors for their contributions, the over 70 referees for their tireless effort and helpful suggestions, Kluwer Academic Publishers for publishing the book, and Alec Resende and Geraldo Veiga for assistance in producing the final camera-ready manuscript.

Last, but not least, we are specially grateful to Ana Viana, for her valuable support and tireless collaboration in the editorial work involving the publication of this book.

Florham Park and Porto, April 2003

Mauricio G. C. Resende
Jorge Pinho de Sousa

Referees

Renata M. Aiex, USA
Antonio Batel Anjo, Portugal
José Elias C. Arroyo, Brazil
Greet Vanden Berghe, Belgium
Mariusz Boryczka, Poland
Edmund Burke, United Kingdom
Maria Eugénia Captivo, Portugal
Maria Antónia Carravilla, Portugal
Carlos Coello Coello, Mexico
Mauro Dell' Amico, Italy
Marco Dorigo, Belgium
Mathias Erghott, New Zealand
Paola Festa, Italy
Paulo França, Brazil
António Miguel Gomes, Portugal
Luís Gouveia, Portugal
Toshihide Ibaraki, Japan
Graham Kendall, United Kingdom
Arne Lokketangen, Norway
Nelson Maculan, Brazil
Rafael Martí, Spain
Nenad Mladenović Yugoslavia
Roberto Montemanni, Switzerland
Heinz Muehlenbein, Germany
José Fernando Oliveira, Portugal
Margarida Pato, Portugal
Stella Porto, USA
Helena Ramalhinho, Spain
Cesar Rego, USA
Celso Ribeiro, Brazil
Andrea Schaerf, Italy
Jorge Pinho de Sousa, Portugal
Jacques Teghem, Belgium
Vincent Valls, Spain
Ana Viana, Portugal
Chris Walshaw, United Kingdom
Horacio Hideki Yanasse, Brazil
Ramon Alvarez-Valdez, Spain
Carlos Hengeller Antunes, Portugal
Barrie M. Baker, United Kingdom
Pedro Borges, Holland
Luciana Buriol, Brazil
Vicente Campos, Spain
Giuliana Carello, Italy
João Claro, Portugal
Alberto Colorni, Italy
Clarisse Dhaenens, France
Richard Eglese, United Kingdom
José Soeiro Ferreira, Portugal
Antonio Fiordaliso, Belgium
Teresa Galvão, Portugal
José Fernando Gonçalves, Portugal
Peter Greistofer, Austria
Andrzej Jaskiewicz, Poland
Andrea Lodi, Italy
Abilio Lucena, Brazil
Vittorio Maniezzo, Italy
Marcelo Medeiros, Brazil
Michele Monaci, Italy
Pablo Moscato, Australia
Gerson Couto de Oliveira, Brazil
Pedro Oliveira, Portugal
João Pedro Pedroso, Portugal
Guenther Raidl, Austria
Colin Reeves, United Kingdom
Mauricio G. C. Resende, USA
Roger Rios, Mexico
Marc Sevaux, France
Thomas Stützle, Germany
Paolo Toth, Italy
Geraldo Veiga, USA
Stefan Voss, Germany
Marino Widmer, Switzerland

To our children:

Alec, Sasha, Pedro, and Mariana.

Metaheuristics: Computer Decision-Making, pp. 1-17
Laurent Alfandari, Agnès Plateau, and Pierre Tolla

1 A PATH RELINKING ALGORITHM FOR THE GENERALIZED ASSIGNMENT PROBLEM

Laurent Alfandari[1], Agnès Plateau[2], and Pierre Tolla[3]

[1]ESSEC
B.P.105 F-95021 Cergy-Pontoise Cedex, France,
alfandari@essec.fr

[2]CEDRIC, CNAM
292 rue Saint-Martin, 75003 Paris, France
aplateau@cnam.fr

[3]LAMSADE, Université Paris IX-Dauphine
Place du Mal De Lattre de Tassigny
75775 Paris Cedex 16, France
tolla@lamsade.dauphine.fr

Abstract: The Generalized Assignment Problem (GAP) consists in finding a maximum-profit assignment of tasks to agents with capacity constraints. In this paper, a path relinking heuristic is proposed for the GAP. The main feature of our path relinking is that both feasible and infeasible solutions are inserted in the reference set of elite solutions, trade-off between feasibility and infeasibility being ruled through a penalty coefficient for infeasibility. Since exploration of the solution space is very sensitive to this penalty coefficient, the coefficient is dynamically updated at each round of combinations so that a balance is kept between feasible and infeasible solutions in the reference set. Numerical experiments reported on classical testbed instances of the OR-library show that the algorithm compares favorably to several other methods in the literature. In particular, more than 95% of the instances in the test-file were solved to optimality with short computation time.
Keywords: Combinatorial optimization, Generalized assignment, Metaheuristics, Path relinking.

1.1 PROBLEM STATEMENT

The Generalized Assignment Problem (GAP) is a much studied NP-hard optimization problem. Given a set of tasks $I = \{1, \ldots, n\}$ and a set of agents $J = \{1, \ldots, m\}$, the GAP consists in assigning each task of I to one agent of J such that the capacity of each agent is not exceeded, while maximizing the total profit of selected assignments. The GAP often appears as a subproblem of many applications such as job scheduling, production planning, storage space allocation, computer and communication network design, facility location, vehicle routing, etc.

The problem is stated as follows. For every task i in I and every agent j in J, c_{ij} is the profit of assigning task i to agent j, a_{ij} is the amount of resource consumed by task i when assigned to agent j, and b_j is the resource availability of agent j. The GAP can formulate as a 0-1 integer linear program (ILP):

$$\max f(X) = \sum_{i=1}^{n} \sum_{j=1}^{m} c_{ij} X_{ij}$$

subject to

$$(ILP) \qquad \sum_{j=1}^{m} X_{ij} = 1 \qquad i = 1, \ldots, n \tag{1.1}$$

$$\sum_{i=1}^{n} a_{ij} X_{ij} \leq b_j \qquad j = 1, \ldots, m \tag{1.2}$$

$$X_{ij} \in \{0, 1\} \qquad i = 1, \ldots, n, \quad j = 1, \ldots, m \tag{1.3}$$

where binary decision variables X_{ij} indicate whether task i is assigned to agent j or not. Constraints (1.1) are called semi-assignment constraints, whereas constraints (1.2) are complicating capacity (or knapsack) constraints. In this paper, another encoding is used for a solution. A solution x is a mapping from I to J, and the set of all mappings $x : I \to J$ is denoted by M. As $x(i) = j \iff X_{ij} = 1$ in the ILP-formulation, the model is restated as finding a mapping $x \in M$ maximizing

$$f(x) = \sum_{i=1}^{n} c_{i,x(i)}$$

$$\text{subject to} \quad \sum_{i \in I : x(i) = j} a_{ij} \leq b_j, \; \forall j = 1, \ldots, m.$$

Since the GAP is NP-hard, heuristic solutions are often wanted for large-scale problems. The paper is organized as follows. Section 1.2 presents a general survey of methods for solving or approximating the GAP. Section 1.4 describes the path relinking algorithm. Numerical experiments are reported in Section 1.4, and Section 1.5 concludes the paper.

1.2 STATE OF THE ART

The GAP has been extensively studied in the literature. The main contributions are reported in the following brief review, separating exact methods and approximate methods.

1.2.1 *Exact methods*

Removing knapsack constraints (1.2) from the GAP formulation provides an easy-to-solve assignment subproblem. This upper bound was exploited by Ross and Soland (1975) within a depth-first branch & bound algorithm. Two papers propose refined branch & bound methods computing upper bounds by Lagrangian relaxation (LR) of either assignment constraints (1.1) or capacity constraints (1.2). These two LR-based methods were respectively developed by Fisher et al. (1986), and Guignard and Rosenwein (1989). Savelsbergh (1997) also proposed an efficient branch & price algorithm solving a set partitioning reformulation of the problem. This reformulation, which was originally introduced in Cattrysse et al. (1994), consists in associating with every agent j in J all subsets of tasks which do not violate the capacity of j. Due to the exponential number of subsets, the problem was solved by column generation.

A comprehensive survey on exact methods for the GAP can be found in Cattrysse and van Wassenhove (1992). Typically, the above methods were able to solve problems with a maximum of one hundred tasks. For very large-scale problems, heuristic methods are needed for obtaining high-quality solutions in short computational time.

1.2.2 *Approximate methods*

A classical reference for greedy heuristics for the GAP is the paper of Martello and Toth (1981), in which they developed a local improvement search applied to a solution constructed in a greedy way so as to minimize an appropriate regret function.

Many other heuristic methods are based on relaxations of the problem. The LP-relaxation was used in the heuristic of Trick (1992) for fixing variables to one and performing swap reassignments. The LP bound was also exploited in the Variable Depth-Search Heuristic by Amini and Racer (1994). The main interest of their paper is that it provides a carefully-designed statistical comparison of several methods. LR-based heuristic methods were implemented as well for finding approximate solutions on large-scale instances, showing their efficiency (Barcia and Jornsten, 1990; Lorena and Narciso, 1996).

Many metaheuristics have been applied to the GAP. An inexhaustive but representative list is comprised of the genetic algorithms by Wilson (1997) and Chu and Beasley (1997), the tabu search algorithms by Osman (1995), Laguna et al. (1995), Yagiura et al. (1999; 1998), Díaz and Fernandez (2001) and Higgins (2001), and a combination of simulated annealing with tabu search by Osman (1995).

Most of the metaheuristics reported above exploit the following principles:

- Use of both shifts and swaps as valid moves for local search (Yagiura et al., 1998; Díaz and Fernandez, 2001; Higgins, 2001); indeed, the single swap neighborhood is restrictive in that it does not allow a change in the number of tasks assigned to each agent. Ejection chains have also been considered in Laguna et al. (1995) and Yagiura et al. (1999).

- Use of a measure of infeasibility representing the global over-capacity of agents (Wilson, 1997; Chu and Beasley, 1997; Laguna et al., 1995; Yagiura et al., 1999; 1998; Díaz and Fernandez, 2001; Higgins, 2001).

- Oscillation between feasibility and infeasibility for a wider diversification of the search (Chu and Beasley, 1997; Laguna et al., 1995; Yagiura et al., 1999; 1998; Díaz and Fernandez, 2001; Higgins, 2001). The balance between feasibility and infeasibility is controlled through an appropriate penalty coefficient for infeasibility. In most papers, infeasible solutions are admitted but the search is guided toward feasibility (Chu and Beasley, 1997; Yagiura et al., 1999; 1998; Laguna et al., 1995; Higgins, 2001), whereas in Díaz and Fernandez (2001), feasible and infeasible solutions are compared through a unique evaluation function so that no preference is put on feasibility when exploring a neighborhood. Also, the penalty term is often dynamically adapted all along the search.

To our knowledge, neither scatter search nor path relinking had been applied to the GAP. Recently, however, a very efficient path relinking algorithm based on ejection chains was proposed by Yagiura et al. (2002). This algorithm obtains the best known values for many test problems of the OR-library.

Our path relinking algorithm for the GAP is described in the following section.

1.3 A PATH RELINKING ALGORITHM FOR THE GAP

1.3.1 *Principles of path relinking*

Path relinking is a generalization of *scatter search* (Glover, 1998a;b), which performs linear combinations of *elite* solutions. It consists in enriching a population of solutions with new ones found on paths linking elite solutions. According to Glover (1998b), the initial endpoint $\underline{x}$ of a path is called the *initiating* solution, whereas the terminal endpoint $\bar{x}$ is the *guiding* solution. The path linking $\underline{x}$ and $\bar{x}$ is a sequence of (feasible or infeasible) solutions

$$\underline{x} = x^0, x^1, x^2, \ldots, x^p = \bar{x} \tag{1.4}$$

where x^{k+1} belongs to a neighborhood of x^k, denoted by $N^{pr}(x^k)$, chosen so as to get closer and closer to (resp. to go farther and farther away from) the structure of the guiding (resp. initiating) solution according to some distance d, i.e. $d(x^{k+1}, \bar{x}) < d(x^k, \bar{x})$ and $d(x^{k+1}, \underline{x}) > d(x^k, \underline{x})$.

The construction of a path can be seen as an intensification process since the common elements between $\underline{x}$ and $\bar{x}$ are never modified throughout the path, as keeping common elements of good solutions helps to generate other good (hopefully, improving) solutions. The best solutions found on paths are then inserted in the population in a dynamic way.

Although principles of path relinking are quite promising, only a few papers in the literature deal with applications of this method to specific combinatorial optimization problems (see, for example, Glover et al. (1994); Laguna et al. (1999); Plateau et al. (2002)).

To adapt path relinking to the generalized assignment problem, two main issues have to be studied: path construction and population management.

Path construction consists in choosing (i) a neighborhood N^{pr}, (ii) a criterion for selecting x^{k+1} in $N^{pr}(x^k)$, (iii) which solution(s) of the path to select for insertion.

Population management comprises the determination of (iv) the initial population, (v) which pairs of solutions to combine, (vi) the stopping criterion of the path relinking algorithm. The two following sections specify the strategies chosen for treating items (i) to (vi).

1.3.2 Path construction

We call $pr(\underline{x}, \bar{x})$ the procedure that constructs the path linking $\underline{x}$ to $\bar{x}$. The intermediate solutions of this path are generated by reassigning only tasks that are not assigned to the same agent in $\underline{x}$ and $\bar{x}$, i.e. tasks in subset $\{i \in I : \underline{x}(i) \neq \bar{x}(i)\}$.

(i) Choice of neighborhood N^{pr}.

The distance d introduced in the previous section to build a neighborhood is defined as follows for the GAP:

$$d(x, x') = |\{i \in I : x(i) \neq x'(i)\}|. \tag{1.5}$$

It represents the number of tasks that are not assigned to the same agent in x and x'. To build a neighborhood N^{pr} for the GAP, we combine a shift neighborhood N^{pr}_{shift} and a swap neighborhood N^{pr}_{swap} defined as follows. Given a current solution x^k, each solution x in $N^{pr}_{shift}(x^k)$ (resp. $N^{pr}_{swap}(x^k)$) differs from x^k by a single (resp. two) new assignment(s) belonging to $\bar{x}$. Using distance formula (5), those two neighborhoods can be expressed by:

$$\begin{aligned} N^{pr}_{shift}(x^k) &= \{x \in M : d(x, x^k) = 1 \wedge d(x, \bar{x}) = d(x^k, \bar{x}) - 1\}, \\ N^{pr}_{swap}(x^k) &= \{x \in M : d(x, x^k) = 2 \wedge d(x, \bar{x}) = d(x^k, \bar{x}) - 2\}. \end{aligned}$$

N^{pr}_{shift} and N^{pr}_{swap} adapt for path relinking the shift and swap neighborhoods usually designed for the GAP:

$$\begin{aligned} N_{shift}(x^k) &= \{x \in M : d(x, x^k) = 1\}, \\ N_{swap}(x^k) &= \{x \in M : d(x, x^k) = 2\}. \end{aligned}$$

The neighborhood N^{pr} is the union $N^{pr}_{shift} \cup N^{pr}_{swap}$. Since M is the set of all mappings $x : I \to J$, a solution in $N^{pr}(x^k)$ is not constrained to respect the capacity constraints, i.e. be feasible.

(ii) Choice of neighbor x^{k+1} in $N^{pr}(x^k)$.

As in several papers in the literature (Chu and Beasley, 1997; Díaz and Fernandez, 2001; Higgins, 2001; Laguna et al., 1995; Yagiura et al., 1999; 1998), our search is not restricted to feasible solutions. To compare feasible and infeasible solutions according to a single criterion, a modified evaluation function is used:

$$v(x,\alpha) = f(x) - \alpha \cdot fitness(x),$$

where *fitness*(x) is a measure of infeasibility representing the average (scaled) over-capacity of agents and defined by

$$fitness(x) = \frac{1}{m}\sum_{j=1}^{m} \max\left(0, \frac{1}{b_j}\sum_{i=1}^{n} a_{ij}x_{ij} - 1\right) \tag{1.6}$$

and $\alpha \geq 0$ is a penalty coefficient for infeasibility. As in Díaz and Fernandez (2001), no priority is given to feasibility. An infeasible solution x' is preferred to a feasible solution x at any time of the search as soon as $v(x',\alpha) > v(x,\alpha)$, i.e. $f(x') - f(x) > \alpha \cdot fitness(x')$. Indeed, restricting the search space to feasible solutions leads to a set of solutions that is both smaller and less diversified. The reason is that tight capacity constraints make the proportion of infeasible assignments much higher than the proportion of feasible assignments. As a consequence, the neighborhood of a feasible solution is almost fully or totally composed of infeasible solutions, so a search allowing infeasibility is more likely to reach this solution. Naturally, the performance of the search depends on the penalty coefficient α. A high valued α forces feasibility but makes the search terminate earlier, whereas a low valued α forces infeasibility and may lead to preferring an infeasible solution to an optimal feasible solution. Therefore, controlling the value of α is crucial for maintaining a balance between feasibility and infeasibility. Coefficient α is initialized as follows. Considering $v(x,\alpha)$, a feasible solution x and an infeasible solution x' have equal value if $f(x') - f(x) = \alpha \cdot fitness(x')$. If x' is such that the capacities of $m-1$ agents are respected and the capacity of the remaining agent is exceeded by one unit, on average $fitness(x') = 1/m \times 1/(\sum_{j=1}^{m} b_j/m) = 1/(\sum_{j=1}^{m} b_j)$ and $f(x') - f(x) = \alpha/(\sum_{j=1}^{m} b_j)$. Therefore, $\alpha/(\sum_{j=1}^{m} b_j)$ represents the minimum objective variation $\Delta f = f(x') - f(x)$ that one is ready to accept for violating one capacity constraint of a feasible solution x by one unit. Δf is initialized to a prefixed value Δf_0 and repeatedly adjusted along the search so that $\alpha = \Delta f/(\sum_{j=1}^{m} b_j)$ is dynamically updated as explained in item (*vii*).

(iii) Selection of the solution to insert.

For each path, a single intermediate (feasible or infeasible) solution, denoted by x^*, is candidate for insertion in the population. x^* is returned by procedure $pr(\underline{x}, \bar{x})$ at the end of the two following steps.

First, $pr(\underline{x}, \bar{x})$ selects the intermediate solution x^k with higher value $v(x^k, \alpha)$. Then, an *ascent* method applied to x^k returns x^*. Given a solution x, $ascent(x)$ iteratively selects the solution x' with higher value $v(x', \alpha)$ in the neighborhood $N = N_{shift} \cup N_{swap}$ of the current solution, until a local optimum is reached. Applying this local search to x^k permits both to improve x^k and to avoid selecting a solution that would lie in the subregion of a local optimum already in the population. Naturally, the output x^* of $ascent(x^k)$ is inserted in the population only if the latter does not contain it.

The complexity of building a path computes simply as follows. The number of solutions generated on a path ranges in $[\lceil d(\underline{x}, \bar{x})/2 - 1\rceil, d(\underline{x}, \bar{x}) - 1]$, i.e. is $O(d(\underline{x}, \bar{x}))$. The complexity of selecting x^{k+1} in $N^{pr}(x^k)$ is $O(d(\underline{x}, \bar{x})^2)$. Therefore, the complexity of computing the whole path from $\underline{x}$ to $\bar{x}$ is $O(d(\underline{x}, \bar{x})^3)$.

1.3.3 Population management

(iv) Initial population

The path relinking algorithm starts with a population $\mathcal{P}_0 = \{x^0, x^1, \ldots, x^m\}$ composed of not necessarily feasible solutions. The first solution x^0 is expected to have a good value, whereas the others are dummy but diversified solutions.

The solution x_0 is derived from the LP-relaxation of (ILP) as follows. First, the fractional optimal solution of the LP-relaxation, denoted by X^*, is computed. Secondly, X^* is rounded to a 0-1 solution x such that constraints (1.1) are satisfied, by setting $x(i) = \text{argmax}_{j \in J} X^*_{ij}$. Finally, x_0 is returned by $ascent(x)$. This solution is a guiding solution for the other initial solutions.

The other m solutions of $\mathcal{P}_0$ are generated in two steps. First, m solutions $y^1, \ldots, y^m$ are constructed so that the distance between each pair of solutions is maximum, i.e. $d(y^k, y^l) = n$ for all $k \neq l$. Suppose that the average number of tasks per agent $\bar{n} = n/m$ is integer (this is the case in the database used for experiments in Section 1.4). Denote by $I_1 = \{1, \ldots, \bar{n}\}$ the first $\bar{n}$ tasks, $I_2 = \{\bar{n}+1, \ldots, 2\bar{n}\}$ the following $\bar{n}$ tasks, etc., i.e.

$$\forall j \in \{1, 2, \ldots, m\},\ I_j = \{(j-1)\bar{n} + 1, \ldots, j \cdot \bar{n}\}$$

For $k = 1, 2, \ldots, m$, y^k is defined by :

$$\forall j \in J,\ \forall i \in I_j,\ y^k(i) = \begin{cases} m - k + j + 1 & \text{if } j < k \\ j - k + 1 & \text{if } j \geq k \end{cases}$$

These m solutions have no assignment in common (if $\bar{n} = n/m$ is not integer, the above process would be applied with $\lfloor \bar{n} \rfloor$ tasks per agent and the remaining tasks would be equally dispatched on the agents).

In the second step, the *ascent* procedure is applied to each y^k to obtain x^k i.e., $x^k = ascent(y^k)$.

The above m solutions are called *dummy* solutions because, before *ascent*, no control is made neither on the objective nor on the fitness of the solutions, which is generally very large and often requires many improvements in *ascent* before reaching a local optimum. However, these solutions are of double interest:

- Before *ascent*, diversification of population is maximum.
- Starting with a medium- or low-quality population in terms of objective function enables us to show the efficiency of path relinking combinations *independently of the quality of the initial population.* Indeed, if the best feasible solution in the original population (before combinations) already has a very low gap to optimality, the efficiency of the algorithm is hardly justified by the combination phase. However, the purpose of this paper is not only to solve a particular problem (the GAP) but also to show in general how path relinking can significantly improve a set of solutions.

(v) Choice of pairs of solutions to combine.

The population is updated in two main phases, a *growing* phase, and a *stabilization* phase.

The *growing* phase consists in increasing the size of the initial population $\mathcal{P}_0$ until it includes at least a prefixed number p_{RS} of solutions. The p_{RS} best solutions of the population will form the initial reference set (RS) of elite solutions. This Reference Set is reached after a r_{RS} number of rounds of combinations. At each round r in $\{1, \dots, r_{RS}\}$, the set $\mathcal{Q}_r$ of new solutions inserted is

$$\mathcal{Q}_r = \{pr(x, x') : (x, x') \in (\mathcal{P}_{r-2} \times \mathcal{Q}_{r-1}) \cup (\mathcal{Q}_{r-1} \times \mathcal{Q}_{r-1}), v(x, \alpha) \leq v(x', \alpha)\}$$

where $\mathcal{P}_{-1} = \emptyset$ and $\mathcal{P}_r$ is the current population, that is

$$\mathcal{P}_r = \bigcup_{t=0}^{r} \mathcal{Q}_t. \tag{1.7}$$

In this way, the last round r_{RS} of the growing phase is

$$r_{RS} = \min_r \{r : |\mathcal{P}_r| \geq p_{RS}\}.$$

Then, at each round, all pairs (x, x') of *(old, new)* and *(new,new)* solutions are combined. In every combination, the guiding solution $\bar{x}$ and the initiating solution $\underline{x}$ are chosen so that $v(\bar{x}, \alpha) \geq v(\underline{x}, \alpha)$. At the end of round r_{RS}, the second phase starts with the reference set composed of the p_{RS} best solutions of $\mathcal{P}_{r_{RS}}$.

In the *stabilization* phase, the size of RS remains equal to p_{RS} until the last round, denoted by r_s. Rounds $r_{RS}+1, r_{RS}+2, \ldots, r_s$ combine each pair (x, x') of (*old,new*) and (*new,new*) solutions as in the growing phase, but from now a new solution x^* output by $pr(x, x')$ is inserted in RS if and only if $v(x^*, \alpha) > v(x^w, \alpha)$, where

$$x^w = \text{argmin}_{x \in RS} v(x, \alpha) \quad (1.8)$$

is the worst solution of RS. In that case, x^w is removed from RS so that exactly p_{RS} solutions are kept in RS. The stopping criterion defining last round r_s is described in the sequel.

(vi) Stopping criterion

Two strategies were tested for stopping the path relinking algorithm:

- Stop the process when no new solution enters RS at the end of a round, i.e. when all solutions x^* generated on paths satisfy $v(x^*, \alpha) \leq v(x^w, \alpha)$ where x^w is defined as in (1.8). This strategy formulates an evolution principle according to which the population dies when it cannot give birth any more to children that outperform their parents.

- Stop the process when a fixed number of rounds *max_r* is reached after round r_{RS}. However, parameter *max_r* is not critical since Strategy 1 automatically stops the search.

A last item in population management is not a general issue of path relinking but is specific to the GAP. It concerns the updating of the infeasibility penalty coefficient α.

(vii) Dynamic adjustment of infeasibility penalty coefficient

Coefficient α is initialized to $\Delta f_0 / (\sum_{j=1}^{m} b_j)$ as mentioned in item (ii), and α is repeatedly adjusted so that the current population has a given threshold of feasible solutions and infeasible solutions. For every r in $\{0, 1, \ldots, r_{RS}\}$, if population $\mathcal{P}_r$ defined in (1.7) does not include any feasible (resp. infeasible) solution, then Δf is changed to $\Delta f + 1$ (resp. $\Delta f - 1$) and *ascent* is applied to each solution of $\mathcal{P}_r$ with new $\alpha = \Delta f / (\sum_{j=1}^{m} b_j)$, until the population includes at least one feasible solution and one infeasible solution.

The same process is applied at each round $r > r_{RS}$, but the threshold for feasibility is increased to a prefixed proportion ρ_0 of feasible solutions in RS. Indeed, at the beginning of the search, a low proportion of feasible solutions is preferable for wider diversification, whereas forcing feasibility and intensification become all the more accurate as the search tends to its end.

1.4 COMPUTATIONAL RESULTS

The path relinking algorithm was tested on testbed instances of the maximization version of the GAP. These test problems of the OR-library are available at `http://mscmga.ms.ic.ac.uk/jeb/orlib/gapinfo.html`. They were used to test the set partitioning heuristic of Cattrysse and van Wassenhove (1992), the simulated annealing heuristic of Cattrysse et al. (1994), the simulated annealing/tabu-search heuristic SA/TS of Osman (1995) and the tabu-search heuristics of Osman (1995), Díaz and Fernandez (2001), and Higgins (2001) denoted respectively by TS1, TS2, and TS3. The test-library contains 60 problems for which optimal value f^* is known. The problem set is composed of 12 series of 5 problems, all 5 problems of a series having the same number of tasks and agents. Precisely,

$$m = 5 \text{ and } n \in \{15, 20, 25, 30\} \text{ in test series 1-4,}$$
$$m = 8 \text{ and } n \in \{24, 32, 40, 48\} \text{ in test series 5-8,}$$
$$m = 10 \text{ and } n \in \{30, 40, 50, 60\} \text{ in test series 9-12.}$$

Therefore, the average number of tasks per agent n/m is in $\{3, 4, 5, 6\}$ for the four series associated with a given m. Values a_{ij} are integers generated from a uniform distribution $U(5, 25)$ and c_{ij} values are integers generated from $U(15, 25)$. Finally, the b_j capacities are set to $(\sum_{i=1}^{n} a_{ij}) \times (0.8/m)$ for every agent j of every problem so that knapsack constraints are tight.

The various parameters of the algorithm were set as follows: $p_{RS} = 25$, $\rho_0 = 0.25$ and, after some tuning, $\Delta f_0 = 4$,. For the small problems of series 1-4 the *ascent* procedure was not applied to the best solution of a path, as when using *ascent* the required size p_{RS} of the reference set was not reached.

Numerical experiments were carried out on a Pentium III computer (550 MHz processor, 256 Mb RAM). The LP-solver used for the initial population was CPLEX 6.6, and the algorithm was programmed in C language. A summary of the results obtained is presented in the following two tables. Table 1.1 presents the average gap from optimum in %, i.e. $100 \times (f^* - f^h)/f^*$, where f^h denotes the heuristic value and f^* is the optimal value, for each heuristic method SA/TS, GA, TS1, TS2 and PR. For all methods except GA and TS2, the algorithm is applied once for each problem, and the average gap is taken over each execution of the 5 problems of a series. For GA and TS2, the algorithm is applied a fixed number of times on each problem because of an alea parameter, and the average gap is computed over all executions of a problem and all problems of a series.

The average gap on the 60 instances is indicated below, as well as the number of problems solved to optimality ((*) for GA and TS2 this number is the number of problems for which at least one execution of the algorithm gave the optimal value).

Table 1.2 presents the average CPU computation times (in seconds) for each series and each method : $\bar{t}_{best}$ is the average CPU time before finding the best

Table 1.1 Gap from optimal (in %) for the 12 test series

test series	SA/TS	TS1	GA	TS2	PR
GAP1	0.00	0.00	0.00	0.00	0.00
GAP2	0.00	0.10	0.00	0.00	0.00
GAP3	0.00	0.00	0.00	0.00	0.00
GAP4	0.00	0.03	0.00	0.00	0.00
GAP5	0.00	0.00	0.00	0.00	0.00
GAP6	0.05	0.03	0.01	0.01	0.00
GAP7	0.02	0.00	0.00	0.00	0.00
GAP8	0.10	0.09	0.05	0.01	0.03
GAP9	0.08	0.06	0.00	0.00	0.00
GAP10	0.14	0.08	0.04	0.03	0.02
GAP11	0.05	0.02	0.00	0.00	0.00
GAP12	0.11	0.04	0.01	0.00	0.00
average gap	0.210	0.070	0.009	0.004	0.004
# optima found	39	45	60*	60*	58

feasible solution and $\bar{t}_{total}$ is the total CPU time of the algorithm. Naturally, CPU-times compare differently since experiments were performed on different machines (VAX-8600 for SA/TS and TS1, Silicon Graphics Indigo R4000 100 MHz for GA, Sun Sparc 10/30 4×100 MHz for TS2, Pentium II 400 MHz for TS3).

The path relinking algorithm solved 58 of the 60 instances. This number is not directly comparable to the 60 instances solved by GA and TS2* since, as mentioned before, the algorithm was applied several times for each problem, 30 times for TS2 where the tenure parameter is an alea parameter. For the two problems not solved by PR, the absolute error between PR and optimum was 1. Although CPU times are not comparable, compared to GA we generally obtain better gaps with short computation times. The average gap over the 60 problems (0.04%) is comparable to TS2, but the average running time for reaching the best value is generally higher for PR. As for TS3, the authors mention in Higgins (2001) that all of the 60 problems were solved within 0.1% from the optimum with less than one second of CPU time. For PR, the average gap to optimality was equal to 0.35 % for initial population, 0.13 % at the end of the *growing* phase and 0.04 % at the end of the search. Therefore, the performance of PR is due mainly to the rounds of combinations rather than to the quality of the initial population. Since tabu search performs well on the GAP, we believe that the results obtained by PR are quite promising when attempting to combine path relinking with tabu search for larger problems.

Table 1.2 CPU times (in seconds) for the 12 test series

test	SA/TS		TS1		GA		TS2		PR	
series	$\bar{t}_{best}$	$\bar{t}_{total}$	$\bar{t}_{best}$	$\bar{t}_{total}$	$\bar{t}_{best}$	$\bar{t}_{total}$	$\bar{t}_{best}$	$\bar{t}_{total}$	$\bar{t}_{best}$	$\bar{t}_{total}$
GAP1	n.a	0.22	n.a	0.73	0.49	72.38	0.04	1.09	0.04	0.31
GAP2	n.a	0.91	n.a	1.44	1.68	79.96	0.04	1.48	0.06	0.28
GAP3	n.a	1.82	n.a	2.85	2.18	85.06	0.08	2.05	0.07	0.32
GAP4	n.a	2.32	n.a	4.86	4.16	92.36	0.13	2.54	0.15	0.63
GAP5	n.a	2.48	n.a	3.97	5.52	100.36	0.22	2.38	0.57	1.48
GAP6	n.a	5.67	n.a	8.85	16.36	130.02	0.41	3.53	1.21	3.28
GAP7	n.a	13.97	n.a	12.72	10.54	130.34	0.35	4.92	1.86	4.92
GAP8	n.a	19.53	n.a	22.99	49.90	184.42	2.38	7.64	2.10	4.64
GAP9	n.a	5.73	n.a	12.19	13.82	129.32	0.77	3.72	1.31	3.65
GAP10	n.a	15.55	n.a	18.62	33.70	167.66	1.07	5.90	3.25	6.36
GAP11	n.a	36.96	n.a	34.46	32.60	193.00	1.53	8.02	3.89	9.23
GAP12	n.a	65.96	n.a	47.07	39.08	260.34	1.83	10.67	6.22	12.65

1.5 CONCLUSION

We have designed a path relinking heuristic for the Generalized Assignment Problem based on the following principles: combination of shifts and swaps for the neighborhood, relaxation of the problem aggregating both the objective function and a measure of infeasibility through a penalty coefficient, and dynamic adjustment of this coefficient. The latter principle has proved efficient, as evidenced in the literature, for controlling oscillation between feasible and infeasible solutions and wider diversification. The PR algorithm was tested on a classical set of 60 problems of the OR-library and compared favorably with three of the other four methods of the literature tested on these problems. For the other tabu search method, PR gave comparable results in terms of average gap to optimality with generally higher CPU-times, however computation times remained quite low. Our PR algorithm has the advantage of being fully deterministic and requires only few parameters – the size of the reference set, the initial penalty coefficient, and the proportion of feasible/infeasible solutions desired in the population. It should give even better results when combined with an efficient heuristic for building the initial population, since ours was intentionally simple in order to demonstrate the efficiency of combinations. For complete validation, experiments will be carried out on larger problem-sets of the OR-library, namely the A-E-type problem-sets. Future work should also focus on strategies for diversifying the initial population and dynamic adjustment of the parameters.

M.M. Amini and M. Racer. A rigorous comparison of alternative solution methods for the generalized assignment problem. *Management Sc.*, 40:868–890, 1994.

P. Barcia and K. Jornsten. Improved lagrangian decomposition : An application to the generalized assignment problem. *European J. of Oper. Res.*, 46: 84–92, 1990.

D.G. Cattrysse, M. Salomon, and L.N. van Wassenhove. A set partitioning heuristic for the generalized assignment problem. *European J. of Oper. Res.*, 72:167–174, 1994.

D.G. Cattrysse and L.N. van Wassenhove. A survey of algorithms for the generalized assignment problem. *European J. of Oper. Res.*, 60:260–272, 1992.

P.C. Chu and J.E. Beasley. A genetic algorithm for the generalised assignment problem. *Comp. Oper. Res.*, 24:17–23, 1997.

J. Díaz and E. Fernandez. A tabu search heuristic for the generalized assignment problem. *European J. of Oper. Res.*, 132:1:22–38, 2001.

M.L. Fisher, R. Jaikumar, and L.N. Van Wassenhove. A multiplier adjustment method for the generalized assignment problem. *Management Sc.*, 32:9:1095–1103, 1986.

F. Glover. Genetic algorithms, evolutionary algorithms and scatter search: Changing tides and untapped potentials. *INFORMS Computer Science Newsletter*, 19:1:7–14, 1998a.

F. Glover. A template for scatter search and path relinking. In J.-K. Hao, E. Lutton, E. Ronald, M. Schoenauer, and D. Snyers, editors, *Artificial Evolution, Lecture Notes in Computer Science*, pages 13–54. Springer, 1998b.

F. Glover, J.P. Kelly, and M. Laguna. Genetic algorithms and tabu search: Hybrids for optimization. *Computers and Oper. Res.*, 22:1:111–134, 1994.

M. Guignard and M. Rosenwein. An improved dual-based algorithm for the generalized assignment problem. *Oper. Res.*, 37:4:658–663, 1989.

A.J. Higgins. A dynamic tabu search for large-scale generalised assignment problems. *Computers and Oper. Res.*, 28:10:1039–1048, 2001.

M. Laguna, J.P. Kelly, J.L. Gonzalez-Velarde, and F. Glover. Tabu search for the multilevel generalized assignment problem. *European J. of Oper. Res.*, 82: 176–189, 1995.

M. Laguna, R. Martí, and V. Campos. Intensification and diversification with elite tabu search solutions for the linear ordering problem. *Computers and OR*, 26:1217–1230, 1999.

L.A.N. Lorena and M.G. Narciso. Relaxation heuristics for a generalized assignment problem. *European J. of Oper. Res.*, 91:600–610, 1996.

S. Martello and P. Toth. An algorithm for the generalized assignment problem. In J.P. Brans, editor, *Oper. Res. '81*, pages 589–603. North Holland, 1981.

I.H. Osman. Heuristics for the generalized assignment problem: Simulated annealing and tabu search approaches. *OR Spektrum*, 17:211–225, 1995.

A. Plateau, D. Tachat, and P. Tolla. A hybrid search combining interior point methods and metaheuristics for 0-1 programming. *Intl. Trans. in Oper. Res.*, 9:6:731–746, 2002.

G.T. Ross and R.M. Soland. A branch and bound algorithm for the generalized assignment problem. *Math. Prog.*, 8:91–103, 1975.

M. Savelsbergh. A branch and cut algorithm for the generalized assignment problem. *Oper. Res.*, 45:6:831–841, 1997.

M. Trick. A linear relaxation heuristic for the generalized assignment problem. *Naval Research Logistics*, 39:137–151, 1992.

J.M. Wilson. A genetic algorithm for the generalised assignment problem. *J. of the Oper. Res. Soc.*, 48:804–809, 1997.

M. Yagiura, T. Ibaraki, and F. Glover. An ejection chain approach for the generalized assignment problem. Technical Report 99013, Department of Applied Mathematics and Physics, Graduate Sch. of Informatics, Kyoto University, 1999.

M. Yagiura, T. Ibaraki., and F. Glover. A path relinking approach for the generalized assignment problem. In *Proc. International Symposium on Scheduling, Japan, June 4-6*, pages 105–108, 2002. To appear in *INFORMS J. on Computing* with title "An Ejection Chain Approach for the Generalized Assignment Problem.".

M. Yagiura, T. Yamaguchi, and T. Ibaraki. A variable depth search algorithm with branching search for the generalized assignment problem. *Optimization Methods and Software*, 10:419–441, 1998.

Metaheuristics: Computer Decision-Making, pp. 19-36
Mousbah Barake, Pierre Chardaire, and Geoff P. McKeown

2 THE PROBE METAHEURISTIC AND ITS APPLICATION TO THE MULTICONSTRAINT KNAPSACK PROBLEM

Mousbah Barake[1], Pierre Chardaire[2], and Geoff P. McKeown[2]

[1]23 Alexandra Road
Colchester, CO3 3DF, UK
mousbah@email.com

[2]School of Information Systems
University of East Anglia
Norwich, NR4 7TJ, UK
pc@sys.uea.ac.uk, gpm@sys.uea.ac.uk

Abstract: A new metaheuristic technique called PROBE is presented. The application of PROBE to the multiconstraint knapsack problem is described. Experimental results obtained using the resulting algorithm are compared with the results obtained by Chu and Beasley using a Genetic Algorithm.
Keywords: Metaheuristic, PROBE, OCTANE, Multiconstraint knapsack problem, Genetic algorithm.

2.1 INTRODUCTION

In this paper, a new population based metaheuristic technique called PROBE is proposed. PROBE (*P*opulation *R*einforced *O*ptimization *B*ased *E*xploration) directs optimization algorithms, general or specific, toward good regions of the search space using some ideas from Genetic Algorithms (GAs) (see Michalwicz (1996) for details on GAs). A disadvantage of a straightforward implementation of the original Genetic Algorithm ideas is that for problems with constraints, there is no guarantee that the cross-over operator will create feasible offspring from feasible parents. One way of dealing with this problem is to devise problem-specific encoding and cross-over operators. Another way is to take the constraints into account by incorporating penalty terms into the fitness function. However, it is difficult with this approach to determine the correct balance between the original objective and the penalty terms. It is therefore often necessary to augment the basic GA procedure with some heuristic, often referred to as a "repair" operator, for transforming an infeasible offspring into a feasible solution. An important property of the (standard n-point) crossover operator used in GAs is that if two parents share a common bit value their offspring inherit it, whatever the position of the cross-over point. In other words, the offspring generated belong to the subspace of solutions obtained by fixing the genes common to both parents. PROBE preserves this property of the GA cross-over operator but then uses an appropriate technique to obtain an improved *feasible* solution in the subspace corresponding to a pair of *feasible* parents.

In this paper, we illustrate the use of PROBE by applying it to the multiconstraint knapsack problem (MKP). The MKP is a useful benchmark problem because many 0-1 integer programming problems may be formulated as instances of the MKP. We have also applied PROBE to the graph bisection problem (GBP) (Chardaire et al., 2003a). Using the GBP as a benchmark problem demonstrates the applicability of PROBE to combinatorial optimization problems.

The rest of the paper is structured as follows. The PROBE metaheuristic is presented in Section 2.2. In Section 2.3, we briefly review other approaches for solving the MKP. The application of PROBE to the MKP is discussed in Section 2.4. Our results are presented in Section 2.5.

2.2 PROBE

PROBE maintains, at each generation g, a population of feasible solutions $S_i^g, i = 1, 2, \ldots, n$. We assume solutions are represented by bit strings. For convenience let $S_{n+1}^g = S_1^g$. The next generation of solutions is obtained as follows. For each $i = 1, \ldots, n$ the solution, S_i^{g+1}, is computed from the pair (S_i^g, S_{i+1}^g) by the sequence of steps in Figure 2.1.

1. Fix *some* of the bits that have the same value in S_i^g and S_{i+1}^g.
2. Fix *some* of the bits, which do not have the same value in S_i^g and S_{i+1}^g, to the value they have in one of the two parent solutions.
3. Find an instantiation of the remaining bits using an appropriate search algorithm.
4. Use the instantiation obtained as a starting point for a local optimizer. The solution obtained is S_i^{g+1}.

Figure 2.1 Generation of solutions in PROBE.

Steps 1 and 2 allow a control of the size of the space searched. Too large a subspace may result in a prohibitive solution time, especially when an exact algorithm is used to search that subspace, while too small a subspace often leads to the return of a replica of the fittest parent. PROBE may use *shrink* and *expand* operations to avoid these situations. A shrink operation fixes additional bits according to the corresponding bit values of one of the parents (step 2). An *expand* operation frees some variables previously fixed (i.e. only *some* variables are fixed in step 1). Extending or shrinking subspaces may be done in many ways. For example, variables may be selected randomly or according to intensification or diversification strategies. The size of the subspace searched may also be fixed. Steps 1 and 2 of PROBE guarantee that the subspace searched in step 3 contains a feasible solution if both parent solutions are feasible. This is because all of the fixed bits are common with at least one of the parents. Thus, at least one parent belongs to the subspace searched.

There is considerable flexibility in the choice of the search algorithm used in step 3. For example, if the subspaces generated in steps 1 and 2 are sufficiently small, we could use an exact algorithm such as Branch-and-Bound (B&B) to determine an optimal solution in a subspace. In other situations, heuristic or other metaheuristic techniques could be used. For example, we have used GRASP and path relinking within PROBE to solve instances of both the MKP and the GBP (Chardaire et al., 2003a;b). A nice feature of PROBE is that when the selected subspace search algorithm is an exact algorithm, a version of PROBE can be implemented to guarantee that the average fitness of the solutions in the pool increases until all solutions have the same fitness.

Proposition 1 *Let F_i^k be the objective value (fitness) of solution i in generation k and let*

$$A^k = \frac{1}{n}\sum_{i=1}^{n} F_i^k$$

be the average fitness of the pool of solutions in generation k. If the following conditions are satisfied:

1. *the subspace search algorithm used by PROBE is an exact algorithm,*
2. *the shrink operation always fixes bits according to the bit values of the fittest parent,*

then $A^{k+1} > A^k$, unless all solutions at generation k have the same fitness.

Proof.

Assume with no loss of generality that $F_1^k \neq F_2^k$ and for convenience let $F_{p+1}^k = F_1^k$.

Case: $F_1^k < F_2^k$.

First, we observe that

$$F_1^{k+1} \geq \max(F_1^k, F_2^k) > F_1^k.$$

Then, from $F_1^{k+1} > F_1^k$ and

$$\begin{aligned} F_i^{k+1} &\geq \max(F_i^k, F_{i+1}^k) \\ &\geq F_i^k, i = 2 \ldots n, \end{aligned}$$

it follows that $A^{k+1} > A^k$.
Case: $F_1^k > F_2^k$.

Now we have

$$F_1^{k+1} \geq \max(F_1^k, F_2^k) > F_2^k.$$

This, together with

$$\begin{aligned} F_i^{k+1} &\geq \max(F_i^k, F_{i+1}^k) \\ &\geq F_{i+1}^k, i = 2 \ldots n, \end{aligned}$$

again implies that $A^{k+1} > A^k$.

■

Corollary 1 *Under the hypotheses of the above proposition, PROBE converges to a population of solutions which all have the same fitness.*

Proposition 2 *Under the hypotheses of Proposition 1, let F^* denote the fitness value to which all of the solutions in the population converge and suppose that the fitness function is integer-valued. Then all solutions in the pool will have converged to F^* after at most $n\left(F^* - A^1\right)$ generations.*

Proof.

The fitness of each solution in the initial population is, on average, A^1. By Proposition 1, the fitness of at least one solution in the pool must improve by at least one unit in each successive generation. Since on average each solution must improve by $F^* - A^1$, the result follows.

■

Practical implementations of PROBE do not necessarily meet the hypotheses of Proposition 1. However, the proposition gives some theoretical foundation to the method.

2.2.1 General-purpose heuristics

Rather than using a general exact algorithm such as B&B for searching the subspace, we could use a customized solver or some heuristic method. One could, for example, use a GRASP to determine values for unfixed variables. It is particularly interesting, however, to try to use general-purpose techniques. If the problem to be solved has a 0-1 integer linear programming formulation, then one possibility could be to use the pivot and complement (*P&C*) heuristic of Balas and Martin (1980). *P&C* consists of two phases: a search phase and an improvement phase. The search phase starts with the optimal solution to the LP relaxation of the integer linear programming problem and performs a sequence of simplex pivot operations in order to satisfy the requirement that all the slack variables are basic. Three types of pivot operation may be performed during the search phase and in addition a rounding and truncating process is used at certain stages in an attempt to find an integer feasible solution. If an integer solution is found during the search phase, the improvement phase is entered during which complements are performed in order to find improved integer solutions.

Aboudi and Jörnsten (1994) consider a number of tabu search algorithms in each of which the *P&C* heuristic is used as a black box, taking a basic feasible solution to the LP relaxation as its input. The tabu procedure can select different basic feasible solutions to input to *P&C*, or it can add additional constraints to cut-off solutions satisfying some tabu condition. An aspiration level goal can be effected by forcing *P&C* to look for a solution whose value is better than some cut-off value. The pivot and complement heuristic could be used in a similar fashion within PROBE.

In this paper, however, we restrict attention to another general-purpose heuristic for 0-1 integer linear programming called OCTANE (OCTAhedral Neighbourhood Enumeration) designed by Balas et al. (2001). A brief description of this heuristic is given in Section 2.4.1

2.2.2 *Diversification*

Diversification strategies can be used within PROBE. For example, when the problem to be solved by PROBE has an integer linear programming (ILP) formulation and the algorithms used within PROBE are applied to this formulation, the following strategy could be used. After searching a subspace, the set of constraints could be augmented by an additional constraint that prohibits the algorithm from searching this subspace again. The schema that defines the subspace can be represented as follows:

$$(\underbrace{1,\ldots,1}_{1\to k},\underbrace{0,\ldots,0}_{k+1\to l},\underbrace{*,\ldots,*}_{l+1\to n}).$$

In this subspace, k bits have been fixed to 1, $l-k$ bits fixed to 0 and the remainder are unfixed. If we complement the 0-1 variables $x_1,\ldots,x_k$ corresponding to the first k positions of the schema, i.e. use the change of variables $y_i = 1 - x_i$ for $i = 1$ to k, then we transform the initial problem and the above schema is correspondingly transformed into

$$(\underbrace{0,\ldots,0}_{1\to l},\underbrace{*,\ldots,*}_{l+1\to n}).$$

The corresponding subspace can be prohibited by adding to the set of constraints for the original problem the additional constraint

$$y_1 + \cdots + y_k + x_{k+1} + \cdots + x_l \geq 1.$$

We may express this constraint in terms of the original variables as follows:

$$-x_1 - \cdots - x_k + x_{k+1} + \cdots + x_l \geq -k + 1.$$

For example, the constraint $x_1 + x_5 \geq 1$ prohibits the subspace $(0 * * * 0)$. In this case, the search of the subspace $(* * 1 * 0)$ would then reduce to the search of the subspace $(1 * 1 * 0)$.

2.3 MULTI-CONSTRAINT KNAPSACK PROBLEM

The multi-knapsack problem (*MKP*) can be formulated either as a maximization problem or as a minimization problem. A maximization instance of the *MKP* has the following form:

$$MKP\_max \begin{cases} \quad \max \quad px & (= Z) \\ \text{subject to} \quad Rx \leq b & \\ \quad x_j \in \{0,1\} & j = 1, \ldots, n. \end{cases}$$

Each of the entries in the $m \times n$ matrix $R = [r_{ij}]$ is a non-negative integer, as is each entry in the m-component right-hand-side vector, b. We assume that $p_j > 0$, and $r_{ij} \leq b_i < \sum_{j=1}^{n} r_{ij}$, for all $1 \leq i \leq m, 1 \leq j \leq n$. Each of the inequality constraints in this model may be regarded as a knapsack constraint. The coefficient r_{ij} then represents the amount of space in the i-th knapsack which would be occupied by the j-th item if that item were selected for inclusion. Many 0-1 integer programming problems can be put into the above form. The *MKP* is applicable to a wide variety of problem areas, including scheduling, capital budgeting, cargo loading, project selection and resource allocation. Instances of the *MKP* also arise as sub-problems in complex models such as database allocation in distributed computer systems (see Shih (1979); Pirkul (1987); Chu and Beasley (1998) for references to application areas).

The *MKP* is a much harder problem than the single constraint knapsack problem (*SKP*). The latter is not strongly NP-hard and effective algorithms are available for solving practical instances to optimality or at least to near-optimality. Many of the methods that have been proposed for solving MKPs involve the solution of SKPs as sub-problems. A review of the *SKP* is given by Martello and Toth (1990).

Since the *MKP* is a 0-1 integer linear programming problem, branch-and-bound approaches may be applied (Shih, 1979; Gavish and Pirkul, 1985). Gavish and Pirkul (1985) investigate a variety of relaxations to provide upper bounds for use in a branch-and-bound algorithm and compare the quality of the bounds generated by these relaxations. They also compare the computational effort required to determine the different bounds. In addition to the standard LP relaxation, they consider Lagrangian relaxation, surrogate relaxations and a combination of Lagrangian and surrogate relaxation.

A Lagrangian relaxation of the *MKP* can be obtained by first selecting one of the knapsack constraints to be active and then relaxing the remainder of these constraints to obtain a single constraint knapsack problem. Gavish and Pirkul use subgradient optimization to derive the Lagrange multipliers.

An alternative to using Lagrangian relaxation is to use a type of relaxation based on the concept of *surrogate duality*, originally proposed by Glover (1975). For a given $\mu \geq 0$, the surrogate problem is defined as follows:

$$MKP_{SR} \begin{cases} \max \quad px & (= Z_S(\mu)) \\ \text{subject to} \quad \mu(Rx - b) \leq 0 & \\ \qquad x_j \in \{0,1\} & j = 1, \ldots, n. \end{cases}$$

Problem MKP_{SR} is a single constraint knapsack problem with left-hand-side coefficient vector μR and right-hand-side vector μb. Ideally, we would like to take as the surrogate multipliers a vector, μ^*, such that

$$Z_S(\mu^*) = \min_{\mu} Z_S(\mu).$$

For such a choice, it has been shown that the bound generated by this relaxation of *MKP* is at least as good as the bounds generated by both the LP relaxation and Lagrangian relaxation. Unfortunately, obtaining optimal surrogate multipliers is difficult in practice. One of the simplest methods of generating good surrogate multipliers is to solve the LP relaxation of the original *MKP* and then to use the resulting shadow price vector (i.e. the optimal dual solution) as μ. Indeed, the bound obtained using this μ in MKP_{SR} is always equal to or tighter than the bound obtained from the LP relaxation.

It is also possible to combine Lagrangian and surrogate relaxation. Given μ and λ, a composite relaxed problem is defined as follows:

$$\text{MKP}_{LSR} \begin{cases} \max \quad px - \lambda(Rx - b) & (= Z_C(\lambda, \mu)) \\ \text{subject to} \quad \mu(Rx - b) \leq 0 & \\ \qquad x_j \in \{0,1\} & j = 1, \ldots, n. \end{cases}$$

This composite relaxation was first proposed by Greenberg and Pierskalla (1970). If λ' and μ' are such that

$$Z_C(\lambda', \mu') = \min_{\lambda, \mu} Z_C(\lambda, \mu),$$

then it is easy to see that $Z_C(\lambda', \mu')$ is at least as good a bound as the best bound obtainable using surrogate relaxation alone, which in turn is at least as good as the best bound obtainable using the Lagrangian relaxation described above. Gavish and Pirkul (1985) use this approach within a B&B algorithm for the *MKP*. They use *utility ratios*, defined for any given vector, μ, of surrogate multipliers by:

$$u_j = \frac{p_j}{\sum_{i=1}^{m} \mu_i r_{ij}}, \qquad j = 1, \ldots, n,$$

to determine the order in which variables are selected for branching on. The branch corresponding to setting the selected variable to 1 is fathomed first. The resulting branch-and-bound algorithm was found to outperform the method proposed by Shih (1979). Pirkul (1987) has also given a heuristic algorithm based on the surrogate relaxation problem and utility ratios. His method of defining utility ratios has been adapted by Chu and Beasley (1998) for use in a repair operator in a GA for the *MKP*.

A number of other researchers have applied metaheuristic approaches to the MKP (see, for example, Aboudi and Jörnsten (1994); Cotta and Troya (1998); Raidl (1998)). An approach based on GRASP is given in Chardaire et al. (2001). The algorithm proposed by Chu and Beasley (1998) appears to be the most successful GA to date for the MKP. In their algorithm, infeasible solutions are "repaired" using a greedy heuristic based on Pirkul's surrogate duality approach (Pirkul, 1987). Because all of the previously used bench-mark problems for the MKP are solved in very short computing times using a modern MIP solver, Chu and Beasley introduced a new set of bench-mark problems in their paper. We have used these problems to test our PROBE approach.

2.4 APPLICATION OF PROBE TO THE MULTI-CONSTRAINT KNAPSACK PROBLEM

Given an instance of the multi-constraint knapsack problem, each solution, $S_i^g, i = 1, \ldots, n$, at each generation, g, is represented by a binary string which corresponds to a 0-1 solution, x, to the given MKP. The fitness is simply the objective value, $Z = px$, associated with S_i^g. PROBE needs to be seeded with a population of feasible solutions, not necessarily distinct. We can either start the algorithm with a population of zero-vector solutions or we can use a random greedy algorithm to initialize it with different feasible solutions. The randomness can be obtained by using a random ordering of the variables in the greedy algorithm. However, there is little difference between these two options in our current implementation because of the nature of the method used to generate solutions from one generation to the next. For this reason, we start the algorithm with zero-vector solutions. In all of our experiments the subspace size is equal to a fixed ratio of the full space size. Expand and shrink operations are performed by selecting variables at random. In step 2 of the algorithm presented in Figure 2.1 the bits fixed in the expand operation are those corresponding to the first of the two parent solutions.

2.4.1 Subspace search

Any algorithm designed to solve the MKP is a candidate algorithm for subspace search as the sub-problems to be solved are themselves smaller instances of the MKP. We have used an approach based on a general heuristic, OCTANE (OCTAhedral NEighbourhood Enumeration), presented in Balas et al. (2001), for solving 0-1 integer programs.

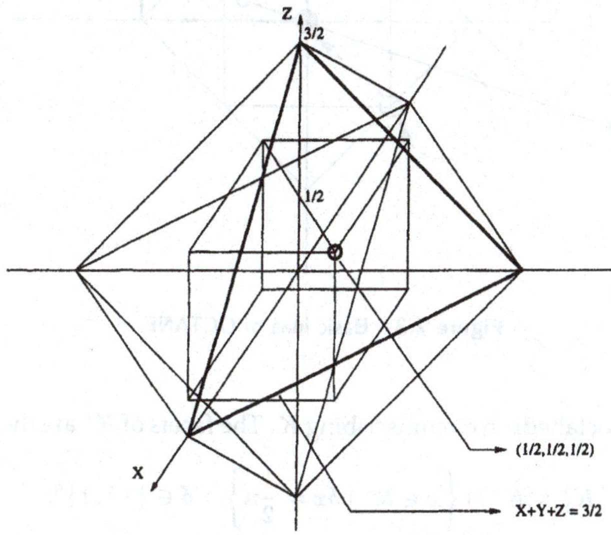

Figure 2.2 Unit cube and circumscribing octahedron.

The ideas underlying OCTANE originate from the work by Balas on intersection cuts in the early 1970s (Balas, 1972). OCTANE selects a fractional solution, y, to the linear programming relaxation of the 0-1 integer program and translates it to obtain the point $x = y - e/2$ where $e = (1, \ldots, 1)$. The point, x, is in the interior of the n-dimensional octahedron that circumscribes the unit hypercube centered at the origin. Each extreme point of the unit hypercube is the center of one facet of the n-dimensional octahedron. For example in three dimensions the point $(1/2, 1/2, 1/2)$ is the centerer of the facet whose supporting hyperplane is $x + y + z = 3/2$ (see Figure 2.2). In n dimensions the unit hypercube centered at the origin is the domain

$$K = \left\{\{x \in \mathbb{R}^n \mid -\frac{e}{2} \leq x \leq \frac{e}{2}\right\},$$

where e denotes the vector of all ones. The domain

$$K^* = \left\{x \in \mathbb{R}^n \mid \delta x \leq \frac{1}{2}n, \delta \in \{-1, 1\}^n\right\}$$

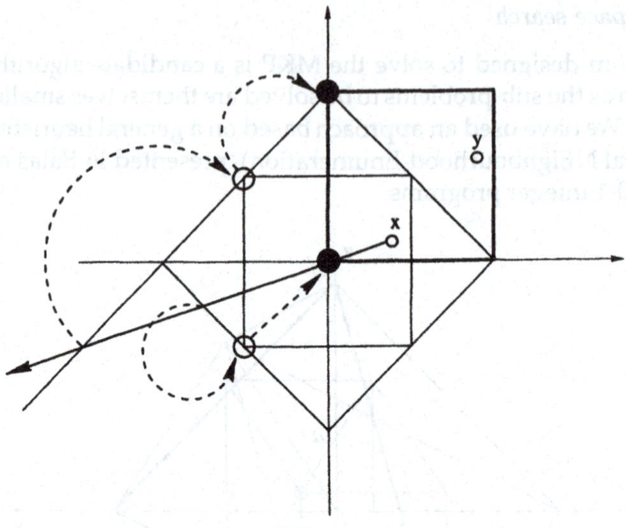

Figure 2.3 Basic idea of OCTANE.

is the regular octahedron circumscribing K. The facets of K^* are the polyhedra

$$K^*_\delta = K^* \cap \left\{ x \in \mathbb{R}^n \mid \delta x = \frac{1}{2}n \right\} \quad \delta \in \{-1,1\}^n.$$

As each facet, K^*_δ, of K^* is characterized by its supporting hyperplane given by $\delta x = \frac{1}{2}n$ we may identify the facets with the normal vectors, δ, of these supporting hyperplanes. The basic idea of OCTANE is illustrated in Figure 2.3.

From the fractional interior point, x, mentioned above a ray is chosen that cuts the extended facets of the n-dimensional octahedron (i.e. the supporting hyperplanes of the facets). There is a one-to-one correspondence between the vertices of $K + \frac{1}{2}e$ (i.e. 0-1 points in $\mathbb{R}^n$ - see the black spots of Figure 2.3), and the facets of K^*, and so the points corresponding to the first k facets intersected by the ray may serve as heuristic solutions to the problem instance. The structure of OCTANE is summarized in Figure 2.4.

The success of the method depends on the fractional point used and the direction of the ray selected. A number of suitable choices are proposed and tested by Balas et al. (2001). OCTANE has been used by Balas et al. to search subspaces corresponding to the nodes in a 0-1 B&B search tree in an attempt to determine good feasible binary solutions. As such, OCTANE is an appropriate algorithm to use when PROBE is specialized to the solution of 0-1 integer programs. The feasibility of the method relies on an efficient method for enumerating the first k facets of the octahedron intersecting with the half-line. An efficient algorithm for facet enumeration is presented by Balas et al. (2001). This algorithm is based on the reverse search paradigm of Avis and Fukada (1991). The resulting algorithm enumerates the first k extended facets in the

1. Solve the LP relaxation of the current problem to get a fractional point, x.
2. Select an appropriate direction, $a \in \mathbb{R}^n$.
3. Determine the first k facets intersecting the half-line

$$\Omega = \{x + \Lambda a | \Lambda \geq 0\}.$$

4. Take the corresponding 0-1 points as heuristic solutions.

Figure 2.4 Basic Structure of OCTANE.

order intersected by the given half-line, Ω, in time $O(kn \log k)$. The first intersected facet is found by applying a sequence of flips of bits from any reachable facet. A facet is said to be *reachable* if $\delta a > 0$, where $\delta \in \{\pm 1\}^n$ represents the facet. Starting with this first intersected facet as the root, a tree is built such that each node in the tree corresponds to a facet not considered previously. To enumerate the first k facets, it is necessary only to expand the first k nodes of this tree. A description of the tree is provided through a function, f, which, given any node in the tree, returns the predecessor of that node. Further details are given by Balas et al. (2001).

In our application of PROBE, OCTANE is used as follows: We compute a fixed number of facets in the subspace to be searched. The fractional feasible point and ray used to initialize the computation in OCTANE are respectively the solution to the LP relaxation of the MKP associated with the subspace and the objective vector of that MKP. Each facet computed gives a 0 - 1 solution vector which when supplemented with the fixed variables provides a solution to the original problem. However, the solution may be feasible or infeasible. When the solution is infeasible a repair operator is used to make it feasible. The repair operator works as follows. The variables are examined in turn according to an ordering, $(\theta_1, \ldots, \theta_n)$. Variables with value 1 are reset to zero until the solution becomes feasible. The ordering used, θ, is defined by

$$i < j \Leftrightarrow ((v(\theta_i) < v(\theta_j)) \vee (v(\theta_i) = v(\theta_j) \wedge \theta_i < \theta_j))$$

where

$$v(j) = p_j / \sum_i r_{ij}.$$

After repairing the infeasible solutions, we apply a limited local optimization (step 4 in Figure 2.1) to all solutions. The variables in a solution are examined in turn according to the ordering, $(\theta_n, \ldots, \theta_1)$ above. Variables with value 0 are set to 1 (one at a time) if the solution remains feasible. This scheme is the same as the one used by Chu and Beasley (1998) except for the ordering used. We have performed a few experiments with the ordering based on the utility ratios of Pirkul's method (Gavish and Pirkul, 1985) and used in (Chu and Beasley, 1998) but this ordering does not give any significant improvement over the one adopted. The best solution is then selected as a member of the next generation.

2.5 EXPERIMENTAL RESULTS

We have tested our algorithm with the 27 problem sets used by Chu and Beasley (1998) and available at OR-Library (Beasley, 1990). There are 10 instances per problem set. Each problem set is characterized by a number, m, of constraints, a number, n, of variables and a *tightness ratio*, t. The vector b of an MKP instance with tightness ratio t is $(b_i = t\sum_j r_{ij})_{i=1,n}$. In other words the closer to 0 the tightness ratio the more constrained the instance.

Our algorithm was applied with the following parameters used for all data sets:

- the population size is initialized to 300,
- the subspace size is equal to one third of the number of variables of the problem,
- the number of facets computed by OCTANE in the subspace search is 100,
- the algorithm stops after 10 generations without improvement of the best solution.

In Tables 2.1, 2.2 we display results for each data set as well as comparisons with Chu & Beasley's results. The group of columns labeled *Relative error, E* in Table 2.1 gives information on the percentage relative error, $E = 100 * (z_{\text{PROBE}} - z_{\text{GA}})/\max(z_{\text{PROBE}}, z_{\text{GA}})$, between the best solution value found by PROBE and the best solution value found by Chu & Beasley's GA. The column labeled *average* (respectively *st. dev, min, max*) represents the average (respectively standard deviation, minimum, max) of E over the 10 instances in the data set. The column labeled b (respectively w) represents the number of times we found a better (respectively worse) solution value than Chu & Beasley. The number of times we found the same value as Chu & Beasley is therefore $10 - b - w$.

CPU times are averaged over the instances of each data set and are given for Chu & Beasley's machine. Although comparing computer times for different types of computer is difficult, we have estimated that the machine used by Chu & Beasley is approximatively 4.7 times slower than a Dec Alpha 500/400. Accordingly, we have scaled our timings by this factor in order to be able to give some indication of relative performance. *total* gives the total execution time of the algorithm whereas *best* gives the CPU time to find the best solution returned by the algorithm.

For data sets with $(m, n) = (5, 100)$ and $(m, n) = (10, 100)$ all solutions found by Chu & Beasley but one were proved to be optimal using an integer programming solver. For those sets we also often found optimal solutions. For problems with a small number of constraints and variables the GA is slightly better than PROBE in terms of quality. For problems with a large number of

Table 2.1 Comparison between PROBE and Chu & Beasley's GA algorithm: Solution quality.

Instance			*Relative error, E*					
m	*n*	*t*	*average*	*st. dev.*	*min*	*max*	*b*	*w*
5	100	0.25	0.0317	0.0637	0.0000	0.2125	0	3
5	100	0.50	0.0195	0.0519	0.0000	0.1739	0	2
5	100	0.75	0.0008	0.0025	0.0000	0.0083	0	1
5	250	0.25	0.0390	0.0365	0.0000	0.1163	0	8
5	250	0.50	0.0076	0.0125	-0.0175	0.0254	1	7
5	250	0.75	0.0027	0.0070	-0.0121	0.0161	1	5
5	500	0.25	0.0187	0.0146	-0.0059	0.0433	1	8
5	500	0.50	0.0062	0.0053	0.0000	0.0151	0	7
5	500	0.75	0.0019	0.0017	0.0000	0.0059	0	7
10	100	0.25	0.0268	0.0708	0.0000	0.2374	0	2
10	100	0.50	0.0213	0.0325	0.0000	0.1019	0	5
10	100	0.75	0.0000	0.0000	0.0000	0.0000	0	0
10	250	0.25	0.0177	0.0773	-0.1652	0.1050	3	6
10	250	0.50	**-0.0044**	0.0293	-0.0572	0.0431	3	4
10	250	0.75	0.0080	0.0110	-0.0091	0.0261	1	6
10	500	0.25	0.0188	0.0299	-0.0125	0.0773	4	6
10	500	0.50	0.0041	0.0114	-0.0271	0.0171	1	8
10	500	0.75	**-0.0025**	0.0071	-0.0189	0.0033	2	4
30	100	0.25	0.0005	0.0014	0.0000	0.0046	0	1
30	100	0.50	**-0.0002**	0.0265	-0.0601	0.0584	1	1
30	100	0.75	**-0.0047**	0.0140	-0.0465	0.0000	1	0
30	250	0.25	**-0.0254**	0.0786	-0.1814	0.0995	4	3
30	250	0.50	**-0.0118**	0.0334	-0.0749	0.0482	4	2
30	250	0.75	**-0.0033**	0.0202	-0.0464	0.0374	5	2
30	500	0.25	**-0.0423**	0.0434	-0.1490	0.0104	8	1
30	500	0.50	**-0.0188**	0.0157	-0.0420	0.0069	8	2
30	500	0.75	**-0.0108**	0.0116	-0.0333	0.0068	8	1

Table 2.2 Comparison between PROBE and Chu & Beasley's GA algorithm: Computing time.

Instance			*CPU PROBE*		*CPU GA*	
m	*n*	*t*	*best*	*total*	*best*	*total*
5	100	0.25	90.2	304.6	9.6	345.9
5	100	0.50	73.3	247.8	23.5	347.3
5	100	0.75	36.2	175.9	26.9	361.7
5	250	0.25	301.8	805.0	50.7	682.0
5	250	0.50	250.5	667.5	276.7	709.4
5	250	0.75	203.9	528.8	195.9	763.3
5	500	0.25	724.9	1682.6	264.6	1271.9
5	500	0.50	557.5	1331.3	391.3	1345.9
5	500	0.75	412.7	1012.7	386.2	1412.6
10	100	0.25	180.4	480.6	97.5	384.1
10	100	0.50	112.1	360.3	97.3	418.9
10	100	0.75	60.4	255.8	16.8	462.6
10	250	0.25	662.0	1318.3	359.0	870.9
10	250	0.50	499.0	1031.4	342.2	931.5
10	250	0.75	269.0	684.8	128.1	1011.2
10	500	0.25	1331.5	2663.2	702.5	1504.9
10	500	0.50	990.3	2062.9	562.2	1728.8
10	500	0.75	693.9	1494.0	937.6	1931.7
30	100	0.25	292.1	888.0	177.4	604.5
30	100	0.50	293.5	782.1	118.0	782.1
30	100	0.75	121.0	499.2	90.1	904.2
30	250	0.25	1613.1	3105.7	582.9	1499.5
30	250	0.50	1252.5	2435.2	901.5	1980.0
30	250	0.75	709.3	1601.6	1059.3	2441.4
30	500	0.25	4098.8	7296.8	1127.2	2437.7
30	500	0.50	3432.0	5955.7	1121.6	3198.9
30	500	0.75	2408.9	4332.0	1903.3	3888.2

constraints, PROBE gives slightly better solutions than the GA but the PROBE computing times are slightly larger than the GA computing times on average. Computing times are however of the same magnitude.

It is interesting to observe that surprisingly enough, for given values of m and n, the more constrained instances (those with the smaller tightness ratio) seem to be less time consuming to solve than the less constrained instances when using Chu & Beasley's algorithm. A common measure of the difficulty of an instance is given by the relative gap between the optimal value of the integer program and the optimal value of its standard linear programming relaxation. According to this measure (see Chu and Beasley (1998)) the difficulty of the instances decreases as the tightness ratio increases. The PROBE computational times and results agree with this measure. A possible explanation may be that the quality of solutions produced by OCTANE depends on the quality of the continuous approximation to the problem. This also indicates that some future refinements such as considering the introduction of polyhedral cuts (knapsack cover inequalities in particular) could be useful to improve the quality of solution obtained.

2.6 CONCLUSIONS

There is very little difference in quality between the results obtained by the GA and PROBE as the maximum gap over the 270 instances is less than 0.3 percent. This is encouraging as the GA seems to be an excellent algorithm for the type of problem solved as shown by Chu & Beasley who report numerical comparisons with several heuristics from the literature (see Chu and Beasley (1998) for more details).

More generally, we can see that two major advantages of PROBE over simple GAs are fast convergence and greater effectiveness in the solution of constrained problems as appropriate search-space algorithms can be used to return feasible solutions. In particular, the difficulty encountered with the penalty methods that are used in GAs to transform a constrained optimization problem into an unconstrained problem is avoided. The speed and rate of convergence of PROBE depends on the heuristics used and on the control of the size of the subspaces. To avoid premature convergence, a fast heuristic might be used at the beginning, then, as the algorithm converges, a more exhaustive search of the subspaces could be performed.

Acknowledgments

This work was supported in part by the UK Defence Evaluation and Research Agency, Malvern, UK.

R. Aboudi and K. Jörnsten. Tabu search for general zero-one integer programs using the pivot and complement heuristic. *ORSA J. Computing*, 6:82–93, 1994.

D. Avis and K. Fukada. A pivoting algorithm for convex hulls and vertex enumerationof arrangements and polyhedra. In *Proceedings of the 7th ACM Symposium on Computational Geometry*, pages 98–104. ACM, 1991.

E. Balas. Integer programming and convex analysis: Intersection cuts from outer polars. *Mathematical Programming*, 2:330–382, 1972.

E. Balas, S. Ceria, M. Dawande, F. Margot, and G. Pataki. OCTANE: a new heuristic for zero-one programs. *Operations Research*, 49:207–225, 2001.

E. Balas and C. H. Martin. Pivot and complement – A heuristic for zero-one programming. *Management Science*, 26:86–96, 1980.

J.E. Beasley. OR-Library: Distributing test problems by electronic mail. *Journal of the Operational Research Society*, 41:1069–1072, 1990. `http://mscmga.ms.ic.ac.uk/info.html`.

P. Chardaire, G. P. McKeown, and J. A. Maki. GRASP algorithms with path relinking and PROBE for the graph bisection problem. Manuscript, 2003a.

P. Chardaire, G. P. McKeown, and J. A. Maki. GRASP algorithms with path relinking and PROBE for the multiconstraint knapsack problem. Manuscript, 2003b.

P. Chardaire, G.P. McKeown, and J. A. Maki. Application of GRASP to the 0-1 multiple knapsack problem. In E. J. W. Boers, editor, *Applications of Evolutionary Computing*, pages 30–39. Springer-Verlag LNCS 2037, 2001.

P. C. Chu and J. E. Beasley. A genetic algorithm for the multidimensional knapsack problem. *Journal of Heuristics*, 4(1):63–86, 1998.

C. Cotta and J. M. Troya. A hybrid genetic algorithm for the 0-1 multiple knapsack problem. In G. D. Smith, N. C. Steele, and R. F. Albrecht, editors, *Artificial neural nets and genetic algorithms 3*, pages 251–255. Springer-Verlag, 1998.

B. Gavish and H. Pirkul. Efficient algorithms for solving multiconstraint zero-one knapsack problems to optimality. *Mathematical Programming*, 31:78–105, 1985.

F. Glover. Surrogate constraint duality in mathematical programming. *Operations Research*, 23(3):434–453, 1975.

H. J. Greenberg and W.P. Pierskalla. Surrogate mathematical programming. *Operations Research*, 18:924–939, 1970.

S. Martello and P. Toth. *Knapsack Problems: Algorithms and Computer Implementations*. Wiley, 1990.

Z. Michalwicz. *Genetic Algorithms + Data Structures = Evolution Programs*. Springer-Verlag, 1996.

H. Pirkul. A heuristic solution procedure for the multiconstraint zero-one knapsack problem. *Naval Research Logistics*, 34:161–172, 1987.

G.R. Raidl. An improved genetic algorithm for the multiconstrained 0-1 knapsack problem. In *Proceedings of the 5th IEEE International Conference on Evolutionary Computation*, pages 207–211. IEEE, 1998.

W. Shih. A branch and bound method for the multiconstraint zero-one knapsack problem. *J. of the Operational Research Society*, 30:369–378, 1979.

Metaheuristics: Computer Decision-Making, pp. 37-63
Alexandre Belloni and Abilio Lucena

3 LAGRANGIAN HEURISTICS FOR THE LINEAR ORDERING PROBLEM

Alexandre Belloni[1] and Abilio Lucena[2]

[1] Instituto de Matemática Pura e Aplicada
Rua Dona Castorina 110, Rio de Janeiro, RJ, 22460-320, Brazil
belloni@impa.br

[2] Departamento de Administração
Universidade Federal do Rio de Janeiro
Av. Pasteur 250, Rio de Janeiro-RJ, 22290-240, Brazil
lucena@openlink.com.br

Abstract: Two heuristics for the Linear Ordering Problem are investigated in this paper. These heuristics are embedded within a Lagrangian Relaxation framework and are started with a construction phase. In this process, some Lagrangian (dual) information is used as an input to guide the construction of initial Linear Orderings. Solutions thus obtained are then submitted to local improvement in an overall procedure that is repeated throughout Subgradient Optimization. Since a very large number of inequalities must be dualized in this application, Relax and Cut is used instead of a straightforward implementation of the Subgradient Method. Heuristics are tested for instances from the literature and also for some new *hard to solve exactly* ones. From the results obtained, one of the proposed heuristics has shown to be competitive with the best in the literature. In particular, it generates optimal solutions for all 79 instances taken from the literature. As a by-product, it also proves optimality for 72 of them.
Keywords: Linear ordering problem, Lagrangian relaxation, Relax and cut.

3.1 INTRODUCTION

A *linear ordering* (or a permutation) of the n elements of a finite set N is a bijective mapping $\pi : \{1, 2, \ldots, n\} \rightarrow N$. For an element $i \in N$ and a linear ordering π, the *position* of i in the ordering is given by $\pi^{-1}(i)$. Assume that a *weight* $d(\pi)$ is associated with π, as follows. For a given set of profits $\{c_{ij} : i, j \in N, i \neq j\}$, every pair of distinct elements $i, j \in N$ must contribute to $d(\pi)$

with either c_{ij} or c_{ji}. If $\pi^{-1}(i) < \pi^{-1}(j)$, i.e. if i precedes j in the ordering, then c_{ij} is incurred. Otherwise, c_{ji} is incurred. The Linear Ordering Problem (LOP) consists in finding an ordering π of N maximizing $d(\pi)$.

LOP can be cast in graph theoretical terms by assuming the elements of N to be the nodes of a complete directed graph $D = (N, A)$. Accordingly, $A = \{(i, j) : i, j \in N, i \neq j\}$ is defined as the arc set and c_{ij} is the profit associated with the use of arc $(i, j) \in A$. A subgraph $D' = (N, A')$, $A' \subseteq A$, denotes a tournament if, for every pair of distinct nodes $i, j \in N$, exactly one of the arcs, (i, j) or (j, i), is contained in A'. An acyclic tournament D' with the largest sum of arc profits implies an optimal LOP solution to N.

LOP has a number of relevant practical applications. Among these, undoubtedly, the most commonly cited one is the *triangulation* of input-output matrices in Economics (see Wessels (1981)). Applications of LOP to Ranking by Paired Comparison, Minimization of Total Completion Time in One-Machine Scheduling, and Chronological Ordering of Pottery Types are cited in Grötschel et al. (1984) where references to additional applications can be found.

An exact solution algorithm for the LOP, based on the use of polyhedral cutting planes, is found in Grötschel et al. (1984). More recently, Mitchell and Borchers (1997) proposed an exact solution algorithm where Linear Programming (LP) relaxations are partially solved by an Interior Point algorithm (complemented, in the end, with the Simplex Method). Families of facet defining inequalities for the LOP polytope can be found in Grötschel et al. (1985), Reinelt (1985), McLennan (1990), Reinelt (1993), Leung and Lee (1994), and Bolotashvili et al. (1999).

Some recent LOP heuristics are presented in Chanas and Kobylánski (1996), Laguna et al. (1999), Campos et al. (1999) and González and Pérez-Brito (2001). The first one uses a sorting through insertion pattern together with permutation reversals. The second one uses Tabu Search (see Glover and Laguna (1997)). Scatter Search (see Glover (1998)) is used in Campos et al. (1999). Finally, a Variable Neighborhood Search (see Mladenović and Hansen (1997)) is used in González and Pérez-Brito (2001).

Two different heuristics for LOP are proposed and computationally tested here. Heuristics are embedded within a Lagrangian Relaxation (LR) framework and are repeatedly called under different inputs generated from Lagrangian relaxation outputs (either the Lagrangian modified profits or the Lagrangian subproblem solutions). As a result, a series of feasible LOP solutions are generated based on information obtained from the LR upper bounds. On implementing such a scheme, a very large number of inequalities would normally have to be dualized (in order to generate Lagrangian LOP upper bounds). This, in turn, usually brings convergence problems to a straightforward implementation of, say the Subgradient Method (SM) of Held et al. (1974) (see Section 3.2 for details). As a result, a Relax and Cut algorithm, in the style proposed in Lucena (1992; 1993), is used instead.

Heuristics proposed in this study have been tested for the 49 instances of LOLIB (see Reinelt (1997)), the 30 instances proposed in Mitchell and Borchers (1997) and 21 new instances introduced here. LOLIB instances originate from the input-output matrices for some European countries. They range in size from 44 up to 60 economic sectors. Instances in Mitchell and Borchers (1997) were randomly generated to resemble input-output matrices. They range in size from 100 up to 250 elements. Finally, the third test set was randomly generated with the aim of obtaining *hard to solve exactly* LOP instances. Instances in this set range in size from 30 up to 500 elements.

This paper is organized as follows. In Section 3.2, Relax and Cut is briefly reviewed. In Section 3.3, Lagrangian Relaxation LOP upper bounds are discussed. Lagrangian based LOP heuristics are described in Section 3.4. In Section 3.5, we attempt to solve to optimality some test instances with Mixed Integer Programming (MIP) solver CPLEX 7.1 (2001). This is done for setting up some reference basis when evaluating the results obtained with the proposed heuristics (see Section 3.6). Finally, some concluding remarks in Section 3.7 close the paper.

3.2 RELAX AND CUT ALGORITHMS

Assume that a (binary 0-1, for simplicity) formulation of an $\mathcal{NP}$-hard combinatorial optimization problem is given. Assume as well that exponentially many inequalities may be included in it. Such a formulation can be generically described as

$$\max\{cx: \ Ax \leq b, \ x \in X\}, \tag{3.1}$$

where $x \in \mathbb{B}^m$ is a vector of variables, $c \in \mathbb{R}^m$, $b \in \mathbb{R}^p$, $A \in \mathbb{R}^{p \times m}$ and, finally, $X \subseteq \mathbb{B}^m$ represents additional families of inequalities that define a *convenient to work with* polytope (such as the spanning tree polytope, the assignment problem polytope, etc). As a result, assume, as it is usual in Lagrangian relaxation, that

$$\max\{cx: \ x \in X\} \tag{3.2}$$

is an easy problem to solve, i.e. it can be solved in polynomial-time. On the other hand, in a way that is unusual for the application of Lagrangian relaxation, let p be exponential in m. In spite of that, assume one wishes to dualize

$$\{a_i x \leq b_i: \ i = 1, 2, \ldots, p\} \tag{3.3}$$

in a Lagrangian fashion and let $\lambda \in \mathbb{R}^p_+$ be the corresponding vector of Lagrangian multipliers. Subgradient Optimization (SO) could then be used to solve

$$\min_{\lambda \in \mathbb{R}^p_+} \left\{ \max\{(c - \lambda A)x + \lambda b : \ x \in X\} \right\}. \tag{3.4}$$

Optimization is typically conducted here in an interactive way with multipliers being updated so that the optimal value of (3.4) is attained. For the sake of completeness, let us briefly review how the Subgradient Method (SM) of Held et al. (1974), as implemented in Fisher (1981), attempts to solve (3.4).

At any given iteration of the SM, for given feasible values of the Lagrangian multipliers λ, let $\overline{x}$ be an optimal solution to

$$\max\{(c - \lambda A)x + \lambda b : \ x \in X\}. \tag{3.5}$$

Denote by z_{ub} the corresponding solution value and let z_{lb} be a known lower bound on (3.1). Additionally, let $g \in \mathbb{R}^p$ be a vector of subgradients associated with the relaxed constraints. Given an optimal solution $\overline{x}$ to (3.5), subgradients g are evaluated as

$$g_i = (b_i - a_i\overline{x}), \quad i = 1, 2, \ldots, p. \tag{3.6}$$

In the literature (see Fisher (1981), for instance) Lagrangian multipliers are usually updated by firstly determining a "step size" θ,

$$\theta = \frac{\alpha(z_{ub} - z_{lb})}{\sum_{i=1,\ldots,p} g_i^2}, \tag{3.7}$$

where α is a real number assuming values in $(0, 2]$. One would then proceed to computing

$$\lambda_i \equiv \max\{0; \lambda_i - \theta g_i\}, \quad i = 1, \ldots, p, \tag{3.8}$$

and move on to the next iteration of the SM.

Under the conditions imposed here, a straightforward implementation of updating formulas (3.6)–(3.8) is not as simple as it might appear. The reason being the exceedingly large number of inequalities that, typically, would have to be dualized (recall that p is assumed to be exponential in m).

For every SM iteration, inequalities (3.3) may be divided into three groups. The first one contains inequalities that are violated by optimal solution $\overline{x}$ to (3.5) and is denoted *group one*. The second group, denoted *group two*, is for those inequalities that have nonzero multipliers currently associated with them. Notice that an inequality may be, simultaneously, in the groups one and two. Finally, the third group, denoted *group three*, consists of the remaining inequalities and evaluating their subgradients would account for most of the computational burden at a SM iteration.

One should notice that inequalities may change groups from one SM iteration to another. It should also be pointed out that the only multipliers that may directly contribute to the Lagrangian profits $(c - \lambda A)$, at any given SM iteration, are the ones associated with inequalities in groups one and two. These inequalities are thus named *active inequalities*. Conversely, we denote inequalities in group three *inactive inequalities*. Finally, it is important to stress that, from (3.8), multipliers for inequalities in group three will not be changing their null values at the end of the current SM iteration.

Clearly, inactive inequalities will not directly contribute to Lagrangian profits (since their multipliers will remain zero valued at the end of the current SM iteration). On the other hand, they do play a decisive role in determining the value of θ. Typically, for our problem, the number of strictly positive subgradients, amongst inequalities in group three, tends to be huge. Consequently, the value of θ would, typically, result extremely small, leaving multiplier values virtually unchanged from iteration to iteration. Bearing this in mind, one may choose to apply (3.6)–(3.8) exclusively to active inequalities. This was first suggested in Lucena (1992; 1993) and is implemented here. That results in a dynamic scheme where the set of active inequalities may continuously change. Notice, in association, that an inequality may become active at one given SM iteration, then become inactive at a subsequent one and become, yet again, active at a later iteration.

One should notice that Relax and Cut, as explained above, can be viewed as an extension, to exponentially many inequalities, of a scheme suggested in Beasley (1993), for updating multipliers.

Relax and Cut is, in a sense, very much akin to cutting planes generation. In the form described above, it has been used for the Steiner Problem in Graphs in Lucena (1992; 1993), for the Edge-Weighted Clique Problem in Hunting et al. (2001), for the Vehicle Routing Problem in Martinhon et al. (2001) and for the Rectangular Partitioning Problem in Calheiros et al. (2003).

The term *Relax and Cut* was coined in Escudero et al. (1994) for an algorithm that, although different from the one in Lucena (1992; 1993), shares a number of points in common with it. In this paper, we use the term Relax and Cut in a broader sense to denote the whole class of LR algorithms where inequalities are dualized *on the fly* (as they become violated at the solution to a LR subproblem). This class therefore encompasses the algorithm in Lucena (1992; 1993) as well as the one in Escudero et al. (1994).

3.3 LAGRANGIAN RELAXATION UPPER BOUNDS

We have used the Linear Integer Programming formulation of LOP found in Grötschel et al. (1984). It involves a set of binary $0-1$ variables $\{x_{ij} : (i,j) \in A\}$, used to impose an ordering precedence between any two distinct elements i and j of N. Accordingly, for any $i, j \in N$, $i \neq j$, whenever i precedes j, then both $x_{ij} = 1$ and $x_{ji} = 0$ must hold. Otherwise, $x_{ij} = 0$ and $x_{ji} = 1$ must hold. Consider, in association, a polyhedral region, $\mathcal{R}$, described as

$$x_{ij} + x_{ji} = 1, \text{ for all } i, j \in N, i \neq j \tag{3.9}$$

$$x_{ij} + x_{jk} + x_{ki} \leq 2, \text{ for all } (i,j), (j,k), (k,i) \in A \tag{3.10}$$

$$0 \leq x_{ij} \leq 1, \text{ for all } (i,j) \in A. \tag{3.11}$$

A valid formulation for LOP (see Grötschel et al. (1984)) is then given by

$$\max \{ \sum_{(i,j)\in A} c_{ij} x_{ij} : x \in \mathcal{R} \cap \mathbb{Z}^{|A|} \}, \tag{3.12}$$

where (3.9) enforces a tournament to be formed while the 3-Cycle inequalities (3.10) ensure the tournament to be cycle free. Attaching multipliers $\lambda_{ijk} \geq 0$ to the inequalities in (3.10) and denoting by $\mathcal{R}_L$ the polyhedral region described by (3.9) and (3.11), a valid upper bound for LOP is given by

$$UB(\lambda) = \max \{ \sum_{(i,j)\in A} c'_{ij} x_{ij} + 2 \times \sum_{i,j,k\in N, i\neq j, i\neq k, j\neq k} \lambda_{ijk} : x \in \mathcal{R}_L \cap \mathbb{Z}^{|A|} \}, \tag{3.13}$$

where $\{c'_{ij} = c_{ij} - \sum_{k\in N\setminus\{i,j\}} \lambda_{ijk} : (i,j) \in A\}$ is the set Lagrangian modified profits. An optimal solution to (3.13) is obtained by setting to one, for every pair $i, j \in N, i \neq j$, that variable, x_{ij} or x_{ji}, with the largest Lagrangian profit (the remaining variable being set to zero). Ties are broken arbitrarily. Denoting by $m = \binom{|N|}{3}$, the number of inequalities in (3.10) to dualize, the best possible Lagrangian bound for this application is

$$\min_{\lambda\in \mathbb{R}^m_+} UB(\lambda) \tag{3.14}$$

and can, in principle, be attained by using Subgradient Optimization methods (see Held et al. (1974) for SM and Bonnans et al. (1997) for Bundle Methods). Notice, however, that an instance of LOP with $|N| = 500$ elements would have as many as 20,708,499 3-Cycle inequalities to dualize. As explained before, explicitly dualizing so many inequalities would jeopardize convergence of, say a straightforward implementation of SM. As a compromise, although the number of inequalities to dualize does not grow exponentially with $|N|$, we propose to use Relax and Cut.

In Section 3.2, a generic description of a Relax and Cut algorithm for Combinatorial Optimization problems was given. In what follows that algorithm is specialized for LOP.

3.3.1 *Relax and cut for LOP*

In our implementation of a Relax and Cut algorithm for LOP, only 3-Cycle inequalities (3.10) are dualized. Preliminary computational results indicated that, at an optimal solution $\overline{x}$ to (3.13), the number of violated 3-Cycle inequalities is typically very large. This appears in agreement with what has been reported for LP based LOP upper bounding algorithms. For these algorithms 3-Cycle inequalities are normally introduced as cutting planes. Faced with too many violated inequalities at a given reoptimization round, only a fraction of these (i.e. the most violated ones) are normally used as cutting planes (see, for instance, Mitchell and Borchers (1997)). In our context, unfortunately, a similar approach can not be pursued since all 3-Cycle inequalities that violate $\overline{x}$ do so by the same amount (i.e. one unit).

For an optimal solution $\overline{x}$ to (3.13), consider all violated 3-Cycle inequalities that currently have a zero valued multiplier associated with them. As pointed out above, a very large number of such inequalities typically exist. To keep down the number of active inequalities, only at most $\eta > 0$ group one inequalities per variable are introduced into the set of active inequalities (as they are identified) at a given SM iteration. Remaining group one inequalities are simply ignored. Acting as suggested above tends to evenly spread active inequalities amongst variables. Overall this appears to be an adequate strategy that does not seem to negatively affect dual bound quality.

Let $\zeta \geq 1$ be the maximum number of iterations allowed for SM. In our computational experiments, the value of parameter α (see Section 3.2) is initially set at 2.0 and is halved after $\lfloor \zeta/20 \rfloor$ iterations without an overall improvement on the best dual bound (upper bound for LOP) so far generated. All other aspects of our LOP Relax and Cut algorithm follow the generic algorithm of Section 3.2.

3.4 LAGRANGIAN BASED HEURISTICS

Two basic heuristics for LOP are introduced next. Their use as stand alone, single pass procedures, is not very attractive. Nevertheless, solution quality tends to be greatly enhanced by using *local search* procedures in addition. This overall scheme, i.e. a basic LOP heuristic followed by local search, tends to be enhanced even further, as we shall see later, when it is allowed to be called more than once (under conveniently modified input profits). In particular, we use dual information (from Lagrangian relaxation) to modify input profits. Accordingly, primal bound generation is incorporated within the LR framework of Section 3.3. In such a scheme, LOP heuristics are called, unless specified otherwise, for every iteration of the SM, either under Lagrangian modified profits or else using, directly, LR solutions as an input. Feasible LOP solutions, thus obtained, are attempted to be improved through local search.

3.4.1 Procedure Position Profit

For any set $S \subset N$, denote by $\overline{S}$ the complement of S over N and compute, for every $j \in \overline{S}$, *position profits* $q_j = \sum_{k \in \overline{S} \setminus \{j\}} c_{jk}$. The basic idea behind the first heuristic, named *Position Profit* (PP), is to use position profits to sequentially build a linear ordering of the elements of N. Accordingly, at step $p = |S| + 1$ of the procedure (note that S is the empty set for $p = 1$), the elements in $S \subset N$ would have been placed in the first $p - 1$ ordering positions. At this stage, that element j, $j \in \overline{S}$, for which $q_j \geq q_k$, for all $k \in \overline{S} \setminus \{j\}$, is to be placed at the p-th ordering position. The step is concluded by setting $S := S \cup \{j\}$. The procedure is finished after $|N|$ steps and has complexity $O(|N|^2)$.

For every call of PP, Lagrangian modified profits are used to compute position profits. Once a feasible solution is generated (from the position profits), one resorts back to the original LOP profits and applies local improvement procedures (see Subsection 3.4.3).

3.4.2 Procedure Node Degree

The second heuristic denoted Node Degree (ND) uses, as input, LR solutions (instead of Lagrangian modified profits). For a given iteration of the SM, let λ be the corresponding set of Lagrangian multipliers and assume $\overline{x}$ to be an optimal solution to (3.13). Clearly, such a solution must be a tournament, though not necessarily an acyclic one. For this tournament compute, for every node $i \in N$, the out degree $\sum_{(i,j) \in A} \overline{x}_{ij}$ (i.e. the number of tournament arcs pointing out of i). Notice that if $\overline{x}$ were an acyclic tournament (and therefore a feasible solution to LOP), exactly one of the nodes in N would have an out degree of $|N| - 1$. In a generalization, exactly one other node in N would have an out degree of $l \in \{0, \ldots, |N| - 2\}$. The basic idea behind this heuristic is then to (linear) order the elements of N in decreasing value of their $\overline{x}$ out-degrees (ties are broken arbitrarily). Heuristic ND has complexity $O(|N|^2)$.

For every call of ND, after a feasible LOP solution is generated from $\overline{x}$, local improvement procedures are applied under the original LOP profits.

3.4.3 Local Search

Let π be a feasible LOP solution. The concept of k-optimality (see Lin and Kerningham (1973)) can be applied over π in an attempt to generate better quality LOP solutions. In a *2-opt* move, two distinct elements of N, say i and j, are to exchange their positions in π, provided that results in an overall increase in ordering weight. In a *3-opt* move, 3 distinct elements of N, say i, j and l, are to exchange their positions in π, if an overall increase in ordering weight results. Given i, j and l, only two possibilities of a 3-opt exchange exist that are not themselves simply 2-opt moves. Clearly, the concept can be generalized into a *k-opt* move, $4 \leq k \leq |N|$, involving k distinct elements of N.

Bearing in mind the high computational cost of either 2-opt or 3-opt, one should consider restricting the associated search neighborhoods, by reducing the number of potential candidates for change. To illustrate the idea, consider a linear ordering π of N, an element $i \in N$ (placed in position $\pi^{-1}(i) = p, 1 \leq p \leq |N|$) and an integral valued parameter $\delta > 0$ (used to reduce, in a controlled way, the size of the search). In a restricted 2-opt move, one searches for candidate elements to exchange (ordering) positions with i in the range of positions defined by $[\max\{1, p - \delta\}, \min\{|N|, p + \delta\}]$. Amongst competing candidates, that element j leading to the largest overall increase in (linear ordering) weight is chosen for the exchange. Search neighborhoods for 3-opt moves, involving element $i \in N$, are defined likewise.

Another improvement procedure, denoted here as *sequence shifts*, has proved to be a better alternative to either 2-opt or 3-opt (in terms of both CPU time and bound quality). Assume that a linear ordering π of N is given and consider an element $i \in N$ such that $\pi^{-1}(i) = p, 1 \leq p \leq |N|$. The basic idea of a sequence shift operation is to find the best position in π to move element i to. As an illustration, assume that an overall increase in ordering weight results if i is moved to, say position $(p + l)$, where $l > 0$. On implementing this move, the sequence of elements in positions $(p + 1)$ through to $(p + l)$ must be shifted one position backwards. Accordingly, if i is moved, instead, to position $(p - l)$, the sequence of elements placed in positions $(p - l)$ through to $(p - 1)$ must be shifted one position forward.

Sequence shifts can be implemented quite efficiently by dividing the associated search neighborhood in two. The first half involves those positions located ahead of p. The second one is for positions placed behind p. Let $d(\pi)$ be the weight of π and consider first the case where one is searching amongst positions located ahead of p. For the simplest possible scenario, assume that element j is currently placed at position $(p+1)$. Moving i one position forward to $(p+1)$ while shifting j one position backwards to p, leads to a new ordering π^1 with a weight $d(\pi^1) = d(\pi) - c_{ij} + c_{ji}$. Consider now the option of moving i, not to position $(p+1)$, but to position $(p+2)$ (where element l is currently assumed to be placed). This action would result in a shift of the sequence of elements placed in positions $(p+1)$ and $(p+2)$, one position backwards (j being moved to p while l is moved to $p+1$). Denoting by π^2 the resulting ordering, notice that $d(\pi^2) = d(\pi^1) - c_{il} + c_{li} = d(\pi) - c_{ij} - c_{il} + c_{ji} + c_{li}$. In general terms, assume that i is to be moved $k \geq 2$ positions ahead, to a position currently occupied by l. The weight $d(\pi^k)$ of the resulting ordering π^k can then be computed as $d(\pi^k) = d(\pi^{k-1}) - c_{il} + c_{li}$. Sequence shifts are thus computed, one at a time, using as input the results from the previous operation. It should be noticed that a procedure, similar to the one outlined above, can be devised for a search involving positions located behind p. Among all possibilities investigated, that position (if any) leading to the largest overall increase in ordering weight is chosen for i to be moved to. Implemented as outlined above, sequence shifts are computationally cheap and no restrictions on the associated search neighborhood need be imposed.

The idea of sequence shifts can be generalized by attempting to move, in one piece, two or more contiguously placed elements of an ordering π. The implementation suggested above for single element moves can be readily adapted for this general case.

A combination of heuristic PP with 2-opt moves has been used in Grötschel et al. (1984). Single element sequence shifts, under a different name, have been used in Campos et al. (1999) and Laguna et al. (1999).

For the computational results in Section 3.6, only sequence shifts are used. Moves involving from one up to seven contiguously placed elements are implemented. The upper limit of seven was reached after some computational experimentation and represents a good trade-off between solution quality and CPU time requirements.

3.5 TEST INSTANCES

To test the proposed heuristics, three different sets of LOP instances are used. As referred above, two of these come from the literature while the third one is introduced in this study. Solver CPLEX 7.1 (2001) is used in an attempt to solve some of these instances to proved optimality. CPLEX 7.1 generates, if necessary, Gomory cuts (see Gomory (1960)), at the root node of the enumeration tree. As a result, the bound thus produced dominates the one given by the LP relaxation of (3.12).

Before moving on to computational experiments, it is important to highlight a detail that is associated with LOP duality gaps. Percentage duality gaps for LOP, as formulated in (3.12), are always no larger than the corresponding duality gaps derived from a reformulation of LOP (to be explained next) obtained directly from (3.12). Duality gaps associated with this reformulation can thus be argued to be more acceptable parameters to measure problem difficulty.

Let LOP be modeled as in (3.12). Notice from (3.9) that variables x_{ji} with $j > i$ can be rewritten as $x_{ji} = 1 - x_{ij}$. In this process, objective function (3.12) becomes

$$\sum_{i,j \in N, i<j} c_{ji} + \sum_{i,j \in N, i<j} (c_{ij} - c_{ji}) x_{ij} \tag{3.15}$$

where

$$\sum_{i,j \in N, i<j} c_{ji} \tag{3.16}$$

is a constant. Typically (3.16) tends to be a very large proportion of the overall optimal LOP solution values. In general, LP based exact solution approaches for LOP use the variable reformulation above. As a result, percentage duality gaps between optimal solution values and their LP relaxation counterparts tend to be masked by constant (3.16). Duality gaps that include (3.16) represent an "underestimate" of the inherent difficulty of solving exactly an LOP instance. It thus appears more appropriate to compute duality gaps that exclude constant (3.16). In this section, duality gaps are computed both ways. This gives a more sound account of the true difficulty of solving a given LOP instance to proved optimality. Instances acknowledged as being difficult to solve exactly are prime candidates for applying approximate solution approaches.

3.5.1 The LOLIB instances

The first test set contains the 49 instances from LOLIB. These instances relate to input-output matrices compiled for the economies of different European countries. For this test set, $44 \leq |N| \leq 60$.

Solver CPLEX 7.1 (2001), under default settings (except for parameter `mingap`) which was lowered to 10^{-6}), was used in an attempt to generate proven optimal solutions for all LOLIB instances. LOP upper bounds are generated from the LP relaxation of (3.12) (all 3-Cycle inequalities included right from the start) reinforced, if necessary, with Gomory cuts. Average results, obtained on a Pentium III machine, running at 933 MHz, are shown on Table 3.1 (CPU times quoted in seconds).

As one may appreciate from the results on Tables 3.1, all LOLIB instances are solved exactly under fairly low CPU times. With the exception of two instances, the remaining ones are solved at the root node of the enumeration tree. Furthermore, the two instances possessing a duality gap (after Gomory cuts were introduced) required only three branch and bound nodes to have

Table 3.1 CPLEX statistics for LOLIB instances: average results

$\|N\|$	b&b nodes	CPU time
48.1	1.08	3.99

optimality proven. As far as the LP relaxation of (3.12) is concerned, only 5 LOLIB instances do not have naturally integral relaxations.

In view of the results quoted above, one may conclude that the LOLIB instances are *quite easy* to solve exactly.

3.5.2 Instances proposed by Mitchell and Borchers

The second test set contains instances randomly generated in Mitchell and Borchers (1997) to resemble input-output matrices. The 30 instances in this set range in size from 100 up to 250 elements. Attempting to solve them exactly, just as done before for the LOLIB instances, although possible, does not seem a very attractive proposition. Model (3.12) would typically involve a large number of 3-Cycle inequalities. In this case, a more appropriate course of action would be to initially exclude all 3-Cycle inequalities from the LP relaxation of (3.12) and only introduce them, as cutting planes, once they become violated. This is done in Mitchell and Borchers (1997) (in a more elaborated way) where an Interior Point LP solver is combined with CPLEX 4.0 to obtain the LP relaxation of (3.12). This combination has proved a much more effective alternative to either the Interior Point algorithm or CPLEX 4.0 acting in isolation.

In what follows, we quote, mostly, the computational results in Mitchell and Borchers (1997). These were obtained on a Sun SPARC 20/71 machine (CPU times in seconds). All instances were solved to proven optimality. Only instances r100b2 and r200d1 do not possess a naturally integral LP relaxation of (3.12). Solution times ranged from 65 seconds for instance r150e0 up to 1666 seconds for instance r200d1. Duality gaps for r100b2 are 0.0043%, including (3.16), and 0.0136%, excluding (3.16). Corresponding values for r200d1 are, respectively, 0.00039% and 0.0014%.

As one may appreciate from the results quoted above, instances in Mitchell and Borchers (1997) do not differ fundamentally from the LOLIB ones. They invariably have very strong LP relaxations of (3.12). Duality gaps exist for only a very small proportion of the instances involved and are not very large. On the other hand, instances in Mitchell and Borchers (1997) have larger dimensions. That fact and also having to deal with highly degenerated LP relaxations of (3.12) constitute the main challenges in solving the Mitchell and Borchers instances to proven optimality. Overall, these instances can be considered much more difficult to solve exactly than the LOLIB ones.

Table 3.2 CPLEX statistics for Euclidean instances

instance	$\|N\|$	optimal	% gap1	% gap2	b&b nodes	CPU time
e30A.mat	30	17490	0.0	0.0	1	1.49
e30B.mat	30	18116	0.0	0.0	1	4.78
e30C.mat	30	25369	0.0	0.0	1	1.31
e50A.mat	50	46439	0.67	3.52	759	70018
e50B.mat	50	$\leq$ 58656	$\leq$ 1.51	$\leq$ 9.7	> 700	> 128346

3.5.3 Randomly generated LOP instances

Based on the LP relaxation bounds obtained in Mitchell and Borchers (1997) and those reported in Table 3.1, duality gaps appear to be almost always equal zero for input-output matrices (or for instances randomly generated to resemble them). By contrast, instances to be introduced next tend to have *large* duality gaps and appear to be *difficult* to solve exactly. This applies even for $|N| \leq 50$. They are generated as follows. One firstly defines the number n of elements in N, an integrally valued perturbation parameter $\epsilon > 0$ and the fraction $\mu \in (0, 1]$ of variables with nonzero profits. Then, n distinct points, within a square of side l (lying on the Euclidean plane), are randomly generated. Each of these points is associated with a different element of N. Euclidean distances are then computed for every pair of points. Let $d(i, j)$ be the (rounded off) Euclidean distance for points associated, respectively, with elements i and j. Profits c_{ij} and c_{ji} are then computed as $c_{ij} = d(i, j) + \epsilon_{ij}$ and $c_{ji} = d(i, j) + \epsilon_{ji}$, where ϵ_{ij} and ϵ_{ji} are randomly generated in the range $[0, \epsilon]$. The fraction $(1 - \mu)$ of variables with zero valued profits is sequentially met as follows. Assume a pair of distinct points i and j is randomly chosen. If fraction $(1-\mu)$ has not yet been reached, profit c_{ij} is set to 0 and the procedure is repeated. Otherwise, a valid LOP instance has been generated.

Test instances are generated by setting $l = 100$, $\epsilon = 30$ and using 3 different sets of values for $(1 - \mu)$. Series A instances have $(1 - \mu) = 0.6$. Series B instances have $(1 - \mu) = 0.4$. Finally, series C instances have $(1 - \mu) = 0.2$.

Once again, solver CPLEX 7.1 (2001) is used in an attempt to generate optimal solutions for the instances described above. Table 3.2 shows the results obtained. Only those instances with $|N| = 30$ were found to be easy to solve exactly. Instances with as little as $|N| = 50$ turned out unexpectedly hard to solve. Experiments for larger instances indicate that difficulty appears to increase with the increase in instance dimension. It is thus reasonable to conclude that (3.12) must be strengthened with additional families of strong valid inequalities if optimality is to be attained for instances with $|N| \geq 50$.

From the computational evidence obtained, one may conclude that LOP instances introduced in this study are *hard to solve exactly*. They therefore constitute a good test bed for LOP heuristics.

3.6 COMPUTATIONAL EXPERIMENTS

Feasible LOP solutions are obtained by firstly defining the maximum number of iterations allowed for the Subgradient Method (SM) for each of the two proposed heuristics. For all test instances with $|N| < 300$, heuristics are called for every Relax and Cut iteration, for runs with ζ fixed at, respectively, 1, 200, and 3000 (see Subsection 3.3.1). One additional run with $\zeta = 400$ was performed for heuristic ND under the instances in Mitchell and Borchers (1997). For instances with $|N| \geq 300$, heuristics are only called whenever an overall improvement on the dual bound occurs during a Relax and Cut run. In all experiments, η (i.e. the maximum number of group one inequalities, dualized at any given SM iteration) is initially fixed at 15 (see yet again Subsection 3.3.1). Once a gap of less than one percent is reached between upper and lower bounds, that value is set to $\eta = 100 + \lfloor \zeta/100 \rfloor$. Obviously, whenever LOP optimality could be proven before the iterations limit is reached, the run was immediately stopped. Given that, for all test instances, profits are integral valued, one possibility for proving optimality is to reach a difference of less than one unit between upper and lower bounds. Another (more rare) possibility arises when, at a given iteration of the SM, an acyclic tournament turns out to be an optimal solution to (3.13). In this case, if the corresponding Lagrangian upper bound equals the tournament weight under the original profits, optimality is again proven.

Tables 3.3 through 3.8 summarize the results obtained. Tables 3.3 and 3.4 are for the LOLIB instances. Tables 3.5 and 3.6 are for the instances introduced in Mitchell and Borchers (1997). Finally, Tables 3.7 and 3.8 are for the instances proposed in this study. For any of these tables, entries give, respectively, the maximum number of SM iterations allowed, the average percentage gap between optimal solution values and best primal bounds generated, number of instances for which optimal feasible solutions are obtained, number of instances for which optimality was proven, and the average CPU times involved (in seconds of a Pentium III 933 MHz machine). All the coding was done in C++. For the test set introduced in this study, optimal solution values are known for only 4 instances. For the remaining instances, some experiments were conducted with heuristic ND, for $\zeta = 20000$, in order to generate *high quality* upper bounds on their corresponding optimal solution values. The upper bounds thus obtained replace optimal solution values, whenever applicable, for the statistics quoted above.

Tables 3.9 through 3.14 provide more detailed statistics on the data from which Tables 3.3 through 3.8 were compiled. For each instance set, tables come in pairs (Tables 3.9 and 3.10 for LOLIB instances, Tables 3.11 and 3.12 for the instances in Mitchell and Borchers (1997), and Tables 3.13 and 3.14 for the instances proposed in this study).

Entries for the first three columns in Tables 3.9, 3.11, and 3.13 give, for each different instance set, respectively, instance identification, number of elements in N, and optimal solution value. Likewise, entries for columns four and five give the difference (in number of units) between optimal solution value and corresponding lower bounds generated, respectively, by heuristics PP and ND, running for 1 SM iteration. The same applies, for columns six and seven, for 200 iterations of SM and, for columns eight and nine, for 3000 SM iterations. The following six columns give the CPU times, in seconds, associated, respectively, with each of the previous six columns. Differently from Tables 3.9 and 3.13, Table 3.11 has two extra columns. The first one is for the the difference between optimal solution value and the corresponding lower bound generated by heuristic ND, running for 400 SM iterations. The second column is for the associated CPU time. The reason for including these additional columns in Table 3.11 is that 400 SM iterations, for the Lagrangian relaxation parameter settings we use, is a good upper bound on the least number of SM iterations required by heuristic ND to generate optimal solutions for every instance in Mitchell and Borchers (1997).

Entries for the first three columns in Tables 3.10, 3.12, and 3.14 carry the same information as their counterparts in Tables 3.9, 3.11, and 3.13. Entries for columns four and five in Tables 3.10, 3.12, and 3.14 give the difference (in number of units) between the Lagrangian relaxation upper bounds generated in association, respectively, with heuristics PP and ND, running for 1 SM iteration, and the corresponding optimal solution value. The same applies for columns six and seven, for 200 SM iterations, and for columns eight and nine, for 3000 iterations. Differently from Tables 3.10 and 3.14, Table 3.12 has one extra column. That column gives the difference (in number of units) between the Lagrangian relaxation upper bound generated in association with heuristic ND, running for 400 SM iterations, and the corresponding optimal solution value.

From the results obtained, the two different heuristics appear to perform very well for the LOP instances considered. However ND appears to have a clear edge over PP (generates better quality LOP solutions in less CPU time). In particular, ND with ζ fixed at 200, generates optimal solutions for all LOLIB instances. For the parameter settings used, $\zeta = 400$ is a very good upper bound on the least number of Relax and Cut iterations necessary to obtain similar results for the instances in Mitchell and Borchers (1997). In relation with the instances introduced in this study, little can be said. Optimal solution values are known for only 4 out of the 21 instances involved. Furthermore, duality gaps appear to be very large and, therefore, Lagrangian relaxation upper bounds ought to be far from their corresponding optimal solution values. For the 4 instances where optimal solutions are known, ND with $\zeta = 3000$ managed to find them all.

Dual bounds generated by Relax and Cut also proved very good. They equaled optimal LOP solution values (rounding off dual bounds, if necessary) for 72 out of the 79 combined LOLIB and Mitchell and Borchers (1997) instances. Furthermore, for the remaining seven instances, a duality gap (between the LP relaxation of (3.12) and the corresponding optimal solution value) is known to exist. Therefore, best possible Lagrangian bounds (3.14), for these instances, are guaranteed to be above their optimal solution values.

Table 3.3 LOLIB instances: PP

iterations	% PP-gap	nb. opts	nb. opt. proofs	CPU time
1	0.07154	17	0	0.03
200	0.01028	41	1	1.50
3000	0.00443	42	37	8.25

Table 3.4 LOLIB instances: ND

iterations	% ND-gap	nb. opts	nb. opt. proofs	CPU time
1	0.09933	15	0	0.03
200	0.00000	49	3	1.43
3000	0.00000	49	44	6.15

In our computational experiments, at most 6, 560 inequalities, out of a maximum of over 34, 220 inequalities in (3.10), were explicitly dualized, at any iteration of Relax and Cut, for any of the LOLIB instances. For the instances in Mitchell and Borchers (1997), the corresponding figures are, respectively, 94, 442 and 2, 576, 000. Finally, for the instances introduced in this study, the figures are 210, 000 and 20, 708, 499. Figures quoted above, for number of dualized 3−Cycle inequalities, for the first two sets of test instances (for which LP relaxation bounds and optimal solution values are known), represent a clear certificate of success for Relax and Cut. This applies since best possible upper and lower bounds were both attained by Relax and Cut (while dualizing only a fraction of all existing 3-Cycles inequalities).

The CPU time advantage of ND over PP, for the LOLIB and Mitchell and Borchers (1997) instances, is easy to explain. That follows from the fact that SM iteration runs are stopped as soon as optimality proofs are attained. From the results in Tables 3.10 and 3.12, it is clear that the good quality feasible solutions returned by ND speed up convergence to optimal Lagrangian dual bounds. These bounds, as pointed out above, equal their corresponding optimal solution values for almost every LOLIB and Mitchell and Borchers (1997) instance. Therefore, typically, less SM iterations are performed for ND, thus resulting in smaller CPU times for that heuristic.

For the test instances introduced in this study, a substantial gap must exist between best possible possible Lagrangian dual bounds and their corresponding optimal solution values. Therefore, for these instances, SM iterations could not be cut short by optimality proofs.

3.7 CONCLUSIONS

Two Lagrangian based LOP heuristics are proposed in this study. Heuristics are embedded within a Lagrangian Relax and Cut framework where some lo-

Table 3.5 Instances from Mitchell and Borchers (1997): PP

iterations	% PP-gap	nb. opts	nb. opt. proofs	CPU time
1	0.01622	0	0	0.17
200	0.00163	9	6	29.73
3000	0.00118	17	17	257.36

Table 3.6 Instances from Mitchell and Borchers (1997): ND

iterations	% ND-gap	nb. opts	nb. opt. proofs	CPU time
1	0.01219	2	0	0.20
200	0.00034	24	10	25.76
400	0.00000	30	21	35.60
3000	0.00000	30	28	54.01

Table 3.7 New instances: PP

iterations	% PP-gap	nb. opts	nb. opt. proofs	CPU time
1	0.55591	1	0	0.42
200	0.13129	1	0	64.86
3000	0.06128	2	2	1009.08

Table 3.8 New instances: ND

iterations	% ND-gap	nb. opts	nb. opt. proofs	CPU time
1	0.63228	0	0	0.39
200	0.08788	2	0	63.95
3000	0.03622	6	2	989.60

Table 3.9 Detailed results for LOLIB instances: feasible solutions (continues on next page).

Instance	$\|N\|$	*Optimal*	*PP-1*	*ND-1*	*PP-2*	*ND-2*	*PP-3*	*ND-3*	*T-PP-1*	*T-ND-1*	*T-PP-2*	*T-ND-2*	*T-PP-3*	*T-ND-3*
be75eec	50	264940	464	177	0	0	0	0	0.04	0.04	1.51	1.42	5.06	4.03
be75np	50	790966	551	3	0	0	0	0	0.04	0.03	1.45	1.53	22.06	19.69
be75oi	50	118159	170	157	6	0	6	0	0.04	0.03	1.81	1.62	22.54	6.08
be75tot	50	1127387	1124	0	0	0	0	0	0.04	0.04	1.43	1.38	3.46	2.98
stabu1	60	422088	3	73	0	0	0	0	0.04	0.05	3.01	2.73	31.11	30.50
stabu2	60	627929	350	952	0	0	0	0	0.05	0.04	2.92	3.13	21.32	18.22
stabu3	60	642050	1791	1894	801	0	801	0	0.05	0.04	2.77	2.79	39.25	11.83
t59b11xx	44	245750	595	595	580	0	15	0	0.04	0.04	0.97	1.04	12.97	10.73
t59d11xx	44	163219	903	725	0	0	0	0	0.04	0.03	1.03	0.96	1.85	1.87
t59f11xx	44	140678	0	0	0	0	0	0	0.04	0.03	0.99	0.97	2.74	2.72
t59i11xx	44	9182291	12869	0	0	0	0	0	0.04	0.03	1.06	0.95	15.94	8.71
t59n11xx	44	25225	0	35	0	0	0	0	0.04	0.03	1.17	1.04	1.27	1.58
t65b11xx	44	411733	347	407	0	0	0	0	0.04	0.03	1.12	0.97	3.57	2.40
t65d11xx	44	283971	634	634	0	0	0	0	0.04	0.03	1.14	1.15	3.71	3.73
t65f11xx	44	254568	25	53	25	0	25	0	0.04	0.04	1.21	1.00	11.23	1.42
t65i11xx	44	16389651	4578	3612	0	0	0	0	0.04	0.04	1.02	1.07	1.79	1.84
t65l11xx	44	18359	0	0	0	0	0	0	0.04	0.04	0.91	0.93	0.81	0.82
t65n11xx	44	38814	61	96	0	0	0	0	0.04	0.04	1.12	0.99	3.48	2.78
t65w11xx	44	166052789	149593	0	0	0	0	0	0.04	0.04	1.05	1.06	2.59	1.48

Table 3.9 Detailed results for LOLIB instances: feasible solutions (continued from previous page).

Instance	$\|N\|$	*Optimal*	*PP-1*	*ND-1*	*PP-2*	*ND-2*	*PP-3*	*ND-3*	*T-PP-1*	*T-ND-1*	*T-PP-2*	*T-ND-2*	*T-PP-3*	*T-ND-3*
t69r11xx	44	865650	0	2943	0	0	0	0	0.04	0.04	1.05	1.01	2.54	2.69
t70b11xx	44	623411	326	326	250	0	250	0	0.04	0.04	0.99	1.02	9.61	1.33
t70d11xn	44	438235	0	251	0	0	0	0	0.04	0.03	1.12	0.97	12.46	9.96
t70d11xx	44	450774	190	496	0	0	0	0	0.04	0.03	1.21	1.16	1.73	1.48
t70f11xx	44	413948	0	1175	0	0	0	0	0.04	0.03	1.13	0.56	0.94	0.54
t70i11xx	44	28267738	20326	1382	0	0	0	0	0.04	0.04	0.93	1.04	1.35	1.23
t70k11xx	44	69796200	0	0	0	0	0	0	0.04	0.04	0.99	0.92	3.50	3.23
t70l11xx	44	28108	0	80	0	0	0	0	0.03	0.04	1.02	1.06	3.02	3.06
t70n11xx	44	63944	78	0	0	0	0	0	0.04	0.04	1.05	0.98	2.23	1.80
t70u11xx	44	27296800	8100	8100	8100	0	8100	0	0.04	0.04	1.03	1.08	10.10	11.41
t70w11xx	44	267807180	0	0	0	0	0	0	0.04	0.04	1.09	0.96	1.06	0.96
t70x11xx	44	343471236	0	276740	0	0	0	0	0.03	0.04	1.12	0.71	0.94	0.71
t74d11xx	44	673346	471	471	0	0	0	0	0.03	0.04	1.23	1.11	1.80	1.66
t75d11xx	44	688601	49	3111	0	0	0	0	0.03	0.03	1.33	1.15	4.31	4.19
t75e11xx	44	3095130	3752	1372	0	0	0	0	0.03	0.04	1.14	1.00	2.58	3.56
t75i11xx	44	72664466	0	0	0	0	0	0	0.04	0.04	1.14	1.11	1.11	1.10
t75k11xx	44	124887	0	0	0	0	0	0	0.03	0.04	1.07	1.03	1.77	1.70
t75n11xx	44	113808	0	0	0	0	0	0	0.03	0.04	1.26	1.10	2.76	2.41
t75u11xx	44	63278034	52959	0	0	0	0	0	0.04	0.04	1.31	1.26	12.72	11.65
tiw56n54	56	112767	8	6	2	0	2	0	0.04	0.04	2.49	2.52	29.57	12.31
tiw56n58	56	154440	127	110	0	0	0	0	0.04	0.04	1.90	1.92	14.90	12.95
tiw56n62	56	217499	0	0	0	0	0	0	0.04	0.04	2.01	1.96	4.93	4.95
tiw56n66	56	277593	418	88	0	0	0	0	0.05	0.04	2.26	2.19	5.09	6.14
tiw56n67	56	277962	591	0	0	0	0	0	0.04	0.04	2.11	2.13	7.51	7.76
tiw56n72	56	462991	82	82	0	0	0	0	0.04	0.03	2.49	2.23	10.92	9.02
tiw56r54	56	127390	29	143	0	0	0	0	0.04	0.04	2.83	2.42	20.00	17.73
tiw56r58	56	160776	0	220	0	0	0	0	0.04	0.04	2.19	2.33	9.08	14.72
tiw56r66	56	256326	156	187	145	0	0	0	0.04	0.04	1.96	2.01	2.51	2.46
tiw56r67	56	270497	0	1376	0	0	0	0	0.04	0.04	2.37	2.15	4.34	3.60
tiw56r72	56	341623	0	0	0	0	0	0	0.04	0.04	2.30	2.27	12.30	11.91

Table 3.10 Detailed results for LOLIB instances: Lagrangian bounds

Instance	$\|N\|$	*Optimal*	*PP-1*	*ND-1*	*PP-2*	*ND-2*	*PP-3*	*ND-3*
be75eec	50	264940	7786	7786	45	27	0	0
be75np	50	790966	25636	25636	508	516	27	27
be75oi	50	118159	2823	2823	111	105	0	0
be75tot	50	1127387	45488	45488	334	510	0	0
stabu1	60	422088	25736	25736	1106	792	6	2
stabu2	60	627929	33382	33382	499	724	0	0
stabu3	60	642050	35383	35383	798	947	0	0
t59b11xx	44	245750	11015	11015	17	54	12	12
t59d11xx	44	163219	6080	6080	36	51	0	0
t59f11xx	44	140678	5668	5668	35	35	0	0
t59i11xx	44	9182291	279139	279139	2486	2579	1	0
t59n11xx	44	25225	873	873	1	6	0	0
t65b11xx	44	411733	17511	17511	146	204	0	0
t65d11xx	44	283971	11902	11902	143	188	0	0
t65f11xx	44	254568	9798	9798	10	56	0	0
t65i11xx	44	16389651	789554	789554	565	860	0	0
t65l11xx	44	18359	219	219	1	1	0	0
t65n11xx	44	38814	1528	1528	18	19	0	0
t65w11xx	44	166052789	6330004	6330004	12719	635	0	0
t69r11xx	44	865650	25082	25082	286	76	0	0
t70b11xx	44	623411	22586	22586	10	128	0	0
t70d11xn	44	438235	16842	16842	84	107	8	8
t70d11xx	44	450774	23570	23570	48	52	0	0
t70f11xx	44	413948	18779	18779	1	0	0	0
t70i11xx	44	28267738	1268664	1268664	4075	1849	0	0
t70k11xx	44	69796200	2327200	2327200	12395	12395	0	0
t70l11xx	44	28108	335	335	3	3	0	0
t70n11xx	44	63944	2034	2034	63	23	0	0
t70u11xx	44	27296800	974800	974800	0	2508	0	0
t70w11xx	44	267807180	9820323	9820323	39419	39419	0	0
t70x11xx	44	343471236	12361817	12361817	93	0	0	0
t74d11xx	44	673346	28990	28990	12	27	0	0
t75d11xx	44	688601	33019	33019	169	140	0	0
t75e11xx	44	3095130	77644	77644	1599	1606	0	0
t75i11xx	44	72664466	3590183	3590183	0	0	0	0
t75k11xx	44	124887	4196	4196	13	13	0	0
t75n11xx	44	113808	3374	3374	20	20	0	0
t75u11xx	44	63278034	2688365	2688365	5612	8593	57	3
tiw56n54	56	112767	4726	4726	63	61	0	0
tiw56n58	56	154440	5598	5598	65	76	0	0
tiw56n62	56	217499	8800	8800	46	46	0	0
tiw56n66	56	277593	10754	10754	55	74	0	0
tiw56n67	56	277962	13005	13005	129	101	0	0
tiw56n72	56	462991	19264	19264	209	276	0	0
tiw56r54	56	127390	5859	5859	123	110	0	0
tiw56r58	56	160776	6356	6356	73	81	0	0
tiw56r66	56	256326	9657	9657	4	10	0	0
tiw56r67	56	270497	12491	12491	111	77	0	0
tiw56r72	56	341623	14627	14627	248	248	0	0

Table 3.11 Detailed results for Mitchell and Borchers (1997) instances: feasible solutions

Instance	*|N|*	*Optimal*	*PP-1*	*ND-1*	*PP-2*	*ND-2*	*PP-3*	*ND-3*	*T-PP-1*	*T-ND-1*	*T-PP-2*	*T-ND-2*	*T-PP-3*	*T-ND-3*	*ND-4*	*T-ND-4*
r100a2	100	197652	43	70	19	0	19	0	0.07	0.09	6.23	5.93	86.46	21.52	0	11.42
r100b2	100	197423	132	93	26	7	26	0	0.07	0.10	6.54	6.03	91.78	80.76	0	11.49
r100c2	100	193952	227	42	6	6	0	0	0.06	0.08	6.89	6.55	58.66	21.98	0	12.43
r100d2	100	196397	92	35	0	0	0	0	0.06	0.08	7.09	6.52	12.32	11.53	0	11.48
r100e2	100	200178	18	0	1	0	1	0	0.06	0.07	6.83	6.26	85.09	15.09	0	11.93
r150a0	150	550666	44	16	0	0	0	0	0.10	0.13	11.52	8.36	11.53	8.42	0	8.35
r150b0	150	554338	47	0	0	0	0	0	0.11	0.14	10.54	8.40	10.62	8.90	0	8.92
r150c0	150	551451	6	30	0	0	0	0	0.11	0.14	9.30	9.27	9.30	9.31	0	9.48
r150d0	150	552772	12	10	1	0	1	0	0.11	0.14	14.28	9.35	206.71	9.38	0	9.32
r150e0	150	554400	4	69	0	0	0	0	0.12	0.14	13.18	10.20	13.07	10.32	0	10.09
r150a1	150	504308	115	127	1	0	0	0	0.11	0.14	17.06	16.20	59.55	25.49	0	25.76
r150b1	150	500841	17	60	2	7	2	0	0.10	0.14	15.52	14.67	220.54	28.78	0	28.10
r150c1	150	500757	210	224	40	0	0	0	0.10	0.13	17.27	16.00	86.24	19.19	0	18.59
r150d1	150	501372	77	158	1	0	0	0	0.10	0.13	16.58	15.24	34.01	22.05	0	21.87
r150e1	150	501422	59	42	0	0	0	0	0.12	0.14	17.25	16.60	34.75	32.01	0	31.97
r200a0	200	989422	33	40	0	0	0	0	0.19	0.24	35.62	31.66	35.29	33.03	0	37.88
r200b0	200	984081	16	66	10	0	0	0	0.20	0.26	33.76	28.57	140.71	28.75	0	28.78
r200c0	200	990568	80	9	5	0	5	0	0.19	0.25	32.15	30.43	484.04	36.66	0	37.80
r200d0	200	989123	43	45	1	0	0	0	0.19	0.26	36.63	31.77	88.09	27.80	0	28.11
r200e0	200	980354	22	22	0	0	0	0	0.20	0.24	34.45	22.43	34.05	22.12	0	22.52
r200a1	200	889222	108	114	9	0	9	0	0.19	0.23	41.17	36.58	591.19	39.56	0	40.36
r200b1	200	893867	54	35	7	3	0	0	0.20	0.25	38.31	35.95	224.76	78.30	0	69.47
r200c1	200	882945	15	105	12	0	12	0	0.19	0.27	38.87	35.07	533.69	73.25	0	68.21
r200d1	200	888562	187	95	32	9	32	0	0.20	0.25	41.44	38.30	603.52	492.95	0	72.94
r200e1	200	883948	64	195	29	6	29	0	0.21	0.25	40.72	36.70	575.97	69.73	0	64.34
r250a0	250	1545431	144	29	2	0	0	0	0.36	0.39	78.19	60.54	129.33	65.32	0	66.34
r250b0	250	1538410	134	126	10	0	7	0	0.33	0.37	69.63	60.50	993.88	145.33	0	119.79
r250c0	250	1534036	68	2	0	0	0	0	0.35	0.42	43.23	41.39	43.32	41.66	0	41.93
r250d0	250	1540117	166	46	12	0	12	0	0.36	0.40	73.42	62.93	1148.38	68.30	0	65.25
r250e0	250	1531709	129	97	7	0	7	0	0.37	0.42	78.27	64.48	1074.17	72.89	0	73.14

Table 3.12 Detailed results for Mitchell and Borchers (1997) instances: Lagrangian bounds

Instance	|*N*|	*Optimal*	*PP-1*	*ND-1*	*PP-2*	*ND-2*	*PP-3*	*ND-3*	*ND-4*
r100a2	100	197652	22391	22391	43	75	0	0	10
r100b2	100	197423	23048	23048	187	204	8	8	45
r100c2	100	193952	22661	22661	124	98	0	0	17
r100d2	100	196397	23702	23702	65	48	0	0	0
r100e2	100	200178	22930	22930	119	121	0	0	5
r150a0	150	550666	30167	30167	0	0	0	0	0
r150b0	150	554338	28737	28737	0	0	0	0	0
r150c0	150	551451	29454	29454	0	0	0	0	0
r150d0	150	552772	29217	29217	0	0	0	0	0
r150e0	150	554400	28776	28776	0	0	0	0	0
r150a1	150	504308	41475	41475	28	30	0	0	0
r150b1	150	500841	42257	42257	27	28	0	0	3
r150c1	150	500757	41328	41328	3	7	0	0	0
r150d1	150	501372	42823	42823	16	21	0	0	0
r150e1	150	501422	42025	42025	19	22	0	0	0
r200a0	200	989422	49964	49964	2	6	0	0	0
r200b0	200	984081	52802	52802	0	0	0	0	0
r200c0	200	990568	50032	50032	5	3	0	0	0
r200d0	200	989123	52369	52369	0	0	0	0	0
r200e0	200	980354	51829	51829	0	0	0	0	0
r200a1	200	889222	75402	75402	26	12	0	0	0
r200b1	200	893867	76370	76370	37	47	0	0	1
r200c1	200	882945	73551	73551	62	42	0	0	4
r200d1	200	888562	75016	75016	170	159	3	3	27
r200e1	200	883948	77113	77113	61	53	0	0	0
r250a0	250	1545431	81723	81723	2	0	0	0	0
r250b0	250	1538410	82871	82871	5	6	0	0	2
r250c0	250	1534036	83790	83790	0	0	0	0	0
r250d0	250	1540117	82128	82128	2	1	0	0	0
r250e0	250	1531709	82063	82063	1	2	0	0	0

Table 3.13 Detailed results for Euclidean instances: feasible solutions

Instance	$\|N\|$	*Optimal*	*PP-1*	*ND-1*	*PP-2*	*ND-2*	*PP-3*	*ND-3*	*T-PP-1*	*T-ND-1*	*T-PP-2*	*T-ND-2*	*T-PP-3*	*T-ND-3*
e30A	30	17490	0	62	0	0	0	0	0.05	0.05	0.76	0.72	6.39	5.76
e30B	30	18116	182	175	25	2	10	0	0.05	0.04	0.67	0.70	8.11	8.15
e30C	30	25369	16	124	13	0	0	0	0.03	0.03	0.81	0.78	4.13	2.03
e50A	50	46439	193	850	8	18	8	0	0.04	0.04	1.61	1.53	22.72	33.68
e50B	50	59019	517	331	167	91	73	0	0.05	0.05	1.59	1.54	23.83	23.34
e50C	50	70806	457	759	57	12	12	0	0.05	0.05	1.62	1.65	40.84	43.60
e70A	70	84351	760	1264	85	123	38	58	0.06	0.06	3.42	3.39	55.87	55.49
e70B	70	110229	736	384	28	110	117	76	0.05	0.05	3.33	3.35	55.77	54.17
e70C	70	138210	350	780	228	168	72	1	0.06	0.06	3.58	3.52	56.86	55.20
e100A	100	164215	1613	1136	544	335	137	139	0.07	0.08	7.23	7.03	120.76	118.35
e100B	100	230155	1547	2051	321	337	241	62	0.07	0.07	6.97	7.40	127.28	125.35
e100C	100	292941	2227	1636	511	259	218	154	0.07	0.07	7.86	7.67	125.49	122.80
e200A	200	615836	5021	4772	1811	820	689	592	0.23	0.22	42.55	40.43	755.39	745.49
e200B	200	870321	4609	3183	1505	873	820	414	0.24	0.27	41.64	41.85	801.16	775.06
e200C	200	1114539	5116	3509	893	1038	160	336	0.24	0.23	46.35	45.88	799.69	782.83
e300A	200	615836	5021	4772	1811	820	689	592	0.24	0.22	44.73	42.72	793.46	782.20
e300B	200	870321	4609	3183	1505	873	820	414	0.24	0.27	42.39	42.09	810.16	787.81
e300C	200	1114539	5116	3509	893	1038	160	336	0.24	0.24	46.83	45.87	807.80	791.70
e500A	500	3678742	16138	7991	2191	4052	4800	2785	2.44	2.10	355.90	346.75	5232.69	5099.40
e500B	500	5345728	11396	11362	4956	3458	2407	0	1.95	2.09	342.24	340.02	5195.70	5002.61
e500C	500	6925960	13142	8927	1459	496	574	1236	2.34	1.95	360.03	358.17	5346.66	5366.64

Table 3.14 Detailed results for Euclidean instances: Lagrangian bounds

Instance	|*N*|	*Optimal*	*PP-1*	*ND-1*	*PP-2*	*ND-2*	*PP-3*	*ND-3*
e30A	30	17490	3163	3163	120	156	0	0
e30B	30	18116	3702	3702	251	226	43	47
e30C	30	25369	2851	2851	37	56	0	0
e50A	50	46439	11235	11235	1404	1463	419	408
e50B	50	59019	12421	12421	1190	1404	633	666
e50C	50	70806	10716	10716	871	791	300	317
e70A	70	84351	24484	24484	4937	4932	1866	1908
e70B	70	110229	26816	26816	4520	4841	2319	2749
e70C	70	138210	20430	20430	2926	2850	1559	1521
e100A	100	164215	52873	52873	13625	13624	6372	5825
e100B	100	230155	59003	59003	12578	12790	6734	6717
e100C	100	292941	45353	45353	9151	9072	5449	5271
e200A	200	615836	244262	244262	88576	85120	44500	44260
e200B	200	870321	264386	264386	89720	89791	50275	51423
e200C	200	1114539	204592	204592	64302	63748	38115	37944
e300A	200	615836	244262	244262	88576	85120	44500	44260
e300B	200	870321	264386	264386	89720	89791	50275	51423
e300C	200	1114539	204592	204592	64302	63748	38115	37944
e500A	500	3678742	1750752	1750752	1266814	1263200	1256497	1257307
e500B	500	5345728	1880330	1880330	1353748	1365235	1337037	1342172
e500C	500	6925960	1431253	1431253	1045085	1037485	1037934	1034727

cal search procedures are also used. Very good quality solutions have been generated for LOP instances associated with input-output matrices. Computational experiments with some new, randomly generated LOP instances (introduced in this paper), appear to indicate these instances are *hard to solve exactly*.

The proposed heuristics present some particular features that may be used to advantage. In particular, dual bounds are generated together with primal ones. Combining these two bounds, an indication of solution quality is obtained. Furthermore, primal bounds appear to benefit from the good quality dual bounds that are generated. In this respect, a point to mention is the fact that additional families of valid LOP inequalities could be incorporated into our Relax and Cut framework. Indeed, as explained in Section 3.2, exponentially many such inequalities may be treated as suitable candidates to Lagrangian dualization. Since a number of different families of facet defining inequalities for the LOP polytope have already been proposed in the literature, a potential exists for further improvements in the dual bounds. Such an improvement, as stressed above, should have a positive impact on primal bound quality.

To our knowledge, the best LOP heuristics in the literature are the ones in Laguna et al. (1999) and González and Pérez-Brito (2001). A comparison of their's with our best heuristic, i.e. ND, must be based on the LOLIB instances, since this is the only common test set for all three procedures. In the comparisons that follow, we use ND results obtained for $\zeta = 200$.

Heuristic ND found optimal solutions for all LOLIB instances. The heuristic in Laguna et al. (1999), found optimal solutions for 47 out of the 49 instances tested. The heuristic in González and Pérez-Brito (2001) found optimal solutions for 44 instances. Average CPU times quoted in Laguna et al. (1999) and González and Pérez-Brito (2001) are, respectively, 0.93 and 1.11 seconds of Pentium based machines running at 166 MHz. Heuristic ND took considerably longer, with an average of 1.43 seconds of a Pentium III based machine running at 933 MHz. In summary ND, produced better quality solutions in higher CPU times.

In an overall assessment, it can be said that ND is competitive with the best in the literature. Furthermore it has the nice feature of producing dual bounds as well primal ones, thus allowing for a measure of solution quality to be available while running the algorithm.

A final point, which is related with future work we intend to do, addresses the results obtained for the instances introduced in this study. To improve the results obtained for these instances, a strengthening of the Lagrangian dual bounds with strong valid inequalities must be carried out. To do so, we plan to investigate the use, within a Relax and Cut framework, of some of the LOP facet defining inequalities cited in Section 3.1. We expect that better quality dual bounds will lead to better quality primal ones.

Acknowledgments

Research supported by Fundação de Amparo a Pesquisa do Estado do Rio de Janeiro (FAPERJ) under grant E26/71.906/00.

J.E. Beasley. Lagrangean relaxation. In C. Reeves, editor, *Modern Heuristics*. Blackwell Scientific Press, 1993.

G. Bolotashvili, M. Kovalev, and G. Girlich. New facets of the linear ordering polytope. *SIAM J. Discrete Math.*, 3:326–336, 1999.

J. F. Bonnans, J.Ch. Gilbert, C. Lemaréchal, and C. Sagastizábal. *Optimisation numérique: aspects théoriques et pratiques*. Springer Verlag, 1997.

F. Calheiros, A. Lucena, and C. de Sousa. Optimal rectangular partitions. *Networks*, 41:51–67, 2003.

V. Campos, M. Laguna, and R. Marti. Scatter search for the linear ordering problem. In D. Come, M. Dorigo, and F. Glover, editors, *New Ideas in Optimization*, pages 331–339. Mc-Graw Hill, 1999.

S. Chanas and P. Kobylánski. A new heuristic algorithm solving the linear ordering problem. *Computational Optimization and Applications*, 6:191–205, 1996.

L. Escudero, M. Guignard, and K. Malik. A lagrangian relax and cut approach for the sequential ordering with precedence constraints. *Annals of Operations Research*, 50:219–237, 1994.

M. L. Fisher. The lagrangian relaxation method for solving integer programming problems. *Mangement Science*, 27:1–18, 1981.

F. Glover. A template for scatter search and path relinking. In J.-K. Hao, E. Lutton, E. Ronald, M. Schoenauer, and D. Snyers, editors, *Artificial Evolution*, Lecture Notes in Computer Science 1363, pages 13–54. Springer, 1998.

F. Glover and M. Laguna. *Tabu Search*. Kluwer Academic Publisher, 1997.

R. E. Gomory. An algorithm for the mixed integer problem. Technical report, The Rand Corporation, 1960.

C. G. González and D. Pérez-Brito. A variable neighborhood search for solving the linear ordering problem. In *Proceedings of MIC'2001-4th Metahruristics International Conference*, pages 181–185, 2001.

M. Grötschel, M. Jünger, and G. Reinelt. A cutting plane algorithm for the linear ordering problem. *Operations Research*, 32:1195–1220, 1984.

M. Grötschel, M. Jünger, and G. Reinelt. Facets of the linear ordering polytope. *Mathematical Programming*, 33:43–60, 1985.

M. Held, P. Wolfe, and H. P. Crowder. Validation of subgradient optimization. *Mathematical Programming*, 6:62–88, 1974.

M. Hunting, U. Faigle, and W. Kern. A lagrangian relaxation approach to the edge-weighted clique problem. *European Journal on Operational Research*, 131: 119–131, 2001.

ILOG, Inc. *CPLEX, version 7.1*, 2001.

M. Laguna, R. Marti, and V. Campos. Intensification and diversification with elite tabu search solutions for the linear ordering problem. *Computers and Operations Research*, 26:1217–1230, 1999.

J. Leung and J. Lee. More facets from fences for linear ordering and acyclic subgraphs polytopes. *Discrete Applied Mathematics*, 50:185–200, 1994.

S. Lin and B. W. Kerningham. An effective heuristic for the traveling salesman problem. *Operations Research*, 21:498–516, 1973.

A. Lucena. Steiner problem in graphs: Lagrangean relaxation and cutting-planes. *COAL Bulletin*, 21:2–8, 1992.

A. Lucena. Tight bounds for the steiner problem in graphs. In *Proceedings of NETFLOW93*, pages 147–154, 1993.

C. Martinhon, A. Lucena, and N. Maculan. Stronger *k*-tree relaxations for the vehicle routing problem. Technical report, PESC-COPPE, Universidade Federal do Rio de Janeiro, 2001. accepted for publication in European Journal on Operational Research.

A. McLennan. Binary stochastic choice. In J. S. Chipman, D. McFadden, and M. K. Richter, editors, *Preferences, Uncertainty and Optimality*. Westview Press, 1990.

J. E. Mitchell and B. Borchers. Solving linear ordering problems with a combined interior point/simplex cutting plane algorithm. Technical report, Rensseler Institute, 1997.

N. Mladenović and P. Hansen. Variable neighborhood search. *Computers and Operations Research*, 24:1097–1100, 1997.

G. Reinelt. *The Linear Ordering Problem: Algorithms and Applications*. Heldermann Verlag, Berlin, 1985.

G. Reinelt. A note on small linear ordering polytope. *Discrete Computational Geometry*, 10:67–78, 1993.

G. Reinelt. Lolib. Internet repository, 1997. http://www.iwr.uni-heildelberg.de/iwr/comopt/soft/LOLIB/LOLIB.html.

H. Wessels. Triagulation und blocktriangulation von input-output-tabellen. Technical report, Deutches Institut für Wirtschaftsforschung, 1981.

Metaheuristics: Computer Decision-Making, pp. 65-90
Regina Berretta, Carlos Cotta, and Pablo Moscato

4 ENHANCING THE PERFORMANCE OF MEMETIC ALGORITHMS BY USING A MATCHING-BASED RECOMBINATION ALGORITHM

Regina Berretta[1], Carlos Cotta[2], and Pablo Moscato[1]

[1]School of Electrical Engineering and Computer Science
Faculty of Engineering and Built Environment
The University of Newcastle
Callaghan, 2308 NSW, Australia
regina@cs.newcastle.edu.au, moscato@cs.newcastle.edu.au

[2]Departamento de Lenguajes y Ciencias de la Computación
ETSI Informática (3.2.49)
Universidad de Málaga, Campus de Teatinos
29071 - Málaga, Spain
ccottap@lcc.uma.es

Abstract: The NUMBER PARTITIONING PROBLEM (MNP) remains as one of the simplest-to-describe yet hardest-to-solve combinatorial optimization problems. In this paper we use the MNP as a surrogate for several related real-world problems, to test new heuristics ideas. To be precise, we study the use of weight-matching techniques to devise *smart* memetic operators. Several options are considered and evaluated for that purpose. The positive computational results indicate that – despite the MNP may be not the best scenario for exploiting these ideas – the proposed operators can be really promising tools for dealing with more complex problems of the same family.
Keywords: Memetic algorithms, Tabu search, Number partitioning problem, Weight matching.

4.1 INTRODUCTION AND MOTIVATION

The MIN NUMBER PARTITIONING problem has been one of the hardest challenges for metaheuristics for at least a decade. It was originally the paper by Johnson et al. (1991) that first identified the problem that *Simulated Annealing* (Kirkpatrick et al., 1983) (SA henceforth), a metaheuristic of pristine prestige among physicists, was having to address this problem. At the beginning of the past decade, this was thought to be a peculiar characteristic since SA was viewed as a powerful method. Currently, although the prestige of SA has somewhat declined, the problem has remained to be an open challenge for other metaheuristics like *Genetic Algorithms* (GAs) (Jones and Beltramo, 1991; Ruml, 1993), SA (Johnson et al., 1991; Sorkin, 1992; Ruml et al., 1996), *problem space local search* (Storer et al., 1996), GRASP (Arguello et al., 1996), or Tabu Search (Glover and Laguna, 1997). The decision version is widely cited as being a conspicuous member of the NP-complete class, one of the "six-essential" NP-complete problems. Moreover, the problem has another source of interest if we also have in mind that is essentially equivalent to finding the ground-state of an infinite range Ising spin-glass system with antiferromagnetic couplings. As a consequence, we can think of this problem as being a *worst-case* scenario (Mertens, 2000) among the tasks of finding the ground-state of a disordered system (Laguna and Laguna, 1995; Ferreira and Fontanari, 1998; Mertens, 1998). Other interesting problems from which reductions to NUMBER PARTITIONING exist are the balancing of rotor blades and cargo loading in aircrafts (Storer, 2001), and the assignment of tasks in low-power application-specific integrated circuits (Kirovski et al., 1999).

We can define the problem as:

Input: A set A of n positive integer numbers $\{a_1, \ldots, a_n\}$.
Question: Is there a partition of A, i.e., two disjoint sets A_1 and A_2 with $A = A_1 \bigcup A_2$, such that

$$\sum_{a_i \in A_1} a_i = \sum_{a_j \in A_2} a_j \ ? \tag{4.1}$$

We will denote this problem as NUMBER PARTITIONING (D), the bracketed D indicating that it is a *decision* problem, i.e., for every *instance* the (unique) answer is either 'Yes' or 'No'. Associated with this problem there is a combinatorial *optimization* search problem or *optimization version* (denoted NUMBER PARTITIONING (O) or MIN NUMBER PARTITIONING — MNP for short). This related problem can be viewed as the task of finding a set $y = \{v_1, \ldots, v_n\}$, where v_i can be either 1 or -1, such that y minimizes the following objective function (a *cost* function in this case):

$$m_P(y, A) = \left| \sum_{i=1}^{n} a_i v_i \right| . \tag{4.2}$$

Though the problem is easy to state, this optimization version is deceptively hard to solve. Several frustrated attempts to classify what makes a problem *"hard"* for GAs have been made and this also applies to the whole field of Evolutionary Computation. Extending this concern to the field of metaheuristics, from a scientific point of view, it is frustrating to see that most results report "successful" applications of a certain technique while many *"negative"* results and failures very seldom reach a published status. This problem is notably an exception, and hence constitutes an ideal battle-ground for testing and comparing different metaheuristics. More precisely, we propose the use of ideas taken from weight-matching to understand some of the associated issues, using MNP as a surrogate for some of the optimization problems mentioned above.

4.2 THE KARMARKAR-KARP HEURISTIC

Johnson et al. (1991) compare the bad performance of SA for MIN NUMBER PARTITIONING with two other problems: MIN GRAPH COLORING and MIN GRAPH PARTITIONING. The authors conclude:

> *"The results for number partitioning were, as expected, decidedly negative, with annealing substantially outperformed by the much faster Karmarkar-Karp algorithm, and even beaten (on a time-equalized basis) by multiple-start local optimization (MSLO)."*

The Karmarkar-Karp (Karmarkar and Karp, 1982) heuristic (KKH) is a constructive heuristic for the MNP that works by marking the two largest numbers to belong to two different subsets, replacing them (in the set of numbers yet to be marked) by their difference, and repeating the process until only one number is left. The remaining number is the value of the resulting partition of the original set. To recover the corresponding partitions as well, this basic scheme must be augmented so as to keep track of the successive groupings performed, finally yielding a tree. The coloring of this tree (done with just two colors as it is straightforward to see) results in the precise partitions produced by the algorithm (see complete pseudocode in Algorithm 4.1).

As an example, Figure 4.1 shows the results for the set of integers {205, 157, 133, 111, 100, 91, 88, 59, 47, 23}. According to the KKH, the numbers 205 and 157 must be assigned to different partitions, and so should 133 and 111. The process is repeated until only one number remains: the value of the partition. If we continue the example, we will have 6 as the weight of the resulting partition.

4.2.1 *A crash introduction to "phase transitions"*

An important optimal algorithm that uses KKH is the Complete Karmarkar Karp (CKK) method proposed by Korf (1998). CKK is an exact anytime algorithm which takes advantage of the KKH.

The CKK algorithm is a hard adversary for metaheuristics, particularly due to a phenomenon that in the Artificial Intelligence literature is generally cited

```
procedure KK()
Input: set A
Output: partition (A1, A2)
begin
    Treelist ← ∅;
    foreach i ∈ A
        T1 ← CreateTree ((i,i));            tree rooted with (i,i)
        Append (TreeList, T1);
    endfor
    while Size(TreeList) ≥ 2
        T1 ← ExtractHighest (TreeList);
        T2 ← ExtractHighest (TreeList);
        (l1, n1) ← Root(T1);
        (l2, n2) ← Root(T2);
        AddBranch (T1, T2);      T2 is inserted as a subtree of T1
        SetRoot (T1, (l1 − l2, n1));
        Append (TreeList, T1);
    endwhile
    T ← ExtractHighest (TreeList);
    (A1, A2) ← ColorPartition(T);
end
```

Algorithm 4.1 The Karmarkar-Karp Algorithm. Trees are constructed, being the nodes pairs (l,n), where l is the label for the node —the value of the partition is represents— and n is an element of A.

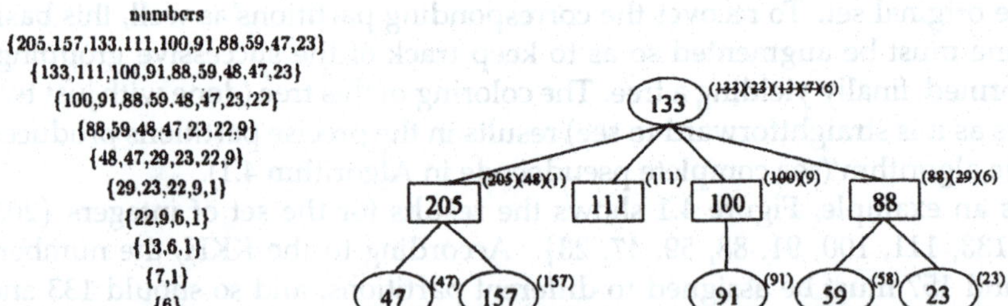

Figure 4.1 Tree provided by the KKH. Oval nodes go to one partition, and rectangular nodes to the other one. The numbers in brackets are the successive labels assigned to each subtree. The final weight of the tree is 6.

as a *"phase transition"* (Cheeseman et al., 1991). i.e., the transition from a "region" in which almost all instances have many solutions to a region in which almost all instances have no solutions, as the constraints become tighter. This feature is sometimes uncovered by exact algorithms, and can be experimentally —and in some situations even theoretically— analyzed using samples of randomly-generated problems (Mitchell et al., 1992; Mammen and Hogg, 1997; Smith and Dyer, 1996).

Regarding the specific case of the MIN NUMBER PARTITIONING problem and in relationship with the CKK algorithm, it uncovers the *"Easy-hard-easy"* type of transition. We will try to give some intuition on this type of transitions by considering an analogous problem. Suppose for a moment that we are given n stones of different weights (with an average weight $\overline{W}$) and that our task is to separate them in two groups of the same weight. For the "large n" limit (assuming we keep a fixed $\overline{W}$), we can analogously think of the "low $\overline{W}/n$" limit, so we can think of the "stones" as grains of sand. It is clear that the task is significantly easier in this case, since the probabilities and combinatorics are playing for us: several optimal solutions may exist, and certainly many suboptimal solutions with costs that are numerically very close to the perfect (optimal) partition value. Also, in the "small n" limit, the problem is also expected to be easy for an exact algorithm, since the search space is greatly reduced. So clearly, there is a relationship between the two magnitudes where the search is expected to be harder.

Recurring once again to the intuition of the reader, certainly we do not expect that the existence of a single optimal solution would characterize the particular scenario that would be the hardest for the exact algorithm under consideration. This is clear from the fact that in the "small n" limit we can easily expect to have a single optimal solution, though the problem would be easy (on average) since it has a small number of possible configurations (2^{n-1}). Thus we can expect that the problem is harder on average when the number of elements is as large as possible yet it still has a single solution.

In this sense, the particular instance of MIN NUMBER PARTITIONING used in Figure 4.1 was used with a purpose in mind. We have chosen it to illustrate the workings of the KK heuristic since this set of $n = 10$ numbers {205,157,133,111,100,91,88,59,47,23} is discussed in page 234 of Papadimitriou (1994). In the author's own words:

> *"...(notice that their sum,* 1014, *is indeed less than* $2^n - 1 = 1023$*). Since there are more subsets of these numbers than there are numbers between* 1 *and* 1014, there must be two different subsets which have the same sum. *In fact, it is easy to see that two* disjoint *subsets must exist that must have the same sum."*

It is thus clear that the concept behind these "phase transitions" is very intuitive. Although it has been probably identified and discussed well before 1991 (Cheeseman et al., 1991), it has very seldom received the attention it deserves when selecting which instances to use in the computational experimentation with heuristic algorithms and metaheuristics.

Curiously, the work of Cheeseman et al. (1991) has been cited a few hundred times in the computing literature, yet most papers in the metaheuristic area do not take into account these facts when selecting the particular instances under study to test their methods. Here we have taken particular care regarding the generation of instances and the behavior of metaheuristics in the three regions which has helped us to understand the characteristics of our methods. It would be interesting to collect evidence on how long the existence of the so-called "phase transitions" (a term borrowed from Physics), has also been reported in the account of experimental studies with exact methods like Branch-and-Bound (Lawler and Wood, 1966) or Branch-and-Cut (Caprara and Fischetti, 1997).

4.2.2 *The IMKK Heuristic*

To our knowledge there has been no other heuristic that would allows us to replace the basic KKH scheme with another fast algorithm. We provide one such an attempt, an iterated constructive heuristic that we tentatively named IMKK for *Iterated Matching and KKH*. For simplicity, we discuss here an implementation of this idea using a greedy matching algorithm.

Suppose we have as input the following partition: $A_1 = \{205, 133, 111, 59, 47\}$ and $A_2 = \{157, 100, 91, 88, 23\}$. Note that the cost associated to this partition is $|555 - 459| = 96$, higher than the one obtained by the KKH for the same instance of Figure 4.1. Now we proceed constructing a weighted bipartite complete graph $G(V, E)$, such that $V = V_1 \cup V_2$ and V_1 and V_2 are two independent sets. We have a vertex in V_1 for each element of A_1 (analogously, we have a vertex in V_2 for each element of A_2). The weight of an edge $(v_i, v_j) \in E$ between two vertices $v_i \in V_1$ and $v_j \in V_2$, is given by $|a_j - a_j|$ of the corresponding numbers $a_i \in A_1$ and $a_j \in A_2$. For the given partition, we can run a greedy algorithm that tries to find a matching of minimum weight. Bending a little the notation, this matching can be written as the following list of paired numbers $M = \{(111, 100), (47, 23), (157, 133), (88, 59), (205, 91)\}$. The weight of such a matching is $11 + 24 + 24 + 29 + 114 = 202$. If we make the heuristic argument that we can assume that the matching can be understood as a constraint that obliges the numbers to be in different partitions, then we can run the KKH on an instance that has half the original size with the numbers $\{11, 24, 24, 29, 114\}$. The whole pseudocode is shown in Algorithm 4.2.

Following the procedure above we obtain the new partition of the original numbers $A_1^{(1)} = \{205, 133, 100, 59, 23\}$ and $A_2^{(1)} = \{157, 111, 91, 88, 47\}$ with an associated cost of $|520 - 494| = 26$, which is still above the one given by the KKH. Iterating the procedure once again, we obtain the matching $M = \{(100, 91), (59, 47), (133, 111), (205, 157), (23, 88)\}$, with weight $9+12+22+48+65 = 156$. Running the KKH on the set of numbers $\{9, 12, 22, 48, 65\}$ we obtain the partition $A_1^{(2)} = \{205, 133, 100, 47, 23\}$ and $A_2^{(2)} = \{157, 111, 91, 88, 59\}$ which has an associated cost of $|508 - 506| = 2$, smaller than the one obtained by the KKH alone.

We leave as an exercise to the reader what happens if we iterate the procedure another time. However, we can anticipate the result: we obtain again the same matching, and as a consequence, the same result for the KKH. In essence,

```
procedure IMKK( )
Input: partition (A_1, A_2)
Output: partition (A'_1, A'_2)
begin
    G ← ∅;
    foreach i ∈ A_1, j ∈ A_2
        G ← G ∪ {(i, j)};
    M ← FindMatching (G);
    A' ← ∅;
    foreach (i, j) ∈ M
        A' ← A' ∪ {|i − j|};
    (A'_1, A'_2) ← KKH (A');
end
```

Algorithm 4.2 A single step of the IMKK Algorithm. This process can be iterated so as to further improve the output partition.

we are trapped in a *"local minimum"*[1] of this procedure. This is certainly interesting and it suggests that a procedure based on repeated application of the KKH and a minimum weight matching algorithm on a complete bipartite graph can be used as an engine for a new type of metaheuristics for this problem. This issue is tackled in the next section.

4.3 RECOMBINATION APPROACHES BASED ON WEIGHT MATCHING

As shown in the previous section, the idea of applying KKH on reduced (via matching) instances of the problem can be a promising mechanism for introducing sensible knowledge about the target problem (MIN NUMBER PARTITIONING in this case) in the search algorithm at hand. While it is still unclear whether its use as a search engine in pure local-search metaheuristics can be useful, it is much more evident that this technique can provide very useful guidelines to model information exchanges in population-based-search metaheuristics such as evolutionary algorithms (EAs). More precisely, the inclusion of problem knowledge into the EA by means of heuristic procedures such as this will result in a memetic algorithm (Moscato, 1989; Moscato and Norman,

[1]Note that actually there is a single solution in the "neighborhood" defined by the procedure described so far. Despite this, it might be possible to somehow define a metaheuristic search scheme, e.g., Tabu Search, on the basis on this neighborhood. This is an interesting yet very substantial topic in itself. For example, a deep study would be required to determine which the solution attributes could be (recall the dramatic difference between "neighboring" solutions in this case). Such a study is clearly out of the scope of the current paper, and will be dealt in subsequent research.

1992) (MA henceforth). A general description of these techniques will be given in Section 4.4. Previously, this section is devoted to discuss the utilization of weight-matching ideas within recombination, a key component in MAs by which a new solution is created by combining information from a set (usually a pair) of existing solutions. In this sense, two main approaches are identified: the use of minimum matchings, and the utilization of balanced matchings.

4.3.1 Minimum-Weight-Matching Recombination

The approach discussed in this subsection is very similar to the plain IMKK algorithm presented before. The pseudocode is shown in Algorithm 4.3.

```
procedure MWMR( )
Input: partitions (A₁, A₂), (A'₁, A'₂)
Output: partition (A''₁, A''₂)
begin
    G₁ ← ∅; G₂ ← ∅;
    foreach i ∈ A₁, j ∈ A₂
        G₁ ← G₁ ∪ {(i, j)};
    foreach i ∈ A'₁, j ∈ A'₂
        G₂ ← G₂ ∪ {(i, j)};
    M ← G₁ ∩ G₂;
    foreach i ∈ A₁ ∪ A₂
        M ← M ∪ {(i, 0)};
    M' ← OrderedList(M);
    (i, j) ← FindFirst (M', max(A₁ ∪ A₂));
    E ← {i, j}; A ← {|i − j|};
    while E ≠ A₁ ∪ A₂
        DeleteAllOccurrences (M', i);
        DeleteAllOccurrences (M', j);
        (i, j) ← GetNextEdge (M', M);          /* (i, j) ∈ M' and
        minimizes max{||i − j| − |k − l|| |(k, l) ∈ M, (k, l) ∉ M'} */
        E ← E ∪ {i, j}; A ← A ∪ {|i − j|};
    endwhile
    (A''₁, A''₂) ← KKH (A);
end
```

Algorithm 4.3 The Minimum Weight-Matching Recombination.

The following example may be useful to illustrate the actual functioning of this pseudocode. Suppose we are given the two parental solutions $S_1 = \{205, 133, 47, 23\}\{157, 111, 100, 91, 88, 59\}$ and $S_2 = \{205, 111, 100\}\{157, 133, 91, 88, 59, 47, 23\}$ having costs of $|408 - 606| = 198$ and $|416 - 598| = 182$; we first start by constructing the graphs associated with them as mentioned above. We then identify common edges, i.e., pairs of numbers being in different partitions in both solutions. Sorting this list according to edge weights yields in this example: (133, 111) = 22, (133, 100) = 33, (205, 157) = 48, (100, 47) = 53, (111, 47) = 64, (100, 23) = 77, (111, 23) = 88, (205, 91) = 114, (205, 88) = 117, and (205, 59) = 146. The next step is augmenting this list with edges connecting each number with a dummy element '0'. The rationale behind this is that it could better to consider a number in isolation rather than taking a bad matching from the parents. In this example, we obtain: (133, 111) = 22, (23, 0) = 23, (133, 100) = 33, (47, 0) = 47, (205, 157) = 48, (100, 47) = 53, (59, 0) = 59, (111, 47) = 64, (100, 23) = 77, (111, 23) = 88, (88, 0) = 88, (91, 0) = 91, (100, 0) = 100, (111, 0) = 111, (205, 91) = 114, (205, 88) = 117, (133, 0) = 133, (205, 59) = 146, (157, 0) = 157, and (205, 0) = 205.

Now, we find the first appearance of the highest number (205), and mark the corresponding edge (205, 157) whose weight is 48. Next, we proceed iteratively by considering the edge that minimizes the largest weight difference with respect to an already marked edge, and whose members (excluding the '0' if it were the case) are not included in any marked edge. This process is repeated until all numbers are included in one marked edge. In our example, edges would be marked in the following order: (205, 157), (0, 47), (0, 59), (133, 100), (0, 23), (0, 88), (0, 91), (0, 111).

The whole process until this point has been devoted to extract some common information from the parents. We now need to use this information to create a new solution. This is done by running the KKH using the edge weights. In our example we obtain the partition {111, 59, 47, 33} {91, 88, 48, 23} that translates into the final partition {157, 133, 111, 59, 47}{205, 100, 91, 88, 23} whose cost is 0.

4.3.2 A recombination algorithm based on finding a balanced matching

After some initial experimentation with the recombination algorithm based on minimum weight matching, we recognized that a variant of the original idea may be more useful for this problem. Again we will resort to our favorite instance example to explain this method.

Using the example discussed in Subsection 4.3.1, we have two parental partitions S_1 = {205, 133, 47, 23} {157, 111, 100, 91, 88, 59} and S_2 = {205, 111, 100} {157, 133, 91, 88, 59, 47, 23} having costs of $|408 - 606| = 198$ and $|416 - 598| = 182$. As we did before, we construct an undirected graph with edge weights given by (133, 111) = 22, (23, 0) = 23, (133, 100) = 33, (47, 0) = 47, (205, 157) = 48, (100, 47) = 53, (59, 0) = 59, (111, 47) = 64, (100, 23) = 77, (111, 23) = 88, (88, 0) = 88, (91, 0) = 91, (100, 0) = 100, (111, 0) = 111, (205, 91) = 114, (205, 88) = 117, (133, 0) = 133, (205, 59) = 146, (157, 0) = 157, and (205, 0) = 205.

Now we can search for the most balanced matching in the graph, that is, a matching that minimizes the absolute difference between the weights of the heaviest and lightest edge in the matching. Instead, to avoid the computational complexity associated to performing this operation, we resort to a simpler heuristic. Since we have 20 edges, we start by considering the tenth edge in the increasing order, which is $(111, 23) = 88$. Selecting this edge means that we have already decided that the numbers 111 and 23 go in different sides of the partition. They are both marked. We next consider the two consecutive edges in the order. Integers 23 and 111 are already marked, so the next edges to consider are $(59, 0) = 59$ and $(88, 0) = 88$. At each step, we will greedily take the edge that minimizes the current imbalance of the matching. In this case $(88, 0) = 88$. We mark 88 and proceed, now with $(59, 0) = 59$ and $(91, 0) = 91$. We select $(91, 0) = 91$, mark 91, and continue, now selecting $(100, 0) = 100$, and so on. We end this procedure when all the integers are marked and we have a matching. We then run the KK heuristic, passing as input the set of weights of the edges. The resulting solution is then used to create the actual partition for the original numbers.

Thus, the main difference with respect to the previous recombination algorithm is the fact that GetNextPair (M', M) returns a pair $(i, j) \in M'$ that minimizes $\max(a, b)$, where

$$a = ||i-j| - \max\{|k-l| \mid (k,l) \in M, (k,l) \notin M'\}| \text{ and } \tag{4.3}$$

$$b = ||i-j| - \min\{|k-l| \mid (k,l) \in M, (k,l) \notin M'\}|, \tag{4.4}$$

rather than simply minimizing

$$\max\{||i-j| - |k-l|| \mid (k,l) \in M, (k,l) \notin M'\}. \tag{4.5}$$

4.4 THE MEMETIC ALGORITHM

As mentioned before, the recombination operators already described will be integrated within a problem-adapted evolutionary algorithm, also known as *hybrid* evolutionary algorithm (Davis, 1991) or memetic algorithm (Moscato, 1989). Besides the utilization of these *smart* recombination operators, MAs use additional mechanisms to include problem-specific knowledge into the search engine. A general view of these mechanisms is provided in Subsection 4.4.1. Subsequently, the particular details for adapting the generic components described in this first subsection to the target problem will be discussed in the next subsections.

4.4.1 Pseudocode for memetic algorithms

The particular MA used in this paper is the so-called Local-Search-based MA, a population-based-search algorithm that intensively uses local search to boost solution quality (see the pseudocode in Algorithm 4.4). This use of local search motivates a view of the process in which each solution is an *agent* that tries to improve its quality, cooperating and competing with other agents in the population (Norman and Moscato, 1991; Slootmaekers et al., 1998).

This generic template can have a huge variety of instantiations. For instance, the FirstPop() function can generate a set of random solutions for the problem at hand, or use a construction heuristic (or a set of them if more than

```
procedure MA( )
begin
  Pop ← FirstPop();
  foreach agent i ∈ Pop
    i ← Local-Search-Engine(i);
    EvaluateSolution(i);
  endfor
  repeat
    for j ← 1 to #recombinations
      S_par ← SelectToMerge (Pop);
      offspring ← Recombine(S_par, x);
      offspring ← Local-Search-Engine(offspring);
      EvaluateSolution(offspring);
      AddInPopulation (offspring, Pop);
    endfor
    for j ← 1 to #mutations
      i ← SelectToMutate (Pop);
      i_m ← Mutate(i);
      i_m ← Local-Search-Engine(i_m);
      EvaluateSolution(i_m);
      AddInPopulation (i_m, Pop);
    endfor
    Pop ← SelectPop(Pop);
    if Converged(Pop)
      Pop ← RestartPop(Pop);
  until termination-condition is satisfied
end
```

Algorithm 4.4 The Local-Search-Based Memetic Algorithm

one is known) to generate a set of good initial solutions. As to the Local-Search-Engine() function, it receives as input a solution and applies an iterative improvement algorithm to it. This improvement algorithm iterates until it is no longer possible to improve the current solution or it is judged not probable to achieve further improvements. Note that in many circumstances, *there is no need* to evaluate the guiding function[2] (using the EvaluateSolution() algorithm) until we got the final solution (the Local-Search-Engine() may be capable of determining whether a certain modification to the current solution is acceptable by using heuristic knowledge about the target function).

After the initial population has been created, at least one *recombination* is done. Some MAs ask all the solutions to be involved in recombinations, so

[2]In this case, the guiding function is $m_P(y, A)$, described in Section 4.1.

#recombinations is fixed and does not need to be specified as a numerical parameter. The SelectToMerge() function is executed by selecting a subset of k solutions to be used as input of the *k-merger*, a k-ary recombination procedure. Some MAs use a random function, while others use some more complex approach. For instance, there are authors who also advocated for the benefits of using a *population structure* for interaction of agents (Gorges-Schleuter, 1989; Mühlenbein, 1991; Gorges-Schleuter, 1991; Moscato, 1993).

A new solution is created by recombining the selected solutions according to the Recombine() function. This can start a variety of procedures, ranging from a single application of any available efficient (i.e., polynomial-time) *k-merger* algorithm up to more complex recombination operators (which means that they need not be efficient) such as the more systematic searches proposed by Aggarwal et al. (1997) as well as the one used in the recent GAs proposed by Balas and Niehaus (1996). Afterwards, it is optimized and added (or not) to the population according to some criteria. Analogously, a number of solutions are subjected to some *mutations*. Sometimes the Mutate() function is implemented as a random process, but this is not the general rule and other forms of mutation are possible. Again, the solution is re-optimized at the end of the mutation process and added or not to the population.

The SelectPop() function will act on the population, having the effect of reducing its size. The selection of this subset is not always determined by the objective or the guiding function. It may be biased by other features of the solutions, like the interest of maximizing some measure of diversity of the selected set. Again, this can be implemented in a variety of ways. The convergence of the population is sometimes decided by reference to a *diversity-crisis*, a measure which indicates, below a certain threshold, that the whole population has very similar configurations. When the population has converged, a RestartPop() function is used. In general, the best solution found so far (or *incumbent solution*) is preserved and a new population is created using some randomized procedure. All the solutions of the population are optimized and evaluated afterwards and the whole process is repeated.

The termination-condition can also be implemented in many ways. It can be a time-expiration or generation-expiration criteria as well as more adaptive procedures, like some dynamic measure of lack of improvement.

4.4.2 Representation and Search Space

In this paper, we have used the *direct representation* (Ruml et al., 1996) which is *complete* since it can generate all possible partitions (the associated search space has 2^n configurations, two for each solution of the problem). In this representation, a partition (A_1, A_2) is represented as a signed bit array $\{a_i \mid 1 \leq i \leq |A_1 \cup A_2|\}$, where $a_i = 1$ if the ith element of $A_1 \cup A_2$ (under an arbitrary but fixed enumeration) belongs to A_1, and $a_i = -1$ otherwise. We will note that any solution is represented by two configurations ($s, s' \in S$). For instance $s = \{1, -1, -1, 1, -1, -1\}$ and $s' = \{-1, 1, 1, -1, 1, 1\}$ both represent the same partition (i.e., the same solution) of six integers.

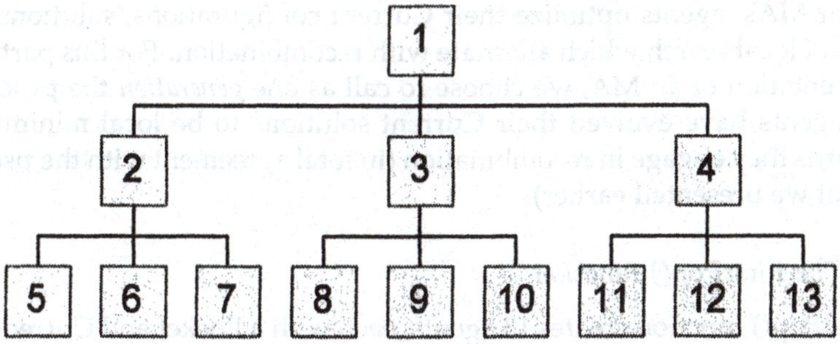

Figure 4.2 Population structure (a complete ternary tree) used in all the MAs in this study on the MIN NUMBER PARTITIONING problem.

Another representation that can be used for this problem is called *ternary direct* since it allows alleles to take one out of three different values, '0', '1', or '−1'. Obviously, the associated search space is also complete since it contains the search space of the direct representation. These two representations were studied using a variety of MAs in (Berretta and Moscato, 1999). We have kept the best MA of that paper (an MA using the direct representation) as benchmark metaheuristic. The aim here would be to see if those results can be improved by other new heuristic ideas for the recombination methods.

4.4.3 Population size and structure

The population has a fixed size of 13 agents, arranged with a neighborhood relationship based on a complete ternary tree of three levels (See Figure 4.2). Initially chosen for historical reasons, this structure has revealed itself as very appropriate with respect to the implementation of *behaviors,* a topic discussed in Subsection 4.4.5.1

Such interaction topology can be interpreted as a variant of *island models* (Tanese, 1989) of evolutionary algorithms, in which each subpopulation has four agents, one *"leader"* node and three *"supporters"*. The latter are one level below in the hierarchy, so agent 1, the root of the tree, is the leader of the top subpopulation and has as supporters agents 2, 3, and 4. Agent 2 has as supporters agents 5, 6, and 7, etc. Note that agents 2, 3, and 4 play both leader and supporter roles.

Each agent of the population is handling two feasible solutions, configurations of the associated search space. One is named *"Pocket"* and the other one *"Current"*. Whenever the Current solution of an agent has a better guiding function than the Pocket solution, they are switched. We can understand the Pocket as playing the role of a "memory" of past good solutions. Another procedure named *PocketPropagation* changes two Pocket solutions if the leader has a Pocket solution which is worse than one of its followers. We only require these two mechanisms to guarantee the flow of better solutions toward the agent at the top of the hierarchy.

In our MAs, agents optimize their Current configurations/solutions with periods of local search which alternate with recombination. For this particular implementation of an MA, we choose to call as one *generation* the process in which agents have evolved their Current solutions to be local minima and afterwards they engage in recombination (in total agreement with the pseudocode that we presented earlier).

4.4.4 *The FirstPop() function*

The FirstPop() function creates 13 agents, each with a Pocket and Current configuration representing a partition (A_1, A_2). All solutions are represented using the direct representation, and are chosen with equal probability.

4.4.5 *The Recombine() procedures*

We have used the complete ternary tree neighborhood topology for interaction between agents. As a consequence, we oblige supporters to recombine with their leaders only. The population structure and this rule suggest that the island models of GAs (taken as a metaphor for evolution with colonization and diffusion) is not the most representative for this MA. Instead, the approach might resemble a hierarchical organization in which "communication of ideas" or "exchange of information" can only occur with the immediate leader member of a subgroup.

At each generation step, the Current solution of each agent is replaced which a new one generated by the Recombine() procedure acting on the Pocket solution of the supporting agent and its leader. For example, the Current of the agent 2, in Figure 4.2, will be replaced by the output of the Recombine() procedure having as input the Pocket of agent 1 and the Pocket of agent 2; the Current of agent 9, will be replaced by the Recombine() using the Pocket of agent 3 and the Pocket of agent 9; etc. Then, all agents except agent 1 will replace its Current solution in one iteration of the generation loop. Regarding the previous pseudo-code, the number *#recombinations* performed at each generation step is 12. Moreover, the SelectToMerge() function can be regarded as *totally constrained*, since each agent always engage in recombination with the same agents.

4.4.5.1 Behaviors. Before a leader recombines its solution with any of its three supporters, a different type of *behavior* is assigned to each one. A behavior can be understood as one extra control parameter in the control set of a given recombination operator (Radcliffe and Surry, 1994b). Each supporter will have different behavior, which will be decided randomly with equal probability. The three types of behaviors used will be described below. To simplify the presentation, let us suppose that the direct representation is being used, and parent P_1 is the leader and the parent P_2 is one of the supporters.

For instance if $P_1 = \{-1,-1,1,-1,1\}$ and $P_2 = \{1,-1,1,1,-1\}$ are the parents, the recombination using the same parents as input and each type of

Table 4.1 Different behaviors in leader/supporter recombination.

behavior	first copied in the *offspring*
rebel	alleles of P_2 which are different from P_1
conciliator	alleles in common to P_1 and P_2
obsequent	alleles of P_1 which are different from P_2

behavior occurs as follows: the 'x' stands for a value that will be decided by algorithms which we will explain at a later stage, then,

rebel	{1 x x 1 -1}
conciliator	{x -1 1 x x}
obsequent	{-1 x x -1 1}

The *conciliator* behavior is an example of a recombination procedure that *respects* the representation (Radcliffe and Surry, 1994a; Radcliffe, 1994) since *"every child it produces contains all the alleles common to its two parents"* (i.e., those in $A \cap B$). It shares the property of being a *respectful* recombination as it is also the case of *uniform crossover*. In this case, since all alleles not in $A \cap B$ have either the value '−1' or '1', the recombination with conciliator behavior is said also to *transmit alleles* since *"each allele in the offspring is present in at least one of its parents"*.

Both the *rebel* and the *obsequent* behaviors do not respect (in the sense postulated by Radcliffe) binary representations since they may exclude allele values in $A \cap B$. Analogously, these behaviors are non-transmitting, since it is allowed to create allele values not present in any parents.

In order to decide the positions in the *offspring* where no allele was chosen (the 'x' marks) we have investigated several different variants. For the direct representation, three types of algorithms have been used: *generalized transmission* (*GT*), *GT with greedy repairing* (*GTgr*), and *GT random seeded - greedy repairing* (*GTrsgr*).

The *GT* decides at random between either '−1' or '1' with equal probability. The *GTgr* uses a greedy algorithm, i.e., a deterministic rule which sequentially decides between '−1' or '1' to minimize the actual imbalance. The order is based in a non-decreasing order of the sequence of not yet decided integers. The *GTrsgr* starts by randomly selecting a gene for which its allele value has to be decided. After giving this gene a values minimizing the imbalance, *GTrsgr* continues in a deterministic way as the *GTgr* does. According to the definitions given in (Moscato, 1999), all of them are crossover operators (can be performed in $O(n \log n)$ time), while the *GT* is the only *blind* one, since it does not use information from the instance of the problem.

To further illustrate the functioning of these "patching" algorithms, consider the instance A = {15,12,10,9,4} and the two parents P_1 and P_2 mentioned above with conciliator behavior. The *offspring* first receives the common allele values, i.e., O = {x,−1,1,x,x}. With *GT*, each 'x' will be decided at random with the same probability (in this case we have the traditional *uniform crossover*. With *GTgr*, we chose O[1]=1, O[4]=−1 and O[5]=−1, in this order. Note that in this case we decided for '1' or '−1' accordingly to the actual absolute difference of the partial partition. With *GTrsgr* we might start by making O[4]=1, followed by O[1]=−1 and O[5]=1.

4.4.6 The Mutate() procedures

We have implemented two types of mutations, *Simple* and *Minimal*. The *Simple* mutation receives as input a solution and for all allele values it decides whether to "flip" it (change from '−1' to '1' or from '1' to '−1') with a fixed probability of 0.1.

The *Minimal* mutation was inspired in the *Binomial Minimal Mutation* as discussed in (Radcliffe and Surry, 1994b). However, we refer to that paper for inspiration only, as a remarkable difference exists. In our case we have taken into account the characteristics of the objective function (to which the label 'minimal' actually refers in this case): alleles will not have its sign changed with the same *a priori* probability used by the *Simple* mutation (i.e., 0.1). Let us suppose allele i has sign $v_i = 1$ and has been selected to be mutated. Let us also suppose we have represented the solutions with the traditional signed bit array, such that the indexing corresponds with a non-increasing order of the integers of the instance A[3]. We then proceed to identify the index of the first allele in the *"left direction"* (referring to the array) j_l and the first allele in the *"right direction"* j_r which have different signs ($v_{jr} = v_{jl} = -1$). Then v_{jl} is the allele value of the smallest integer which is higher or equal than integer a_i and assigned to a different partition. Analogously, v_{jr} is the allele of the largest integer which is smaller or equal to integer a_i but in a different partition. Then we select which one of j_r and j_l minimizes $|a_i - a_j|$. We then swap its allele value with the value of v_i.

We can then say, analogously to the recombination procedures, that the *Simple* mutation is a blind operator while the *Minimal* mutation is not since it uses information of the objective function (Moscato, 1999; Moscato and Cotta, 2003).

4.4.7 The Local-Search-Engine() procedures

We have implemented two different types of local search algorithms. One was called *GreedyImprovement* and the other one was called *TabuImprovement*. For both of them, the input is one solution, and the output is a solution with the same or better objective function value $m_P(y, x)$.

[3]E.g., this means that the allele value of v_1 corresponds to the partition assigned to the largest integer in A (integer a_1).

Using the *GreedyImprovement* we start by selecting one allele i at random. Then, as we did with the minimal mutation, we identify the position of the first allele in the *"right" direction* (j_r) and the position of the first allele in the *"left" direction* (j_l) such that $v_{j_r} = v_{j_l} \neq v_i$ (it may be possible to find only one satisfying the condition). We decide to swap the values of v_i with either v_{jl} or v_{jr} if the objective function value ($m_P(y,x)$) is reduced. If both can do that, we chose the one that causes the largest reduction. We repeat this process until

$$(FailuresTries - SuccessTries) > MaxNumberOfTries.$$

Where $SuccessTries$ is the number of tries that causes better $m_P(y,x)$ value and $FailuresTries$ count the opposite. We have used several values of the parameter $MaxNumberOfTries$ and we will comment on it later.

The *TabuImprovement* uses the same idea of *GreedyImprovement*. The difference is that in this case we use a basic Tabu Search metaheuristic (Glover and Laguna, 1997) inside of the *GreedyImprovement*. Each swap done between i and j, a tabu matrix ($TABU$) of integers stores, in the position $TABU[i][j]$, the number of trials that the swap between i and j will be 'tabu'. In *TabuImprovement* we can allow swaps which do not reduce the value of $m_P(y,x)$, but the output of this procedure is the configuration with the best $m_P(y,x)$ found. We repeat this process until

$$(FailuresTries - SuccessTries) > MaxNumberOfTries.$$

In this case, the variable $SuccessTries$ is the number of tries that could be done, i.e., the number of times a try was not tabu. In addition, we use a simple *aspiration criteria* (Glover and Laguna, 1997). If a swap causes a improvement in the objective function that has never been reached before, we do this swap, even it was declared 'tabu'.

4.4.8 RestartPop()

During the evolution of the population, agents recombine only within its subpopulation. More specifically, an agent only recombines its Pocket solution with the one that its leader has. A *"diversity crisis"* can then happen when three supporters of the same subpopulation all have very similar configurations. Obviously, it is necessary to provide a precise definition of *similarity* to formally establish when a diversity crisis is taking place.

The criterion we used to define a diversity crisis is the following: we select at random 20% of the allele values of the Current solution of the three supporters; if the three solutions have the same values in these alleles, then a diversity crisis happened. When a diversity crisis is detected, the leader will not recombine with its three supporters, but the three supporters will recombine with supporters which belong to different subpopulations. For example, in general, agent 2 recombines with its three supporters, i.e., agents 5, 6, and 7. Whenever a diversity crisis is detected (i.e., agents 5, 6, and 7 have similar configurations), the recombination will be done between {5, 6, and 7} and {8, 9, and 10} or between {5, 6, and 7} and {11, 12, and 13}. Three recombinations will modify the Current solutions of agents 5, 6, and 7; i.e., if it was chosen {5, 6, and 7} and {8, 9, and 10} (this second subpopulation was randomly selected), the recombination pairs can be: 5 with 9, 6 with 10, and 7 with 8. The assignments are chosen at random too. As a consequence, we may get some extra diversity.

4.4.9 Lack of improvement

We have also investigated a rule for lack of improvement. In the context of this problem, the results have been very promising. We have introduced a very simple rule: if after three consecutive generations there is no improvement of the incumbent solution in the population (i.e., if the Pocket solution of the agent at the top of the hierarchy has not been updated), we save the solution (for the off-line assessment of the algorithm), we eliminate it from pocket 0 and we set it to a randomly created solution.

4.5 TABU SEARCH

The Tabu Search heuristic implemented in Berretta and Moscato (1999) was used in that paper as a kind of "background" operator. It was introduced rather for being in charge of diversifying the search than for providing improved solutions. However, it used the *"minimum"* neighborhood, meaning that it is analogous to the *minimum* mutation as explained in that chapter. This is a severe drawback since it means that, given a solution, we only move to other solutions that have the same number of elements in each side of the partition. This said, it is the case that a more powerful neighborhood must be used to improve the search, allowing to change the cardinalities of the two subsets.

Given the good results of this Tabu Search implementation, but aiming to provide a larger neighborhood that would free it from being "cardinality constrained", we have added another Tabu Search procedure that interleaves with the former one. We have named it "exhaustive": for each allele value, we compute the cost of reversing its sign, i.e., to move the associated integer to the other side of the partition; we perform the change that maximizes the reduction in cost, or the one that minimizes the increment if no decreasing change is available.

These two search strategies are alternatively executed during a single Tabu Search individual optimization step. Since the "exhaustive" method requires more computer time, we make it run for a shorter period than the other one. More precisely, TS runs for k steps using the minimum neighborhood, and for $k/10$ steps using the exhaustive neighborhood, where k is one tenth of the total number of allowed TS steps.

Note that both Tabu Search strategies have their own different attribute lists. For the "minimum" TS strategy, an attribute is a pair (i, j) while for the the "exhaustive" scheme it is just the i that corresponds to the integer than has changed. The time an attribute remains tabu is the same as in Berretta and Moscato (1999), i.e., a uniformly distributed integer random number in $[1, n]$.

4.6 COMPUTATIONAL RESULTS

We have studied three groups of instances. In each group the integers have 10, 12, and 14 decimal digits respectively. We also note that the instances with 10 decimal digits where the same as the one used in Berretta and Moscato (1999) to help us evaluate the benefits, if any, of the new ideas tested in this paper. Again, we have taken extreme care with the generation of the instances, and the best approach has been to generate each decimal digit of each integer of each instance uniformly at random to avoid any spurious correlations. In each group we have 100 instances, with $n = 15, 25, 35, \ldots, 105$ integers. This means that we have 10 independently generated instances for each value of n. Table 4.2 shows the results of the KKH on these instances. Results for the complete KK algorithm are also included for the smallest test instances.

Table 4.2 The performance of CKK and KKH on the three sets of instances (CKK could only be run on instances with 10 digits).

n	CKK (D=10)	KKH (D=10)	KKH (D=12)	KKH (D=14)
15	1785469	62529719	11143286419	284759439645
25	3161	2087226	409925198	27573832748
35	3	422345	44169859	7557306861
45	1	329255	20000541	3308094310
55	1	97390	7841909	833784155
65	1	15250	2541069	81996483
75	1	20488	464585	205571875
85	0	3386	1249715	23506635
95	1	3122	290572	45466008
105	0	1656	261532	21748182
mean (≥ 35)	1	111612	9602473	1509684313

We have first run a Tabu Search algorithm for 10^9 iterations. It is important to remark that this Tabu Search algorithm is different to the one implemented in Berretta and Moscato (1999). This implementation obtains better results than the one that uses the *"minimum"* neighborhood described in (Berretta and Moscato, 1999). The results for this improved Tabu Search algorithm are shown in Table 4.3.

The memetic algorithms have been run for 5,000 generations, which approximately corresponds to the same time employed by the Tabu method. The results obtained are shown in Tables 4.4, 4.5, and 4.6. With MA-behavior, we denote the best memetic algorithm that has been reported in Berretta and Moscato (1999), i.e., it uses behavior-based recombination but the same restart mechanism that has been described in this paper. With MA-matching, we

Table 4.3 The performance of Tabu Search on the three sets of instances.

n	Tabu (D=10)	Tabu (D=12)	Tabu (D=14)
15	1785469	190858470	18097490259
25	3161	243962	24534199
35	11	926	209194
45	11	771	57018
55	6	1027	42114
65	12	825	50144
75	6	478	77951
85	6	602	52960
95	6	301	34633
105	4	328	26581
mean ($\geq$ 35)	8	657	68824

Table 4.4 The performance of the four MA variants tested on instances with 10 decimal digits. The best results are shown in boldface.

	Propagate *after* mutation		Propagate *before* mutation	
n	MA-behavior	MA-matching	MA-behavior	MA-matching
15	**1785469**	**1785469**	**1785469**	**1785469**
25	**3161**	**3161**	**3161**	**3161**
35	**9**	16	20	12
45	7	7	7	**5**
55	5	11	7	**2**
65	5	**4**	**4**	**4**
75	**1**	4	3	2
85	5	6	**2**	4
95	**3**	5	**3**	**3**
105	3	3	3	**2**
mean ($\geq$ 35)	5	7	6	**4**

denote the same memetic algorithm that uses restart, but with the difference that the new recombination algorithm based on finding a balanced matching was used. A variant of each of these MAs has been considered: if, after a recombination, the resulting solution is better than its pocket solution then it is immediately substituted. Note that in the original algorithm we might first introduce a mutation, before the actualization of pockets is tested. We have observed that this small change in the algorithm resulted in a better performance of the balanced matching recombination method, since in this case the creation of a better offspring solution —i.e., better than its parents— seems to happen more frequently than in the behavior-based approach.

From the inspection of these tables we can conclude that the new Tabu Search method has certainly improved over the one presented in Berretta and Moscato (1999). It is now able to solve instances with 10 decimal digits and $n = 15$ and $n = 25$. However, that is not the case for the 12- and 14-digit

Table 4.5 The performance of the four MA variants on instances with 12 decimal digits. The best results are shown in boldface.

	Propagate *after* mutation		Propagate *before* mutation	
n	MA-behavior	MA-matching	MA-behavior	MA-matching
15	**190858470**	**190858470**	**190858470**	**190858470**
25	**243962**	**243962**	284877	**243962**
35	2355	1667	1026	**875**
45	545	665	1524	**284**
55	672	646	563	**296**
65	837	715	**269**	360
75	696	433	**393**	443
85	465	501	361	**347**
95	540	388	**201**	219
105	**271**	604	370	478
mean ($\geq$ 35)	798	702	588	**413**

Table 4.6 The performance of the four MA variants on instances with 14 decimal digits. The best results are shown in boldface.

	Propagate *after* mutation		Propagate *before* mutation	
n	MA-behavior	MA-matching	MA-behavior	MA-matching
15	**18097490259**	**18097490259**	**18097490259**	**18097490259**
25	**22076988**	**22076988**	**22076988**	**22076988**
35	64093	181249	**58548**	86491
45	158327	39284	95513	**30868**
55	64027	97166	**36719**	50294
65	53855	67961	66396	**45443**
75	**34295**	55204	45654	40678
85	**25623**	37645	58598	35035
95	40189	26460	39309	**21692**
105	29501	34264	37409	**19731**
mean ($\geq$ 35)	58739	67404	54768	**41279**

instances where it is still defeated by some of the new memetic counterparts. The general improvement over the previous methods is clear.

4.7 DISCUSSION AND FUTURE WORK

The computational results obtained are consistent with our *a priori* analysis of weight-matching recombination: on average, the MA with the matching-based recombination (plus the new pocket-propagation strategy) tends to be the best or the second best approach in almost all groups of instances tested. Nevertheless, we believe that a point of caution has to be introduced.

We have developed, at a great effort, a new recombination algorithm based on finding an almost perfectly balanced matching. This has resulted in a new memetic algorithm, which can be generalized to introduce the concept of behaviors, and that has already improved over the best methods we have previously developed for this problem. However, a simple rule that we have introduced in the previous scheme had a significant impact on the final performance. Moreover, we believe that —for this particular problem and this particular type of instances— the performance gain obtained when using a population-based approach does not compensate the cost of coding such a complex method as an MA if we have some other alternatives, like the CKK method or a simple Tabu Search scheme. In our opinion, this is an important conclusion that raises once again the *"killing flies with a gun"* issue. Knowing that no technique is better than any other one in an absolute sense (a crucial fact that the "No Free Lunch" Theorem (Wolpert and Macready, 1997) popularized, but that antedates this result), interdisciplinarity and cross-breeding reveal themselves as the necessary strategies for selecting the appropriate technique for a given problem.

In any case, the really positive results of this type of "balanced matching recombination" indicate that it is a useful and promising idea, that deserves to be further explored in the future. As mentioned in the introduction, there are several real world problems that can be assimilated to the MNP. The enhanced capabilities of Tabu-based MAs endowed with this recombination scheme can be crucial for tackling these problems.

Bibliography

C.C. Aggarwal, J.B. Orlin, and R.P. Tai. Optimized crossover for the independent set problem. *Operations Research*, 45(2):226–234, 1997.

M.F. Arguello, T.A. Feo, and O. Goldschmidt. Randomized methods for the number partitioning problem. *Computers & Operations Research*, 23(2):103–111, 1996.

E. Balas and W. Niehaus. Finding large cliques in arbitrary graphs by bipartite matching. In D.S. Johnson and M.A. Trick, editors, *Cliques, Coloring, and Satisfiability: Second DIMACS Implementation Challenge*, volume DIMACS 26, pages 29–51. American Mathematical Society, 1996.

R. Berretta and P. Moscato. The number partitioning problem: An open challenge for evolutionary computation? In D. Corne, M. Dorigo, and F. Glover, editors, *New Ideas in Optimization*, pages 261–278. McGraw-Hill, 1999.

A. Caprara and M. Fischetti. Branch-and-cut algorithms. In M. Dell'Amico, F. Maffioli, and S. Martello, editors, *Annotated bibliographies in combinatorial optimization*, pages 45 – 63. John Wiley and Sons, Chichester, 1997.

P. Cheeseman, B. Kanefsky, and W.M. Taylor. Where the Really Hard Problems Are. In *Proceedings of the Twelfth International Joint Conference on Artificial Intelligence, IJCAI-91, Sydney, Australia*, pages 331–337, 1991.

L. Davis. *Handbook of Genetic Algorithms*. Van Nostrand Reinhold, New York NY, 1991.

F.F. Ferreira and J.F. Fontanari. Probabilistic analysis of the number partitioning problem. *Journal of Physics A: Math. Gen*, pages 3417–3428, 1998.

F. Glover and M. Laguna. *Tabu Search*. Kluwer Academic Publishers, Norwell, Massachusetts, USA, 1997.

M. Gorges-Schleuter. ASPARAGOS: An asynchronous parallel genetic optimization strategy. In J. David Schaffer, editor, *Proceedings of the Third International Conference on Genetic Algorithms*, pages 422–427, San Mateo, CA, 1989. Morgan Kaufmann Publishers.

M. Gorges-Schleuter. Explicit Parallelism of Genetic Algorithms through Population Structures. In H.-P. Schwefel and R. Männer, editors, *Parallel Problem Solving from Nature I*, volume 496 of *Lecture Notes in Computer Science*, pages 150–159. Springer-Verlag, Berlin, Germany, 1991.

D.S. Johnson, C. R. Aragon, L. A. McGeoch, and C. Schevon. Optimization by simulated annealing: An experimental evaluation; Part II: Graph coloring and number partitioning. *Operations Research*, 39(3):378–406, 1991.

D.R. Jones and M.A. Beltramo. Solving partitioning problems with genetic algorithms. In R.K Belew and L.B. Booker, editors, *Proceedings of the Fourth International Conference on Genetic Algorithms*, pages 442–449, San Mateo, CA, 1991. Morgan Kaufmann.

N. Karmarkar and R.M. Karp. The differencing method of set partitioning. Report UCB/CSD 82/113, University of California, Berkeley, CA, 1982.

S. Kirkpatrick, C.D. Gelatt Jr., and M.P. Vecchi. Optimization by simmulated annealing. *Science*, 220(4598):671–680, 1983.

D. Kirovski, M. Ercegovac, and M. Potkonjak. Low-power behavioral synthesis optimization using multiple-precision arithmetic. In *ACM-IEEE Design Automation Conference*, pages 568–573. ACM Press, 1999.

R. Korf. A complete anytime algorithm for number partitioning. *Artificial Intelligence*, 106:181–203, 1998.

M. Laguna and P. Laguna. Applying Tabu Search to the 2-dimensional Ising spin glass. *International Journal of Modern Physics C - Physics and Computers*, 6(1):11–23, 1995.

E.L. Lawler and D.E. Wood. Branch and bounds methods: A survey. *Operations Research*, 4(4):669–719, 1966.

D.L. Mammen and T. Hogg. A new look at the easy-hard-easy pattern of combinatorial search difficulty. *Journal of Artificial Intelligence Research*, 7:47–66, 1997.

S. Mertens. Phase transition in the number partitioning problem. *Physical Review Letters*, 81(20):4281–4284, 1998.

S. Mertens. Random costs in combinatorial optimization. *Physical Review Letters*, 84(6):1347–1350, 2000.

D.G. Mitchell, B. Selman, and H.J. Levesque. Hard and easy distributions for SAT problems. In P. Rosenbloom and P. Szolovits, editors, *Proceedings of the*

Tenth National Conference on Artificial Intelligence, pages 459–465, Menlo Park, California, 1992. AAAI Press.

P. Moscato. On Evolution, Search, Optimization, Genetic Algorithms and Martial Arts: Towards Memetic Algorithms. Technical Report Caltech Concurrent Computation Program, Report. 826, California Institute of Technology, Pasadena, California, USA, 1989.

P. Moscato. An Introduction to Population Approaches for Optimization and Hierarchical Objective Functions: The Role of Tabu Search. *Annals of Operations Research*, 41(1-4):85–121, 1993.

P. Moscato. Memetic algorithms: A short introduction. In D. Corne, M. Dorigo, and F. Glover, editors, *New Ideas in Optimization*, pages 219–234. McGraw-Hill, 1999.

P. Moscato and C. Cotta. A gentle introduction to memetic algorithms. In F. Glover and G. Kochenberger, editors, *Handbook of Metaheuristics*, pages 105–144. Kluwer Academic Publishers, Boston MA, 2003.

P. Moscato and M. G. Norman. A Memetic Approach for the Traveling Salesman Problem Implementation of a Computational Ecology for Combinatorial Optimization on Message-Passing Systems. In M. Valero, E. Onate, M. Jane, J. L. Larriba, and B. Suarez, editors, *Parallel Computing and Transputer Applications*, pages 177–186, Amsterdam, 1992. IOS Press.

H. Mühlenbein. Evolution in Time and Space – The Parallel Genetic Algorithm. In Gregory J.E. Rawlins, editor, *Foundations of Genetic Algorithms*, pages 316–337, San Mateo, CA, 1991. Morgan Kaufmann Publishers.

M.G. Norman and P. Moscato. A competitive and cooperative approach to complex combinatorial search. In *Proceedings of the 20th Informatics and Operations Research Meeting*, pages 3.15–3.29, Buenos Aires, 1991.

C.H. Papadimitriou. *Computational Complexity*. Addison-Wesley, 1994.

N.J. Radcliffe. The algebra of genetic algorithms. *Annals of Mathematics and Artificial Intelligence*, 10:339–384, 1994.

N.J. Radcliffe and P.D. Surry. Fitness Variance of Formae and Performance Prediction. In L.D. Whitley and M.D. Vose, editors, *Proceedings of the Third Workshop on Foundations of Genetic Algorithms*, pages 51–72, San Francisco, 1994a. Morgan Kaufmann.

N.J. Radcliffe and P.D. Surry. Formal Memetic Algorithms. In T. Fogarty, editor, *Evolutionary Computing: AISB Workshop*, volume 865 of *Lecture Notes in Computer Science*, pages 1–16. Springer-Verlag, Berlin, 1994b.

W. Ruml. Stochastic approximation algorithms for number partitioning. Technical Report TR-17-93, Harvard University, Cambridge, MA, USA, 1993. available via ftp://das-ftp.harvard.edu/techreports/tr-17-93.ps.gz.

W. Ruml, J.T. Ngo, J. Marks, and S.M. Shieber. Easily searched encodings for number partitioning. *Journal of Optimization Theory and Applications*, 89(2): 251–291, 1996.

R. Slootmaekers, H. Van Wulpen, and W. Joosen. Modelling genetic search agents with a concurrent object-oriented language. In P. Sloot, M. Bubak, and B. Hertzberger, editors, *High-Performance Computing and Networking*, volume 1401 of *Lecture Notes in Computer Science*, pages 843–853. Springer, Berlin, 1998.

B.M. Smith and M.E. Dyer. Locating the phase transition in binary constraint satisfaction. *Artificial Intelligence*, 81(1–2):155–181, 1996.

G. Sorkin. *Theory and Practice of Simulated Annealing on Special Energy Landscapes*. Ph.d. thesis, University of California at Berkeley, Berkeley, CA, 1992.

R.H. Storer. Number partitioning and rotor balancing. In *Talk at the INFORMS Conference, Optimization Techniques Track, TD15.2*, 2001.

R.H. Storer, S.W. Flanders, and S.D. Wu. Problem space local search for number partitioning. *Annals of Operations Research*, 63:465–487, 1996.

R. Tanese. Distributed genetic algorithms. In J.D. Schaffer, editor, *Proceedings of the Third International Conference on Genetic Algorithms*, pages 434–439, San Mateo, CA, 1989. Morgan Kaufmann.

D.H. Wolpert and W.G. Macready. No free lunch theorems for optimization. *IEEE Transactions on Evolutionary Computation*, 1(1):67–82, 1997.

Metaheuristics: Computer Decision-Making, pp. 91-125
Urszula Boryczka and Mariusz Boryczka

5 MULTI-CAST ANT COLONY SYSTEM FOR THE BUS ROUTING PROBLEM

Urszula Boryczka[1] and Mariusz Boryczka[1]

[1] Institute of Computer Science
University of Silesia
Bedzińska 39, 41-200 Sosnowiec, Poland
uboryczk@us.edu.pl, boryczka@us.edu.pl

Abstract: MCACS-BRP, a new Ant Colony Optimization (ACO) based approach to solve the Bus Routing Problem is presented. MCACS is an extension of ACO, where two hierarchically connected casts of ants optimize two different objective functions. In MCACS-BRP, ants collaborate using information about the best results obtained in the particular cast. Experiments with real data from the Municipal Public Transport Union of the Upper Silesian Industrial District (KZK GOP) show that MCACS-BRP is worth further experiments and extensions.
Keywords: Ant colony system, Bus routing problem, Multi-cast ant colony system, Self-organization, Cooperation.

5.1 INTRODUCTION

This paper investigates a routing problem in a public transportation network, more precisely, the problem of finding an optimal journey plan in the Silesian Bus Network, given a timetable. The following issues are discussed: a graph model of a bus network that takes into account changes of bus lines, and algorithms based on a Multi-Cast Ant Colony Optimization, which finds an optimal connection by different optimality criteria given.

In our previous research, we analyzed an Ant System (AS) for the Bus Routing Problem (BRP). Numerical experiments showed that the evaluation function applied caused many difficulties and system performance was unsatisfactory. We now try to improve the system, using the idea of Multi-Cast Ant Colony System (MCACS).

Multi-Cast Ant Colony Optimization is a new method for BRP, where two hierarchically connected casts of ants solve the BRP by analyzing two optimization criteria. Moreover, miscellaneous constraints may be imposed on the communication media between the casts of ants and on the rules used by the casts.

Several scheduling problems in a transportation network have been extensively presented in the literature, e.g. Daduna and Wren (1988); Dias and e Cunha (2000); Engelhardt-Funke and Kolonko (2000); Freling (2000); Kwan et al. (2000) and Meilton (2000). There are a few commercial products aimed at train or bus scheduling and crew rostering, used or introduced a few years ago. For instance, the British BUSMAN, the German HOT, and the Canadian HASTUS (Rousseau, 2000; 1985; Schrijver, 1993; Serna and Bonrostro, 2000). However, much less attention has been paid to the problem of finding an optimal route in a transportation network, although some working systems are known. In Europe, the Dutch Journey Planner is probably the most advanced (Tulp, 1993). Unfortunately, technical information about existing systems, considering their commercial value, is not freely available. The purpose of the paper is, at least partly, to fill the gap (especially in Poland).

In our paper, we try to gain an insight into the routing problem based on our investigation and experience gained during the experiments with the Silesian bus network in the Upper Silesian district in Poland. Below, we state the problem. In Section 5.2, we analyze shortest path problems in a transportation model. In Section 5.3, we describe the bus routing problem. In Section 5.4, we present different Ant Colony Optimization algorithms and different rules for pheromone-updating and node transition. In Section 5.5, we focus on the architecture of the MCACS and relations and methods of communication between the casts of ants in MCACS. In that section, we formally describe algorithms used in Multi-Cast Ant Colony System for Bus Routing Problem (MCACS-BRP). Finally, in Section 5.6, we shortly describe our experiments with a computer implementation of MCACS-BRP.

5.2 SHORTEST PATH PROBLEMS IN TRANSPORTATION MODELS

Since the end of the 1950s, shortest path problems have been among the most studied network flow optimization problems. One of the most interesting application fields is transportation. In many transportation problems, in fact, shortest path problems of different kinds need to be solved. These include both classical problems, e.g. to determine shortest paths (under various measures, such as length or cost) between given origin/destination pairs in a certain area, and also non-standard variants, e.g. to compute shortest paths either under additional constraints or on specially structured graphs.

Shortest path routines are never absent from any computer code used in transportation analysis and planning. This explains why, together with the increasing role of large scale mathematical models, interest in efficient shortest path algorithms has been growing, and numerous algorithms have been proposed.

Interesting problems that frequently arise in transportation applications are the so-called *dynamic shortest path problems,* where the factor "time" is taken into consideration. Applications concern, for example, street networks, real-time intelligent vehicles, and so on (see, for instance, Kaufman and Smith (1993); Palma et al. (1993), and Nachtigall (1995)).

Given a directed graph $G = (N, A)$, in dynamic problems, a travel time or delay $d_{ij}(t)$ is associated with each arc (i, j) with the following meaning: if t is the (nonnegative) departure time from node i, then $t + d_{ij}(t)$ is the arrival time at node j. In addition to the delay, a time-dependent cost $c_{ij}(t)$ is generally associated with (i, j) as well. This is the cost of traveling from i to j through (i, j) starting at time t. Furthermore, there is the possibility of waiting at the nodes. In particular, a (unit time) waiting cost $w_i(t)$ can be associated with each node i, which gives the (unit time) cost of waiting at i at time t.

Different models have been defined and analyzed in the literature depending on the properties of the delay functions (e.g. continuous or discrete), on the possibility of waiting at the nodes (e.g. no waiting, waiting at each node, waiting only at the root node), and on the choice of the departure time from the root node (in particular, dynamic shortest paths for a fixed departure time or for all possible departure times). As far as the models based on continuous delay functions are concerned, we refer to Orda and Rom (1990) and to Orda and Rom (1991). Here, we shall limit ourselves to the case of discrete models, i.e. we assume that the time variable t can vary in the discrete set $T = \{t_1, t_2, \ldots, t_q\}$, and that, when applied to a time instant $t \in T$, each delay function $d_{ij}(t)$ is such that $t + d_{ij}(t) \in T$. This is not a loss of generality in the context under consideration, since in the transportation field, time "discretization" is generally performed.

In many transportation problems, paths are characterized by two or more attributes: length, cost, time, risk of congestion, and so on. In some cases, by combining the original attributes, the different attributes can be condensed into a new unique attribute, the "generalized cost". In these cases, paths with a minimum generalized cost are sought. However, in some applications, the condensation of the attributes is either impossible or discouraged, and direct approaches must be followed to find optimal solutions. For example, if we want to find all the non-dominated paths under the two criteria "cost" and "time". It is easy to show that a unique optimal path could not exist in this situation, and that even a large set of non-dominated paths might represent the "target" solution set. For the sake of simplicity, let us consider the bicriterion shortest path problem. Given the graph $G = (N, A)$, assume that two measures c_{ij} and w_{ij} are associated with each arc $(i, j) \in A$, which hereafter will be called the cost and the weight of (i, j), respectively. Let P be the set of the arcs in a path from a given origin to a given destination, and let $C(P)$ and $W(P)$, i.e. the cost and the weight of P, denote two functions of the costs and the weights, respectively, of the arcs belonging to P:

$$C(P) = f(\{c_{ij} : (i, j) \in P\}), \qquad W(P) = g(\{w_{ij} : (i, j) \in P\}).$$

Let Π denote the family of all the feasible subsets of arcs, i.e. the arcs in the paths from the given origin to the given destination. $P \in \Pi$ is a dominated path if there exists a path $P' \in \Pi$ such that $C(P) \geq C(P')$, $W(P) \geq W(P')$, and at least one strict inequality holds. Otherwise, P is said to be non-dominated (or efficient). The bicriterion shortest path problem is to find the sub-family $\Pi_E \subseteq \Pi$ formed by all the efficient paths from the given origin to the given destination. The general bicriterion shortest path problem is NP-hard. Hansen (1980) studied when the problem can be solved in polynomial time, depending on the properties of functions f and g.

Mote et al. (1991) proposed a parametric algorithm, based on linear programming relaxation, for determining all the efficient paths. Warburton (1984) and Warburton (1987) described exact and approximate algorithms for the problem solution. In particular, Warburton (1987) proposed some fully polynomial approximation schema for approximating all the efficient paths. Generalizations of fully polynomial approximation schema for multicriteria shortest path problems are suggested in Safer and Orlin (1995).

Transit services in urban areas are frequently organized so that passengers do not know the exact arrival time of the buses at the stops, but only the frequency of each bus line. As shown in Nguyen and Pallottino (1988), the behavior of passengers cannot be represented by a classical path, since when passengers wait at a certain stop to start (or continue) their trip toward their destination, passengers select "a priori" a subset of lines from the lines serving that stop, which is called the attractive set. When passengers wait at a stop, they do not board buses serving a non-attractive line, but rather board the first bus to arrive at that stop that is serving an attractive line.

If we want to draw the passenger's "a priori" strategy defined by the attractive sets, we should consider a set of subpaths leaving the same origin node and arriving at subsequent stops along the attractive sets. Then, we should consider other subpaths outgoing from the new stops and use the attractive sets, and so on, until the destination node. Each path from the origin to the destination using only attractive lines is in fact an "alternative" for the passenger's trip. The partial graph defined by all these alternatives is a "stream" of paths from the origin to the destination of the trip, called a hyperpath. Depending on the arrivals of the buses at the stops, the passenger will travel along exactly one of the paths of the stream, with a certain probability.

In the case of equilibrium transit assignment, it is possible to formulate the problem in terms of hypergraphs, and based on this formulation, solve the problem by iteratively computing "shortest hyperpaths" (Gallo et al., 1992).

5.3 BUS ROUTING PROBLEM

The transportation system is one of the basic components of an urban area's social, economic, and physical structure. One of the major challenges faced today is to ensure that cities have operationally and economically efficient public transportation system, which enhances their environment, reduces congestion, and conserves energy. Therefore, operation of buses and particularly that of urban bus routes is one potential area of study. Effective design of transit routes and service frequencies can decrease the overall cost of providing transit service, which is generally comprised of user costs and operator costs. Below, we present some observations concerning the role of the Municipal Transport Union of the Upper Silesian Industrial District.

The Silesian Region covers an area of over twelve thousand square kilometers and is inhabited by 4.8 million people (over 12% of the population of Poland). The population density stands at over 398 people per square kilometer, with the national average of 123 people per square kilometer. There are 68 cities in the region, 22 of which are large metropolitan centers. The Katowice agglomeration, with 2.5 million inhabitants, is the most central and urbanized part of the region. The Municipal Public Transport Union of the Upper Silesian Industrial District (KZK GOP) was incorporated in October 1991. Statutorily, the role of the KZK GOP is to organize public transport according to local needs of the union member municipalities.

In July 1993, the KZK GOP took over the role of organizing the bus transport system. The Union has been developing in a very dynamic way. It has doubled the number of its member municipalities in the 9 years of its existence. Currently, 23 municipalities participate in the Union and 10 of them are the main cities of the region. The Union's operating area covers almost entirely the Katowice agglomeràtion and some neighboring areas having a total area standing at 1.4 thousand square kilometers, inhabited by over two million people.

The statutory task of the KZK GOP is to organize urban transport in the Katowice agglomeration. The mission of the Union is the effective rendering of transport services for citizens of the agglomeration by applying market mechanisms and complying to the public transport policy realized by local self-governments that constitute the Union. Presently, many projects has been concerned with application of new optimization algorithms to the mentioned bus routing problem.

In this paper, a simplified version of the transportation network reduced to 1210 bus-stops and 250 bus-lines is analyzed. Figure 5.1 presents an exemplary part of this network, where two tours from bus-stop 857 to bus-stop 939 are shown:

tour 1: 857 → 864 → 872 → 879 → 928 → 937 → 938 → 939

tour 2: 857 → 864 → 872 → 879 → 880 → 919 → 930 → 939

The Bus Routing Problem is a classical network optimization problem, which has received special attention of researchers since the 1950s. However, one objective function may not be sufficient to describe real world problems. In fact, in different fields of science, problems arise that are described as a multiobjective optimal path problem, since we need to deal with a lot of information about the same solution. With this information we can define several objectives that we want to optimize. This problem, in the context of shortest path, has received the attention of some researchers (Climaco and Martins, 1980;

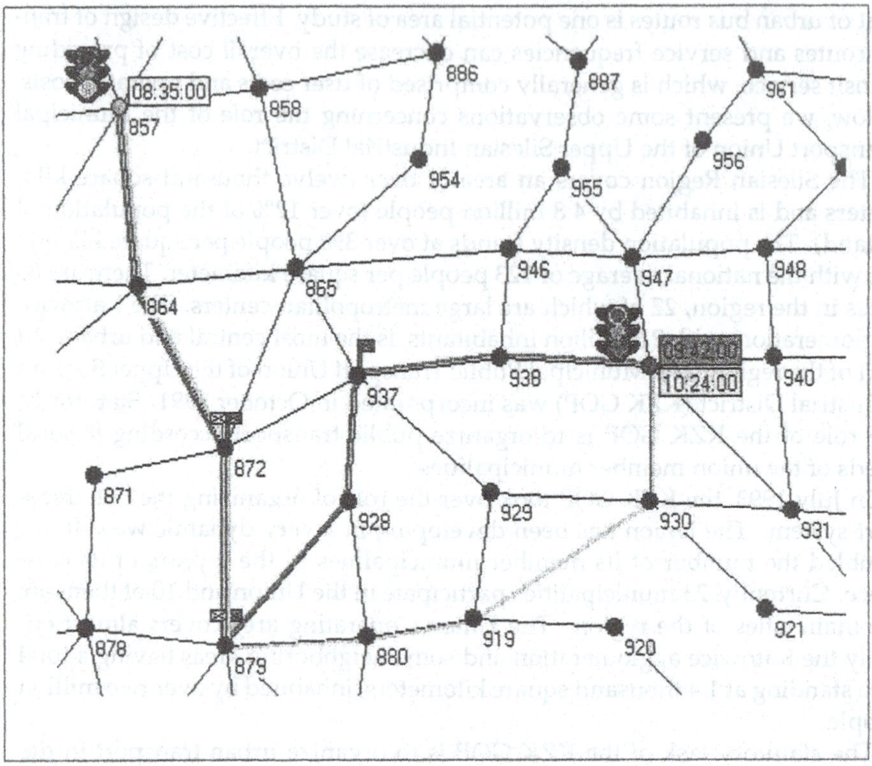

Figure 5.1 An example of the bus network.

1982; Hansen, 1980; Martins, 1984). Despite the importance of this problem, few papers have been written on this subject and the greater part of them refer only the bicriterion path problem.

In the multiobjective optimal path problem, a conflict usually exists among the objectives, i.e. the optimal solution according some objective can be different than the optimal solution according another one. In this situation, an optimal solution does not exists in the usual sense. We use an order relation to define an optimal solution for this relation, which usually is not unique and it is called a non-dominated solution. It is important to point out that, in this case, there are several non-dominated solutions and we must not restrict ourselves to computing only one of them. To overcome this problem, we can associate weights with each objective and define a linear combination of these objectives to obtain a real function that we can optimize in the usual way. This is called a parameterization of the multiobjective optimal path problem. However, this is not a convenient way of solving this problem, since non-dominated solutions, which are not optimal solutions of any parameterization, can exist. Therefore, if we compute all non-dominated solutions, the choice of the best solution has to be done by someone with a knowledge of the real world problem. BRP is a version of time-dependent, dynamic, bicriterion shortest path problem, analyzed for the bus network.

Let us describe the problem of representation of our transportation network more precisely. We have a directed network or a di-multigraph $G = (N, A)$, where $N = 1, \ldots, n$ is a set of nodes (bus-stops) and $A = (i, j)$ is a finite set of directed edges (arcs joining nodes). Parallel edges are allowed. Each edge $(i, j) \in A$ carries two attributes denoted by (c_{ij}, t_{ij}). For simplicity, assume that c_{ij} is the cost using edge (i, j) and t_{ij} is the travel time from node i to node j. The travel time matrix is asymmetric, i.e. $(t_{ij} \neq t_{ji})$. Each node x carries a waiting time w_x. The objective is to find a "shortest" path from a particular node, the origin node $s \in N$, to another particular node, the destination node $e \in N$.

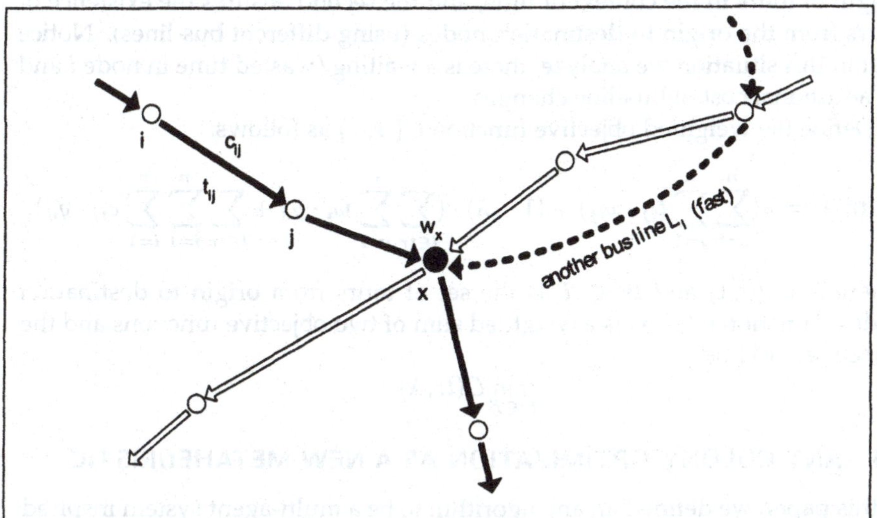

Figure 5.2 An example of different bus-lines connecting the same pair of bus-stops.

Figure 5.2 presents two bus-lines connecting the same pair of bus-stops. In this version of BRP, we must take into account the time t_{ij} needed to pass from bus-stop i to bus-stop j, its cost c_{ij}, and the expected (waiting) time w_i in the case of changes inside a route.

The problem of finding an optimal route in a transportation network in relation to a bus network can be formulated as follows. We are given a set of points (bus-stops) and bus-lines, consisting of a set of edges (every one with a specific length), connecting pairs of these points. Several bus-lines can connect a pair of points. There is a set of bus-lines (each bus-line is defined by a list of places where the bus stops), times of arrivals (a_i), and times of departures (d_i). Each bus-line is assigned a type (e.g. slow, fast) that defines a special ticket price (c_{ij}).

Define binary decision variables as follows:

$$x_{ij} = \begin{cases} 1, & \text{if edge } (i,j) \text{ is used in the path,} \\ 0 & \text{otherwise,} \end{cases}$$

$$y_{il} = \begin{cases} 1, & \text{if the new bus-line } l \text{ is chosen in node } i, \\ 0 & \text{otherwise.} \end{cases}$$

If $x_{ij} = 1$, then $d_i + t_{ij} = d_j$, for all $(i,j) \in A$, where d_i and d_j are the departure times from nodes i and j, respectively. If $y_{il} = 1$, then $a_i + y_{il} \cdot w_i = d_i$ and $a_i + y_{il} \cdot w_i + t_{ij} = d_j$. These equations represent the time constraints, i.e. the time wasted during the bus-line changes.

There are two types of constraints. The first constraint ensures that the length of tours in the context of time, and the second ensures the existence of tours from the origin to destination nodes (using different bus-lines). Notice that in this situation we analyze, there is a waiting/wasted time in node i and an additional cost of bus-line changes.

Define the weighted objective function $C(tr, \lambda)$ as follows:

$$C(tr, \lambda) = \lambda(\sum_{i=1}^{n}\sum_{j=1}^{n} t_{ij} \cdot x_{ij}) + (1 - \lambda) \cdot (\sum_{l \in tr}\sum_{i=1}^{n} w_i \cdot y_{il} + \sum_{l \in tr}\sum_{j=1}^{n}\sum_{i=1}^{n} c_{ij} \cdot y_{il}),$$

where $\lambda \in (0,1)$ and $tr \in T$ is the set of tours from origin to destination nodes. Function $C(tr, \lambda)$ is a weighted sum of two objective functions and the objective is to find:

$$\min_{tr \in T} C(tr, \lambda).$$

5.4 ANT COLONY OPTIMIZATION AS A NEW METAHEURISTIC

In this paper, we defined an ant algorithm to be a multi-agent system inspired by the observation of some real ant colony behavior exploiting the stigmergic communication paradigm. The optimization algorithm that we propose in this paper was inspired by previous work on ant systems and, in general, by the term *stigmergy*. This phenomenon was first introduced by Grasse (1959) and Grasse (1984).

The last two decades have been highlighted by the development and the improvement of approximate solution methods for combinatorial optimization problems, usually called heuristics and metaheuristics. In the context of combinatorial optimization, the term heuristic is used as a contrast to methods that are guaranteed to find a global optimum, such as branch and bound or dynamic programming. A heuristic is defined by Reeves (1995) as a technique which seeks good (i.e. near-optimal) solutions at a reasonable computational cost without being able to guarantee either feasibility or optimality, or even in many cases to state how close to optimality a particular feasible solution is. Often heuristics are problem-specific, so that a method which works for one problem cannot be used to solve a different one. In contrast, metaheuristics are powerful techniques applicable generally to a large number of problems. A metaheuristic refers to an iterative master strategy that guides and modifies the operations of subordinate heuristics by combining intelligently different concepts for exploring and exploiting the search space (Glover and Laguna, 1997; Osman and Laporte, 1996). A metaheuristic may manipulate a complete (or incomplete) single solution or a collection of solutions at each iteration. The family of metaheuristics includes, but is not limited to, Constraint Logic Programming, Genetic Algorithms, Evolutionary Methods, Neural Networks, Simulated Annealing, Tabu Search, Non-monotonic Search Strategies, Scatter Search, and their hybrids. The success of these methods is due to the capacity of such techniques to solve in practice some hard combinatorial problems.

Traditional and well-established heuristic optimization techniques, such as Random Search, Local Search (Boffey, 1995), or the class of Greedy Heuristics (GH) (Boffey, 1995), may be considered as metaheuristics.

Considering that greedy heuristics are available for most practical optimization problems and often produce good results, it seems, in these cases, less expensive with regard to development costs to further improve their solution quality by extending them to repetitive procedures than to replace them by iterative heuristics which follow completely different optimization strategies. So it seems desirable to have a constructive and repetitive metaheuristics including GH as a special (boundary) case.

An essential step in this direction was the development of Ant System (AS) by Dorigo (1992); Dorigo et al. (1991); Dorigo and Gambardella (1996). An AS is a type of heuristic inspired by analogies to the foraging behavior of real ant colonies, which has proved to work successfully in a series of experimental studies. Diverse modifications of AS have been applied to many types of discrete optimization problems and have produced very satisfactory results (Dorigo et al., 1999). Recently, the approach has been extended by Dorigo and Caro (1999) to a full discrete optimization metaheuristic, called the Ant Colony Optimization (ACO) metaheuristic.

AS, which was the first ACO algorithm (Colorni et al., 1991; Dorigo and Gambardella, 1997a; Dorigo et al., 1991) was designated as a set of three ant algorithms differing in the way the pheromone trail was updated by ants. Their names were: ant-density, ant-quantity, and ant-cycle. A number of algorithms, including the metaheuristics, were inspired by ant-cycle, the best performing of the ant algorithms (Figure 5.3).

The Ant Colony System (ACS) algorithm was introduced by Dorigo and Gambardella (1996; 1997b) to improve the performance of Ant System (Dorigo and Gambardella, 1997b; Gambardella and Dorigo, 1997), which allowed good solutions to be found within a reasonable time for small size problems only. The ACS is based on three modifications of Ant System:

- a different node transition rule,
- a different pheromone trail updating rule,

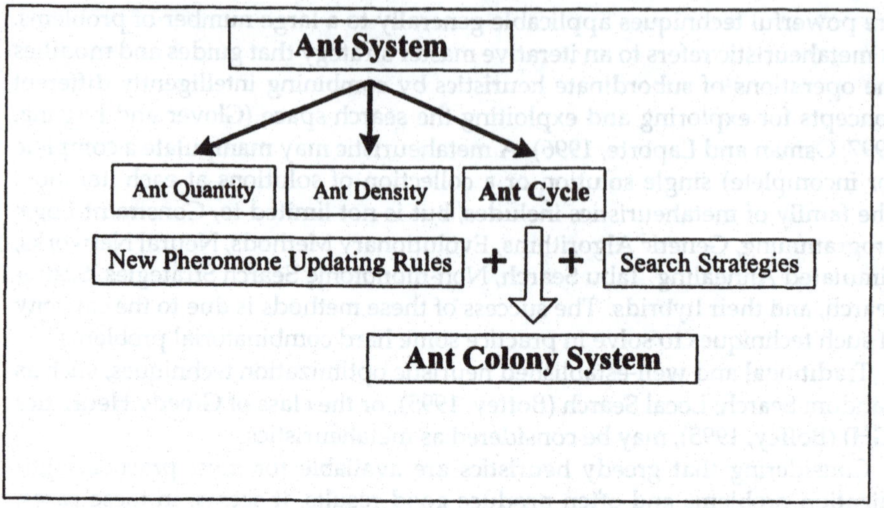

Figure 5.3 Descent of ACS.

- the use of local and global pheromone updating rules (to favor exploration).

The node transition rule is modified to allow explicitly for exploration. An ant k in city i chooses the city j to move to following the rule:

$$j = \begin{cases} \text{argmax}_{u \in J_i^k} \{[\tau_{iu}(t)] \cdot [\eta_{iu}]^\beta\} & \text{if } q \leq q_0 \\ J & \text{if } q > q_0 \end{cases}$$

where q is a random variable uniformly distributed over $[0, 1]$, q_0 is a tuneable parameter ($0 \leq q_0 \leq 1$), and $J \in J_i^k$ is a city that is chosen randomly according to a probability

$$p_{iJ}^k(t) = \frac{\tau_{iJ}(t) \cdot [\eta_{iJ}]^\beta}{\sum_{l \in J_i^k} [\tau_{il}(t)] \cdot [\eta_{il}]^\beta},$$

which is similar to the transition probability used by Ant System. We see, therefore, that the ACS transition rule is identical to the Ant System's when $q > q_0$, and is different when $q \leq q_0$. More precisely, $q \leq q_0$ corresponds to an exploitation of the knowledge available about the problem, that is, the heuristic knowledge about distances between cities and the learned knowledge memorized in the form of pheromone trails, whereas $q > q_0$ favors more exploration.

In Ant System, all ants are allowed to deposit pheromone after completing their tours. In contrast, in the ACS, only the ant that generated the best tour since the beginning of the trail is allowed to globally update the concentrations of pheromone on the branches. The updating rule is

$$\tau_{ij}(t+n) = (1-\rho)\cdot\tau_{ij}(t) + \rho\cdot\Delta\tau_{ij}(t,t+n),$$

where (i,j) is the edge belonging to T^+, the best tour since the beginning of the trail, ρ is a parameter governing pheromone decay, and

$$\Delta\tau_{ij}(t,t+n) = \frac{1}{L^+},$$

where L^+ is the length of the T^+.

The local update is performed as follows. When performing a tour, if ant k is in city i and selects city $j \in J_i^k$ to move to, the pheromone concentration on edge (i,j) is updated by the following formula:

$$\tau_{ij}(t+1) = (1-\rho)\cdot\tau_{ij}(t) + \rho\cdot\tau_0.$$

The value of τ_0 is identical to the initial value of pheromone trails. It was empirically determined that setting $\tau_0 = (n \cdot L_{nn})^{-1}$, where n is the number of cities, and L_{nn} is the length of a tour produced by the nearest neighbor heuristic, produces good results (Dorigo and Gambardella, 1996; Gambardella and Dorigo, 1997).

The ACO metaheuristic has been successfully applied to many discrete optimization problems, as listed in Table 5.1 (Bonabeau et al., 1999). ACO algorithms result to be competitive with the best available heuristic approaches. In particular, results obtained by the application of ACO algorithms to the TSP are very encouraging. They are often better than those obtained using other general purpose heuristics like Evolutionary Computation or Simulated Annealing. Also, when adding local search procedures based on 3-opt to ACO algorithms, the quality of the results is close to that obtainable by other state-of-the-art methods. ACO algorithms are currently one of the best performing heuristics available for the particularly important class of quadratic assignment problems which model real world problems. AntNet, an ACO algorithm for routing in packet switched networks, outperformed a number of state-of-the-art routing algorithms for a set of benchmark problems. AntNet-FA, an extension of AntNet for connection oriented network routing problems, also shows competitive performance. HAS-SOP, an ACO algorithm coupled to a local search routine, has improved many of the best known results on a wide set of benchmark instances of the sequential ordering problem (SOP), i.e. the problem of finding the shortest Hamiltonian path on a graph which satisfies a set of precedence constraints on the order in which cities are visited. ACO algorithms have also been applied to a number of other discrete optimization problems like the shortest common supersequence problem, the multiple knapsack, single machine total tardiness, and others, with very promising results.

Multiple Ant Colony System for Vehicle Routing Problem with Time Windows (MACS-VRPTW) is a new ACO based approach to solve vehicle routing problems with time windows. To adapt ACS for multiple objectives in VRPTW, the idea is to define two colonies, each dedicated to the optimization of a different objective function. In MACS-VRPTW, the number of tours

Table 5.1 Current applications of ACO algorithms.

Problem	Algorithm	Main references
Traveling salesman	AS	(Dorigo et al., 1991)
		(Dorigo, 1992)
		(Dorigo et al., 1996)
	Ant-Q	(Gambardella and Dorigo, 1995)
	ACS and ACS-3-opt	(Dorigo and Gambardella, 1997b)
		(Dorigo and Gambardella, 1997a)
		(Gambardella and Dorigo, 1996)
	AS	(Stützle and Hoos, 1997b)
		(Stützle and Hoos, 1997a)
	ASrank	(Bullnheimer et al., 1997b)
Quadratic assignment	AS-QAP	(Maniezzo et al., 1994)
	HAS-QAP	(Gambardella et al., 1997)
		(Gambardella et al., 1999b)
	MM AS-QAP	(Stüzle and Hoos, 1998)
	ANTS-QAP	(Maniezzo, 1998)
	AS-QAP	(Maniezzo and Colorni, 1999)
Scheduling problems	AS-JSP	(Colorni et al., 1994)
	AS-FSP	(Stützle, 1997)
	ACS-SMTTP	(Bauer et al., 1999)
	ACS-SMTWTP	(den Besten et al., 1999)
Vehicle routing	AS-VRP	(Bullnheimer and Strauss, 1996)
		(Bullnheimer et al., 1997a)
		(Bullnheimer et al., 1998)
	HAS-VRP	(Gambardella et al., 1999a)
Connection-oriented network routing	ABC	(Schoonderwoerd et al., 1996)
		(Schoonderwoerd et al., 1997)
	ASGA	(White et al., 1998)
	AntNet-FS	(DiCaro and Dorigo, 1998c)
	ABC-smart ants	(Bonabeau et al., 1998)
Connection-less network routing	AntNet and AntNet-FA	(DiCaro and Dorigo, 1998a)
		(DiCaro and Dorigo, 1998b)
		(DiCaro and Dorigo, 1998d)
	Regular ants	(Subramanian et al., 1997)
	CAF	(Heusse et al., 1998)
	ABC-backward	(van der Put, 1998)

Table 5.1 (continued)

Problem name	Algorithm	Main references
Sequential ordering	HAS-SOP	(Gambardella and Dorigo, 1997)
Graph coloring	ANTCOL	(Costa and Hertz, 1997)
Shortest common supersequence	AS-SCS	(Michel and Middendorf, 1998) (Michel and Middendorf, 1999)
Frequency assignment	ANTS-FAP	(Maniezzo and Carbonaro, 1998) (Maniezzo and Colorni, 2000)
Generalized assignment	MMAS-GAP	(Lourenço and Serra, 1998)
Multiple knapsack	AS-MKP	(Leguizamón and Michalewicz, 1999)
Optical networks routing	ACO-VWP	(Varela and Sinclair, 1999)
Redundancy allocation	ACO-RAP	(Liang and Smith, 1999)

(or vehicles) and the total travel time are minimized simultaneously by two ACS based colonies. They are ACS-VEI and ACS-TIME. The goal of the first colony is to diminish the number of vehicles used, whereas the second colony optimizes the feasible solutions found by ACS-VEI. Each colony uses its own pheromone trails, but collaborates with the other by sharing information about the best results obtained so far. MACS-VRPTW is shown to be competitive with the best existing methods both in terms of solution quality and computation time. Moreover, MACS-VRPTW improves good solutions for a number of problem instances in the literature.

5.5 MULTI-CAST ANT COLONY SYSTEM

Multi-Cast Ant Colony System were firstly analyzed for VRPTW (Gambardella et al., 1999a). We suggest the new idea of two casts of ants, each dedicated to the optimization of a different objective function. Our approach is motivated by aspect of the real life of ants. We consider this new system because of analogy of ant self-organization and holism theory.

There are many types of ants with very different styles of predation. However, they all have one thing in common. Ants play an important part in nature's population control. Several ant species have been observed making invasions of other colonies, be it other ants or even termite colonies. Ants have attacked termite colonies with a high degree of strategy. It has been experimentally shown that some ant species will attack termite colony more than others. The termite species *Captotermes formosanus* is known to have a frontal gland with fluid in it. As the ants attack the termite mounds, fighting between the soldiers of the two species commences. The ants are able to learn that the termites had a chemical defense, i.e. the fluid produced in the frontal glands. Thus, they did not attack that particular species of termites as much as the other mounds of termites (Reticulitermes). The C. formosanus soldiers ejected a sticky fluid from their glands and therefore had a higher survival rate than the Reticulitermes. The learning capacity of ants is fascinating.

Each ant is genetically identical (identical genotypes), all originating from the same queen, but many ant casts that differ physically (different phenotypes) make up the colony. As the colony functions as a unit, natural selection "tests" the colony, while the ant casts are the agents of the colony. Just as our hands serve as tools for our whole body, the worker ants serve the whole colony. If, through a genetic change, a queen produces workers with unusually small jaws, so that the workers are unable to collect the usual food, then the whole colony fails. Genetic changes or variations originate with the queen. Mutations that improve the performance of one ant cast, improves the survival of the whole colony.

It is obvious that complex cooperative societies exist only when their members are close relatives. In most animal species cooperation is either limited to very small groups or is absent altogether. Among the few animals that cooperate in large groups are social insects like bees, ants, and termites, and the

Naked Mole Rat, a subterranean African rodent. These are the main reasons of creation such distributed but connected by the common goal system.

MCACS is organized with a hierarchy of casts of artificial ants designated to optimize the tours in the bus network according two different groups of optimality criteria. The first cast of ants minimizes the total length of the tour, while the second one minimizes the number of bus changes. Cooperation between these two casts belonging to the same colony is performed by the pheromone medium. It is similar to a real colony of ants, where the system of casts works well for the common goal – the development of the nest as a structure. The architecture of the MCACS for our problem is presented in Figure 5.4.

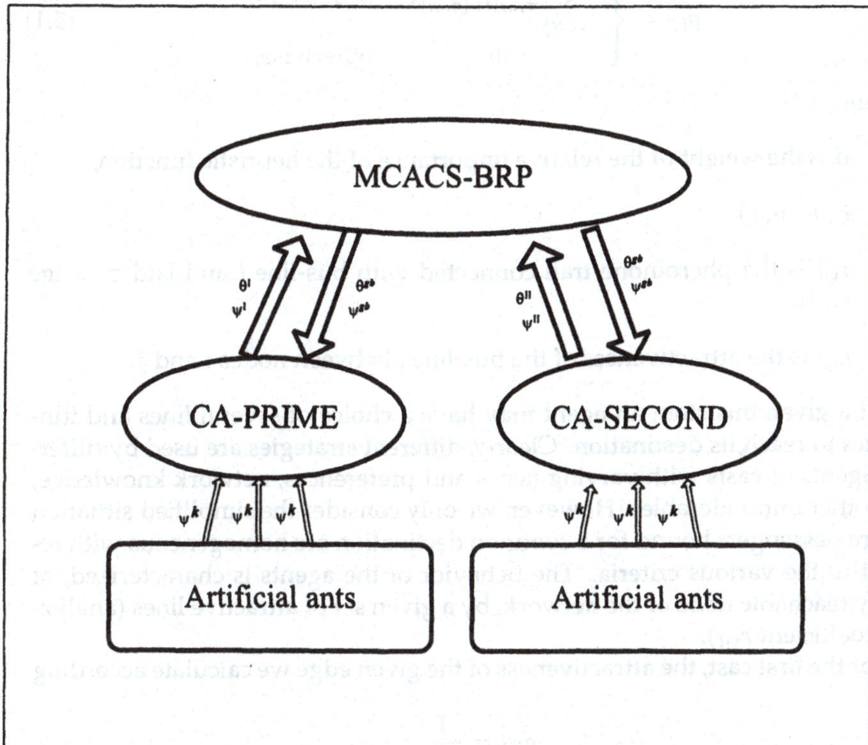

Figure 5.4 Architecture of MCACS-BRP.

Basically, in MCACS-BRP both casts optimize their functions using coordination of their activities. The first cast, CA-PRIME, minimizes the total length of tours, whereas the goal of the second cast, CA-SECOND, is to try to decrease the number of bus-changes. Furthermore, each cast uses its own table of the pheromone, but casts can influence each other by the pheromone. The collaboration between casts concerns sharing information about the best-known solutions obtained so far. More precisely, two casts search the same solution space, but minimize different functions. It is our expectation that the double pheromone updating process will improve the performance of the system searching for the best solutions for two objective functions.

The node transition rules in both casts are similar and may be calculated as follows: When ant k is located in node i, it chooses the next node j randomly from the set of nodes N_i^k (not yet visited). The probabilistic rule used to construct a tour is the following: With probability q_0 a node with the highest value of $\tau_{ijl}(t) \cdot [\eta_{ijl}]\beta$, where $j \in N_i^k$ is chosen (exploitation), while with probability $(1 - q_0)$, node j is chosen with measure $p_{ijl}(t)$ proportional to $\tau_{ijl}(t) \cdot [\eta_{ijl}]^\beta, j \in N_i^k$ (exploration):

$$p_{ijl} = \begin{cases} \frac{\tau_{ijl}(t)\cdot[\eta_{ijl}]^\beta}{\sum_{n\in N_i^k} \tau_{inl}(t)\cdot[\eta_{inl}]^\beta}, & \text{if } j \in N_i^k, \\ 0 & \text{otherwise,} \end{cases} \tag{5.1}$$

where

- β is the weight of the relative importance of the heuristic function,
- $q_0 \in (0,1)$,
- τ_{ijl} is the pheromone trail connected with bus-line l and laid on edge (i,j),
- η_{ijl} is the attractiveness of the bus-line l between nodes i and j.

At a given bus-stop, an agent may have a choice of several lines and itineraries to reach its destination. Clearly, different strategies are used by different agents of casts with varying tastes and preferences, network knowledge, and other imponderables. However, we only consider the simplified situation where passengers bound for a common destination are homogeneous with respect to the various criteria. The behavior of the agents is characterized, at every reachable node of the network, by a given set of attractive lines (analyzing coefficient r_{ijl}).

For the first cast, the attractiveness of the given edge we calculate according to

$$\eta_{ijl} = \frac{1}{\delta_{ijl}},$$

where δ_{ijl} is a length of the edge connected with the bus line l. For CA-SECOND, the attractiveness is represented by

$$\eta_{ijl} = \frac{1}{\delta_{ijl}} \cdot r_{ijl}$$

and

$$r_{ijl} = \begin{cases} 1 & \text{for } l \notin L_d \wedge l_n \neq l_{n-1} \\ 2 & \text{for } l \notin L_d \wedge l_n = l_{n-1} \\ 3 & \text{for } l \in L_d, \end{cases}$$

where r_{ijl} is a coefficient determining the range of bus-lines and l_n is the bus-line connected with node n.

We assume that L_d is a set of bus-lines including the destination point. In this case, the value of 3 is assigned to the bus-line belonging to L_d. When the tour is created without any changes and does not belong to L_d, we assign the value of 2 to appropriate bus line. Finally, the value of 1 is assigned to the bus-line when there are changes and this line does not belong to L_d.

The best result computed by each cast is used in the global pheromone-updating rule where, additionally, the number of changes, C_ψ^{gb} and C_θ^{gb}, is considered:

- CA-PRIME:

$$\tau_{ijl}(t+n) = (1-\rho)\cdot\tau_{ijl}(t) + \rho\cdot\frac{1}{J_\psi^{gb}\cdot(C_\psi^{gb}+1)} \quad \forall i,j,l \in \psi^{gb}$$

- CA-SECOND:

$$\tau_{ijl}(t+n) = (1-\rho)\cdot\tau_{ijl}(t) + \rho\cdot\frac{1}{J_\theta^{gb}\cdot(C_\theta^{gb}+1)} \quad \forall i,j,l \in \theta^{gb}$$

where

J_ψ^{gb} – the length of the shortest tour for CA-PRIME,

J_θ^{gb} – the length of the best tour for CA-SECOND,

C_ψ^{gb} – the number of changes in the shortest tour,

C_θ^{gb} – the number of changes in the best tour for CA-SECOND,

ψ^{gb} – the shortest tour obtained so far,

θ^{gb} – the tour with the minimal number of changes obtained so far.

The local updating rule is used by each ant of the casts according to

$$\tau_{ijl}(t+1) = (1-\rho)\cdot\tau_{ijl}(t) + \rho\cdot\tau_0$$

where

$$\tau_0 = \frac{1}{s\cdot J_\psi^h\cdot(C_\psi^{gb}+1)}$$

is an initial value and is the length of the initial solution obtained by the greedy algorithm, and s is the number of bus-stops.

In MCACS-BRP, both time and cost are optimized simultaneously by two casts of ants: CA-PRIME and CA-SECOND. Casts use separate pheromone trails, and consequently need to analyze two graphs and two data structures. Casts exchange information via variables ψ and θ. ψ is used to store the best solution found so far and is updated each time by one of the casts.

The initial procedures for both casts have two parameters: the set of bus-lines L_d and a Boolean value: *range*. CA-PRIME exploits an original ACS algorithm and its goal is to create a route as short as possible. Each route is created by calling the procedure *CalculateRoute*. When $\psi^i, i = 1, 2, 3, \ldots, m_{cast}$ have been computed, they are compared with variables ψ and θ managed by MCACS-BRP. Every improving solution is sent to MCACS-BRP. Solution ψ is improving when the travel time is less then ψ^{gb}. Solution θ is improving when the number of changes is less than θ^{gb}. After the creation of the solution, the global updating of the pheromone trail is performed for two different solutions ψ and θ. This improves the performance of the algorithm.

Cast CA-SECOND minimizes time and cost of the route. For this cast of ants the Boolean parameter has value *true*. It means that the heuristic function uses information about the range of lines. When we obtain all of the solutions, the comparison between ψ^{gb} and θ^{gb} is performed. MCACS-BRP obtains the result and stores it in ψ^{gb}.

```
procedure MCACS-BRP( )
begin
   ψ^gb ← initial solution for cast CA-PRIME;          Initialization
   θ^gb ← initial solution for cast CA-SECOND;
   L_d ← lines connected with the destination bus stop;
   initialize CA-PRIME( L_d, ψ^gb );
   initialize CA-SECOND( L_d, ψ^gb );
   repeat
      run CA-PRIME( ψ, θ );                              main loop
      run CA-SECOND( ψ, θ );
      wait for an improved solution ψ or solution θ from CA-
      PRIME or CA-SECOND;
      ψ^gb ← ψ;
      θ^gb ← θ;
   until stopping criterion is satisfied
end
```

Algorithm 5.1 The MCACS-BRP algorithm.

Algorithm 5.2 presents a high-level description of CA-PRIME_and_CA-SECOND procedure representing the performance of the casts. and Algorithm 5.3 describe other algorithms used: the main procedure, controlling the MCACS-BRP algorithm and the procedure called from algorithm ACS, calculating the route from a given origin point to a given destination point of the bus network.

5.6 EXPERIMENTS AND RESULTS

The experimental part of this research consisted of three major phases. First, an efficient and effective set of parameters settings was determined. Then, the

procedure ACS($Proc, \psi, \theta$)
begin
  foreach ant k
    { Create solution ψ^k }
    if $Proc$ = CA-PRIME
      $\psi^k \leftarrow$ CalculateRoute(k, *false*);
    else
      $psi^k \leftarrow$ CalculateRoute(k, *true*);
    $\theta^k := \psi^k$;
  endfor
  { Update the best solution if it is improved }
  if $\exists_k \psi^k$ is feasible and $J_\psi^k < J_\psi^{gb}$
    $\psi := \psi^k$;
  if $\exists_k \theta^k$ is feasible and $C_\psi^k < C_\theta^{gb}$
    $\theta := \theta^k$;
  { Perform global updating rule }
  $\tau_{ijl}(t+n) \leftarrow (1-\rho) \cdot \tau_{ijl}(t) + \frac{\rho}{J_\psi^{gb} \cdot (C_\psi^{gb}+1)}, \forall (i,j,l) \in \psi^{gb}$;
  if $\theta^{gb} \neq \psi^{gb}$
    $\tau_{ijl}(t+n) \leftarrow (1-\rho) \cdot \tau_{ijl}(t) + \frac{\rho}{J_\theta^{gb} \cdot (C_\theta^{gb}+1)}, \forall (i,j,l) \in \theta^{gb}$;
end

Algorithm 5.2 ACS working as CA-PRIME or CA-SECOND cast.

MCACS-BRP algorithm was run several times using this setting to compare the behavior of two casts. Finally, the results concerning the number of bus changes and travel time for different tours were examined.

It is known that the good results obtained by the algorithm depend strongly on the parameters affecting the computational formula (5.1). In the case of the presented algorithm, β, ρ, q, and m_{cast} had to be determined. We analyzed the same parameters for casts: CA-PRIME and CA-SECOND. As the starting point for the determination of the role of parameter β, the following values of other parameters were chosen: $q = 0.9$, $\rho = 0.2$, and $m_{cast} = 30$. The number of iterations was equal to 30.

We adjusted experimentally the value of parameter β, observing two histograms: Figure 5.5 and Figure 5.6. The experimental results showed that the value 3 for this parameter is the best one (from two mentioned points of view).

Both the number of bus-changes and the travel time have the same kind (tendency) of fluctuation. Generally, increasing the value of β to 10 caused a significant decrease in solution quality. Experimental observation has shown the similarity between two values of parameter β: 2 and 3. This comparison is clearer from the histograms: Figure 5.5 and Figure 5.6. These values pro-

```
procedure CalculateRoute(k, range )
begin
    $\psi^k$ ← origin bus-stop;
    $i$ ← destination bus-stop;                          destination point
    CurrTime ← Starting time of the tour;
    repeat
        { Starting from node $i$, calculate set $N_i^k$ of feasible bus-
        stops which are not visited yet by ant $k$ }
        GetStops( $i$, CurrTime, $N_i^k$ );
        { $\forall j \in N_i^k$ compute attractiveness $\eta_{ijl}$ as follows: }
        foreach $N_i^k$
            $DriveTime_{ijl} \leftarrow t_{ijl} + w_{il}$;
            if range
                $Distance_{ijl} :\leftarrow DriveTime_{ijl} \cdot r_{ijl}$;
            else
                $Distance_{ijl} \leftarrow DriveTime_{ijl}$;
            $\eta_{ijl} \leftarrow 1/Distance_{ijl}$;
        endfor
        { Choose randomly next node $j$ using $\eta_{ijl}$ in exploita-
        tion and exploration mechanisms }
        $\psi^k \leftarrow \psi^k + j$;
        { Local pheromone updating rule: }
        $\tau_{ijl}(t+1) \leftarrow (1-\rho) \cdot \tau_{ijl}(t) + \rho \cdot \tau_0$;
        $i \leftarrow j$;
        $CurrTime \leftarrow CurrTime + DriveTime_{ijl}$;
    until $j$ = destination or $Count(\psi^k)$ =MAXCOUNT

    { Check if solution is feasible }
    if $j \in \psi^k$
        $\psi^k$ is feasible;
    else
        $\psi^k$ is not feasible;
end
```

Algorithm 5.3 Procedure calculating a route from origin to destination.

duce the same results or very close to each other. Therefore, we may use two different values: 2 and 3, for cast CA-PRIME and CA-SECOND, respectively.

Next, an experiment was carried out to find the best value for parameter ρ. Let us mention that all the initial values of the parameters were chosen in the same way. Trying to understand a mechanism of the searching process in MCACS-BRP, changes of evaporation of the pheromone trail during the performance of the algorithm were studied. We considered two types of

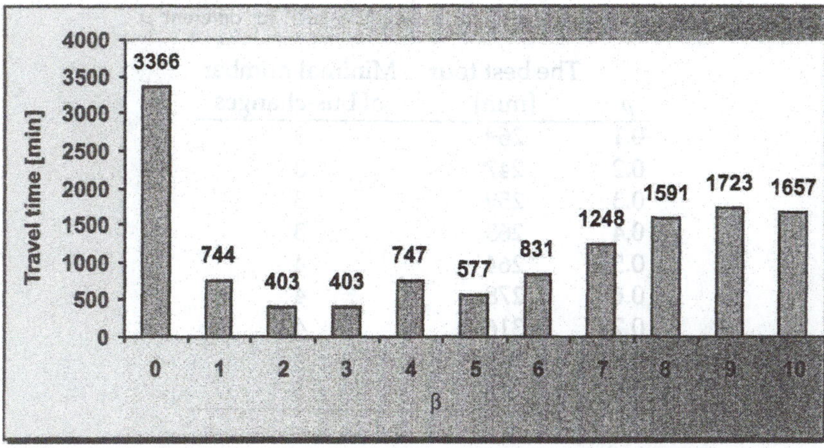

Figure 5.5 Travel time of the best tour for different β.

histograms: the travel time for the best tour and the number of bus changes (similar to the first experiment).

To compare the results found by MCACS-BRP for different values of parameter ρ and demonstrate the characteristics of solutions, we propose two histograms (Figure 5.7 and Figure 5.8). For $\rho \in < 0.1; 0.4 >$, performance of the algorithm is better than for other values of ρ. Results obtained by MCACS-BRP (travel time) clearly shows that the best value was achieved for ρ equal to 0.2, so we favor this value. The effect of increasing the value of ρ is to increase the travel time.

Table 5.2 Results found by MCACS-BRP for different β.

β	The best tour [min]	Minimal number of bus-changes
0	3366	46
1	744	14
2	403	8
3	403	7
4	747	12
5	577	17
6	831	13
7	1248	19
8	1591	25
9	1723	24
10	1657	30

Table 5.3 Results found by MCACS-BRP for different ρ.

ρ	The best tour [min]	Minimal number of bus-changes
0.1	269	3
0.2	247	3
0.3	259	3
0.4	263	3
0.5	264	4
0.6	278	4
0.7	316	4
0.8	342	5
0.9	323	6

After the analysis of the results of the mentioned two experiments, values of the parameters were fixed. Using these values, the algorithm gives better solutions. The values are:

- $\beta = 2$,
- $\rho = 0.2$,
- $q = 0.9$,
- $m_{cast} = 30$

for both casts.

We also analyzed the global parameters, such as:

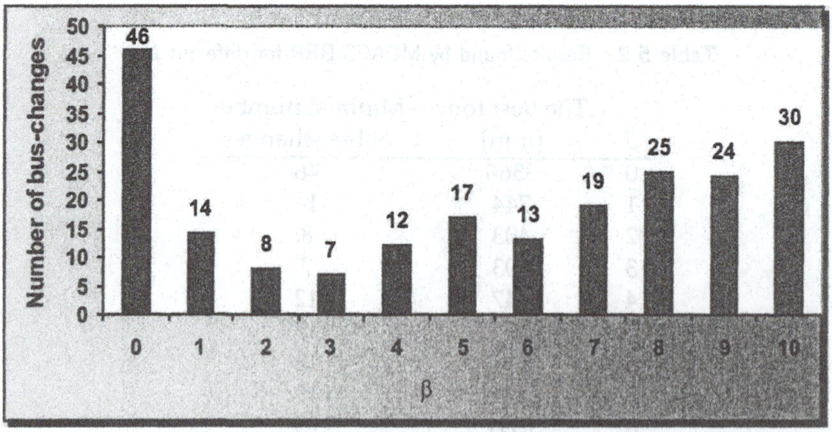

Figure 5.6 Number of bus-changes for different β.

Table 5.4 Best results for casts and MCACS-BRP.

Tour no.	The best tour CA-PRIME		The best tour CA-SECOND		The best tour MCACS-BRP	
	Travel time	Number of changes	Travel time	Number of changes	Travel time	Number of changes
1	442	9	397	5	264	4
2	409	7	409	4	293	4
3	325	4	325	3	263	3
4	231	6	275	6	231	4
5	286	5	275	5	249	5

- the average number of the bus-stops on the tour is 120,
- the number of iterations is 30,
- the starting time is 08:00h.

We report our primary computational results showing the efficiency of MCACS for BRP. The simplified version of the Silesian Bus Network analyzed consists of 120 bus-lines with about 1000 bus-stops. Experiments were made by executing 5 runs for each origin and destination point. The best solution was chosen for each cast and for MCACS and the results are reported in Table 5.4 and Table 5.5.

Table 5.4 shows the best solutions obtained by the casts and MCACS and the number of bus-line changes in different tours. This characterizes that MCACS-BRP is able to produce tours with a short travel time and a small number of bus-line changes. Figure 5.9 and Figure 5.10 report the comparison of two parameters of the tour mentioned above (for casts and for the

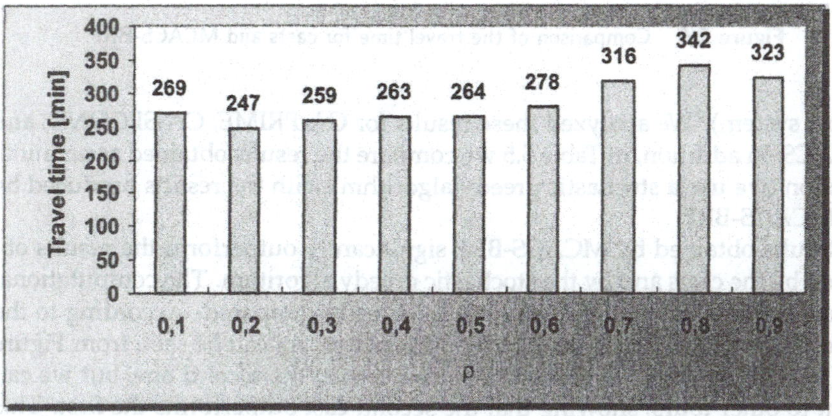

Figure 5.7 Travel time of the best tour for different ρ.

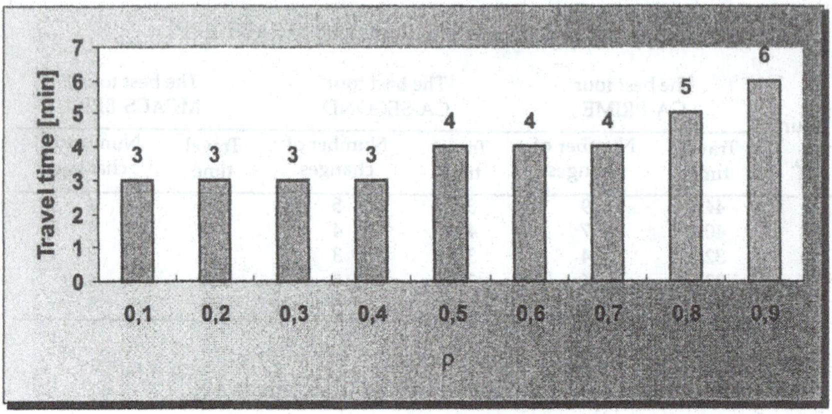

Figure 5.8 Number of bus-changes for different ρ.

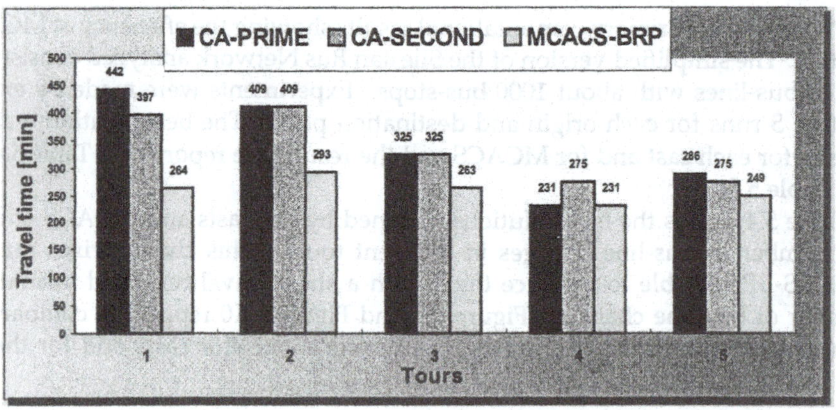

Figure 5.9 Comparison of the travel time for casts and MCACS-BRP.

whole system). We analyzed these results for CA-PRIME, CA-SECOND, and MCACS. In addition, in Table 5.5 we compare the results obtained as an initial solution (we use a stochastic greedy algorithm) with the results produced by the MCACS-BRP.

Results obtained by MCACS-BRP significantly outperform the results obtained by the casts and by the stochastic greedy algorithm. The computational results for two different casts cannot be directly compared. According to the tables, we cannot determine which cast is better. As can be seen from Figure 5.9 (tour 4), the first cast finds better results than the second one, but we can point to other tours, showing that the second cast outperforms the first. This property was exploited in our system. When one cast cannot manage to find a good result, the other one helps it using a new heuristic in its search process.

Although the tables and figures do not give us the complete results for every origin and destination point, nevertheless we may derive some conclusions. It may be possible to say that results obtained by MCACS-BRP increase by 25% the results obtained by CA-PRIME. There was a slight increase of results obtained by CA-SECOND.

Now, we want to analyze different characteristics representing the special abilities of two casts, especially different parameters connected with characteristics of the casts. The problem of cooperation between two casts is undoubtedly worth further analysis, so we decided to analyze different methods of governing the pheromone trails in context of reinforcement and punishment.

To measure the performance of the adaptive, dynamic versions of MCACS described above, test runs were performed using a selection of prespecified values of parameters:

- $\alpha = \rho = \{0.1, 0.2\}$,

- $\beta = \{2, 4, 6, 8\}$,

- $q_0 = \rho = \{0.5, 0.6, 0.7\}$,

- $m_{cast} = 30$,

- number of iterations = 50.

Global parameters were not changed.

Up to now, we considered only one variant of values (unvarying within two casts). To test whether the values have some influence on the results, we applied various sets of parameters and strategies concerning the pheromone trails. Different methods of updating pheromone trails were analyzed. In

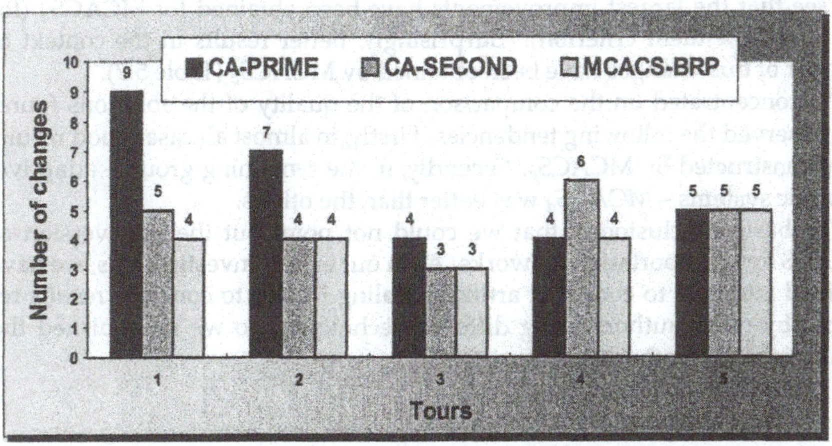

Figure 5.10 Comparison of the number of changes for tours obtained by casts and MCACS-BRP.

Table 5.5 Comparison between MCACS-BRP and the initial solution.

Bus-stop		Time of the tour [min]		No. of bus-changes	
Origin	Destination	MCACS-BRP	Initial solution	MCACS-BRP	Initial solution
1037	353	205	267	3	8
2	1188	403	486	7	12
877	1071	282	339	5	12
870	1039	341	376	5	9
857	939	73	110	2	5
900	819	234	262	3	12
853	808	277	277	6	8
97	445	118	141	2	7
164	752	263	271	3	8
1151	525	140	205	3	6
2	540	264	414	3	14

$MCACS_c$, we used constant values of parameters α, ρ, β and q_0. On the other hand, in $MCACS_d$ we differentiated two casts. Finally, in $MCACS_p$, we punished "bad" behavior of ants in the cast with worse results, and we rewarded "good" results reinforcing the pheromone trail by additional value of pheromone.

We concentrated on the previously analyzed sets of tours and we generated new efficient solutions for each tour. The results are given in Table 5.6 and Table 5.7. As can be seen, three new versions of MCACS did not substantially influence the results. The percentage errors were computed comparing $MCACS_p$ and $MCACS_+$ with $MCACS_d$ (for $MCACS_p$ and $MCACS_+$) and comparing $MCACS_d$ with $MCACS_c$ (for $MCACS_d$). From Table 5.6, one can see that the largest improvements have been obtained for $MCACS_d$ (for the time-dependent criterion). Surprisingly, better results in the context of number of bus-changes have been obtained by $MCACS_c$ (Table 5.7).

We concentrated on the comparison of the quality of the solutions found and observed the following tendencies. Firstly, in almost all cases good results were constructed by $MCACS_c$. Secondly, in the remaining group – adaptive, dynamic systems – $MCACS_d$ was better than the others.

The basic conclusion is that we could not point out the best version of MCACS for transportation networks. As in our earlier investigations, we have avoided attempts to construct artificial scaling factors to compare results reported by other authors using different techniques, so we relinquished the comparison with other metaheuristics.

5.7 CONCLUSIONS

With the number of algorithms implemented, and the computational results obtained, many real-life problems can now be modeled with more than one objective. This leads to a more realistic representation of the problem. Most

Table 5.6 Comparison between different versions of MCACS (time-dependent criterion).

Bus-stop (org.-dest.)	Time of the tour [min]						
	$MCACS_c$	$MCACS_d$	err.	$MCACS_p$	err.	$MCACS_+$	err.
1037-353	205	265	29.0	326	-16.1	234	13.2
877-1071	282	249	-11.7	342	-27.1	249	0.0
870-1039	341	333	-2.3	333	0.0	333	0.0
857-939	73	54	-26.0	54	0.0	54	0.0
900-819	234	174	-25.6	174	0.0	174	0.0
853-808	277	321	15.8	321	0.0	321	0.0
97-445	118	77	-34.6	77	0.0	70	10.0
164-752	263	263	0.0	263	0.0	263	0.0
1151-525	140	147	5.0	153	-3.9	153	-3.9
2-540	264	309	17.0	294	5.1	294	5.1

Table 5.7 Comparison between different versions of MCACS (number of bus-changes criterion).

Bus-stop (org.-dest.)	Number of bus-changes			
	$MCACS_c$	$MCACS_d$	$MCACS_p$	$MCACS_+$
1037-353	3	6	9	5
877-1071	5	5	7	5
870-1039	5	6	7	5
857-939	2	2	2	2
900-819	3	4	3	3
853-808	6	4	5	6
97-445	2	2	2	3
164-752	3	3	4	4
1151-525	3	3	5	5
2-540	3	6	3	3

of the algorithms discussed can be easily modified to handle more than two objectives, making even more sophisticated models applicable.

The idea of MCACS can be exploited not only for the Bus Routing Problem. In fact, there are other transportation problems which need to model a user behavior in which the exact itinerary is not completely defined. In these cases, casts can model this kind of adaptive user choice quite well, in terms of a stream of alternative paths in addition to the ideal path.

The presented algorithm contains elements of several different ant optimization approaches found in literature. The obtained results show that the algorithm is effective in finding good solutions. MCACS-BRP is a new metaheuristic for optimizing multiple objective functions. The basic idea of coordination and self-organization between two casts belonging to the same colony really helped ants to find the good tours for different origin and destination points within the transportation network. MCACS-BRP shows that it is worth further experiments and extensions. The results obtained during our experiments tend to be more attractive than those, which were proposed in our previous version (Boryczka, 1999).

To sum up, MCACS is similar to MACS and in this paper we have adopted this idea to the Bus Routing Problem. This conception is really useful for multiple objective functions and our experiments have confirmed this assumption. MCACS is shown to be interesting among many modern metaheuristics in term of different parameters evaluating solutions of the more complicated problems.

Though no comparison to other metaheuristics has been carried out, the good results justify the selection of ACO for the BRP. Future research will aim at such a comparison.

A. Bauer, B. Bullnheimer, R.F. Hartl, and C. Strauss. An Ant Colony Optimization approach for the single machine total tardiness problem. In *Proceedings of the 1999 Congress on Evolutionary Computation*, pages 1445–1450, Piscataway, NJ, 1999. IEEE Press.

B. Boffey. Multiobjective routing problems. *Top*, 3(2):167–220, 1995.

E. Bonabeau, M. Dorigo, and G. Theraulaz. *Swarm Intelligence. From Natural to Artificial Systems*. Oxford University Press, 1999.

E. Bonabeau, F. Henaux, S. Guérin, D. Snyers, P. Kuntz, and G. Théraulaz. Routing in telecommunication networks with "Smart" ant-like agents telecommunication applications. In *Proceedings of IATA'98, Second Int. Workshop on Intelligent Agents for Telecommunication Applications, Lectures Notes in AI vol. 1437*. Springer Verlag, 1998.

U. Boryczka. Ant Colony System and Bus Routing Problem. In *Proceedings of CIMCA'99*, Vienna, February 1999.

B. Bullnheimer, R. F. Hartl, and C. Strauss. An improved Ant System algorithm for the Vehicle Routing Problem. Technical Report POM–10/97, Institute of Management Science, University of Vienna, 1997a.

B. Bullnheimer, R. F. Hartl, and C. Strauss. A new rank?based version of the Ant System: A computational study. Technical report, Institute of Management Science, University of Vienna, 1997b.

B. Bullnheimer, R. F. Hartl, and C. Strauss. Applying the Ant System to the Vehicle Routing Problem. In S. Martello In I. H. Osman S. Voß and C. Roucairol, editors, *Meta?Heuristics: Advances and Trends in Local Search Paradigms for Optimization*, pages 109–120, Kluwer Academics, 1998.

B. Bullnheimer and C. Strauss. Tourenplanung mit dem Ant System. Technical report, Instituts für Betriebwirtschaftslehre, Universität Wien, 1996.

J. C. N. Climaco and E. Q. V. Martins. On the determination of the nondominated paths in a multiobjective network problem. In *Proceedings of V Symposiöum über Operations Research*, pages 255–258, Köln, 1980.

J. C. N. Climaco and E. Q. V. Martins. A bicriterion shortest path algorithm. *European Journal of Operational Research*, 11:399–404, 1982.

A. Colorni, M. Dorigo, and V. Maniezzo. Distributed optimization by ant colonies. In F. Vavala and P. Bourgine, editors, *Proceedings First Europ. Conference on Artificial Life*, pages 134–142, Cambridge, 1991. MIT Press.

A. Colorni, M. Dorigo, V. Maniezzo, and M. Trubian. Ant system for job–shop scheduling. *Belgian Journal of Operations Research, Statistics and Computer Science (JORBEL)*, 34:39–53, 1994.

D. Costa and A. Hertz. Ants can colour graphs. *Journal of the Operational Research Society*, 48:295–305, 1997.

J. R. Daduna and A. Wren. *Computer–Aided Transit Scheduling*. Springer–Verlag, 1988.

M. den Besten, T. Stützle, and M. Dorigo. Scheduling single machines by ants. Technical Report 99–16, IRIDIA, Université Libre de Bruxelles, Belgium, 1999.

T. Galvão Dias and J. Falcão e Cunha. Evaluating DSS for operational planning in public transport systems: Ten years of experience with GIST system. In *Proceedings of CASPT 2000*, Berlin, Germany, June 2000.

G. DiCaro and M. Dorigo. AntNet: A mobile agents approach to adaptive routing. Technical report, IRIDIA, Université Libre de Bruxelles, 1998a.

G. DiCaro and M. Dorigo. AntNet: Distributed stigmergetic control for communications networks. *Journal of Artificial Intelligence Research (JAIR)*, 9:317–365, 1998b.

G. DiCaro and M. Dorigo. Extending AntNet for best–effort quality–of–service routing. In Unpublished presentation at *ANTS'98 — From Ant Colonies to Artificial Ants: First International Workshop on Ant Colony Optimization*, October 15–16, 1998c.

G. DiCaro and M. Dorigo. Two ant colony algorithms for best–effort routing in datagram networks. In *Proceedings of the Tenth IASTED International Conference on Parallel and Distributed Computing and Systems (PDCS'98)*, pages 541–546, IASTED/ACTA Press, 1998d.

M. Dorigo. *Optimization, Learning and Natural Algorithms (in Italian)*. PhD thesis, Dipartimento di Elettronica, Politecnico di Milano, IT, 1992.

M. Dorigo and G. Di Caro. The ant colony optimization meta–heuristic. In D. Corne, M. Dorigo, and F. Glover, editors, *New Ideas in Optimization*. McGraw–Hill, London, UK, 1999.

M. Dorigo, G. DiCaro, and L. Gambardella. Ant algorithms for distributed discrete optimization. *Artif. Life*, 5(2):137–172, 1999.

M. Dorigo and L. M. Gambardella. A study of some properties of Ant–Q. In *Proceedings of Fourth International Conference on Parallel Problem Solving from Nature, PPSNIV*, pages 656–665, Berlin, 1996. Springer–Verlag.

M. Dorigo and L. M. Gambardella. Ant colonies for the Traveling Salesman Problem. *Biosystems*, 43:73–81, 1997a.

M. Dorigo and L. M. Gambardella. Ant Colony System: A cooperative learning approach to the Traveling Salesman Problem. *IEEE Trans. Evol. Comp.*, 1: 53–66, 1997b.

M. Dorigo, V. Maniezzo, and A. Colorni. Positive feedback as a search strategy. Technical Report 91–016, Politechnico di Milano, Italy, 1991.

M. Dorigo, V. Maniezzo, and A. Colorni. The Ant System: Optimization by a colony of cooperating agents. *IEEE Trans. Syst. Man. Cybern.*, B26:29–41, 1996.

O. Engelhardt-Funke and M. Kolonko. Cost–Benefit–Analysis of investments into railway networks with randomly perturbed operations. In *Proceedings of CASPT 2000*, Berlin, Germany, June 2000.

R. Freling. Scheduling Train Crews. A case study for the dutch railways. In *Proceedings of CASPT 2000*, Berlin, Germany, June 2000.

G. Gallo, G. Longo, S. Nguyen, and S. Pallottino. Directed hypergraphs and applications. *Discrete Applied Mathematics*, 40:177–201, 1992.

L. M. Gambardella and M. Dorigo. Ant–Q. A reinforcement learning approach to the Traveling Salesman Problem. In *Proceedings of Twelfth International Conference on Machine Learning*, pages 252–260, Palo Alto, CA, 1995. Morgan Kaufman.

L. M. Gambardella and M. Dorigo. Solving symmetric and asymmetric TSPs by ant colonies. In *Proceedings of the IEEE Conference on Evolutionary Computation, ICEC96*, pages 622–627. IEEE Press, 1996.

L. M. Gambardella and M. Dorigo. HAS–SOP: Hybrid Ant System for the Sequential Ordering Problem. Technical Report 11, IDSIA, 1997.

L. M. Gambardella, E. Taillard, and G. Agazzi. MACS–VRPTW: A Multiple Ant Colony System for vehicle routing problems with time windows. Technical Report 06–99, IDSIA, Lugano, Switzerland, 1999a.

L. M. Gambardella, E. D. Taillard, and M. Dorigo. Ant colonies for the QAP. Technical Report 4–97, IDSIA, Lugano, Switzerland, 1997.

L. M. Gambardella, E. D. Taillard, and M. Dorigo. Ant colonies for the QAP. *Journal of the Operational Research Society (JORS)*, 50(2):167–176, 1999b.

F. Glover and M. Laguna. *Tabu Search*. Kluwer Academic Publishers, Dordrecht, 1997.

P.-P. Grasse. La reconstruction du nid et les coordinations inter–individuelles chez bellicositermes natalensis et cubitermes sp. La theorie de la stigmerie. *Insects Soc.*, 6:41–80, 1959.

P.-P. Grasse. *Termitologia*, volume II. Paris, Masson, 1984.

P. Hansen. Bicriterion path problems. In G. Fandel and T. Gal, editors, *Multicriteria decision making: theory and applications, Lecture Notes in Economics and Mathematical Systems 177*, pages 236–245. Springer, Heidelberg, 1980.

M. Heusse, S. Guérin, D. Snyers, and P. Kuntz. Adaptive agent–driven routing and load balancing in communication networks. Technical Report RR–98001–IASC, Départment Intelligence Artificielle et Sciences Cognitives, ENST Bretagne, 1998.

D. E. Kaufman and R. L. Smith. Fastest paths in time–dependent networks for intelligent vehicle–highway systems applications. *IVHS Journal*, 1(1):1–11, 1993.

A. S. K Kwan, R. S. K. Kwan, M. E. Parker, and A. Wren. Proving the versatility of automatic driver scheduling on difficult train & bus problems. In *Proceedings of CASPT 2000*, Berlin, Germany, June 2000.

G. Leguizamón and Z. Michalewicz. A new version of Ant System for subset problems. In *Proceedings of the 1999 Congress on Evolutionary Computation*, pages 1459–1464, Piscataway, NJ, 1999. IEEE Press.

Y.-C. Liang and A. E. Smith. An Ant System approach to redundancy allocation. In *Proceedings of the 1999 Congress on Evolutionary Computation*, pages 1478–1484, Piscataway, NJ, 1999. IEEE Press.

H. Ramalhinho Lourenço and D. Serra. Adaptive approach heuristics for the generalized assignment problem. Technical Report EWP Series No. 304, Department of Economics and Management, Universitat Pompeu Fabra, Barcelona, 1998.

V. Maniezzo. Exact and approximate nondeterministic tree–search procedures for the quadratic assignment problem. Technical Report CSR 98–1, C. L. In Scienze dell'Informazione, Università di Bologna, sede di Cesena, Italy, 1998.

V. Maniezzo and A. Carbonaro. An ANTS heuristic for the frequency assignment problem. Technical Report CSR 98–4, Scienze dell'Informazione, Università di Bologna, Sede di Cesena, Italy, 1998.

V. Maniezzo and A. Colorni. The Ant System applied to the Quadratic Assignment Problem. *IEEE Trans. Knowledge and Data Engineering*, 1999.

V. Maniezzo and A. Colorni. An ANTS heuristic for the frequency assignment problem. *Future Generation Computer Systems*, 16:927–935, 2000.

V. Maniezzo, A. Colorni, and M. Dorigo. The Ant System applied to the Quadratic Assignment Problem. Technical Report 94-?-28, IRIDIA, Université Libre de Bruxelles, Belgium, 1994.

E. Q. V. Martins. On a Multicriteria Shortest Path Problem. *European Journal of Operational Research*, 16:236–245, 1984.

M. Meilton. Selecting and implementing a computer aided scheduling system for a large bus company. In *Proceedings of CASPT 2000*, Berlin, Germany, June 2000.

R. Michel and M. Middendorf. An island model based Ant System with lookahead for the Shortest Supersequence Problem. In A. E. Eiben, T. Back, M. Schoenauer, and H.-P. Schwefel, editors, *Proceedings of PPSN–V, Fifth International Conference on Parallel Problem Solving from Nature*, pages 692–701. Springer–Verlag, 1998.

R. Michel and M. Middendorf. An ACO algorithm for the Shortest Common Supersequence Problem. In D. Corne, M. Dorigo, and F. Glover, editors, *New Methods in Optimisation*. McGraw–Hill, 1999.

J. Mote, I. Murthy, and D. Olson. A parametric approach to solving bicriterion shortest path problems. *European Journal of Operational Research*, 53:81–92, 1991.

K. Nachtigall. Time depending shortest–path problems with applications to railway networks. *European Journal of Operational Research*, 83(1):154–166, 1995.

S. Nguyen and S. Pallottino. Equilibrium traffic assignment for large scale transit networks. *European Journal of Operational Research*, 37:176–186, 1988.

A. Orda and R. Rom. Shortest–path and minimum–delay algorithms in network with time–dependent edge length. *Journal of the ACM*, 37(3):607–625, 1990.

A. Orda and R. Rom. Minimum weight paths in time–dependent network. *Networks*, 21(3):295–320, 1991.

I. Osman and G. Laporte. Metaheuristics: A bibliography. *Annals of Operations Research*, 63:513–623, 1996.

A. De Palma, P. Hansen, and M. Labbé. Commuters' paths with penalties for early or late arrival times. *Transportation Science*, 24(4):276–286, 1993.

C. Reeves. Modern heuristic techniques for combinatorial problems. In *Advanced Topics in Computer Science*. McGrawHill, London, 1995.

J.-M. Rousseau. *Computer Scheduling of Public Transport 2*. North Holland, 1985.

J.-M. Rousseau. Scheduling regional transportation with HASTUS. In *Proceedings of CASPT 2000*, Berlin, Germany, June 2000.

H. M. Safer and J. B. Orlin. Fast approximation schemes for multi-criteria combinatorial optimization. Technical Report 3756–95, Sloan School of Management, Massachusetts Institute of Technology, 1995.

R. Schoonderwoerd, O. Holland, and J. Bruten. Ant-like agents for load balancing in telecommunications networks. In *Proceedings of the First International Conference on Autonomous Agents*, pages 209–216. ACM Press, 1997.

R. Schoonderwoerd, O. Holland, J. Bruten, and L. Rothkrantz. Ant-based load balancing in telecommunications networks. *Adaptive Behavior*, 5(2):169–207, 1996.

A. Schrijver. Minimum circulation of railway stock. *CWI Quarterly*, 6:205–217, 1993.

C. R. Delgado Serna and J. Pacheco Bonrostro. MINMAX vehicle routing problems: Application to school transport in the province of Burgos (Spain). In *Proceedings of CASPT 2000*, Berlin, Germany, June 2000.

T. Stützle. An ant approach to the Flow Shop Problem. Technical Report AIDA–97–07, FG Intellektik, FB Informatik, TH Darmstadt, September 1997.

T. Stützle and H. Hoos. Improvements on the Ant System: Introducing MAX-MIN Ant System. In *In Proceedings of the International Conference on Artificial Neural Networks and Genetic Algorithms*, pages 245–249, Wien, 1997a. Springer Verlag.

T. Stützle and H. Hoos. The MAX–MIN Ant System and Local Search for the Traveling Salesman Problem. In T. Baeck, Z. Michalewicz, and X. Yao, editors, *Proceedings of IEEE–ICEC–EPS'97, IEEE International Conference on Evolutionary Computation and Evolutionary Programming Conference*, pages 309–314. IEEE Press, 1997b.

T. Stüzle and H. Hoos. MAX–MIN Ant System and Local Search for combinatorial optimisation problems. In *Proceedings of the Second International conference on Metaheuristics MIC'97*, Dordrecht, 1998. Kluwer Academic.

D. Subramanian, P. Druschel, and J. Chen. Ants and Reinforcement Learning: A case study in routing in dynamic networks. In *Proceedings of IJCAI–97, International Joint Conference on Artificial Intelligence*. Morgan Kaufmann, 1997.

T. Tulp. *CVI: Bilder of Dutch Public Transport Information System*. Personal Communication, 1993.

R. van der Put. Routing in the faxfactory using mobile agents. Technical Report R&D–SV–98–276, KPN Research, 1998.

G. Navarro Varela and M. C. Sinclair. Ant Colony Optimisation for virtual-wavelength–path routing and wavelength allocation. In *Proceedings of the 1999 Congress on Evolutionary Computation*, pages 1809–1816, Piscataway, NJ, 1999. IEEE Press.

A. R. Warburton. Bicriterion shortest path problems. Technical Report 84–27, University of Ottawa, 1984.

A. R. Warburton. Approximation of pareto optima in multiple–objective shortest–path problems. *Operations Research*, 35:70–79, 1987.

T. White, B. Pagurek, and F. Oppacher. Connection management using adaptive mobile agents. In H.R. Arabnia, editor, *Proceedings of the International Conference on Parallel and Distributed Processing Techniques and Applications (PDPTA'98)*, pages 802–809. CSREA Press, 1998.

Metaheuristics: Computer Decision-Making, pp. 127-151
Domingo Ortiz-Boyer, César Harvás-Martínez, and José Muñon-Pérez

6 STUDY OF GENETIC ALGORITHMS WITH CROSSOVER BASED ON CONFIDENCE INTERVALS AS AN ALTERNATIVE TO CLASSICAL LEAST SQUARES ESTIMATION METHODS FOR NONLINEAR MODELS

Domingo Ortiz-Boyer[1], César Harvás-Martínez[1],
and José Muñoz-Pérez[2]

[1]Department of Computer Science
University of Córdoba
Building 2, Campus of Rabanales s/n
E-14071 Córdoba, Spain
ma1orbod@uco.es, ma1hemac@uco.es

[2]Department of Languages and Computer Science,
University of Málaga
Technological Complex, Teatinos Campus
E-29071 Málaga, Spain
munozp@lcc.uma.es

Abstract: Genetic algorithms are optimization techniques especially useful in functions whose nonlinearity makes an analytical optimization impossible. This kind of functions appear when using least squares estimators in nonlinear regression problems. Least squares optimizers in general, and the Levenberg-Marquardt method in particular, are iterative methods especially designed to solve this kind of problems, but the results depend on both the features of the problem and the closeness to the optimum of the starting point. In this paper we study the least squares estimator and the optimization methods that are based on it. Then we analyze those features of real-coded genetic algorithms that can be useful in the context of nonlinear regression. Special attention will be devoted to the crossover operator, and a new operator based on confidence intervals will be proposed. This crossover provides an equilibrium between exploration and exploitation of the search space, which is very adequate for this kind of problems. To analyze the fitness and robustness of the proposed crossover operator, we will use three complex nonlinear regression problems with search domains of different amplitudes and compare its performance with that of other crossover operators and with the Levenberg-Marquardt method using a multi-start scheme.

Keywords: Real coded genetic algorithms, Nonlinear regression, Confidence interval based crossover.

6.1 INTRODUCTION

Most processes in the real world follow a nonlinear model. This explains the scientific community's interest in the development of new methods for the estimation of parameters of nonlinear models (nonlinear Regression, NLR, methods) for such processes. Existing methods are based on the estimation of the parameters through the optimization of a cost function. So the squared sum of errors is minimized, the likelihood function is maximized or some a-posteriori density function is maximized depending on the method.

Different nonlinear techniques for parameter estimation impose, to a certain extent, some conditions such as the target function continuity, a specific identical distribution for the residuals, compact domains for the parameters, etc. Unlike linear models, only convex nonlinear equations have analytical solution, which makes it necessary to resort to iterative optimization methods. Those methods impose additional constraints to guarantee the convergence to, at least, a local optimum. This causes many estimation methods to perform very well on certain kinds of problems, and fail in other types not conforming to the imposed conditions (Midi, 1999).

In many real models, these conditions are either not met or very difficult to test. For this reason, it is usual to choose a method with a good performance in most models. The scientific community seems to agree that the Levenberg-Marquardt method, or any of its modifications, is the most versatile. In fact, this is the standard nonlinear regression method in most statistical packages. Nevertheless, any investigator who has used the method knows that when the residuals are large or the starting point is far from the optimum the method can converge to a sub-optimal local minimum.

Different global optimization techniques have been applied to nonlinear regression problems. Simulated annealing is one of the most popular in the computational biology community and other application fields. Simulated annealing is a probabilistic approach introduced by Kirkpatrick et al. (1983; 1987). It works by emulating the annealing process of a solid. Simulated annealing creates solutions in the neighborhood of the current solution. If the new solution improves the current solution it is accepted, otherwise it is accepted according to an exponentially decreasing probability function of the temperature.

Other representative technique is the classical and exponential tunneling described by Levy and Gómez (1985); Levy and Montalvo (1985). It is an iterative procedure that consists of two phases. The first phase involves finding a local minimizer via a local minimization technique, and the second phase, tunneling, finds a new point with a function value no greater than the previous minimum found. The exponential tunneling function is preferred over the classical tunneling function because it can create a pole at a local minimizer independent of the precision of the local minimum found (Barron and Gómez, 1991).

An alternative, studied in this article, is the use of genetic algorithms (GA). Genetic algorithms in general, and more specifically Real Coded Genetic Algorithms (RCGA's), are parallel stochastic search algorithms, very robust, and widely used in optimization problems. This makes them a promising alternative to classical NLR methods. One of the first proposals was made by Johnson and Husbands (1990), where a binary GA is applied to the estimation of the parameters of a model. In Kargupta and Smith (1991), a system to evolve polynomial networks is proposed. Billings and Mao (1996) propose an identification algorithm based on a rational model and a genetic algorithm. Other evolutionary techniques for system modeling were proposed in Nissesen and Koivisto (1996), where a GA is applied to the identification of Volterra's series.

The crossover operator plays a central role in GAs. In fact it may be considered to be one of the algorithm's defining features, and it is one of the components to be borne in mind to improve the GA behavior (Liepins and Vose, 1992). In this context, it is fundamental the capacity of crossover operators to solve the balance between the exploration and exploitation of the search space, which is associated to the intervals determined by the bounds of the domain of the genes and by the corresponding alleles of the parents chosen for crossing (Herrera et al., 1998).

In this article, a new parameter estimation method is proposed, using a new crossover operator based on confidence intervals. This crossover uses information from the best individuals of the population, and achieves an equilibrium between exploration and exploitation that makes it very efficient and robust for the resolution of this kind of problems, especially when compared to other crossover operators in RCGAs. To validate the methods based on RCGAs, we have used NLR problems taken from the Statistical Reference Datasets Project [1] (STRDP) developed by the faculty of the Statistical Engineering Division and the Mathematical and Computational Sciences Division within the Information Technology Laboratory of the National Institute of Standards and Technology Gaithersburg, USA.

[1]Can be consulted at `http://www.nist.gov/itl/div898/strn/nls`

6.2 LEAST SQUARES ESTIMATOR

Let $(\mathbf{x}_i, y_i)$, $i = 1, 2, ..., m$, be m observations, linked through a nonlinear function f, defined as

$$y_i = f(\mathbf{x}_i; \beta^*) + \epsilon_i \quad (i = 1, 2, ..., m), \tag{6.1}$$

where y_i is the dependent variable, $\mathbf{x}_i$ is the q-dimensional vector of independent variables for observation i ($\mathbf{x}_i \in \mathcal{X} \subset \Re^q$), $\mathbf{x}$ is the observations vector, β^* are real values of the p-dimensional parameter vector β ($\beta \in \boldsymbol{\beta} \subset \Re^p$), and ϵ_i is the random error.

If the following conditions apply

1. ϵ_i are independent variables with identical distributions, with zero mean and $\sigma^2 > 0$ variance,
2. function $f(\mathbf{x}; \beta)$ is continuous in $\mathcal{X} \times \boldsymbol{\beta}$,
3. $\mathcal{X} \subset \Re^q$ and $\boldsymbol{\beta} \subset \Re^p$, are closed and bounded, i.e. are compact,

then the least squares estimator of β^*, denoted as $\hat{\beta}$, minimizes the sum of squared error or sum of squared residuals:

$$S(\beta) = \sum_{i=1}^{m} [y_i - f(\mathbf{x}_i; \beta)]^2 \tag{6.2}$$

If, additionally, the following points are true

4. observations $\mathbf{x}$ are such that $H_m(\mathbf{x}) \rightarrow H(\mathbf{x})$ with probability 1, where $H_m(\mathbf{x})$ is the empirical distribution function of $\mathbf{x}$ and $H(\mathbf{x})$ is a theoretical distribution function,
5. if $f(\mathbf{x}; \beta) = f(\mathbf{x}; \beta^*)$ with probability 1, then $\beta = \beta^*$,

then the least squares estimator is strongly consistent, being asymptotically normal if

6. β^* is an interior point of $\boldsymbol{\beta}$,
7. functions $\partial f(\mathbf{x}; \beta)/\partial\beta_j$ and $\partial^2 f(\mathbf{x}; \beta)/(\partial\beta_j \partial\beta_s)$ $(j, s = 1, 2, ..., p)$ are continuous in $\mathcal{X} \times \boldsymbol{\beta}$,
8. matrix $\boldsymbol{\Omega} = w_{js}(\beta^*)$, where

$$w_{js}(\beta^*) = \int \frac{\partial f(\mathbf{x}; \beta^*)}{\partial\beta_j} \frac{\partial f(\mathbf{x}; \beta^*)}{\partial\beta_s} dH(\mathbf{x}) \tag{6.3}$$

   is not singular.

We should point out that even if $\hat{\beta}$ is asymptotically normal, this does not generally imply it is asymptotically efficiency. For this to happen, it should have to meet

9. ϵ_i are normally distributed.

More on the above conditions can be found in Malinvaud (1970), Gallant (1987), and Seber and Wild (1989).

If ϵ_i follows a $N(0, \sigma^2)$ distribution, the least squares estimator $\hat{\beta}$ is the maximum likelihood estimator.

6.3 OPTIMIZATION METHODS

It has been pointed before that the different parameter estimation methods for nonlinear models try to optimize a function defined as either a sum of squared residuals, a likelihood function, a linear approximation of the model, a sum of the absolute value of residuals, or an a-posteriori density function.

Let us suppose this function to be $S(\beta)$, as defined in equation (6.2). By setting the partial derivatives of $S(\beta)$ with respect to β_j, $j = 1, 2, ..., p$, to zero, we obtain a system of *p normal equations*, whose solutions are the least squares estimators $\hat{\beta}$ of β^*. Each normal equation has the form

$$\frac{\partial S(\beta)}{\partial \beta_j} = -\sum_{i=1}^{n} [y_i - f(\mathbf{x}_i; \beta)] \left[\frac{\partial f(\mathbf{x}_i; \beta)}{\partial \beta_j}\right] = 0. \tag{6.4}$$

Unlike linear models, the partial derivatives of a nonlinear model are functions of the parameters. The residual equations are nonlinear and, in general, an analytical solution can not be obtained.

If we consider a nonlinear response model of the form

$$y_i = \beta_1 e^{\beta_2 x_i} + \epsilon_i, \tag{6.5}$$

then the partial derivatives of the model with respect to the parameters are

$$\frac{\partial f(\mathbf{x}_i; \beta)}{\partial \beta_1} = \frac{\partial \left(\beta_1 e^{\beta_2 x_i}\right)}{\partial \beta_1} = e^{\beta_2 x_i}, \tag{6.6}$$

$$\frac{\partial f(\mathbf{x}_i; \beta)}{\partial \beta_2} = \frac{\partial \left(\beta_1 e^{\beta_2 x_i}\right)}{\partial \beta_2} = \beta_1 x_i e^{\beta_2 x_i}. \tag{6.7}$$

The two normal equations of this model would be

$$\sum_{i=1}^{n} \left(y_i - \beta_1 e^{\beta_2 x_i}\right) \left(e^{\beta_2 x_i}\right) = 0 \tag{6.8}$$

$$\sum_{i=1}^{n} \left(y_i - \beta_1 e^{\beta_2 x_i}\right) \left(\beta_1 x_i e^{\beta_2 x_i}\right) = 0 \tag{6.9}$$

Even for simple models such as this one, normal equations do not have an analytical solution. Iterative numerical methods are used to obtain solutions. These methods need an initial solution or initial values for the parameters

$\hat{\beta}^{(0)}$. From this initial solution, the sum of squared error is reduced in each iteration until the improvement for each step is small enough, i.e. until the method converges to a solution. The choice of the starting point is of dramatic importance, as the convergence of the process (to a global optimum) depends on this point and on the rest of the method's conditions.

6.4 LEAST SQUARES OPTIMIZERS

The features of least squares estimators, detailed in Section 6.2, make this method the most widely used for the estimation of nonlinear model parameters. Although the least squares optimization problem can be solved using any standard optimization method, specialized methods have been designed that exploit the structure (sum of squares) of the function to optimize. These methods are based on a modification of Newton's method introduced by Gauss in 1809, known as the Gauss-Newton Method (GN).

6.4.1 Gauss-Newton method

Gauss-Newton method uses a Taylor expansion of $f(\mathbf{x}_i;\beta)$ from the values $\beta^{(a)}$ to obtain a linear approximation of the model in a region near β^* in the a-th iteration.

$$f(\mathbf{x}_i;\beta) \approx f(\mathbf{x}_i;\beta^{(a)}) + \sum_{j=1}^{p} \left(\frac{\partial f(\mathbf{x}_i;\beta^{(a)})}{\partial \beta_j} \right) (\beta_j - \beta_j^{(a)}) \tag{6.10}$$

or

$$\mathbf{f}(\beta) \approx \mathbf{f}(\beta^{(a)}) + \mathbf{F}(\beta^{(a)})(\beta - \beta^{(a)}) \tag{6.11}$$

where $\mathbf{F}(\beta^{(a)})$ is the matrix of $m \times p$ partial derivatives, evaluated for $\beta^{(a)}$ and m data $\mathbf{x}_i$, defined as:

$$\mathbf{F}(\beta^{(a)}) = \begin{bmatrix} \frac{\partial f(\mathbf{x}_1;\beta^{(a)})}{\partial \beta_1} & \frac{\partial f(\mathbf{x}_1;\beta^{(a)})}{\partial \beta_2} & \cdots & \frac{\partial f(\mathbf{x}_1;\beta^{(a)})}{\partial \beta_p} \\ \frac{\partial f(\mathbf{x}_2;\beta^{(a)})}{\partial \beta_1} & \frac{\partial f(\mathbf{x}_2;\beta^{(a)})}{\partial \beta_2} & \cdots & \frac{\partial f(\mathbf{x}_2;\beta^{(a)})}{\partial \beta_p} \\ \vdots & \vdots & \vdots & \vdots \\ \frac{\partial f(\mathbf{x}_m;\beta^{(a)})}{\partial \beta_1} & \frac{\partial f(\mathbf{x}_m;\beta^{(a)})}{\partial \beta_2} & \cdots & \frac{\partial f(\mathbf{x}_m;\beta^{(a)})}{\partial \beta_p} \end{bmatrix} \tag{6.12}$$

This way, the nonlinear minimization problem approaches in each iteration a linear minimization problem on the least squares function,

$$\textbf{Minimize } S(\beta^{(a)}) = \sum_{i=1}^{m} \left[y_i - \mathbf{f}(\beta^{(a)}) + \mathbf{F}(\beta^{(a)})(\beta - \beta^{(a)}) \right]^2 . \tag{6.13}$$

Its solution provides an estimation of the parameter variation in each iteration.

$$\beta - \beta^{(a)} = (\mathbf{F}(\beta^{(a)})\mathbf{F}(\beta^{(a)})^{-1}([\mathbf{F}(\beta^{(a)})(\mathbf{y} - \mathbf{f}(\beta^{(a)}))] \tag{6.14}$$

The Gauss-Newton algorithm defines

$$\beta^{(a+1)} = \beta^{(a)} + \delta^{(a)} \tag{6.15}$$

where

$$\delta^{(a)} = (\mathbf{F}(\beta^{(a)})\mathbf{F}(\beta^{(a)})^{-1}) \, [\mathbf{F}(\beta^{(a)})(\mathbf{y} - \mathbf{f}(\beta^{(a)}))]. \tag{6.16}$$

Substituting $\mathbf{y} - \mathbf{f}(\beta^{(a)}) = \mathbf{r}(\beta^{(a)})$, the gradient and the Hessian matrix can be calculated as

$$g(\beta) = \frac{\partial \mathbf{S}(\beta^{(a)})}{\partial \beta} = 2\sum_{i=1}^{m} r_i(\beta^{(a)}) \frac{\partial r_i(\beta^{(a)})}{\partial \beta} = 2\mathbf{r}(\beta^{(a)})\mathbf{J}(\beta^{(a)}) \tag{6.17}$$

$$\begin{aligned} \mathbf{H} = \frac{\partial^2 \mathbf{S}(\beta^{(a)})}{\partial^2 \beta} &= 2\textstyle\sum_{i=1}^{m} \frac{\partial^2 r_i(\beta^{(a)})}{\partial^2 \beta} + 2\sum_{i=1}^{m} r_i(\beta^{(a)}) \frac{\partial^2 r_i(\beta^{(a)})}{2\partial \beta} \\ &= 2(\mathbf{J}(\beta^{(a)})\mathbf{J}(\beta^{(a)}) + \mathbf{A}(\beta^{(a)})) \end{aligned} \tag{6.18}$$

where

$$\mathbf{J}(\beta^{(a)}) = \frac{\partial r_i(\beta^{(a)})}{\partial \beta} = -F(\beta^{(a)}), \tag{6.19}$$

$$\mathbf{A}(\beta^{(a)})) = \sum_{i=1}^{m} r_i(\beta^{(a)}) \frac{\partial^2 r_i(\beta^{(a)})}{\partial^2 \beta}. \tag{6.20}$$

The increment for the Newton method $\delta^{(a)}$ would be

$$\begin{aligned} \delta^{(a)} &= -\mathbf{H}(\beta^{(a)})^{-1}\mathbf{g}(\beta)^{(a)} \\ &= -(\mathbf{J}(\beta^{(a)})\mathbf{J}(\beta^{(a)}) + \mathbf{A}(\beta^{(a)}))^{-1}\mathbf{r}(\beta^{(a)})\mathbf{J}(\beta^{(a)}), \end{aligned} \tag{6.21}$$

whereas for the Gauss-Newton method it is

$$\delta^{(a)} = -(\mathbf{J}(\beta^{(a)})\mathbf{J}(\beta^{(a)}))^{-1}\mathbf{r}(\beta^{(a)})\mathbf{J}(\beta^{(a)}). \tag{6.22}$$

The advantage of the Gauss-Newton method over Newton's method is that only the first derivative of $r_i(\beta)$ is required to calculate $\mathbf{J}$. So, each iteration is less time and memory consuming, as it is not necessary to keep the Hessian matrix. Moreover, the Gauss-Newton method improves the performance of methods that only use the first derivative to minimize the sum of squared residuals.

If we compare (6.20) and (6.22), we can see that Gauss-Newton and Newton methods are very similar when $\mathbf{A}(\hat{\beta})$ is small compared to $\mathbf{J}(\hat{\beta})\mathbf{J}(\hat{\beta})$, and they are identical when $\mathbf{A}(\hat{\beta}) = 0$. According to equation (6.20) $\mathbf{A}(\hat{\beta}) = 0$ when one of the following applies: $\mathbf{r}(\hat{\beta}) = 0$, i.e., the nonlinear regression model, perfectly fits the data; or $\partial^2 r_i(\beta)/\partial^2 \beta = 0$, which implies that $r_i(\beta)$, so $f(\mathbf{x}_i; \beta)$, is linear with respect to β in the vicinity of $\hat{\beta}$. We can interpret $\mathbf{A}(\hat{\beta})$ as a measure of the size of the residual and of the nonlinearity of the problem.

When the point $\beta^{(1)}$ is close enough to $\hat{\beta}$ and $\mathbf{A}(\hat{\beta})$ is small compared to $\mathbf{J}(\hat{\beta})\mathbf{J}(\hat{\beta})$, it can be proved that the Gauss-Newton method converges linearly (Dennis and Schnabel, 1983). But when these conditions are not met the method may not converge, or may converge very slowly. To avoid this, different methods based on Gauss-Newton method have been proposed. Among them is the *Levenberg-Marquardt method* (Marquardt, 1963).

6.4.1.1 Levenberg-Marquardt method. Marquardt, using Levenberg's idea (Levenberg, 1944), describes the method known as Levenberg-Marquardt (Marquardt, 1963). Based on Gauss-Newton, it defines the increment as

$$\delta^{(a)} = -(\mathbf{J}(\beta^{(a)})\mathbf{J}(\beta^{(a)}) + \eta^{(a)}\mathbf{D}^{(a)})^{-1}\mathbf{r}(\beta^{(a)})\mathbf{J}(\beta^{(a)}), \tag{6.23}$$

where $\mathbf{D}^{(a)}$ is a diagonal matrix with positive elements in the diagonal. To simplify, usually $\mathbf{D}^{(a)} = \mathbf{I}_p$. Another common choice, to make the method insensitive to rescaling β, is to assign the same values of the diagonal elements in $\mathbf{J}(\beta^{(a)})\mathbf{J}(\beta^{(a)})$.

When $\mathbf{D}^{(a)} = \mathbf{I}_p$, the direction of the Levenberg-Marquardt method is an interpolation between the direction of Gauss-Newton, when $\eta^{(a)} \to 0$, and the direction of the maximum slope method, when $\eta^{(a)} \to \infty$. To avoid the problems in both methods, Levenberg-Marquardt's algorithm defines a policy for assigning values to η. Initially it is assigned a small positive value, for instance $\eta^{(1)}$ =0.01. If, in the a-th iteration, the increase $\delta^{(a)}$ reduces $S(\beta)$, then $\beta^{(a+1)} = \beta^{(a)} + \delta^{(a)}$ and η is divided by a factor, for instance $\eta^{(a+1)} = \eta(a)/10$, so that in the next iteration the increase will be greater and the method will converge faster. This way the algorithm resembles Gauss-Newton. If in the a-th iteration the increase $\delta^{(a)}$ does not reduce $S(\beta)$, then $\eta^{(a)}$ is progressively increased by a factor, for instance $\eta^{(a)} = 10\eta^{(a)}$, and $\delta^{(a)}$ is recalculated until $S(\beta)$ is reduced. This way, the increase is reduced and the algorithm converges in the direction of maximum slope.

Some modern Levenberg-Marquardt algorithms use a method using confidence regions to update $\eta^{(a)}$ (Moré, 1977), others incorporate a linear search (Nash, 1977), or a partial linear search (Dennis, 1977). Dennis and Schnabel (1983) recommend Moré's implementation. In practice, Levenberg-Marquardt algorithms, in general, have proved to be both robust and fast. These two features makes them the reference nonlinear regression methods. Nevertheless, for problems with large residuals, the linear convergence can be very slow or the algorithm may not converge at all (Seber and Wild, 1989).

6.5 GENETIC ALGORITHMS

Genetic algorithms are iterative probabilistic search algorithms that maintain a *population of individuals*, *chromosomes* or *strings of numbers*,

$$P(t) = \{\beta^{1(t)}, \beta^{2(t)}, ..., \beta^{i(t)}, ..., \beta^{N(t)}\}$$

in iteration or *generation* t. Each individual represents a potential solution to the problem being solved. This solution is represented by a complex data structure, or *genotype*, β. Each field in this structure represents a variable or characteristic of the problem, called *gene*. Each gene can take different states or values called *alleles*. The meaning of a particular chromosome is called *phenotype* and is defined by the user according to the nature of the problem. Each solution $\beta^{i(t)}$ is evaluated to obtain a measure of its *fitness*. In iteration $t+1$ a new population is obtained by the *selection* of the best individuals. Some members of the new population are subject to various alterations by the application of *genetic operators* that generate new solutions. There are unary operators or *mutators*, that create new individuals by making small changes in existing individuals, and higher degree operators called *crossovers*, which create new individuals by combining parts of two or more individuals. After a number of generations the algorithm converges to a solution which, in case the GA is well designed, will be reasonably close to the optimum. The general structure of a genetic algorithm can be seen in Algorithm 6.1.

procedure GENETICALGORITHM( )
$t \leftarrow 0$;
Initialize $P(t)$;
Evaluate $P(t)$;
while stopping condition is not satisfied
  $t \leftarrow t+1$;
  Select $P(t)$ from $P(t-1)$;
  Apply operators to $P(t)$;
  Evaluate $P(t)$;
endwhile

Algorithm 6.1 Structure of a genetic algorithm

Genetic algorithms are stochastic multiple, general purpose, blind search algorithms (Goldberg, 1989). They are probabilistic, but not random, algorithms, as they combine direct search and random search. They do not need any specific knowledge of the problem, so the search is based exclusively on the values of an objective function (this is why they are qualified as "blind"). Other methods, based on the characteristics of the problem, fail when the necessary information is not available or is difficult to obtain. The multiplicity of the search in a GA is inherent to its parallel nature: a GA works with a population of candidate solutions which are evolved until an optimal solution is reached. This characteristic, not present in other methods using only one point of search, reduces the chance of finding a local (i.e. false) optimum. Genetic algorithms are general-purpose because the only knowledge needed to apply them is the expression of the objective function, which is easily obtained. This makes GAs applicable to virtually any problem.

6.5.1 *The encoding problem*

The simplicity of binary encoding traditionally used in genetic algorithms makes theoretical studies possible. Therefore, Holland (1975) postulated the *schema theory*, which explains the good behavior of genetic algorithms. The binary encoding presents certain difficulties when using continuous multidimensional search spaces, that require high numerical precision. In these scenarios, binary encoding can limit the window through which a genetic algorithm perceives its world (Koza, 1992). To avoid this we can increase the size of the binary code. For instance, in a problem with 100 variables in the range [-500, 500] needing a precision of six decimal places, a binary code of 3000 bits would be needed, thus generating a search space of 10^{1000}. This makes the genetic algorithm perform very poorly (Michalewicz, 1992).

Although theoretical studies demonstrate that binary encoding is one of the most adequate, simple and theoretically manageable (Goldberg, 1991), the good properties of genetic algorithms are not due to the use of bit strings (Antonisse, 1989; Radcliffe, 1992). For this reason, other representations have been used which are more adequate to each particular problem. One of the most important is *real encoding*, which seems very natural in optimization problems with continuous domains, where each gene represents a problem variable. Now the precision of the solution depends only on the computing system used for the simulation of the genetic algorithm. The values of the genes are kept inside the valid intervals, and the operators must take this constraint into account. This kind of genetic algorithms are known as Real Coded genetic algorithms (RCGAs). RCGAs have been extensively used in numerical optimization problems with continuous domains (Herrera et al., 1995). In 1991 the first theoretical studies appeared, which justified the power of RCGAs (Wright, 1991; Goldberg, 1991).

6.5.2 *Balance between exploration and exploitation*

Genetic algorithms are general purpose search methods whose operators establish a very adequate equilibrium between exploration and exploitation. This has motivated their extended use for the resolution of a variety of highly complex combinatorial and engineering problems with hard constraints, only solvable by approximating their optimal values (Grefenstette, 1987; Bennett et al., 1991).

The mutation operator randomly modifies one or more genes in a chromosome with a given probability, thus increasing the structural diversity of the population. Therefore, it is clearly an exploratory operator, which restores the genetic material lost during the selection phase and explores new solutions avoiding the premature convergence of the genetic algorithm to a local optimum. This way, the probability of reaching a given point in the search space is never zero.

The crossover operator combines the characteristics of two or more parents to generate better offsprings. It is based on the idea that the exchange of

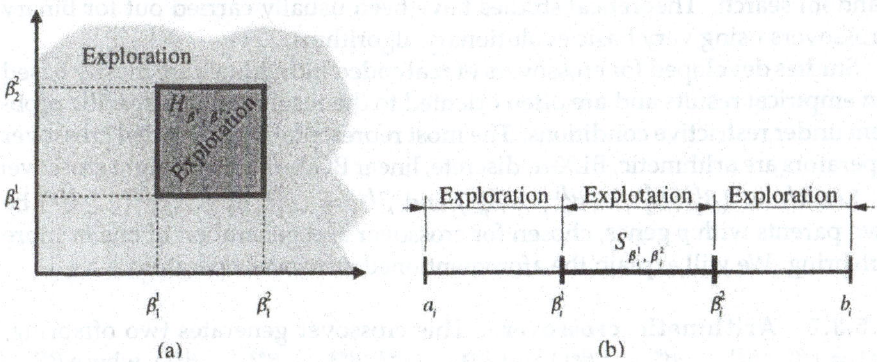

Figure 6.1 (a) Hypercube defined by the two first genes of the parents; (b) Representation of the segment defined by the i-th genes of two chromosomes

information between good chromosomes will generate even better offsprings. The effect of the crossover operator can be studied from two different points of view: at the chromosome level and at the gene level. The effect of the crossover operator at the chromosome level can be considered in a geometric way. Given two parents $\beta^1 = (\beta_1^1, \beta_2^1, ..., \beta_i^1, ..., \beta_p^1)$ and $\beta^2 = (\beta_1^2, \beta_2^2, ..., \beta_i^2, ..., \beta_p^2)$ with p genes, we denote by $H_{\beta^1\beta^2}$ the hypercube defined by the two first genes (Figure 6.1a). At the gene level the representation would be linear, defining in this case a segment or interval $S_{\beta_i^1,\beta_i^2}$ for each pair of genes (Figure 6.1b). Most crossover operators generate individuals in the exploitation regions $S_{\beta_i^1,\beta_i^2}$ or $H_{\beta^1\beta^2}$. In this way, the crossover operator implements a depth search or exploitation, leaving the breadth search or exploration to the mutation operator. This policy, intuitively very natural, makes the population converge to inner values from the search space, producing a fast decrement in the population diversity which could lead to a premature convergence to a non-optimal solution. Studies on the BLX-α crossover (Eshelman and Schaffer, 1993) confirm the good performance of those crossover operators that also generate individuals in the exploration zone. These operators avoid the loss of diversity and the premature convergence to inner points of the search space, but the generation of new individuals in the exploration region could slow the search process. For this reason the crossover operator should establish an adequate balance between exploration and exploitation from the values of the parents' genes (alleles) and generate offspring in the exploration and exploitation regions with an adequate balance.

6.5.3 Crossover operators

Since the first applications of genetic algorithms, much research has focused on the study of the crossover operator, perhaps the most innovative, considering that the mutation operator inherits most of the theoretical aspects of random search. Theoretical studies have been usually carried out for binary crossovers using very basic evolutionary algorithms.

Studies developed for crossovers of real coded individuals are mostly based on empirical results and are often oriented to the resolution of a specific problem under restrictive conditions. The most representative real coded crossover operators are arithmetic, BLX-α, discrete, linear BGA, flat and Wright crossover.

Let $\beta^{f_1} = (\beta_1^{f_1}, \beta_2^{f_1}, ..., \beta_i^{f_1}, ..., \beta_p^{f_1})$ and $\beta^{f_2} = (\beta_1^{f_2}, \beta_2^{f_2}, ..., \beta_i^{f_2}, ..., \beta_p^{f_2})$ be two parents with p genes, chosen for crossover and generation of one or more offspring. We will explain the aforementioned crossover operators.

6.5.3.1 Arithmetic crossover. This crossover generates two offspring, $\beta^{s_1} = (\beta_1^{s_1}, \beta_2^{s_1}, ..., \beta_i^{s_1}, ..., \beta_p^{s_1})$ and $\beta^{s_2} = (\beta_1^{s_2}, \beta_2^{s_2}, ..., \beta_i^{s_2}, ..., \beta_p^{s_2})$, where $\beta_i^{s_1} = \lambda\beta_i^{f_1} + (1-\lambda)\beta_i^{f_2}$ and $\beta_i^{s_2} = \lambda\beta_i^{f_2} + (1-\lambda)\beta_i^{f_1}$, where λ is a constant (Michalewicz, 1992). This kind of crossover tends to generate solutions close to the center of the search space.

6.5.3.2 Flat crossover. This crossover generate only one child $\beta^s = (\beta_1^s, \beta_2^s, ..., \beta_i^s, ..., \beta_p^s)$, such that β_i^s is a random value chosen in the interval $[\beta_i^{f_1}, \beta_i^{f_2}]$ (Radcliffe, 1991). It is a clearly exploitative crossover, both at the gene and at the chromosome level.

6.5.3.3 BLX-α crossover. This crossover generates one offspring, $\beta^s = (\beta_1^s, \beta_2^s, ..., \beta_i^s, ..., \beta_p^s)$ where β_i^s is chosen randomly in the interval $[\beta_{min} - I \cdot \alpha, \beta_{max} + I \cdot \alpha]$, with $\beta_{max} = max(\beta_i^{f_1}, \beta_i^{f_2})$, $\beta_{min} = min(\beta_i^{f_1}, \beta_i^{f_2})$ and $I = \beta_{max} - \beta_{min}$ (Eshelman and Schaffer, 1993). BLX-0.0 crossover is identical to flat crossover. For $\alpha = 0.5$, the probability of the children's genes taking values inside the intervals defined by the parents' genes is identical to their probability of taking values outside this region. So, depending on the value of α this crossover can be more or less exploitative. In experiments carried out by Herrera et al. (1998) for different values of α, it can be seen that the best results are obtained for $\alpha = 0.5$.

6.5.3.4 Wright crossover. Considering that β^{f_1} is the parent with best fitness, one offspring is generated $\beta^s = (\beta_1^s, \beta_2^s, ..., \beta_i^s, ..., \beta_p^s)$, where $\beta_i^s = \tau \cdot (\beta_i^{f_1} - \beta_i^{f_2}) + \beta_i^{f_1}$, τ being a random number in the interval [0, 1] (Wright, 1991). This is a highly exploratory crossover, as the descendants are always out of the hypercube formed by the parents. Even so, this operator performs a directed search in the direction of the fitness function optimum.

6.5.3.5 Linear BGA crossover. Under the same considerations as above, $\beta_i^s = \beta_i^{f_1} \pm rang_i.\gamma.\Lambda$, where $\Lambda = (\beta_i^{f_2} - \beta_i^{f_1})/(\|\beta^{f_1} - \beta^{f_2}\|)$. The "-" sign is chosen with probability 0.9. Normally, $rang_i = 0.5 \cdot (b_i - a_i)$ and $\gamma = \sum_{k=0}^{15} \alpha_k 2^{-k}$, where $\alpha_i \in \{0, 1\}$ is randomly generated with $p(\alpha_i = 1) = \frac{1}{16}$ (Schlierkamp-Voosen, 1994), and a_i and b_i are the lower and upper limits of each gene's domain. This operator is based on Mühlenbein's (Mühlebein and Schlierkamp-Voosen, 1993) and it performs a search around the best individual, with an amplitude depending both on the distance between the genes' values and on the distance between the chromosomes. The high probability of the sign being negative makes it a very exploitative crossover and most descendants are inside the hypercube defined by the parents.

6.5.3.6 Discrete crossover. This crossover generates one offspring $\beta^s = (\beta_1^s, \beta_2^s, \ldots, \beta_i^s, \ldots, \beta_p^s)$, where β_i^s is chosen randomly from the set $\{\beta_i^{f_1}, \beta_i^{f_2}\}$ (Mühlebein and Schlierkamp-Voosen, 1993). This crossover generates descendants in the extremes of the hypercube defined by their parents. It is similar to one-point crossover, apart from the fact that in this case the mixing of the genes is random. As in the one-point crossover, the mutation operator is the only one able to change each gene's possible values.

6.6 CROSSOVER BASED ON CONFIDENCE INTERVALS

This operator relies on the exploitation ability associated with a population parameter belonging to a confidence interval, and on the exploration ability associated with it not belonging to such interval.

Let β be the set of the N individuals constituting the population, $\beta^+ \subset \beta$ the set of the n best individuals, in terms of fitness, and $\beta^f = (\beta_1^f, \beta_2^f, \ldots, beta_i^f, \ldots \beta_p^f) \in \beta$ a chromosome to be crossed. If we consider that the genes of the chromosomes of β^+ are independent random variables following a normal distribution $N(\mu_i, \sigma_i^2)$ (and the same for each gene in the chromosomes of β), we can consider three new individuals associated to the sample of the best n individuals $\beta^+ = \{\beta^1, \beta^2, ..., \beta^j, ..., \beta^n\}$: the ones formed by all the lower limits CILL[2], upper limits CIUL[3] and means CIM[4] of each gene confidence interval. Their definition would be

$$\begin{aligned} CILL &= (CILL_1, CILL_2, \ldots, CILL_i, \ldots CILL_p) \\ CIUL &= (CIUL_1, CIUL_2, \ldots, CIUL_i, \ldots CIUL_p) \\ CIM &= (CIM_1, CIM_2, \ldots, CIM_i, \ldots CIM_p) \end{aligned} \tag{6.24}$$

where

$$CILL_i = \overline{\beta_i} - t_{n-1}(\alpha/2)\frac{\overline{S}_{\beta_i}}{\sqrt{n}}; \; CIUL_i = \overline{\beta_i} + t_{n-1}(\alpha/2)\frac{\overline{S}_{\beta_i}}{\sqrt{n}}; \; CIM_i = \overline{\beta_i} \tag{6.25}$$

where n is the number of best individuals in the population, $\overline{\beta_i}$ the mean of each gene of the n best individuals, $i = 1, 2, \ldots, p$, $\overline{S}_{\beta_i}$ the standard quaside-

[2]Confidence Interval Lower Limit
[3]Confidence Interval Upper Limit
[4]Confidence Interval Mean

viation, $t_{n-1}(\alpha/2)$ the value of student's t with $n-1$ degrees of freedom, and $1-\alpha$ the confidence coefficient for the interval.

The CILL and CIUL individuals divide each gene's domain, D_i, into three subintervals I_i^L, I_i^{CI} and I_i^R (Figure 6.2), such that

$$D_i \equiv I_i^L \cup I_i^{CI} \cup I_i^R \tag{6.26}$$

$$I_i^L \equiv [a_i, CILL_i);\; I_i^{CI} \equiv [CILL_i, CIUL_i];\; I_i^R \equiv (CIUL_i, b_i]$$

where a_i and b_i the lower and upper limits of the domain D_i. The interval I_i^{CI} is a confidence interval built from the n best individuals in the population, under the hypothesis that they are distributed following a normal distribution and they have a probability $1-\alpha$ of their genes' values belonging to that interval (the exploitation interval), and a probability α of their genes' values belonging to intervals I_i^L and I_i^R (exploration interval).

If we consider $\alpha = 0.5$ we shall establish an equilibrium between the probability of exploration and that of exploitation. In addition, if n increases, for a fixed α, the amplitude of the interval will diminish and we shall have a greater precision in the exploitation level, as the confidence interval will be smaller and more selective. Analogously, if $\overline{S}_{\beta_i}$ is reduced, which is usual as the GA converges, the amplitude of the interval diminishes and the precision in the level of exploitation will increase. Thus, it is necessary to assign values to both α and n that set the adequate balance between exploration and exploitation for each kind of problem.

The crossover operator proposed in this work will create, from the individual $\beta^f \in \beta$, the individuals CILL, CIUL and CIM, and their fitness, a single offspring β^s in the following way:

- If $\beta_i^f \in I_i^L$ then, if the fitness of β^f is higher than the fitness of CILL then $\beta_i^s = r(\beta_i^f - CILL_i) + \beta_i^f$, else $\beta_i^s = r(CILL_i - \beta_i^f) + CILL_i$.

- If $\beta_i^f \in I_i^{CI}$ then, if the fitness of β^f is higher than the fitness of CIM then $\beta_i^s = r(\beta_i^f - CIM_i) + \beta_i^f$, else $\beta_i^s = r(CIM_i - \beta_i^f) + CIM_i$; Figure 6.3b.

- If $\beta_i^f \in I_i^R$ then, if the fitness of β^f is higher than the fitness of CIUL then $\beta_i^s = r(\beta_i^f - CIUL_i) + \beta_i^f$, else $\beta_i^s = r(CIUL_i - \beta_i^f) + CIUL_i$. This last case can be graphically seen in Figure 6.2.

where r is a random number in the interval [0,1].

From this definition we can deduce that the genes of the offspring always take values toward the best parent, β^f, and one of CILL, CIUL or CIM, depending on the interval to which β^f belongs. If β^f is far from the other parent, the offspring will probably suffer an important change, and vice versa. Obviously, the first circumstance will appear mainly in the early stages of the evolution, and the second one in the last stages.

The underlying idea in this crossover operator is taken from the real world, where many species evolve within highly hierarchical groups, in which only

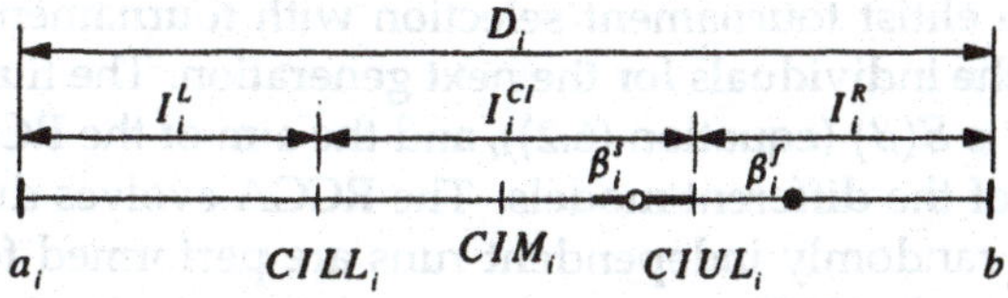

Figure 6.2 Crossover based on confidence intervals

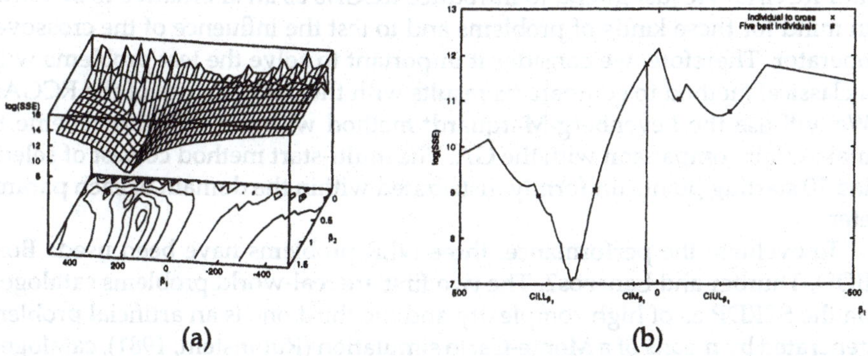

Figure 6.3 (a) log(SSE) surface for BoxBOD where β_1 and β_2 represent the model's parameters; (b) Cut of BoxBOD parallel to β_1 showing graphically the philosophy of confidence interval crossover

a elite of individuals mate with the rest of the population, transmitting their genetic heritage to the descendants of the group.

From the statistical viewpoint, the three generated individuals, CILL, CIUL, and CIM, extract characteristics of centralization and dispersion of the population from the fittest individuals. The proposed multiparent crossover is based on the idea that crossing individuals generated from the characteristics of the best individuals will create better offspring, provided we can find in each problem the localization and dispersion statistics, obtain its exact distribution and build a confidence interval by setting the optimal values of n and $1 - \alpha$.

6.7 EXPERIMENTS

The RCGA used has a constant population size of 100 individuals, from which 10 percent are mutated using a non-uniform mutation with parameter b equal to 5 (Michalewicz, 1992). The probability of mutating a gene in the selected individual is 0.5. Sixty percent of the population will be subject to crossover using the following operators: confidence interval crossover ($n = 5, 1-\alpha = 0.90$), arithmetic crossover ($\lambda = 0.5$), BLX-α crossover ($\alpha = 0.5$), discrete crossover, linear BGA crossover ($rang = 0.5$), flat crossover and Wright's crossover. A uniform selection method is used to choose the individuals for crossover and mutation, and a elitist tournament selection with tournament size 2 will be used to choose the individuals for the next generation. The fitness function of each individual is $S(\beta)$ (equation (6.2)), and the aim of the RCGA is to obtain the parameters of the different models. The RCGA evolves during 5000 generations and 10 randomly independent runs are performed for each type of RCGA (Herrera et al., 1998).

The objective of these experiments is not to make an exhaustive comparison among classical least squares estimation methods for nonlinear models and RCGAs. We just intend to introduce RCGAs as an alternative to be borne in mind for these kinds of problems and to test the influence of the crossover operator. Therefore, we consider it important to solve the test problems with a classical method to compare its results with the performance of the RCGAs. We will use the Levenberg-Marquardt method with a multi-start scheme to make a fair comparison with the GA. The multi-start method consist of selecting 10 starting points uniformly distributed within the domain of each parameter.

To evaluate the performance, three NLR problems have been used: BoxBOD, Thurber and Lanczos2. The two first are real-world problems cataloged on the STRDP as of high complexity and the third one is an artificial problem generated by means of a Monte-Carlo simulation (Rubinstein, 1981), cataloged as average complexity.

6.7.1 BoxBOD problem

This problem is described in Box et al. (1978). The response variable y is biochemical oxygen demand (BOD) in mg/l, and the predictor variable x is incubation time in days (Figure 6.4a). To model this problem we have six observations and an exponential model is used, with two parameters β_1 and β_2:

$$f_1 = y = \beta_1(1 - e^{-\beta_2 x}) + \epsilon \tag{6.27}$$

The minimum value of $S(\beta^*)$ or sum of squared error (SSE) is 1168.008 for values $\beta_1^* = 213.80$ and $\beta_2^* = 0.54$. In Figure 6.3a we can see that the response surface $S(\beta)$ of BoxBOD is multimodal and there exist a very high epistasis among its parameters.

6.7.2 Thurber problem

This problem's data are the result of a NIST (National Institute of Standards and Technology) study involving semiconductor electron mobility. The response variable y is a measure of electron mobility, and the predictor variable x is the natural log of the density (Figure 6.5a). We have 36 samples that must be adjusted to the following rational model:

$$f_2 = y = \frac{\beta_1 + \beta_2 x + \beta_3 x^2 + \beta_4 x^3}{1 + \beta_5 x + \beta_6 x_2 + \beta_7 x^3} + \epsilon \tag{6.28}$$

The optimum value of the sum of squared error for this problem obtained when $\beta^* = (1288.13, 1491.07, 583.23, 75.41, 0.96, 0.39, 0.04)$ is 5642.708. As with BoxBOD the response surface of this model is highly multimodal and very epistatic, making for a complex problem.

6.7.3 *Lanczos2 problem*

This problem is taken from an example discussed in Lanczos (1956). We have 24 points generated with six digits of accuracy using the six parameters function (Figure 6.6a):

$$f_3 = y = 0.0951e^{-x} + 0.8607e^{-3x} + 1.5576e^{-5x} \tag{6.29}$$

The optimal value of the sum of squared error for this problem is 2.22E-11. The problem is cataloged on the STRDP as of average complexity, being its main difficulty the fine tuning of the solution.

6.7.4 *Domains*

To establish performance and robustness differences among the different crossover operators when the search space amplitude varies, three domains are defined for each problem. We have avoided that the optimal parameter value is centered in its domain, as this could favor the convergence of some crossover methods.

For the BoxBOD problem, the domains are:

- wide: $D_1^{f_1} = \{[10, 1000], [0.01, 1]\}$,
- medium: $D_2^{f_1} = \{[100, 1000], [0.1, 1]\}$,
- small: $D_3^{f_1} = \{[1, 327], [0.3, 1]\}$.

For the Thurber problem, the domains are:

- wide: $D_1^{f_2} = \{[100, 10000], [100, 10000], [10, 1000], [1, 100], [0.01, 1], [0.01, 1], [0.001, 0.1]\}$,
- medium: $D_2^{f_2} = \{[1000, 10000], [1000, 10000], [100, 1000], [10, 100], [0.1, 1], [0.1, 1], [0.01, 0.1]\}$,
- small: $D_3^{f_2} = \{[1000, 1300], [1000, 1500], [400, 667], [40, 76], [0.7, 1], [0.3, 0.4], [0.03, 0.05]\}$.

Finally, for the Lanczos2 problem, the domains are:

- wide: $D_1^{f_3} = \{[0, 100], [0, 100], [0, 100], [0, 100], [0, 100], [0, 100]\}$,
- medium: $D_2^{f_3} = \{[0, 10], [0, 10], [0, 10], [0, 10], [0, 10], [0, 10]\}$,
- small: $D_3^{f_3} = \{[0, 1.2], [0.3, 1.7], [0, 5.6], [0, 5.5], [0, 6.5], [0, 7.6]\}$.

6.8 RESULTS

As can be seen on Table 6.1, for the BoxBOD problem, all crossover operators converge to the optimal solution. For the three domains, the speed of convergence is high (see Figure 6.4bcd), although the confidence interval based crossover is the best one. The reason is that, even though this problem is qualified as complex due to its epistasis and its multimodal character (Figure 6.3a), the fact that it has only two parameters makes it very easy to be tackled by any kind of RCGA, even using wide domains. The Levenberg-Marquartd method converges in all the experiments to a less precise solution. In our experiments, the initial values obtained have always been close to the optimum, but if they are chosen not so close to the optimum the algorithm fails. For instance, with starting values ($\beta_1 = 1, \beta_2 = 0.5$) this method converges to a SSE value of 9771.5, far from the global optimum.

Table 6.2 shows the results for the Thurber problem. For these problems, the Levenberg-Marquartd method is very sensitive to the initial values. Therefore, in some runs it achieves values very close to the optimum, and in other runs it converges to values far from the optimum. This effect is observed in the mean values and in the standard deviation. The RCGA are also sensitive to the starting point but to a lesser degree. The use of the confidence interval crossover decreases such fluctuations as we reduce the domain of the parameters. The mean values obtained with this crossover are the best and most stable ones. For the small domain interval it achieved the value closest to the optimum in $\beta = (1288.13, 1491.07, 583.23, 75.41, 0.96, 0.39, 0.04)$.

Figures 6.5b, 6.5c, and 6.5d show the speeds of convergence for Thurber's wide, medium and small domains, respectively. It can be seen that the confidence interval based crossover is the fastest in all cases. From these results it can be deduced that this problem is very complex for a RCGA due to its epistasis, multi-modality and the fact that the number of dimensions of the search space is large. This makes that a slight increment in the domain of the parameters affects decisively the complexity of the problem, often making it unfit for classical NLR methods.

For the Lanczos2 problems, the best results are also obtained with a RCGA using the confidence interval based crossover (see Table 6.3). This points out the adequacy of this operator when a high precision of the solution is needed. The operator allows the RCGA to carry out a fine-tuning of the solution, being this feature distinctive of the local search methods.

As we can see in Figures 6.6b, 6.6c, and 6.6d, at the beginning of the evolution the confidence interval based crossover is not always the one that converges fastest. However, from generation 2500 on, it achieves the best results in all the three cases. The reason probably has to do with the reduced complexity of this problem. Some crossovers can converge faster at the beginning but the are unable to fine tune the obtained solution.

Table 6.1 Average (AVG) and Standard Deviation (STD) of SSE for BoxBOD problem, with domains of different amplitude over 10 runs for Levenberg-Marquardt algorithm and for each RCGA using different crossovers.

	$D_1^{f_1}$		$D_2^{f_1}$		$D_3^{f_1}$	
	AVG	STD	AVG	STD	AVG	STD
Levenberg-M.	1168.010	0.00E+00	1168.010	0.00E+00	1168.010	0.00E+00
Confidence Interval	1168.008	2.03E-05	1168.008	2.03E-05	1168.008	2.03E-05
Arithmetic	1168.008	2.03E-05	1168.008	2.03E-05	1168.008	2.03E-05
BLX-α	1168.008	2.03E-05	1168.008	2.49E-05	1168.008	2.03E-05
Discrete	1168.008	2.49E-05	1168.008	2.03E-05	1168.008	2.03E-05
Linear BGA	1168.008	1.44E-05	1168.008	2.88E-05	1168.008	2.03E-05
Flat	1168.008	2.03E-05	1168.008	2.03E-05	1168.008	2.03E-05
Wright	1168.008	2.03E-05	1168.008	2.03E-05	1168.008	2.03E-05

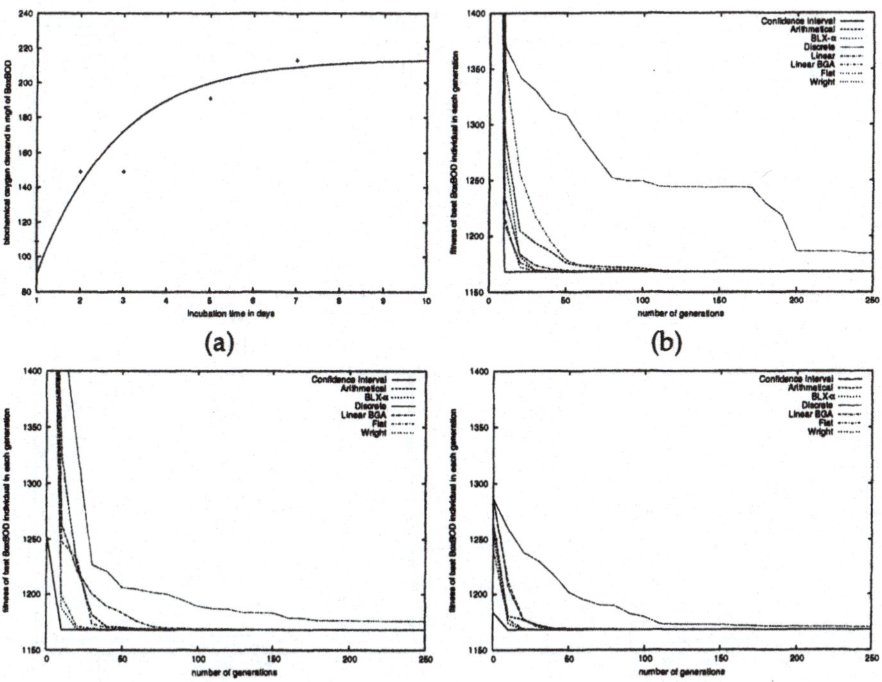

Figure 6.4 (a) Nonlinear regression model for BoxBOD; (b) Average fitness of the best individual in 10 runs for BoxBOD using different crossovers for $D_1^{f_1}$ domain; (c) Idem b) for $D_2^{f_1}$ domain; (d) Idem b) for $D_3^{f_1}$ domain

Table 6.2 Average (AVG) and Standard Deviation (STD) of SSE for Thurber problem, with domains of different amplitude over 10 runs for Levenberg-Marquardt algorithm and for each RCGA using different crossovers.

	$D_1^{f_2}$		$D_2^{f_2}$		$D_3^{f_2}$	
	AVG	STD	AVG	STD	AVG	STD
Levenberg-M.	8.78E+09	2.62E+10	6.42E+09	2.03E+10	3.16E+10	8.65E+10
Confidence Interval	6061.835	6.85E+02	5659.5861	1.64E+01	5646.115	6.53E+00
Arithmetic	28600.199	8.68E+03	23557.185	7.00E+03	7298.986	5.55E+02
BLX-α	7839.888	2.33E+03	7597.951	1.38E+03	6833.509	3.61E+02
Discrete	63698.269	2.07E+04	22539.227	1.61E+04	7401.607	1.24E+03
Linear BGA	44889.829	2.10E+04	21707.524	1.69E+04	7525.315	1.52E+03
Flat	7245.728	1.28E+03	6285.199	4.69E+02	6796.092	4.15E+02
Wright	8157.669	4.37E+02	6646.249	9.23E+02	5652.669	2.76E+01

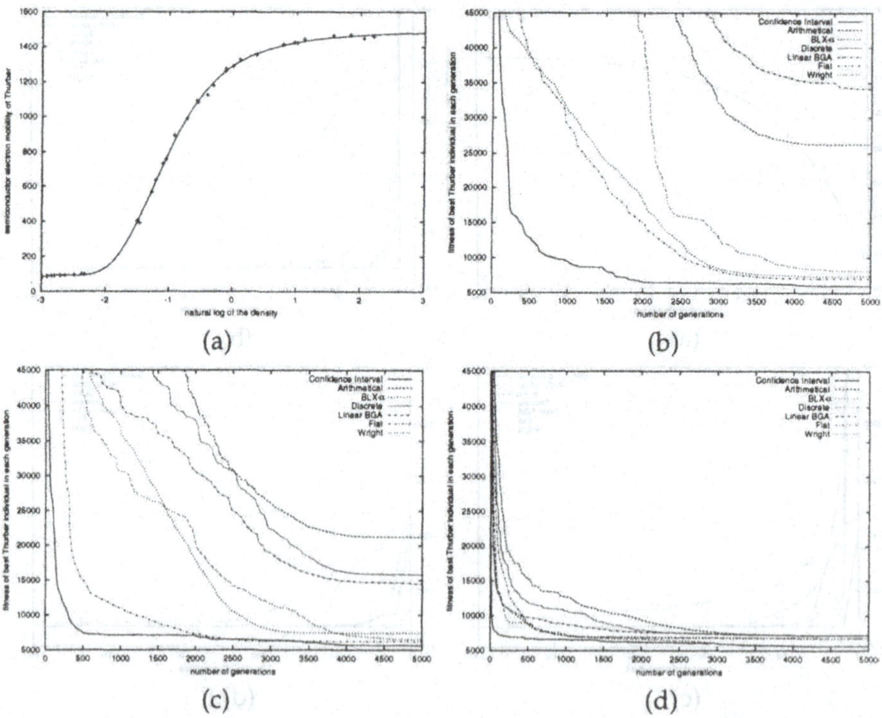

Figure 6.5 (a) Nonlinear regression model for Thurber; (b) Average fitness of the best individual in 10 runs for Thurber using different crossovers for $D_1^{f_2}$ domain; (c) Idem b) for $D_2^{f_2}$ domain; (d) Idem b) for $D_3^{f_2}$ domain

Table 6.3 Average (AVG) and Standard Deviation (STD) of SSE for Lanczos2 problem, with domains of different amplitude over 10 runs for Levenberg-Marquardt algorithm and for each RCGA using different crossovers.

	$D_1^{f_3}$		$D_2^{f_3}$		$D_3^{f_3}$	
	AVG	STD	AVG	STD	AVG	STD
Levenberg-M.	2.15E-06	2.27E-06	1.72E-06	2.22E-06	2.15E-06	2.27E-06
Confidence Interval	1.77E-06	1.11E-06	1.49E-06	1.47E-06	3.38E-07	6.52E-07
Arithmetic	8.03E-03	7.18E-03	2.32E-04	1.06E-04	3.68E-05	3.42E-05
BLX-α	7.92E-04	8.11E-04	1.67E-04	1.76E-04	1.11E-05	1.93E-05
Discrete	1.03E-03	1.34E-03	3.99E-04	5.28E-04	5.08E-05	1.07E-04
Linear BGA	7.45E-05	5.96E-05	9.37E-06	1.00E-05	1.47E-05	1.95E-05
Flat	1.42E-03	6.87E-04	1.11E-04	1.03E-04	5.27E-06	6.36E-06
Wright	2.34E-06	1.31E-06	2.41E-06	1.94E-06	4.63E-06	7.63E-06

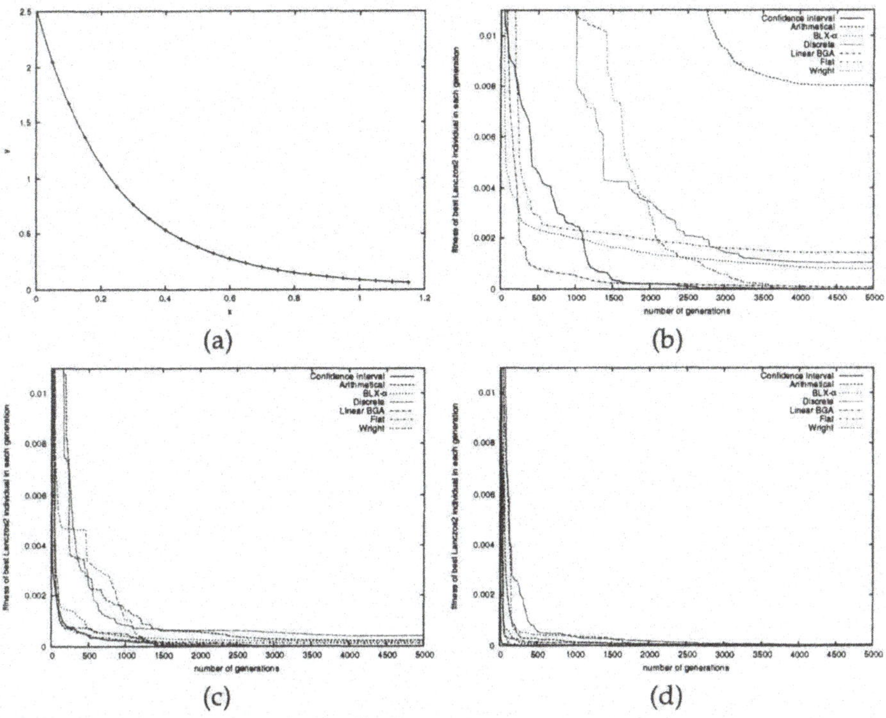

Figure 6.6 (a) Nonlinear regression model for Lanczos2; (b) Average fitness of the best individual in 10 runs for Lanczos2 using different crossovers for $D_1^{f_2}$ domain; (c) Idem b) for $D_2^{f_2}$ domain; (d) Idem b) for $D_3^{f_2}$ domain

6.9 CONCLUSIONS

We make two concluding remarks.

First, RCGAs using the confidence interval based crossover operator are an alternative to classical parameter estimation methods, specially in problems where there is no a priori information that allows to choose an initial solution close to the optimum.

Second, when the search space is expanded, RCGAs using the confidence interval based crossover operator, which balances the exploration and exploitation of the search space, perform better that the classical methods for the three problems under study.

Acknowledgments

This research has been supported in part by projects ALI98-0676-CO-02 and TIC2001-2577 of the Spanish CICYT.

Bibliography

J. Antonisse. A new interpretation of schema notation that overturns the binary encoding constraint. In J. David Schaffer, editor, *Third International Conference on Genetic Algorithms*, pages 86–91, San Mateo, 1989. Morgan Kaufmann.

C. Barron and S. Gómez. The exponential tunneling method. Technical report, IIMAS-UNAM, 1991.

K. Bennett, M. C. Ferris, and Y. E. Ioannidis. A genetic algorithm for database query optimization. In *Fourth International Conference on Genetic Algorithms*, pages 400–407, San Mateo, CA, 1991. Morgan Kaufmann.

S. A. Billings and K. Z. Mao. Structure detection for non-linear rational models using genetic algorithms. Technical Report 634, Department of Automatic Control and Systems Engineering, University of Sheffield, U. K., 1996.

G. P. Box, W. G. Hunter, and J. S. Hunter. *Statistics for Experimenters*. Wiley, New York, 1978. pp. 483-487.

J. E. J. Dennis. Non-linear least squares and equations. In D. A. H. Jacobs, editor, *The State of the Art in Numerical Analysis*, pages 269–312, London, 1977. Academic Press.

J. E. J. Dennis and R. B. Schnabel. *Numerical Methods for Unconstrained Optimization and Nonlinear Equations*. Prentice-Hall, New York, 1983.

L. J. Eshelman and J. D. Schaffer. Real-coded genetic algorithms and interval-schemata. In L. Darrell Whitley, editor, *Foundation of Genetic Algorithms 2*, pages 187C3.3.7:1–C3.3.7:8.–202, San Mateo, 1993. Morgan Kaufmann.

A. R. Gallant. *Nonlinear Statistical Models*. Wiley, New York, 1987.

D. E. Goldberg. *Genetic Algorithms in Search, Optimization, and Machine Learning*. Addison-Wesley, New York, 1989.

D. E. Goldberg. Real-coded genetic algorithms, virtual alphabets, and blocking. *Complex Systems*, 5:139–167, 1991.

J. J. Grefenstette. *Incorporating Problem Specific Knowledge into Genetic Algorithms*. Morgan Kaufmann, San Mateo, CA, 1987.

F. Herrera, M. Lozano, and J. L. Verdegay. Tunning fuzzy logic controllers by genetic algorithms. *International Journal of Approximate Reasoning*, 12:299–315, 1995.

F. Herrera, M. Lozano, and J. L. Verdegay. Tackling real-coded genetic algorithms: Operators and tools for behavioural analysis. *Artificial Intelligence Review*, pages 265–319, 1998. Kluwer Academic Publishers. Printed in Netherlands.

J. H. Holland. *Adaptation in natural and artificial systems*. The University of Michigan Press, Ann Arbor, MI, 1975.

T. Johnson and P. Husbands. System identification using genetic algorithms. In *Parallel Problem Solving from Nature*, volume 496 of *Lecture Notes in Computer Science*, pages 85–89, Berlin, 1990. Springer-Verlag.

H. Kargupta and R. E. Smith. System identification with evolving polynomial networks. In *Fourth International Conference on Genetic Algorithms*, pages 370–376. Morgan Kaufmann, 1991.

S. Kirkpatrick, C. D. Gelatt, and M. P. Vecchi. Optimization by simulated annealing. *Science*, 220:671–680, 1983.

S. Kirkpatrick, C. D. Gelatt, and M. P. Vecchi. Optimization by simulated annealing. In M. A. Fischler and O. Firschein, editors, *Readings in Computer Vision: Issues, Problems, Principles, and Paradigms*, pages 606–615. Kaufmann, Los Altos, CA., 1987.

J. R. Koza. *Genetic Programming*. The MIT Press, 1992.

C. Lanczos. *Applied Analysis. Englewood Cliffs*. Prentice Hall, Englewood Cliffs, NJ, 1956.

K. Levenberg. A method for the solution of certain problems in least squares. *Quart. Appl. Math.*, 2:164–168, 1944.

A. V. Levy and S. Gómez. The tunneling method applied to global optimization. In *Society for Industrial and Applied Mathematics (SIAM)*, pages 213–244, Philadelphia, PA, 1985.

A. V. Levy and A. Montalvo. The tunneling algorithm for the global minimization of functions. *SIAM Journal on Scientific and Statistical Computing*, 6 (1):15–29, 1985.

G. E. Liepins and M. D. Vose. Characterizing crossover in genetic algorithms. *Annals of Mathematics and Artificial Intelligence*, 5:27–34, 1992.

E. Malinvaud. The consistency of nonlinear regression. *Ann. Math. Stat.*, 41: 956–969, 1970.

D. W. Marquardt. An algorithm for least-squares estimation of nonlinear parameters. *Journal of the American Statistical Association*, 75:87–91, 1963.

H. Mühlebein and D. Schlierkamp-Voosen. Predictive models for breeder genetic algorithm i. continuous parameter optimization. *Evolutionary Computation*, 1:25–49, 1993.

Z. Michalewicz. *Genetic Algorithms + Data Structures = Evolution Programs*. Springer-Verlag, New York, 1992.

H. Midi. Preliminary estimators for robust non-linear regression estimation. *Journal of Applied Statistics*, 26(5):591–600, 1999.

J. J. Moré. Levenberg-marquardt algorithm: Implementation and theory. In G. A. Watson, editor, *Lecture Notes in Mathematics*, number 630 in Numerical Analysis, pages 105–116, Berlin, 1977. Springer-Verlag.

J. C. Nash. Minimizing a nonlinear sum of squares function on a small computer. *J. Inst. Math. Appl.*, 19:231–237, 1977.

A. S. Nissesen and H. Koivisto. Identification of multivariate volterra series using genetic algorithms". In J.T. Alander, editor, *Second Nordic Workshop on Genetic Algorithms and Their Applications*, pages 151–161, University of Vaasa, Finland, 1996.

N. J. Radcliffe. Equivalence class analysis of genetic algorithms. *Complex Systems*, 2(5):183–205, 1991.

N. J. Radcliffe. Non-linear genetic representations. In R. Männer and B. Manderick, editors, *Second International Conference on Parallel Problem Solving from Nature*, pages 259–268, Amsterdam, 1992. Elsevier Science Publishers.

R. Y. Rubinstein. *Simulation and the Monte Carlo Method*. Wiley series in probability and mathematical statistics. John Wiley & Sons, 1981.

D. Schlierkamp-Voosen. Strategy adaptation by competition. In *Second European Congress on Intelligent Techniques and Soft Computing*, pages 1270–1274, 1994.

G. A. F. Seber and C. J. Wild. *Non linear regression*. Wiley, 1989.

A. Wright. Genetic algorithms for real parameter optimization. In G. J. E. Rawlin, editor, *Foundations of Genetic Algorithms 1*, pages 205–218, San Mateo, 1991. Morgan Kaufmann.

Metaheuristics: Computer Decision-Making, pp. 153-172
Edmund Burke, Patrick De Causmaecker, Sanja Petrovic, and Greet Vanden Berghe

7 VARIABLE NEIGHBORHOOD SEARCH FOR NURSE ROSTERING PROBLEMS

Edmund Burke[1], Patrick De Causmaecker[2], Sanja Petrovic[1], and Greet Vanden Berghe[2]

[1]School of Computer Science & IT
University of Nottingham
Jubilee Campus, Nottingham, NG8 1BB, UK
ekb@cs.nott.ac.uk, sxp@cs.nott.ac.uk

[2]KaHo St.-Lieven, Information Technology
Gebr. Desmetstraat 1, 9000 Gent, Belgium
patrick.decausmaecker@kahosl.be, greetvb@kahosl.be

Abstract: Nurse rostering problems consist of assigning varying tasks, represented as shift types, to hospital personnel with different skills and work regulations. The goal is to satisfy as many soft constraints and personal preferences as possible while constructing a schedule which meets the required personnel coverage of the hospital over a predefined planning period. Real-world situations are often so constrained that finding a good quality solution requires advanced heuristics to keep the calculation time down. The nurse rostering search algorithms discussed in this paper are not aimed at specific hospitals. On the contrary, the intention is that such algorithms should be applicable across the whole sector. Escaping from local optima can be very hard for the metaheuristics because of the broad variety of constraints. In this paper, we present a variable neighborhood search approach. Hidden parts of the solution space become accessible by applying appropriate problem specific neighborhoods. The method allows for a better exploration of the search space, by combining shortsighted neighborhoods, and very greedy ones. Experiments demonstrate how heuristics and neighborhoods can be assembled for finding good quality schedules within a short amount of calculation time.
Keywords: Variable neighborhood search, Nurse rostering.

7.1 INTRODUCTION

Developing nurse rostering algorithms for real-world problems often involves very generic heuristics to deal with widely varying hospital customs and requirements. Unlike many other personnel planning environments, in which cyclical schedules cover nearly all the needs, nurse scheduling requires much more flexible algorithms to deal with varying patient needs, round the clock coverage, a large number of shift types, different skill categories, flexible work regulations, and personal preferences.

This paper deals with real-world personnel scheduling problems in Belgian hospitals. Such problems are particularly complex because there are very few legal restrictions on the problem definition. The characteristics of the constraints are so different that the solution space seems to consist of a hilly landscape with very deep and narrow valleys, in which good quality solutions are very hard to find. Some constraints refer to particular duties of the nurses while other constraints restrict consecutive shifts, days, weekends, etc. The cost function used by the metaheuristics sums the penalties of the violations of each of the soft constraints (Section 7.2).

The paper is organized as follows. The description of the nurse scheduling problem is given in Section 7.2. Different neighborhoods for the search heuristics are defined in Section 7.3. In Section 7.4, we explain how the heuristics can be combined in order to reach results which might remain behind big barriers when using single neighborhood strategies. A few ideas to combine, repair and restart different heuristics applying several neighborhoods are explored. The results of the developed variable neighborhood algorithms are discussed in Section 7.5, and we conclude in Section 7.6.

7.2 PROBLEM DESCRIPTION

7.2.1 *Nurse Rostering*

The nurse scheduling problem tackled in this paper is based on the formulation considered by *Plane*, a commercial nurse rostering package developed in co-operation with Impakt[1]. The software, which is now implemented in many Belgian hospitals, makes use of a hybrid Tabu Search algorithm (Burke et al., 1999). Since the system has to address widely varying hospital requirements, the schedulers are allowed considerable freedom with respect to defining the needs and customs of the ward, people, and the constraints on the schedule. Both personal constraints and statutory regulations are implemented. All the constraints are modifiable and they are evaluated with respect to a cost parameter set by the user. The cost function is evaluated separately for every personnel member of the ward. Summing the values for the separate personnel members' schedules gives the value of the entire schedule (see also Burke et al. (2001a)). Although the problem definition is user-defined to a large ex-

[1] Impakt N.V., Dendermondsesteenweg 42, B-9000 Gent

tent, the scheduling algorithms have to be efficient in different settings. The heuristics, which are currently implemented in Plane, are very successful with respect to solution quality and calculation time. Experiments with other approaches, such as Memetic Algorithms in Burke et al. (2001b), often produce better results at the expense of calculation time. The research in this paper attempts to make more use of problem specific characteristics to dynamically change the search heuristics and their neighborhoods in order to overcome some typical drawbacks of metaheuristics for NP hard problems.

7.2.2 *Formulation*

The nurse scheduling problem consists of assigning personnel requirements (expressed in terms of shift types) to qualified personnel over a certain planning period. Note that shift types, qualifications, requirements, and the planning period are defined by the users. Typical shift types are Morning, Day, and Night shifts but users of Plane are not restricted to this limited number of types. New shift types require the setting of a number of features. For example, a start and end time, and recommended breaks. Qualifications reflect the skills of the personnel. They normally depend on experience, level of education, etc. Some people can be replaced by personnel members from a different skill category. However, unlike in some other personnel planning systems, Plane does not automatically allow higher qualified personnel to replace less experienced people. Indeed, in practice, higher qualified nurses are usually reluctant to undertake duties more normally assigned to less qualified staff.

7.2.2.1 Hard constraints. Hard constraints are those which must never be violated. In Plane, the personnel requirements are hard constraints and we call all the solutions satisfying them 'feasible' solutions. Any solution must provide a sufficient number of qualified personnel at any time during the entire planning period. The problems are defined so that this sufficient number lies between a range, which represents the minimum and maximum numbers that are acceptable.

7.2.2.2 Soft constraints. Soft constraints can be, but are preferably not, violated. Of course, in practice it is impossible to completely satisfy all soft constraints and, during the search process, the algorithms aim at minimizing the number of violations of these constraints (taking cost parameters into account). The previous planning period influences the evaluation of some constraints (e.g. minimum and maximum number of consecutive working days, etc). Some relaxations also have to be made when people take illness leave or have a holiday period. It would be beyond the scope of this paper to explain and fully discuss the large group of soft constraints, and the entire evaluation procedure in detail. The interested reader can see Burke et al. (2001a) and Berghe (2001) for more details. We have extracted a set of important soft constraints, which we think have applications beyond Belgian hospital wards and these are briefly presented below.

7.2.2.3 Hospital constraints. These constraints are rules which hold for the entire organization. They are often based on legal regulations which dominate the local habits and preferences of each ward. Examples are:

- Minimum time between two assignments: wards can make the constraint locally more stringent by increasing the required free time before and after particular shift types.

- Allow the use of an alternative skill category in certain situations: a personnel member should not undertake jobs which do not belong to his/her main skill category.

7.2.2.4 Constraints defined by the work regulation. These constraints are modifiable by the users of the software and are not always applied to every personal schedule in the problem. Each personnel member has a contract with the hospital, called the work regulation or work agreement, which sets a group of soft constraints. Typical examples are full time personnel contracts, half time and night nurses. In practice, many hospitals allow for a personal work agreement per nurse. This enables them to formulate personal constraints such as 'Every Friday morning should be free', 'Work a weekend every two weeks', etc. When defining the work regulation, any of the following constraints can be defined or made idle.

- Maximum number of assignments in the planning period: Holiday periods or illness leave are taken into account.

- Minimum/Maximum number of consecutive days: Special rules hold when the work schedule includes days off, illness, etc.

- Minimum/Maximum number of hours worked.

- Minimum/Maximum number of consecutive free days.

- Maximum number of assignments per day of the week: There is a possibility to set different numbers for different days.

- Maximum number of assignments for each shift type: The constraint can be defined per shift type.

- Maximum number of a shift type per week: This constraint can also be set per shift type.

- Number of consecutive shift types (can be different per shift type).

- Assign two free days after night shifts: Night shifts are all the shift types which start before and end after midnight.

- Assign complete weekends: Single assignments in weekends are to be avoided. Plane allows for extended definitions of weekends (e.g. Friday till Sunday).

- Assign identical shift types during the weekend.
- Maximum number of consecutive working weekends.
- Maximum number of working weekends in a 4-week period.
- Maximum number of assignments on bank holidays: This constraint is generally evaluated over a longer period than the short term planning horizon.
- Restriction on the succession of shift types: Some shift sequences have to be avoided on consecutive days.
- Patterns enable the formulation of specific cyclic constraints: Several pattern elements (such as free day, working day, assignment of special shift types, etc) can be combined to construct both flexible and very rigid patterns of a user definable length. The start dates of patterns can be set per personnel member in order to enable a good overall distribution of work and free time.
- Balancing the workload among personnel, working time, weekend work, night shifts, etc: Tasks should be divided evenly among nurses.

7.2.2.5 Personal constraints. Personnel managers or head nurses can allow individual personnel members to take a day off or to schedule their holiday period. Nurses sometimes ask to carry out a specific shift on a certain day. These requests can be formulated with:

- Days off; shifts off.
- Requested assignments.

Other personal constraints are those in which the schedules of different nurses are linked with each other. Some junior nurses, for example, can only work when their supervisor is at work. On the other hand it is often the case that we specifically do not want two particular people to work together. This could be for a variety of reasons including: bad use of resources, e.g. two head nurses working together, husband and wife who want to maximize the time that they look after their children themselves, two people who do not like each other, etc. Such constraints can be represented as follows:

- Tutorship (people not allowed to work alone)
- People should not work together

The cost function is the motor of the search heuristics, but since it does not interpret the problem characteristics, the algorithms are quite blind to certain improvements. While improving the schedule with respect to one constraint, it might make the solution much worse with respect to others.

Some of the constraints are of particular importance for the research presented in this paper. We have constructed neighborhoods in order to especially satisfy a number of these constraints (see Section 7.3.2).

7.3 VARIABLE NEIGHBORHOOD SEARCH

7.3.1 *Variable neighborhood search heuristics*

Glover and Laguna (1997) suggest diversification strategies to efficiently explore the search space of combinatorial problems. It is often the only possible way to reach regions behind barriers in the landscape of solutions. Variable neighborhood search (Mladenović and Hansen, 1997) combines local search heuristics, which stop in local optima, and neighborhood changes to escape from these local optima. The approach is applicable in combination with metaheuristic algorithms as a diversification strategy for the local search.

Variable neighborhood search has been applied to several NP hard problems such as the traveling salesman problem, the location-allocation problem, the clustering problem, the bilinear programming problem with bilinear constraints (Hansen and Mladenović, 1999; Mladenović and Hansen, 1997), the linear ordering problem (Gonzáles and Pérez-Brito, 2001), scheduling problems (Davidović et al., 2001; den Besten and Stützle, 2001), vehicle routing (Crispim and Brandão, 2001), the p-median (Crainic et al., 2001; Hansen and Mladenović, 1999), the MAX-CUT (Festa et al., 2002), and the k-cardinality tree problem (Mladenović and Urošević, 2001). Different approaches exist for selecting neighborhoods and for going from one neighborhood to another. It is often recommended to *shake* the solution, i.e. to randomly swap to a solution in the neighborhood of the current one.

For the approach presented in this paper, a set of neighborhood structures, which use specific information about the problem, is defined. When the search heuristic fails to improve the solution (within a certain amount of time or a number of iterations), the algorithm dynamically chooses a different neighborhood.

7.3.2 *Neighborhoods for the nurse rostering problem*

In the hybridization algorithms discussed in Burke et al. (1999), three different neighborhoods are introduced. We expand this group with sets of new neighborhoods related to soft constraints, and greedier neighborhoods, that are inspired by manual scheduling processes. Hard constraints are never violated during the course of the rostering process; shifts are never assigned to unskilled personnel members, nor is an excess or shortage in working personnel allowed at any time during the planning period. To guarantee the satisfaction of the hard constraints, shifts will only be moved to another person's schedule on the same day. The moves are not allowed unless the person is qualified to perform the duty and provided this person is not yet assigned to the same shift.

7.3.2.1 Single shift-day (D). The simplest neighborhood of a schedule includes all the feasible solutions that differ in the position of one scheduled shift. It is the basic neighborhood for all the metaheuristic approaches executed on the problem described in Section 7.2. Note that the position refers to

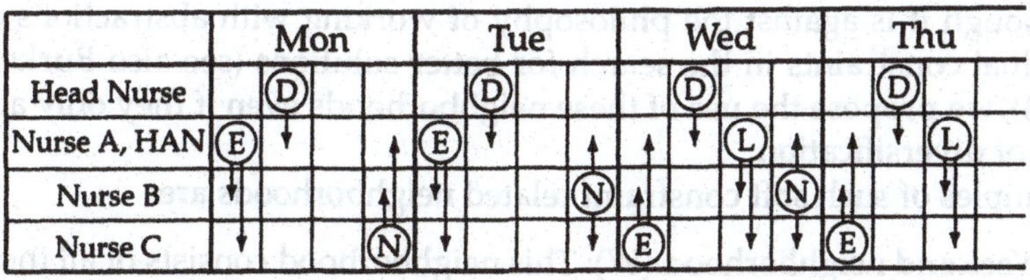

Mon
Tue
Wed
Thu
Head Nurse
Nurse A, HAN
Nurse B
Nurse C
D
E
N
L

Figure 7.1 Possible moves in a 'single shift-day' neighborhood. Each day is divided into columns that represent the shifts. Shifts are Early (E), Day (D), Late (L), and Night (N).

the personnel member whose schedule the assigned shift belongs to. The single shift-day neighborhood considers the solutions in the nearest environment of the current solution. In order to create the neighborhood, it suffices to consider all the allowed moves of a scheduled shift from the personal schedule, which contains the shift type, to another person's schedule which is empty for that shift type on the same day. The solution corresponding to that move does not violate the hard constraints, provided the 2nd person is qualified to work that shift type. Figure 7.1 presents the moves which are allowed in the single shift-day neighborhood. A very small ward, consisting of one head nurse and three regular nurses is presented. One of the regular nurses has the head nurse skill as an alternative. This will be the person to replace the head nurse during absence. A very small part of a realistic planning period is shown and there are four shift types: Early (E), Day (D), Late (L), and Night (N). Arrows demonstrate the possible moves in the neighborhood. Note that the head nurse's Day shifts cannot be moved into the schedules of Nurse B and C because that would violate the hard constraint on skills. Also, shifts from the regular nurses (Nurses A, B, and C) cannot be moved to the head nurse's schedule because to do so would be to violate a hard constraint. Shifts cannot be moved horizontally in the schedule either because that would disturb the equilibrium between required personnel members and scheduled ones.

7.3.2.2 Soft constraint related neighborhoods. The neighborhoods introduced in this section are not comparable to the others because they perceive the landscape of the solution space in a different way. While searching schedules, which better satisfy a particular soft constraint, the algorithms are blind to the overall quality of the result. This is one of the main reasons why such neighborhoods are not applied in the final phase of a search algorithm.

The inspiration for developing these neighborhoods comes from real-world suggestions from hospital planners (customers of Plane). Looking at an automatically generated schedule, they can point out shortcomings with respect to some sensitive constraints. Solving these problems does not necessarily lead the search to a very interesting part of the solution space but it generally does not hinder the search either.

Although it is against the philosophy of working with abstractions of the individual constraints in the search for better solutions (see also Burke et al. (2001a)), we propose the use of these neighborhoods even if they only act as a means of diversification.

Examples of such soft constraint related neighborhoods are:

- **Weekend neighborhood (W)**: This neighborhood consists of all the solutions differing from the current solution by the assignment of one shift on a day of the weekend. This weekend neighborhood is of importance only in the case where the constraint of 'complete weekends' is applied to at least one of the personnel members. The weekend neighborhood is empty if the constraint on complete weekends is completely satisfied. If it is not completely satisfied, all the personal schedules which are subject to the complete weekends constraint have one violation of this constraint less than the current solution.

- **Overtime - Undertime neighborhood (OU)**: This neighborhood only considers moving shifts from people with overtime to people with undertime. An extension of this neighborhood includes all the moves which do not increase the sum of the overtime and undertime violations in the schedule.

- **Alternative qualifications neighborhood (AQ)**: Experienced people have the authority to carry out work for other skill categories in order to replace absent personnel members. It is better for the quality of a schedule when the number of replacements is low. This neighborhood consists of all schedules which have one less assignment (which involves a replacement).

- **Personal requests neighborhood (PR)**: The soft constraint on personal requests has a modifiable cost parameter, like all other soft constraints. In many circumstances, the result of the scheduling algorithm will violate a few of these constraints. The nature of the cost function (which sums the violations on soft constraints) guarantees a solution, which is not biased toward solving a particular constraint. Nurses can be very sensitive about their personal request for a certain shift or day off. This particular neighborhood has been developed to search for solutions which satisfy the personal requests. By moving from one solution to another in this personal requests neighborhood, the size of the neighborhood should decrease. Ideal schedules without penalties for personal requests have this neighborhood empty.

- **The most violated constraint neighborhood (MV)**: The modular nature of the cost function allows for isolating constraints. This neighborhood pays more attention to moves affecting a particular constraint, namely the constraint that is violated to the highest extent. We consider the number of violations per constraint in every personal schedule in order to determine the most violated constraint. The neighborhood itself contains all the solutions of the simplest neighborhood (single shift-day) but the evaluation function temporarily takes a higher value for the cost parameter of the most violated constraint.

7.3.2.3 Swapping large sections of personal schedules. The user interface of Plane allows hospital schedulers to change the schedule manually. Their manipulations often have the aim of creating schedules, which are visually more satisfying. This inspired us to design this category of neighborhood, in which we try to imitate very common real-world manipulations of schedules.

Unlike the previous group of neighborhoods, in which neighboring solutions only differ in the position of one single shift type, this set of neighborhoods looks at schedules which differ considerably from the original solution. Re-allocating larger groups of shifts at once is often less harmful for the quality of a schedule than moving single shifts around. The drawback of applying this category of neighborhoods is that the number of neighboring solutions is very large, and thus so is the calculation time. Examples are:

- **Shuffle** neighborhood (SH), see also Burke et al. (1999): The "shuffle" environment considers switches from a part of the worst personal schedule, in terms of the evaluation function, with any other schedule. Instead of moving duties (as in the simple single shift-day neighborhood), all the duties, which are scheduled in a period from one day to a number of days equal to half the planning period, are switched between the person with the worst schedule and another person in the ward. All possible feasible shuffles during the planning period are considered (see Figure 7.2).

- **Greedy shuffling** neighborhood (GS), see also Burke et al. (1999): The greedy shuffling environment is comparable to the shuffle environment, but it is much bigger. It consists of all possible shuffles between any set of two people in the schedule. We call this shuffling greedy because it is very large and very time-consuming to evaluate.

- **Core shuffle** neighborhood (CS): Compared to the shuffle neighborhood, we apply an extra shuffle, moving an internal part of the shuffle section (see Figure 7.3). The core shuffle neighborhood considers two consecutive swaps between a pair of personal schedules at a time. In the first phase, a move from the greedy shuffling neighborhood is performed. Within the swapped time interval of that move, a new time interval, also consisting of full days, is swapped back in the second phase. The second interval must start at least one day after the beginning of the first time interval and end at least one day before the other ends.

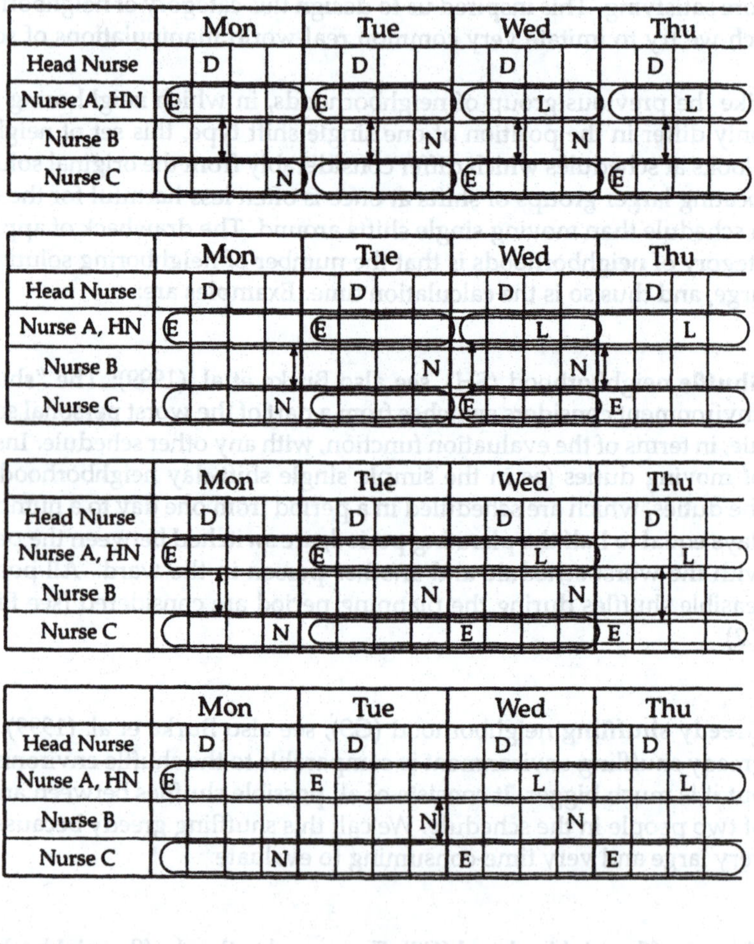

Figure 7.2 Possible moves in a 'shuffle' neighborhood between the personal schedules of Nurse A and Nurse C. For clarity, the moves are presented on 4 instances of the schedule.

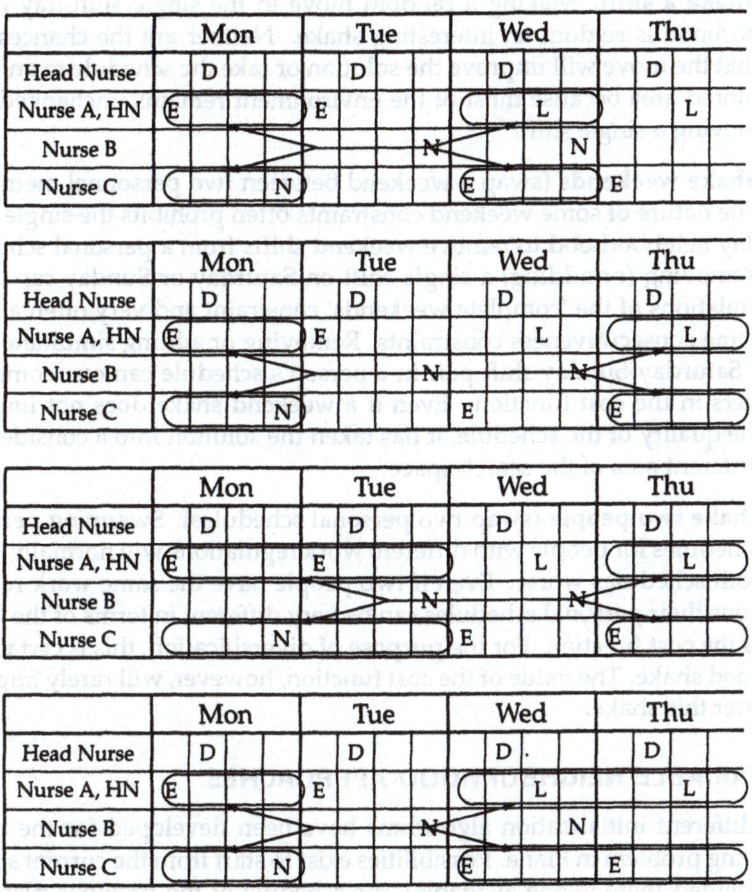

Figure 7.3 Examples of moves between Nurse A and Nurse C in the core shuffle neighborhood.

7.3.3 *Shaking the solution*

Shaking allows the algorithm to explore the solution space in a random manner. It is defined as the move to a random element of the neighborhood. Some moves within the 'soft constraint neighborhoods' can act as shakes. They do not generally improve the overall quality of the schedule but they provide a different viewpoint in the search space. Examples of shakes are:

- **Shake a shift:** Making a random move in the single shift-day neighborhood is seldom an interesting shake. Neither are the chances high that the move will improve the solution or take the schedule to an unexplored area because most of the environment remains unchanged after moving a single shift.

- **Shake weekends** (swap a weekend between two personnel members): The nature of some weekend constraints often prohibits the single shift-day neighborhood to remove weekend shifts from a personal schedule. Removing (or adding) a single shift on Saturday or Sunday can create violations of the 'complete weekends' constraint and very often also on some consecutiveness constraints. Removing or adding simultaneously a Saturday-Sunday shift pair in a person's schedule can overcome barriers in the cost function. Even if a weekend shake does not improve the quality of the schedule, it has taken the solution into a considerably different area of the search space.

- **Shake two people** (swap two personal schedules): Swapping personal schedules for people with different work regulations will normally make both schedules worse. Even if two people have the same work regulations their personal schedules can be very different in terms of the value of the cost function. For the purpose of diversification, this is certainly a good shake. The value of the cost function, however, will rarely improve after this shake.

7.4 VARIABLE NEIGHBORHOOD APPROACHES

Three different initialization algorithms have been developed for the nurse scheduling problem in Plane. Possibilities exist to start from the current schedule (provided there is one available), the schedule of the previous planning period or an empty schedule. Shifts are added and/or removed randomly in order to satisfy the personnel demands for every skill category, thus satisfying the hard constraints. In Burke et al. (1999) and Burke et al. (2001b), the procedure of creating a feasible initial solution is explained in more detail. For the Variable Neighborhood Search research presented in this paper, we always choose the last initialization option, in which all the required shifts are assigned at random.

The nurse rostering problem is solved by scheduling each skill category separately. We opted for applying two different search algorithms to this problem, namely Steepest Descent and Tabu Search. For both algorithms, the

Table 7.1 Selection of pairs of neighborhoods and heuristics for the test results of Table 7.2.

Neighborhood	Steepest Descent	Tabu Search
single shift-day		x
personal requests	x	
weekend		x
worst constraint		x
shuffle	x	
greedy shuffle	x	
core shuffle	x	

decision for a move is made at random out of the set of equally good solutions. Several layouts have been implemented for swapping between these algorithms, combining different neighborhoods from the set defined in Section 7.3.2. The experiments which have been carried out for this paper, however, have fixed algorithm-neighborhood pairs. The combinations are presented in Table 7.1. Previous research indicated that it is better to combine the single shift-day neighborhood with Tabu Search than with Steepest Descent. As long as there are improving moves, the search in shuffle neighborhoods continues. We deliberately combine these neighborhoods with Steepest Descent because they generally reach good quality local optima (provided they are explored after a search in a smaller size neighborhood has been applied first). Applying Tabu Search in these neighborhoods would increase the calculation time enormously. Experiments have also been carried out with 'shaking' the neighborhood (Section 7.3.1). Shaking did not contribute to finding good quality schedules and we decided not to keep it in the algorithm. Rather, most of the neighborhoods of the soft constraint class (2nd category in Section 7.3.2), have the effect of shakes since they search improvements for a partial set of soft constraints only. Searching in these neighborhoods can be seen as a process which reduces the contribution of a particular soft constraint to the overall cost function.

The Steepest Descent algorithm (obviously) stops when the neighborhood contains no better solution than the current one. The stop criterion for the Tabu Search algorithm is defined as a number of iterations without improvement. The number depends on the problem dimensions (number of people, number of shift types, length of the planning period, etc). When the stop criterion for a heuristic and neighborhood combination is reached, the heuristic switches to another neighborhood, or to the other search algorithm.

Starting from an initial solution, local search is applied in the first neighborhood. If the local optimum thus found is better than the current best solution, the algorithm moves there and continues the search in the first neighborhood; otherwise, it employs the next neighborhood and applies the corresponding search method. The algorithm stops when the search in the last neighborhood does not lead to an improvement. Algorithm 7.1 presents a schematic overview of the procedure.

```
procedure VNS( )
initialize:
Select a set of (algorithm, neighborhoods) pairs (neighbor-
hoods N_k, k = 1...k_max);
Set success rate success_k;
Define a local stopping criterion;
Construct an initial feasible solution x;
search:
Set k ← 1;
while k ≤ k_max
    if success_k ≥ 1
        Explore the neighborhood until local stop criterion is
        met;
        if best solution (x') is better than the current solution x
            x ← x';
            k ← 1;
        endif
    else
        success_k ← success_k − 1;
        k ← k + 1;
    endif
endwhile
```

Algorithm 7.1 Pseudo code for the variable neighborhood approach.

Neighborhoods can be applied in many different orders, which affect the result of the search considerably. The best approach is to develop algorithms which explore neighborhoods with increasing size. Whenever a neighborhood generates a better solution, the algorithm starts over from the first (finest) neighborhood, which is generally also the least time consuming. As explained in earlier papers, e.g. Burke et al. (1999), it is always beneficial for use in practice when, after the most greedy step, no finer neighborhood is explored. The same holds for the core shuffle neighborhood as it brings the solution into a 'finalized shape'. The nature of the GS and CS environments leads to solutions, which are judged positively by the schedulers. Additional moves which do not worsen the quality might bring the solution into a new area which encourages hospital planners to carry out some manual exploration. The greedy shuffling, and (even more so) the core shuffling neighborhood, are not recommended for very large problems when the calculation time is limited. Exploring the entire neighborhood is (for both approaches) an extremely intensive task.

The soft constraint related neighborhoods are not equally interesting for every type of problem. We therefore developed a method to avoid those neighborhoods which never contribute to better solutions. The probability of selecting a particular soft constraint neighborhood will change during the course of the calculations, depending on the results produced by that neighborhood. We introduce a parameter $success_k$ which is decreased by 1 each time that neighborhood k does not lead to better solutions (see also Algorithm 7.1). When $success_k$ is less than 1, the neighborhood k will not be applied in later iterations. The single shift-day neighborhood initially has a very high value for $success_k$ whereas soft constraint related neighborhoods preferably start with smaller values (1 or 2).

When changing the neighborhood, it is possible to start from the most recent solution reached in the previous algorithm-neighborhood combination; or from the overall best solution found. It appeared from our experiments that the best solution is always a recommendable start position.

Since the approach of this paper has been applied to real-world problems, we cannot ignore the calculation time. The test data sets are complex and large, and hospital schedulers expect a schedule to be generated within a reasonable amount of calculation time.

7.5 TEST RESULTS

Experiments have been carried out on real-world data with different combinations of the neighborhoods defined in Section 7.3.2. Depending upon the nature of the test data (whether certain soft constraints are applied or not, whether the corresponding cost parameter is high, etc) the effect of the neighborhoods corresponding to soft constraints is completely different.

In Table 7.2, the test results on a rather simple real-world problem are presented for a variety of algorithms. The scheduled ward consists of 20 personnel members, 6 shift types and very stringent soft constraints for which simultaneous satisfaction can never produce a feasible schedule. The combination of applied neighborhoods is denoted by the abbreviations in the column 'algorithm'. All the abbreviations stand for the neighborhoods which can be found in Section 7.3.2. To summarize, the neighborhoods used in the test example are: single shift-day (D), weekend (W), most violated constraint (MV), shuffle (SH), greedy shuffle (GS), and core shuffle (CS). Columns 2–4 present restart options:

- r repeat 'large section' neighborhoods with the best solution found after the global stop criterion is reached for the first time
- rsD restart from the first neighborhood with the best solution found after the global stop criterion is reached for the first time
- rs identical to rsD but skip the single shift-day neighborhood

Figure 7.4 schematically presents the scenario for one of the test algorithms. The algorithm applies four different neighborhoods: the single shift-day neigh-

Table 7.2 Test results for algorithms combining different neighborhoods in the search. x denotes which the restart options in the algorithms are.

Algorithm	r	rsD	rs	Result	Time
D CS	x		x	572	23'19"
D W CS	x		x	572	23'13"
D SH CS				527	21'16"
D W GS	x			572	8'09"
D W MW GS (Figure 7.4)	x	x		572	9'14"
D PR W SH GS	x			572	8'12"
D PR W SH GS	x	x		527	7'34"
D PR W SH GS	x		x	527	7'13"
D PR W SH CS	x		x	572	21'05"
D W SH GS CS	x		x	527	11'16"
D W SH GS				602	3'30"
D W SH GS	x			573	4'09"
D W SH GS	x	x		573	4'09"
D W SH GS	x		x	587	3'41"
D PR W SH GS CS				527	9'18"
D PR W SH GS CS	x		x	527	13'24"

borhood D, two soft constraint related neighborhoods W and MW, and one large section neighborhood GS. The smallest box shows the nearest neighborhood whereas bigger peripheries of the neighborhoods are represented by bigger boxes. Note that D and W are explored with Tabu Search while in the other 2 neighborhoods Steepest Descent is applied. The numbers 1-8 explain the order in which the neighborhoods are passed through. After the first exploration of GS was finished, the numbers 4-6 demonstrate how the option rsD is applied. It makes the search restart from the smallest neighborhood. Number 7 indicates option r, for which GS is applied at the end of the search, starting from the best solution found.

The Result is the value of the cost function, i.e. the weighted sum of the violations of soft constraints, summed over all the personnel members of the ward. The calculation time was recorded on an IBM RS6000 PowerPC. It is presented in the column Time.

Many algorithms reached a solution with value 527, which is the best cost function value found. However, the solutions are all different. Some constraints cannot be satisfied and their violation appears in all the solutions, all be it in the schedule of different people. It is remarkable that the category of algorithms, which reached the value 527, all make use of the greedy and/or core shuffle neighborhood. This confirms the conclusions made in Burke et al. (1999) and Burke et al. (2001b) that larger scale swaps are very useful at the end of the search.

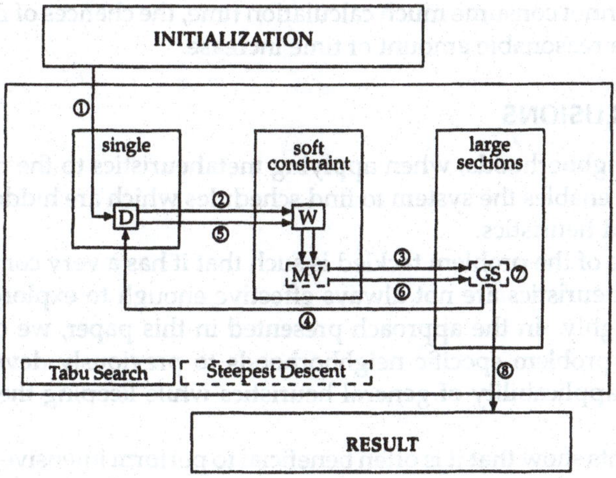

Figure 7.4 Diagram of the scenario for the algorithm D W MV GS with options r and rsD.

The order in which neighborhoods are explored is very important. It would be a waste of effort to use a greedy neighborhood to improve the randomly obtained initial solution. Greedy algorithms require a lot of time to explore the entire search space and they would improve the schedule in a very slow manner. The single shift-day neighborhood is rather small and quickly brings the initial solution into an area with acceptable quality. It is also remarkable that a combination of the single shift-day (D) and the core shuffle (CS) neighborhood alone is not interesting at all. The solution quality is not impressive and the calculation time is very high. With the D neighborhood only, the search stops in a solution which is the result of shift swaps only. The CS neighborhood finds many changes which make the solution better and therefore requires a lot of calculation time. When applying the CS neighborhood after a series of smaller scale neighborhoods (but larger than single shift-day) the possible improvements are smaller, and so is the calculation time.

Some medium scale (soft constraint related) neighborhoods have been used in the algorithms: weekend, personal requests and most violated constraint. Searches in these neighborhoods are rather considered as swaps than as real improvement steps. They do not contribute by generating overall better solutions but they act as a diversification in the search. If the soft constraint neighborhood does not contribute it is eliminated by the *success* variable. The soft constraint related neighborhoods are not necessarily developed to solve particular soft constraints but rather to explore parts of the solution space in which these constraints are satisfied. Later iterations might take the solution back to a schedule with a reduced number of violations for that constraint. Since the problem specific neighborhoods, which the soft constraint neighborhoods are, cannot consume much calculation time, the chances of finding good solutions in a reasonable amount of time increase.

7.6 CONCLUSIONS

Changing neighborhoods, when applying metaheuristics to the nurse rostering problem, enables the system to find schedules which are hidden for single neighborhood heuristics.

The nature of the problem tackled is such that it has a very complex search space. Metaheuristics are not always effective enough to explore the search space thoroughly. In the approach presented in this paper, we demonstrate how adding problem specific neighborhoods to previously developed ones increase the applicability of general heuristics while keeping the calculation time down.

Experiments show that it is often beneficial to perform intensive local search in the immediate surroundings of a schedule. After reaching a local optimum, we recommend the exploration of wider environments.

Several algorithms reach results with equally good qualities. The fastest among these is D PR W SH GS, with the single shift-day, two soft constraint related neighborhoods, and the shuffle and greedy shuffle as large section neighborhoods. When the stop criterion is reached for the first time, the algorithm passes through all the stages again, except the smallest neighborhood one (rs option). The algorithm ends with exploring the GS neighborhood proceeding from the best solution found. The second runner up is nearly the same algorithm but it restarts with the rsD instead of the rs option. It does not find a better solution although it explores an extra neighborhood after restarting (which explains the longer calculation time). The worst algorithm in terms of quality is D W SH GS, but it is the fastest. Algorithms with a long calculation time do not necessarily produce good results (see D CS and D W CS).

We developed a method to organize the changes of neighborhoods, and to choose particular soft constraint-related neighborhoods which are beneficial in the approach, resulting in schedules with a low value of the overall cost function. It is often more beneficial to apply simple heuristics such as Steepest Descent, with a variety of neighborhoods than to use sophisticated heuristics which are blind to large parts of the search space.

G. Vanden Berghe. Soft constraints in the nurse rostering problem. *http://www.cs.nott.ac.uk/~gvb/constraints.ps*, 2001.

E.K. Burke, P. De Causmaecker, and G. Vanden Berghe. A hybrid tabu search algorithm for the nurse rostering problem. *X. Yao et al. (Eds.): Simulated Evolution and Learning 1998, Lecture Notes in Artificial Intelligence*, 1585:187–194, 1999.

E.K. Burke, P. De Causmaecker, S. Petrovic, and G. Vanden Berghe. Fitness evaluation for nurse scheduling problems. *Proceedings of Congress on Evolutionary Computation, CEC2001, IEEE Press*, pages 1139–1146, 2001a.

E.K. Burke, P. Cowling, P. De Causmaecker, and G. Vanden Berghe. A memetic approach to the nurse rostering problem. *Applied Intelligence, Special issue on Simulated Evolution and Learning*, 15:199–214, 2001b.

T.G. Crainic, M. Gendreau, P. Hansen, N. Hoeb, and N. Mladenović. Parallel variable neighbourhood search for the p-median. In J. P. de Sousa, editor, *Proceedings of the 4th Metaheuristics International Conference, Porto, MIC 2001*, pages 595–599, 2001.

J. Crispim and J. Brandão. Reactive tabu search and variable neighbourhood descent applied to the vehicle routing problem with backhauls. In J. P. de Sousa, editor, *Proceedings of the 4th Metaheuristics International Conference, Porto, MIC 2001*, pages 631–636, 2001.

T. Davidović, P. Hansen, and N. Mladenović. Variable neighborhood search for multiprocessor scheduling problem with communication delays. In J. P. de Sousa, editor, *Proceedings of the 4th Metaheuristics International Conference, Porto, MIC 2001*, pages 737–741, 2001.

M. den Besten and T. Stützle. Neighborhoods revisited: An experimental investigation into the effectiveness of variable neighborhood descent for scheduling. In J. P. de Sousa, editor, *Proceedings of the 4th Metaheuristics International Conference, Porto, MIC 2001*, pages 545–549, 2001.

P. Festa, P.M. Pardalos, M.G.C. Resende, and C.C. Ribeiro. Randomized heuristics for the MAX-CUT problem. *Optimization Methods and Software*, 7:1033–1058, 2002.

F. Glover and M. Laguna. *Tabu Search*. Kluwer, 1997.

C.G. Gonzáles and D. Pérez-Brito. A variable neighbourhood search for solving the linear ordering problem. In J. P. de Sousa, editor, *Proceedings of the 4th Metaheuristics International Conference, Porto, MIC 2001*, pages 181–185, 2001.

P. Hansen and N. Mladenović. An introduction to variable neighborhood search. In S. Voss, S. Martello, I.H. Osman, and C. Roucairol, editors, *Advances and Trends in Local Search Paradighms for Optimization*, pages 433–358. Kluwer Academic Publishers, 1999.

N. Mladenović and P. Hansen. Variable neighborhood search. *Computers & Operations Research*, 24:1097–1100, 1997.

N. Mladenović and D. Urošević. Variable neighbourhood search for the k-cardinality tree. In J. P. de Sousa, editor, *Proceedings of the 4th Metaheuristics International Conference, Porto, MIC 2001*, pages 743–747, 2001.

Metaheuristics: Computer Decision-Making, pp. 173-186
Marco P. Carrasco, and Margarida V. Pato

8 A POTTS NEURAL NETWORK HEURISTIC FOR THE CLASS/TEACHER TIMETABLING PROBLEM

Marco P. Carrasco[1,3] and Margarida V. Pato[2,3]

[1]Escola Superior de Gestão, Hotelaria e Turismo
University of Algarve
8501 859 Portimão, Portugal
pcarras@ualg.pt

[2]Instituto Superior de Economia e Gestão
Technical University of Lisbon
1200 781 Lisboa, Portugal
mpato@iseg.utl.pt

[3]Centro de Investigação Operacional
University of Lisbon
Campo Grande, 1749 Lisboa, Portugal
centro.io@fc.ul.pt

Abstract: This paper describes the application of a neural network metaheuristic to a real timetabling problem, the Class/Teacher Timetabling Problem (CTTP). This problem is known to be a complex, highly constrained optimization problem, thus exhibiting limitations to be solved using classical optimization methods. For this reason, many metaheuristics have been proposed to tackle real CTTP instances. Artificial neural networks have, during the last decade, shown some interesting optimization capabilities supported mainly by the seminal work of Hopfield and Tank. By extending this approach, the current paper proposes a Potts neural network heuristic for the CTTP. Computational tests taken with real instances yield promising results, which suggest that this Potts neural heuristic is an effective method to the solving of this class of timetabling problem.

Keywords: School timetabling, Metaheuristics, Artificial neural networks, Hopfield neural network.

8.1 INTRODUCTION

Timetabling problems emerge in many variations, depending on the nature of the assignments involved (exam timetabling, course timetabling, school timetabling and others), as well as the characteristics of the institution and the set of constraints and objectives assumed (Schaerf, 1999). In this paper, one focuses on a particular school timetabling issue: the Class/Teacher Timetabling Problem (CTTP) found in schools with rigid curricula, as described by Werra (1985), to be addressed in Section 8.2.

The fact that most timetabling problems have for some time been known to belong to the class of NP-hard problems (Even et al., 1976), limits the approach involving exact algorithms. Hence, over the last two decades, several heuristics, specially metaheuristics ranging from tabu search (Costa, 1994), to genetic algorithms (Carrasco and Pato, 2001) and simulated annealing (Abramson, 1982) – have been used with relative success to solve difficult timetabling problems (Burke and Erben, 2001; Burke and Carter, 1998; Burke and Ross, 1996).

Meanwhile, artificial neural networks have received particular attention from the Operational Research community since the pioneering work of Hopfield and Tank on the Traveling Salesman Problem (Hopfield and Tank, 1985). Though this initial work enjoyed limited success, it aroused an interest in the use of neural network-based metaheuristics to solve combinatorial optimization problems. This interest is evident from the increasing number of papers that address the subject (Smith, 1999).

As for the specific application of neural networks for timetabling construction purposes only a few papers exist. For example, Mausser et al. (1996) address the scheduling of interviews and present results for small randomly generated instances, and Tsuchiya and Takefuji (1997) describe parallel neural networks to solve meeting schedule problems with positive empirical results for distinct size instances. Kovacic (1993), focusing on the CTTP and reporting favorable results for real instances, uses a Markovian neural network to tackle a small instance. Similarly, Pellerin and Herault (1994) use a neural network with back lateral inhibitions to solve a particular CTTP problem.

In this context, the present paper investigates an extension of a neural network metaheuristic to solve a real, highly dimensioned CTTP. The main motivation of this paper is to propose an alternative approach to several metaheuristics already applied to this problem.

There are two main approaches to tackle optimization problems when using artificial neural networks: the Hopfield-Tank neural network (HTNN) and the self-organizing neural network. The present work focuses on a generalization of the HTNN proposed by Peterson and Söderberg (1983) which provides a more direct encoding to solve general combinatorial optimization problems, as is the case of some timetabling problems (Gislén et al., 1989). Such a mean field network heuristic, based on multi-state Potts neurons, is described in Section 8.3 for the CTTP. Section 8.4, provides some results and considerations regarding the computational application of the above metaheuristic to real CTTPs and comments follow in Section 8.5.

8.2 THE CLASS/TEACHER TIMETABLING PROBLEM

The CTTP addressed in this paper is based on the problem already described in Carrasco and Pato (2001) and can be defined as the constraint satisfaction problem relative to the scheduling of a batch of lessons over a weekly set of time periods, using suitable rooms, while satisfying a wide range of constraints. In this context, a lesson is a prior assignment of one or more rigid classes of students to one teacher and one subject. The set of constraints included in the CTTP is characterized as follows.

1. The curricula of all classes must be respected, i.e. all lessons assigned to each class are fully scheduled.

2. Each teacher and class is assigned to no more than one lesson and one room within one time period. This is a hard constraint common to all timetabling problems.

3. Lessons can be of different but fixed durations, expressed as a multiple of the basic time period.

4. For pedagogical reasons, each subject is scheduled to be given no more than once a day.

5. Rooms are suitable for the lessons assigned. Each lesson may require a specific type of room, in terms of seating or special resources. For example, computer-related lessons traditionally require a room with computers and specific software.

6. The teaching shift for each class is respected. In some schools there are teaching shifts that must always be satisfied; such is the case when scheduling lessons for day or evening classes.

7. Each class and teacher must have a lunch break. This constraint imposes a free time period, with a minimum duration of one hour, to be used for the lunch activity.

8. The unavailability periods for each class and each teacher is respected. In fact, classes and, far more often, teachers have periods of unavailability arising from periodic professional or academic commitments.

These conditions define the most important features regarding our CTTP. However, different particularities can emerge from other real timetabling problems and be implemented within the model to be presented in the next section. Note that, although the CTTP described assume the structure of a constraint satisfaction problem, the neural method presented in this paper uses an optimization problem to force the maximum satisfaction of the CTTP constraint set.

8.3 THE POTTS NEURAL NETWORK MODEL

For optimization purposes, the seminal work of Hopfield and Tank (1985) presents a HTTP characterized by a fully interconnected network with N binary neurons, whose basic features will now be described. Neuron i has a net input value U_i and an output value or state V_i. The input value U_i results from the weighted sum of output values arising from other neurons, added to a negative bias current I_i:

$$U_i = \sum_{j \neq i}^{N} w_{ji} V_j + I_i, \tag{8.1}$$

where the weight, corresponding to the synapse connection, from a different neuron j to neuron i is given as a parameter w_{ji}. The state V_i, which is bounded by zero and one, is calculated from U_i by the activation function $g_i(U_i)$, traditionally a sigmoid function:

$$V_i = g_i(U_i) = \frac{1}{2}\left(1 + \tanh\left(\frac{U_i}{T}\right)\right), \tag{8.2}$$

where T is a temperature parameter determining the steepness.

Hopfield and Tank (1985) proved that, through the use of the difference equation

$$U_i^{(t)} = U_i^{(t-1)}(1 - \Delta t) + \Delta t \left(\sum_{j \neq i}^{N} w_{ji} V_j^{(t-1)} + I_i\right) \tag{8.3}$$

for a symmetrical weight matrix ($w_{ij} = w_{ji}$) and a proper cooling scheme, the neural network converges to a local minimum of the following energy function

$$E(\overline{V}) = -\frac{1}{2} \sum_{i=1}^{N} \sum_{j=1}^{N} w_{ij} V_i V_j - \sum_{i=1}^{N} I_i V_i. \tag{8.4}$$

Combinatorial optimization problems can be tackled by using this HTNN model. To map each problem, one must construct an appropriate quadratic energy function, whose minimum value corresponds to the optimum of the optimization problem. Typically this function is defined on the basis of the constraints of the problem in the form of a penalized sum, added to the objective expression, thus requiring additional care in the balance between constraints and objective terms. Note that the formulation proposed here for the CTTP assumes the nature of a constraint satisfaction problem. It could therefore be translated through an energy function, embedding the constraint-penalizing terms alone.

To improve the basic HTNN for the CTTP, the two final state Hopfield neurons were replaced by Potts multi-state neurons. In addition, a factorization scheme proposed by Gislén et al. (1992) was applied, and, as a result, the number of neurons needed was significantly reduced. Simultaneously, a specific neural energy function for this particular CTTP was constructed on the basis of neural network factorization.

Assuming that L lessons are to be scheduled (involving prior assignments of T teachers to C classes), in P identical time periods and R rooms, a direct formalization (Carrasco, 1995) uses $L \times P \times R$ binary variables as follows:

$$x_{lpr} = \begin{cases} 1 & \text{if lesson } l \text{ occurs during time period } p \text{ in room } r \\ 0 & \text{otherwise.} \end{cases}$$

Hence, within a direct Hopfield and Tank neural encoding, each of these variables is associated with a generic neuron, thus requiring $L \times P \times R$ neurons to represent a CTTP solution.

Whereas, when using multi-state neurons and factorization, the neural network requires only $L \times P$ lesson-period neurons, with states given by V^x_{lp}, and $L \times R$ lesson-room neurons, corresponding to final states V^y_{lr}. Furthermore, this encoding is far more efficient than the previous one as it allows direct implementation of some of the CTTP constraints referred to in Section 8.2, more precisely constraints 5, 6 and 8, through an initial freezing of the neurons involved in appropriate states. For example, when considering constraint 5, if room $r1$ cannot support a specific lesson $l1$, the corresponding neuron state $V^y_{l1,r1}$ is set to 0 and excluded from the updating procedure.

Moreover, CTTP constraint 1 is automatically satisfied by the Potts neural formalism expressed by equations (8.5) because they force the activation of only one neuron from a specific group of neurons:

$$\sum_{p=1}^{P} V^x_{lp} = 1\,, \sum_{r=1}^{R} V^y_{lr} = 1\,, \tag{8.5}$$

where V^x_{lp} and V^y_{lr} represent the states of the lesson-period neuron (l, p) and the lesson-room neuron (l, r), respectively. The remaining CTTP constraints are used to define the energy function.

Constraints 2 and 3 are implemented simultaneously, using penalty terms E_1 and E_2 that follow. First, to prevent room occupation conflicts, E_1 is defined:

$$E_1 = \frac{1}{2} \sum_{p=1}^{P} \sum_{r=1}^{R} \sum_{\substack{l,l'=1 \\ l \neq l'}}^{L} V^y_{lr} V^y_{l'r} S_{lp} S_{l'p}\,, \tag{8.6}$$

where $S_{lp} = \sum_{k=0}^{Dur(l)-1} V^x_{l,p-k}$ and $Dur(l)$ gives the duration of lesson l.

In addition, to avoid class and teacher time conflicts, the energy component E_2 is stated as

$$E_2 = \frac{1}{2} \sum_{p=1}^{P} \sum_{\substack{l,l'=1 \\ l \neq l'}}^{L} Q_{ll'}^{teca} S_{lp} S_{l'p} , \tag{8.7}$$

where

$$Q_{ll'}^{teca} = \begin{cases} 2 & \text{if lessons } l \text{ and } l' \text{ share the same teacher and class} \\ 1 & \text{if lessons } l \text{ and } l' \text{ share the same teacher or class} \\ 0 & \text{otherwise.} \end{cases}$$

Energy term E_3 attempts to avoid scheduling lessons of the same subject on the same day (constraint 4):

$$E_3 = \frac{1}{2} \sum_{c=1}^{C} \sum_{\substack{l,l' \in LClass(c) \\ l \neq l'}} Q_{ll'}^{sub} \sum_{d=1}^{D} \left(\sum_{p \in TDay(d)} S_{lp} \right) \left(\sum_{p \in TDay(d)} S_{l'p} \right), \tag{8.8}$$

where $LClass(c)$ is the set of lessons attended by class c, $TDay(d)$ represents the set of time-periods of day d, and

$$Q_{ll'}^{sub} = \begin{cases} 1 & \text{if lessons } l \text{ and } l' \text{ belong to the same subject} \\ 0 & \text{otherwise.} \end{cases}$$

To enforce satisfaction of constraint 7, concerning the existence of lunch breaks, two auxiliary sets of variables are defined for both classes and teachers:

$$A_{cp}^{cla} = \sum_{l \in LClass(c)} S_{lp} , \quad A_{tp}^{tea} = \sum_{l \in LTeacher(t)} S_{lp} , \tag{8.9}$$

where $LTeacher(t)$ is the set of lessons taught by teacher t.

Using these aggregated variables, the energy term E_4 ensures that a daily lunch break is created within the set of time periods corresponding to $TLunchF(d)$ and $TLunchL(d)$, as follows:

$$E_4 = \frac{1}{2} \sum_{d=1}^{D} \left[\sum_{c=1}^{C} \left(\prod_{p=TLunchF(d)}^{TLunchL(d)} A_{cp}^{cla} \right) + \sum_{t=1}^{T} \left(\prod_{p=TLunchF(d)}^{TLunchL(d)} A_{tp}^{tea} \right) \right]. \tag{8.10}$$

Finally, the complete Potts neural network energy function for the CTTP is given by

$$E(\overline{V}^x, \overline{V}^y) = E_1 + E_2 + E_3 + E_4. \tag{8.11}$$

Minimization of the customized energy function (8.11) leads to final neural network states corresponding to low-conflict configurations. As the total penalization associated with the activation of any pair of neurons is symmetric, the corresponding weights matrix is symmetric. Hence, simulation of the dynamics of the neural network toward a final minimum configuration is obtained by iterating equations (8.12) and (8.13) according to a mean-field annealing method, controlled by a temperature parameter T' (Bout and Miller, 1989; Peterson and Söderberg, 1983):

$$V_{lp}^x = \frac{e^{U_{lp}^x}}{\sum_{p'=1}^{P} e^{U_{lp'}^x}}, \qquad \text{with } U_{lp}^x = -\frac{1}{T'}\frac{\partial E}{\partial V_{lp}^x} \tag{8.12}$$

$$V_{lr}^y = \frac{e^{U_{lr}^y}}{\sum_{r'=1}^{R} e^{U_{lr'}^y}}, \qquad \text{with } U_{lr}^y = -\frac{1}{T'}\frac{\partial E}{\partial V_{lr}^y}. \tag{8.13}$$

8.4 ALGORITHM SPECIFICATIONS AND RESULTS

To approximate the minimum of the Potts neural network energy function (8.11), constructed over the V_{lr}^y and V_{lp}^x neuron states, the respective neural sub-networks are iterated according to the generic algorithmic procedure in Algorithm 8.1.

$T'_0 = 1$;
$MaxIter = 100$;
Initialize all neuron states;
for $Iter = 0$ **to** $MaxIter$
  if $NSum^x = \frac{1}{N^x}\sum_{i=1}^{N^x} V_i^2 < 0.9$
    Randomly perform one update per neuron of lesson-room sub-network by using equations (8.13) with T'_{iter};
  if $NSum^y = \frac{1}{N^y}\sum_{i=1}^{N^y} V_i^2 < 0.9$
    Randomly perform one update per neuron of lesson-room sub-network by using equations (8.13) with T'_{iter};
  $T'_{iter+1} = T'_{iter} \times 0.95$;
endfor
Discretize all neuron states;

Algorithm 8.1 Pott neural network algorithm

The algorithm starts by defining the initial values of the main parameters. The temperature T'_0 $(= 1)$ was determined by trial and error. Generally, the solution configuration is formed around a critical temperature level T'_c, so it is important to ensure that the initial temperature is set to a higher level than T'_c. In addition, a maximum number of iterations is given to act as a stopping criterion, $MaxIter = 100$.

Then both neural sub-networks are initiated with a small random perturbation around the equilibrium state value for each lesson-room and lesson-period neuron, respectively. The starting equilibrium values are calculated by inverting the number of active neurons in each row of matrices whose elements are the state values V^x_{lp} or V^y_{lr}. Note that, as mentioned above, some of the CTTP constraints, namely constraints 5, 6 and 8, are directly implemented through activation or deactivation of the corresponding neurons, in the initialization phase of the procedure.

The main iterative step is executed while the maximum number of iterations $MaxIter$ is not exceeded. On the assumption that N^x and N^y represent the total number of neurons within the lesson-period and the lesson-room neural sub-networks, for each of the sub-networks two measures of global saturation of the neuron output values, defined as $NSum^x$ and $NSum^y$, are checked to ensure they are below 0.9. This occurs whenever the neurons globally assume non-discrete values, thus requiring more iterations to define their final discrete output values (1 or 0). The main loop consists in sequentially updating the neuron states V^x_{lp} and V^y_{lr} using equations (8.12) and (8.13) according to a random, asynchronous process, for a specific temperature level T'_{Iter}. Then the temperature level and the iteration counter are updated.

The final step forces discretization, constrained to equations (8.5), of all neurons not in state 0 or 1, thus allowing direct extraction of the binary solution for the CTTP.

The above-described Potts neural network heuristic was tested in several timetabling situations and applied to real CTTPs, namely the first semester of the last academic year timetabling problems of the Escola Superior de Gestão, Hotelaria e Turismo of the Algarve University, located in two distinct cities, Portimão and Faro.

Table 8.1 briefly shows the main characteristics of these two instances.

Table 8.1 Characteristics of the instances.

Problem instance	*Lessons*	*Time periods*	*Rooms*	*Teachers*	*Classes*	*Time-slot occupation index*(*)
ESGHT-PTM	173	50	9	51	32	48.6
ESGHT-ALL	568	50	27	107	92	49.4

(*) $\frac{L}{P \times R} \times 100$

The last column of Table 8.1 presents the time-slot occupation index for the instances tackled. This index provides a measure of the difficulty level of the timetabling issues, by dividing the number of lessons to be scheduled by the potential number of time-slots (rooms × time periods). As observed, the degree of occupancy for both cases is about 49%, thus revealing some intense room occupation. Here, the basic time period is 90 minutes and the lesson length may vary from 1 to 2 consecutive time periods.

Table 8.2 presents the number of neurons of each type required to map the corresponding CTTP instance through a Potts neural network model.

Using the above algorithm, a series of 20 experiments for each case was performed, each starting from distinct initial state values for the neurons. In all

Table 8.2 Neural network dimensions.

Problem instance	*Number of neurons* Lesson-room V_{lr}^{y}	Lesson-period V_{lp}^{x}	*Total*
ESGHT-PTM	1557	8650	10207
ESGHT-ALL	15336	28400	43736

runs tested so far, both neural sub-networks converged to stable state configurations in less than 68 iterations. The summarized results are presented below in Table 8.3.

Table 8.3 Results after 20 runs.

Problem instance	*Potts neural network heuristic*				
	Energy value			*Execution (average)*	
	Minimum	*Average*	*% Feasible*	*Time (s.)*	*Iterations*
ESGHT-PTM	0	0.25	75%	1251	64.5
ESGHT-ALL	0	0.65	55%	6583	58

As for the quality of the solutions, expressed precisely by the energy function in terms of minimum and average number of conflicts per solution (second and third columns of Table 8.3), the neural heuristic proved to be able to find very low-conflict timetables. Note that the majority of runs for each instance obtained conflict-free timetables, hence corresponding to a significant percentage of feasible solutions for these real CTTPs (75% and 55%, respectively).

Figure 8.1 shows the evolution of the results obtained with the Potts neural heuristic on a typical run of the algorithm for the ESGHT-ALL instance. The Potts neural network energy value, expressing conflicts occurring in the CTTP constraints, is represented on the left axis. The horizontal axis displays the iteration counter and the right axis is a scale for both neuron saturation and temperature.

As may be observed, the algorithm took about 66 iterations to find a stable configuration corresponding to a zero conflict set of timetables which is a feasible CTTP solution, as seen by the energy function plot. From iterations 1 to 20, the neural system evolved slowly while the temperature decreased rapidly. Between iterations 20 to 40, when T' was below 0.1, the system dynamics became unstable and neurons started to define their final states, thus decreasing the neural energy value. The evolution of the neuron saturation levels on the lesson-room and lesson-period sub-networks, given by $NSum^x$ and $NSum^y$, is also graphically represented in Figure 8.1. Note that the lesson-room neu-

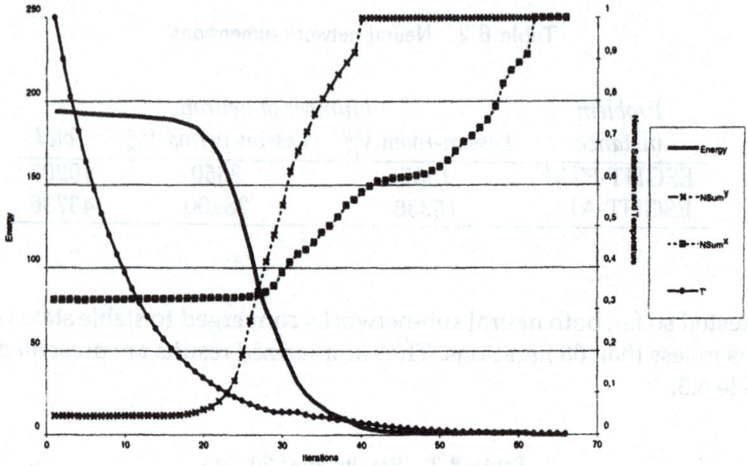

Figure 8.1 Potts neural algorithm evolution on a typical run.

rons had already completely defined their states in iteration 40, while the lesson-period neurons only determined their final configurations at iteration 62.

Finally, it should be mentioned that the algorithm was implemented using Delphi 5 and, on average, took less than 110 minutes on a Pentium III 800 MHz with 128 MBytes of RAM to complete a full run, for the largest case, ESGHT-ALL. It should be also noted that the manual construction of an equivalent solution for this instance involved one week's work for a three-person team.

8.5 FINAL COMMENTS

This paper described an extension of the Hopfield and Tank neural network, namely a factorized Potts neural network, to model a difficult constraint satisfaction problem, a real Class/Teacher Timetabling Problem (CTTP). Then, the corresponding neural algorithm features were presented and computational experiments on real instances were discussed.

The main contribution of this paper consists in presenting a neural network model as a basis of an alternative heuristic to construct feasible solutions for real CTTP. In fact, the computing results suggest that the Potts neural network heuristic is an effective approach for timetabling generation, both in terms of computational execution time and solution quality. When compared with a genetic algorithm for a similar problem, the neural model showed a competitive computing time (Carrasco and Pato, 2001).

It should be noted that, as CTTP problems are highly diverse in real situations, resolution methods are often customized for each timetabling problem or instance, which does not allow fair comparison between alternative methods. For this reason a comparison study is currently being performed using a set of CTTP instances for benchmark evaluation between this neural approach and the above mentioned genetic algorithm.

Other instances of this type of CTTP are being tackled, featuring different timetabling constraints, along with a definition of respective optimization-oriented energy terms for the neural network. Nevertheless, several possibilities of enhancing this Potts neural network heuristic are being studied, namely by combining this methodology with other metaheuristics.

Acknowledgments

This paper is part of the first author's Ph.D. research.

Bibliography

D. Abramson. Constructing school timetables using simulated annealing: Sequential and parallel algorithms. *Management Science*, 37(1):98–113, 1982.

D. Van Den Bout and T. Miller. Improving the performance of the hopfield-tank neural network through normalization and annealing. *Biological Cybernetics*, 62:123–139, 1989.

E. Burke and M. Carter, editors. *The Practice and Theory of Automated Timetabling II*, volume 1408. LNCS Springer Verlag, 1998.

E. Burke and E. Erben, editors. *The Practice and Theory of Automated Timetabling III*, volume 2079. LNCS Springer Verlag, 2001.

E. Burke and P. Ross, editors. *The Practice and Theory of Automated Timetabling*, volume 1153. LNCS Springer Verlag, 1996.

M. Carrasco. Redes neuronais na elaboração de horários escolares. Master's thesis, Instituto Superior de Economia e Gestão - Universidade Técnica de Lisboa, 1995.

M. Carrasco and M. Pato. A multiobjective genetic algorithm for the class/teacher timetabling problem. *The Practice and Theory of Automated Timetabling III*, 2079:3–17, 2001.

D. Costa. A tabu search algorithm for computing an operational timetable. *European Journal of Operational Research*, 76:98–110, 1994.

S. Even, A. Itai, and A. Shamir. On the complexity of timetabling and multicommodity flow problems. *SIAM Journal of Computation*, 5(4):691–703, 1976.

L. Gislén, C. Peterson, and B. Söderberg. Teachers and classes with neural networks. *International Journal of Neural Systems*, 1:167–176, 1989.

L. Gislén, C. Peterson, and B. Söderberg. Complex scheduling with Potts neural networks. *Neural Computation*, 4:805–831, 1992.

J. Hopfield and D. Tank. Neural computation of decisions in optimization problems. *Biological Cybernetics*, 52:141–152, 1985.

M. Kovacic. Timetable construction with a Markovian neural network. *European Journal of Operational Research*, 69(1):92–96, 1993.

H. Mausser, M. Magazine, and J. Moore. Application of an annealed neural network to a timetabling problem. *INFORMS Journal on Computing*, 2(8): 103–117, Spring 1996.

D. Pellerin and J. Herault. Scheduling with neural networks: Application to timetabling construction. *Neurocomputing*, 4(6):419–442, 1994.

C. Peterson and B. Söderberg. A new method for mapping optimization problems onto neural networks. *International Journal of Neural Systems*, 3(1):3–22, 1983.

A. Schaerf. A survey of automated timetabling. *Artificial Intelligence Review*, 13:87–127, 1999.

K. Smith. Neural networks for combinatorial optimization: A review of more than a decade of research. *INFORMS Journal on Computing*, 11:15–34, 1999.

K. Tsuchiya and Y. Takefuji. A neural network parallel algorithm for meeting scheduling problems. *Applied Intelligence*, 3(7):205–213, 1997.

D. De Werra. An introduction to timetabling. *European Journal of Operations Research*, 19:151–162, 1985.

Metaheuristics: Computer Decision-Making, pp. 187-216
Maria João Cortinhal and Maria Eugénia Captivo

9 GENETIC ALGORITHMS FOR THE SINGLE SOURCE CAPACITATED LOCATION PROBLEM

Maria João Cortinhal[1] and Maria Eugénia Captivo[1]

[1]ISCTE-CIO
Av. Forças Armadas, 1649-026
Lisboa, Portugal
mscc@iscte.pt

[2]DEIO-CIO
Faculdade de Ciências da Universidade de Lisboa
Edifício C2 - Campo Grande
1749-016 Lisboa, Portugal
mecaptivo@fc.ul.pt

Abstract: The single source capacitated location problem is considered. Given a set of potential locations and the plant capacities, it must be decided where and how many plants must be open and which clients must be assigned to each open plant. Genetic algorithms that use different methodologies for handling constraints are described and tested. Computational experiments on different sets of problems are presented.
Keywords: Capacitated facility location, Genetic algorithms, Search methods.

9.1 INTRODUCTION

In the Single Source Capacitated Location Problem (SSCLP) a set $I = \{1, .., n\}$ of possible plant locations, each with a maximum capacity a_i, $i \in I$, and a set $J = \{1, .., m\}$ of customers, each with a demand b_j, $j \in J$, are given. One must choose a subset of plants and to fully assign each customer to one of the chosen plants, in such a way that the total cost is minimized, and the plant capacities are not exceeded. Total cost is composed by fixed costs, f_i, $i \in I$, whenever plant i is open, and by the cost of assigning each customer to one plant, $c_{ij}, i \in I, j \in J$. Considering $x_{ij} = 1$, if customer j is assigned to plant i and $x_{ij} = 0$, otherwise and $y_i = 1$, if plant i is open and $y_i = 0$, otherwise , the SSCLP can be formulated as follows:

$$\min \sum_{i=1}^{n} \sum_{j=1}^{m} c_{ij} x_{ij} + \sum_{i=1}^{n} f_i y_i \tag{9.1}$$

subject to

$$\sum_{i=1}^{n} x_{ij} = 1, \qquad j = 1, \ldots, m, \tag{9.2}$$

$$\sum_{j=1}^{m} b_j x_{ij} \leq a_i y_i, \qquad i = 1, \ldots, n, \tag{9.3}$$

$$x_{ij} \in \{0,1\}, \qquad i = 1, \ldots, n, j = 1, \ldots, m, \tag{9.4}$$

$$y_i \in \{0,1\}, \qquad i = 1, \ldots, n. \tag{9.5}$$

Constraints (9.2) stipulate that each customer must be fully assigned to a single plant. Constraints (9.3) are the plant capacity constraints, and also ensure that a customer can only be assigned to an open plant. Constraints (9.4) and (9.5) are the integrality constraints.

The SSCLP is a combinatorial optimization problem that belongs to the class of NP-hard problems. Several authors have studied the problem. In Neebe and Rao (1983) the problem is formulated as a set partitioning problem and solved by a column generation branch and bound procedure. Bounds were obtained using linear relaxation. They observe that the tree search is not very large due to the high probability of having integer solutions with the linear program. Computational results are given for a set of problems, supplied by DeMaio and Roveda (1971), and for two randomly generated sets, with a maximum of 25 facilities and 40 customers.

Barceló and Casanovas (1984) propose a Lagrangean heuristic for the problem in which the customer assignment constraints are relaxed. The heuristic for finding feasible solutions, consists of two phases: one for the plant selection and the other for the clients' assignment.

Klincewicz and Luss (1986) describe a Lagrangean heuristic for the SSCLP, where they dualize the capacity constraints. Feasible solutions for the relaxed problems were obtained by the dual ascent method of Erlenkotter (1978). An initial add heuristic and a final adjustment heuristic were used. Computational results are given for a set of 24 test problems, taken from Kuenh and Hamburger (1963), with n=25 or 26, m=50, and where demands greater than 3700 were set to 3700.

In Darby-Dowman and Lewis (1988), the same Lagrangean relaxation is used to establish some relationships between fixed and assignment costs, allowing the identification of problems for which the optimal solution to the relaxed problem is guaranteed not to be a feasible solution to the original problem.

Beasley (1993) presents a framework to develop Lagrangean heuristics for location problems, where capacity constraints and customer assignment constraints are dualized. For the SSCLP, an allocation cost procedure was used to calculate the cost of one set of open facilities. Computational results are given for a set of 24 problems, taken from Klincewicz and Luss (1986).

Sridharan (1993) describes a Lagrangean heuristic for the problem that improves and extends the one presented by Nauss (1978) for the Capacitated Plant Location Problem where customer assignment constraints are dualized. Computational experiments were made over a set of randomly generated problems, and a set of problems given in Guignard and Kim (1983).

In Agar and Salhi (1998), a general framework, similar to the one presented by Beasley (1993), for solving location problems including SSCLP, is described. The main differences can be found in the generation of feasible solutions. They report computational results obtained with some instances taken from Beasley (1990) and other test problems, with $100 \leq n = m \leq 600$, which were constructed based on the standard data sets given in Christofides et al. (1979).

Hindi and Pienkosz (1999) describe a heuristic that combines Lagrangean relaxation with restricted neighborhood search for the SSCLP. They dualize customer assignment constraints and use subgradient optimization. For finding feasible solutions a heuristic procedure with three phases is used. Computational results are given for a set of 36 problems, taken from Beasley (1990), and a set of 33 problems used in Barceló et al. (1991) and in Delmaire et al. (1997).

Holmberg et al. (1999) present a Lagrangean heuristic and a branch and bound to the problem. In both, they dualize customer assignment constraints, and use subgradient optimization. Upper bounds were obtained by a primal heuristic that solves a sequence of related matching problems using Lagrangian's solutions as starting points. Computational results are given for 4 sets of randomly generated test problems, with $10 \leq n \leq 30$ and $50 \leq m \leq 200$. Comparisons between the results obtained with branch and bound, Lagrangean heuristic and with CPLEX (1994) are made.

In Cortinhal and Captivo (2001), we proposed a heuristic that combines Lagrangean relaxation with search methods. Customer assignment constraints are dualized and subgradient optimization is used. We compare three methods for finding feasible solutions: a Lagrangean heuristic, a local search, and a tabu search. Computational results are given for the same set of instances presented here and compared with other results published, for some of the instances tested.

The SSCLP as many other optimization problems, is a constrained problem. So, a relevant issue is how to deal with constraints and infeasible solutions when developing a genetic algorithm.

Most of the methodologies for handling constraints presented are problem dependent. So, questions like "Should we construct special genetic operators, that disable the creation of infeasible solutions? If not, how to deal with infeasible solutions?" arise. In our experimental study, four approaches that use different methodologies for handling constraints were implemented and tested.

Section 9.2 describes all the algorithms proposed. Computational experiments are presented in Section 9.3, and Section 9.4 contains the conclusions.

9.2 GENETIC ALGORITHMS

All the algorithms proposed here are genetic algorithms and use the same representation. Three of them also use the same general structure. So, we will start by introducing some general concepts.

9.2.1 Representation

The first step in designing a genetic algorithm for a particular problem is to devise a suitable representation. For the SSCLP the usual 0-1 binary representation is not suitable since it needs binary strings of length mxn and, at the same time, it does not ensure that assignment and capacity constraints are satisfied. Due to this fact, we used a representation where each chromosome is a m-dimensional vector of integers in the set $\{1,..,n\}$: the integer value of the j-th position indicates the plant where customer j is assigned. So, for example, if $m = 8$ and $n = 3$, the chromosome $\langle 2,1,3,3,1,1,2,1\rangle$ represents one solution for the problem where customers 2, 5, 6 and 8 are assigned to plant 1, customers 1 and 7 to plant 2 and, customers 3 and 4 to plant 3. We note, however, that this representation may also lead to infeasible solutions for the problem, since it does not ensure that capacity constraints are satisfied. So, denoting by

$$s_i = \begin{cases} 0 & \text{if } \sum_{J_i^*} b_j < a_i \text{ where } J_i^* = \{\text{clients assigned to plant i}\} \\ |a_i - \sum_{J_i^*} b_j| & \text{otherwise.} \end{cases}$$

We will say, from now on, that one solution is feasible if and only if $s_i = 0$ $(\forall i = 1,..,n)$. Consequently, the degree of infeasibility of one solution x is given by $u(x) = \sum_{i=1}^{n} s_i$.

9.2.2 Pseudocode

The general pseudocode for the first three algorithms is shown in Algorithm 9.1. This pseudocode refers to five procedures: `Init_Pop()`, `Cros_Sol()`, `Mut_Sol()`, `Eval_Sol()` and `Imp_Sol()`. The procedures `Init_Pop()`, that initializes the population, `Cros_Sol` and `Mut_Sol()`, used to perform the genetic operators, and `Imp_Sol()`, that tries to improve the quality of one given solution, are algorithm dependent, and will be explained later in detail. The

```
procedure GENERAL()
Initialize Pop using Init_Pop();
foreach Individual z ∈ Pop
    z ← Imp_sol(z);
    Evaluate z using Eval_Sol(z);
endfor
for t = 1 to MaxGen
    for k = 1 to MaxCross
        Select two individuals z1 and z2;
        New_z ← Cros_Sol(z1, z2);
        New_z ← Mut_Sol(New_z);
        Evaluate New_z using Eval_Sol(New_z);
        New_z ← Imp_Sol(New_z);
    endfor
    z ← New_z;
endfor
```

Algorithm 9.1 General pseudocode for the first three algorithms.

procedure `Eval_Sol()` is defined in the same way for all the algorithms and evaluates the quality of the chromosome, by measuring two different values: the cost, $c(x)$, and the degree of infeasibility, $u(x)$ of the solution.

9.2.3 First approach

In our first approach, an algorithm that tries to avoid the creation of infeasible solutions was developed. For this purpose, specialized genetic operators were considered. The `Imp_Sol()` procedure was not used and the cost $c(x)$ is set to ∞, each time that $u(x) > 0$. Generational replacement and proportional selection were used.

9.2.3.1 Init_Pop procedure. Each chromosome of the initial population is obtained using the procedure given in Algorithm 9.2.

The stopping criterion is satisfied if a feasible solution is found or the number of tries is 50.

9.2.3.2 Cros_Sol and Mut_Sol procedures. Crossover operators combine the information contained in two chromosomes as a way to generate a new one. In this approach, since we were interested in preserving feasibility, usual crossovers as one point and uniform crossover, were not suitable. For this reason, two crossover operators, `Cross1` and `Cross2`, were developed and are described in Algorithms 9.3 and 9.4, respectively.

Since we can change one solution either by modifying customer assignments or by modifying the set of open plants, and consequently customer as-

```
procedure INIT_POP( )
while stopping criterion not satisfied
    Generate a permutation of customers P_β ← {β_1, β_2, ...,
    β_i, ..., β_n};
    foreach plant i, i ∈ I
        J_i ← {j ∈ J : customer j is assigned to plant i};
        Δ_i ← a_i − Σ_{J_i} b_j;
    endfor
    foreach customer α_j, j = 1, ..., m
        Assign customer α_j to the first plant β_i, such that Δ_{β_i} −
        b_{α_j} ≥ 0. If such plant does not exist, then assign α_j to
        β_1;
        Update Δ_β;
    endfor
endwhile
```

Algorithm 9.2 Initial population generation.

signments, we have decided to use two mutation operators. In the first one, Mut1, couples of customers assigned to different plants, are randomly chosen and there assignment is exchanged. In the second, Mut2, open plants are mutated. These operators are applied with some probability, previously defined, to each chromosome. We note, however, that for both of them, changes are performed only if the resulting chromosome represents a feasible solution. Mutation operators Mut1 and Mut2 are described in Algorithms 9.5 and 9.6, respectively.

9.2.4 *Second approach*

In this case, instead of avoiding the creation of infeasible solutions, by the use of special genetic operators, the procedure Imp_Sol, was used as a way of obtaining better chromosomes.

As in the first approach, generational replacement was used. For parent selection and since the Imp_Sol does not guarantee feasibility, we decided to treat infeasible and feasible solutions the same way, i.e. only by their cost $c(x)$.

To initialize the population, the Init_Pop procedure is used: for each chromosome, allele values are randomly chosen in the set $\{1, \ldots, n\}$.

The Imp_Sol procedure is applied to each chromosome, immediately after the Eval_sol procedure, and consists of a local search where moves are defined as shift or swap. A shift transfers the assignment of one customer to another open plant. A swap exchanges assignment between two customers. This local search is a descent local method, with the additional requirement that the entire neighborhood, at each iteration, must be exploited. So, the best

procedure Cross1( )
Initialize the child chromosome:

$$Child[j] \leftarrow \begin{cases} Father1[j] & \text{if } Father1[j] = Father2[j]; \\ 0 & \text{otherwise;} \end{cases}$$

Choose randomly one of the parent chromosomes: *Father*1;
Choose randomly a number $r \in [0, m]$;
repeat
  Set $\alpha \leftarrow Father1[r]$;
  Set $A \leftarrow \{j \mid Father1[j] = \alpha\}$;
  foreach $j \in A$ such that $Child[j] = 0$
    if j can be assigned to α without violating the capacity of plant α)
      $Child[j] \leftarrow \alpha$ and update the capacity of plant α;
    else
      Set $\beta \leftarrow Father2[j]$ and $B \leftarrow \{l \mid Father2[l] = \beta\}$;
      foreach $l \in B \mid Child[l] = 0$
        $Child[l] \leftarrow \beta$;
        Update the capacity of plant β;
      endfor
    endif
  endfor
  Set $\alpha \leftarrow Father2[r]$;
  Set $A \leftarrow \{j \mid Father2[j] = \alpha\}$;
  foreach $j \in A$ such that $Child[j] = 0$
    if j can be assigned to α without violating the capacity of plant α)
      $Child[j] \leftarrow \alpha$ and update the capacity of plant α;
    else
      Set $\beta \leftarrow Father2[j]$ and $B \leftarrow \{l \mid Father2[l] = \beta\}$;
      foreach $l \in B \mid Child[l] = 0$
        $Child[l] \leftarrow \beta$;
        Update the capacity of plant β;
      endfor
    endif
  endfor
until $\nexists\, j$ such that $Child[j] = 0$

Algorithm 9.3 Crossover operator Cross1. With a very low probability, this operator can create infeasible solutions.

move is the one, over all possible moves, that most improves the current solu-

procedure Cross2()
Choose randomly one of the parent chromosomes: *Father*1;
Generate a permutation P of the set $\{1, 2, \ldots, m\}$;
foreach $j \in J$
  $Child[j] \leftarrow Father1[j]$;
endfor
foreach element $P(j)$, $j = 1, \ldots, , m$
  Replace in the child chromosome the value of the position $P(j)$ by $Father2[P[j]]$ if such assignment does not violate the capacity constraints.;
endfor

Algorithm 9.4 Crossover operator Cross2.

procedure MUT1()
Randomly generate a permutation of customers $\{\alpha_1, \ldots, \alpha_m\}$;
foreach customer $j = 1, \ldots, m$
  if customers j and α_j are assigned to different plants
    Swap their assignment;
endfor

Algorithm 9.5 Mutation operator Mut1.

procedure MUT2()
Randomly select one closed plant β and one open plant α;
$A \leftarrow \{j \mid j$ is assigned to plant $\alpha\}$;
foreach customer $j \in A$
  Assign j to plant β;
endfor

Algorithm 9.6 Mutation operator Mut2.

tion. Since solutions can be infeasible, improving moves are moves that lead to:

- Feasible solutions with smaller costs, if the current solution is feasible;
- Solutions with a smaller degree of infeasibility, if the current solution is infeasible.

Procedure Cros_Sol uses one point and uniform crossovers, while procedure Mut_Sol is similar to the one previously described: the same mutation operators were used but without the constraint that feasible solutions must be obtained.

9.2.5 *Third approach*

This approach is very similar to the second one and was motivated by some computational results obtained. The main differences are:

- Steady state replacement was used instead of generational replacement;
- Procedure `Imp_Sol` only uses swap moves, and is applied when more than 50 iterations were executed after the last update of the best solution;
- The use of heuristic `ClosPlant`, that tries to close some open plants and is applied only to feasible solutions that improve the best feasible solution already found. Heuristic `ClosPlant` is described in Algortihm 9.7.

procedure CLOSPLANT()
repeat
  Define $J_i \leftarrow \{j \in J$ such that j is assigned to plant $i\}$;
  Define $O \leftarrow \{i \in I$ such that $J_i \neq \emptyset\}$;
  foreach $i_1, i_2 \in O$
    if $\sum_{j \in J_{i_1}} b_j < a_{i_2} - \sum_{j \in J_{i_2}} b_j$
      Assign customers $j \in J_{i_1}$ to plant i_2;
  endfor
until no more plants can be closed

Algorithm 9.7 Heuristic `ClosPlant`.

9.2.6 *Fourth approach*

This approach is completely different from the ones previously described, since it uses a structured population with only 13 individuals. This procedure was developed based on evolutionary algorithms proposed for the number partitioning problem, presented by Berretta and Moscato (1999).

The population structure is a complete ternary tree with three levels, where each individual handles two chromosomes: the *Pocket* and the *Current*. In this structure, a hierarchy relation is maintained between the quality of the solutions represented by *Pocket* and *Current* chromosomes: *Pocket* chromosomes represents always better solutions than *Current* chromosomes. There is also maintained a relation between *Pocket* chromosomes: the *Pocket* chromosome of one leader represent a better solution than the *Pocket* chromosome of its subordinates. Therefore, the solution contained in the *Pocket* chromosome

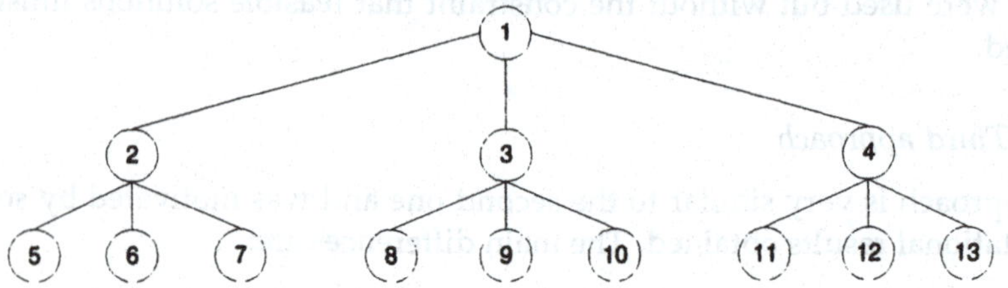

of individual 1 is the best solution contained in *Pocket* chromosomes, the solution contained in the *Pocket* chromosome of individual 2 is better than the solutions contained in *Pocket* chromosomes of individual 5, 6 and 7, and so on.

The pseudocode for this approach is given in Algorithm 9.8, where six procedures are cited: `Init_Pop()`, `Eval_Sol()`, `Swap_Pop()`, `Cros_Sol()`, `Mut_Sol()`, and `Imp_Sol`. Procedures `Init_Pop()`, that initializes the population, `Cros_Sol()` and `Mut_Sol()` used to perform the genetic operators and `Imp_Sol` that tries to improve the quality of one given solution will be explained latter in detail. The procedure `Eval_Sol()` is the same as mentioned in Subsection 9.2.2. The `Swap_Pop()` procedure is only used to compare the quality of the chromosomes and to switch them if necessary, in order to maintain the hierarchy relation between the *Current* and *Pocket* chromosomes. The `Divers_Pop` function measures the diversity of the population. In our algorithm, we established that the population looses diversity when the *Pocket* chromosomes of the leaders in the second level of the tree, have all the same genotype, in 25% of randomly chosen positions and K is a given parameter.

9.2.6.1 Init_Pop procedure. The initial population is obtained by one Lagrangean heuristic based on the Lagrangean relaxation of the demand constraints and adding the surrogate constraint:

$$\sum_{i=1}^{n} a_i y_i \geq \sum_{j=1}^{m} b_j.$$

At the end of each subgradient iteration, a solution to the relaxed problem is obtained. In this solution, a set of plants, $A \subseteq I$ is open, with a total capacity larger or equal than total demand, but where the set J of customers is partitioned in three subsets:

$$J_{AF} = \{j \in J : \sum_{i=1}^{n} x_{ij} = 1\},$$

$$J_{PAF} = \{j \in J : \sum_{i=1}^{n} x_{ij} > 1\},$$

$$J_{WAF} = \{j \in J : \sum_{i=1}^{n} x_{ij} = 0\},$$

```
procedure FOURTHAPPROACH( )
Initialize Pocket and Current using Init_Pop();
for i = 1 to 13
    Evaluate Pocket(i) using Eval_Sol();
    Evaluate Current(i) using Eval_Sol();
endfor
Call Swap_Pop();
for t = 1 to MaxGen
    for i = 2 to 4
        New_Current(i) ← Cros_Sol(Pocket(i), Pocket(1));
    endfor
endfor
if Divers_Pop
    for i = 2 to 4
        for j = 3i − 1 to 3i + 1
            New_Current(j) ← Cros_Sol(Pocket(j), Pocket(i));
        endfor
    endfor
else
    foreach subpopulation P_i of level 3
        Randomly select another population P_j such that j ≠ i
        of level 3;
    endfor
    foreach individual i ∈ P_i
        Randomly select an individual j ∈ P_j;
        New_Current(i) ← Cros_Sol(Pocket(i), Pocket(j));
    endfor
endif
for i = 2 to 13
    New_Current(i) ← Mut_Sol(New_Current(i));
    Evaluate New_Current(i) using Eval_Sol();
endfor
Current ← New_Current;
Call Swap_Pop();
if t mod K = 0
    for i = 1 to 13
        Imp_Sol(Pocket(i));
    endfor
endif
```

Algorithm 9.8 Pseudocode for fourth approach.

such that $J_{AF} \cap J_{PAF} \cap J_{WAF} = \emptyset$ and $J_{AF} \cup J_{PAF} \cup J_{WAF} = J$.

A heuristic procedure is used to search for feasible solutions to the problem. This procedure is described in Algorithm 9.9.

```
procedure SEARCH()
foreach customer j ∈ J_PAF
    Let I_j = {i ∈ A : x_ij = 1};
    c_{i1 j} = min_{i∈I_j} c_ij,                 Assign customer j to plant i1
endfor
foreach plant i ∈ A
    Let J_i ← {j ∈ J | x_ij = 1};
    Let Δ_i ← {a_i − Σ_{J_i} b_j};
endfor
Sort the customers j ∈ J_WAF, in descending order of demand;
foreach customer j ∈ J_WAF, in sorted order
    if plant i1 : c_{i1 j} = min_{i∈A:Δ_i−b_j≥0} c_ij exists
        Assign j to it;
    else
        Assign j to open plant i1 such that c_{i1 j} = min_{i∈A} c_ij;
    endif
    Update J_{i_i} and Δ_{i1};
endfor
```

Algorithm 9.9 Heuristic to search for feasible solutions.

The initial population will consist of the best 26 solutions found during 500 iterations of the subgradient optimization.

9.2.6.2 Imp_Sol procedure.

`Imp_Sol()` is a tabu search procedure that changes the assignment of the clients. For this purpose only two kinds of moves are allowed:

- *MI*: Switches the assignment of one client to another plant;
- *MII*: Swaps the assignment between two clients that are assigned to different open plants.

At each iteration, *MI* moves are applied only to one randomly chosen client. Once the best *MI* move is found, *MII* moves are performed only between clients that belong to plants involved in the best move *MI*. Since our population can contain infeasible solutions, the best move is the one that leads to:

- Feasible solutions with smaller costs, if the current solution is feasible;

- Solutions with a smaller degree of infeasibility, if the current solution is infeasible.

As these moves are performed, tabu restrictions are employed to prevent moving back to previous solutions. In our procedure a static tabu list with dimension 7 was used.

9.2.6.3 Procedures Cros_Sol and Mut_Sol. We used three types of crossover operators. The one point (**Cross1**), the uniform (**Cross2**), and one, which we call **Cross3**. **Cross3** is described in Algortihm 9.10.

procedure CROSS3( )
Initialize the child chromosome:

$$Child[j] = \begin{cases} Father1[j] & \text{if } Father1[j] = Father2[j]; \\ 0 & \text{otherwise;} \end{cases}$$

while $\exists\, j$ such that $Child[j] = 0$
  Choose randomly one position r such that $Child[r] = 0$;
  if $Child[r] = Father1[r]$ produces a less degree of infeasibility or the same degree of infeasibility and a better cost than $Child[r] = Father2[r]$
    Set $Child[r] \leftarrow Father1[r]$;
  else
    Set $Child[r] \leftarrow Father2[r]$;
  endif
endwhile

Algorithm 9.10 Crossover operator **Cross3**.

Two mutation operators were used: **Mut1** and **Mut2**. **Mut1** swaps the assignment between two randomly chosen customers and is applied, with a certain probability, to each allele. On the other hand, **Mut2** switches the assignment of the customers that are assigned to one randomly chosen open plant to another randomly chosen close plant. This mutation operator is applied with a given probability, only one time for each chromosome.

9.3 COMPUTATIONAL RESULTS

The algorithms were written in Fortran and run on a 233 MHz Pentium II computer. Computational results were obtained using available test problems and randomly generated problems.

Some available test problems were taken from OR-Library (Beasley, 1990). Since some of these problems are infeasible for the single source version of the problems, some modifications were made (Klincewicz and Luss, 1986). For small-medium size problems, a total of 61 instances with 50 customers and 16,

25, or 50 potential plant locations were tested. Large size problems, with 1000 customers and 100 potential plant locations were also used for testing. The other test problems, available at `www.ups.es/catala/webs/webs.html`, were already used by other researchers, namely by Hindi and Pienkosz (1999).

Randomly generated problems were obtained according to different methodologies. The first set was generated according to our own methodology, with the objective of showing how the results can be influenced by the data structure. These instances were generated in the following way. Client demands were drawn from a uniform distribution on the range $[5, 35]$ and plant capacities in the same way but in the range $[\lfloor 5mq/n \rfloor, \lfloor 35mq/n \rfloor]$, where $\lfloor x \rfloor$ represent the greatest integer less or equal to x, n the number of potential plant locations, m the number of customers and q the ratio between total supply and total demand. Fixed costs are defined as $f_i = U[0, 90] + U[100, 110]\sqrt{a_i}$, where $U[x, y]$ represents an uniform distribution in the range $[x, y]$. For assignment costs, a network of density d and $(m + n)$ vertices was used. The cost of each edge was drawn from a uniform distribution in the range $[0, 100]$. This matrix was submitted to the Floyd (1962) algorithm and we worked with a sub-matrix with n rows and m columns.

A total of 72 classes of problems were tested, with parameters:

- $n \times m$: $8 \times 25, 25 \times 25, 33 \times 50, 50 \times 50$
- q: 1.2, 1.3, 1.4, 1.5, 3.0, 5.0
- d: 25%, 50%, 75%.

We have also randomly generated a total of 5 classes of problems, with higher dimensions. For these classes the parameters used were:

- $n \times m$: $25 \times 50, 25 \times 100, 25 \times 250, 25 \times 500, 50 \times 100, 50 \times 250, 50 \times 500$
- q: 5.0
- d: 50

To test the effect of varying the relation between fixed and assignment costs, a new set of instances was considered. This new set was obtained multiplying the assignments costs by the demand of each client. To simplify, from now on, this set will be denoted by set O_2, while the previous one will be denoted by O_1.

For each class of test problems, five instances were developed.

The other sets were randomly generated according to the methodologies of Pirkul (1987), Filho and Galvão (1998) and Sridharan (1993).

In the following tables, the maximum, mean and minimum deviations in percentage were calculated as $100 \cdot (UB - LB)/LB$, where UB is the best upper bound obtained and LB is the best known lower bound. The maximum, mean and minimum time, are expressed in seconds.

Our first computational experiments were made using the first approach with proportional selection. And, as we suspected, the following conclusions were obtained:

- It was impossible to find feasible solutions and to run the algorithm for some highly constrained instances ($q = 1.2$ and $q = 1.3$);
- The algorithm converges very quickly so it needs very high mutation's probability;
- The algorithm is very time consuming and does not give good results.

Some of these conclusions can be observed in Tables 9.1 and 9.2, that show the average results obtained for set O_1 and for some OR-Library problems. These results were obtained with **Cross2**, population size set to 50, MaxGen set to 1000 iterations and probabilities of crossover and mutation, respectively equal to 0.9 and 0.25.

Table 9.1 Results for the set O_1 with the first approach.

Problem	*Deviation(%)*			*Time(s)*		
	Max	*Mean*	*Min*	*Max*	*Mean*	*Min*
8x25	3.78	0.47	0.00	11.00	8.72	6.00
25x25	8.55	1.95	0.01	20.00	12.00	8.00
33x50	7.59	1.98	0.08	7039.00	132.67	25.00
50x50	8.01	2.50	0.65	1802.00	78.08	31.00
25x50	2.99	1.73	1.09	44.00	42.80	42.00
25x100	3.41	2.53	1.94	123.00	122.00	121.00
25x250	4.38	3.66	1.71	616.00	595.20	560.00
25x500	8.40	6.81	3.98	2206.00	2171.00	2064.00
50x100	5.47	4.51	3.34	167.00	165.60	163.00
50x250	7.76	5.99	2.91	714.00	698.60	670.00
50x500	8.27	7.74	6.96	2419.00	2410.40	2402.00

Table 9.2 Results for some OR-Library problems with the first approach.

Problem	*Deviation(%)*			*Time(s)*		
	Max	*Mean*	*Min*	*Max*	*Mean*	*Min*
16x50	17.65	11.12	2.97	37.00	32.90	30.00
25x50	32.53	12.06	2.97	39.00	33.19	30.00
50x50	37.27	24.14	5.08	59.00	54.95	53.00
c-100x1000	105.13	91.65	76.90	10321.00	10210.75	10072.00

Our next experiments were made using the second approach. Several experiments were made, as a way of comparing results obtained with uniform and one point crossover operators and also tournament and proportional selection. Analyzing the results, some conclusions can be established:

- It was not possible to obtain feasible solutions to some problems ($q = 1.2$, 1.3 and $n \times m = 50 \times 50, 33 \times 50$);

- On average, one point crossover gives better results than uniform crossover;
- On average, tournament selection gives better results than proportional selection;
- The algorithm is very time consuming and does not give good results.

In Table 9.3, some results are presented. These results were obtained with a population size of 50 individuals and 2000 iterations. One point crossover and tournament selection were used, and probability of crossover and mutation were set to 0.9 and 0.01, respectively.

Table 9.3 Results for larger instances of problem set O_1 with the second approach.

	Deviation(%)			*Time(s)*		
Problem	*Max*	*Mean*	*Min*	*Max*	*Mean*	*Min*
25x50	7.28	4.12	1.13	74.00	65.00	54.00
25x100	12.03	7.72	4.34	385.00	333.80	280.00
25x250	7.42	6.37	5.35	4584.00	4217.60	3781.00
25x500	8.68	5.69	2.82	35432.00	30172.60	22801.00
50x100	14.83	10.74	4.89	509.00	412.60	371.00
50x250	20.15	13.53	8.56	21024.00	7821.40	4233.00
50x500	9.14	6.23	3.30	37435.00	29199.80	20185.00

Our next experiments were made with the third approach. All the results that are going to be presented, were obtained with a population of 100 individuals and 5000 iterations. For mutation the following probability values were used:

- Mut1: $p_{mut1} = 0.01$;
- Mut2: p_{mut2} is set to 0.01, and increased by 0.01 until the value of 0.1 is reached.

In Table 9.4, we give the average deviations in percentage, D(%), and the number of infeasible solutions (NIS) obtained when different criteria for parent selection and different types of crossover were used. So, for parent selection we used proportional and binary tournament. For each one of them, we also used two different rules for comparing solutions: with and without ranking. Without ranking means that the solution x_1 is better than x_2 if $c(x_1) < c(x_2)$, and with ranking means that x_1 is better than x_2 if

$$\begin{cases} c(x_1) < c(x_2) & \text{and both are feasible} \\ u(x_1) < u(x_2) & \text{otherwise.} \end{cases}$$

Analyzing Table 9.4 we can conclude that:

Table 9.4 Comparing the average deviations and the number of infeasible solutions obtained with the third approach.

	One point crossover				*Uniform Crossover*			
	Proportional		*Tournament*		*Proportional*		*Tournament*	
	D(%)	*NIS*	*D(%)*	*NIS*	*D(%)*	*NIS*	*D(%)*	*NIS*
Without ranking								
Set O1								
8x25	4.18	0	3.31	0	3.69	0	3.69	0
25x25	3.72	1	4.61	0	3.98	3	4.63	1
33x50	5.85	2	7.61	1	7.62	3	9.03	3
50x50	5.81	4	7.12	2	6.78	5	7.84	3
Set O2								
8x25	1.02	0	1.57	0	1.12	0	1.44	0
25x25	2.13	0	3.20	0	1.93	1	3.57	0
33x50	5.24	3	7.81	3	5.71	2	7.96	3
50x50	5.55	7	8.62	7	5.06	6	8.46	7
With ranking								
Set O1								
8x25	5.18	0	5.69	0	4.77	0	4.18	0
25x25	3.58	0	3.86	0	3.18	1	3.19	0
33x50	6.97	0	6.82	2	5.60	3	5.52	1
50x50	5.84	4	6.10	3	5.25	3	5.29	2
Set O2								
8x25	0.88	0	0.95	0	0.88	0	0.79	0
25x25	1.80	0	1.87	0	2.06	0	1.89	0
33x50	5.26	2	5.22	2	5.47	1	5.32	2
50x50	5.29	5	5.07	6	5.01	7	5.09	5

- In general, better results were obtained with ranking;
- The number of infeasible solutions obtained is higher for set O_2, namely when $m = 50$;
- The number of infeasible solutions increases with the dimension of the problem. The same conclusion cannot be establish for the average deviation in percentage.

Now we present the results obtained with the fourth approach. All the results were obtained with the following parameters:

- 5000 iterations
- The stopping criterion for tabu search is 100 iterations without improvement
- Ttabu search is only used in every 100 iterations

- Probability mutation values:

$$\begin{cases} p_{mut1} = 0.01 & \text{for Mut1} \\ p_{mut2} = 0.1 & \text{for Mut2} \end{cases}$$

Table 9.5 Average deviations in percentage for our problems with the fourth approach.

	Cross1			*Cross2*			*Cross3*		
	Max	*Mean*	*Min*	*Max*	*Mean*	*Min*	*Max*	*Mean*	*Min*
Set O1									
8x25	0.79	0.20	0.00	0.79	0.20	0.00	0.79	0.20	0.00
25x25	4.35	1.87	0.20	3.81	1.67	0.20	4.47	1.91	0.21
33x50	5.15	2.53	0.89	3.95	2.22	1.16	4.43	2.35	1.37
50x50	6.57	2.92	1.06	5.34	2.28	1.04	4.67	2.22	0.89
Set O2									
8x25	1.46	0.38	0.00	1.43	0.37	0.00	1.44	0.39	0.00
25x25	4.33	0.88	0.03	4.16	0.86	0.03	4.33	0.09	0.03
33x50	2.58	0.86	0.00	2.24	0.74	0.01	3.00	0.85	0.00
50x50	5.32	0.65	0.51	4.56	0.48	0.30	4.84	0.55	0.40

Table 9.6 Average times for our problems with the fourth approach.

	Cross1			*Cross2*			*Cross3*		
	Max	*Mean*	*Min*	*Max*	*Mean*	*Min*	*Max*	*Mean*	*Min*
Set O1									
8x25	2.83	1.52	0.00	3.50	1.80	0.00	4.00	2.00	0.00
25x25	5.50	4.19	2.67	6.33	4.76	3.00	7.17	5.63	3.83
33x50	10.00	8.49	6.83	11.50	9.88	8.00	13.50	11.56	9.00
50x50	16.67	11.91	9.67	18.33	13.29	11.33	20.67	15.41	12.50
Set O2									
8x25	2.67	1.59	0.17	3.50	1.98	0.33	3.50	2.14	0.17
25x25	4.50	3.40	1.67	5.17	3.98	2.00	6.00	4.38	2.00
33x50	8.67	7.53	6.17	10.50	8.88	7.17	11.33	9.67	7.17
50x50	11.17	7.89	6.83	12.50	9.22	7.67	13.83	10.11	8.50

Analyzing Table 9.5 and Table 9.6 we can conclude that :

- For set O_1, Cross2 gives better average deviations than Cross1. Except for 50x50 problems in the set O_1, the same conclusion can be established between Cross2 and Cross3;
- For set O_2, Cross2 gives the best average deviations;
- The average time increases with the dimension of the problem.

In Table 9.7, we can see the average deviations for each class of our problems tested, when Cross2 is used. In that table, results are given for each pair of parameters tested (d_q) and for each dimension (nxm). The letters A, B and C represent $d = 25\%$, 50%, and 75% and the numbers 1.2, 1.3, 1.4, 1.5, 3.0 and 5.0 represent the values of q.

Analyzing Table 9.7 we can conclude that it does not exist any kind of relation between parameters q and d and the quality of the mean deviation in percentage. We can also observe that, generally, the mean deviations obtained for set O_2 are smaller than the deviations obtained for O_1.

Comparing the average deviations (1.59 and 0.61) and average times (7.43 and 6.01) obtained for sets O_1 and O_2, with the average deviations (0.95 and 0.98) and the average times (42.73 and 44.08) reported in Cortinhal and Captivo (2001), we can conclude that this procedure has performed well for set O_2.

The results obtained for our larger instances are given in Table 9.8 and Table 9.9.

In these tables, we can observe that it is not possible to conclude which crossover is the best operator, since the deviations obtained are very similar. This fact can be confirmed by the values obtained for the mean value of the average deviations (1.50, 1.49, and 1.51) and the mean value of the average times (45.69, 53.29, and 66.01) for Cross1, Cross2, and Cross3, respectively.

Since the crossover operator, Cross2 gave slightly better average deviations (Table 9.5 and Table 9.8) and almost the same average times (Table 9.6 and Table 9.9) for our problems. From this point on, we will restrict our presentation to computational results obtained with Cross2.

Comparing the mean values for the average deviations (0.21 and 3.84) and average times (52.18 and 54.40) obtained for sets O_1 and O_2, using Cross2, with the mean values for the average deviations (0.15 and 1.95) and the average times (694.74 and 755.06) reported in Cortinhal and Captivo (2001), we can conclude that this procedure gives worst average deviations.

In Tables 9.10 to 9.16, computational results are given for each other set of problems tested.

Comparing the average deviation and average time obtained for small-medium size OR-Library problems (Table 9.10), with the average deviation (0.08) and the average time (42.93) reported in Cortinhal and Captivo (2001), we can conclude that this procedure has performed well. When comparing the average deviation obtained for large size problems (Table 9.11) (0.72) with the average deviations reported in Hindi and Pienkosz (1999) which is 0.92, and reported in Cortinhal and Captivo (2001), which is 0.93, we can conclude that our procedure performed well for these instances.

For the Filho and Galvão problems (Table 9.12), in some groups of problems, there exist a large difference between maximum and minimum deviations obtained. However, this fact does not seem to be related with the parameters values. Mean CPU time increases as m increases, but is not influenced by the n, $W1$, or $W2$ parameters.

Table 9.7 Average deviations in percentage for our set O_1 and O_2 with Cross2.

	8x25			*25x25*			*33x50*			*50x50*		
	Max	*Mean*	*Min*	*Max*	*Mean*	*Min*	*Max*	*Mean*	*Min*	*Max*	*Mean*	*Min*
Set O1												
A1.2	0.22	0.11	0.00	9.94	2.93	0.00	6.73	4.05	2.79	6.81	3.54	0.71
A1.3	0.68	0.27	0.06	3.24	1.54	0.00	4.41	3.40	2.19	8.32	3.85	2.03
A1.4	0.28	0.08	0.01	4.20	2.46	0.52	4.38	2.93	1.96	5.29	3.67	1.36
A1.5	1.24	0.30	0.00	3.97	2.36	0.60	3.74	2.88	1.22	3.31	2.71	2.34
A3.0	0.02	0.01	0.00	1.17	0.64	0.09	0.49	0.25	0.04	7.82	2.46	0.25
A5.0	2.32	0.65	0.00	0.04	0.04	0.01	0.34	0.15	0.00	0.47	0.18	0.01
B1.2	1.14	0.37	0.00	3.47	1.18	0.00	9.11	4.49	2.91	7.08	3.30	1.58
B1.3	1.43	0.38	0.00	8.60	3.91	0.26	6.25	4.03	3.11	7.19	3.84	1.38
B1.4	0.46	0.23	0.01	4.16	1.46	0.05	4.84	3.81	3.17	3.54	3.35	1.02
B1.5	0.48	0.14	0.01	5.51	3.71	2.97	3.23	2.44	0.56	2.47	2.32	2.16
B3.0	0.11	0.03	0.01	0.92	0.23	0.00	0.19	0.06	0.00	4.49	1.91	0.10
B5.0	0.02	0.01	0.00	0.22	0.05	0.00	0.09	0.03	0.00	0.04	0.03	0.02
C1.2	1.97	0.42	0.00	5.57	3.93	2.86	4.26	3.13	1.90	3.20	2.58	1.98
C1.3	0.58	0.21	0.01	2.78	2.18	1.34	4.29	3.75	3.11	2.61	2.12	1.75
C1.4	0.68	0.21	0.01	4.91	1.43	0.00	4.25	3.04	1.75	4.46	3.06	2.41
C1.5	0.26	0.07	0.00	3.10	1.85	0.12	2.79	1.29	0.14	2.44	2.06	1.37
C3.0	0.12	0.03	0.00	0.24	0.10	0.01	1.18	0.28	0.04	4.14	1.05	0.07
C5.0	0.00	0.00	0.00	0.10	0.04	0.01	0.11	0.03	0.00	0.37	0.09	0.01
Set O2												
A1.2	1.03	0.28	0.00	0.24	0.09	0.00	4.87	1.25	0.00	12.79	3.28	0.07
A1.3	0.39	0.11	0.00	0.85	0.31	0.00	1.96	0.67	0.01	2.29	0.88	0.00
A1.4	1.41	0.36	0.00	1.30	0.44	0.00	0.59	0.33	0.00	0.85	0.45	0.01
A1.5	2.34	0.96	0.16	0.73	0.49	0.07	2.37	0.77	0.00	4.61	2.85	1.61
A3.0	0.85	0.17	0.00	1.34	0.69	0.11	1.87	0.86	0.02	4.05	2.03	0.49
A5.0	0.00	0.00	0.00	1.50	0.79	0.00	0.50	0.27	0.00	2.79	2.00	1.00
B1.2	3.95	0.91	0.00	8.59	1.86	0.02	3.73	1.61	0.00	4.43	1.01	0.00
B1.3	1.39	0.52	0.00	7.51	1.82	0.00	2.71	0.74	0.04	12.78	2.69	0.05
B1.4	0.58	0.12	0.00	0.65	0.21	0.00	1.66	1.17	0.54	0.21	0.13	0.00
B1.5	0.63	0.23	0.01	2.57	0.79	0.03	2.92	1.15	0.25	2.68	1.99	1.03
B3.0	0.54	0.22	0.01	3.57	1.78	0.54	1.46	0.48	0.12	3.08	1.15	0.42
B5.0	0.87	0.18	0.00	2.09	0.65	0.01	0.99	0.27	0.00	4.09	1.56	0.31
C1.2	1.27	0.37	0.02	3.25	0.79	0.00	0.99	0.59	0.13	0.84	0.40	0.00
C1.3	1.62	0.56	0.00	0.78	0.52	0.25	1.03	0.44	0.00	2.49	1.04	0.07
C1.4	1.40	0.52	0.00	0.53	0.19	0.00	1.62	0.56	0.00	0.91	0.58	0.19
C1.5	0.24	0.12	0.01	4.17	2.60	0.35	3.52	1.84	0.66	3.85	2.98	1.53
C3.0	3.97	1.01	0.00	1.54	0.75	0.14	0.83	0.37	0.00	3.25	2.41	1.89
C5.0	0.06	0.01	0.00	1.94	0.83	0.00	0.08	0.02	0.00	2.26	1.01	0.01

Comparing the mean average deviation and the standard deviation, presented in Filho and Galvão (1998) (0.43 and 1.12) and in Cortinhal and Captivo (2001) (0.01 and 0.01), we can say that our approach performed better than the one presented in Filho and Galvão (1998). Concerning CPU times, direct comparisons with Filho and Galvão (1998) are not possible, since different computers were used. However, we can observe that our standard deviation (2.70) is smaller than the one reported by them (9.07).

Table 9.8 Average deviations in percentage for our higher dimension problems with the fourth approach.

	Cross1			*Cross2*			*Cross3*		
	Max	*Mean*	*Min*	*Max*	*Mean*	*Min*	*Max*	*Mean*	*Min*
Set O1									
b25x50	1.02	0.31	0.01	1.02	0.31	0.01	1.02	0.31	0.01
b25x100	0.37	0.14	0.02	0.37	0.14	0.02	0.37	0.14	0.02
b25x250	0.22	0.06	0.00	0.22	0.06	0.00	0.22	0.06	0.00
b25x500	0.47	0.23	0.05	0.48	0.23	0.05	0.48	0.24	0.05
b50x50	0.06	0.04	0.02	0.04	0.00	0.02	0.04	0.03	0.02
b50x100	0.97	0.44	0.15	0.97	0.44	0.15	0.97	0.44	0.15
b50x250	0.48	0.29	0.08	0.48	0.29	0.08	0.48	0.29	0.08
b50x500	0.36	0.20	0.05	0.37	0.20	0.05	0.37	0.20	0.05
Set O2									
b25x50	1.95	0.76	0.08	1.95	0.76	0.08	1.95	0.74	0.08
b25x100	2.22	1.66	1.07	2.22	1.62	0.87	2.22	1.73	1.44
b25x250	2.30	1.28	0.30	2.30	1.28	0.29	2.30	1.27	0.26
b25x500	1.81	0.86	0.05	1.81	0.87	0.03	1.81	0.87	0.06
b50x50	3.59	2.31	1.56	2.79	2.00	1.00	2.79	2.33	1.66
b50x100	7.31	4.85	2.63	7.54	5.05	2.63	7.54	5.05	2.63
b50x250	6.69	5.46	3.25	6.69	5.46	3.25	6.69	5.46	3.25
b50x500	5.44	5.05	4.70	5.44	5.05	4.70	5.44	5.05	4.70

For Pirkul problems, presented in Table 9.13, we can conclude that the deviations seem to be influenced by some parameters values, namely by W_2. The average time seem to be influenced by m, increases as m increases, but not by n, $W1$ or $W2$ parameters.

It is not possible to make direct comparisons, with Pirkul (1987) results reported for this set of problems, since deviations are calculated in a different way. However comparing with the average and the standard deviation reported in Cortinhal and Captivo (2001), whose values are respectively 0.01 and 0.01, we can conclude that our algorithm did not perform well for this set of instances.

For the Shridharan problems, as we can see in Table 9.14, our procedure performed very well. Comparing them, with the results presented in Sridharan (1993) and in Cortinhal and Captivo (2001) we can observe that except for

Table 9.9 Average times in percentage for our higher dimension problems with the fourth approach.

	Cross1			*Cross2*			*Cross3*		
	Max	*Mean*	*Min*	*Max*	*Mean*	*Min*	*Max*	*Mean*	*Min*
Set O1									
b25x50	11.00	9.40	5.00	15.00	11.20	4.00	18.00	15.00	5.00
b25x100	19.00	12.40	4.00	19.00	12.40	4.00	26.00	16.80	4.00
b25x250	42.00	21.40	6.00	52.00	25.00	6.00	53.00	25.40	6.00
b25x500	115.00	94.49	14.00	137.00	109.80	16.00	137.00	109.60	15.00
b50x50	20.00	14.80	10.00	22.00	16.00	9.00	23.00	16.40	10.00
b50x100	50.00	32.00	18.00	53.00	35.60	21.00	65.00	42.60	26.00
b50x250	81.00	61.20	48.00	90.00	72.20	61.00	101.00	83.80	62.00
b50x500	149.00	118.40	27.00	168.00	135.20	27.00	208.00	150.60	26.00
Set O2									
b25x50	10.00	10.00	10.00	14.00	12.40	12.00	15.00	14.20	13.00
b25x100	16.00	15.80	15.00	21.00	19.40	18.00	23.00	22.40	22.00
b25x250	40.00	39.40	39.00	53.00	51.40	50.00	68.00	57.80	51.00
b25x500	108.00	106.60	106.00	134.00	128.60	123.00	194.00	152.00	126.00
b50x50	8.00	8.00	8.00	10.00	9.60	9.00	11.00	10.60	10.00
b50x100	18.00	18.00	18.00	21.00	21.00	21.00	30.00	29.00	28.00
b50x250	46.00	45.20	45.00	55.00	54.00	53.00	87.00	84.60	81.00
b50x500	129.00	124.00	119.00	141.00	138.80	136.00	244.00	225.40	203.00

8x25 problems, it was possible to improve almost all the deviations already presented.

The next set of test problems, presented in Table 9.15, is also available at www.upc/es/catala/webs/webs.html, and contains instances with 30 potential plant locations and four different values for the number of costumers.

Comparing with the results presented in Cortinhal and Captivo (2001), we can conclude that the average and standard deviation (1.75 and 2.11) are significantly worst than the values reported there, which are 0.15 and 0.22. Concerning CPU times, this approach gives better results.

The final set tested, was also used by Hindi and Pienkosz (1999), Delmaire et al. (1997) and Barceló et al. (1991) and is available at `www.upc.es/catala/-webs/webs.html`. The results obtained with our approach are presented in Table 9.16.

In Table 9.16, we can observe that the quality of the average deviations did not seem to have any kind of relation with the dimension of the problem. They seem to be very dependent of the instance itself. Comparing the average deviation (1.95) with the one obtained by Hindi and Pienkosz (1999) (0.56) and by us in Cortinhal and Captivo (2001) (0.66) we can conclude that our algorithm performed worse.

Table 9.10 Results for small-medium size OR-Library problems with the fourth approach.

Problem	*Dev(%)*	*Time(s)*	*Problem*	*Dev(%)*	*Time(s)*
cap41-5	0.00	1.00	cap51	0.17	5.00
cap41-6	0.00	6.00	cap61	0.00	1.00
cap41-7	0.00	1.00	cap62	0.00	0.00
cap42-5	0.00	1.00	cap63	0.01	6.00
cap42-6	0.03	6.00	cap64	0.05	7.00
cap42-7	0.00	1.00	cap71	0.00	0.00
cap43-5	0.00	1.00	cap72	0.00	1.00
cap43-6	0.06	5.00	cap73	0.00	0.00
cap43-7	0.00	1.00	cap74	0.00	0.00
cap44-5	0.17	6.00	cap91	0.00	1.00
cap44-6	0.25	6.00	cap92	0.00	0.00
cap44-7	0.64	5.00	cap93	0.22	6.00
cap81-5	0.02	6.00	cap94	0.21	6.00
cap81-6	0.00	1.00	cap101	0.00	0.00
cap81-7	0.00	1.00	cap102	0.00	1.00
cap82-5	0.00	1.00	cap103	0.00	1.00
cap82-6	0.12	6.00	cap104	0.00	0.00
cap82-7	0.00	1.00	cap121	0.00	2.00
cap83-5	0.00	1.00	cap122	0.00	1.00
cap83-6	0.31	6.00	cap123	0.02	7.00
cap83-7	0.17	6.00	cap124	0.19	8.00
cap84-5	0.55	7.00	cap131	0.00	0.00
cap84-6	0.70	6.00	cap132	0.00	0.00
cap84-7	0.18	6.00	cap133	0.00	2.00
cap111-5	0.02	7.00	cap134	0.00	1.00
cap111-6	0.00	1.00			
cap111-7	0.00	1.00			
cap112-5	0.00	1.00			
cap112-6	0.12	8.00			
cap112-7	0.00	0.00			
cap113-5	0.00	2.00			
cap113-6	0.10	8.00			
cap113-7	0.00	0.00			
cap114-5	0.19	8.00			
cap114-6	0.20	8.00			
cap114-7	0.07	8.00			
Std.Dev				0.15	2.98
Average				0.08	3.20

Analyzing each table, we can conclude that the fourth approach is better than the others since it leads to better deviations and gives feasible solutions

Table 9.11 Computational results for large size OR-Library problems with the fourth approach.

Problem	*Dev(%)*	*Time(s)*
CapA1	0.28	336.00
CapA2	0.12	305.00
CapA3	0.29	369.00
CapA4	0.00	45.00
CapB1	0.01	298.00
CapB2	2.87	381.00
CapB3	2.04	375.00
CapB4	0.47	296.00
CapC1	0.70	435.00
CapC2	0.65	478.00
CapC3	1.15	466.00
CapC4	0.09	460.00
Std.Dev	0.89	117.48
Average	0.72	353.67

to all the problems tested. However, when we compare these deviations with the deviations obtained using tabu search, reported in Cortinhal and Captivo (2001), we can conclude that this approach is not as good as the other, for the majority of the problems tested.

9.4 CONCLUSIONS

In this paper, we have studied several genetic algorithms for the SSCLP. The computational results show that genetic algorithms only performed well when the initial population was obtained using a Lagrangean heuristic. By our study we can conclude that genetic algorithms are not good methods for solving this problem.

On one hand, is very difficult to define what is a good solution for the problem and consequently to find good crossover operators. On the other hand, when usual crossover operators are used and the problem is very constrained, it can become very difficult to find feasible solutions. However it seems to exist some problems, with very special structure, like those available at OR-Library, for which this approach can perform very well.

Table 9.12 Computational results for Filho and Galvão problems with the fourth approach.

Parameters			*Dev(%)*			*Time(s)*		
nxm	*W1*	*W2*	*Max*	*Mean*	*Min*	*Max*	*Mean*	*Min*
10x40	200	2	0.49	0.06	0.00	6.00	1.40	0.00
10x40	200	4	4.44	0.40	0.00	7.00	0.90	0.00
10x40	400	2	3.24	0.63	0.00	7.00	2.50	0.00
10x40	400	4	1.40	0.14	0.00	7.00	0.80	0.00
10x60	200	2	4.38	0.61	0.00	10.00	3.90	0.00
10x60	200	4	0.37	0.04	0.00	10.00	1.30	0.00
10x60	400	2	0.30	0.09	0.00	9.00	2.70	0.00
10x60	400	4	3.45	0.38	0.00	9.00	1.90	0.00
10x80	200	2	4.97	0.50	0.00	11.00	1.70	0.00
10x80	200	4	1.10	0.21	0.00	12.00	3.80	0.00
10x80	400	2	14.03	1.56	0.00	13.00	5.20	0.00
10x80	400	4	0.57	0.15	0.00	13.00	4.00	0.00
10x100	200	2	3.12	0.70	0.00	22.00	9.70	0.00
10x100	200	4	1.22	0.24	0.00	19.00	6.90	0.00
10x100	400	2	3.49	0.43	0.00	18.00	7.50	0.00
10x100	400	4	2.88	1.01	0.00	17.00	10.60	0.00
20x40	200	2	0.28	0.03	0.00	8.00	1.80	0.00
20x40	200	4	0.48	0.05	0.00	7.00	2.20	0.00
20x40	400	2	0.00	0.00	0.00	1.00	0.40	0.00
20x40	400	4	0.29	0.05	0.00	8.00	3.10	0.00
20x60	200	2	2.21	0.22	0.00	10.00	2.40	0.00
20x60	200	4	0.86	0.13	0.00	9.00	2.10	0.00
20x60	400	2	3.73	0.78	0.00	10.00	5.80	0.00
20x60	400	4	3.84	1.15	0.00	10.00	7.50	0.00
20x80	200	2	0.14	0.03	0.00	9.00	4.60	0.00
20x80	200	4	1.30	0.13	0.00	12.00	3.80	0.00
20x80	400	2	5.84	1.11	0.00	12.00	4.20	0.00
20x80	400	4	3.01	0.32	0.00	12.00	5.10	0.00
20x100	200	2	7.95	1.10	0.00	16.00	5.20	00.0
20x100	200	4	1.16	0.33	0.00	17.00	8.30	0.00
20x100	400	2	0.00	0.00	0.00	1.00	0.40	0.00
20x100	400	4	2.10	0.30	0.00	16.00	5.20	0.00
Std.Dev				0.41				2.70
Average				0.40				3.97

Table 9.13 Computational results for Pirkul problems with the fourth approach.

Parameters			*Dev(%)*			*Time(s)*		
nxm	*W1*	*W2*	*Max*	*Mean*	*Min*	*Max*	*Mean*	*Min*
10x40	200	2	1.70	0.32	0.00	5.00	1.50	0.00
10x40	200	4	0.21	0.02	0.00	4.00	0.60	0.00
10x40	400	2	2.98	0.58	0.00	4.00	1.30	0.00
10x40	400	4	2.38	0.50	0.00	5.00	2.90	0.00
10x60	200	2	2.34	0.47	0.00	6.00	2.60	0.00
10x60	200	4	2.02	0.31	0.00	6.00	2.50	0.00
10x60	400	2	4.72	0.50	0.00	6.00	2.10	0.00
10x60	400	4	2.30	0.75	0.00	6.00	3.70	0.00
10x80	200	2	13.81	2.18	0.00	11.00	5.80	0.00
10x80	200	4	2.01	0.30	0.00	8.00	1.90	0.00
10x80	400	2	14.02	2.04	0.00	8.00	4.70	0.00
10x80	400	4	0.97	0.23	0.00	8.00	3.30	0.00
10x100	200	2	2.90	0.78	0.00	10.00	4.50	0.00
10x100	200	4	1.99	0.53	0.00	10.00	4.30	0.00
10x100	400	2	1.36	0.24	0.00	10.00	3.40	0.00
10x100	400	4	1.42	0.31	0.00	10.00	4.00	0.00
20x40	200	2	4.19	0.48	0.00	5.00	2.10	0.00
20x40	200	4	1.33	0.37	0.00	5.00	2.20	0.00
20x40	400	2	27.51	5.12	0.00	5.00	2.00	0.00
20x40	400	4	0.62	0.10	0.00	5.00	1.40	0.00
20x60	200	2	4.99	0.52	0.00	7.00	1.90	0.00
20x60	200	4	0.33	0.03	0.00	6.00	0.80	0.00
20x60	400	2	1.49	0.28	0.00	7.00	3.10	0.00
20x60	400	4	1.69	0.23	0.00	6.00	1.70	0.00
20x80	200	2	3.79	0.87	0.00	9.00	4.60	0.00
20x80	200	4	2.32	0.79	0.00	9.00	5.30	0.00
20x80	400	2	33.09	5.64	0.00	9.00	4.50	0.00
20x80	400	4	20.42	2.12	0.00	9.00	3.10	0.00
20x100	200	2	13.81	2.18	0.00	11.00	5.80	0.00
20x100	200	4	0.19	0.02	0.00	11.00	2.10	0.00
20x100	400	2	29.00	5.64	0.00	11.00	5.70	0.00
20x100	400	4	15.36	2.18	0.00	11.00	6.80	0.00
Std.Dev				1.57				1.62
Average				1.14				3.19

Table 9.14 Computational results for Sridharan problems with the fourth approach.

Problem	*Dev(%)*			*Time(s)*		
	Max	*Mean*	*Min*	*Max*	*Mean*	*Min*
Srid8x25	0.03	0.02	0.00	1.00	0.20	0.00
Srid16x25	0.05	0.02	0.00	7.00	3.20	1.00
Srid25x25	0.01	0.00	0.00	7.00	2.60	1.00
Srid16x50	0.02	0.01	0.00	1.00	1.00	1.00
Srid33x50	0.03	0.02	0.00	10.00	4.40	2.00
Srid50x50	0.00	0.00	0.00	14.00	9.60	4.00

Table 9.15 Computational results for problems available at www.ups/es/catala/webs/webs.html with the fourth approach.

Problem	*nxm*	*Dev(%)*	*Time(s)*
p34	30x60	2.25	16
p35	30x60	0.16	18
p36	30x60	2.93	18
p37	30x60	0.83	19
p38	30x60	0.69	18
p39	30x60	1.13	21
p40	30x60	5.80	22
p41	30x60	0.00	3
p42	30x75	1.09	20
p43	30x75	1.84	21
p44	30x75	9.78	23
p45	30x75	2.39	21
p46	30x75	1.42	23
p47	30x75	0.47	22
p48	30x75	1.31	22
p49	30x90	0.45	22
p50	30x90	2.05	22
p51	30x09	1.12	21
p52	30x90	1.21	25
p53	30x90	0.56	25
p54	30x90	1.31	24
p55	30x90	2.66	27
p56	30x90	0.57	26
p57	30x90	0.00	3
Std.Dev		2.11	5.90
Average		1.75	20.08

Table 9.16 Computational results for Barceló et al. problems with the fourth approach.

Problem	*nxm*	*Dev(%)*	*Time(s)*
p1	10x20	3.07	10
p2	10x20	1.18	9
p3	10x20	0.20	9
p4	10x20	1.45	10
p5	10x20	0.53	9
p6	10x20	1.06	9
p7	15x30	5.17	12
p8	15x30	1.08	12
p9	15x30	2.89	10
p10	15x30	0.05	12
p11	15x30	3.39	12
p12	15x30	0.96	11
p13	15x30	1.28	11
p14	15x30	4.65	13
p15	15x30	0.44	11
p16	15x30	0.29	12
p17	15x30	1.00	11
p18	20x40	0.12	14
p19	20x40	1.45	14
p20	20x40	0.41	15
p21	20x40	5.00	15
p22	20x40	3.68	12
p23	20x40	0.28	13
p24	20x49	0.48	13
p25	20x40	0.00	1
p26	20x50	6.59	16
p27	20x50	0.46	15
p28	20x50	0.47	14
p29	20x50	0.27	15
p30	20x50	5.54	17
p31	20x50	4.66	16
p32	20x50	0.59	14
p33	20x50	5.57	18
Std.Dev		2.03	3.16
Average		1.95	12.27

Bibliography

M.C. Agar and S. Salhi. Lagrangean heuristics applied to a variety of large scale capacitated plant location problems. *Journal of the Operational Research Society*, 49:1072–1084, 1998.

J. Barceló and J. Casanovas. A heuristic Lagrangean algorithm for the capacitated plant location problem. *European Journal of Operational Research*, 15: 212–226, 1984.

J. Barceló, E. Fernandez, and K. Jornsten. Computational results from a new relaxation algorithm for the capacitated plant location problem. *European Journal of Operational Research*, 15:38–45, 1991.

J.E. Beasley. Or-library: Distributing test problems by electronic mail. *Journal of the Operational Research Society*, 41:1069–1072, 1990.

J.E. Beasley. Lagrangean heuristics for location problems. *European Journal of Operational Research*, 65:383–309, 1993.

R. Berretta and P. Moscato. The number partitionating problem: An open challenge for evolutionary computation? In M. Dorigo D.Corn and F. Glover, editors, *New ideias in Optimization*, pages 261–278. Mc Graw Hill Companies ISBN 007 709506 5, 1999.

N. Christofides, A. Mingozzi, and P. Toth. *Combinatorial Optimization.* Wiley, Chichester, 1979.

M.J. Cortinhal and M.E. Captivo. Upper and lower bounds for the single source capacitated location problem. Accepted for publication in the special issue of European Journal of Operational Research dedicated to the EURO Winter Institute XVIII, 2001.

K. Darby-Dowman and H.S. Lewis. Lagrangean relaxation and the single source capacitated facility location problem. *Journal of the Operational Research Society*, 39:1035–1040, 1988.

H. Delmaire, J.A. Diaz, E. Fernandez, and M.Ortega. Comparing new heuristics for the pure integer capacitated location problem. Research report DR97/10, Universidad Politecnica de Catalunya, Spain, 1997.

O. DeMaio and C.A. Roveda. An all zero-one algorithm for a certain class of transportation problems. *Operations Research*, 19:1406–1418, 1971.

D. Erlenkotter. A dual-based procedure for the uncapacitated facility location. *Operations Research*, 26:992–1009, 1978.

V.J.M.F. Filho and R.D. Galvão. A tabu search heuristic for the concentrator location problem. *Location Science*, 6:189–209, 1998.

R.W. Floyd. Algorithm 97: Sorthest path. *Communications of the ACM*, 5:345, 1962.

M. Guignard and S. Kim. A strong Lagrangean relaxation for the capacitated plant location problems. Report 56, Department of Statistics, University of Pennsylvania, 1983.

K.S. Hindi and K. Pienkosz. Efficient solution of large scale, single source, capacitated plant location problems. *Journal of the Operational Research Society*, 50:268–274, 1999.

J.K. Holmberg, M. Ronnqvist, and D. Yuan. An exact algorithm for the capacitated facility location problems with single sourcing. *European Journal of Operational Research*, 113:544–559, 1999.

CPLEX Optimization Inc. *Using the Cplex Callable Library, version 3.0*, 1994.

J.G. Klincewicz and H. Luss. A Lagrangean relaxation heuristic for the capacitated facility location with single-source constraints. *Journal of the Operational Research Society*, 37:495–500, 1986.

A. Kuenh and M.J. Hamburger. A heuristic program for locating warwhouses. *Management Science*, 9:643–666, 1963.

R.M. Nauss. An improved algorithm for the capacitated facility location problem. *Journal of the Operational Research Society*, 29:1195–1201, 1978.

A.E. Neebe and M.R. Rao. An algorithm for the fixed-charge assignment users to sources problem. *Journal of the Operational Research Society*, 34:1107–1113, 1983.

H. Pirkul. Efficient algorithms for the capacitated concentrator problem. *Computers and Operations Research*, 14:197–208, 1987.

R. Sridharan. A Lagrangean heuristic for the capacitated plant problem with single source constraints. *European Journal of Operational Research*, 66:305–312, 1993.

Metaheuristics: Computer Decision-Making, pp. 217-236
Lino Costa and Pedro Oliveira

10 AN ELITIST GENETIC ALGORITHM FOR MULTIOBJECTIVE OPTIMIZATION

Lino Costa[1] and Pedro Oliveira[1]

[1]Departamento de Produção e Sistemas
Universidade do Minho
Campus de Gualtar
4710 Braga, Portugal
lac@dps.uminho.pt, pno@dps.uminho.pt

Abstract: Solving multiobjective engineering problems is, in general, a difficult task. In spite of the success of many approaches, elitism has emerged has an effective way of improving the performance of algorithms. In this paper, a new elitist scheme, by which it is possible to control the size of the elite population, as well as the concentration of points in the approximation to the Pareto-optimal set, is introduced. This new scheme is tested on several multiobjective problems and, it proves to lead to a good compromise between computational time and size of the elite population.
Keywords: Genetic algorithms, Multiobjective optimization, Elitism.

10.1 INTRODUCTION

In general, solving multiobjective (engineering) problems is a difficult task since the objectives conflict across a high-dimensional problem space. In these problems, where, usually, there is no single optimal solution, the interaction of multiple objectives gives rise to a set of efficient solutions, known as the Pareto-optimal solutions.

During the past decade, Genetic Algorithms (GAs)(Goldberg, 1989) were extended to tackle this class of problems, such as the work of Schaffer (1985), Fonseca and Fleming (1995), Horn et al. (1994), Srinivas and Deb (1994) and Zitzler and Thiele (1998). These multiobjective approaches explore some features of GAs to tackle these problems, in particular:

- as GAs work with populations of candidate solutions, they can, in principle, find multiple potential Pareto-optimal solutions in a single run;
- GAs, using some diversity-preserving mechanisms, can find widely different potential Pareto-optimal solutions.

In spite of the success of the application of these approaches to several multiobjective problems, some authors (Zitzler et al., 2000; Veldhuizen and Lamont, 2000) suggested that elitism can speed up the performance of GAs and also prevent the loss of good solutions found during the search process. In this paper, a GA with a new elitist approach is proposed to solve multiobjective optimization problems.

Section 10.2 introduces some elementary definitions concerning multiobjective optimization. In Section 10.3, the elitist GA is described. Next, the results of the application to several problems are presented. Finally, some conclusions and future work are addressed.

10.2 MULTIOBJECTIVE OPTIMIZATION

Usually, in multiobjective optimization there is no single optimum and, due to that, there are several solutions (referred to as Pareto-optimal solutions) which are not easily comparable. A multiobjective minimization problem with n decision variables and m objectives can be formulated as $\min y = f(x) = (f_1(x), \ldots, f_m(x))$ where $x = (x_1, ..., x_n) \in X$ and $y = (y_1, \ldots, y_m) \in Y$, without loss of generality. For a multiobjective minimization problem, a solution a is said to dominate a solution b, if and only if, $\forall i \in \{1, \ldots, m\} : f_i(a) \leq f_i(b)$ and $\exists j \in \{1, \ldots, m\} : f_j(a) < f_j(b)$. A solution a is said to be non-dominated regarding a set $X' \subseteq X$ if and only if, there is no solution in X' which dominates a. The solution a is Pareto-optimal if and only if a is non-dominated regarding X.

The set of all non-dominated solutions constitutes the Pareto-optimal set. Thus, the main feature for a multiobjective algorithm is to find a good and balanced approximation to the Pareto-optimal set.

10.3 AN ELITIST GA

One of the first evolutionary algorithms proposed for multiobjective optimization was the NSGA by Srinivas and Deb (1994). This approach differs from conventional GAs with respect to the selection operator emphasizing the non-domination of solutions. Non-domination is tested at each generation in the selection phase, thus defining an approximation to the Pareto-optimal set. Crossover and mutation operators remain as usual. On the other hand, a sharing method is used to distribute the solutions in the population over the Pareto-optimal region.

The multiobjective GA here considered is, in terms of fitness sharing, similar to the non-domination sorting algorithm proposed by Deb and Goldberg (1989). Thus, for each generation, all non-dominated solutions will constitute the first front. To these solutions a fitness value equal to the population size (P) is assigned. To maintain diversity, a sharing scheme is then applied to the fitness values of these solutions (Deb and Goldberg, 1989). However, all distances are measured in the objective space. Thus, the fitness value of each solution is divided by a quantity, called niche count, proportional to the number of solutions having a distance inferior to a parameter, the σ_{share}. The parameter σ_{share} is the maximum distance between solutions necessary to form a well distributed front. This parameter must be set carefully to obtain a well distributed set of solutions approximating the Pareto-optimal set. Thereafter, the solutions of the first front are ignored temporarily, and the rest of solutions are processed. To the second level of non-dominated solutions is assigned a fitness value equal to the lowest computed fitness value from the solutions in the first front minus 1. Next, the fitness value of each solution in the second front is divided by the respective niche count value. This process is repeated until all solutions are assigned a fitness value. This fitness assignment process will emphasize the non-domination of solutions, since the fitness values of all solutions in the first front will have a value superior to all the fitness values of solutions in the second front, and so on. Moreover, the co-existence of multiple non-dominated solutions is encouraged by the sharing scheme.

10.3.1 Elitism with a secondary population

The elitist technique is based on a separate population, the secondary population (SP) composed of all (or a part of) potential Pareto-optimal solutions found so far during the search process. In this sense, SP is completely independent of the main population and, at the end of the entire search, it contains the set of all non-dominated solutions generated.

A parameter Θ is introduced to control the elitism level. This parameter states the maximum number of non-dominated solutions of SP, the so-called elite, that will be introduced in the main population. These non-dominated solutions will effectively participate in the search process. If the number of solutions in SP (n_{SP}) is greater or equal than Θ, then Θ non-dominated solutions are randomly selected from SP to constitute the elite. Otherwise, only n_{SP} non-dominated solutions are selected from SP to constitute the elite.

In its simplest form, for all generations the new potential Pareto-optimal solutions found are stored in SP. The SP update implies the determination of Pareto optimality of all solutions stored so far, to eliminate those that became dominated. As the size of SP grows, the time to complete this operation may become significant. Therefore, to prevent the growing computation times, in general, a maximum SP size is imposed. Thus, the algorithm consists on, for all generations, to store, in SP, each potential Pareto-optimal solution x_{nd} found in the main population if:

1. all solutions in SP are different from x_{nd};
2. none of the solutions in SP dominates x_{nd}.

Afterward, all solutions in SP that became dominated are eliminated.

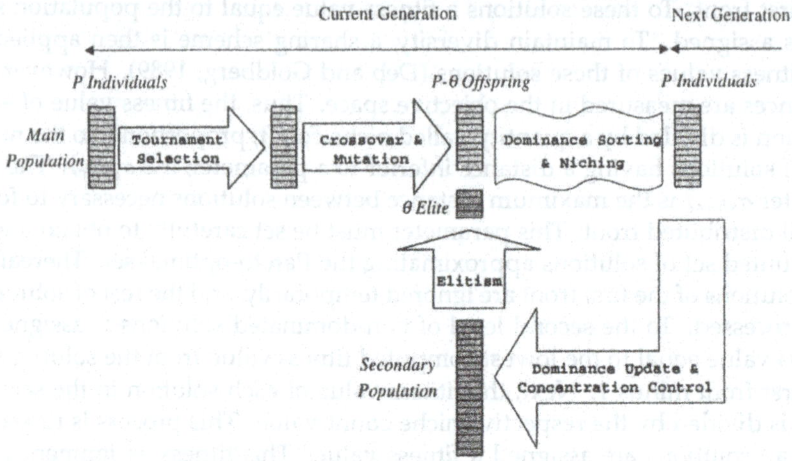

Figure 10.1 Outline of the algorithm

Figure 10.1 presents the outline of the algorithm. Tournament selection is applied to the P individuals of the main population. A new population is formed by $P - \theta$ offspring generated by crossover and mutation and θ members selected from SP. Afterward, the non-dominance sorting and niching are performed. Finally, the new non-dominated solutions found are processed according to the SP update scheme.

To illustrate this algorithm, let us consider the following multiobjective problem:

$$\min f_1(x_1, x_2) = (x_1 + x_2)^2 \tag{10.1}$$
$$\min f_2(x_1, x_2) = (x_1 x_2 - 2)^2.$$

The two variables were represented by 32 bits strings. In the experiment, a population size of 100, a tournament selection, a two point crossover with probability 0.7, a uniform mutation with probability 0.001 and a sigma share of 0.05 were considered. For illustrative purposes of the importance of a secondary population, Figure 10.2 shows the non-dominated (potential Pareto-optimal) solutions, after 200 generations, in the main and secondary populations for this problem. In this figure, it is possible to observe that a significant number of the potential Pareto-optimal solutions in the main population is dominated by solutions in the secondary population. This highlights the importance of the use of a secondary population in preventing the loss of good solutions. Furthermore, the distribution of the solutions in the main population, in the objective space, is not very uniform.

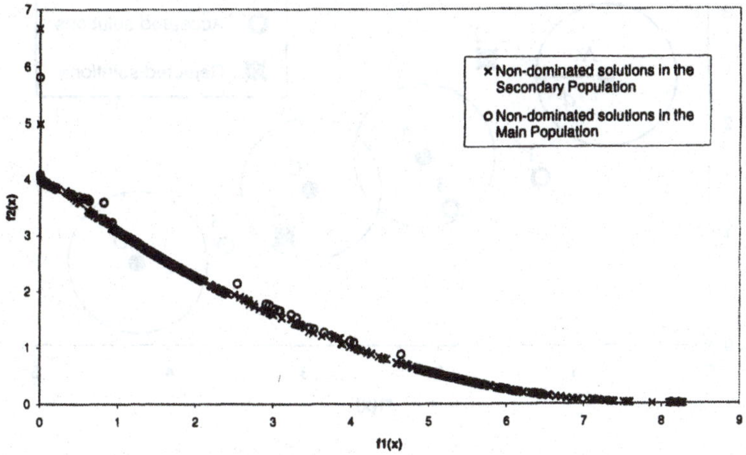

Figure 10.2 Non-dominated solutions in the secondary and main populations after 200 generations

10.3.2 Controlling the size of the secondary population

As mentioned, as the size of SP increases, the execution time and memory requirements also increase. So, it is convenient to keep relatively small sizes of SP. In this sense, the previous algorithm can be modified accordingly. A new parameter *d* is introduced, stating the minimum desirable distance in objective space between potential Pareto-optimal solutions in SP. So, the algorithm is modified by the introduction of the following step:

3. the distance from x_{nd} to any of the solutions in SP which are not dominated by x_{nd} is greater than *d* (euclidean distance measured in the objective space).

Figure 10.3 illustrates the acceptance and rejection of new solutions in SP. At each generation, non-dominated solutions in the main population are candidate to being inserted in SP. In Figure 10.3, the solutions *A* to *D* are the solutions in SP (non-dominated solutions found so far). Several possible cases are illustrated by the candidate solutions 1 to 5 to be included in SP (for simplicity, all cases will be considered individually, so, for all cases, SP will contain the same solutions *A* to *D*). Solution 1 is accepted because it is not dominated by any of the solutions present in SP (solutions *A* to *D*) and the distance between solution 1 and *A* is superior to *d*. After inserting solution 1 in SP, solution *B* is removed from SP because it is dominated by solution 1. Solution 2 is obviously rejected because it is dominated by solution *A*. Solution 3 is accepted because it dominates solution *B* (in spite of the distance between solutions being inferior to *d*). Solutions *B* and *C* are removed from SP after inserting solution 3 in SP because they are dominated by solution 3. However, solution

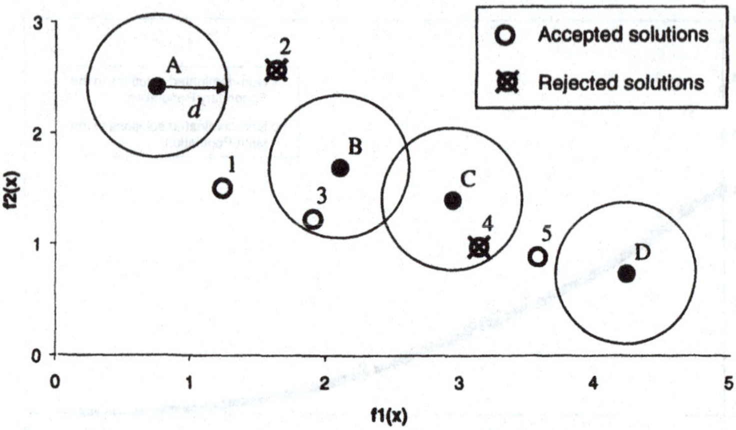

Figure 10.3 Example of acceptance or rejection of new solutions in SP

4 is not accepted in spite of being a non-dominated solution. This is due to the fact that solution C is not dominated by solution 4 but the distance between solution 4 and C is inferior to d. Finally, the non-dominated solution 5 is accepted.

Thus, parameter d defines a region of indifference with respect to non-domination. The order by which the candidate solutions are inserted in SP influences the approximation to the Pareto-optimal front. However, this influence can be minimized by successive runs with decreasing values of parameter d, thus defining approximations to the front with increasing precision.

Figure 10.4 presents the non-dominated (potential Pareto-optimal) solutions in SP after 200 generations for $d = 0.1$ and $d = 0.5$. The solutions, in this figure, are widely distributed in the objective space according to the value of d.

With respect to the computational effort required by the algorithm, it should be noted that the number of function evaluations would be the same with or without secondary population and proportional to the size of the main population (P). The use of a secondary population implies an additional effort resulting from the non-dominance testing of the archived solutions. The complexity of the algorithm without secondary population is $O(mP^2)$. Considering a limited size (A) of the secondary population, the algorithm must execute the non-dominance testing of order $O(mAP)$, which translates into an overall complexity of $O(mAP)$. This complexity depends on the relative size of the two populations, i.e., whether the secondary population size is greater or lesser than P. As an example, the computational times for problem (10.1) are 2 sec, 13 sec, 7 min 42 sec, respectively, without secondary population, with a limited secondary population of size 1500 and archiving all non-dominated solutions from every generation (with 5000 non-dominated solutions in the secondary population).

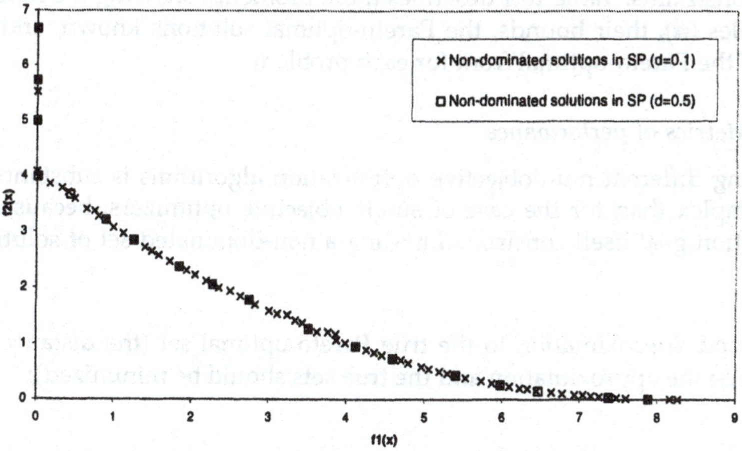

Figure 10.4 Non-dominated solutions in the secondary population for distances of 0.1 and 0.5 after 200 generations

Bearing in mind the previous examples and figures, the elitist strategy presented shows the following main advantages, when compared with the results from the main population:

- the set of all non-dominated solutions in SP constitutes a far better approach to the Pareto-optimal set;
- the solutions in SP clearly present a balanced distribution along the Pareto front;
- parameter *d* allows the definition of the concentration of points all along the Pareto front;
- the size of SP is small when compared to an approach that keeps all non-dominated solutions found during the search;
- the additional computational time required is negligible considering the quality of the results when compared with just one main population; moreover, the execution time is much lower than the time required to maintain all non-dominated points in SP.

10.4 RESULTS

10.4.1 Test problems

The multiobjective problems were chosen from a number of significant past studies in the area (Schaffer, 1985; Fonseca and Fleming, 1998; Poloni, 1997; Kursawe, 1990; Zitzler et al., 2000). All problems have two objective functions and no constraints. Table 10.1 describes these problems, showing the number of variables (n), their bounds, the Pareto-optimal solutions known, and the nature of the Pareto-optimal front for each problem.

10.4.2 Metrics of performance

Comparing different multiobjective optimization algorithms is substantially more complex than for the case of single objective optimizers, because the optimization goal itself consists on finding a non-dominated set of solutions that is:

- a good approximation to the true Pareto-optimal set (the distance between the approximation and the true sets should be minimized);

- a well distributed set in the objective space (in most cases uniform).

Several attempts can be found in the literature to express the above statements by means of quantitative metrics. The metric here considered is described by Knowles and Corne (2000) and is based on a statistical method proposed by Fonseca and Fleming (1995). The basic idea is to compare the resulting approximations to the Pareto surface from different algorithms by defining attainment surfaces through joining lines. Figure 10.5 illustrates the definition of the attainment surfaces for the resulting approximations to the Pareto front for two algorithms A and B (corresponding to the attainment surfaces A and B, respectively). The lines L1 to L4 are considered to determine which algorithm outperforms the other in different zones of the objective space. For instance, lines L1 and L2 intersect the A attainment surface in a point closer to the origin than the B attainment surface. The opposite occurs for lines L2 and L3. An attainment surface divides the objective space into two regions: one containing solutions that are non-dominated by the solutions returned by the algorithm, and another containing all the points that are dominated. A collection of sampling lines that intersect the attainments surfaces is considered to determine which algorithm outperforms or it is outperformed (for each sampling line, an algorithm outperforms another if the intersection of the sampling line with its attainment surface is closer to the origin). Then, for several executions of the algorithms, a statistical test based on the Mann-Whitney rank-sum test is applied to the previous collected data. The results of a comparison can be presented in a pair $[a, b]$, where a is the percentage of the objective space on which algorithm A was found statistically superior to B, and b gives the similar percentage for algorithm B. Thus, a is the percentage of the objective space where algorithm A is 'unbeaten' and, b is the percentage of the objective space where algorithm B is 'unbeaten'. So, typically, if $a \approx b \approx 100\%$ then the algorithms A and B have similar results. For all results presented in the paper the statistical significance is at the 5% level and 1000 sampling lines were used.

Table 10.1 Multiobjective problems

Problem	Objective functions	Solutions
SCH ($n = 1$) $x \in [-10^3, 10^3]$	$f_1(x) = x^2$ $f_2(x) = (x-2)^2$	$x \in [0, 2]$ convex
FON ($n = 3$) $x_i \in [-4, 4]$ $i = 1, ..., n$	$f_1(x) = 1 - \exp(-\sum_{i=1}^{3}(x_i - \frac{1}{\sqrt{3}})^2)$ $f_2(x) = 1 - \exp(-\sum_{i=1}^{3}(x_i + \frac{1}{\sqrt{3}})^2)$	$x_1 = x_2 = x_3$ $\in [-\frac{1}{\sqrt{3}}, \frac{1}{\sqrt{3}}]$ nonconvex
POL ($n = 2$) $x_i \in [-\pi, \pi]$ $i = 1, ..., n$	$f_1(x) = [1 + (A_1 - B_1)^2 + (A_2 - B_2)^2]$ $f_2(x) = [(x_1 + 3)^2 + (x_2 + 1)^2]$ $A_1 = 0.5 \sin 1 - 2 \cos 1 + \sin 2 - 1.5 \cos 2$ $A_2 = 1.5 \sin 1 - \cos 1 + 2 \sin 2 - 0.5 \cos 2$ $B_1 = 0.5 \sin x_1 - 2 \cos x_1 + \sin x_2 - 1.5 \cos x_2$ $B_2 = 1.5 \sin x_1 - \cos x_1 + 2 \sin x_2 - 0.5 \cos x_2$	nonconvex, disconnected
KUR ($n = 3$) $x_i \in [-5, 5]$ $i = 1, ..., n$	$f_1(x) = \sum_{i=1}^{n-1}(-10 \exp(-0.2\sqrt{x_i^2 + x_{i+1}^2}))$ $f_2(x) = \sum_{i=1}^{n}(\|x_i\|^{0.8} + 5 \sin x_i^3)$	nonconvex
ZDT1 ($n = 30$) $x_i \in [0, 1]$ $i = 1, ..., n$	$f_1(x) = x_1$ $f_2(x) = g(x)[1 - \sqrt{x_1/g(x)}]$ $g(x) = 1 + 9(\sum_{i=2}^{n} x_i)/(n-1)$	$x_1 \in [0, 1]$ $x_i = 0$ $i = 2, ..., n$ convex
ZDT2 ($n = 30$) $x_i \in [0, 1]$ $i = 1, ..., n$	$f_1(x) = x_1$ $f_2(x) = g(x)[1 - (x_1/g(x))^2]$ $g(x) = 1 + 9(\sum_{i=2}^{n} x_i)/(n-1)$	$x_1 \in [0, 1]$ $x_i = 0$ $i = 2, ..., n$ nonconvex
ZDT3 ($n = 30$) $x_i \in [0, 1]$ $i = 1, ..., n$	$f_1(x) = x_1$ $f_2(x) = g(x)[1 - \sqrt{x_1/g(x)} - \frac{x_1}{g(x)} \sin(10\pi x_1)]$ $g(x) = 1 + 9(\sum_{i=2}^{n} x_i)/(n-1)$	$x_1 \in [0, 1]$ $x_i = 0$ $i = 2, ..., n$ convex, disconnected
ZDT4 ($n = 10$) $x_1 \in [0, 1]$ $x_i \in [-5, 5]$ $i = 2, ..., n$	$f_1(x) = x_1$ $f_2(x) = g(x)[1 - \sqrt{x_1/g(x)}]$ $g(x) = 1 + 10(n-1) + \sum_{i=2}^{n}[x_i^2 - 10 \cos(4\pi x_i)]$	$x_1 \in [0, 1]$ $x_i = 0$ $i = 2, ..., n$ nonconvex
ZDT6 ($n = 10$) $x_i \in [0, 1]$ $i = 1, ..., n$	$f_1(x) = 1 - \exp(-4x_1) \sin^6(4\pi x_1)$ $f_2(x) = g(x)[1 - (f_1(x)/g(x))^2]$ $g(x) = 1 + 9[(\sum_{i=2}^{n} x_i)/(n-1)]^{0.25}$	$x_1 \in [0, 1]$ $x_i = 0$ $i = 2, ..., n$ nonconvex, nonuniformly spaced

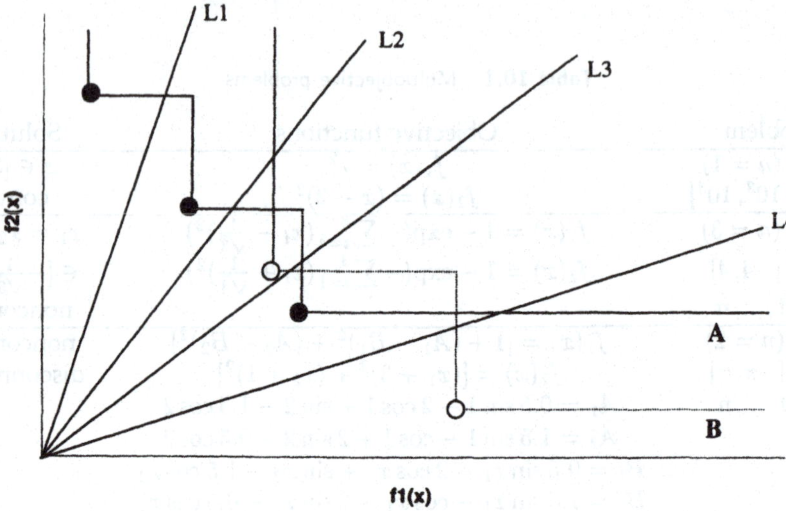

Figure 10.5 Sampling the Pareto front using attainment surfaces

10.4.3 Elitist GA parameters

The Elitist GA was applied to each problem with standard values for the parameters (no effort was made in finding the best parameter setting for each problem) as presented in Table 10.2.

For each problem, the Elitist GA was executed 20 times and, for each run, the set of all non-dominated solutions generated during the entire search was taken as the outcome of one optimization run (off-line performance). Independent of the problem, the stopping criterion was to terminate the execution after 250 generations. All problem variables were encoded using a binary string of 30 bits. A tournament selection, a two point crossover and a uniform mutation were adopted. The crossover probability was, for all problems, 0.7. The mutation probability was given by $1/b$ where b is the binary string length ($b = 30n$ bits where n is the number of variables of the problem). The value of sigma share was kept constant for all problems as indicated in Table 10.2. All distances were measured in the objective space. Several scenarios were considered for studding the influence of Θ (the elitism level) and d parameters.

10.4.4 Influence of elitism, the Θ and d parameters

Table 10.3 presents the results obtained for the ZDT1 problem with $d = 0$ for different values of Θ. The d parameter was fixed equal to 0 to guarantee that in SP all non-dominated solutions found during the search are present. This table shows that for increasing values of Θ there is a degradation of the performance of the algorithm, due to the lack of diversity in the main population.

Table 10.2 Elitist GA parameters

Parameter	Value
Number of generations	250
Population size	100
Crossover rate	0.7
Mutation rate	$1/b$
σ_{share}	0.027

Table 10.3 Influence of Θ parameter (ZDT1 problem, $d = 0$)

ZDT1	$\Theta = 5$	$\Theta = 10$	$\Theta = 15$	$\Theta = 20$	$\Theta = 25$	$\Theta = 30$
$\Theta = 0$	[90.5,100]	[46.6,100]	[96.0,100]	[71.2,100]	[94.6,100]	[71.6,100]
$\Theta = 5$	-	[87.0,100]	[100,99.4]	[98.8,100]	[100,100]	[97.4,92.1]
$\Theta = 10$	-	-	[100,40.3]	[100,100]	[100,70.3]	[100,93.8]
$\Theta = 15$	-	-	-	[84.3,100]	[100,100]	[79.6,99.9]
$\Theta = 20$	-	-	-	-	[84.3,100]	[100,100]
$\Theta = 25$	-	-	-	-	-	[96.0,97.2]

Table 10.4 Influence of parameter d (ZDT1 problem, $\Theta = 0$)

ZDT1	$d = 0.014$	$d = 0.027$	$d = 0.054$	$d = 0.108$
$d = 0$	[100,100]	[100,100]	[100,99.3]	[100,59.5]
$d = 0.014$	-	[100,100]	[100,100]	[100,65.1]
$d = 0.027$	-	-	[100,99.9]	[100,73.8]
$d = 0.054$	-	-	-	[100,89.5]

Although parameter Θ is the number of elite solutions used in the search, since the main population comprises 100 solutions, Θ can be perceived as a percentage. Consistently, the best results were obtained with 10% of elitism ($\Theta = 10$).

Parameter d allows the control of the concentration of non-dominated solutions in SP and, consequently, leads to a good distribution of these solutions in the objective space. Table 10.4 shows all pairwise comparisons for the Elitist GA applied to the ZDT1 problem. Here, different values of d were considered ($d = 0$, $d = 0.014$, $d = 0.027$, $d = 0.054$ and $d = 0.108$) and $\Theta = 0$. Figure 10.6 presents the maximum, average, and minimum number of non-dominated solutions in SP after 250 generations for the different values of d considered for the ZDT1 problem.

For values of $d = 0.014$ and $d = 0.027$, there are no significant differences in the results (this can be also considered for $d = 0.054$). On the other hand, from Figure 10.6 it is clear the significant reduction on the average number of non-dominated solutions in SP, when d increases. Thus, the importance of controlling the value of d is now evident, since it is possible to obtain a

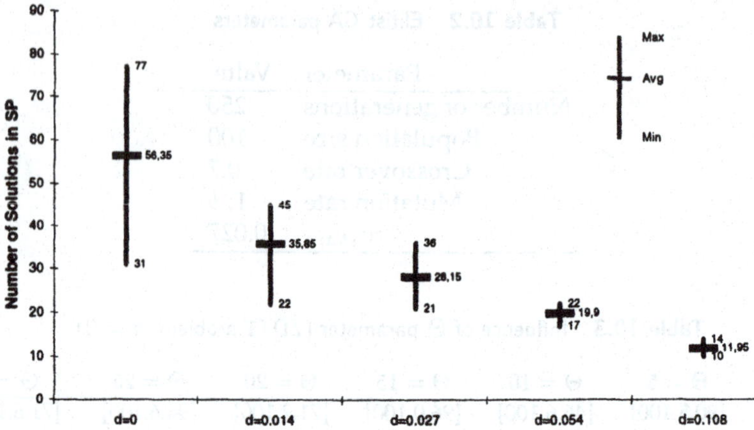

Figure 10.6 Number of non-dominated solutions in SP after 250 generations versus parameter d (ZDT1 problem)

Table 10.5 Interaction between Θ and d parameters (ZDT1 problem)

		$\Theta = 0$		$\Theta = 10$		
ZDT1		$d = 0.014$	$d = 0.027$	$d = 0$	$d = 0.014$	$d = 0.027$
$\Theta = 0$	$d = 0$	[100,100]	[100,100]	[46.6,100]	[51.7,100]	[89.0,100]
	$d = 0.014$	-	[100,100]	[42.9,100]	[49.7,100]	[86.6,100]
	$d = 0.027$	-	-	[38.0,100]	[43.6,100]	[78.2,100]
$\Theta = 10$	$d = 0$	-	-	-	[100,100]	[100,87.6]
	$d = 0.014$	-	-	-	-	[100,94.4]

convenient set of non-dominated solutions with a significant reduction on the effort to maintain SP, in terms of computation and memory storage.

However, it is important to note that the impact of parameter d on the definition of a set of non-dominated solutions in SP depends largely on the multiobjective problem being solved and, specially, on the nature of the Pareto-optimal front of the problem. Thus, the value of d must be set carefully and values close or inferior to the typical values suggested for σ_{share} produce, in general, good results (when all distances are measured in the objective space).

Table 10.5 presents the results for different combinations of parameters Θ and d for the ZDT1 problem. As before, it is clear the advantage of using convenient levels of elitism. However, when elitism is used, the effect of the d parameter is more effective than expected. This is due to the fact that the non-dominated solutions in SP are used to guide the search. Thus, in this case the value of d must be carefully chosen.

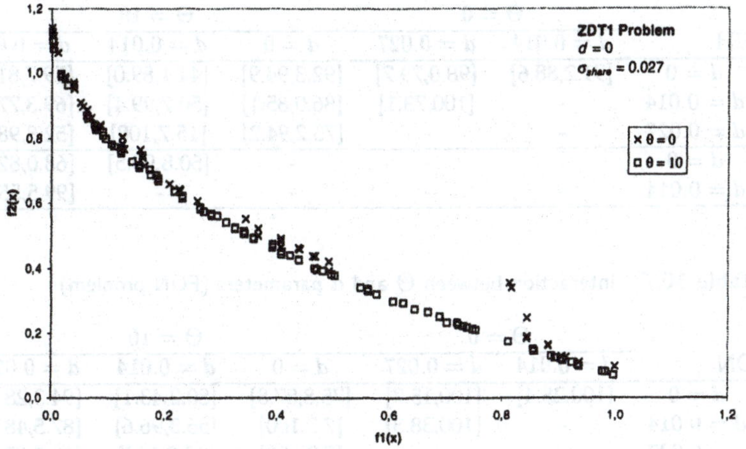

Figure 10.7 Non-dominated solutions in SP after 250 generations for ZDT1 problem with $\Theta = 0$ and $\Theta = 10$

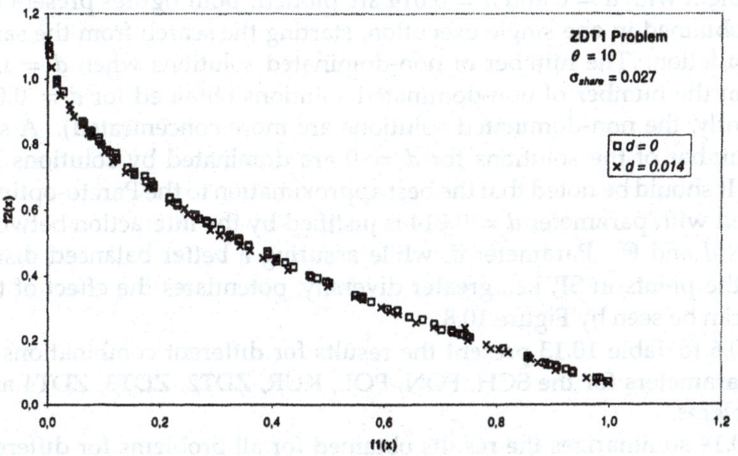

Figure 10.8 Non-dominated solutions in SP after 250 generations for ZDT1 problem with $d = 0$ and $d = 0.014$

Table 10.6 Interaction between Θ and d parameters (SCH problem)

		$\Theta = 0$			$\Theta = 10$	
SCH		$d = 0.014$	$d = 0.027$	$d = 0$	$d = 0.014$	$d = 0.027$
	$d = 0$	[91.2,88.6]	[98.0,79.7]	[92.3,94.9]	[44.1,89.0]	[59.7,81.0]
$\Theta = 0$	$d = 0.014$	-	[100,73.1]	[86.0,85.1]	[50.7,99.4]	[69.3,77.0]
	$d = 0.027$	-	-	[75.2,94.2]	[15.7,100]	[52.8,98.4]
$\Theta = 10$	$d = 0$	-	-	-	[50.8,90.5]	[68.0,82.0]
	$d = 0.014$	-	-	-	-	[99.5,55.7]

Table 10.7 Interaction between Θ and d parameters (FON problem)

		$\Theta = 0$			$\Theta = 10$	
FON		$d = 0.014$	$d = 0.027$	$d = 0$	$d = 0.014$	$d = 0.027$
	$d = 0$	[100,28.4]	[100,15.7]	[75.8,87.8]	[80.2,43.1]	[94.3,28.3]
$\Theta = 0$	$d = 0.014$	-	[100,38.3]	[7.3,100]	[56.3,96.6]	[87.5,48.7]
	$d = 0.027$	-	-	[5.3,100]	[10.8,100]	[68.5,97.1]
$\Theta = 10$	$d = 0$	-	-	-	[93.0,27.9]	[100,13.3]
	$d = 0.014$	-	-	-	-	[100,15.9]

Figure 10.7 presents the non-dominated solutions in SP after 250 generations for the ZDT1 problem with $\Theta = 0$ and $\Theta = 10$. It is clear that the best approximation to Pareto-optimal front is achieved with elitism ($\Theta = 10$). In Figure 10.8, the non-dominated solutions in SP after 250 generations for the ZDT1 problem with $d = 0$ and $d = 0.014$ are plotted. Both figures present the solutions obtained in one single execution, starting the search from the same initial population. The number of non-dominated solutions when $d = 0$ is higher than the number of non-dominated solutions obtained for $d = 0.014$ (consequently, the non-dominated solutions are more concentrated). A significant number of the solutions for $d = 0$ are dominated by solutions for $d = 0.014$. It should be noted that the best approximation to the Pareto-optimal set obtained with parameter $d = 0.014$ is justified by the interaction between parameters d and Θ. Parameter d, while assuring a better balanced distribution of the points in SP, i.e., greater diversity, potentiates the effect of the elitism as can be seen by Figure 10.8.

Table 10.6 to Table 10.13 present the results for different combinations of Θ and d parameters for the SCH, FON, POL, KUR, ZDT2, ZDT3, ZDT4 and ZDT6 problems.

Table 10.14 summarizes the results obtained for all problems for different combinations of parameters Θ and d. The values in the table, for each combination of parameters Θ and d, are the median percentage of the objective space that is "unbeaten" when compared with the remaining combinations of parameters Θ and d. Again, it is clear the advantage of using elitism. In a global manner, the best combination of parameters Θ and d was $\Theta = 10$ and $d = 0.014$.

Table 10.8 Interaction between Θ and d parameters (POL problem)

		$\Theta=0$		$\Theta=10$		
POL		$d=0.014$	$d=0.027$	$d=0$	$d=0.014$	$d=0.027$
$\Theta=0$	$d=0$	[100,94.2]	[100,90.4]	[97.7,77.3]	[75.1,95.9]	[55.8,92.2]
	$d=0.014$	-	[100,95.0]	[93.0,99.9]	[73.8,99.9]	[55.3,96.4]
	$d=0.027$	-	-	[86.2,99.9]	[66.7,100]	[51.7,99.9]
$\Theta=10$	$d=0$	-	-	-	[72.1,95.7]	[74.8,86.5]
	$d=0.014$	-	-	-	-	[77.8,83.9]

Table 10.9 Interaction between Θ and d parameters (KUR problem)

		$\Theta=0$		$\Theta=10$		
KUR		$d=0.014$	$d=0.027$	$d=0$	$d=0.014$	$d=0.027$
$\Theta=0$	$d=0$	[100,83.6]	[100,69.6]	[84.5,66.8]	[63.6,83.9]	[75.3,75.9]
	$d=0.014$	-	[100,84.0]	[75.0,74.8]	[55.3,97.9]	[72.6,84.4]
	$d=0.027$	-	-	[63.7,82.6]	[34.9,100]	[58.1,97.6]
$\Theta=10$	$d=0$	-	-	-	[54.2,93.7]	[67.3,75.2]
	$d=0.014$	-	-	-	-	[99.2,68.5]

Table 10.10 Interaction between Θ and d parameters (ZDT2 problem)

		$\Theta=0$		$\Theta=10$		
ZDT2		$d=0.014$	$d=0.027$	$d=0$	$d=0.014$	$d=0.027$
$\Theta=0$	$d=0$	[100,100]	[100,100]	[100,100]	[100,100]	[100,100]
	$d=0.014$	-	[100,100]	[100,100]	[100,100]	[99.8,100]
	$d=0.027$	-	-	[100,100]	[100,100]	[99.4,100]
$\Theta=10$	$d=0$	-	-	-	[100,100]	[100,100]
	$d=0.014$	-	-	-	-	[100,100]

Table 10.11 Interaction between Θ and d parameters (ZDT3 problem)

		$\Theta=0$		$\Theta=10$		
ZDT3		$d=0.014$	$d=0.027$	$d=0$	$d=0.014$	$d=0.027$
$\Theta=0$	$d=0$	[100,100]	[100,100]	[76.6,100]	[23.2,100]	[83.4,100]
	$d=0.014$	-	[100,100]	[74.5,100]	[20.7,100]	[80.8,100]
	$d=0.027$	-	-	[71.6,100]	[18.6,100]	[77.4,100]
$\Theta=10$	$d=0$	-	-	-	[52.3,100]	[99.5,95.2]
	$d=0.014$	-	-	-	-	[100,78.2]

Table 10.12 Interaction between Θ and d parameters (ZDT4 problem)

		Θ = 0		Θ = 10		
ZDT4		$d = 0.014$	$d = 0.027$	$d = 0$	$d = 0.014$	$d = 0.027$
	$d = 0$	[100,100]	[100,100]	[100,100]	[100,100]	[100,100]
Θ = 0	$d = 0.014$	-	[100,100]	[100,100]	[100,100]	[99.9,100]
	$d = 0.027$	-	-	[100,100]	[100,100]	[100,100]
Θ = 10	$d = 0$	-	-	-	[100,100]	[100,100]
	$d = 0.014$	-	-	-	-	[100,100]

Table 10.13 Interaction between Θ and d parameters (ZDT6 problem)

		Θ = 0		Θ = 10		
ZDT6		$d = 0.014$	$d = 0.027$	$d = 0$	$d = 0.014$	$d = 0.027$
	$d = 0$	[100,100]	[100,100]	[32.5,100]	[30.4,100]	[38.1,100]
Θ = 0	$d = 0.014$	-	[100,100]	[32.5,100]	[30.4,100]	[38.1,100]
	$d = 0.027$	-	-	[37.0,100]	[35.1,100]	[37.9,100]
Θ = 10	$d = 0$	-	-	-	[100,100]	[100,100]
	$d = 0.014$	-	-	-	-	[100,100]

Table 10.14 Global performance

	Θ = 0			Θ = 10		
Problem	$d = 0$	$d = 0.014$	$d = 0.027$	$d = 0$	$d = 0.014$	$d = 0.027$
SCH	91.2	85.1	94.2	85.1	99.4	81.0
FON	94.3	56.3	15.7	87.8	96.6	28.3
POL	97.7	93.0	86.2	77.3	95.9	92.2
KUR	84.5	75.0	63.7	67.3	97.9	75.9
ZDT1	89.0	86.6	78.2	100.0	100.0	100.0
ZDT2	100.0	100.0	100.0	100.0	100.0	100.0
ZDT3	83.4	80.8	77.4	100.0	100.0	100.0
ZDT4	100.0	100.0	100.0	100.0	100.0	100.0
ZDT6	38.1	38.1	37.9	100.0	100.0	100.0

10.5 CONCLUSIONS AND FUTURE WORK

In real multiobjective problems, the Pareto-optimal front is not usually known. Thus, any algorithm designed to search for this front must assure a uniform distribution of the solutions all along the Pareto front, but also, to prevent premature convergence.

In this work, an elitist genetic algorithm for multiobjective optimization was presented. This new approach, tested on several problems, produces a well balanced distribution of the solutions on the Pareto front. Furthermore, the level of elitism and the concentration of points can be controlled by two parameters and, for reasonable values, the increase on computational time is negligible.

Elitism proves to be useful in the approximation to the Pareto-optimal set. However, the control of elitism was yet not deeply well studied, and this work shows how the level of elitism can influence the performance of the algorithms. Furthermore, a compromise between a large secondary population and a well balanced distribution can be reached through a parameter that controls the density of points in the approximation set to the Pareto-optimal front. Future work will extend the study of these parameters to other multiobjective algorithms.

Bibliography

K. Deb and D. Goldberg. An investigation of niche and species formation in genetic function optimization. In *Proceedings of the Third International Conference on Genetic Algorithms*, pages 42–50, USA, 1989.

C.M. Fonseca and P.J. Fleming. On the performance assessment and comparison of stochastic multiobjective optimizers. In H.-M. Voigt, W. Ebeling, I. Rechenberg, and H.-P. Schwefel, editors, *Parallel Problem Solving from Nature IV*, pages 584–593. Springer, 1995.

C.M. Fonseca and P.J. Fleming. Multiobjective optimization and multiple constraint handling with evolutionary algorithms-part ii: Application example. *IEEE Transactions on Systems, Man, and Cybernetics: Part A: Systems and Humans*, 28(1):38–47, 1998.

D.E. Goldberg. *Genetic Algorithms in Search, Optimization, and Machine Learning*. Addison-Wesley, Reading, Massachusetts, 1989.

J. Horn, N. Nafploitis, and D. Goldberg. A niched pareto genetic algorithm for multi-objective optimization. In *Proceedings of the First IEEE Conference on Evolutionary Computation*, pages 82–87, 1994.

J.D. Knowles and D.W. Corne. Approximating the nondominated front using the pareto archived evolution strategy. *Evolutionary Computation*, 8(2):149–172, 2000.

F. Kursawe. A variant of evolution strategies for vector optimization. In H.-P. Schwefel and R. Manner, editors, *Parallel Problem Solving from Nature*, pages 193–197. Springer, 1990.

C. Poloni. *Genetic Algorithms in engineering and computer science*, chapter Hybrid GA for multiobjective aerodynamic shape optimization, pages 397–414. G. Winter, J. Periaux, M. Galan, and P. Puesta, Ed. Hillsdale, 1997.

J.D. Schaffer. Multiple objective optimization with vector evaluated genetic algorithms. In J.J. Grefensttete, editor, *Proceedings of the First International Conference on Genetic Algorithms*, pages 93–100. Ed. Hillsdale, 1985.

N. Srinivas and K. Deb. Multi-objective function optimization using nondominated sorting genetic algorithms. *Evolutionary Computation*, 2(3):221–248, 1994.

D.A. Van Veldhuizen and G.B. Lamont. Multiobjective evolutionary algorithms: Analysing the state-of-the-art. *Evolutionary Computation*, 8(2):125–147, 2000.

E. Zitzler, K. Deb, and L. Thiele. Comparison of multiobjective evolutionary algorithms: Empirical results. *Evolutionary Computation*, 8:173–195, 2000.

E. Zitzler and L. Thiele. Multiobjective optimization using evolutionary algorithms a comparative case study. In *Parallel Problem Solving from Nature V*, pages 292–301, 1998.

Metaheuristics: Computer Decision-Making, pp. 237-256
Raphaël Dorne and Christos Voudouris

11 HSF: THE IOPT'S FRAMEWORK TO EASILY DESIGN METAHEURISTIC METHODS

Raphaël Dorne[1] and Christos Voudouris[1]

[1] Intelligent Complex Systems Research Group
BTexact Technologies
Adastral Park, PP12/MLB1, Martlesham Heath
Ipswich, IP5 3RE
Suffolk, United Kingdom
raphael.dorne@bt.com, chris.voudouris@bt.com

Abstract: The *Heuristic Search Framework* (*HSF*) is a Java object-oriented framework allowing to easily implement single solution algorithms such as *Local Search*, population-based algorithms such as *Genetic Algorithms*, and hybrid methods being a combination of the two. The main idea in *HSF* is to break down any of these heuristic algorithms into a plurality of constituent parts. Thereafter, a user can use this library of parts to build existing or new algorithms. The main motivation behind *HSF* is to provide a "well-designed" framework dedicated to heuristic methods in order to offer representation of existing methods and to retain flexibility to build new ones. In addition, the use of the infra-structure of the framework avoid the need to re-implement parts that have already been incorporated in *HSF* and reduces the code necessary to extend existing components.
Keywords: Heuristic search framework, Local search, Evolutionary algorithms, Hybrid algorithms, iOpt.

11.1 INTRODUCTION

For some years now, metaheuristic methods have demonstrated their ability to tackle large-scale optimization problems. Up to now, several software frameworks that allow to build more easily this family of methods, have been implemented. Some of them are either dedicated to Local Search such as *EasyLocal++* (Gaspero and Schaerf, 2000), *Localizer* (Michel and Hentenryck, 1999),

LocalSearch framework (Andreatta et al., 1998), *Templar* (Jones et al., 1998), *HotFrame* (Fink and Voß, 2002) or to Evolutionary Computation such as *EOS* (Bonsma et al., 2000), *EASEA* (Collet et al., 2000). These usually tend to provide templates with the user having to define, for each problem addressed, the move operators and/or evolutionary operators with the need to construct tedious and hard-coded move evaluation mechanisms. Furthermore, sometimes the user has to implement most of the structure of the algorithm and to code the flow of execution going through such an algorithm at runtime. As a result, most of these frameworks are limited in the choice of techniques provided. Therefore, it appears difficult using existing frameworks to quickly model efficient metaheuristics such as Hybrid Methods, combining evolutionary algorithms with local search methods which require more generalization and flexibility.

The main motivation behind the *Heuristic Search Framework* is to develop a Java object-oriented framework for building and combining population and single solution algorithms. Such a framework should provide infra-structure and design well-adapted to heuristic methods offering generalization, flexibility and efficiency in order to tackle academic problems and real applications. In addition, using a "well-designed" framework makes it much easier for the developer to add extensions, to factor out common functionality, to enable interoperability, and to improve software maintenance and reliability (Taligent Inc., 1993). All these points are particularly important in heuristic search methods, which tend to reuse the same set of procedures and concepts such as initial solution generation, iterative improvement, tabu restriction, etc.

11.2 HEURISTIC SEARCH FRAMEWORK

The Heuristic Search Framework (*HSF*) was created as a generic framework for the family of optimization techniques known as Heuristic Search. The motivation here is to provide representation of existing methods, retain flexibility to build new ones and use generalization to avoid re-implementing concepts that have already been incorporated.

In *HSF*, we make use of the fact that many Heuristic Search algorithms can be broken down into a plurality of constituent parts, and that at least some of these parts are common between algorithms. For example, algorithm *A* could be considered as comprising parts *a*, *b*, and *c* while algorithm *B* could be considered as comprising *a*, *c*, and *d*. While with most existing systems, algorithms *A* and *B* would be considered as independent algorithms; we identify constituent parts, then look to see which of these parts are already available within the framework. Such a constituent part is called *Search Component* (*SC*) and these components can be considered to comprise a "library" of parts. *HSF* offers to the user the capability to build a complete and complex algorithm by connecting several parts together.

In addition, a user can add/create new parts to specialize an algorithm. For example, if, with algorithm *B*, parts *a*, *b*, and *c* already exist in the framework, but part *d* does not, a user can create part *d*. This part is then added to the framework and becomes available for future use by other algorithms.

Three main concepts are the basis of *HSF*: ***HeuristicSearch, HeuristicSolution*** and ***HeuristicProblem***.

- ***HeuristicSearch*** is the concept used to define a search algorithm.
- ***HeuristicSolution*** is the solution representation of an optimization problem manipulated inside *HSF*. At the present moment, a *Vector of Variables* and a *Set of Sequences* are readily available.
- ***HeuristicProblem*** is the interface between an optimization problem and *HSF*.

In *HSF*, a valid ***HeuristicSearch*** instance is a tree of search components. Figure 11.1 shows a class hierarchy of the main search components necessary to build the core of a local search method.

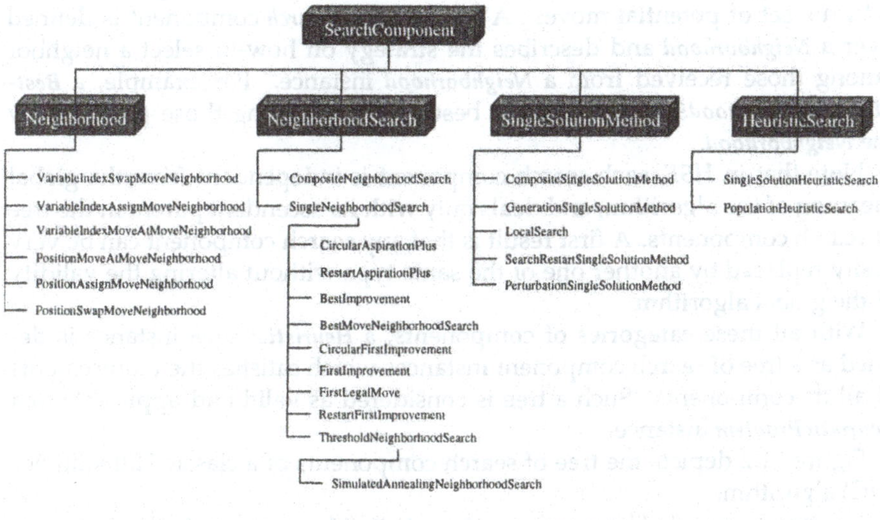

Figure 11.1 Search Components for single solution heuristic methods in *HSF*

To ensure a framework-oriented programming, *HSF* provides the flow of control between the search components, therefore the developer's specific components will wait for a call from *HSF*. This is a significant benefit as the developer does not have to be concerned with the details of the implementation and can focus his attention on the specific parts of his component.

For each search component, we define a role and requirement that must be satisfied to guarantee the correct behavior of the component. For example, the role of a ***HeuristicSearch*** component is to apply a heuristic algorithm to solve a ***HeuristicProblem***. Its requirement is only to take a ***HeuristicProblem*** as input. According to such a generic description, each main component can be extended to specialize the behavior of its parent component. Thus the role of a

SingleSolutionHeuristicSearch instance is to apply a single solution algorithm to solve a ***HeuristicProblem***. It requires a ***SingleSolutionMethod*** instance and returns an (optimized) ***HeuristicSolution***.

In the same manner, the concept of a ***SingleSolutionMethod*** is a search component taking as input a ***HeuristicSolution***, modifies it using its own process and returns as output a new ***HeuristicSolution***. Under a ***SingleSolutionMethod***, we find ***GenerationSingleSolutionMethod*** to generate initial solution, ***CompositeSingleSolutionMethod*** to execute several single solution methods in sequence, ***LocalSearch*** to run a local search iterative process, ***SearchRestartSingleSolutionMethod*** to restart a search and ***PerturbationSingleSolutionMethod*** to perturb a solution.

In particular, a ***LocalSearch*** component (within the ***SingleSolutionMethod***) manipulates a ***NeighborhoodSearch*** to realize the iterative improvement process. In its turn, a ***NeighborhoodSearch*** is using a ***Neighborhood*** component. In *HSF*, a ***Neighborhood*** is almost equivalent to the classic definition of a *Neighborhood* concept for a Local Search. Given a ***HeuristicSolution***, a *HSF* ***Neighborhood*** will return a set of potential moves. A ***NeighborhoodSearch*** component is defined over a ***Neighborhood*** and describes the strategy on how-to select a neighbor among those received from a ***Neighborhood*** instance. For example, a ***BestMoveNeighborhoodSearch*** selects the best neighbor among those provided by the ***Neighborhood***.

Note that in *HSF*, each search component is independent from the global meaning of the algorithm, and deals only with its ascendant parent in the tree of search components. A first result is that any search component can be very easily replaced by another one of the same type without altering the validity of the global algorithm.

With all these categories of components, a ***HeuristicSearch*** instance is defined as a tree of search component instances which satisfies the requirements of all its components. Such a tree is considered as valid and applicable to a ***HeuristicProblem*** instance.

Figure 11.2 depicts the tree of search components of a classic Hill-Climber (HC) algorithm.

The tree is executed by passing a ***HeuristicProblem*** instance to the root component. Then the execution process follows the specific communication protocol defined within each search component. Most of the time, such a process goes through the search components from the left to the right following a depth-first style traversal algorithm.

In Figure 11.2, we can see the flow of data exchanged between components during the execution phase of an algorithmic tree. Different types of data are exchanged: ***HeuristicProblem, HeuristicSolution*** or ***Move*** instances. Note that certain components will iterate on a subset or all of their children. Iterative components (*e.g.* ***NeighborhoodSearch, LocalSearch***) stop when at least one of their local stopping conditions or one stopping condition of their ascendant components is satisfied. For example, the ***LocalSearch*** component of Figure 11.2 may stop when its limit on the number of iterations is reached or

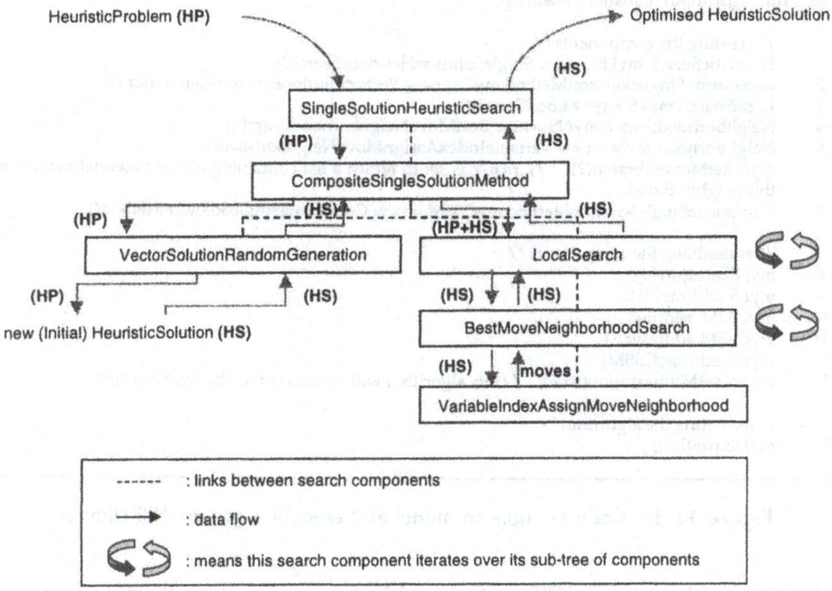

Figure 11.2 Tree of Search Components for an Hill-Climber

when the time limit of one of its ascendant components is exceeded. Finally, the global search process finishes when one of the stopping conditions of the root component is satisfied, and then an optimized solution is returned.

The following algorithm in pseudo-code is equivalent to the tree of Figure 11.2 (related components are given in parenthesis).

1. generate initial solution s (*VectorSolutionRandomGeneration*)
2. generate a set of potential moves $M(s)$ using the *Neighborhood* instance (*VariableIndexAssignMoveNeighborhood*)
3. select the best move m in $M(s)$ (all moves are evaluated here) (*BestMoveNeighborhoodSearch*)
4. perform move m on s (*LocalSearch*)
5. while stopping condition not satisfied, go back to step 2 (*LocalSearch*)

Figure 11.3 shows the corresponding code to create and execute the generic Hill-climber of Figure 11.2 on a given heuristic problem. This code is complete and confirms the simplicity of use of *HSF*.

HSF being a sound and complete framework, the user does not have to specify anything in the code of Figure 11.3 to express how the components should interact together. All these interactions have been already coded in the infra-structure of *HSF* within the role/behavior of each category of components.

```
void runAlgorithm(HeuristicProblem hp)
{
    /* creating the components */
    HeuristicSearch myHS = new SingleSolutionHeuristicSearch();
    GenerationSingleSolutionMethod myG = new VectorSolutionRandomGeneration();
    LocalSearch myLS = new LocalSearch();
    NeighborhoodSearch myNS = new BestMoveNeighborhoodSearch();
    Neighborhood myN = new VariableIndexAssignMoveNeighborhood();
    myN.setMovesVersion(2);   // myN is set to return a list containing all the potential moves of
    this neighborhood
    CompositeSingleSolutionMethod myCSSM = new CompositeSingleSolutionMethod();

    /* assembling the components */
    myNS.add(myN);
    myLS.add(myNS);
    myCSSM.add(myG);
    myCSSM.add(myLS);
    myHS.add(myCSSM);
    myHS.setMinimization(true);   //this algorithm will minimize the objective function

    /* executing the algorithm */
    myHS.run(hp);
}
```

Figure 11.3 Code example to model and execute a generic Hill-climber

Another objective of *HSF* is to be able to represent well-known existing methods. Next paragraphs present how methods such as Simulated Annealing, Tabu Search and Hybrid Methods can be modeled using the framework.

11.2.1 Simulated Annealing

A Simulated Annealing (SA) algorithm can be seen as a classic HC with a specific neighborhood search mechanism, where a neighbor is selected if it verifies a specific predicate involving the current solution s and the neighbor under consideration s'. The following search components have been added to the framework to model the more general family of Threshold methods which includes Simulated Annealing:

- ***ThresholdNeighborhoodSearch*** extends ***NeighborhoodSearch***. It describes a specific Neighborhood strategy, which goes through all potential neighbors and selects the first one satisfying the predicate ***isAccepted(Move move)***.

- ***SimulatedAnnealingNeighborhoodSearch***, which extends ***ThresholdNeighborhoodSearch***, overrides only the predicate ***isAccepted(Move move)*** to redefine it as the classic Simulated Annealing acceptance criterion using Temperature, Cooling program, etc.

With the addition of only two new components, a Threshold method has been included in *HSF*. As a result all existing ***Neighborhood*** instances are automatically available to any Threshold strategy. Note that the classic Simulated Annealing iterative process combines a ***SimulatedAnnealingNeighborhoodSearch*** component with a ***Neighborhood*** instance providing moves at random (*cf.* Figure 11.4). These additions give an example of the simple process of extending *HSF* to cover a broader range of heuristic search methods.

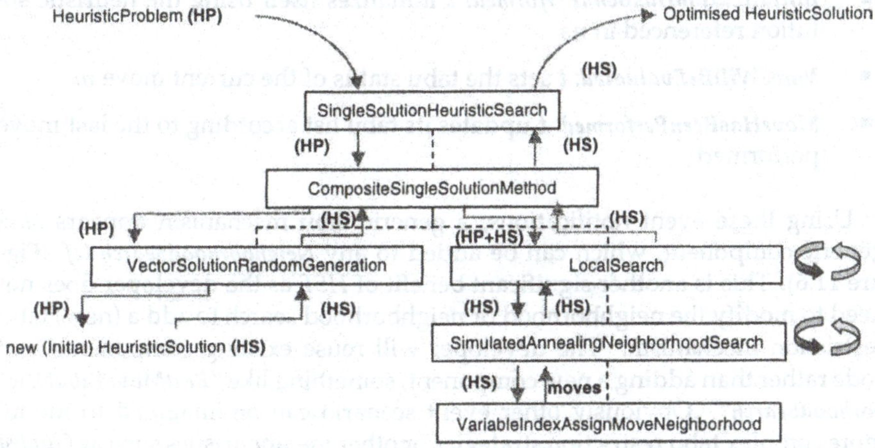

Figure 11.4 Tree of Search Components for a Simulated Annealing

With regard to the implementation, only two lines of the code of Figure 11.3 need to be replaced to obtain a ready-to-use Simulated Annealing:

```
line 4.   NeighborhoodSearch myNS = new SimulatedAnnealingNeighborhoodSearch();
line 6.   myN.setMovesVersion(0);          // myN is set to return moves of this neighborhood at
          random
```

Figure 11.5 Code to be replaced for modeling a generic Simulated Annealing

11.2.2 *Tabu search*

Another feature included in *HSF* is event notification and handling facilities. A ***SearchComponent*** can send events to, or listen for events from other components of *HSF* and then interact with them. Examples of event notification can be illustrated by considering a Tabu Search (TS) method. To implement a tabu mechanism, we need to listen for events sent by a ***NeighborhoodSearch*** (*NS*) instance:

- ***InitializationHasBeenPerformed***: sent when a *NS* has been initialized,
- ***MoveWillBeEvaluated***: sent when a move is going to be evaluated in a *NS* instance,
- ***MoveHasBeenPerformed***: sent when a move has been performed in a *NS* instance.

TS has a very specific behavior which is to set "tabu" restrictions to prevent the last configurations visited for a certain number of iterations. In *HSF*, a

Tabu mechanism is a search component that can be added as a listener to any *NS* instance. A tabu instance *t* reacts as events are sent from a *NS* instance *ns*:

- ***InitializationHasBennPerformed***: *t* initializes itself using the heuristic solution referenced in *ns*.
- ***MoveWillBeEvaluated***: *t* sets the tabu status of the current move *m*.
- ***MoveHasBeenPerformed***: *t* updates its tabu list according to the last move performed.

Using these event notifications, a generic tabu mechanism appears as a generic component, which can be added to any ***NeighborhoodSearch*** (*cf.* Figure 11.6). This is another significant benefit of *HSF* as the developer does not need to modify the neighborhood or neighborhood search to add a (new) tabu restriction mechanism. The developer will reuse existing components and code rather than adding a new component, something like "***BestMoveTabuNeighborhoodSearch***". Obviously other event scenario can be imagined to model more complex tabu restriction strategies or other metaheuristics such as *Guided Local Search* or *Perturbation Methods*. Furthermore mechanisms such as Aspiration Criterion or dynamic modifications of the objective function (used in Guided Local Search) are implemented in *HSF* using similar event scenario.

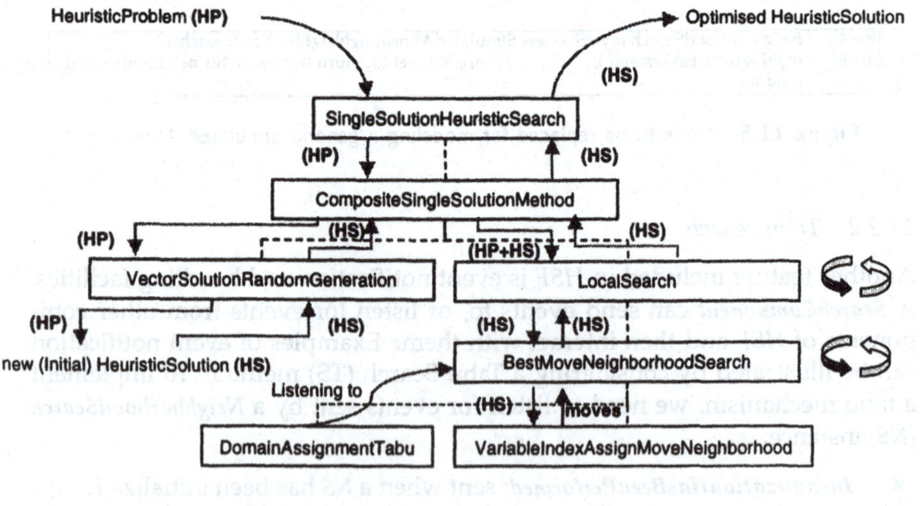

Figure 11.6 Tree of Search Components for a Tabu Search

With regard to the code, a generic tabu mechanism is available in *HSF* which sets tabu the last assignments made to decision variables. To change the initial hill-climbing algorithm into a tabu search, only one line is added for attaching a tabu mechanism on the neighborhood search component:

```
after line 4.    myNS.add(new DomainAssignmentTabu());
```

Figure 11.7 Code to be added in Figure 11.3 for modeling a generic Tabu Search

This line specifies only which ***NeighborhoodSearch*** instance, the tabu component is going to listen to. Again, the user does not have to implement the interaction between these two components except if he wants to implement a new event scenario not already included in *HSF*.

11.2.3 Hybrid Methods

For a population heuristic search, the goal is similarly to identify the pertinent principles governing the search algorithms, this time specifically related to population-based methods. A ***PopulationMethod*** is a similar concept to ***SingleSolutionMethod*** but taking as input a ***HeuristicPopulation*** and returning as output another ***HeuristicPopulation*** modified by operators such as Crossover, Mutation, etc. Figure 11.8 shows the main components for population-based methods.

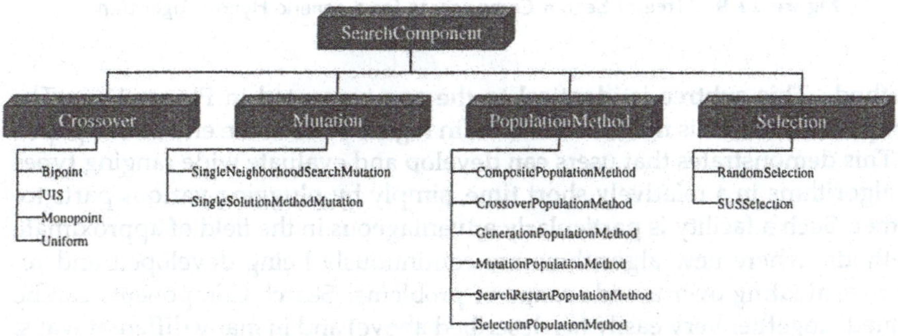

Figure 11.8 Search Components for population-based methods

In particular, a mutation is generically described as an operator that locally modifies a ***HeuristicSolution*** to generate a new ***HeuristicSolution***. The flexibility of *HSF* allows the *Search Components* defined for a ***SingleSolutionHeuristicSearch*** to be reused here. A ***NeighborhoodSearch***, or a ***SingleSolutionMethod***, can be plugged as a mutation into a ***PopulationHeuristicSearch***. Thus confirming the reusability and flexibility of *HSF* as any novel ***NeighborhoodSearch*** is automatically available inside any mutation operator. Figure 11.9 shows an example where a Tabu Search is combined with a classic *Genetic Algorithm* to obtain a hybrid algorithm.

As *HSF* breaks down an algorithm into parts, to create the hybrid algorithm of Figure 11.9, we only need to plug a subtree of components representing the iterative process of a Tabu Search as the mutation operator of the overall

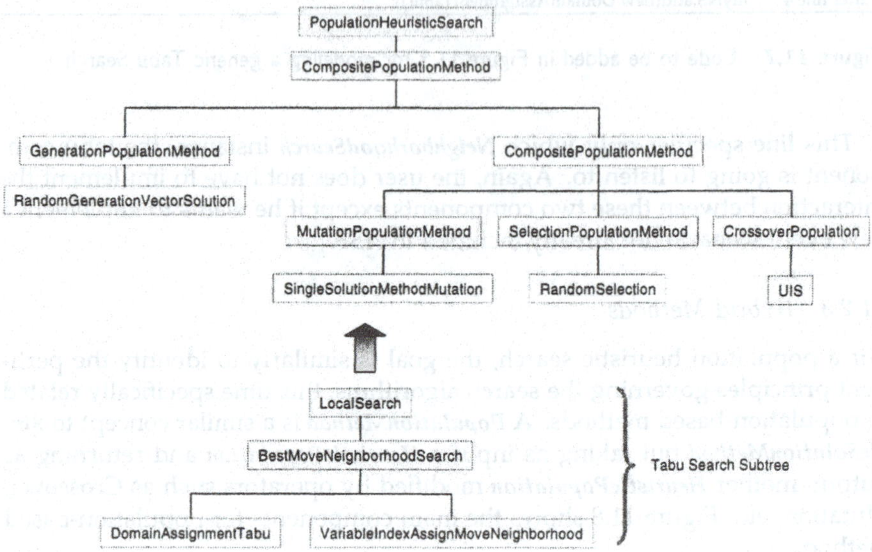

Figure 11.9 Tree of Search Components for a generic Hybrid Algorithm

method. This subtree is identical to the one presented in Figure 11.6. The complete code of this method is shown in Figure 11.11 at the end of this paper.

This demonstrates that users can develop and evaluate wide ranging types of algorithms in a relatively short time, simply by plugging various parts together. Such a facility is particularly advantageous in the field of approximate methods, where new algorithms are continuously being developed and require evaluating over a wide range of problems. Search Components can be "glued" together very easily (as described above) and in many different ways, to create many different types of algorithms. These algorithms can then be applied to a problem, so that the performance of the different algorithms can be evaluated very easily. In addition, using the same framework to evaluate the performances of different heuristic searches allows a fair comparison more independent from the implementation and the machine used.

11.3 APPLYING HSF TO CONSTRAINED OPTIMIZATION PROBLEMS

The *HSF* framework is part of a complete project called *iOpt* for *Intelligent Optimization Toolkit* created by BTexact Technologies. Typically, *HSF* interacts with both an Invariant library (IL) and a framework for Problem Modeling (*PMF*).

The IL is a set of functions providing a comprehensive mathematical algebra allowing definition of relations between variables, and performing incremental updates when the value of a variable is changed. This library is suitable for heuristic search methods, where we want to quickly evaluate the results of a change to one or more decision variables on the objective function.

The *PMF* is a high level framework based on invariant library where the user can define a Constrained Optimization Problem (*COP*) with decision variables, constraints and objective function. More precisely, *PMF* offers the flexibility to use Boolean, Integer, Real, String and Object basic data types and also to define a domain of values for each of them. For more details with regard to integrating *HSF* and *PMF*, the reader can refer to Voudouris and Dorne (2002).

To illustrate the combination of *PMF* with *HSF*, we consider a classic Graph Coloring problem. Given a graph $G(V, E)$ to color in k colors, we obtain the following problem description using *COP* format (X, D, C, F) where:

- X = set of integer decision variables representing the $|V|$ nodes of the Graph G.
- D = domain $[1..k]$ for each variable
- $C = \{c_l | c_l = \{x_i \neq x_j\}\}$ for each edge between node i and node j contained in E
- F = minimize the number of colors of a complete[1] coloring of the graph

HSF can easily provide search trees for HC, SA, TS and Hybrid Genetic Algorithms methods (*cf.* Figures 11.2, 11.4, 11.6 and 11.9). These default methods do not need to use any specific information from the Graph Coloring problem. They only assume two things:

Firstly, a *Vector Solution* is used to represent a potential solution of the problem, (valid as in a *COP* a set of variables can be represented by a vector of variables).

Secondly, they assume that a generic neighborhood modifying the values assigned to variables is appropriate to tackle this family of problems (valid as in a COP, the objective is to find the best assignment of values to variables).

Thereafter *HSF* is capable, through its default ***HeuristicProblem*** interface, to access a problem instance from *PMF*, get the type and the domain of the decision variables, and read the objective value for any potential solution. Assuming a *Vector Solution* representation, *HSF* can build an initial ***HeuristicSolution*** instance and then apply ***NeighborhoodSearch***, ***Mutation*** and ***Crossover***. With regard to the Tabu Search, we assume a tabu mechanism, which sets "tabu" the last assignments made.

Therefore, this set of default methods is readily available. *HSF* combined with *PMF* offers a complete and integrated solving system based on Heuristic Search methods for any COP. In similar ways, *HSF* can be combined with *PMF* to address more specific classes of problem such as Machine Scheduling, Vehicle Routing, etc. (*cf.* Voudouris et al. (2001)).

---
[1] A complete coloring is a coloring of all the nodes satisfying all the difference constraints.

11.4 EXPERIMENTS ON GRAPH COLORING

In this section, we have carried out experiments[2] on a subset of well-known random graphs provided by Johnson et al. (1991). To develop efficient methods to tackle this problem, we start from the default methods and specialize these algorithms by plugging some specific components into them. First of all, to obtain good results on the Graph Coloring, we would better consider this problem as a series of *k-coloring* satisfaction problems[3], define a neighborhood (***GraphColoringNeighborhood***) considering only nodes having conflicts[4] and use a more efficient coloring-specific crossover named *UIS* (Dorne and Hao, 1998). With *HSF*, we only need to create these components, and to include them into the tree of search components. Figure 11.10 shows a new algorithm obtained from the Tabu Search of Figure 11.6.

In *HSF*, we followed a "framework-oriented" approach where a special effort has been made to get a sound and complete class hierarchy covering most of the main principles found in the literature on Heuristic Search methods. By providing such an infra-structure, *HSF* dramatically reduces the amount of code that the developer has to implement, test and debug. The developer writes only the code that extends or specializes existing components to suit the specific needs of his algorithm. As we did for the ***GraphColoringNeighborhood*** a special neighborhood for the Graph Coloring problem extending a ***VariableIndexAssignMoveNeighborhood***)).

Note that by adding new components, they are automatically at the disposal of any other algorithm. For example, the ***GraphColoringNeighborhood*** component is now available for any other local search algorithm to use in similar problems. This confirms again the high reusability of the *HSF* framework. The other local search algorithms experimented in this section, SA and HC, are obtained by adding the same set of components to their respective trees of search components. In addition, transforming a Graph Coloring problem into a sequence of *k*-coloring problems in *HSF* is a very easy task. This requires only one specific component ***ReduceRestart***, reducing the number of available colors when a complete *k*-coloring is found, and to link it into the tree of components as a new step (*i.e.* after the Local Search process, we include the restart process). This is done using a ***CompositeSingleSolutionMethod*** instance which has the capability to iterate on a list of Single Solution Methods such as the ***LocalSearch*** and ***ReduceRestart*** instances. Again, this validates the high flexibility of *HSF*. Other frameworks using templates are not able to model such a transformation as they cannot dynamically connect components or change the algorithm configuration at runtime. To add this new type of algorithm with a template-based framework, the user should have to implement a new

[2]More experiments on applying *HSF* on instances of the Vehicle Routing Problem are reported in Voudouris and Dorne (2002).

[3]A *k*-coloring satisfaction problem comes down to find a complete coloring with *k*-colors.

[4]A node in conflict is a node having a difference constraint not satisfied with at least one of its neighbor nodes.

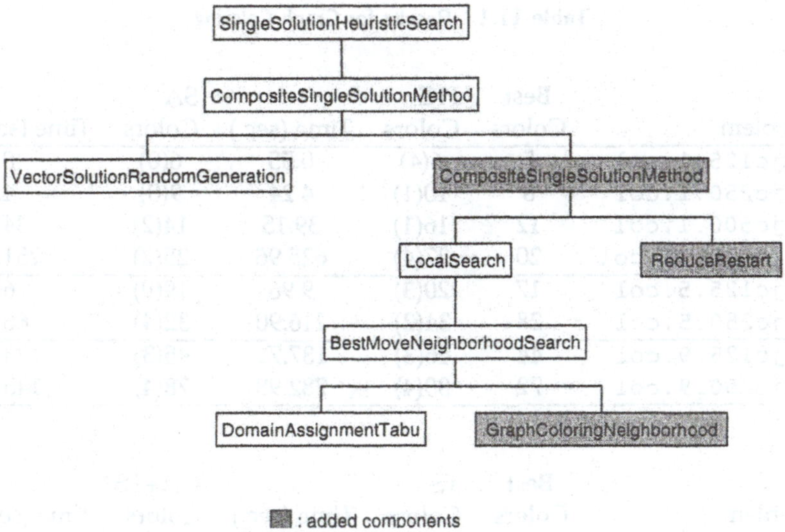

Figure 11.10 Tabu Search specialized to Graph Coloring

template accepting more parameters to deal with the reduce restart part of the algorithm and then describe in the code how these components work together.

With regard to our evolutionary algorithm GA+TS[5], we plugged the Tabu Search iterative process of Figure 11.6 as a mutation operator, replaced ***Uniform*** crossover by ***UIS*** crossover and *Stochastic Universal Sampling (SUS)* selection (Baker, 1987) (***SUSSelection***) by a ***RandomSelection*** instance. This latter selection ranks only the population in a random order to avoid crossing same pairs of individuals. As our population size is small (only 10 individuals), we cannot use a more intensive selection such as *SUS*, that will lead to a premature convergence of the population.

The machine used for these tests is a PC Dell Precision 330 with a Pentium IV 1.7GHz with 1Gb memory running Windows2000. The results in the Table 11.1 show the best known coloring and for each method, the best number of colors found, the percentage of success over 5 runs, the time required to get such a coloring with a time limit of 20 minutes for local search algorithms and one hour for the hybrid algorithm per run[6]. For example, on the first row, the two first cells report that Hill-Climber has found one time over 5 runs (4 failed attempts) a coloring of 6 colors for `dsjc125.1.col` instance, and it takes on average 0.55 seconds to do so.

[5]Corresponding to a Genetic Algorithms using a Tabu Search algorithm as mutation operator.
[6]The time limit is expressed here in real time.

Table 11.1 Results for Graph Coloring

Problem	Best Colors	HC Colors	HC Time (sec.)	SA Colors	SA Time (sec.)
dsjc125.1.col	5	6(4)	0.55	6(0)	0.32
dsjc250.1.col	8	10(1)	4.24	9(0)	4.39
dsjc500.1.col	12	16(1)	39.15	14(2)	34.20
dsjc1000.1.col	20	27(4)	625.96	25(2)	251.49
dsjc125.5.col	17	20(3)	9.96	19(0)	6.90
dsjc250.5.col	28	34(2)	216.90	32(4)	66.84
dsjc125.9.col	44	46(3)	187.71	45(3)	23.25
dsjc250.9.col	72	80(4)	782.98	78(4)	145.27

Problem	Best Colors	TS Colors	TS Time (sec.)	GA+TS Colors	GA+TS Time (sec.)
dsjc125.1.col	5	5(0)	22.27	5(0)	122.68
dsjc250.1.col	8	8(2)	467.98	9(0)	227.58
dsjc500.1.col	12	13(4)	808.20	15(1)	792.70
dsjc1000.1.col	20	24(4)	1142.07	26(0)	1886.40
dsjc125.5.col	17	18(0)	156.47	18(1)	761.15
dsjc250.5.col	28	31(1)	866.60	36(4)	1195.82
dsjc125.9.col	44	44(1)	665.78	46(0)	786.91
dsjc250.9.col	72	79(4)	950.81	79(2)	3347.33

According to previous results published in the literature, TS and GA+TS are supposed to be the best algorithms among those tested. Thus, as expected, TS appears to provide in general better colorings with less colors than the other methods. But, with regard to the Hybrid Algorithm GA+TS, theses tests are not in favor of this method as the time limit stops most often its search process too early. However, for small instances this algorithm is capable to find good results as well. With regard to *HSF*, in terms of the solution quality, the results reported here are equivalent or sometimes better than those found by similar frameworks as *Localizer* (Michel and Hentenryck, 1999) or *EasyLocal++* (Gaspero and Schaerf, 2000). In terms of speed, *HSF* appears to be slower but still competitive with *Localizer* especially if we consider that *HSF* has been implemented in Java. Note that both frameworks use a different Graph Coloring Model representation although in both cases is an Invariant-based one. The speed of the algorithms is mainly dependent on the time necessary to propagate changes into the invariant network. Thus we believe the *Localizer*'s model updates more efficiently the network when the size of the GC instances is increasing. *EasyLocal++* remains clearly faster as it is less generic and declarative framework than both *HSF* and *Localizer*.

In terms of memory size, even if *HSF* is a generic framework implemented in Java, this framework is capable to run all these families of algorithms on all theses instances. For bigger instances such as dsjc250.9.col with 250 nodes and an edge density of 90% corresponding to 55794 difference constraints, *HSF* requires less than 40Mb running GA+TS.

With regard to the development time, each algorithm requires at most half page[7] of code as we only connect different search components together. For new components, one page of code was necessary to implement each specific component such as ***UIS*** crossover and ***ReduceRestart***. Therefore, including one page of code needed for the Graph Coloring model, we obtain around 250 lines of code to provide a complete solution with 5 different families of algorithms. This requires about three working days to implement. Obviously, the development time depends on the *HSF* knowledge of the developer but plugging together different components and creating independent components remains an easy task.

In comparison to other systems such as *EasyLocal++*, *HotFrame*, *Localizer* (Voß and Woodruff, 2002), where a user must respectively understand a complex class hierarchy, manipulate templates or learn a specific language to be capable of modeling his own algorithm, *HSF* appears to be clearly easier to use. Indeed, a developer only have to add his problem-specific components following their respective communication protocols. Then to plug them into an existing generic tree of search components for local searches or population algorithms. It is these "in-wires" inter-connections between search components that provides the architectural model and design of *HSF* to developers and free them to master the whole framework to create complex and efficient algorithms.

Future work is likely to be the development of a graphic user interface allowing to create a complete algorithm using a drag-and-drop interface. This GUI will allow a user to develop algorithms without typing a line of code if all search components required are available. Given an optimization problem *P*, a user could issue a request to community of users accessing a database through Internet to find out if any specific search components have been already developed for *P*. In the case of an affirmative answer, the user will combine classic and specific components to graphically build the algorithm extending the knowledge of the community.

11.5 CONCLUSION

This paper has presented *HSF*, a Java-based innovative framework, which proposes a new way to design Heuristic Search algorithms by identifying fundamental concepts already defined in the literature. Those concepts are imple-

[7]One page is considered here as 50 lines of code.

mented as basic components, and thereafter they are plugged together to obtain a Heuristic Search algorithm. This approach allows to easily implement a wide range of well-known algorithms in a short time and facilitate the integration of future algorithms by quickly and easily adding required new components. *HSF* is a very good foundation particularly for future visual tools which would allow the full synthesis of an algorithm without typing a single line of code.

Moreover, combined with *PMF*, *HSF* is capable to automatically provide a broad range of default algorithms for COPs as shown in section 11.3. By its flexibility, generality and efficiency, *HSF* offers an innovative complete solving system based on Heuristic Search algorithms. This is particularly suitable for Heuristic Methods since for a given optimization problem, we still cannot determine in advance which method will be the most effective. Thus, *HSF* can be seen as the next tool for Heuristic Search that can be used to evaluate different families of methods and select the one that suits best the problem.

Finally, *HSF* is part of a complete toolkit *iOpt* dedicated to tackle combinatorial optimization problems. *iOpt* is currently used for developing in-house applications for the *BT Group* in the areas of resource planning & scheduling as well as network optimization problems. Future plans for the toolkit include the release of a first version for external use and also industrialization of the system.

```
static HeuristicSearch createGeneticAlgorithms()
{
    /* creating components for the generation part of the algorithm */
    GenerationSingleSolutionMethod myRSSG = new VectorSolutionRandomGeneration();
    GenerationPopulationMethod myPGM = new GenerationPopulationMethod(10, myRSSG);

    /* creating components for the mutation */
    BestMoveNeighborhoodSearch myRMNS = new BestMoveNeighborhoodSearch();
    DomainAssignmentTabu tabu = new DomainAssignmentTabu();
    myRMNS.setTabu(tabu);
    VariableIndexAssignMoveNeighborhood myNeighborhood = new VariableIndexAssign-
    MoveNeighborhood();
    myNeighborhood.setMovesVersion(2);          // myNeighborhood is set to return a list
    containing all the potential moves of this neighborhood
    myRMNS.setNeighborhood(myNeighborhood);
    LocalSearch myLS = new LocalSearch();
    myLS.setNeighborhoodSearch(myRMNS);
    myLS.setMaxMovesPerformed(100);          // limits the number of moves to be per-
    formed
    myLS.setMaxMovesEvaluated(10000);          // limits the number of moves to be evalu-
    ated
    SingleSolutionMethodMutation myM = new SingleSolutionMethodMutation(myLS);
    MutationPopulationMethod myMPM = new MutationPopulationMethod(myM);

    /* creating components for the crossover */
    CrossoverPopulationMethod myCRPM = new CrossoverPopulationMethod(new Uni-
    form());

    /* creating components for the selection */
    SelectionPopulationMethod mySPM = new SelectionPopulationMethod(new SUSSelec-
    tion());

    /* final assembling */
    CompositePopulationMethod myCPM = new CompositePopulationMethod();
    CompositePopulationMethod myCPM1 = new CompositePopulationMethod();

    /* assembling left part composed of selection+mutation+crossover */
    myCPM1.addMethod(myCRPM);
    myCPM1.addMethod(myMPM);
    myCPM1.addMethod(mySPM);
    myCPM1.setMaxIterations(Long.MAX_VALUE);          // no limitation on the number of
    iterations

    /* assembling left and right parts of the algorithm */
    myCPM.addMethod(myPGM);          // left part
    myCPM.addMethod(myCPM1);          // right part
    myCPM.setMaxIterations(1);          // maximum number of iterations for this compo-
    nent

    /* finalizing assembling */
    PopulationHeuristicSearch myPHS = new PopulationHeuristicSearch();
    myPHS.setPopulationMethod(myCPM);
    myPHS.setSearchTitle("Hybrid Algorithm: TS+GA");          // assign a name to the algo-
    rithm
    myPHS.setMinimization(true);          // this algorithm will minimize the objective func-
    tion
    return myPHS;
}
```

Figure 11.11 Code sample to create a generic hybrid method

Bibliography

A. Andreatta, S. Carvalho, and C. Ribeiro. An object-oriented framework for local search heuristics. In *Proceedings of TOOLS USA'98*, pages 33–45, 1998.

J.E. Baker. Reducing bias and inefficiency in the selection algorithm. In John J Grefenstette, editor, *2nd International Conference on Genetic Algorithms*, pages 14–21. Lawrence Erlbaum Associates, 1987.

E. Bonsma, M. Shackleton, and R. Shipman. EOS: an evolutionary and ecosystem research platform. *BT Technology Journal*, 18:24–31, 2000.

P. Collet, E. Lutton, M. Schoenauer, and J. Louchet. Take it EASEA. In Marc Schoenauer, Kalyanmoy Deb, Günter Rudolph, Xin Yao, Evelyne Lutton, Juan Julian Merelo, and Hans-Paul Schwefel, editors, *Parallel Problem Solving from Nature – PPSN VI*, pages 891–901, Berlin, 2000. Springer.

R. Dorne and J.K. Hao. A new genetic local search algorithm for graph coloring. In Agoston E. Eiben, Thomas Bäck, Marc Schoenauer, and Hans-Paul Schwefel, editors, *Parallel Problem Solving from Nature – PPSN V*, pages 745–754, Berlin, 1998. Springer.

A. Fink and S. Voß. Hotframe: A heuristic optimization framework. In S. Voß and D. Woodruff, editors, *Optimization Software Class Libraries*, OR/CS Interfaces Series, pages 81–154. Kluwer Academic Publishers, Boston, 2002.

L.D. Di Gaspero and A. Schaerf. Easylocal++: An object-oriented framework for flexible design of local search algorithms. Technical Report UDMI/13/2000/RR, Università degli Studi di Udine, 2000.

D.S. Johnson, C.R. Aragon, L.A. McGeoch, and C. Schevon. Optimization by simulated annealing: An experimental evaluation; part ii, graph coloring and number partitioning. *Operations Research*, 39(3):378–406, 1991.

M. Jones, G. McKeown, and V. Rayward-Smith. Templar: An object oriented framework for distributed combinatorial optimization. In *UNICOM Seminar on Modern Heuristics for Decision Support*, 1998.

L. Michel and P. Van Hentenryck. Localizer: A modeling language for local search. *INFORMS Journal of Computing*, 11:1–14, July 1999.

Taligent Inc. Leveraging object-oriented frameworks. *A Taligent White Paper*, 1993.

S. Voß and D. Woodruff, editors. *Optimization Software Class Libraries*. OR/CS Interfaces Series. Kluwer Academic Publishers, Boston, 2002.

C. Voudouris and R. Dorne. Integrating heuristic search and one-way constraints in the iopt toolkit. In S. Voß and D. Woodruff, editors, *Optimization Software Class Libraries*, OR/CS Interfaces Series, pages 177–191. Kluwer Academic Publishers, Boston, 2002.

C. Voudouris, R. Dorne, D. Lesaint, and A. Liret. iOpt: A software toolkit for heuristic search methods. In Springer-Verlag, editor, *7th International Conference on Principles and Practice of Constraint Programming (CP2001)*, Paphos, Cyprus, 2001.

Metaheuristics: Computer Decision-Making, pp. 257-278
Zvi Drezner and George A. Marcoulides

12 A DISTANCE-BASED SELECTION OF PARENTS IN GENETIC ALGORITHMS

Zvi Drezner[1] and George A. Marcoulides[1]

[1]Department of Information Systems and Decision Sciences
California State University-Fullerton
Fullerton,CA 92834 USA
zdrezner@fullerton.edu, gmarcoulides@fullerton.edu

Abstract: In this paper we propose an improvement to the widely used metaheuristic genetic algorithm. We suggest a change in the way parents are selected. The method is based on examining the similarity of parents selected for mating. Computational comparisons for solving the quadratic assignment problem using a hybrid genetic algorithm demonstrate the effectiveness of the method. This conclusion is examined statistically. We also report extensive computational results of solving the quadratic assignment problem Tho150. The best variant found the best known solution 8 times out of 20 replications. The average value of the objective function was 0.001% over the best known solution. Run time for this variant is about 18 hours per replication. When run time is increased to about two days per replication the best known solution was found 7 times out of 10 replications with the other three results each being 0.001% over the best known.
Keywords: Genetic algorithms, Parent selection, Quadratic Assignment Problem.

12.1 INTRODUCTION

Genetic algorithms, first suggested by Holland (1975), have recently become very popular metaheuristic methods for solving optimization problems. Borrowed from the natural sciences and Darwin's law of natural selection and survival of the fittest, genetic algorithms are based on the premise that successful matching of parents will tend to produce better, improved offspring (for a review see Goldberg (1989) and Salhi (1998)).

Genetic algorithms are based on several components that require special attention for each specific problem. Exploiting the structure of the specific problem at hand based on the following issues can determine the success of the genetic algorithm.

- How to code a solution as a chromosome comprised of genes? Normally every variable is defined as a gene.

- The fitness function which determines whether one solution is better than another. Normally, the objective function can serve as a fitness function.

- The way that parents are selected for mating. This is the topic of the present paper.

- The way that two parents produce an offspring. The common approach is to have a crossover point so that all the genes on one side of the crossover point are taken from the first parent and all the other genes are taken from the second parent.

- What is the rule for entering offspring to the population and removing existing population members? Most algorithms start with a fixed number of randomly generated solutions and maintain a constant population count. Consequently, the number of offspring introduced into the population is equal to the number of existing population members that are removed from the population.

- When to stop the procedure? In many algorithms the number of generations is predetermined.

- It is recognized that increased diversity among population members usually results in better solutions. Many genetic algorithms perform mutations on a regular basis. A mutation means that for one population member (or the just created offspring) a randomly selected gene's value is arbitrarily (or randomly) changed to a different value.

- When the process stops, the best population member is the solution. However, usually all population members are "good", and a user may choose an inferior solution from the population that meets his/her needs better.

There are various schemes for the selection of parents. We propose to select parents that are quite different to each other. We believe that this will enhance the diversity of the population and consequently tend to result in better solutions. For example, assigning gender to population members and restricting mating to opposite sexes has also been shown to increase diversity and result in better solutions (Drezner and Drezner, 2003). To select parents in our approach we use a Hamming distance metric. The Hamming distance metric between two parents is defined as the number of genes that are different in the chromosomes of the two parents (Grimaldi, 1998; Marcoulides, 2001). We randomly select one parent, and choose K potential mates (where K is a parameter of the approach). The Hamming distance between each potential mate and the first parent is calculated, and the furthest one is selected for mating.

It is interesting to note that the use of a Hamming distance metric to select parents can be used in any genetic algorithm or a hybrid-genetic algorithm. For every genetic algorithm there is a counterpart which incorporates the distance-based parent selection approach.

The paper is organized as follows. In the next section we outline the hybrid-genetic algorithm. In Section 12.3, we describe the quadratic assignment problem that is used for comparison, and describe the particular genetic algorithm used for the comparison. In Sections 12.4 and 12.5, we present the computational and statistical results. We conclude with a summary of the results and suggestions for future research.

12.2 THE HYBRID GENETIC ALGORITHM

A genetic algorithm produces offspring by mating parents and attempting to improve the population make-up by replacing existing population members with superior offspring. A hybrid genetic algorithm, sometimes called a memetic algorithm (Moscato, 2002), incorporates some improvement heuristic on every offspring before considering its inclusion into the population. A typical genetic algorithm is outlined in Algorithm 12.1:

procedure GA( )

(1) A chromosome is defined for each solution. The chromosome is composed of genes which are usually 0-1 variables. Each chromosome defines a "fitness function" which is usually the value of the objective function.

(2) A population of solutions is generated and maintained.

(3) At each generation, pairs of parents are selected and merged to produce offspring.

(4) Some offspring enter the population and replace existing population members.

(5) Generations are repeated until a stopping criterion is met. The best solution in the final population is the result of the algorithm.

Algorithm 12.1 A typical genetic algorithm.

The specific hybrid genetic algorithm (without the consideration of a Hamming distance metric) tested in this paper for comparison purposes has the structure shown in Algorithm 12.2:

procedure HGA( )
(1) A starting population of size P is randomly selected, and the post-merging procedure is applied on each starting population member.
(2) **for** $t = 1$ **to** G
(3) Randomly select two population members (each population member has an equal chance of being selected, regardless of the objective function value) and merge them to produce an offspring;
(4) Apply the post merging procedure on the merged solution, possibly improving it;
(5) **if** offspring objective function value is at least as good as the objective function value of the worst population member
(6) **if** offspring is diiferent from all population members
(7) Offspring replaces the worst population member;
(8) **endif**
(9) **endif**
(10) **endfor**

Algorithm 12.2 A hybrid genetic algorithm.

In this paper, we propose to modify Step (3) by randomly selecting the first parent, and randomly selecting K potential candidates for the second parent. The Hamming distance between the first parent and each of the potential candidates is calculated and the furthest one is selected as the second parent. Note that in the common genetic algorithm, $K = 1$.

12.3 THE QUADRATIC ASSIGNMENT PROBLEM

We tested our approach on the solution of the quadratic assignment problem. The quadratic assignment problem is considered to be one of the most difficult optimization problems to solve optimally. The problem is defined as follows.

A set of n possible sites are given and n facilities are to be located on these sites, one facility at a site. Let c_{ij} be the cost per unit distance between facilities i and j and d_{ij} be the distance between sites i and j. The cost f to be minimized over all possible permutations, calculated for an assignment of facility i to site $p(i)$ for $i = 1, \ldots, n$, is:

$$f = \sum_{i=1}^{n} \sum_{j=1}^{n} c_{ij} d_{p(i)p(j)} \tag{12.1}$$

Optimal algorithms can solve relatively small problems ($n \leq 36$). Nug30, Kra30b, Tho30 were solved by Anstreicher et al. (2002); Kra30a by Hahn and Krarup (2001); Ste36a by Brixius and Anstreicher (2001); Ste36b and Ste36c by Nyström (1999). Consequently, considerable effort has been devoted to constructing heuristic algorithms. The first heuristic algorithm proposed for this problem was CRAFT (Armour and Buffa, 1963) which is a descent heuristic. More recent algorithms use metaheuristics such as tabu search (Battiti and Tecchiolli, 1994; Skorin-Kapov, 1990; Taillard, 1991), simulated annealing (Burkard and Rendl, 1984; Wilhelm and Ward, 1987; Connoly, 1990), simulated jumping (Amin, 1999), genetic algorithms (Ahuja et al., 2000; Fleurent and Ferland, 1994; Tate and Smith, 1995; Drezner, 2003a;b), ant colony search (Gambardella et al., 1999), or specially designed heuristics (Drezner, 2002; Li et al., 1994). For a complete discussion and list of references see Burkard (1990); Çela (1998); Rendl (2002); Taillard (1995).

For the implementation of the genetic algorithm for the solution of the quadratic assignment problem, each solution (chromosome) is defined by the facility assigned to sites #1, #2, ..., #n. The Hamming distance between two solutions is the *number* of facilities located at *different* sites. We define the merging procedure and the post-merging procedure used in this paper. These are also described in Drezner (2003a;b).

12.3.1 The merging procedure

The merging procedure used is the "cohesive merging procedure" (Drezner, 2003a) shown in Algorithm 12.3.

12.3.2 The post merging procedure

We use the "short" concentric tabu search. The concentric tabu search was first presented in Drezner (2002) and was used as a post merging procedure in hybrid genetic algorithms (Drezner, 2003a;b). The short version is used in Drezner (2003b).

One iteration of the concentric tabu search is very similar to the variable neighborhood search (Mladenović and Hansen, 1997; Hansen and Mladenović, 2001). The search is performed in "rings" around the center solution, proceeding from one ring to a larger one, and so on, until the maximum possible radius is obtained. A starting solution is selected as the center solution. The same Hamming distance Δp is defined for each solution (permutation p of the center solution). The distance Δp is the number of facilities in p that are not in their center solution site. The tabu list consists of solutions that are not further from the center solution, forcing the search away from the center solution. For each Δp we keep three solutions: sol_0, sol_1, and sol_2. sol_0 is the best encountered solution with distance Δp. Similarly, sol_1 and sol_2 are the best encountered solutions whose distance is $\Delta p + 1$, and $\Delta p + 2$, respectively. The depth of the search is set to $d \leq n$. A description of one iteration of short concentric tabu search is given in Algorithm 12.4.

procedure COHESIVEMERGING( )
(1) **for** *PivotSite* = 1 **to** *n*
(2) Calculate median distance from *PivotSite* to all sites;
(3) Sites which are closer than the median to the *PivotSite* are assigned the facility from the first parent;
(4) All other sites are assigned a facility from the second parent;
(5) Create list of unassigned facilities;
(6) All facilities from the second parent that are assigned twice are replaced with an unassigned facility;
(7) Calculate the value of the objective function for the merged solution.
(8) **endfor**
(9) The best of the *n* merged permutations (i.e. the one with the best value of the objective function) is selected for a post merging procedure;
(10) **return** offspring

Algorithm 12.3 The cohesive merging procedure.

12.3.2.1 The concentric tabu search with a given number of levels. Let L be the number of levels. In our experiments we use $L = 6$ for $n \leq 100$ problems. The concentric tabu search is described in Algorithm 12.5.

12.3.2.2 Specific parameters. We use a population size $P = 100$ for problems with $n \leq 100$ and various values of P for problem Tho150. We use $G = \max\{1000, 20n\}$ generations for problems with $n \leq 100$ problems and $G = 50P$ for problem Tho150.

12.4 COMPUTATIONAL RESULTS FOR PROBLEMS OF DIMENSION NOT GREATER THAN 100

All algorithms were programmed in FORTRAN PowerStation 4.0 and conducted on a 600MHz notebook 7200 Toshiba Portege computer.

We tested the algorithms on the same set of 29 problems tested in Drezner (2003b)[1]. These are all the symmetric problems with unknown optimum (although optimality was recently established for Nug30 by Anstreicher et al. (2002) and for Kra30a by Hahn and Krarup (2001)). Each problem was run 10 times from randomly generated starting solutions. Twelve of the 29 problems

[1]Downloaded from `http://www.mim.du.dk/~sk/qaplib`.

Table 12.1 Results for problems with $n \leq 100$ using $K = 1, 2, 3$.

Problem	Best Known	$K=1$ (1)	$K=1$ (2)	$K=1$ (3)	$K=2$ (1)	$K=2$ (2)	$K=2$ (3)	$K=3$ (1)	$K=3$ (2)	$K=3$ (3)
Ste36a	9526	7	0.031	0.70	10	0	0.71	7	0.031	0.74
Ste36c	8239.11	7	0.039	0.68	9	0.013	0.71	5	0.035	0.72
Tho40	240516	4	0.010	1.15	2	0.012	1.19	2	0.016	1.21
Sko49	23386	9	0.005	2.17	8	0.010	2.25	7	0.015	2.29
Wil50	48816	8	0.003	2.30	10	0	2.38	9	0.002	2.43
Sko56	34458	8	0.002	3.94	8	0.002	4.18	8	0.003	4.48
Sko64	48498	10	0	7.17	10	0	7.96	10	0	8.24
Sko72	66256	3	0.014	10.59	4	0.015	11.76	7	0.002	12.05
Sko81	90998	4	0.015	17.50	6	0.004	18.80	2	0.015	20.64
Sko90	115534	5	0.007	28.90	4	0.008	34.33	4	0.009	34.78
Sko100a	152002	2	0.025	45.47	4	0.014	57.38	3	0.031	58.96
Sko100b	153890	6	0.007	46.18	8	0.001	53.48	6	0.007	57.97
Sko100c	147862	2	0.009	44.15	2	0.001	52.35	0	0.002	59.35
Sko100d	149576	1	0.051	48.08	3	0.039	57.88	2	0.041	57.76
Sko100e	149150	10	0	43.68	10	0	46.98	4	0.005	56.22
Sko100f	149036	0	0.019	44.75	3	0.026	58.36	1	0.036	65.12
Wil100	273038	4	0.001	45.44	3	0.002	52.79	4	0.002	55.71
Average		5.29	0.014	23.11	6.12	0.009	27.26	4.76	0.015	29.33

(1) Number of times out of 10 replications that best known solution obtained
(2) Percentage of average value over the best known solution
(3) Average time in minutes per replication

procedure SHORTCONCENTRICTS-ITERATION( )

(1) Set $\Delta p \leftarrow 0$. The starting solution sol_0 is the center solution and the best found solution.
(2) All pair exchanges of sol_0 are evaluated.
(3) **if** the exchanged solution is better than the best solution found, the best solution is updated and the rest of the exchanges are evaluated.
(4)   **if** the distance of an exchanged solution is Δp or lower, it is in the tabu list. It is ignored and the rest of the exchanges are evaluated.
(5)   **if** its distance is $\Delta p + 1$ or $\Delta p + 2$, sol_1 or sol_2 are updated, if necessary.
(6) **endif**
(7) **if** a new best found solution is found by scanning all the exchanges of sol_0
(8)   The starting (center) solution is set to the new best found solution;
(9)   Go to Step (1);
(10) **else**
(11) $sol_0 \leftarrow sol_1$, $sol_1 \leftarrow sol_2$, and sol_2 is emptied;
(12)   Set $\Delta p \leftarrow \Delta p + 1$;
(13)   **if** $\Delta p = d + 1$ **then** stop the iteration
(14)   **else** go to Step (2).
(15) **endif**

Algorithm 12.4 Description of one iteration of short concentric tabu search.

can be classified as "easy" problems (i.e., all variants of the genetic algorithms were able to find the best known solution in all runs (Drezner, 2003a). The easy problems are Kra30a, Kra30b Krarup and Pruzan (1978), Nug30 (Nugent et al., 1968), Tho30 (Thonemann and Bölte, 1994), Ste36b (Steinberg, 1961), Esc32a, Esc32b, Esc32c, Esc32d, Esc32h (Eschermann and Wunderlich, 1990), Sko42 (Skorin-Kapov, 1990), and Esc64a (Eschermann and Wunderlich, 1990). As such, the results for these twelve "easy" problems are not reported in this paper. Only results for the remaining 17 problems are reported and averaged. These problems are: Ste36a, Ste36c (Steinberg, 1961), Tho40 (Thonemann and Bölte, 1994), Sko49 (Skorin-Kapov, 1990), Wil50 (Wilhelm and Ward, 1987), Sko56, Sko64 (for which the best known solution was found for all runs in this paper), Sko72, Sko81, Sko90, Sko100a-f (Skorin-Kapov, 1990), and Wil100 (Wilhelm and Ward, 1987).

```
procedure CONCENTRICTS( )
(1)  Select a random center solution. It is also the best
     found solution.
(2)  Set a counter c ← 0.
(3)  repeat
(4)     Select d randomly in [0.3n, 0.9n]. Perform an itera-
        tion on the center solution.
(5)     if the iteration improved the best found solution
(6)        Go to Step (2).;
(7)     else
(8)        Advance the counter c ← c + 1;
(9)        if c ≤ L and is odd
(10)          Use the best solution with depth d as the
              new center solution;
(11)          Go to Step (4);
(12)       endif
(13)       if c ≤ L and is even
(14)          Use the best solution found throughout the
              scan (the previous center solution is not con-
              sidered) as the new center solution;
(15)          Go to Step (4);
(16)       endif
(17)    endif
(18) until c = L + 1
(19) return best found solution
```

Algorithm 12.5 The concentric tabu search.

12.4.1 Statistical analysis

The computational results are summarized in Table 12.1. Analysis of variance (ANOVA) procedures were examined to determine if there were any significant differences between the $K = 1$, $K = 2$, and $K = 3$ approaches to selecting parents. A two-factor ANOVA indicated that there were significant differences between the three parent selection approaches [$F_{2,491} = 4.12, p < 0.0168$]. *Post hoc* pairwise comparisons using Neuman-Keuls and LSD procedures between the three approaches revealed that the $K = 2$ is the best approach, followed by $K = 1$ and $K = 3$ [$F_{1,491} = 3.325, p < .01$]. The $K = 1$ and $K = 3$ approaches were found not to be statistically significantly different to each other. It is also interesting to note that the $K = 1$ and $K = 3$ approaches each have an instance in which the best known solution was never found, while the $K = 2$ approach found the best known solution at least twice out of ten replications for every problem. There were also four instances (in addition to the twelve "easy" problems not reported in the paper) in which the $K = 2$

Table 12.2 Comparison with Ahuja et al. (2000).

Problem	Best Known	$K=2$ (1)	$K=2$ (2)	Ahuja et al. (2000) (1)	Ahuja et al. (2000) (2)
Ste36a	9526	0.000	0.71	0.270	11.83
Tho40	240516	0.012	1.19	0.320	15.97
Sko49	23386	0.010	2.25	0.210	20.87
Wil50	48816	0.000	2.38	0.070	35.25
Sko56	34458	0.002	4.18	0.020	49.60
Sko64	48498	0.000	7.96	0.220	63.14
Sko72	66256	0.015	11.76	0.290	84.63
Sko81	90998	0.004	18.80	0.200	182.74
Sko90	115534	0.008	34.33	0.270	211.63
Sko100a	152002	0.014	57.38	0.210	276.80
Sko100b	153890	0.001	53.48	0.140	245.49
Sko100c	147862	0.001	52.35	0.200	338.57
Sko100d	149576	0.039	57.88	0.170	338.37
Sko100e	149150	0.000	46.98	0.240	352.12
Sko100f	149036	0.026	58.36	0.290	357.98
Wil100	273038	0.002	52.79	0.200	342.40
Average		0.008	28.92	0.207	182.96

(1) Percentage of average over the best known
(2) Average time in minutes per replication

approach found the best known solution in all 10 runs. There are only two such instances for $K = 1$, and one such instance for $K = 3$. Run times increase slightly with increased K because the parents are less similar to one another and offspring tend to require more post-merging iterations.

It is clear that increasing the value of K does not necessarily improve the performance of the algorithm. We considered the possibility that for $K = 3$ the number of generations is not large enough and the procedure is stopped prematurely before convergence. We therefore experimented also with a 30% increase in the number of generations and got the same performance. As expected, the performance of all three versions ($K = 1, 2, 3$) improved, but $K = 2$ was still clearly the best one. This variant, of course, requires longer computer times and thus is not recommended. We conclude that it is beneficial to select a distant second parent, but one should not overdo it. It is clear that selecting the further of two candidates works very well for solving the quadratic assignment problem.

12.4.2 Comparison with Ahuja et al. (2000)

It should be noted that the results reported in this study are quite comparable (but statistically superior) to the best available genetic algorithm presented in Drezner (2003b) with comparable run times (for $K = 1$ they are identical to those reported in Drezner (2003b)). Nevertheless, the results presented in this paper are much better than those reported in other similar studies. For example, in Table 12.2 we compare our results with those reported in Ahuja et al. (2000), which represents a typical performance of most algorithms. Ahuja et al. (2000) solved each problem once and therefore we compare our average over the best known solution with their result. Ahuja et al. (2000) did not obtain the best known solution for two of the problems that we considered "easy", because the best known solution was found in all runs by all three of our parent selection approaches. These are Nug30 and Sko42, which Ahuja et al. (2000) obtained solutions of 0.070%, and 0.250% over the best known, respectively. Of the remaining 17, Ahuja et al. (2000) do not report results for Ste36c, which leaves 16 problems for further comparison. Run times in Ahuja et al. (2000) are reported for GA-1, and for GA-3 (their best variant) they state that run times are about double those of GA-1. We therefore report double the times of GA-1. Note that Ahuja et al. (2000) used an HP-6000 platform which is comparable or faster than our notebook computer.

By examining Table 12.2 we conclude that

- The percentage over the best known solution is, on the average, more than 25 times better using our procedure.
- Run times are, on the average, more than six times faster using our method (assuming the same computer performance) .
- There are four instances (in addition to the twelve "easy" problems not reported here) that our procedure found the best known solution in all ten runs, while Ahuja et al. (2000) do not report even one case in which the best known solution was found.

12.5 COMPUTATIONAL RESULTS FOR THO150 PROBLEM

Thonemann and Bölte (1993) introduced the Tho150 problem. The best known solution for this problem was improved over the years as follows:

8134056	Thonemann and Bölte (1994)
8133864	Cung et al. (1997)
8133642	Taillard and Gambardella (1997)
8133484	Amin (1999)
8133398	Misevičius (2001)

In this paper we found the best known solution of 8133398 in 22 replications.

The program was run on several computers due to the extensive computational effort (many months) required for all experiments with Tho150. All times in this section are reported for a 2.8GHz Pentium 4 computer. We established a run time ratio of 0.286 between the 600MHz Pentium 3 computer and the 2.8GHz Pentium 4 computer. All times on the 600MHz computer were multiplied by this factor.

We first solved the problem 480 times using $L = 6$. All six combinations of $K = 1, 2, 3$ and $P = 100, 150$ were replicated 80 times each. The results are summarized in Table 12.3.

We found that $K = 3$ is best for Tho150. We performed a two-way ANOVA with replications using 480 values of the objective function. We used three columns (one for each value of K) and two rows (One for each value of P) leading to six cells with 80 values in each cell. The columns were significantly different ($p = 3.45 \times 10^{-8}$). Using $P = 150$ was superior to using $P = 100$ ($p = 0.004$). When the $K = 2$ results were compared with the $K = 3$ results (A total of 320 replications in four cells of 80 observations in each), then even though the $K = 3$ variant was better, its superiority was not statistically significant ($p = 0.38$). The superiority of $P = 150$ over $P = 100$ was statistically significant ($p = 0.009$).

We were somewhat disappointed that the best known solution of 8133398 (Misevičius, 2001) was missed in these 480 replications. The best solution found in these replications was 8133520 which is 0.001% higher. We did not even obtain the solution 8133484 reported by Amin (1999). We therefore used $K = 3$ and $P = 200$ for 120 replications. The results are depicted in Table 12.4. The results are better than those variants reported in Table 12.3 but we still did not find a solution better than 8133520.

We then experimented with increasing the number of levels (we used $K = 3$). This time we ran each variant for 20 replications. The results are summarized in Table 12.5. These results are much better than those obtained using $L = 6$. The quality of the solutions generally improves with an increase in L and P with moderate increase in run times. The best known solution of 8133398 was obtained 15 times.

It is interesting to note that even though no solution between 8133398 and 8133520 was found, the final population included solutions (which are local minima even for $L = 30$) between these two values. We encountered no local minima with a value of the objective function between 8133520 and 8133642. However, we encountered six local minima below 8133520. These were 8133398, 8133414, 8133476, 8133484, 8133492, and 8133500. Each solution has four identical representations which are mirror images of one another. We select the mirror image that leads to the closest distances (i.e., the minimum number of facilities in different sites) between the various solutions. The distances between these solutions are depicted in Table 12.6. Note that the other three mirror images of the solution 8133520 are at distances 146, 150, and 150 from the selected best known solution.

Note that two pairs of local minima are different by an exchange of only four facilities. For example, by exchanging the facilities 7, 137, 81, and 15 to the sites of facilities 137, 15, 7, and 81, respectively, the solution 8133476 is improved to 8133398. We observe a clear cluster of four solutions 8133398, 8133414, 8133476, and 8133492. A second cluster of two solutions 8133484 and 8133500 is a little farther away, and the solution 8133520 is by itself "in a different region of the solution space".

Table 12.3 Using $L = 6$ and $P = 100, 150$ with 80 replications for each case.

Values Below 8134000		$P = 100$			$P = 150$		
		$K = 1$	$K = 2$	$K = 3$	$K = 1$	$K = 2$	$K = 3$
8133520	0.001%		1	4	2	3	10
8133642	0.003%	4	8	14	6	19	19
8133702	0.004%	1	4	2	2	3	
8133712	0.004%						1
8133718	0.004%			1			1
8133760	0.004%	4	4	5	4	6	4
8133862	0.006%		1				1
8133864	0.006%	12	17	8	15	17	14
8133914	0.006%			2			3
8133926	0.006%			1			
8133936	0.007%						3
8133946	0.007%	3	3	3	3	3	3
8133954	0.007%		1	2	2		
8133956	0.007%			1			
8133984	0.007%		1		1	1	
Total (out of 80)		24	40	43	35	52	59
Average (all 80)		8135595	8134734	8134599	8135129	8134345	8134228
%		0.027%	0.016%	0.015%	0.021%	0.012%	0.010%
Median (all 80)		8134409	8133994	8133950	8134063	8133864	8133864
%		0.012%	0.007%	0.007%	0.008%	0.006%	0.006%
Time (min/run)		84.82	97.92	111.74	131.01	153.79	167.02

Table 12.4 Results for 120 replications using $K = 3$, $L = 6$, and $P = 200$.

Value	Count	Value	Count	Value	Count	Value	Count
8133520	28	8133864	15	8134048	1	8135118	1
8133642	35	8133914	1	8134066	1	8135310	1
8133700	1	8133932	1	8134080	2	8135380	1
8133702	3	8133946	6	8134158	1	8135386	1
8133712	2	8133954	1	8134210	1	8135468	1
8133718	1	8133984	1	8134224	1	8135780	1
8133760	5	8134008	1	8134802	1	8135786	1
8133862	1	8134038	1	8134940	1	8136046	1

Average: 8133858 (0.006%)
Median: 8133642 (0.003%)
Time per run: 248.07 minutes

We depicted the solutions by applying the transformation proposed in Marcoulides and Drezner (1993) (see Figure 12.1). This procedure was suggested in Drezner and Marcoulides (2003) to monitor the progression of the population structure in genetic algorithms. The top part of the figure shows a scatter diagram of the seven solutions. It shows clearly the clustering described above by examining Table 12.6. Since the solution 8133520 is far from the other solutions and may distort the spatial pattern of the other solutions, we repeated the calculation of the scatter diagram just for the other six solutions. The result is depicted at the bottom of Figure 12.1. It clearly shows the two clusters with the cluster of two being more dispersed than the cluster of four.

As a final experiment we used $P = 300$, $L = 50$, and $K = 3$ which requires about two days (2832 minutes) of computer time. Out of 10 replications the best known solution 8133398 was found seven times, and the solution 8133520 three times. In one replication (seed=77) 282 offspring (out of 15,000) had an objective function of 8133520 or lower and 19 of them were distinct and were part of the final population. The worst final population member had an objective of 8134368. The counts are:

Table 12.5 Using $L = 12, 18, 24, 30$ and $P = 150, 200$ with 20 replications each.

Values	$P = 150$				$P = 200$			
	$L = 12$	$L = 18$	$L = 24$	$L = 30$	$L = 12$	$L = 18$	$L = 24$	$L = 30$
8133398	1	3	2			1		8
8133520	4	5	7	11	6	8	17	8
8133642	9	7	6	6	9	9	3	3
8133760	1		1					
8133862			1					
8133864	3	1	2	3	1	1		
8133914					1			
8133936		1						
8133946	1	1	1		1			1
8134030						1		
8134078	1							
8134208					1			
8134288					1			
8135046		1						
8135414		1						
Average	8133682	8133775	8133629	8133608	8133706	8133611	8133538	8133511
%	0.003%	0.005%	0.003%	0.003%	0.004%	0.003%	0.002%	0.001%
Median	8133642	8133642	8133642	8133520	8133642	8133642	8133520	8133520
%	0.003%	0.003%	0.003%	0.001%	0.003%	0.003%	0.001%	0.001%
Time	299.68	470.28	646.78	757.00	466.21	663.36	858.67	1069.89

Table 12.6 Distances between local minima.

	8133398	8133414	8133476	8133484	8133492	8133500	8133520
8133398	0	6	4	27	10	33	117
8133414	6	0	10	33	4	27	117
8133476	4	10	0	31	6	37	119
8133484	27	33	31	0	37	6	122
8133492	10	4	6	37	0	31	119
8133500	33	27	37	6	31	0	122
8133520	117	117	119	122	119	122	0

Objective	(1)	(2)	(3)
8133398	116	3	170
8133414	11	3	19
8133476	22	3	44
8133484	88	3	161
8133492	3	2	3
8133500	8	3	13
8133520	34	2	296
Total	282	19	706

(1) offspring in seed=77.
(2) distinct offspring in seed=77.
(3) offspring in all 10 replications.

12.6 SUMMARY

In this paper we proposed to select for mating parents that are not similar to one another. This increases the diversity of the population, and the homogenizing effect of the genetic algorithm is reduced. Genetic algorithms normally terminate when the population becomes homogeneous, and no improvement in the population is likely. This is because any pair of parents is very similar to one another and offspring generated are either very similar to their parents, or identical to an existing population member. By increased diversity, this phase of the algorithm is delayed to later generations and better solutions are expected.

We found that by randomly selecting one parent, randomly selecting two or three candidates for the second parent, and selecting the furthest one as a second parent, produces the best results on a set of quadratic assignment problems. As future research we propose to apply the distance-based parent selection approach to other genetic algorithms as well.

Acknowledgments

This research was completed while the first author was visiting the Graduate School of Management, University of California, Irvine (UCI), and the second author was visiting the Division of Social Research Methodology, University of California, Los Angeles (UCLA).

The FORTRAN program is available for download [2].

[2]http://business.fullerton.edu/zdrezner

Figure 12.1 Distribution of local minima.

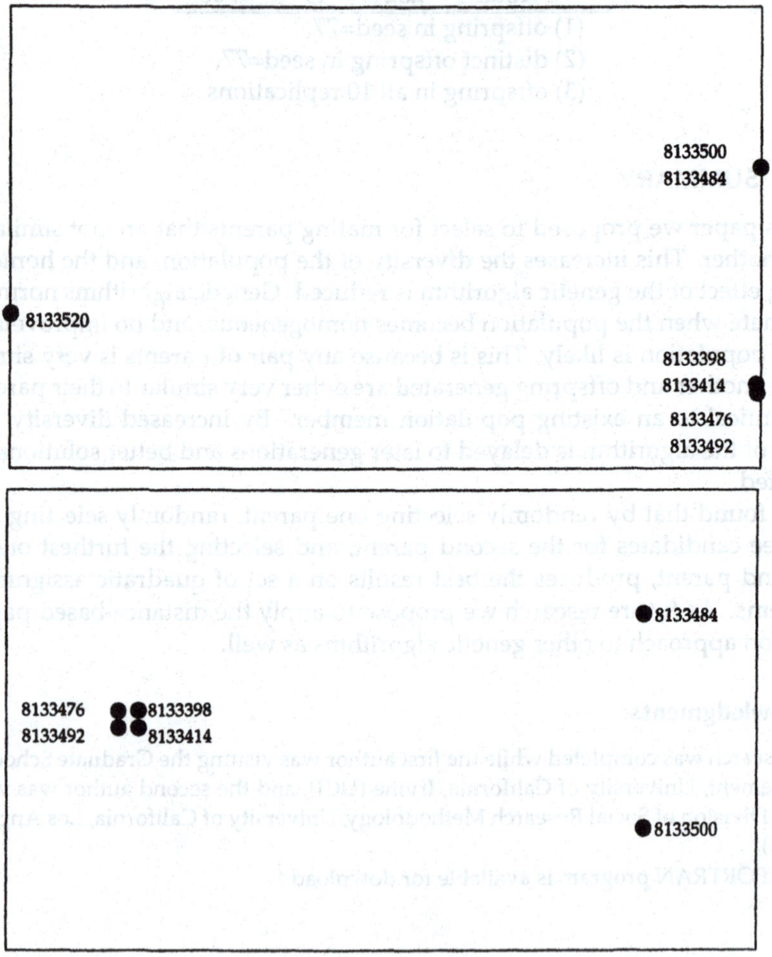

Bibliography

R.K. Ahuja, J.B. Orlin, and A. Tiwari. A descent genetic algorithm for the quadratic assignment problem. *Computers and Operations Research*, 27:917–934, 2000.

S. Amin. Simulated jumping. *Annals of Operations Research*, 84:23–38, 1999.

K. Anstreicher, N. Brixius, J.-P. Gaux, and J. Linderoth. Solving large quadratic assignment problems on computational grids. *Mathematical Programming*, 91:563–588, 2002.

G.C. Armour and E.S. Buffa. A heuristic algorithm and simulation approach to relative location of facilities. *Management Science*, 9:294–309, 1963.

R. Battiti and G. Tecchiolli. The reactive tabu search. *ORSA Journal on Computing*, 6:126–140, 1994.

N.W. Brixius and K.M. Anstreicher. The Steinberg wiring problem. Working paper, The University of Iowa, 2001.

R.E. Burkard. Locations with spatial interactions: The quadratic assignment problem. In P.B. Mirchandani and R.L. Francis, editors, *Discrete Location Theory*. Wiley, Berlin, 1990.

R.E. Burkard and F. Rendl. A thermodynamically motivated simulation procedure for combinatorial optimization problems. *European Journal of Operational Research*, 17:169–174, 1984.

E. Çela. *The Quadratic Assignment Problem: Theory and Algorithms*. Kluwer Academic Publishers, Dordrecht, 1998.

D.T. Connoly. An improved annealing scheme for the QAP. *European Journal of Operational Research*, 46:93–100, 1990.

V.-D. Cung, T. Mautor T., P. Michelon, and A. Tavares A. A scatter search based approach for the quadratic assignment problem. In *Proceedings of the*

IEEE International Conference on Evolutionary Computation and Evolutionary Programming (ICEC'97), pages 165–170, Indianapolis, 1997.

T. Drezner and Z. Drezner. Gender-specific genetic algorithms. under review, 2003.

Z. Drezner. Heuristic algorithms for the solution of the quadratic assignment problem. *Journal of Applied Mathematics and Decision Sciences*, 6:163–173, 2002.

Z. Drezner. A new genetic algorithm for the quadratic assignment problem. *INFORMS Journal on Computing*, 2003a.

Z. Drezner. Robust heuristic algorithms for the quadratic assignment problem. Under review, 2003b.

Z. Drezner and G.A. Marcoulides. Mapping the convergence of genetic algorithms. under review, 2003.

B. Eschermann and H.J. Wunderlich. Optimized synthesis of self-testable finite state machines. In *20th International Symposium on Fault-Tolerant Computing (FFTCS 20)*, Newcastle upon Tyne, 1990.

C. Fleurent and J.A. Ferland. Genetic hybrids for the quadratic assignment problem. In P. Pardalos and H. Wolkowicz, editors, *Quadratic Assignment and Related Problems*, volume 16 of *DIMACS Series in Discrete Mathematics and Theoretical Computer Science*, pages 173–187. American Mathematical Society, 1994.

L. Gambardella, E. Taillard, and M. Dorigo. Ant colonies for the quadratic assignment problem. *Journal of the Operational Research Society*, 50:167–176, 1999.

D.E. Goldberg. *Genetic Algorithms in Search, Optimization and Machine Learning*. Addison-Wesley, Wokingham, England, 1989.

R.P. Grimaldi. *Discrete and Combinatorial Mathematics: An Applied Introduction*. Addison-Wesley, Wokingham, England, 1998.

P.M. Hahn and J. Krarup. A hospital facility problem finally solved. *The Journal of Intelligent Manufacturing*, 12:487–496, 2001.

P. Hansen and N. Mladenović. Variable neighborhood search: Principles and applications. *European Journal of Operational Research*, 130:449 – 467, 2001.

J.H. Holland. *Adaptation in Natural and Artificial Systems*. University of Michigan Press, Ann Arbor, 1975.

J. Krarup and P.M. Pruzan. Computer-aided layout design. *Mathematical Programming Study*, 9:75–94, 1978.

Y. Li, P.M. Pardalos, and M.G.C. Resende. A greedy randomized adaptive search procedure for the quadratic assignment problem. In P. Pardalos and H. Wolkowicz, editors, *Quadratic Assignment and Related Problems*, volume 16 of *DIMACS Series in Discrete Mathematics and Theoretical Computer Science*, pages 237–261. American Mathematical Society, 1994.

G.A. Marcoulides. A eugenic algorithm for condcuting specification searches in structural equation modeling, 2001. Paper presented at the annual meeting of the Society of Multivariate Experimental Psychology, Monterey, CA.

G.A. Marcoulides and Z. Drezner. A procedure for transforming points in multi-dimensional space to two-dimensional. *Educational and Psychological Measurement*, 53:933–940, 1993.

A. Misevičius. An efficient simulated annealing algorithm for the quadratic assignment problem. Technical report, Kaunas University of Technology, 2001. working paper.

N. Mladenović and P. Hansen. Variable neighborhood search. *Computers and Operations Research*, 24:1097–1100, 1997.

P. Moscato. Memetic algorithms. In P.M. Pardalos and M.G.C. Resende, editors, *Handbook of Applied Optimization*, pages 157–167. Oxford University Press, Oxford, U.K., 2002.

C.E. Nugent, T.E. Vollman, and T. Ruml. An experimental comparison of techniques for the assignment of facilities to locations. *Operations Research*, 16: 150–173, 1968.

M. Nyström. Solving certain large instances of the quadratic assignment problem: Steinberg's examples. Technical report, California Institute of Technology, 1999. Working paper.

F. Rendl. The quadratic assignment problem. In Z. Drezner and H. Hamacher, editors, *Facility Location: Applications and Theory*. Springer, Berlin, 2002.

S. Salhi. Heuristic search methods. In G.A. Marcoulides, editor, *Modern Methods for Business Research*. Lawrence Erlbaum Associates, Mahwah, NJ., 1998.

J. Skorin-Kapov. Tabu search applied to the quadratic assignment problem. *ORSA Journal on Computing*, 2:33–45, 1990.

L. Steinberg. The backboard wiring problem: a placement algorithm. *SIAM Review*, 3:37–50, 1961.

E.D. Taillard. Robust tabu search for the quadratic assignment problem. *Parallel Computing*, 17:443–455, 1991.

E.D. Taillard. Comparison of iterative searches for the quadratic assignment problem. *Location Science*, 3:87–105, 1995.

E.D. Taillard and L.M. Gambardella. Adaptive memories for the quadratic assingnment problem. Technical report, IDSIA Lugano, Switzerland, 1997. Research report.

D.M. Tate and A.E. Smith. A genetic approach to the quadratic assignment problem. *Computers and Operations Research*, 22:73–83, 1995.

U.W. Thonemann and A. Bölte. Optimizing simulated annealing schedules with genetic programming. Technical report, Lehrstuhl für Betriebswirtschaftslehre, Insbes. Produktionswirtschaft, Universität Paderborn, Germany, 1993. Working paper.

U.W. Thonemann and A. Bölte. An improved simulated annealing algorithm for the quadratic assignment problem. Technical report, School of Business, Department of Production and Operations Research, University of Paderborn, Germany, 1994. Working paper.

M.R. Wilhelm and T.L. Ward. Solving quadratic assignment problems by simulated annealing. *IIE Transactions*, 19:107–119, 1987.

Metaheuristics: Computer Decision-Making, pp. 279-300
Peter Greistorfer

13 EXPERIMENTAL POOL DESIGN: INPUT, OUTPUT AND COMBINATION STRATEGIES FOR SCATTER SEARCH

Peter Greistorfer[1]

[1]Institut für Industrie und Fertigungswirtschaft
Karl-Franzens-Universität
Graz, Austria
peter.greistorfer@kfunigraz.ac.at

Abstract: We investigate several versions of a tabu scatter search heuristic to solve permutation type optimization problems. The focus lies on the design of the three main components which are comprised in every pool-oriented method. These components are the input and output procedures, which are responsible for pool maintenance and determine the transfer of elite solutions, and a solution combination method which must effectively combine a set of elite solutions. We propose several methods for each of these three components and evaluate their combinations as heuristic design variants on a sample set of capacitated Chinese postman instances. Descriptive results are discussed in detail and supported by the testing of a statistical hypothesis.
Keywords: Pool method, Scatter search, Metaheuristics design, Statistical evaluation.

13.1 INTRODUCTION

Many metaheuristic search methods are, in one form or another, based on a set of high quality solutions. Besides the most prominent member of this group, the genetic algorithms (GAs), there is also a number of interesting methods which build upon this paradigm. The most important ones are the scatter search (SCS), path relinking variants and mere restart techniques, as they are often used in connection with a tabu search (TS) framework. All these methods maintain a collection of so-called elite solutions, which permanently or from time to time, are accessed within an algorithm's run in order to pursue an intensification or a diversification purpose. In the literature such a set is designated as *population* (GA) and as *reference set* (SCS), respectively. In this study we prefer using the generalizing term *pool* which can be simply understood as an algorithmic data structure that enables the storage of a solution set. We restrict ourselves to non-GA architectures, i.e. to the development of strategies which are basically designed for operations in phenotypical solution representations (Reeves, 1993).

Our focus is on problems which can be fully or partially modeled using permutation strings. Many of them can be found in the area of routing problems, but are not restricted to it. Problems which can be fully addressed by permutation encoding are *uncapacitated* problems, e.g. the traveling salesman problem (TSP), the Chinese postman problem (CPP) and the quadratic (and linear) assignment problem. If a first-level constraint is added to such an unrestricted problem, which is usually a capacity restriction, permutation encoding can still be used if the underlying solution representation is extended by means that enable the consideration of clusters. In this case pointers may be used to mark the beginning and the end of a route in a permutation, thus transferring a TSP to a (node-oriented) vehicle routing problem (VRP). Setting up a capacitated problem like this and assuming an appropriate cost function still allows the application of generic permutation neighborhoods if such constraints are additionally observed while checking potential neighborhood solutions.

The theme of this paper is to theoretically work out relevant pool strategies in the context of SCS and to evaluate them by means of computational comparisons. The focus lies on the pool building process (input strategies), on taking elite solutions from that pool (output methods) and on how these can be effectively combined (combination methods). Test results refer to a sample of arc routing instances for the so-called capacitated CPP (CCPP). However, the context-independent encoding used allows generalizing the algorithmic findings for a number of different settings.

In Section 13.2, we shortly discuss the main components of a pool method and address some general aspects of controlling pool transfers. Section 13.3 summarizes a set of procedures which are alternative components for an existing pool method (Greistorfer, 2003). Different input and output strategies as well as solution combination methods (SCMs) have been implemented, tested and compared with each other. Section 13.4 is a presentation and detailed analysis of the computational results obtained for the CCPP case. We conclude with a summary and an outlook in Section 13.5.

13.2 MAIN COMPONENTS IN POOL DESIGN

A *neighborhood* is an algorithmic component that constructs a set of solutions. Multi-solution neighborhoods form the classical pool-oriented design (GA and SCS), whereas single-solution neighborhoods have become more frequently used in hybrid pool methods only recently (Rochat and Taillard, 1995; Voß, 1995; Greistorfer, 1998; Bastos and Ribeiro, 2002). Basic neighborhood operations are *add* and *drop* as well as *exchange* and *insert* for a phenotypical solution presentation and genetic *crossover* for the genotypical case. SCMs are often based on linear programming (LP) relaxations (Cung et al., 1997; Laguna, 1997; Greistorfer, 2003).

The most common *selection* schemes are best improvement and first improvement cost criteria or, more generally, *best admissibility*, which includes the application of memory. To counteract a loss of quality in limited neighborhoods, several strategies have been proposed which rely on similar basic ideas (e.g. see the so-called restricted candidate list in GRASP (Feo and Resende, 1989; 1995) or aspiration-plus and the elite candidate list setting (Glover and Laguna, 1997)). Further distinctions with respect to choice criteria are made in Section 13.3.2.

The purpose of every input strategy is to decide whether potential *pool* candidates should be treated as high quality information carriers or not, while the purpose of the output strategy is to ensure the guidance of a desired search trajectory (intensification and diversification). The main aspect is the definition of *quality* which has both cost and structural properties. Cost is always measured by an objective function, while structural properties are reflected by values of decision variables. If structural properties are included in transfer decisions, then mechanisms are asked for which are able to define a sort of relatedness.

Relatedness, expressing vicinity or diversity, is often measured by a distance function. Well-known concepts are the Minkowsky metric (including the rectangular and Euclidean measure), Camberra and Chebychev metric, chi-square distance, Mahalanobis distance, correlation coefficients (Kendall, Spearman) and the Hamming distance (see Wilson and Martinez (1997)). A special type of *binary* distance is incorporated within identity checks, which play an important role, e.g. when checking for repetitions in TS (Woodruff and Zemel, 1993) or when avoiding duplications within an SCS (Campos et al., 1999). In addition to these hashing estimates, some authors also propose full duplication checks (Glover, 1998). Another way to directly avoid duplications is the checking for move redundancy in a neighborhood generation mechanism (Greistorfer and Voß, 2001). A different area of duplication avoidance covers special TS memory strategies, e.g. the reverse elimination method (Glover, 1990; Dammeyer and Voß, 1993) or elements of the reactive TS (Battiti and Tecchiolli, 1994).

After cost and structure considerations, there are some additional decisions to be made on both sides of a pool. The number of solutions which are subject to an input or output operation is mainly determined by computational restrictions (time and storage). Furthermore, every input requires the removal of a solution from the pool. This is often done by applying an inverse quality measure, e.g. worst-cost.

Generally, the management of pool transfers gives rise to a wide field of activities, straightforward and advanced, to vary the architecture of a standard pool method. A framework for these is given in Greistorfer and Voß (2001), some specific suggestions are explored in the next section.

13.3 APPLYING POOL STRATEGIES

The basis of all computational experiments is a hybrid procedure, called *tabu scatter search* (TSCS), and was introduced in Greistorfer (2003) for the solving of the the CCPP. A compound neighborhood is the central component of a TS, which maintains a pool of elite solutions to be used in an LP clustering SCM, i.e. the assignment of edges to a certain vehicle. After the generation of a new solution this SCM outcome is again returned to the TS improvement process.

For the encoding of a solution a record vector of length *number-of-edges* is used. All first record attributes store the permutation in which the customers are serviced, while the routes are stored as second edge attributes. Assuming that c_p is the customer number and r_p the route number, then a solution of a problem with dimension n, where n is the number of customers, is written as $S = ((c_1, r_1), \ldots, (c_p, r_p), \ldots, (c_n, r_n))$. It should be added that these storage conventions algorithmically entail a cost function which, assisted by shortest paths, links the customers between one another and to the depot node. In their entity the values in those two record fields are also addressed as C and R, respectively. Further, we define a position function $p(c_p) = p$, while also using the vector style $p(C)$.

The next three subsections describe the distinct algorithmic procedures with regard to the input and output sides of the pool and to the SCM used. The original TSCS approaches are denoted by I_0, O_0 and M_0. All in all, we study four input strategies, $I_{0,\ldots,4}$, four output strategies $O_{0,\ldots,4}$ and three SCMs, namely $M_{0,\ldots,2}$.

13.3.1 *Input strategies*

For all input strategies we follow the well-known worst-replacement approach that cancels the current worst solution from the pool whenever an input is made.

13.3.1.1 Input Strategy I_0. We start our input considerations with a straightforward setting as shown in Table 13.1. The quality aspect in I_0 is simply restricted to the cost dimension of a solution.

Table 13.1 Input strategy I_0 ("good" cost concept).

- A new best-cost solution is always fed into the pool and, moreover, a solution, identified as "good", is accepted, where
  - good means that its cost is within a percentage range of the current best-cost solution.

A solution is deemed to be elite if its cost qualifies, which is indicated by means of a percentage deviation from the currently known best-cost solution. Experiments with different percentage levels showed that this range should not be too small because then it appears to counteract the aim of having a set of *diverse* elite solutions. This range was set to 20%. A consequence of using I_0 is a relatively high transfer rate into the pool, which may support diversification on the one hand. But this tendency may also be overestimated since solutions can be replaced which have a better cost value than the new arrival. Another obvious disadvantage is the fact that this strategy does not pay attention to any structural aspect.

13.3.1.2 Input Strategy I_1. The next input strategy is I_1, which tries to overcome the simple cost aspect by means of examining the logical structure of a solution. This is done by a full comparison of all solution vector elements with regard to all structural attributes which are considered. Before we outline the conditions which have to be fulfilled for such a comparison, a summary of I_1 is presented in Table 13.2.

Table 13.2 Input strategy I_1 (cost, duplication check).

- A new best-cost solution is always fed into the pool and, moreover, a solution is added to the pool
  - if its cost is better than the worst-cost solution contained in the pool and no full duplicate is found after detecting cost identity.

Basically, I_1 is much more restrictive. In contrast to I_0 it does not displace solutions which, in terms of cost, are better than the new candidate. At any time, I_0 keeps the pool holding a number of diverse least-cost solutions which were found during the past search. As depicted in Table 13.2, the first indicator of a duplicate is the detection of a cost identity between the solution to be included and any other current member of the pool. The problem is that structural real-world properties have to be adequately mapped into algorithmic data structures. This is especially the case if properties of symmetry are or have to be allowed for a solution representation. Cost symmetry, which is assumed for the CCPP, gives a higher degree of freedom in data constellations and may therefore complicate a safe identification of duplicates. (Strongly related is the fact that *sets*, e.g. a set of routes, are computationally more likely to be treated as sequences.) The following situation exemplifies the above said:

$$S_1 = ((5,1),(4,1),(6,1),(2,2),(1,2),(3,2),(8,3),(7,3))$$
$$S_2 = ((7,1),(8,1),(6,2),(4,2),(5,2),(3,3),(1,3),(2,3))$$

Scanning either one of the two or both record fields (*customer* or *route*) from left to right and comparing their values in the solutions S_1 and S_2 would yield *different* solutions and thus a wrong result. To achieve a representation that sufficiently supports an unequivocal identification of duplicates, all records are rearranged to satisfy the following primary and secondary sorting keys which both apply to the first field of a record (customer number): (1) increasing numbers of the first element of each substring and (2) the first element number is smaller than the last element's number in each substring. After the rearrangement all entries in the second fields (route numbers) are reset and filled with increasing numbers in accordance with the respective substring (route). If S_1 and S_2 are sorted in this way, yielding $S_3 = ((2,1),(1,1),(3,1),(5,2),(4,2),(6,2),(7,3),(8,3))$, it becomes more apparent that both describe the same fact. This sorting is always performed whenever a solution enters the pool or if an unsorted trial candidate is going to be compared with a pool member. Sorting is also needed for the next two input strategies.

13.3.1.3 Input Strategy I_2. Glover (1998) introduced input strategy I_2 as the so-called *reference set update method.* It maintains a cost-sorted array of elite solutions which all are diverse, evaluated either in the sense of their (cost) or with respect to their structure. In comparison to I_2, full duplication checking is complemented by a preceding hash test.

Table 13.3 Input strategy I_2 (cost, hashing, duplication check).

- A new best-cost solution is always fed into the pool and, moreover, a solution is added to the pool
  - if its cost is better than the worst-cost solution contained in the pool and no (full) duplicate is found after detecting a
    - cost identity, and a
    - hashed identity.

As shown in Table 13.3, a full duplication check is only performed if an identity of the respective hash values indicates a potential solution identity. Hash values are calculated from the first and second attribute values of a solution. For the following two solutions Table 13.4 specifies the calculations using the new example instance S_4:

$$S_3 = ((2,1),(1,1),(3,1),(5,2),(4,2),(6,2),(7,3),(8,3))$$
$$S_4 = ((2,1),(1,1),(3,1),(5,2),(4,2),(6,2),(7,2),(8,3))$$

For the first field numbers (e.g. customers) the hash function suggested is $h_1(S) = \sum_p pc_p^2$ and for the second field numbers (e.g. routes) it is $h_2(S) = \sum_p pr_p^2$ (compare Campos et al. (2001)).

S_3 and S_4 are different solutions with the same hash value on the first attribute but a different hash value on the second attribute. Since $h_2(S_3) \neq$

Table 13.4 Example hashing with two attributes.

$p(\cdot)$	1	2	3	4	5	6	7	8	sum
$h_1(S_3)$	4	2	27	100	80	216	343	512	1284
$h_2(S_3)$	1	2	3	16	20	24	63	72	201
$h_1(S_4)$	4	2	27	100	80	216	343	512	1284
$h_2(S_4)$	1	2	3	16	20	24	28	72	166

$h_2(S_4)$, these two solutions are (correctly) considered to be different. As already mentioned, hashing requires ordered structures, i.e. the solutions to be judged have to be sorted (as done in our example).

13.3.1.4 Input Strategy I_3. The last input strategy to be introduced, I_3, is an extreme variant of I_2. To test the influence of the hashing phase with respect to a subsequent full duplication check, I_3 only makes use of a hashed comparison. This also allows the identification of duplicates, although with a slight degree of error, but it considerably saves computational running time.

Table 13.5 Input strategy I_3 (cost, hashing).

- A new best-cost solution is always fed into the pool and, moreover, a solution is added to the pool
  - if its cost is better than the worst-cost solution contained in the pool and no hashed identity is found after detecting
    - cost identity.

13.3.2 Output strategies

We present four output methods and start again with the setting as it was originally used in the TSCS (Greistorfer, 2003).

13.3.2.1 Output Strategy O_0. Selection method O_0 is parameterized by an upper bound for the maximum number of solutions to be combined and it only refers to random sampling without any further information utilization.

13.3.2.2 Output Strategy O_1. In O_1 advantages are drawn from recording the number of times a solution currently present in the pool has been se-

Table 13.6 Output strategy O_0 (random approach).

- Randomly select a random number of different solutions from the pool.

lected for a combination process. This counter is set to zero by O_1 whenever a new solution is accepted and inserted into the pool.

Table 13.7 Output strategy O_1 (frequency memory).

- Frequency counts record the selection of a solution.
- Select those solutions for combination which have not been used before or have been used rarely.

Keeping frequency counts in that way widely avoids the double-use of subsets and introduces a memory effect in the selection process. In this context it must be noted that the TSCS does not provide an explicit algorithmic structure to *securely* avoid a duplicate subset selection on the output side. A corresponding method, called the *subset generation method*, was introduced in the SCS template of Glover (1998).

13.3.2.3 Output Strategies O_2 and O_3. Apart from random selection and frequency based approaches, it is reasonable to utilize structural properties of the solutions in the pool to decide for a convenient subset choice. Structural relations can be detected if solutions are evaluated according to their *distance* to each other. A distance function evaluates a candidate with regard to its vicinity to the current *best-cost* solution in the pool. These distances, which are stored for every pool member, are used to add structural information in the selection process. We suggest two distance involving output strategies, which are called O_2 and O_3.

As in the sorting algorithm in Section 13.3.1.2, the distance function $d(.)$ is based on the first and second attributes of each solution element. The following explanatory example is based on solutions S_3 and S_5:

$$S_3 = ((2,1),(1,1),(3,1),(5,2),(4,2),(6,2),(7,3),(8,3))$$
$$S_5 = ((1,1),(2,1),(3,2),(5,2),(4,2),(6,2),(7,3),(8,3))$$

Rows 2 and 3 in Table 13.8 give the positions of the customers, C, in the corresponding solutions. The absolute differences for these positions and for the route numbers, δ_p and δ_r, are given below. As shown, the sum of the absolute differences of customer positions plus the sum of the absolute differences of

Table 13.8 Example distance between two solutions.

C	1	2	3	4	5	6	7	8	sum
$p(C_3)$	2	1	3	5	4	6	7	8	
$p(C_5)$	1	2	3	5	4	6	7	8	
δ_p	1	1	0	0	0	0	0	0	2
δ_R	0	0	1	0	0	0	0	0	1
$d(S_3, S_5)$									3

route numbers gives a dissimilarity measure where a high value corresponds to a high distance between the two solutions. Using the idea of distance can be motivated from two angles. On the one hand, it can be argued that worse solutions need not have similar inferior cost neighbors; on the other hand distant solutions may have almost the same good cost. The following two extreme distance involving output strategies are derived from that assumption and are summarized in Table 13.9.

Table 13.9 Output strategies O_2 (min-distance) and O_3 (max-distance).

- Choose a random number of solutions which have the smallest (largest) distance to the best solution.

In O_2 (smallest distance) the best solution is always included in the combination process, which realizes the argument that good solutions may be found in the structural vicinity of the best solution. Hence, this output strategy pursues an intensification aspect. By contrast, O_3 (largest distance) hardly includes the best solution, which, however, is implicitly used by spanning some distant quality solutions around it. So this selection strategy can be understood as a max-min selection approach.

13.3.3 Solution combination Methods

In this subsection three different SCMs are described. In the beginning, we shortly summarize method M_0, which is an LP approach and then two new methods are introduced, which can be directly applied to the case of pure permutation problems and which are extended to handle an additional capacity constraint, e.g., for the solution of the CCPP (or standard VRP). Complementary information can be found in the study of Campos et al. (2001).

13.3.3.1 Combination strategy M_0. The SCM called M_0 was originally proposed in Greistorfer (2003). It is an LP application making use of a classical transportation problem (TPP) model. Given a set of elite solutions $S_1, \ldots, S_{n_c}$, the TPP coefficients a_{ij} depict the number of times a customer i is assigned to a route j in this pool subset. The coefficient matrix can be interpreted as an *assignment frequency matrix* (AFM), being the linear combination of the n_c individual assignment matrices. Unit demands are a consequence of the need that every customer has to be serviced by one vehicle and the supply of a route is approximated as the average number of customers that can be serviced by a single vehicle. A dummy column picks up the oversupply. Maximizing this TPP results in desirable customer-route assignments while simultaneously minimizing the Euclidean distance to the (unfeasible) combination of the initial trial points, namely AFM. Although the outcome of this SCM can be directly used, it is clearly improvable since the optimal clusters provided (sets) do not imply any guidance on how the vehicle routes (sequences) should be formed. Therefore, a greedy sequencing heuristic, GSH, is used to put the customers of all routes into a cost-convenient order. Table 13.10 contains a summary of SCM M_0.

Table 13.10 Solution combination method M_0 (LP-combination).

- Solve the max-TPP:
  - derive the assignment frequencies
  - set up the demand row, the supply and dummy columns
  - determine optimal assignment coefficients
- Use this optimal solution to build a preliminary routing schedule and the sequencing heuristic GSH to form final routes.

13.3.3.2 Combination strategy M_1. The solution construction in M_1 is based on average customer numbers which are derived from all positions in the set of customer permutations. In doing so, a *position votes for its element*. In the following example we refer to a pair of solutions, S_6 and S_7, containing customer arrays C_6 and C_7 as the first field value of the record of each element. (The elements C_{avg} in the example are actually not averages since the division by 2 is not an algorithmic need.) C'_{avg} is the vector of sorted average customers and $p(C'_{avg})$ gives their original position in the unsorted array C_{avg}. According to these positions the adjusted customers of C'_{avg} (the first three 8's become 1, 2, 3; 10 becomes 4 etc.) receive their final position as shown in the C_c row of Table 13.11.

The disadvantage of procedure M_1 is a comparatively higher degree of freedom (and, hence, arbitrariness) if average elements are identical, which is the case in the example, e.g., for $C'_{avg}(i)$ with $i = 1, 2, 3$. However, this more rarely

Table 13.11 Example solution combination method M_1 (average customers).

$p(.)$ and C	1	2	3	4	5	6	7	8	9	10	11	12
C_6	11	5	12	7	6	4	3	2	1	10	9	8
C_7	11	6	9	3	2	10	5	12	7	8	1	4
C_{avg}	22	11	21	10	8	14	8	14	8	18	10	12
C'_{avg}	8	8	8	10	10	11	12	14	14	18	21	22
$p(C'_{avg})$	5	7	9	4	11	2	12	6	8	10	3	1
C_c	12	6	11	4	1	8	2	9	3	10	5	7

occurs if the new solution is composed of a higher number of originals (which is only 2 in this example). Since the new permutation does not reflect the second attribute (e.g. TSP (CPP) obtained, but VRP (CCPP) wanted), feasibility and cost quality are obtained by heuristic clustering according to capacity followed by the greedy improvement procedure GSH which finds least-cost route sequences. The heuristic clustering procedure splits the permutation sequence C_c into a set of route clusters (Greistorfer, 2003). Table 13.13 in the next subsection gives a summary of this SCM together with that of M_2.

13.3.3.3 Combination strategy M_2. In analogy to M_1, the dual approach is used for solution combination method M_2. Here averages are not taken for customers, but for their positions in the sequence of the original permutations. The arrangement idea is that an *element votes for its position*. Table 13.12 describes an application of combination operator M_2. The row $p_{avg}(C)$ shows the average positions of each customer (expressed as sums as above); e.g. $20 = 9 + 11$ or $18 = 6 + 12$. After the adjustment of the sorted positions p'_{avg} (2, 7, 9,... become 1, 2, 3,...), the combined solution is obtained by assigning those customers to C_c who correspond to the unsorted average positions (1, 2, 3,... receive 11, 6, 5,...). The central aspects of combination methods M_2 and M_1 are given in Table 13.13.

13.4 COMPUTATIONAL EXPERIENCE

To gain reliable statements from the computational results, we decided to use a total testing model based on statistical inference. In other words, the set of all 4+4+3 methods ($I_{0,\dots,3}$, $O_{0,\dots,3}$ and $M_{0,\dots,2}$) described in Section 13.3 was subject to a complete crosswise check. The number of all possible *configurations* $K = (i, o, m)$ is $4 \cdot 4 \cdot 3 = 48$, where (i, o, m) means the TSCS with its variable components I_i, O_o and M_m. Each configuration was run over a test set of 50 CCPP instances. This set contains literature examples, e.g. all 23 DeArmon files, and own examples (Hertz et al., 2000; Greistorfer, 2003). To increase statistical significance the whole set of configurations was run twice, the second time starting from a different initial random population.

A single instance run within a given configuration covers 5,000 iterations (or less in the case of reaching an optimal lower bound solution earlier). All runs were performed on a PIV-1.7 GHz under Windows 2000. The running time for a configuration was approximately 20 minutes. The quality of a configuration is evaluated by the average best-cost objective function value calculated from the 50 best solutions.

Configurations are aggregated and combined into several *groups* to evaluate the individual designs of the respective input, output, or SCM procedures. In testing I_3, for instance, all configurations $(3, o, m)$ with $o = 0\dots, 3$ and $m = 0, \dots, 2$ are grouped together. Ensuring that I_3 is tested over all possible output-SCM pairs thus eliminates all effects of a specific output and combination setting and therefore allows a comparison with other input strategies which are grouped in a similar manner. The specific groupings for the evalua-

Table 13.12 Example solution combination method M_2 (average positions).

$p(.)$ and C	1	2	3	4	5	6	7	8	9	10	11	12
C_6	11	5	12	7	6	4	3	2	1	10	9	8
C_7	11	6	9	3	2	10	5	12	7	8	1	4
$p_{avg}(C)$	20	13	11	18	9	7	13	22	14	16	2	11
p'_{avg}	2	7	9	11	11	13	13	14	16	18	20	22
C_c	11	6	5	3	12	2	7	9	10	4	1	8

Table 13.13 Solution combination methods M_1 (average customers) and M_2 (average positions).

- Build a combination of solutions to obtain a relaxed (uncapacitated) solution
  - M_1: Average customers are assigned to positions to keep the deviation from the original customer numbers as small as possible.
  - M_2: Average positions determine the customer sequence to keep the deviation from the original positions as small as possible.
- Include the capacity constraint by heuristic fifo-clustering.
- Improve the solution with the GSH.

Table 13.14 Descriptive results ranked for all types of strategies.

rank	Input, output, combination method		Avg. Cost	Std. Dev.
1	I_3	**cost, hashing**	9072.8333	42.0379
2	I_1	cost, duplication check	9072.8750	41.9610
2	I_2	cost, hashing, duplication check	9072.8750	41.9610
4	I_0	*"good" cost concept*	9075.4167	40.3279
1	O_0	random approach	9064.5000	37.3118
2	O_1	frequency memory	9068.7917	34.2979
3	O_3	**max-distance**	9070.7917	47.5111
4	O_2	*min-distance*	9089.9167	41.2035
1	M_0	**LP-combination**	9021.5938	13.3631
2	M_1	average customers	9098.7188	17.9106
3	M_2	*average positions*	9100.1875	21.4948

tion of the remaining input strategies, the output strategies and the combination methods can be given in a similar manner.

13.4.1 *Descriptive results*

Table 13.14 shows the *ranks* of all methods (the best is 1) and gives the average cost as well as its standard deviation within a given group. From the input strategies results in Table 13.14 it becomes clear that option I_0 is an inferior alternative, which means that the inclusion of structural properties is an indispensable need. Checking the solution structures for identity by means of full duplication checks and/or hashing can increase the algorithmic performance of the TSCS. Obviously, however, it should be emphasized that the similarity of the numerical values underlying the input ranking does not support the assumption that the differences in input design, especially $I_{1,...,3}$, are a valid indicator for deviating results.

The output procedure that works best is the random procedure described as O_0. Having in mind the well-reasoned argumentation regarding advanced output strategies, this is a surprising result which needs to be interpreted in more detail. Random output is a good choice if no other specific technique, either on the input side or for the SCM, is put into action. Although, there is no output strategy testing *without* these mandatory components, the impacts resulting from them are *neutralized* on the average by the grouping of the test designs. This is exactly the reason why O_0 is the best alternative: it causes a positive diversification effect in the absence of another influencing search guidance. The second best output version is strategy O_1. It is a sound argument that any type of memory can improve a heuristic procedure, although the frequency memory implemented cannot keep up with the random choice of O_0. An explanation for this behavior is that these frequency counts are limited in time during which a solution is kept in the pool. Comparing the inferior min-distance strategy O_2 to the max-distance choice in O_3, one can realize the benefit of defining search regions between solutions which are rather distinct from each other while still providing enough quality as they are elite solutions. This nicely coincides with the theory of SCS, which has the aim to span (convex) regions between the trial points to be combined.

The most profitable position of the LP SCM strategy is in line with the comments made for output method O_3. Good linear combinations formed from subsets of elite solutions are more likely to be found if the regions from which they are derived are set up by reference points that are sufficiently diverse. And, obviously, exact methods outperform simple heuristics in combining such a complex set of trial points. We note that the average positions, realized by M_2, do not perform better, as one might have intuitively assumed. It is difficult to give reasons for such a behavior. But, again anticipating some following statistical test results, we can state that this difference to M_1 is not significant.

To finish the discussion on the descriptive results, we specify the overall worst and best configuration, K^- and K^*, found within the set of 48 configurations. The corresponding entries in Table 13.14 are printed in italic and bold typeface. The worst choice, $K^- = (0,2,2)$, is characterized by ignoring structural properties (input side) and/or by suffering from a high coherence in the solution subsets to be combined (output side). In the best configuration, $K^* = (3,3,0)$, however, duplication checking (done without the hash interface) and the interplay between the max-distance output strategy and the LP-combination ensure the best possible heuristic design. Again note that, in contrast to the individual testing of the output methods, the random output strategy loses its position against O_3 in K^* as a result of the positive influence of (non-averaged) effects on the input side (I_3) and as a result of a well-performing combination method (M_0) on the output side.

13.4.2 Hypothesis testing

To include and analyze the *amount* of cost differences within the various configurations, we use the software package SPSS (1999). The primary question is, to what extent a specific input, output or combination variant is able to perform significantly better than another variant. A corresponding hypothesis is $H_0 : \mu_1 = \mu_2 = \ldots = \mu_m$, where $m = 3$ or 4 is the number of design variants. It examines whether the samples represented by the results of the m configuration groups, e.g. $I_{0,\ldots,3}$, come from populations whose means $\mu_{0,\ldots,3}$ are identical. The statistical model is a univariate repeated measures *analysis of variance* (ANOVA). The only dependent variable is the average cost, which is influenced by a number of independent variables (factors). In our context these independent variables are the design variants whose results are summarized in the input, output and combination groups. To evaluate the various designs in the input, output and SCM cases, three ANOVAs with an α-level of 0.05 were run over the respective configuration groups.

13.4.2.1 Input strategies. The non-significance of all input variants was already assumed. Neither the univariate nor the multivariate standard tests indicate any difference between the input strategies introduced. The values of the F-distribution and corresponding significance levels are given as $F = 0.273$ and $sig = 0.606$ (univariate) and $F = 0.517$ and $sig = 0.603$ (multivariate). This needs some interpretation.

The first explanatory argument is again the isolated measuring of the input designs, which means that there is no significant difference in the choice for a specific input method if the influence of the other design options (output and combination) is neutralized. On the contrary, the overall best configuration, K^*, works with the non-trivial input setting I_3. Secondly, non-significance does not mean ineffectiveness or even the confirmation that H_0 is *true*. Nevertheless, it can be deduced in any case that H_0 could not be *rejected* in the present investigation because it is limited in many respects, e.g. by properties of the data set, by a specific heuristic parametrization or by a number of (unknown) factors which were not explicitly controlled in the test environment. And, thirdly, of course reasons *can* be found to underline this statistical result. One is that $I_1, \ldots, I_3$ work very similarly. They prevent duplicates effectively. Test runs in which hashing was simultaneously inspected by full duplication checks revealed that the (empirical) probability of a hashing error is smaller than 0.2%. Another reason may be that the functionality provided by the input strategies is not sufficiently exploited by the other variable components. The usefulness of input strategies might, for instance, be better utilized if more combinations are performed on the output side or if the induced pool diversity is combined with a higher degree of diversification in the selection part, a task which is perfectly realized within the dynamic variant of the subset generation method (Glover, 1998).

Table 13.15 Pairwise comparison of output strategies.

O_i	O_j	Mean Diff. (i-j)	Std. Error	*sig*	95% Confidence Interval Lower Bound	Upper Bound
0	1	-4.292	5.178	0.416	-15.004	6.420
	2	**-25.417**	3.546	0.000	-32.752	-18.082
	3	-6.292	5.617	0.274	-17.910	5.327
1	0	4.292	5.178	0.416	- 6.420	15.004
	2	**-21.125**	4.563	0.000	-30.565	-11.658
	3	-2.000	5.269	0.708	-12.899	8.899
2	0	**25.417**	3.546	0.000	18.082	32.752
	1	**21.125**	4.563	0.000	11.685	30.565
	3	**19.125**	3.435	0.000	12.018	36.232
3	0	6.292	5.617	0.274	- 5.327	17.910
	1	2.000	5.269	0.708	- 8.899	12.899
	2	**-19.125**	3.435	0.000	-26.232	-12.018

13.4.2.2 Output strategies. The significance discussion appears in a completely different light when output strategies or different SCMs are looked upon. We continue our analysis with the selection procedures $O_0, \ldots, O_3$. Here univariate and multivariate significance levels statistically indicate output method differences. The corresponding values are $F = 11.576$ (univariate) and $F = 35.901$ (multivariate), both with a $sig < 0.001$. Therefore, it makes sense to interpret the so-called estimated marginal means which are provided by SPSS as pairwise comparisons between the means of the within-subjects effects. Table 13.15 summarizes these results for the output strategies. It also lists the mean difference between any two output variants, the standard error and gives a $(1 - \alpha)$-confidence interval for the differences of the second column.

For output strategy O_2, which is the min-distance approach, the bad descriptive results can be confirmed. In comparison with all its competitors the mean of O_2 is significantly ($sig < 0.001$) greater, namely with up to 25.417 units of cost. Hence it can be said that the best individual choice is O_0 and the worst one O_2. As for the max-distance strategy in 0_3, there does not seem to be any significant inferiority compared with the random approach ($sig = 0.274$), but it underlines its significant superiority over O_2. The relationship between O_3 and O_1 is not significant nor is the mean difference of importance.

13.4.2.3 Combination methods . The pairwise comparisons for the combination methods M_0 to M_2, i.e. the SCMs designated as LP approach, as average customers strategy and as average positions strategy, can be given without a table. The analysis clearly reveals ($sig < 0.001$) that the LP approach is the most desirable choice for an effective SCM. It improves the cost by an amount of 77.125 and 78.594 if compared with the competitors M_1 and M_2. While there is nothing worth mentioning as regards significant relations between M_1 and M_2, the individual choice for M_0 additionally emphasizes its high usefulness in the collective optimal design $K^* = (3, 3, 0)$.

13.5 SUMMARY AND OUTLOOK

For several reasons, pool design has its attractions for a number of heuristic methods. Using direct, phenotypical encoding, these methods have become known in literature as path relinking, scatter search or as restart methods in their manifold variants. The natural assumption that solutions qualify due to inherent properties that go beyond the cost dimension leads to the effort to identify strategies which are able to effectively support the search process by utilizing structural information that can be derived from a pool of elite solutions.

According to our understanding, a pool has its meaning as an algorithmic component which is able to store a predefined number of high quality solutions. At the first sight, the opportunities connected with this black box approach are clear, however, in fact they do open a wide range of questions where answers cannot be found easily. The general scope of a pool is to provide a collection of restart points from which the search can be newly initialized. A restart point is a certain member of the pool or still has to be generated from several pool solutions using a kind of *solution combination method* (SCM). Apart from the possibilities in SCM design, the main considerations are to decide, depending on what a solution should be added to the pool and in which way solution subsets should be taken from it.

In this study we proposed a variety of input and output strategies as well as different kinds of SCMs, which were implemented as design variants in an existing hybrid tabu scatter search method. All methods were tested using capacitated Chinese postman instances. It should, however, be mentioned that the permutation encoding proposed is context-independent so that our experience could be successfully applied to other problems.

There is every indication that even if simple algorithmic modifications are used, the tuning of a pool process can have a clearly positive effect. For the present framework of application, these effects tend to be smaller on the input side of the pool than on the output side, on which the inclusion of strongly diversifying strategies was found to be useful. This can be performed either by random choice or by a selection criterion that makes use of maximizing the distances to the best solution in the pool. The main contribution comes from an LP-based SCM which provides optimized permutation clusters. All results have been tested and appear to be statistically significant.

Supported by the present findings, future effort will be put into the refinement of heuristic components and into the development of new strategies. In this respect, introducing a type of long-term memory which can guide pool transfers, will certainly be of interest. Memory considerations could also be useful for an SCM whose solution generation makes use of past circumstances.

Another challenge would be the task to explore and develop heuristic design variants in order to gain running time, while simultaneously keeping a given solution quality standard. Finally, encouraged by the present study, we believe that a strengthened employment of statistical tools and corresponding ideas can certainly contribute to finding improvements in metaheuristics design.

Bibliography

M. P. Bastos and C. C. Ribeiro. Reactive tabu search with path-relinking for the Steiner problem in graphs. In C. C. Ribeiro and P. Hansen, editors, *Essays and Surveys in Metaheuristics*, pages 39–58. Kluwer Academic Publishers, Boston, 2002.

R. Battiti and G. Tecchiolli. The reactive tabu search. *ORSA J. on Computing*, 6 (2):126–140, 1994.

V. Campos, F. Glover, M. Laguna, and R. Martí. An experimental evaluation of a scatter search for the linear ordering problem. *J. of Global Optimization*, 21:397–414, 2001.

V. Campos, M. Laguna, and R. Martí. Scatter search for the linear ordering problem. In D. Corne, M. Dorigo, and F. Glover, editors, *New Ideas in Optimization*, pages 331–339. McGraw-Hill, London, 1999.

V.-D. Cung, T. Mautor, P. Michelon, and A. Tavares. A scatter search based approach for the quadratic assignment problem. In T. Bäck, Z. Michalewicz, and X. Yao, editors, *Proceedings of IEEE-ICEC-EPS'97, IEEE International Conference on Evolutionary Computation and Evolutionary Programming Conference*, pages 165–170, Indianapolis, USA, April 1997.

F. Dammeyer and S. Voß. Dynamic tabu list management using the reverse elimination method. *Annals of Ops Res.*, 41:31–46, 1993.

T. A. Feo and M. G. C. Resende. A probabilistic heuristic for a computationally difficult set covering problem. *Ops Res. Letters*, 8:67–71, 1989.

T. A. Feo and M. G. C. Resende. Greedy randomized adaptive search procedures. *J. of Global Optimization*, 6:109–133, 1995.

F. Glover. Tabu search – part II. *ORSA J. on Computing*, 2:4–32, 1990.

F. Glover. A template for scatter search and path relinking. In J.-K. Hao, E. Lutton, E. Ronald, M. Schoenauer, and D. Snyers, editors, *Artificial Evolution*.

Lecture Notes in Computer Science, Vol. 1363, pages 3–51. Springer, Heidelberg, 1998.

F. Glover and M. Laguna. *Tabu Search.* Kluwer Academic Publishers, Boston, 1997.

P. Greistorfer. Hybrid genetic tabu search for a cyclic scheduling problem. In S. Voß, S. Martello, I. H. Osman, and C. Roucairol, editors, *Meta-Heuristics: Advances and Trends in Local Search Paradigms for Optimization*, pages 213–229. Kluwer Academic Publishers, Boston, 1998.

P. Greistorfer. A tabu scatter search metaheuristic for the arc routing problem. *Computers and Industrial Engineering*, 44(2):249–266, 2003.

P. Greistorfer and S. Voß. Controlled pool maintenance in combinatorial optimization. In *Conference on Adaptive Memory and Evolution: Tabu Search and Scatter Search*, 2001.

A. Hertz, G. Laporte, and M. Mittaz. A tabu search heuristic for the capacitated arc routing problem. *Ops Res.*, 48(1, Jan-Feb):129–135, 2000.

M. Laguna. Optimization of complex systems with OptQuest. *http://www.decisioneering.com/articles/article_index.html*, 1997.

C. R. Reeves. Genetic algorithms. In C. R. Reeves, editor, *Modern heuristic techniques for combinatorial problems*, pages 151–196. Blackwell Scientific Publications, 1993.

Y. Rochat and É. Taillard. Probabilistic diversification and intensification in local search for vehicle routing. *Journal of Heuristics*, 1(1):147–167, 1995.

SPSS. Version 10.0, SPSS GmbH Software. München, 1999.

S. Voß. Solving quadratic assignment problems using the reverse elimination method. In S. G. Nash and A. Sofer, editors, *The Impact of Emerging Technologies on Computer Science and Operations Research*, pages 281–296. Kluwer, Dordrecht, 1995.

D. R. Wilson and T. R. Martinez. Improved heterogeneous distance functions. *J. of Artificial Intelligence Research*, 6:1–34, 1997.

D. L. Woodruff and E. Zemel. Hashing vectors for tabu search. *Ann. Ops Res.*, 41:123–137, 1993.

Metaheuristics: Computer Decision-Making, pp.301-324
Baris Güyagüler and Roland N. Horne

14 EVOLUTIONARY PROXY TUNING FOR EXPENSIVE EVALUATION FUNCTIONS: A REAL-CASE APPLICATION TO PETROLEUM RESERVOIR OPTIMIZATION

Baris Güyagüler[1] and Roland N. Horne[2]

[1] ChevronTexaco, EPTC
bguy@chevrontexaco.com

[2] Petroleum Engineering Department
Stanford University
Stanford, CA 94305-2220
horne@stanford.edu

Abstract: Decisions have to be made at every level of petroleum reservoir development. For many cases, optimal decisions are dependent on many nonlinearly correlated parameters, which makes intuitive judgement difficult. In such cases automated optimization is an option. Decisions should be based on the most relevant and accurate tools available. For the well placement problem a numerical simulator that computes the movement and interaction of subsurface fluids is the most accurate tool available to engineers. However, numerical simulation is most often computationally expensive making direct optimization prohibitive in terms of CPU requirements. To overcome the computational infeasibility, one can try to utilize mathematical proxies (surrogates) to replace numerical simulators. Although these proxies are very cheap to compute, they often require an initial investment in computation for calibration purposes. The magnitude of this initial computational investment is unclear. Also the calibration points, that are used to calibrate the proxy, are chosen synchronously; that is, the choice of a particular point to be simulated is independent of the others even though in real life the choice of later *experiments* would be based on the experience of earlier *observations*. In this study, an approach is proposed which employs direct optimization and proxy approaches simultaneously. The Genetic Algorithm (GA) forms the basis of the approach. The proxy ought to evolve *intelligently* as the GA iterates. This work investigated the design of a composite and adaptive algorithm, and tested its effectiveness in a range of artificial test problems and real field cases. Kriging was considered as the proxy. The polytope method was also utilized to help with local search. The composite algorithm was applied to the highly non-linear problem of an offshore Gulf of Mexico hydrocarbon reservoir development and significant improvement in efficiency was observed.

Keywords: Genetic algorithm, Hybrid algorithm, Expensive evaluation, Proxy function, Petroleum reservoir.

14.1 STATEMENT OF THE PROBLEM

The decision of the configuration of oil wells to drill during the development stages of a petroleum reservoir is critical to the success of the overall recovery from the oil field. It is intuitive to base this decision on simulation models which represent the subsurface physics of fluids and rocks. These simulation models are generally very detailed numerical models (Aziz and Settari, 1979) which are computationally demanding and a single simulation may take hours or even days. Since optimization requires many repeated calls to the evaluation function, it is often infeasible to use a numerical model to evaluate the well configurations.

An alternative to using the numerical simulator as the evaluation function is to use a calibrated proxy (substitute) function. This proxy would generally be a mathematical function that would be anticipated to approximate the numerical model and would possess little or no notion of the physics of the petroleum reservoir. In other words, given a well configuration, the proxy function would give a similar result to that obtained by numerical simulation in a shorter computational time without explicit consideration of the physics underlying the evaluation process. The calibration of the proxy would be done using instances obtained by carrying out numerical simulation with different well configurations.

To utilize proxies as a method one would make an initial computational investment by carrying out numerical simulation with selected well configurations, calibrating a proxy with these configurations and finally using this proxy for optimization. However the initial number of well configurations to simulate is unclear. Also the training points are chosen synchronously and generally there is no rational basis for the choice of initial configurations.

Direct optimization (Rian and Hage (1994); Beckner and Song (1995); Bittencourt and Horne (1997); Güyagüler and Gümrah (1999)) and use of proxies (Rogers et al. (1994); Aanonsen et al. (1995); Pan and Horne (1998); Stoisits et al. (1999); Centilmen et al. (1999); Johnson and Rogers (2001)) have been examined for the well placement problem in the petroleum literature. However the proposed approaches carry both the advantages and the disadvantages of either direct optimization or proxy approaches.

Table 14.1 Problem definition.

Index name	Scheme
Objective	To maximize oil field performance by determining the best well locations and pumping rates
Objective Function	Cumulative oil production obtained at the end of a run, or the Net Present Value (NPV) calculated from the numerical model output and the costs associated with the project
Decision variables	Location of proposed wells as (i, j) indices of grid location on the numerical model and pumping rates as discretized values within a specified range.
Data Structure	Integer strings for exact mapping arranged linearly representing (i, j) indices and pumping rates
Constraints	Wells must be conformed within the numerical grid, all wells should be within active grid blocks and the specified pumping rates range should be honored

To avoid the problems associated with direct optimization using numerical simulation as the evaluation function and the utilization of the proxy approach as a standalone method, a hybrid algorithm is proposed which iteratively evolves a proxy within the simple Genetic Algorithm (GA) context.

The ordinary kriging system (Matheron, 1965) is used as the proxy function in this study. Kriging has favorable properties such as being data exact and easy extension to multiple-dimensions.

In addition to iterative proxy calibration the polytope algorithm (Nelder and Mead, 1965) is utilized to resolve possible local refinement problems associated with GAs.

The problem as defined in this study is summarized in Table 14.1. This is a highly nonlinear combinatorial problem and cannot be expected to be solved in polynomial time. Other techniques exist, such as Simulated Annealing or simple GAs, which are able to deliver reasonable results for nonlinear combinatorial problems. However the special case for this problem is that the evaluation function is very expensive (in the order of hours or tens of hours) thus there is a limit to the number of evaluations one can make. In this study, considerable computational time is spent on the optimization algorithm itself to minimize the number of calls to the evaluation function.

14.2 ALGORITHM COMPONENTS

14.2.1 *Genetic Algorithm*

GAs are evolutionary algorithms that have been developed by Holland (1975). The central theme of GAs is robustness, the balance between exploitation and exploration. Such a balance is necessary for survival of species in different environments. GAs employ natural selection (selection based on fitness) as well as crossover for information exchange and mutation to introduce further variety (randomness) into search. Details of GAs can be found in the literature (Goldberg, 1989).

14.2.2 *Kriging*

Kriging is an interpolation algorithm assuming correlation among data points (Matheron, 1965). Kriging has been used widely in earth sciences, particularly in geostatistics. Geostatistical applications are usually limited to three dimensions, however the algorithm can be extended to as many dimensions as necessary (n_d), thus kriging can be used for interpolation of multivariate functions.

Ordinary kriging, which is the most widely used kriging method, was used in this study. Ordinary kriging is used to estimate a value at a point of a region for which the variogram is known, without prior knowledge about the mean. Furthermore, ordinary kriging implicitly evaluates the mean in the case of a moving neighborhood.

Suppose the objective function value at point x_0 in Figure 14.1 is to be estimated with the n data points. Combining function values at the data points linearly with weights λ_i (Wackernagel, 1998):

$$Z^*(x_0) = \sum_{i=1}^{n} \lambda_i Z(x_i) \tag{14.1}$$

The weights should be constrained to add up to one. In ordinary kriging, this constraint is imposed by Lagrange formalism, with the introduction of the Lagrange parameter μ. In simple kriging, which is the fundamental kriging algorithm, there is no need for such a constraint since the weights are made to sum up to one by giving the weight λ_{n+1} to the mean, hence in simple kriging, prior knowledge of the mean is necessary.

The data are assumed to be part of a realization of an intrinsic random function $Z(x)$ with the variogram $\gamma(h)$. In a geostatistical study, the variogram that has a best fit to the data would be chosen. Here in the context of optimization a boundless variogram or a variogram with a long range is necessary. Also Gaussian behavior of the variogram near the origin is desired to prevent non-smooth estimation surface with artifacts resembling *circus tents* at data locations. Variograms with Gaussian behavior at the origin ensure smooth estimations since estimations very close to a data point will have almost per-

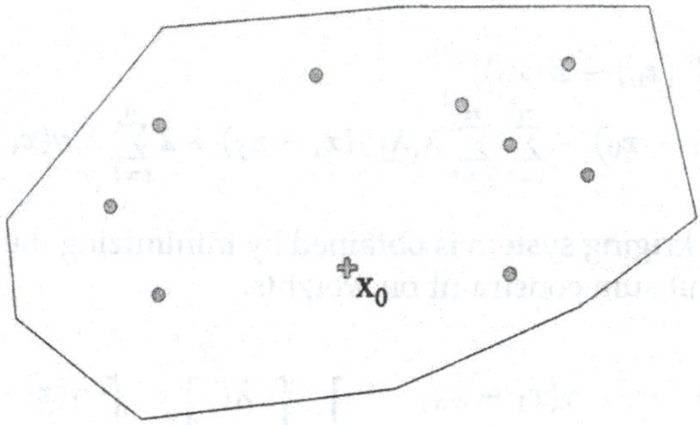

Figure 14.1 A two-dimensional parameter space with irregularly spaced data (circles) and a location where value is to be estimated (cross).

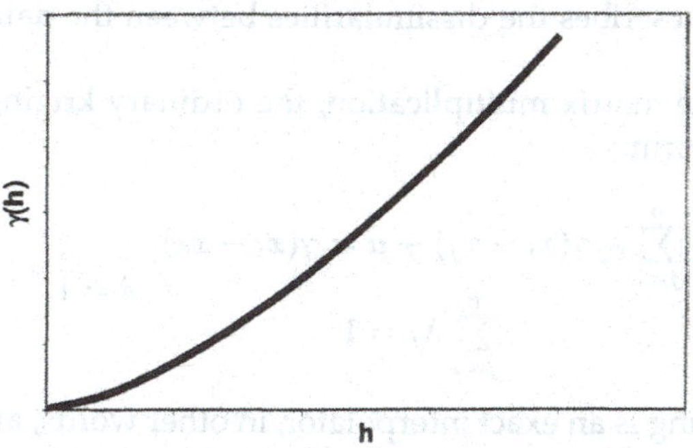

Figure 14.2 The power variogram with exponent of 1.5.

fect correlation with the value at the data point. The power variogram with exponent of 1.5 was used in this study:

$$\gamma(h) = h^{1.5} \tag{14.2}$$

Equation (14.2) is visualized in Figure 14.2. The power variogram is not bounded (i.e. does not have a practical range) and behaves near-Gaussian near the origin.

The unbiasedness is assured with unit sum of the weights:

$$\sum_{i=1}^{n} \lambda_i \mathrm{E}[Z(x_i) - Z(x_0)] = 0 \tag{14.3}$$

The estimation variance is given by:

$$\begin{aligned} \sigma_{\mathrm{E}}^2 &= \mathrm{E}\left[(Z^*(x_0) - Z(x_0))^2\right] \\ \sigma_{\mathrm{E}}^2 &= -\gamma(x_0 - x_0) - \sum_{i=1}^{n}\sum_{j=1}^{n} \lambda_i\lambda_j\gamma(x_i - x_j) + 2\sum_{i=1}^{n} \lambda_i\gamma(x_i - x_0) \end{aligned} \tag{14.4}$$

The ordinary kriging system is obtained by minimizing the estimation variance with the unit sum constraint on weights:

$$\begin{bmatrix} \gamma(x_1 - x_1) & \cdots & \gamma(x_1 - x_n) & 1 \\ \vdots & \ddots & \vdots & \vdots \\ \gamma(x_n - x_1) & \cdots & \gamma(x_n - x_n) & 1 \\ 1 & \cdots & 1 & 0 \end{bmatrix} \cdot \begin{bmatrix} \lambda_1 \\ \vdots \\ \lambda_n \\ \mu \end{bmatrix} = \begin{bmatrix} \gamma(x_1 - x_0) \\ \vdots \\ \gamma(x_n - x_0) \\ 1 \end{bmatrix} \tag{14.5}$$

The left hand side describes the dissimilarities among data points while the right hand side describes the dissimilarities between the data points and the estimation point.

Performing the matrix multiplication, the ordinary kriging system can be rewritten in the form:

$$\begin{cases} \sum_{j=1}^{n} \lambda_j\gamma(x_i - x_j) + \mu = \gamma(x_i - x_0) \\ \sum_{j=1}^{n} \lambda_j = 1 \end{cases} \qquad i = 1, \ldots, n \tag{14.6}$$

Ordinary kriging is an exact interpolator, in other words, at the point of the data, the estimated value is equal to the data value.

If all data are used for each estimation point, only the right hand side vector in Equation (14.5) is modified, while the left hand side matrix remains unchanged. Thus given a data configuration, the left hand side matrix in Equation (14.5) has to be inverted only once. After the inversion, every estimation requires a single matrix multiplication. Also regardless of the number of dimensions of the problem (n_d) the left hand side matrix in Equation (14.5) has the size $n + 1$, n being the number of data points.

14.2.3 Polytope Method

The polytope method is a direct search method which relies on function value comparison (Nelder and Mead, 1965). It is basically a greedy hill-climber that does not require derivative information. Details of the polytope algorithm can be found in literature (Gill et al., 1981).

14.2.4 Hybridization

In this study GA was hybridized with the proxy approach and the polytope method. The proxy was improved iteratively at each generation. The resulting algorithm is illustrated in Algorithm 14.1.

```
procedure HybridAlgorithm( )
begin
    repeat
        Select individuals;
        Apply GA operators: crossover and mutation;
        Apply PM to n + 1 points in population;
        Create proxy function from database;
        Optimize with proxy function from database;
        Verify results with actual evaluation function;
    until Stopping criteria is met
    return best solution;
end
```

Algorithm 14.1 The hybrid algorithm utilizing Genetic Algorithm (GA), Polytope Method (PM) and the proxy approach.

The finite-difference numerical simulation is used as the evaluation function in this study. The finite-difference formulation of flow equations divides the physical space into discrete intervals thus the GA population consists of linearly arranged strings representing problem variables such as well location as the (i, j) indices of the coordinates of an injection well on the finite-difference grid and the injection rate of water from the well. The GA operators of crossover and mutation are applied to all the individuals in the population with specified probabilities. At the end of each generation a polytope is constructed from the fittest individual ever encountered and the n_d closest individuals to this individual, n_d being the dimensions of the problem.

To integrate the proxy method, a database of visited points is constructed and updated whenever a function evaluation is made. This database is used to construct the proxy, which replaces the actual evaluation function. Using this proxy, estimations are made for the unevaluated points. The maximum point of the proxy representation of the evaluation function is found by a recursive call to the hybrid with the proxy replacing the numerical model as the evaluation function. This proxy maximum is then evaluated by the actual evaluation function for verification, and if better than the best solution, the worst solution in the generation is replaced. The proxy method requires only one function evaluation per generation.

An additional error analysis step may be employed when using the kriging proxy to further ensure robustness. The kriging framework enables the computation of the estimation variance (Equation (14.4)). The location in the search space with the maximum estimation variance can be interpreted as the location with the poorest kriging estimation. Thus making an additional simulation at this location ensures that the proxy is not missing a significant optimal region, making the overall hybrid algorithm more robust.

Table 14.2 Well indexing schemes.

Index name	Scheme
Row	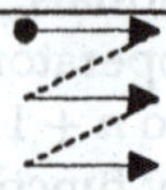
Column	
Snake row	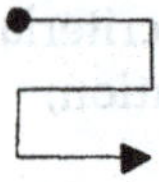
Snake column	
Spiral	
Hilbert curve	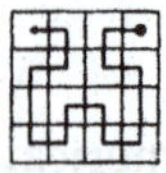

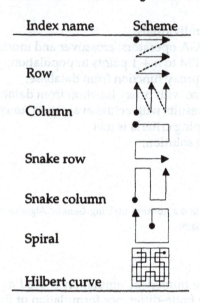

The composite algorithm is referred to as the Hybrid Genetic Algorithm (HGA).

14.3 DATA STRUCTURES

The representation of the solution vectors for the well placement problem is discussed in this section. Different types of indexing schemes are investigated. The way the well-blocks are indexed is important since it determines the behavior of the evaluation function. Bittencourt and Horne (1997) indexed active cells only arguing that (i, j) indexing is not suitable because the optimization algorithm could place wells in inactive regions.

A synthetic petroleum reservoir with grid dimensions of 32×32 was constructed. Several types of well-block indexing were investigated for the injector placement problem using this synthetic case. Well indexing types investigated are given in Table 14.2.

The response surface of the 32×32 problem was generated through exhaustive simulation. The objective of optimization was to maximize cumulative oil production. The cumulative oil production surface given by exhaustive runs is shown in Figure 14.3.

The indexing schemes in Table 14.2 map the two-dimensional surface (Figure 14.3) into a single dimension. The resulting one-dimensional functions are given in Figure 14.4.

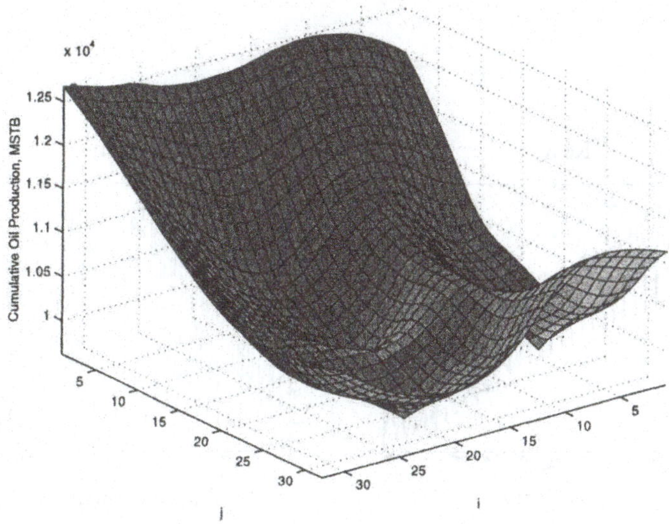

Figure 14.3 Cumulative oil production surface for the single injection well placement problem for the 32×32 synthetic case obtained by exhaustive simulation.

All of the indexing schemes except the Hilbert curve are easy to generate. The Hilbert curve was first described by David Hilbert in 1892. The Hilbert curve is a space-filling curve that visits every point in a square grid with a size of any power of 2. The useful feature of Hilbert curves is that two points close to each other on the Hilbert curve are also close in the two-dimensional space. The Hilbert curve mapping results in a smaller number of large scale local optima but the one-dimensional function has a lot of small scale noise (Figure 14.4). The Hilbert curve has links to the area of fractals and the generation of the Hilbert curve requires a recursive algorithm. The one limitation in our case is that the Hilbert curve mapping works only for square grids with a size of any power of 2. The 32×32 Hilbert curve, also referred to as the level 5 Hilbert curve ($2^5 = 32$), is given in Figure 14.5;

After mapping with the given indexing schemes, the smooth two-dimensional function (32×32) given in Figure 14.3 is replaced by noisy one-dimensional functions (1024×1) given in Figure 14.4. This artificial noise is problematic from an optimization point of view. The HGA is applied to the indexed one-dimensional functions (Figure 14.4) and the results are given in Figure 14.6.

It is seen from Figure 14.6 that (i, j) indexing of wells is most suitable for optimization algorithms for the well placement problem since other kinds of indexing may introduce artificial noise due to the discontinuities in the indexed search space. Oddly, the promising Hilbert curve indexing scheme results in the worst optimization performance which implies that small-scale noise is more of a hindrance to optimization with GAs than large-scale noise.

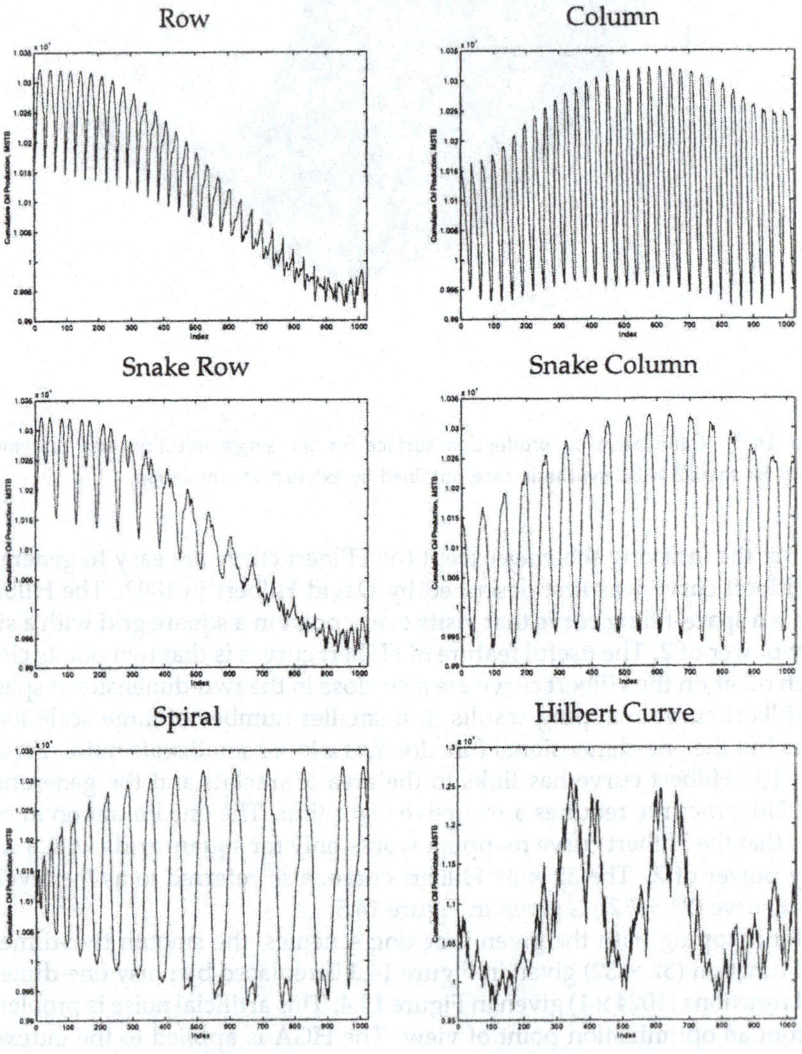

Figure 14.4 One dimensional objective functions generated by mapping from the 32×32 synthetic injector placement problem by different well indexing schemes.

Figure 14.5 The Hilbert curve used to map the 32×32 two-dimensional surface into a single-dimension.

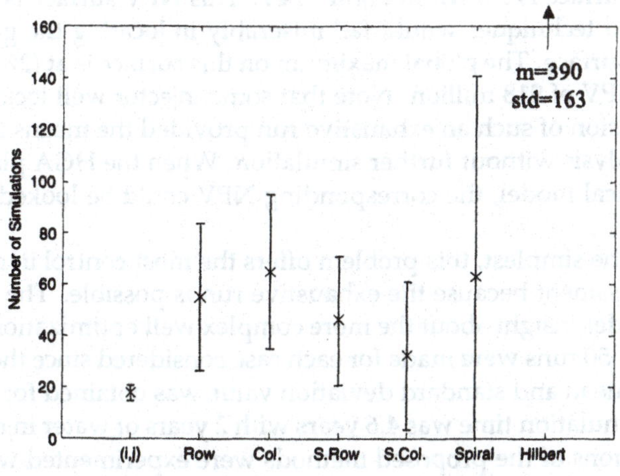

Figure 14.6 Comparison of the efficiency of the HGA for different types of indexing; crosses indicate the means and the error bars indicate the standard deviations of 50 runs.

14.4 REAL-WORLD APPLICATION

The hybrid algorithm was applied to the Pompano field, which is an offshore development in the Gulf of Mexico. The operating companies decided to carry out water injection from an infill well after 2.6 years of production. The optimum location and rate of water injection wells were sought. The evaluation function used was the Pompano numerical reservoir model which is a full field-scale model with 40,000 cells with dimensions of 40 × 40 × 25 having 7,533 active cells.

Each individual in the GA generation represents the (i, j) location of injectors and their injection rates. Integer chromosomes were utilized to achieve exact mapping. In order to obtain the fitness value for an individual, a numerical simulation of the reservoir is carried out with the existing producer wells and the injector wells represented by the individual. The resulting flow history goes through an economic analysis to obtain the Net Present Value (NPV) of the proposed well configuration and the injection rates.

14.4.1 Controlled Experimentation

The HGA was used first to optimize the location of one injection well injecting water with constant rate. Fixing the pumping rate decreased the size of the search space and the relatively small search space enabled the exhaustive simulation of every possible well location. The exhaustive run for this problem was made by carrying out a simulation at every possible active cell in the numerical model. The corresponding incremental NPVs from the no-injection case for each well location were calculated from the simulation output. The parameters for NPV calculations are given in Table 14.3. The resulting incremental NPV surface is shown in Figure 14.7. This NPV surface is very noisy. Gradient based techniques would fail miserably in locating the global maximum for this surface. The global maximum on this surface is at (27,14) with an incremental NPV of $18 million. Note that some injector well locations result in loss. Possession of such an exhaustive run provided the means to carry out sensitivity analysis without further simulation. When the HGA placed a well on the numerical model, the corresponding NPV could be looked up from a table.

While it is the simplest, this problem offers the most control in overall performance assessment because the exhaustive run is possible. The exhaustive run also provides insight about the more complex well optimization problems of similar sort. 50 runs were made for each case considered since the process is stochastic. A mean and standard deviation value was obtained for each case.

The total simulation time was 4.6 years with 2 years of water injection. Various combinations of the proposed methods were experimented with for this problem. The number of simulations consumed to correctly locate the optimum well location are given in Table 14.4 and visualized in Figure 14.8. It is observed that ordinary GA performs poorly with an average of 184.2 simulations. Utilization of the polytope method with the GA improves the search

Table 14.3 Parameters for net present value calculations.

Parameter	Value
Discount rate, %	10.0
Oil price, $/bbl	25
Gas price, $/MSCF	3
Water handling cost, $/bbl	1
Operation cost, $/day	40,000
Well cost, $/well	20,000,000
Capital expenditures, $	500,000,000

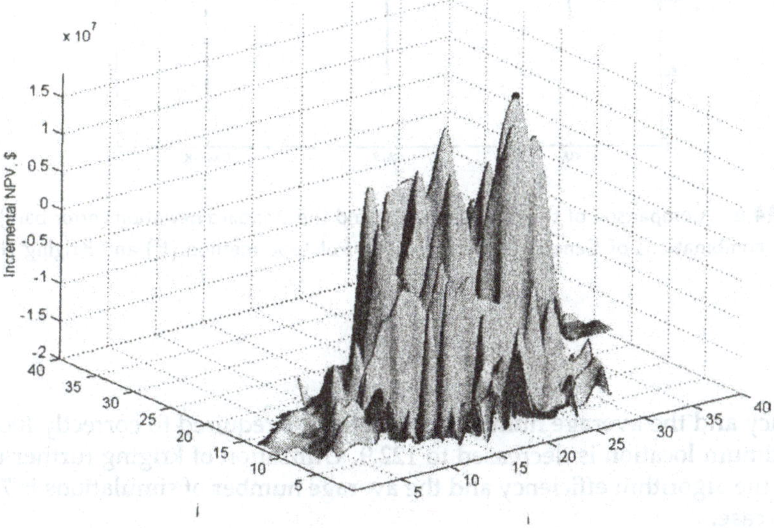

Figure 14.7 The incremental net present value surface for the single injector well placement problem generated by exhaustive simulations.

Table 14.4 Means (m) and standard deviations (σ) of number of simulations to correctly locate global optimum for different algorithms.

	GA	GA+P	GA+P+K
m	184.2	122.9	72.8
σ	96.47	72.95	46.35

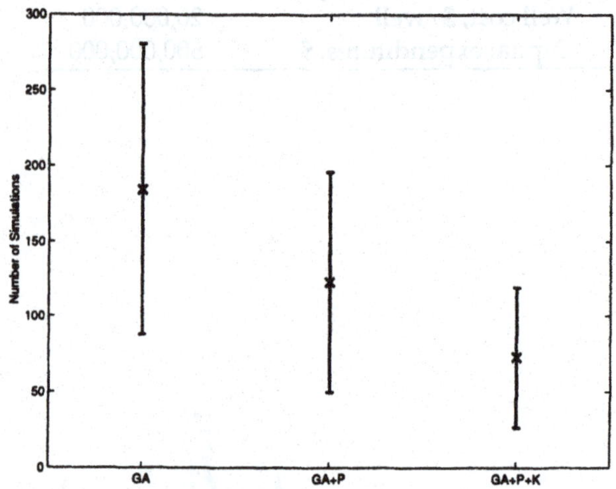

Figure 14.8 Comparison of the mean (crosses) and the standard deviation (error bars) for different combinations of Genetic Algorithm (GA), Polytope method (P) and Kriging (K).

efficiency and the average number of simulations required to correctly locate the optimum location is decreased to 122.9. Utilization of kriging further improves the algorithm efficiency and the average number of simulations is 72.8 for this case.

The kriging proxy was iteratively evolved through the generations. The evolution of the kriging estimate of the NPV surface for one of the runs using GA, polytope method and the kriging proxy is shown in Figure 14.9. The actual surface is also given in Figure 14.9 which is a zoomed view of the exhaustive dataset at around the global optimum. It is seen that as generations progress and more points are visited, the kriging estimates get better and eventually at generation 6, the algorithm correctly locates the optimum well location resulting in the highest NPV.

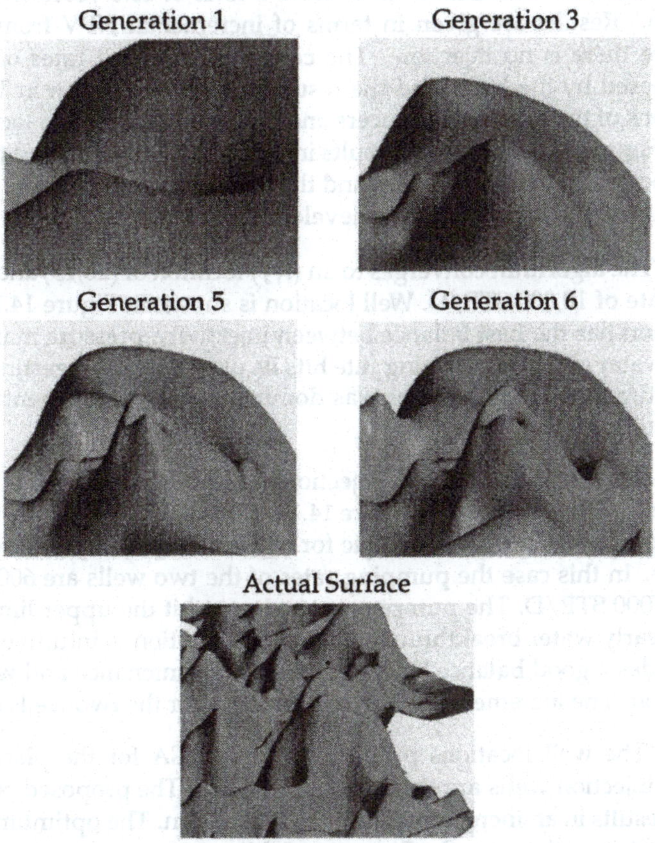

Figure 14.9 Evolution of the kriging estimation for the single injector placement problem, the black dot is the location of the true global optimum.

14.4.2 Optimization of Multiple Well Locations and Pumping Rates

The configuration and pumping rate of up to four wells was optimized using the HGA. In these cases it is not feasible to carry out exhaustive runs since the search space size is very large and a very large number of simulations would be necessary. The pumping rate was discretized into ten intervals from 1000 STB/D to 10,000 STB/D. Search space sizes for the problems considered are shown in Table 14.5. Runs were allowed to iterate for 200 generations before termination. Simulation times were a total of 10.6 years with 8 years of injection. Results are given in terms of incremental NPV from the base case, where there is no injection. The configurations and rates of injection wells proposed by the HGA and the resulting NPVs are given in Table 14.6. The locations of the existing producers and the proposed injector locations are plotted in Figure 14.10. From the results in Table 14.6, it is observed that water injection increases profit in all cases and that three injectors with the proposed configuration is the most profitable development plan.

1 Injector The algorithm converges to an (i, j) location of (26,13) and a pumping rate of 10,000 STB/D. Well location is shown in Figure 14.10.a. This location has the best balance between injectivity, pressure maintenance and water cut. The pumping rate hits its upper limit suggesting that the pressure maintenance factor was dominating. The incremental NPV is $80 million.

2 Injectors The locations for two injection wells proposed by the HGA for highest NPV are shown in Figure 14.10.b. These locations have favorable rock properties and are strategic for better sweep and pressure maintenance. In this case the pumping rates of the two wells are 6000 STB/D and 7000 STB/D. The pumping rates do not hit the upper limit to prevent early water breakthrough. The HGA solution is intuitive in that it provides a good balance between pressure maintenance and water production. The incremental NPV is $87 million for the two wells case.

3 Injectors The well locations proposed by the HGA for the placement of three injection wells are shown Figure 14.10.c. The proposed configuration results in an incremental NPV of $92 million. The optimum number of injectors is three for the Pompano field.

4 Injectors For the four well placement problem, the HGA places the wells as shown in Figure 14.10.d. Introduction of the additional well does not pay off and the resulting incremental NPV of $81 million is less than the two and three well scenarios.

In all the cases considered the HGA proposed intuitive solutions. A reservoir engineer would not oppose the well locations and pumping rates proposed by the HGA. It should be noted that, different from a human being, the optimization procedure is able to evaluate all the effects of hundreds of factors in a straightforward and precise manner. Some of these factors are rock and

Table 14.5 Search space sizes for the problems considered for the Pompano field.

Number of Injectors	Search space size
1	8,700
2	37,801,500
3	109,372,340,000
4	237,064,546,950,000

Figure 14.10 Optimum well locations of one to four injectors for the Pompano field single injector location optimization problem.

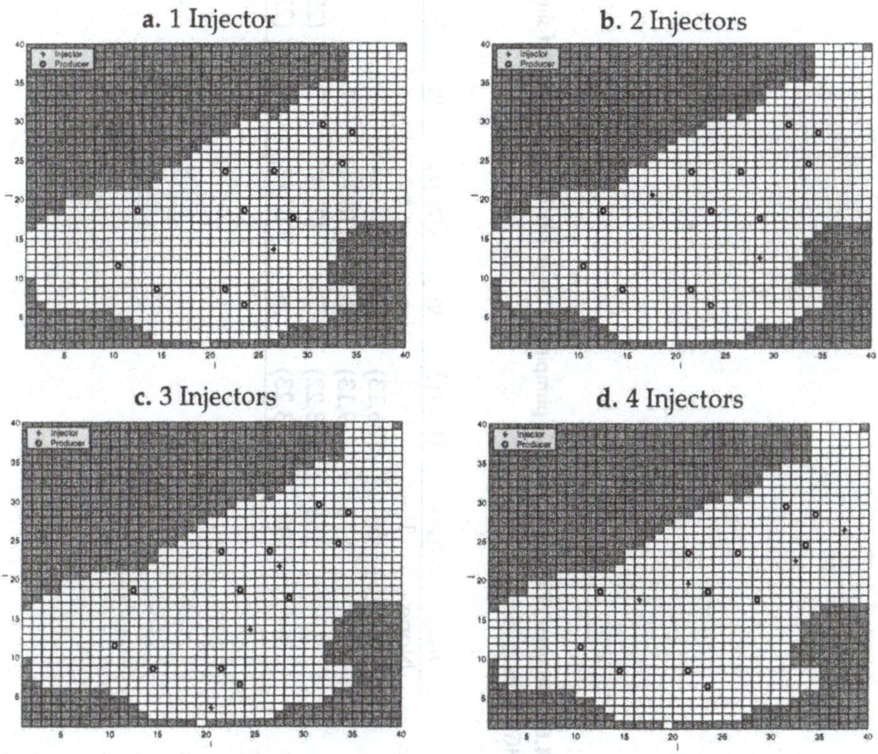

Table 14.6 Number of injectors (n_{inj}), pumping rates in MSTB/D, number of simulations (n_{sim}) and the resulting incremental NPVs in MM$ for the HGA runs.

n_{inj}	n_{sim}	(i_1, j_1)	q_1	(i_2, j_2)	q_2	(i_3, j_3)	q_3	(i_4, j_4)	q_4	NPV
None	1	-	-	-	-	-	-	-	-	0
1	53	(26,13)	10	-	-	-	-	-	-	80
2	173	(29,13)	6	(18,21)	7	-	-	-	-	87
3	156	(28,22)	7	(21,4)	5	(25,14)	8	-	-	92
4	92	(33,23)	6	(17,18)	9	(22,20)	7	(38,27)	4	81

fluid properties, physics of flow through porous media, economic parameters, etc. Most of these factors have nonlinear and implicit effects on the objective function, which are hard to evaluate manually. In particular it should be noted that adding more injectors changes the optimal location of earlier injectors; this phenomenon would prevent the application of successive or sequential optimization methods such as dynamic programming.

14.5 GA PARAMETER DEPENDENCE

The GA parameters used were based on rule-of-thumb values from the literature. For binary alphabets, Goldberg (1989) suggests an appropriate population size and De Jong (1975) suggests appropriate GA operator probabilities based on experimentation over problems of different nature:

Population size - same as total string length, $n = l$

Crossover probability - $p_c = 0.6$ works well for a wide range of problems

Mutation probability - inverse of population size, $p_m = 1/n$

For the Pompano problem, integer data structures were used. A population size of 12 was used since the length of the equivalent binary string would have been 12. Since De Jong suggests a mutation probability of 1/12 per bit the equivalent of this for integer data structures is 0.5 mutation probability.

The GA parameter dependence of optimization performance was also investigated. Sensitivity of the GA performance to mutation and crossover probabilities was determined by varying these probabilities from 0.0 to 1.0 with 0.01 increments. The mean and variance of the necessary number of simulation runs consumed by the GA to locate the global optimum for the single injector placement problem (Figure 14.7) was determined by carrying out 100 GA optimizations for each mutation and crossover probability pairs.

The dependency of the GA to mutation and crossover probability is given in Figure 14.11 which shows the average number of simulations that was required to locate the global optimum. The total number of GA optimizations made to construct Figure 14.11 were:

$$(101\ p_m \text{ values}) \times (101\ p_c \text{ values}) \times (100 \text{ GA runs per } p_c - p_m \text{ pair}) = 1{,}020{,}100 \text{ optimization runs,}$$

where p_m is the mutation probability and p_c is the crossover probability.

The variance of the number of simulations that the GA consumed to locate the global optimum is given in Figure 14.12. In each of the 1,020,100 optimization runs carried out, the GA was allowed to consume at most 1000 simulations, otherwise the run was terminated after 1000 generations.

The result acquired from Figure 14.11 and Figure 14.12 is rather surprising: *GA optimization performance does not depend on crossover probability for the single injector placement problem for the Pompano field.* However this result is reasonable when one evaluates the single injector placement problem more carefully.

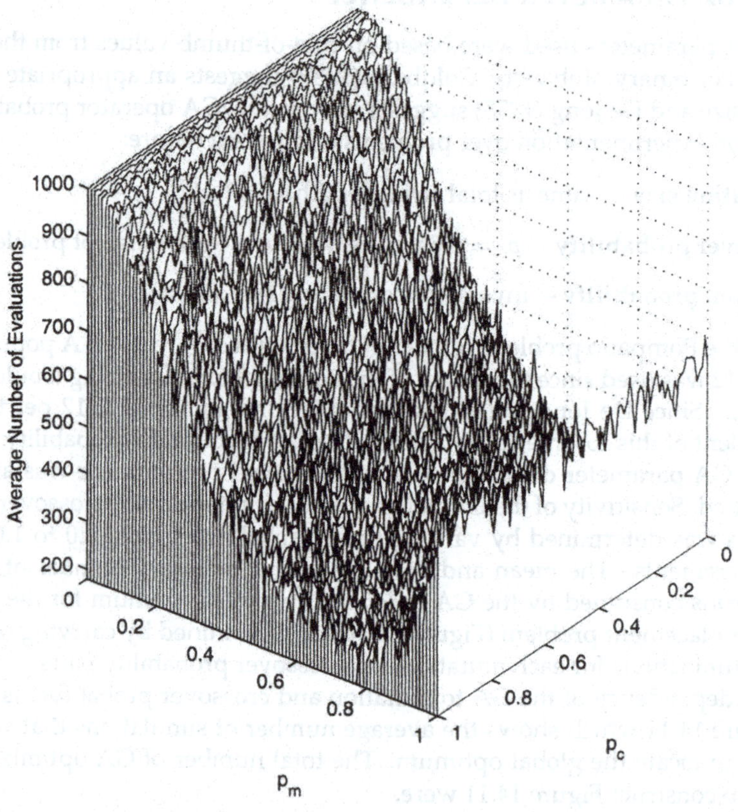

Figure 14.11 The average number of simulations that the GA consumed to locate the global optimum for different values of mutation probability (p_m) and crossover probability (p_c).

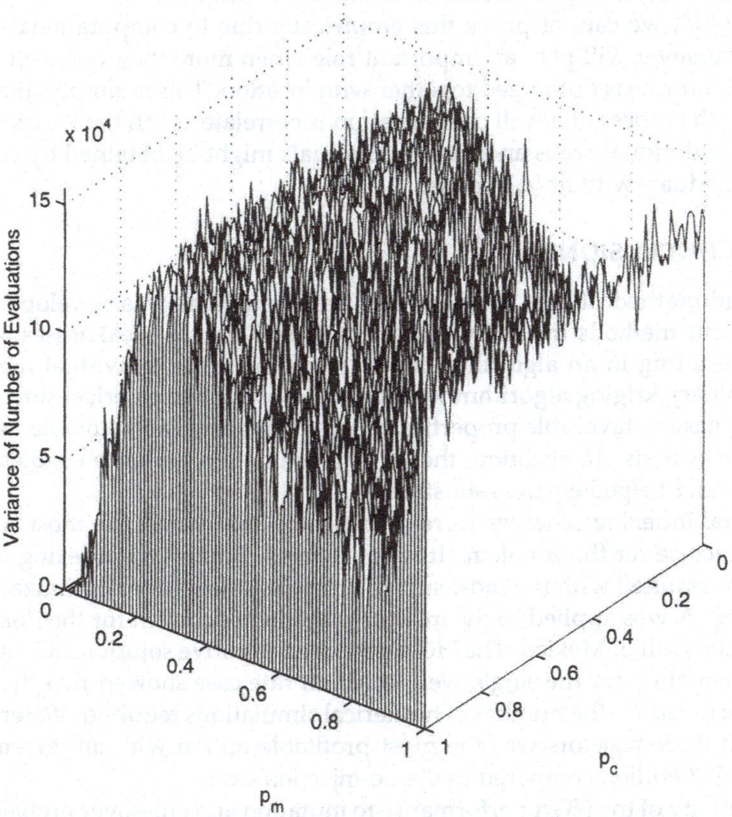

Figure 14.12 The variance of the number of simulations that the GA consumed to locate the global optimum for different values of mutation probability (p_m) and crossover probability (p_c).

Optimized parameters are the i and j location of the proposed injector and there is no insight suggesting that a particular i or j value contributes to the fitness of an individual. In other words the combination of i and j is what makes an individual fit, i or j alone is not correlated with the fitness of a given well configuration. This can also be observed by looking at the search space (Figure 14.7).

Also the mutation probability obtained with the method suggested by De Jong ($p_m = 0.5$) is observed to be a reasonable value.

Although we cannot prove this empirically due to computational limitations, crossover will play an important role when more than one well is considered and rate is optimized together with location. This is simply due to the fact that the fitness of a well configuration is correlated with the locations and rates of individual wells and fitter individuals might be obtained by combining individuals with fit (i, j) pairs.

14.6 CONCLUSIONS

A hybrid method of GA, polytope and proxy methods was developed. The component methods in the hybrid compensated for the weaknesses of each other, resulting in an algorithm that outperformed the individual methods. The ordinary kriging algorithm was used as a proxy to numerical simulation. Kriging has the favorable properties of being data-exact and suitable for multiple dimensions. In addition, the kriging estimation variance enabled error analysis and helped ensure robustness.

Several indexing schemes were investigated to find out the most suitable data structure for the problem. It was discovered that (i, j) indexing of well locations resulted with the most suitable response surface for optimization.

The HGA was applied to the injector placement problem for the Pompano field in the Gulf of Mexico. The HGA proposed intuitive solutions. Controlled experimentation for the single well, constant rate case showed that the HGA was able to reduce the number of numerical simulations required. Waterflooding with three injectors was the most profitable option with an incremental NPV of $92 million compared to the no-injection case.

Sensitivity of the HGA performance to mutation and crossover probabilities were also examined. It was observed that algorithm performance had no correlation with crossover probability for the single injector placement problem for the Pompano field. However it is perceived that crossover will play an important role for problems with higher dimensions.

Acknowledgments

The authors acknowledge the financial support from SUPRI-D Research Consortium on Innovation in Well Testing and SUPRI-B Research Consortium on Reservoir Simulation and also the permission from BP-Amoco and the DeepLook consortium to publish information from the Pompano field.

Bibliography

S. I. Aanonsen, A. L. Eide, L. Holden, and J. O. Aasen. Optimizing reservoir performance under uncertainty with application to well location. *In Proceedings 1995 SPE Annual Technical Conference & Exhibition, SPE 30710*, 1995.

K. Aziz and A. Settari. *Petroleum reservoir simulation*. Applied Science Publishers, 1979.

B. L. Beckner and X. Song. Field development planning using simulated annealing - optimal economic well scheduling and placement. *In Proceedings 1995 SPE Annual Technical Conference & Exhibition, SPE 30650*, 1995.

A. C. Bittencourt and R. N. Horne. Reservoir development and design optimization. *In Proceedings 1997 SPE Annual Technical Conference & Exhibition, SPE 38895*, 1997.

A. Centilmen, T. Ertekin, and A. S. Grader. Applications of neural networks in multiwell field development. *In Proceedings 1999 SPE Annual Technical Conference & Exhibition, SPE 56433*, 1999.

K. A. De Jong. *An Analysis of the Behavior of a Class of Genetic Adaptive Systems*. PhD thesis, University of Michigan, 1975. Dissertation Abstracts International 36(10), 514B.

P. E. Gill, W. Murray, , and M. H. Wright. *Practical Optimization*. Academic Press, London, UK, 1981.

D. E. Goldberg. *Genetic Algorithms in Search, Optimization, and Machine Learning*. Addison-Wesley, Reading, MA, 1989.

B. Güyagüler and F. Gümrah. Comparison of genetic algorithm with linear programming for the optimization of an underground storage field. *IN SITU*, pages 131–150, 23(2) 1999.

J. H. Holland. *Adaptation in Natural and Artificial Systems*. The University of Michigan Press, Ann Harbor, 1975.

V. M. Johnson and L. L. Rogers. Applying soft computing methods to improve the computational tractability of a subsurface simulation-optimization problem. *Journal of Petroleum Science and Engineering*, special issue on Soft Computing 2001.

G. Matheron. *Les Variables Régionalisées et leur Estimation, une Application de la Theorie de Fonctions Aleatoires aux Sciences de la Nature.* Paris, Masson et Cie, 1965.

J.A. Nelder and R. Mead. A simplex method for function minimization. *Computer Journal*, pages 308–313, July 1965.

Y. Pan and R. N. Horne. Improved methods for multivariate optimization of field development scheduling and well placement design. *In Proceedings 1998 SPE Annual Technical Conference & Exhibition, SPE 49055*, 1998.

D. T. Rian and A. Hage. Automatic optimization of well locations in a north sea fractured chalk reservoir using a front tracking reservoir simulator. *In Proceedings 1994 SPE International Petroleum & Exhibition of Mexico SPE 28716*, 1994.

L. L. Rogers, , and F. U. Dowla. Optimal groundwater remediation using artificial neural networks with parallel solute transport. *Water Resources Research*, pages 457–481, 30(2) 1994.

R. F. Stoisits, K. D. Crawford, D. J. MacAllister, A. S. Lawal, , and D. O. Ogbe. Production optimization at the kuparuk river field utilizing neural networks and genetic algorithms. *In Proceedings 1999 SPE Mid-Continent Operations Symposium, SPE 52177*, 1999.

H. Wackernagel. *Multivariate Geostatistics.* Springer-Verlag, Berlin Heidelberg, 1998.

Metaheuristics: Computer Decision-Making, pp. 325-345
Jean-Philippe Hamiez and Jin-Kao Hao

15 AN ANALYSIS OF SOLUTION PROPERTIES OF THE GRAPH COLORING PROBLEM

Jean-Philippe Hamiez[1] and Jin-Kao Hao[2]

[1] LGI2P, École des Mines d'Alès (site EERIE)
Parc Scientifique Georges Besse
F-30035 Nîmes CEDEX 01, France
Jean-Philippe.Hamiez@ema.fr

[2] LERIA, Université d'Angers (U.F.R. des Sciences)
2, Boulevard Lavoisier
F-49045 Angers CEDEX 01, France
Jin-Kao.Hao@univ-angers.fr

Abstract: This paper concerns the analysis of solution properties of the Graph Coloring Problem. For this purpose, we introduce a property based on the notion of *representative sets* which are sets of vertices that are always colored the same in a set of solutions. Experimental results on well-studied DIMACS graphs show that many of them contain such sets and give interesting information about the diversity of the solutions. We also show how such an analysis may be used to improve a tabu search algorithm.
Keywords: Graph coloring, Solution analysis, Representative sets, Tabu search.

15.1 INTRODUCTION

Given an undirected graph $G = (V, E)$ with a vertex set V and an edge set E, the goal of the graph coloring problem (GCP for short) is to find a color assignment to every vertex in V such that any pair of adjacent (or connected) vertices receive different colors, and the total number of colors required for the feasible color assignment be minimized (*assignment approach*). The smallest color size corresponds to the *chromatic number* $\chi(G)$ of graph G. The GCP can also be defined more formally as follow: partition V into a minimal number of

k subsets $V_1, \ldots, V_k$ such that $\forall u \in V_i(u \in V, 1 \leq i \leq k), \nexists v \in V_i, v \in V/(u, v) \in E$ (*partition approach*).

Many practical problems, such as timetable construction (Leighton, 1979) or frequency assignment (Gamst, 1986), can be mapped into a GCP. Graph coloring is also a classic constraint satisfaction problem (Tsang, 1993) with applications to many problems in Artificial Intelligence.

GCP in an arbitrary graph is a well-known *NP-complete* problem (Garey and Johnson, 1979) and only small problem instances can be solved exactly within a reasonable amount of time in the general case (Dubois and De Werra, 1993). It is also hard even to approximate the chromatic number of a graph. In Lund and Yannakakis (1994), it is proved that for some $\epsilon > 0$, approximating the chromatic number within a factor of $|V|^\epsilon$ is NP-hard. Indeed, one of the best known approximation algorithm (Halldórsson, 1993) provides an extremely poor performance guarantee[1] of $O(|V|(\log\log|V|)^2/(\log|V|)^3)$.

The reported heuristics for the solution of the graph coloring problem range from greedy constructive methods, such as DSATUR (Brélaz, 1979) and the Recursive Largest First algorithm (Leighton, 1979), to sophisticated hybrid strategies like HCA (Dorne and Hao, 1998a; Galinier and Hao, 1999) and those proposed by Morgenstern (1996) or Funabiki and Higashino (2000). The last ones are among the most efficient approaches for GCP. We may also mention local search metaheuristics, e.g., simulated annealing (Johnson et al., 1991) or tabu search (Hertz and De Werra, 1987; Fleurent and Ferland, 1996b; Dorne and Hao, 1998b), pure genetic algorithms (Davis, 1991), neural network attempts (Jagota, 1996), and scatter search (Hamiez and Hao, 2002). One can find more methods in Johnson and Trick (1996).

Although most authors in the graph coloring community try to explain why their approach gives good results on some graphs and poor ones on others (mainly by analyzing the behavior of the algorithm according to different options and settings), few studies are available concerning analysis of properties of GCP solutions. Moreover, to our knowledge, no algorithm exploiting such properties exists for the GCP.

Analysis of solution properties may help explain the behavior of some algorithms on a particular graph. Such an analysis can also be used to develop new algorithms relying on such properties. In this paper, we are concerned with studying a particular property that may be called *representative sets* (RS for short). Informally, a RS for a graph is composed of a set of vertices shared by its solutions. More formally, given a set C_k of legal k-colorings (solutions with exactly k colors) of a graph G, $\{v_1, \ldots, v_m\}(m > 1), v_j \in V(1 \leq j \leq |V|)$, is a RS if $\forall c_i \in C_k(1 \leq i \leq |C_k|), col_i(v_1) = \cdots = col_i(v_m)$, with $col_i(v_j)$ being the color of vertex v_j in solution c_i.

A RS may coincide exactly with a complete color class in which case the RS is called a *complete representative set* (CRS). It may also correspond simply to

[1]The performance guarantee is the maximum ratio, taken over all inputs, of the color size over the chromatic number.

a subset of color classes in which case the RS is called *partial representative set* (PRS). The formal definitions of CRS and PRS are given later in Section 15.3. Like the notion of "backbone" for the satisfiability problem (Monasson et al., 1999), the RS reveals an invariant of solutions for the graph coloring problem.

The paper begins by recalling some existing analysis schemes (Section 15.2). Then, we describe the analysis schemes we propose (Section 15.3). Experimental results on a set of well-studied DIMACS graphs are presented in Section 15.4. A simple algorithm using analysis information is then introduced (Section 15.5), showing a better performance. Some concluding remarks and future studies are discussed in the last section.

15.2 SOME EXISTING ANALYSIS SCHEMES

Before reporting any previous studies done in the field of graph coloring analysis, and so the usefulness of such research works, let us first recall a few notions and definitions used later on.

- A *critical subgraph* is a subgraph which is uncolorable with a given number k of colors but which becomes k-colorable if any edge is removed from it.
- The set of *critical edges* is the set of edges that occur in every critical subgraph.
- A $G_{n,p}$ *random graph* denotes a graph with n vertices, where p is the probability that there exists an edge between any pair of vertices.
- *Phase transition* separates over-constrained problems (probability of finding a solution near 0) from under-constrained instances (probability near 1). See, e.g., Cheeseman et al. (1991).

One recent study we found about topological analysis of GCP solutions is due to Culberson and Gent (1999). They define *frozen same pairs* as pairs of vertices that are always in the same color class, formally, $(u, v) \in V \times V, u \neq v$, is a frozen same pair if $\forall c_i \in C_k, col_i(u) = col_i(v)$ where $col_i(u)$ is the color of vertex u in solution c_i. Note that this definition is a particular case of representative sets introduced in this paper. They reported results for 4-coloring random graphs, 3-coloring random graphs and 3-coloring triangle-free graphs using a backtracking program. One of their most interesting conclusions is that, at phase transition, the coloring hardness relies on the large size of *critical subgraphs*, which are not easily checkable to confirm uncolorability quickly; the size of critical subgraphs evolving with the number of vertices. Furthermore, existence of *critical edges*, that are difficult to check explicitly, makes the search of solutions harder, "suggesting that hardness at phase transitions is an algorithm independent property". A similar study can be found in Galinier (1999).

Hertz et al. (1994) developed the first topological analysis for the k-coloring problem. They investigated in particular the number and distribution of *lo-*

cal optima of the k-coloring problem. All local optima of small random graphs ($|V| \leq 20$) were enumerated, up to a permutation of the colors, using a branch-and-bound procedure. The authors observed that the percentage of local optima which are global ones increases with the value of k. Then, the probability of reaching a local optimum which is a global one is extremely small for $k < \chi(G)$ (over-constrained problem, no valid k-coloring). Furthermore, when k is increasing from $\chi(G)$, "most local optima which are not global have a small number of conflicts". Finally, using additional statistical measures, they also explain the performance of a tabu algorithm (Glover and Laguna, 1997) according to the evolution of k around $\chi(G)$. If k is a few units above $\chi(G)$, tabu search seems effective since the landscape contains few *valleys* (or *plateaus*). It produces poor results when $k \leq \chi(G)$ since a small proportion of local optima are global ones.

Yokoo (1997) also analyzed the search space landscape of the 3-coloring problem. Analyzes were performed using a descent algorithm[2] (with restart) over small random instances. He reported results by varying the edge density ($|E|/|V|$): probability of satisfiability, number of solutions, number of local minima, ... The main objective was to clarify the cause of a paradoxical phenomenon. For incomplete algorithms, problems are easier beyond the phase transition region (few solutions) than problems in the phase transition region (although there are more solutions). One of the main results showed that, while increasing the edge density, the number of local minima decreases due to plateaus of small size. Thus, more paths lead to solutions.

The performance of several hybrid algorithms developed for the graph coloring problem are examined by Fleurent and Ferland (1996a). The authors reported comparative results for $G_{n,p}$ random graphs and Leighton graphs (Leighton, 1979). They mainly study the effect of three string-based crossovers (*1-point*, *2-point* and *uniform* (Davis, 1991)) on the quality (number of conflicts) of solutions. For a 450-node Leighton graph, they also give an interesting analysis, from a topological point of view, of the effect of these crossovers on the entropy of a population. Recall that the entropy measure evaluates the diversity of a population[3]. Hence, it can be used to monitor the convergence of a population or to provide information on the behavior of population-based algorithms using different options and settings. They note that "the uniform crossover operator converges more slowly, but ultimately gives better results". In other words, the uniform crossover seems to insure a certain level of diversity.

Walsh (1999) studied graphs with a *small world* topology, i.e., graphs in which vertices are highly clustered yet the path length between them is small (Watts and Strogatz, 1998). He demonstrated that register allocation graphs have a small world topology and observed similar results with other DIMACS

[2]The algorithm moves iteratively to a better neighboring configuration, i.e., with fewer conflicts, until a local minimum is reached.

[3]See Welsh (1988), e.g., for discussions about interesting properties of a measure based on entropy.

benchmark graphs. He also used a DSATUR-based backtracking algorithm (Brélaz, 1979) to color graphs with a small world topology generated according to the model proposed by Watts and Strogatz (1998)[4]. Results showed that the cost of solving such graphs has a heavy-tailed distribution. He suspected then that "problems with a small world topology can be difficult to color since local decisions quickly propagate globally". Finally, to combat this heavy-tailed distribution, he tried the strategy of randomization and restart (Gomes et al., 1998). As in other studies, this technique appeared particularly effective to eliminate these heavy tails. See Walsh (2001), for extended results on these graphs. Analysis are also reported for other non-uniform graphs generated using alternative models, namely, *ultrametric* graphs (Hogg, 1996) and *power law* graphs (Barabási and Albert, 1999).

15.3 EXTENDED ANALYSIS SCHEMES

In this section, we present two analysis schemes based on the notion of representative sets. The first scheme is designed to extract complete representative sets (CRS), i.e., complete color classes shared by a set of given k-colorings. The second one aims at extracting partially shared representative sets (PRS) of color classes. These two schemes may be considered as two applications of the same idea at two different levels. To implement these schemes, we use extensively binary tree techniques.

15.3.1 Extracting complete representative sets: surface analysis

15.3.1.1 Surface analysis: definition and example. Recall that the vertices of a representative set (RS) $\{v_1, \ldots, v_m\} (m > 1), v_j \in V (1 \leq j \leq |V|)$ are always colored the same in a set of solutions, i.e., $\forall c_i \in C_k (1 \leq i \leq |C_k|), col_i(v_1) = \cdots = col_i(v_m)$, with C_k a set of legal k-colorings of a graph G and $col_i(v_j)$, the color of vertex v_j in solution c_i.

To be a complete representative set (CRS), the set $\{v_1, \ldots, v_m\}$ must coincide exactly with a color class for each of the given k-colorings in C_k. In other words, the color class $V_j^i \subseteq V (1 \leq j \leq k, 1 \leq i \leq |C_k|)$ from solution c_i is a complete RS if $\forall c_p \in C_k, p \neq i, \exists V_q^p \in c_p (V_q^p \subseteq V) / V_q^p = V_j^i$.

To explain the underlining idea, we will use the example shown in Figure 15.2 (left) where the sample set $C_k (k = 4)$ comprises three solutions c_1, c_2 and $c_3 (|C_k| = 3)$ for the sample input graph of 10 vertices (numbered from 1 to 10) shown in Figure 15.1. In this example, one complete color class {1,2,3,4} appears in all the three solutions. Thus we have only one CRS.

In what follows, we use this example to explain in detail the technique used to identify and extract such a CRS from a given set of k-colorings.

[4]Starting from a regular graph, randomly rewire each edge with probability p. As p increases from 0, the graph develops a small world topology.

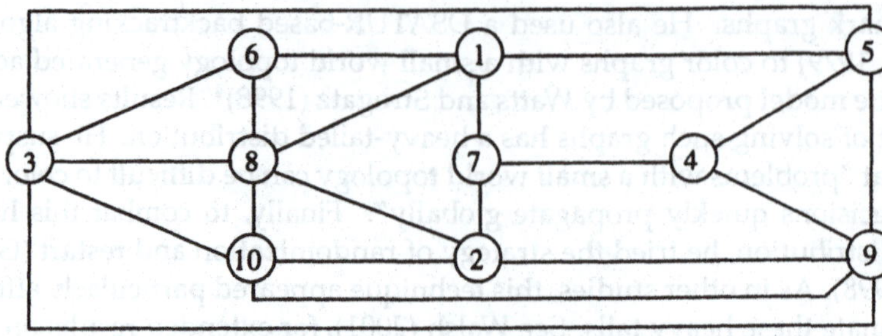

Figure 15.1 A sample input graph.

15.3.1.2 Surface analysis: a binary tree based technique. The core technique for extracting CRS is a representation of the given solutions as a binary tree. In such a tree, each node corresponds to a vertex in the given graph. Figure 15.2 (right) shows such a representation for the solutions c_1, c_2 and c_3.

From a logical point of view, two steps are necessary to carry out a surface analysis: building the binary tree and then extracting the CRS from the tree. From a practical point of view, these two steps are merged such that CRS are successively identified and extracted while building the tree. Let us now see how this happens.

Let *root vertices* be the nodes that can be reached from the root node of the tree (included) following right edges ({1, 5, 7, 8, 10} in our sample). In such a structure, a path exists from a *root vertex* v_r to all nodes v_i reached following left edges, if v_r and all v_i belong to the same color class of a solution. For instance, the path 7 → 8 → 9 exists since vertices 7, 8 and 9 are in the same color class (V_4) in solution c_2.

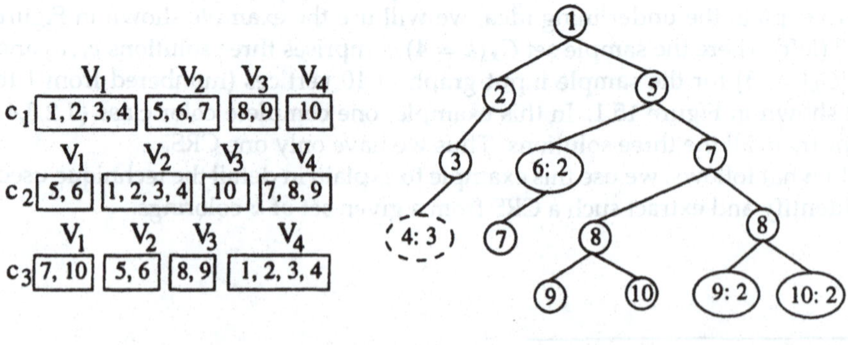

Figure 15.2 Illustration of the surface analysis.

When building the tree[5], right edges are added (and, so, root vertices) to include all the color classes of C_k that are not already in the structure. They are also used to represent color classes issued from the same root vertex. This is the case in Figure 15.2 for the path $7(\rightarrow 8) \rightarrow 10$ since the root vertex 7 appears in $V_4 = \{7, 8, 9\}$ of c_2 and in $V_1 = \{7, 10\}$ of c_3. If one adds a fourth solution containing, for instance, the color class $\{7, 9\}$, then node 8 will still be the left node of 7, 9 will be inserted as a right node of 8 and 10 will become the right node of 9 (instead of 8).

Let *representative nodes* be the particular nodes which contain additional information, e.g. *6: 2* in the tree of Figure 15.2. The first number is a vertex of the studied graph and the second (let us call it *FREQ*, for "frequency") shows the number of times the set composed of vertices which belong to a path from a root vertex to a representative node covers entirely a color class within all studied solutions. For instance, *6:* 2 means "the color class $\{5, 6\}$ appeared two times (in c_2 and c_3) as a complete color class over the studied solutions $\{c_1, c_2, c_3\}$". Thus, CRS can directly be retrieved by following paths from root vertices to all representative nodes which have a frequency equal to the number of solutions studied ($|C_k|$). There is only one CRS $\{1, 2, 3, 4\}$ in our illustration corresponding to the path $1 \rightarrow 2 \rightarrow 3 \rightarrow 4$ (tree node with a dashed line style).

Algorithm 15.1 gives an outline of the procedure we used to build the surface analysis structure (denoted by *SA_STRUCT*). $first(V_j^i)$ refers to the first vertex of V_j^i. "*min_FREQ*" is an integer parameter ($1 \leq min\_FREQ \leq |C_k|$) used to restrict the output to CRS which have a minimal frequency. For instance, it has been set to 1 to generate the structure illustrated in Figure 15.2 (right drawing). Notice that *min_FREQ* influences the size of *SA_STRUCT*.

Finally, let us say that the order the solutions are examined has no incidence on the identification of CRS. Indeed, the core treatment of Algorithm 15.1 consists in comparing each color class from any solution with color classes of the other solutions (line 3), the order of these comparisons being irrelevant. For instance, Figure 15.2 (right) has been generated following the order permutation $\pi_1 = \{c_1, c_2, c_3\}$, but $\pi_2 = \{c_2, c_3, c_1\}$ will give the same result.

One must pay attention to the structures storing the representative sets and the solutions since all the color classes from all the solutions are first checked for existence in *SA_STRUCT* (line 2). If a color class V_j^i is already in the tree, it is just ignored since it has already been considered. Insertion (line 4) may also be easy and fast. We implemented the *SA_STRUCT* and solutions storage by means of binary trees because they are well convenient for searching and inserting, but balanced trees (*B-Trees*) may also be of great interest.

15.3.1.3 Time and space complexity. Algorithm 15.1 has an $O(|V|^3)$ worst time complexity, weighted by $|C_k|^2$:

[5]Note that the tree is ordered for fast searching and inserting operations.

```
ALGORITHM 1()
SA_STRUCT ← ∅;
for i = 1 to |C_k|
1    for j = 1 to k
2        if V_j^i ∉ SA_STRUCT(V_j^i ∈ c_i)
             counter ← 1;     /*counts the frequency of V_j^i in C_k*/
3            for i' = i + 1 to |C_k|
                 if V_j^i = V^{i'}_{col_{i'}(first(V_j^i))}
                     counter ← counter + 1;
             endfor
             if counter ≥ min_FREQ
4                SA_STRUCT ← SA_STRUCT ∪ V_j^i;
                 FREQ_{V_j^i} ← counter;
             endif
         endif
     endfor
endfor
```

Algorithm 15.1 Building the surface analysis structure

- If $k = 1$ (one color class) then *SA_STRUCT* only contains one CRS (which is V) and $FREQ_V = |C_k|$. This leads to a simplified algorithm running in O(1). Remark that this case corresponds to null graphs ($|E| = 0$).
- If $k = |V|$ then *SA_STRUCT* contains exactly $|V|$ CRS, each one being composed of a single vertex with frequency $|C_k|$. Here again the algorithm is reduced and runs in O(1). In this case, the input graph $G = (V, E)$ is complete ($|E| = |V|(|V| - 1)/2$) if $k = \chi(G)$.
- The worst case is the most commonly encountered and occurs when $1 < k < |V|$. Lines 1–3 require here a maximum of $|V|$ elementary operations each, leading to an overall time complexity of $O(|V|^3)$. Note that insertions (line 4) can easily be achieved in O(1) by storing the possible right place when searching (line 2).

Recall that Algorithm 15.1 aims at counting the frequency of each color class in C_k. So, assuming that all color classes occur exactly once and $min\_FREQ = 1$ (worst case), *SA_STRUCT* contains exactly $k|C_k|$ items. This leads to an $O(|V|)$ worst space complexity, up to a factor of $|C_k|$.

15.3.2 Extracting partial representative sets: depth analysis

15.3.2.1 Definition and example. The previous analysis scheme allows us to identify complete representative sets; we will see later that many DIMACS benchmark graphs do contain such sets. However, this analysis scheme is interested only in complete color classes and is blind of subsets of color classes. Indeed, if there is no entire color class shared by all the studied solutions, surface analysis cannot give useful information.

For instance, in Figure 15.2 (left), the set of vertices {5, 6} is shared by the given colorings. However, it represents only a *subset* of the color class V_2 of the solution c_1 though it corresponds to a complete color class in c_2 (V_1) and in c_3 (V_2). Surface analysis will miss such a partial representative set (PRS).

This section presents the *depth analysis* which is designed to extract PRS. This scheme allows us to discover a finer and deeper information compared to surface analysis.

Before giving the description of the technique used for depth analysis, let us define formally the notion of partial representative set. Given a set C_k of legal k-colorings of a graph G, the set of vertices $\{v_1, \ldots, v_m\}, v_i \in V (1 < m \leq |V|)$, is a PRS if:

1. $\exists (V_a^q, V_b^{q'}) / \{v_1, \ldots, v_m\} = V_a^q \cap V_b^{q'} (V_a^q \in c_q, a \in [1..k], c_q \in C_k, V_b^{q'} \in c_{q'}, b \in [1..k], c_{q'} \in C_k, q \neq q')$ and

2. $\forall c_p \in C_k (1 \leq p \leq |C_k|), \exists V_j^p \in c_p (j \in [1..k]) / \{v_1, \ldots, v_m\} \subseteq V_j^p$.

15.3.2.2 Depth analysis: the technique. The technique used to carry out a depth analysis is quite similar to that of a surface analysis. The main difference is that partial representative sets are not required to cover an entire color class; they can be **subsets** of color classes. Here, the tree structure also stores **intersections** of color classes. In this case, the "frequency" counts the number of **inclusions** of any intersection into color classes.

Figure 15.3 gives an illustration of this *depth analysis,* with respect to the graph of Figure 15.1. Like for complete representative sets, partial representative sets can directly be extracted by following paths from root vertices to all representative nodes which have a frequency equal to the number of solutions studied. For our example, there are three PRS: {1, 2, 3, 4}, {5, 6}, and {8, 9}.

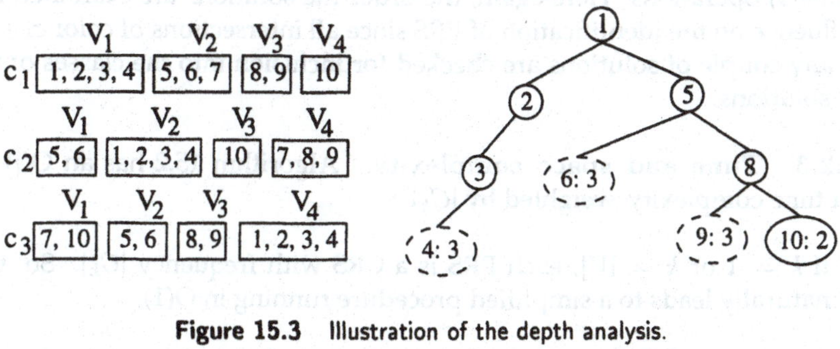

Figure 15.3 Illustration of the depth analysis.

Algorithm 15.2 gives an outline of the procedure we used to build the tree structure of the depth analysis (denoted by *DA_STRUCT*). Some practical details are voluntarily omitted for a greater readability. Let us say simply that

```
ALGORITHM 2()
DA_STRUCT ← ∅;
for i = i to |C_k| − 1
  for j = i to k
    for i' = i + 1 to |C_k|
      for j' = i to k
1       if V_j^i ∩ V_{j'}^{i'} ∉ DA_STRUCT(V_j^i ∈ c_i, V_{j'}^{i'} ∈ c_{i'})
          counter ← 1; /*counts the frequency of V_j^i ∩ V_{j'}^{i'}
          in C_k*/
2         for i'' = i + 1 to |C_k|
            if (V_j^i ∩ V_{j'}^{i'}) ⊆ V^{i''}_{col_{i''}(first(V_j^i ∩ V_{j'}^{i'}))}
              counter ← counter + 1;
          endfor
          if counter ≥ min_FREQ
                                  /*2 ≤ min_FREQ ≤ |C_k|*/
3           DA_STRUCT ← DA_STRUCT ∪ (V_j^i ∩
            V_{j'}^{i'});
            FREQ_{V_j^i ∩ V_{j'}^{i'}} ← counter;
          endif
        endif
      endfor
    endfor
  endfor
endfor
```

Algorithm 15.2 Building the depth analysis structure

special data structures are necessary to allow fast searching (line 1) and inserting (line 3) operations. Here again, the order the solutions are examined has no influence on the identification of PRS since all intersections of color classes from any couple of solutions are checked for inclusion into the classes of the other solutions.

15.3.2.3 Time and space complexity. Algorithm 15.2 has an $O(|V|^4)$ worst time complexity, weighted by $|C_k|^3$:

- If $k = 1$ or $k = |V|$, each PRS is a CRS with frequency $|C_k|$. So, this naturally leads to a simplified procedure running in $O(1)$.

- for $1 < k < |V|$, the additional cost of Algorithm 15.2, compared to the time complexity of Algorithm 15.1 (Section 15.3.1.3), comes from line 2 (which requires $|C_k||V|$ elementary operations at most), leading to a time complexity $O(|V|^4)$.

The space complexity of *DA_STRUCT* only relies on the number of times the test in line 1 is true (assuming also that $V_j^i \cap V_{j'}^{i'} \neq \emptyset$). So, suppose that it is always true and $min\_FREQ = 2$ (worst case).

From a theoretical point of view, this case generates a maximum of $\#PRS$ different sets to be included in the depth analysis structure, with:

$$\#PRS = \frac{k|C_k|(k|C_k|-1)}{2} - |C_k|\frac{k(k-1)}{2} \tag{15.1}$$

$\#PRS$ can be retrieved by building a complete graph $G' = (V', E')$ with $|V'| = k|C_k|$ vertices, each vertex $v_j^i \in V'$ representing the color class V_j^i in c_i. Each edge $(v_j^i, v_{j'}^{i'}) \in E'$ means that $V_j^i \cap V_{j'}^{i'} \notin DA\_STRUCT$ (and $V_j^i \cap V_{j'}^{i'} \neq \emptyset$). At this stage $|E'| = k|C_k|(k|C_k| - 1)/2$ (first part of Equation 15.1).

We must remove all $(v_j^i, v_{j'}^i)$ edges from E' since $V_j^i \cap V_{j'}^i = \emptyset$ (V_j^i and $V_{j'}^i$ belong to the same solution c_i): there are $k|C_k|(k-1)/2$ (second part of the equation).

Equation 15.1 leads to an $O(|V|^2)$ worst space complexity, up to a factor of $|C_k|^2$.

15.4 COMPUTATIONAL RESULTS

15.4.1 Generic tabu search

To generate solutions to be studied, we use the generic tabu search (GTS) algorithm from Dorne and Hao (1998b). GTS is a general algorithm designed to solve several coloring problems (graph coloring, T-coloring and set T-coloring). We recall here the main components of GTS and the general procedure (see Algorithm 15.3).

Initial configuration A *DSATUR-based* greedy algorithm (Brélaz, 1979) is used by GTS to generate initial configurations.

Configuration regeneration The regeneration aims at producing a (k-1)-coloring from a k-coloring. The nodes in the last color class k are given a new color from $[1..k-1]$ in such a way that the number of conflicting nodes[6] over the graph is minimized.

Searching for proper coloring Beginning from a regenerated conflicting configuration, GTS iteratively makes best *1-moves*, changing the color of a conflicting node to another one, until achieving a proper coloring. "Best moves" are those which minimize the difference between the cost (the number of conflicting nodes) of the configuration before the move is made and the cost of the configuration after the move is performed[7]. A tabu move leading to a configuration better than the best configuration found so far is always accepted (*aspiration criterion*).

[6] A node $u \in V_i$ is said to be conflicting if $\exists v \in V_i / (u, v) \in E$.

[7] If there are multiple best 1-moves, one is chosen randomly.

ALGORITHM 3()
$TL \leftarrow \emptyset$; /*Initialize the tabu list TL to empty*/
Generate the initial configuration c using DSATUR;
Regenerate c with one color less;
while Stopping condition not satisfied
    $c^* \leftarrow c$;
    while $f(c^*) > 0$ and stopping condition not satisfied
        Update c by performing a best 1-move $m(u, V_{old}, V_{new})$;
        $TL \leftarrow TL \cup (u, V_{old})$;
        if $f(c) < f(c^*)$
            $c^* \leftarrow c$;
    endwhile
    if $f(c^*) = 0$
        Regenerate c with one color less;
endwhile

Algorithm 15.3 Generic tabu search

The tabu tenure l is dynamically computed using the following formula:

$$l = \alpha \times f(c) + random(g) \tag{15.2}$$

where $f(c)$ stands for the number of conflicting edges in configuration c. $random(g)$ is a function which returns an integer value uniformly chosen in $[1..g]$. α weights the number of conflicting edges. A move m can be characterized by a triplet (u, V_{old}, V_{new}), $u \in V$, V_{old} and V_{new} being, respectively, the previous and the new colors of u. So, when a move m is performed, assigning u to the color class V_{old} is forbidden for the next l iterations by introducing the (u, V_{old}) couple in the tabu list TL.

Each time a solution is found with k colors, a new configuration is regenerated from this solution with $k-1$ colors and the search process starts over again. The algorithm stops when an optimal or a k-coloring (k fixed) is obtained or when a maximum number of moves have been carried out without finding a solution.

15.4.2 Analysis results

In this section, we give the results of *surface analysis* (Section 15.3.1) and *depth analysis* (Section 15.3.2) for a set of the DIMACS benchmark graphs[8]. To generate solutions, we use the GTSD procedure which is our implementation of

[8]Available via anonymous FTP from *FTP://dimacs.rutgers.edu/pub/challenge/graph/ benchmarks/*.

the above GTS algorithm[9]. GTSD is initiated with a *DSATUR* algorithm. For each graph, we indicate if the set of solutions found by GTSD contains *complete representative sets* or *partial representative sets*.

Settings The α and g parameters used for computing the dynamic tabu tenure l (Equation 15.2) were empirically determined, respectively 2 and 10 at most for all graphs. A maximum of 10 million moves were allowed for the search process.

Table 15.1 Some results of analysis on DIMACS benchmark graphs using GTSD.

Characteristics of the graphs				*Results of analysis*			
Name	$\lvert V\rvert$	$\lvert E\rvert$	χ	*k*	*FREQ. (%)*	*#RS*	$\overline{\lvert RS\rvert}$
school1	385	19095	14	14	100 (CRS)	1	29
school1_nsh	352	14612	14	14	50 (CRS)	3	23.3
dsjr500.1	500	3555	12	12	20 (CRS)	1	4
dsjr500.1c	500	121275	84	85	30 (CRS)	17	6
r125.1	125	209	5	5	20 (CRS)	3	5
r125.1c	125	7501	46	46	100 (CRS)	37	2.8
r125.5	125	3838	36	36	90 (CRS)	1	4
r250.1	250	867	8	8	50 (CRS)	1	3
r250.1c	250	30227	64	64	100 (CRS)	31	3.9
r250.5	250	14849	65	66	100 (CRS)	14	3.6
r1000.1	1000	14378	20	20	30 (CRS)	1	3
r1000.1c	1000	485090	≥ 90	98	100 (CRS)	18	10.4
le450_15a	450	8168	15	15	0 (CRS)	0	0
le450_15b	450	8169	15	15	0 (CRS)	0	0
le450_15c	450	16680	15	15	100 (CRS)	1	30
le450_15d	450	16750	15	15	40 (CRS)	5	30
mulsol.i.1	197	3925	49	49	100 (CRS)	24	1.5
flat300_20_0	300	21375	20	20	100 (CRS)	20	15
flat300_26_0	300	21633	26	26	100 (CRS)	26	11.5
flat300_28_0	300	21695	28	32	100 (PRS)	1	3
flat1000_50_0	1000	245000	50	50	100 (CRS)	50	20
flat1000_60_0	1000	245830	60	60	100 (CRS)	60	16.7
dsjc125.5	125	3891	≥ 10	17	100 (CRS)	1	8
dsjc250.5	250	15668	≥ 11	28	100 (PRS)	1	2
dsjc500.5	500	62624	≥ 13	50	50 (PRS)	2	2

Columns 1–4 in Table 15.1 show for each graph, its name, the number of nodes and edges and its chromatic number (or its best known lower bound) respectively. The last four columns show the results of the analysis: best coloring found by GTSD after 10 million moves, percentage of solutions containing at least one RS, number of RS and mean RS size. Entries with a "100 %" pattern in the "FREQ. (%)" field indicate that a RS has been found. If no RS is found in *all* the studied solutions (entries < 100 %), we give the number of solutions

[9]GTSD is coded in C (CC compiler with –O5 option) and executed on a Sun Ultra 1 (256 RAM, 143 MHz).

(in percentage) that contain a RS. Analyzes were performed over five to ten solutions generated by GTSD.

From Table 15.1 we can make several remarks. First, 12 out of the 25 studied graphs have complete representative sets. Note that most of them have a significant number of RS (compared to k) with an average total size (#RS $\times$ $\overline{|RS|}$) going from almost 20 % to 100 % of $|V|$, although these graphs belong to quite different families.

Second, for a few graphs, when no CRS is found, or with a low percentage, partial representative sets are sometimes identified. Nevertheless, ten graphs have less than 60 % of RS, meaning that solutions are quite different for these graphs. Especially, the le450_15a and le450_15b graphs have no color class in common.

Third, for flat graphs, except flat300_28_0 for which GTSD finds no optimal solution, the analysis reveals that all solutions were identical. This may be due to some structural properties of these graphs and we believe that other flat graphs, optimally colored, may have the same property[10]. It is possible that only one solution (equivalent solutions by permutation of colors are excluded) exists for each of these graphs.

15.5 BOOSTING THE PERFORMANCE OF GTSD BY EXPLOITING THE ANALYSIS RESULTS

We show below a simple way of boosting the performance of the above GTSD algorithm by exploiting the results of solution analysis (Figure 15.4 illustrates the idea).

Our main motivations are twofold. First, it is always easier to color a graph with $k + \epsilon$ colors (integer $\epsilon > 0$) than with k colors[11]. In other words, GTSD needs a less number of moves to find $(k + \epsilon)$-colorings than to reach solutions with k colors. Secondly, we believe that solutions with $k + \epsilon$ colors share common characteristics (representative sets) with k-colorings.

15.5.1 Using analysis results to build an initial configuration

Recall that the initial configuration of GTSD is built using the DSATUR greedy algorithm. Now, we modify the way the GTSD procedure is initiated as follows.

For a given graph, we first generate a set of five to ten solutions with $k + \epsilon$ colors (integer $\epsilon > 0$). These solutions are analyzed to extract RS. Then the extracted RS are sorted in decreasing order of appearance percentage (frequency). Finally, a configuration with only k colors is constructed as follows.

---

[10]The designer of these graphs had inadvertently left out a *feature* whose omission made these graphs significantly easier to solve than he had previously imagined. He also stated that "they could be optimally colored by a simple greedy algorithm using a *particular natural ordering*". See Jagota (1996).

[11]There are more alternative to color any vertex since ϵ additional colors are available.

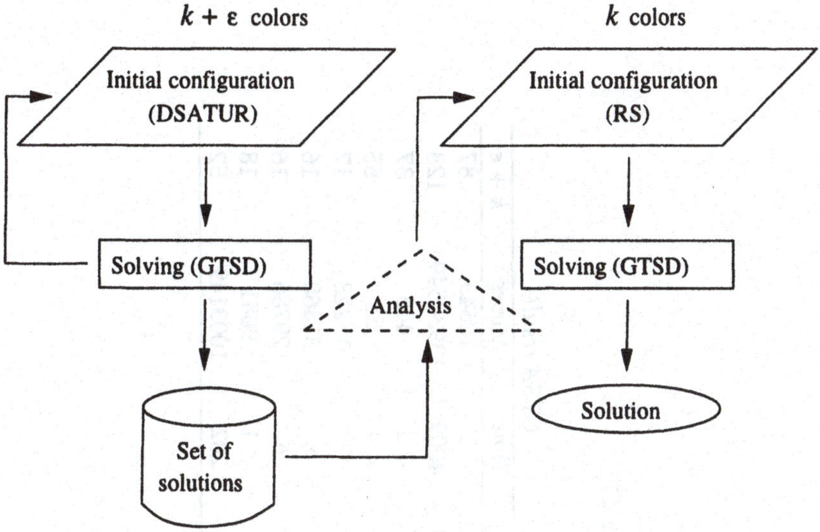

Figure 15.4 Overall view of the GTSA global process.

Color classes are filled one by one with these sorted RS, beginning with the representative set most frequently encountered and ending when all the vertices are included in the k color classes or when no more representative set is available. In the latter case, free vertices are given a color in $[1..k]$ such that the conflicts over the graph are minimized.

For the presentation purpose, we call GTSA (A for Analysis) the overall procedure using this special initialization and including the search for a valid k-coloring.

15.5.2 Computational results of GTSA

Table 15.2 gives results of GTSA. Times (in seconds) and the number of moves include the generation of solutions with $k + \epsilon$ colors, the analysis process, the above initialization step and the search for a proper k-coloring (within 10 million moves). All reported results are averaged over five to ten runs. They were obtained using the same α and g parameters (Equation 15.2) for GTSA and GTSD, namely 2 and 10 at most (respectively) for all graphs.

From Table 15.2, we notice that solution analysis remarkably speeds up the search. Indeed, all reported results of GTSA are better than those of GTSD in terms of resolution speed. For some graphs, the resolution times needed to find a solution is divided by a factor of ten. GTSA is even able to find much better colorings for the dsjr500.5 graph in terms of solution quality (using 124 colors instead of 127). Finally, for the le450_15c and le450_15d graphs, an optimal coloring with 15 colors is found directly by constructing the initial configuration using the results of solution analysis with 16 colors. This last re-

Table 15.2 Using solution analysis to improve GTS.

Graph		*GTSD results*			*GTSA results*			
Name	χ	*k*	*Time*	*Moves*	*k*	*Time*	*Moves*	$k + \epsilon$
dsjr500.1c	84	85	278	88972	85	13	10643	87
dsjr500.5	122	127	3	376	*124*	4009	13930549	128
r125.5	36	36	2	25736	36	< 1	406	37
r250.1c	64	64	< 1	252	64	< 1	54	65
le450_15a	15	15	13	84848	15	2	15228	17
le450_15c	15	15	71	412719	15	6	31460	16
le450_15d	15	15	14	75269	15	6	29786	16
dsjc125.5	≥ 10	17	15	242968	17	< 1	9081	18
dsjc500.5	≥ 13	50	5585	16532498	50	282	1003149	52

mark suggests that these graphs have representative sets not only for a set of solutions with k colors but also for solutions with different number of colors.

15.6 CONCLUSIONS AND DISCUSSIONS

In this paper, we have introduced the notion of *representative sets* to characterize an intrinsic property of solutions for the graph coloring problem. We have also presented two practical schemes allowing for the extraction of *complete representative sets* and *partial representative sets* from a set of given solutions. Thanks to binary tree techniques, we avoid the permutation problem of coloring solutions.

Analyzes have been carried out on a set of well-studied DIMACS graphs. We observed that a large number of these graphs contain representative sets, sometimes quite numerous with consequent total size. One may suspect some links between the existence, number and size, of representative sets and special topological structure of the graphs. However, more evidence is needed to confirm or refute such a hypothesis. The analysis also reveals that some graphs have quite different solutions while others share common coloring information.

We have also used the analysis result to improve a tabu search algorithm. This is achieved by building a special initial configuration with k-colors from solutions with $k + \epsilon$ (integer $\epsilon > 0$) colors. We observed that this simply technique greatly speeds up the initial search algorithm. There are certainly other possibilities to integrate such analysis results in a search algorithm.

Representative sets may also be useful in the context of population-based coloring algorithms to measure the diversity of the configurations of a population, leading to a new stopping criterion for this kind of algorithm. Similarly, representative sets can be used with non-deterministic coloring algorithms running on a single configuration to measure diversity of solutions found within multiple executions.

To carry out the analysis proposed in this paper, one needs a (large) set of solutions. These solutions may be difficult to obtain when the problem instance is hard to solve. One possibility to get around this problem would be to relax the requirement for legal solutions. In this case, the analysis may be based on improper colorings (with a few conflicts). Such colorings can be obtained more easily and more quickly by a tabu (or any other search) algorithm with a limited number of iterations or a faster (say greedy) algorithm. Of course, the results produced in such a way will be less accurate. Nevertheless such a treatment constitutes a fast approximate analysis and may still give useful information about the solutions.

Acknowledgments

We would like to thank gratefully the reviewers of the paper for their questions and comments, which help to improve the quality of the paper. This work was partially supported by the Sino-French Joint Laboratory in Computer Science, Control and Applied Mathematics (LIAMA) and the Sino-French Advanced Research Program (PRA).

Bibliography

A.L. Barabási and R. Albert. Emergence of scaling in random networks. *Science*, 286:509–512, 1999.

D. Brélaz. New methods to color the vertices of a graph. *Communications of the ACM*, 22(4):251–256, 1979.

P. Cheeseman, B. Kanefsky, and W.M. Taylor. Where the really hard problems are. In J. Mylopoulos and R. Reiter, editors, *Proceedings of the Twelfth International Joint Conference on Artificial Intelligence*, volume 1, pages 331–337. Morgan Kaufmann Publishers, 1991.

J.C. Culberson and I.P. Gent. Well out of reach: Why hard problems are hard. Technical Report APES-13-1999, APES Research Group, 1999. *http://www.dcs.st-and.ac.uk/~apes/reports/apes-13-1999.ps.gz.*

L. Davis, editor. *Handbook of Genetic Algorithms*. Van Nostrand Reinhold, 1991.

R. Dorne and J.K. Hao. A new genetic local search algorithm for graph coloring. In A.E. Eiben, T. Bäck, M. Schoenauer, and H.P. Schwefel, editors, *Parallel Problem Solving from Nature - PPSN V*, volume 1498 of *Lecture Notes in Computer Science*, pages 745–754. Springer-Verlag, 1998a.

R. Dorne and J.K. Hao. Tabu search for graph coloring, T-colorings and set T-colorings. In S. Voss, S. Martello, I.H. Osman, and C. Roucairol, editors, *Meta-Heuristics: Advances and Trends in Local Search Paradigms for Optimization*, pages 77–92. Kluwer Academic Publishers, 1998b.

N. Dubois and D. De Werra. EPCOT: an efficient procedure for coloring optimally with tabu search. *Computers and Mathematics with Applications*, 25 (10/11):35–45, 1993.

C. Fleurent and J.A. Ferland. Genetic and hybrid algorithms for graph coloring. *Annals of Operations Research*, 63:437–461, 1996a.

C. Fleurent and J.A. Ferland. Object-oriented implementation of heuristic search methods for graph coloring, maximum clique, and satisfiability. In D.S. Johnson and M.A. Trick, editors, *Cliques, Coloring, and Satisfiability*, volume 26 of *DIMACS Series in Discrete Mathematics and Theoretical Computer Science*, pages 619–652. American Mathematical Society, 1996b.

N. Funabiki and T. Higashino. A minimal-state processing search algorithm for graph coloring problems. *IEICE Transactions on Fundamentals*, E83-A(7): 1420–1430, 2000.

P. Galinier. *Étude des métaheuristiques pour la résolution du problème de satisfaction de contraintes et de la coloration de graphes*. PhD thesis, University of Montpellier II, France, 1999.

P. Galinier and J.K. Hao. Hybrid evolutionary algorithms for graph coloring. *Journal of Combinatorial Optimization*, 3(4):379–397, 1999.

A. Gamst. Some lower bounds for a class of frequency assignment problems. *IEEE Transactions on Vehicular Technology*, 35(1):8–14, 1986.

M.R. Garey and D.S. Johnson. *Computers and Intractability: A Guide to the Theory of NP-Completness*. W.H. Freeman and Company, 1979.

F. Glover and M. Laguna. *Tabu Search*. Kluwer Academic Publishers, 1997.

C.P. Gomes, B. Selman, and H.A. Kautz. Boosting combinatorial search through randomization. In C. Rich, J. Mostow, B.G. Buchanan, and R. Uthurusamy, editors, *Proceedings of the Fifteenth National Conference on Artificial Intelligence and Tenth Conference on Innovative Applications of Artificial Intelligence*, pages 431–437. AAAI Press / MIT Press, 1998.

M.M. Halldórsson. A still better performance guarantee for approximate graph coloring. *Information Processing Letters*, 45(1):19–23, 1993.

J.P. Hamiez and J.K. Hao. Scatter search for graph coloring. In P. Collet, E. Lutton, M. Schoenauer, C. Fonlupt, and J.K. Hao, editors, *Artificial Evolution*, volume 2310 of *Lecture Notes in Computer Science*, pages 168–179. Springer-Verlag, 2002.

A. Hertz and D. De Werra. Using tabu search techniques for graph coloring. *Computing*, 39:345–351, 1987.

A. Hertz, B. Jaumard, and M. Poggi De Aragão. Local optima topology for the *k*-coloring problem. *Discrete Applied Mathematics*, 49(1-3):257–280, 1994.

T. Hogg. Refining the phase transition in combinatorial search. *Artificial Intelligence*, 81(1-2):127–154, 1996.

A. Jagota. An adaptive, multiple restarts neural network algorithm for graph coloring. *European Journal of Operational Research*, 93(2):257–270, 1996.

D.S. Johnson, C.R. Aragon, L.A. McGeoch, and C. Schevon. Optimization by simulated annealing: an experimental evaluation; Part II, Graph coloring and number partitioning. *Operations Research*, 39(3):378–406, 1991.

D.S. Johnson and M.A. Trick, editors. *Cliques, Coloring, and Satisfiability*, volume 26 of *DIMACS Series in Discrete Mathematics and Theoretical Computer Science*. American Mathematical Society, 1996.

F.T. Leighton. A graph coloring algorithm for large scheduling problems. *Journal of Research of the National Bureau of Standards*, 84:489–506, 1979.

C. Lund and M. Yannakakis. On the hardness of approximating minimization problems. *Journal of the ACM*, 41(5):960–981, 1994.

R. Monasson, R. Zecchina, S. Kirkpatrick, B. Selman, and L. Troyansky. Determining computational complexity from characteristic 'phase transitions'. *Nature*, 400(8):133–137, 1999.

C.A. Morgenstern. Distributed coloration neighborhood search. In D.S. Johnson and M.A. Trick, editors, *Cliques, Coloring, and Satisfiability*, volume 26 of *DIMACS Series in Discrete Mathematics and Theoretical Computer Science*, pages 335–357. American Mathematical Society, 1996.

E.P.K. Tsang. *Foundations of Constraint Satisfaction*. Academic Press, 1993.

T. Walsh. Search in a small world. In T. Dean, editor, *Proceedings of the Sixteenth International Joint Conference on Artificial Intelligence*, volume 2, pages 1172–1177. Morgan Kaufmann Publishers, 1999.

T. Walsh. Search on high degree graphs. In B. Nebel, editor, *Proceedings of the Seventeenth International Joint Conference on Artificial Intelligence*, volume 1, pages 266–271. Morgan Kaufmann Publishers, 2001.

D.J. Watts and S.H. Strogatz. Collective dynamics of 'small-world' networks. *Nature*, 393:440–442, 1998.

D. Welsh. *Codes and Cryptography*. Oxford University Press, 1988.

M. Yokoo. Why adding more constraints makes a problem easier for hill-climbing algorithms: analyzing landscapes of CSPs. In G. Smolka, editor, *Principles and Practice of Constraint Programming - CP97*, volume 1330 of *Lecture Notes in Computer Science*, pages 356–370. Springer-Verlag, 1997.

Metaheuristics: Computer Decision-Making, pp. 347-367
B. de la Iglesia, J. J. Wesselink, V. J. Rayward-Smith, J. Dicks, I. N. Roberts,
V. Robert, and T. Boekhout

16 DEVELOPING CLASSIFICATION TECHNIQUES FROM BIOLOGICAL DATABASES USING SIMULATED ANNEALING

B. de la Iglesia[1], J. J. Wesselink[1], V. J. Rayward-Smith[1], J. Dicks,[2], I. N. Roberts[3], V. Robert[4], and T. Boekhout[4]

[1]School of Information Systems
University of East Anglia
Norwich, England
bli@sys.uea.ac.uk, jjw@sys.uea.ac.uk, vjrs@sys.uea.ac.uk

[2]John Innes Centre
Norwich Research Park
Colney, Norwich, England
jo.dicks@bbsrc.ax.uk

[3]Institute of Food Research
Norwich Research Park
Colney, Norwich, England
ian.roberts@bbsrc.ax.uk

[4]Centraalbureau voor Schimmelcultures
Utrecht, The Netherlands
robert@cbs.knaw.nl, boekhout@cbs.knaw.nl

Abstract: This paper describes new approaches to classification/identification of biological data. It is expected that the work may be extensible to other domains such as the medical domain or fault diagnostic problems. Organisms are often classified according to the value of tests which are used for measuring some characteristic of the organism. When selecting a suitable test set it is important to choose one of minimum cost. Equally, when classification models are constructed for the posterior identification of unnamed individuals it is important to produce optimal models in terms of identification performance and cost. In this paper, we first describe the problem of selecting an economic test set for classification. We develop a criterion for differentiation of organisms which may encompass fuzzy differentiability. Then, we describe the problem of using batches of tests sequentially for identification of unknown organisms, and we explore the problem of constructing the best sequence of batches of tests in terms of cost and identification performance. We discuss how metaheuristic algorithms may be used in the solution of these problems. We also present an application of the above to the problem of yeast classification and identification.

Keywords: Classification, Identification, Minimum test set (MTS), Heuristic techniques.

16.1 INTRODUCTION

Heuristic algorithms for mining large databases are being adapted to enable discriminatory analysis to be performed on biological data, accelerating the progress in understanding biological diversity and its industrial implications. A range of knowledge discovery algorithms are being applied to yeast characteristics data, providing new research leads and decision making tools. The research presented here is part of a project funded by the BBSRC which involves the curation and data mining analysis of yeast species and strain data, including DNA data for 700+ yeast species. There is special industrial interest in the investigation of yeast species capable of causing food spoilage, including emerging spoilage food agents.

The initial phases of the project, which cover some of the work reported here, have focused on understanding and improving the current methods for identification and classification in the biological domain. The new methods being developed should lead to faster, cheaper and more reliable classification and identification. Yeast data is being used as an initial case study, but it is expected that the approach can be extended to other taxa in the biological domain, and possibly to other domains such as medical diagnosis.

The problem of identification/classification in the biological domain has received attention in the literature for decades (Pankhurst, 1975; Payne, 1991; Payne and Thompson, 1989; Willcox and Lapage, 1972). According to Payne (1993), the aim is to find a model for identifying taxa (i.e. species, genera, populations, etc.) whose properties can be defined in terms of tests each of which has only a finite number of possible results. The term test is used to provide a general terminology. For example, in botanical identification each test may take the form of determining which state of a particular character is exhibited by the specimen being identified (e.g. the test "color of petals" may have values "red", "blue","yellow"). In yeast, a test may represent the ability of a specimen to use various compounds for growth (e.g. the test "D-Xylose Growth" may have results "positive", "negative" and "equivocal"). The problem of identification is not only confined to the biological domain, though. In electronic fault diagnosis, a test may involve seeing whether a particular set of components contains a fault, and in medical diagnosis, a test may contain recorded signs or symptoms for a patient.

Traditionally in biology, diagnostic(or identification) keys or diagnostic tables were constructed and used for identification, see for example (Payne and Preece, 1981; Payne, 1992). A diagnostic key is a type of decision tree, for those familiar with Knowledge Discovery in Databases (KDD). A diagnostic table is simply a table of results listing the outcome of each test or experiment for each organism. To produce diagnostic keys or tables, it is sometimes necessary to find a minimum characteristic set (or test set) containing enough characteristics to discriminate all organisms. Payne (1991) describes an algorithm which can be used to find all irredundant test sets (i.e. those containing no tests which can be omitted without making any organisms unidentifiable). The Minimum Test Set (MTS) or the test set of minimum cost (if costs are given) capable of classifying all organisms can be chosen from the list of irredundant test sets, as test sets with redundant tests could not be minimum. However, Payne's algorithm has exponential worse case complexity. In fact, the problem of finding a Minimum Test Set (MTS) to differentiate organisms is NP-hard according to Garey and Johnson (1979). For tests with binary outcomes, finding the optimal decision tree is NP-hard (Hyafil and Rivest, 1976). It follows that, constructing a key with the minimum expected cost of identification is NP-hard. A heuristic approach to key construction, very similar to the approach used for constructing decision trees (Quinlan, 1993; 1996), has been used in practice. The approach is based on selecting sequentially a test that "best" divides the organisms into sets. Intuitively, the best test may be that which more nearly splits the taxa into groups of equal size. However, if equivocal or variable responses to tests are present the choice is not so simple. Various selection criteria for the tests have been used and evaluated (Payne, 1981; Payne and Dixon, 1984; Payne and Preece, 1981; Payne and Thompson, 1989). Programs such as Genkey (Payne, 1993) can be used to produce diagnostic tables and diagnostic keys.

The problem of choosing a classification or identification model is closely related to that of choosing a classification model in data mining. There are also some parallels with Feature Subset Selection (FSS) (Debuse and Rayward-Smith, 1997; Liu et al., 1998; Kononenko, 1994; Hall, 1998), which is another tasks of the pre-processing stage in the KDD process (Debuse et al., 2000). The objective of FSS is to isolate a small subset of highly discriminant features and this has some parallels with finding a MTS.

This paper will first report how it may be possible to use metaheuristics, such as Simulated Annealing previously used for FSS and data mining (de la Iglesia et al., 1996; de la Iglesia and Rayward-Smith, 2001), to the problem of finding a MTS. To do this, we first model the problem as a combinatorial optimization problem. The modeling involves establishing a method that allows organisms to be distinguished by some fuzzy criteria, when a crisp differentiation is not possible. The objective of the metaheuristic algorithm will be to choose the classification model which allows for maximum differentiability of organisms, while minimizing the total number of characteristics (or the total cost of characteristics, if this is given) required for classification. The problem of delivering a model that will allow for "economic" identification of unnamed organisms using batches of tests will also be discussed. For identification, the objective is to choose a model which will maximize the chances of delivering an identity for an unknown organism while minimizing the total cost of the identification process.

Organisms may be assigned a weight, so that if it is crucial to identify a particular organism within the model induced, the weights can be adjusted to give this a priority. The weight may for example be used to indicate known frequency of occurrence, in which case the model will represent minimizing expected cost of classification/identification. Alternatively, a relatively rare organism, but one that is of particular importance, may be given a high weight to ensure its identification. The approach has immediate application to the determination of the tests necessary to identify types of yeast, such as food spoilage agents, which are of great industrial interest.

16.2 DIFFERENTIABILITY OF ORGANISM

The data will consist of a number of records which contain a series of characteristics with their values for different organisms. Characteristics recorded may have an associated cost value. Organisms may have a weight. It is worth noting that with the available data it may not be possible to obtain a perfect classifier (i.e. one that is capable of distinguishing all available organisms); also some characteristics may be redundant to the classification. In this context, the definition of an optimal model may vary but will incorporate the idea of "economy" of classification/identification.

More formally, let us consider a set of characteristics

$$P = \{t_1, t_2, \ldots, t_n\}$$

which applies to each organism $s \in S$ (S is the set of all known organisms to be classified). Each t_i may have an associated cost value $c(t_i) \in \Re^+$. Each organism may have a weight, $w(s) \in]0,1[$ which may, for example, account for the frequency with which they occur. Each organism is initially defined by a set of values for each characteristic in P; the value of characteristic i for organism s is given by $s[t_i] \in \mathrm{Dom}_i$.

We define an organism s as *differentiable* if, for a given set of characteristics, $Q \subseteq P$, the combination of values for s, $s[Q]$, is unique in S, i.e. $s[Q] = s'[Q] \Rightarrow s = s'$

In some applications, uniqueness is not strong enough for differentiability since some values, although distinct are not significantly different, e.g. yeast experts consider a <positive> and a <positive, weak, delayed> response to a growth test as equivocal. Hence, in some cases the concept of differentiability is *fuzzy*. To store the perceived difference between two test responses, we use a differentiability function. This function records the difference value assigned to each pair of characteristic values

$$diff_i \in \text{Dom}_i \times \text{Dom}_i \rightarrow [0, 1].$$

If $diff_i$ only takes a value of 0 or 1, then we have *crisp* differentiability of organisms, otherwise we have *fuzzy* differentiability. *diff* will denote the n-tuple of differences

$$(diff_1, diff_2, \ldots, diff_n).$$

Associated with any set of characteristics, $Q \subseteq P$, any *diff*, and any $s \in S$, is a measure $\delta(s, Q, diff)$ where $\delta(s, Q, diff)$ is equal to zero if s is not differentiable from all other species in S according to tests in Q and it is one otherwise. To compute the *total differentiability* of an organism, $\delta(s, Q, diff)$, we must first compute the differentiability between that organism and each of the other organisms. The differentiability between two organisms $s \in S$ and $s' \in S$ in $Q = \{t_{\lambda 1}, t_{\lambda 2}, \ldots, t_{\lambda n}\}$ is computed as

$$\begin{aligned}\delta(s, s', Q, diff) = \quad & diff_{\lambda 1}(s[t_{\lambda 1}], s'[t_{\lambda 1}]) \\ * \quad & diff_{\lambda 2}(s[t_{\lambda 2}], s'[t_{\lambda 2}]) \\ * \quad & \ldots * diff_{\lambda n}(s[t_{\lambda n}], s'[t_{\lambda n}]).\end{aligned}$$

The operator $*$ needs to satisfy certain properties. For any $x, y, z \in [0, 1]$, $*$ is an operator $[0, 1]^2 \rightarrow [0, 1]$ that must satisfy

$$\begin{array}{ll} x * y = y * x & \text{(commutative)} \\ x * (y * z) = (x * y) * z & \text{(associative)} \\ x_1 \leq x_2 \text{ implies } x_1 * y \leq x_2 * y & \text{(non-decreasing)} \\ x * 1 = 1; x * 0 = x & \\ x * y \geq x & \end{array}$$

i.e. $*$ is a t-conorm (Zimmermann, 1991). One example of a t-conorm which could be used for $*$ is the max-operator. Another possibility is to define $*$ by

$$x * y = x + y - xy.$$

Once the differentiability between each pair of organisms,

$$\delta(s, s_i, Q, diff) : \forall s_i \in S, s_i \neq s$$

is calculated then the total differentiability of an organism, $\delta(s, Q, diff)$ is obtained as

$$\begin{array}{ll} \delta(s, Q, diff) = 1, & \text{if } \min\{\delta(s, s_i, Q, diff) : s_i \in S,\ s_i \neq s\} \geq M \\ \delta(s, Q, diff) = 0, & \text{otherwise,} \end{array}$$

where M is a threshold parameter which can be adjusted to allow different degrees of differentiability. According to the previous definitions of $*$, if M is set to 1, then the criterion for differentiability will be non-fuzzy, whereas if M is set to a value less than 1, then the criterion will be fuzzy.

If $\delta(s, Q, diff)$ is equal to 1 we say that organism s is uniquely identifiable in S according to the tests in Q and to the differentiability matrix *diff*.

16.3 CHOOSING A TEST SET FOR CLASSIFICATION

The previous calculations allow us to compute the differentiability of any organism with respect to the rest of organisms in S, according to any subset of characteristics, $Q \subseteq P$, and using a differentiability function, *diff*, which can encompass fuzzy or crisp differentiability criteria.

If the problem to be solved is to find the Minimum Test Set (MTS) capable of classifying the organisms of maximum combined weight (equivalent to the maximum number of organisms when weights are not used), then we must choose the subset of characteristics $Q \subseteq P$ that minimizes

$$\sum_{t \in Q} c(t),$$

subject to the following constraint

$$\sum_{\forall s \in S} \delta(s, Q, diff) \times w(s) = \sum_{\forall s \in S} \delta(s, P, diff) \times w(s).$$

However, we may be able to approach this by solving a wider problem. It may be possible, and perhaps desirable in some situations, to trade-off the unique identification of some organisms for a lower cost of the test set Q. If the criterion of cost is set against the criterion of performance of the model, in terms of the number of identifications achievable, then the problem becomes a Pareto optimization or multi-optimization problem (Fonseca and Fleming, 1995; Osyczka, 1985). We then want to choose the subset of characteristics, $Q \subseteq P$, that maximizes

$$\sum_{\forall s \in S} \delta(s, Q, diff) \times w(s),$$

and minimizes

$$\sum_{t \in Q} c(t).$$

This problem will have a number of non-dominated solutions, in Pareto optimization terms, which form the set of Pareto optimal solutions.

Modern heuristics are widely used in Pareto optimization (Horn and Nafpliotis, 1994; Parks and Miller, 1998; Srinivas and Deb, 1994). Often in multi-objective problems, objectives are artificially combined into a scalar function, for example by the use of simple weighted sums (Jakob et al., 1992). Other approaches treat objectives separately, for example Schaffer (1985) uses different objectives to select sub-populations in a genetic algorithm and then merges and shuffles all sub-populations continuing the process of mating in a normal way but monitoring the population for non-dominated solutions. Approaches using Pareto-based genetic algorithms (Goldberg, 1989) assign equal probability of reproduction to all non-dominated individuals in the population. Recent developments have outperformed previous approaches. For example, the Pareto-based algorithm NSGA II (Deb et al., 2000) (based on the sorting of solutions according to the concept of non-domination) has been shown to be efficient and effective at finding good approximations to the Pareto optimal front.

An alternative approach to performing Pareto optimization may be to use a bound on the total cost of the characteristics contained in Q, i.e.

$$\sum_{t \in Q} c(t) \leq B,$$

where B is a cost bound. If we start, for example, with a high value of B and repeat the search for decreasing values of B then we may be able to sample solutions from the Pareto optimal set using a simple metaheuristic. In fact, a single solution obtained in this way may be all that is required in some practical applications. This is true of many Pareto optimality problems in which a single compromise solution is usually sought. We can perform this optimization using simulated annealing, and we present a possible implementation in Section 16.5.

Whatever way the problem is solved, the resulting test set for the organisms identified could be used as a diagnostic table, which can be used for identification. When trying to identify a new organism, the test set values would be checked to try to find one row of test results matching the values of the organism being identified. A test set of minimum cost can also be used to construct an identification model.

16.4 SELECTING A MODEL FOR IDENTIFICATION USING BATCHES OF TESTS

We denote by identification the process of finding the classification of an unknown organism. Note that, in yeast identification for example, it is common practice to perform all tests necessary for classification, i.e. all the tests that form part of the MTS, in advance of the identification process. However, it is possible that identification of some organisms can be delivered by doing a batch of the tests in Q, with the rest of the tests performed if the results of the first batch do not produce an identification. This may be particularly suitable if some tests are difficult to perform simultaneously. The first batch could be constrained to contain tests that can be performed simultaneously in an economic way, with other tests being done if the results of the first batch prove inconclusive. As a hypothetical example, to identify a food spoilage yeasts it may be possible to perform only some fermentation tests which may be cheaply performed simultaneously.

16.4.1 Problem formulation for tests without equivocal responses

Any set of characteristics, Q, partitions the organisms into classes or groups of organisms indistinguishable by Q. In the model described in the previous section, the ideal is to produce a cheap set where the associated classes each contain at most one organism as this would deliver a perfect classification. Now, we consider a sequence of tests.

Instead of selecting all tests, the new approach involves selecting an initial batch of experiments $B \subseteq Q$. The initial batch would produce some classes of organisms, some with more than one element, and those could be further partitioned by an additional set of tests in the next stage, repeating this process until a unique organism is identified, or the tests are exhausted. This approach is applicable when there are no equivocal tests values in the database (i.e. no unknown, variable, or other equivocal results are recorded for any tests). As long as tests are unequivocal, then classes form a partition of S. However, if there are tests with equivocal outcomes, then organisms may appear in more than one class and therefore S cannot be partitioned into disjoint groups.

Let T be the tree induced by some experimental procedure. Each internal node in T represents a batch of experiments, and each leaf node represents a partitioning of S. If a leaf node comprises a singleton set, $\{s\}$, then we set $\delta(s,T) = 1$, otherwise $\delta(s,T) = 0$. We denote by cost(s) the cost of all the experiments performed at each node on the path from the root to the leaf containing s. For identification, we seek a tree, T, such that

$$\sum_{s \in S} \delta(s,T) * w(s)$$

is maximized subject to

$$\sum_{s \in S} \{\text{cost}(s) : \delta(s,T) \neq 0\} \times w(s) \leq B$$

for some bound B.

We have modeled the batch problem only for unequivocal tests. Let us examine the procedure for constructing trees. Since branches represent combinations of tests values and there are k possible outcomes to each test, if we choose m tests to form part of a batch, B, there are potentially m^k combinations of test values. An approach which tried to construct branches of the tree by examining each possible combination of test values would not be practical for large values of m and k. Naturally, some of the combination of test values may not exist in the real data.

Instead of examining all branches, we could examine only those combinations of test values that have support in the data. To do this we could perform the following procedure. First we would start by comparing each organism with every other organism to decide if each pair are differentiable according to the tests in batch B. Pairs of organisms that cannot be differentiated need to be placed in the same group, i.e. they need to follow the same branch in the tree. This is fairly straightforward procedure for tests with unequivocal responses.

Table 16.1 A simple example with variable or unknown results

Organism	t_1	t_2	t_3
s_i	+	+	V
s_j	+	V	-
s_k	V	+	+

16.4.2 Problem formulation for tests with equivocal responses

When we have to cater for equivocal outcomes, and this is realistically the case for most biological applications, the previous approach cannot be used. Suppose that we are going to build a tree, as presented above, for a set of tests with equivocal responses. When some of the tests considered have equivocal outcomes an organism may follow multiple branches in the tree, and therefore appear in a number of classes of indistinguishable organisms. For example, suppose that we have three organisms s_i, s_j and s_k with the values shown on Table 16.1 for tests t_1 to t_3. In this case, s_i cannot be distinguished from s_j or from s_k according to the batch of tests $B = \{t_1, t_2, t_3\}$, but s_j can be distinguished from s_k. In this context,therefore, $s_i[B] = s_j[B]$ and $s_j[B] = s_k[B]$ does not imply $s_i[B] = s_k[B]$. Hence, if we group those organisms according to the results of the batch B, there will be two separate groups, one containing s_i and s_j, and another one containing s_i and s_k. Note that s_j and s_k should not be placed in the same group. If at the next stage of tree building, the node containing s_i and s_j is partitioned by another batch of tests, then organism s_i may appear as distinguishable. However, if the node containing s_i and s_k cannot be further partitioned by any batch of tests, then s_i is not identifiable in the tree. The previous modeling approach could not be used in this circumstances.

Furthermore, at the end of the initial comparison of pairs of organisms, the output is a list of pairs of organisms that are indistinguishable according to the tests in B. These pairs need to be considered for merging with other pairs of indistinguishable organisms to form the final groups or classes. For tests with unequivocal responses, this is fairly straightforward because in this case it would follow that if $s_i[B] = s_j[B]$ and $s_i[B] = s_k[B]$ then $s_j[B] = s_k[B]$, and hence s_i, s_j and s_k must be placed together in a node. In the case of organisms with equivocal responses, however, the problem is much harder to solve. If we represent the organisms as nodes in an undirected graph, and the pairs as edges joining the nodes then we can convert this into a graph problem. For example, suppose there is a set of organisms $S = \{s_1, \cdots, s_9\}$ which produces the following list of pairs of indistinguishable organisms for some batch of tests B with equivocal responses:

$$(s_1, s_6), (s_2, s_4), (s_5, s_8), (s_5, s_9), (s_6, s_9), (s_8, s_9).$$

The corresponding graph appears in Figure 16.1. We note that each maximal

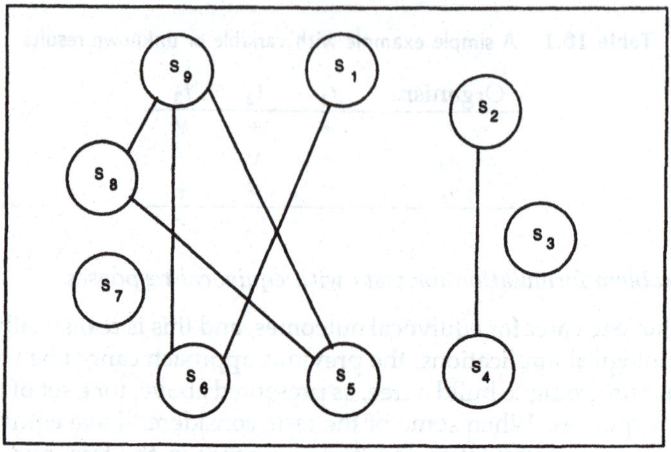

Figure 16.1 An undirected graph used to represent indistinguishable organisms

clique in the graph (i.e. each clique that is not a proper subset of any other clique) represents one of the final groupings of indistinguishable organisms:

- Each unconnected organism (a clique of size 1) represents a distinguishable organism.
- Two nodes connected to one another but with no other common neighbor (a clique of size 2) represent a group with two organisms, for example s_6 cannot be distinguished from s_1 or s_9, but s_1 can be distinguished from s_9 hence s_1 and s_9 must form a group on their own.
- s_5, s_8 and s_9 can be merged together as none of them can be distinguished from the others, and in the graph they form a maximal clique of size 3.

Hence the problem of finding how organisms are grouped into classes of indistinguishable organisms according to batch B when the batch contains equivocal tests can be solved by finding all the maximal cliques in a graph in which each node represents an organism and each edge represents a pairing of indistinguishable organisms. Unfortunately, this could be a rather hard problem to solve! In the worst case, there is an exponential number of maximal cliques in a graph. To prove this, consider the complement of the graph comprising $n/2$ cliques of size 2. This has $2^{n/2}$ maximal cliques.

Finding the maximum clique (i.e. the maximal clique that has the maximum cardinality) is an NP-hard problem (Karp, 1972). The maximal clique problem is even harder, in fact intractable for non-trivial problems, as the solution size may grow exponentially. This may signal the fact that the problem is not defined realistically.

16.4.3 Alternative "tractable" formulation

For tests with equivocal outcomes, it is necessary to find a different problem formulation. We can constrain ourselves to finding an initial batch of experiments to differentiate between some organisms, and then finding a second batch which would differentiate between the remaining organisms. Of course, in doing this, we loose the ability to determine later batches of tests depending on the outcome of the first batch of tests. However, the following is a more tractable problem. Let us have a solution consisting of a set of batches $B_1, B_2, \ldots, B_k$. The application of batch B_i will produce a partition R_i consisting of all unidentified organisms. To evaluate a solution of this type, all organisms will be compared according to tests included in the first batch B_1. Then all those not identified by the first batch will form R_1 and will be compared according to tests in batch B_2, etc. Note that the costs of the first batch $cost(B_1)$ has to be added to the identification of all those organisms identified by the second batch, etc. We denote by p_i the total weight of organisms (or total number of organisms if weights are not used) identifiable in partition R_i using batch B_i, i.e.

$$p_i = \sum_{s \in R_i} \delta(s, \bigcup_{k=1}^{i} B_k, diff) * w(s),$$

where $\delta(s, B_i, diff)$ is equal to 1 is s is identifiable in partition R_i according to tests in B_i and 0 otherwise. The total cost of the solution can be calculated as follows:

$$\text{Total cost}\{B_1, B_2, \ldots B_k\} = \sum_{i=1}^{k} \left(\sum_{j=1}^{i} \text{cost}(B_j) \right) \times p_i.$$

Similar to previous formulations, we need to maximize the total weight (or number) of identifiable organisms according to the set of batches, i.e. maximize

$$\sum_{\forall s \in S} \delta(s, \{B_1, \ldots, B_n\}, diff) \times w(s),$$

and minimizes the total cost as calculated in the previous equation. We can again approach this Pareto optimization problem by setting a bound on the total cost of the solution. Alternatively, we may add a large cost penalty for each organism that is not identified with the subsequent batches of tests, and simply concentrate on minimizing the total cost of the solution.

16.5 SIMULATED ANNEALING IMPLEMENTATION

A general purpose Simulated Annealing toolkit, such as *SAmson* (Mann, 1996), can be used as the platform for the implementation of the solution to the problem of finding the MTS. SAmson is an easy to use environment for the development of optimization problems using Simulated Annealing. The SAmson toolkit consists of an intuitive interface including binary, integer and floating point representations. Adjustable parameters include the initial temperature, various cooling schedules, and different conditions for temperature equilibria. Various neighborhood operators are implemented, and user defined operators can easily be incorporated. There is also a combined, random and adaptive method for neighborhood operator selection. The main code in SAmson is implemented in the C programming language.

Sampling of the Pareto optimal set can also be achieved with SAmson, by use of a cost bound. Sampling of the Pareto optimal set should also deliver an approximation to the optimal MTS.

The implementation for this problem can be approached as follows:

- A binary alphabet can be used to represent a solution. Each position represents a characteristic. A 1 in position i implies that characteristic i is included in the model; 0 implies that characteristic i is excluded. This establishes the set of characteristics Q

- A solution is initialized by randomly setting some bits to 1, and checking that the total cost, $\sum_{t \in Q} c(t)$ is less than or equal to a user defined bound B. If the total cost exceeds the bound the initial solution can be modified or reconstructed until a satisfactory initial solution is reached. Alternatively, a penalty can be applied at the time of evaluation, proportional to the cost of the solution. In the case where costs of characteristics are unknown, it is only necessary to set less than or equal to B bits in the initially solution.

16.5.1 *Neighborhood structures*

A new solution is reached by including a new characteristic previously excluded, excluding a characteristic previously included, or both. For each new solution, it is necessary to check that the bound on the total cost of the set Q is not exceeded. If the bound is exceeded, two approaches can be used:

- In the first more restrictive approach, the neighborhood operator would reconstruct any solution that exceeds the total cost bound, so that any solution produced is guaranteed to be a valid solution in terms of the cost bound B. This may lead to the neighborhood operator slowing down the algorithm.

- In the second approach, the neighborhood operator would be allowed to produce invalid solutions, in terms of the bound B. The evaluation function would penalize invalid solutions to ensure that the search is encouraged in a different direction.

To evaluate a solution, the total number of differentiable organisms using the characteristic set Q is computed, using the differentiability criterion described in Section 16.2. If penalties are to be applied to invalid solutions, a penalty which is related to the cost of the characteristic set Q, if this exceeds bound B, can be added at this stage.

Experiments to choose suitable SA parameters such as cooling schedule, initial temperature, etc, will have to be performed to establish adequate parameter values.

The previous implementation can be changed by modifying the solution representation. For this approach, a solution can be represented by an array of n integers, where n is the total number of tests. The values of the array will be in the range [0... Max. number of batches]. Note that a value of 0 would mean that a test is excluded from all batches, hence this approach will also solve the previous problem. For each position in the solution string the corresponding array value represents the first batch number in which the test is included.

The initialization process will randomly set each position in the solution array to a value in the range [0... Max. number of batches].

The neighborhood operator will change one of the positions in the solution array to a different value. For example, a test previously excluded, may not be include in batch 2.

The evaluation of a solution will be performed by looking at all organisms and computing the number (or combined weight) of organisms identified by the first batch; the process will continue by selecting the subset of organisms still unidentified, and computing the number (or combined weight) of organisms identified by the second batch, etc. At the end of this process the total cost of the solution will be calculated as described previously. A penalty may be added for each unidentified organism.

The algorithm would be set to minimize the total cost of the solution generated.

16.6 COMPUTATIONAL EXPERIMENTS AND RESULTS

Motivation for this work was the need to investigate the classification of yeasts and in particular the characteristics of food spoilage yeasts not shared by other yeasts.

There are 10 species which have been identified by the experts as food spoilers (Pitt and Hocking, 1997), hence yeasts associated with food spoilage represent a minority of species. Nevertheless, they may cause very significant losses to the food and beverages industry, and therefore the isolation of characteristics that distinguish them from other yeast species is an important exercise.

Some of the yeast data available was donated by the Centraalbureau voor Schimmelcultures (CBS) [1]. The CBS maintains the largest collection of living fungi in the world. Their yeast database contains data on more than 4,500 strains kept in the Netherlands, and on about 1,250 strains from the IGC collection in Oeiras, Portugal. They also have data at the species level, for 745 species.

[1]The CBS is located in Utrecht. Details can be found at http://www.cbs.knaw.nl

Table 16.2 Differentiability matrix for the yeast

Responses	–	+	others
-	0	1	0
+	1	0	0
others	0	0	0

The CBS data stored for each yeast, at species or at strain level, consists of what we will name as *conventional* characteristics used to classify the yeasts. There are 96 characteristics recorded for each yeast species and they include microscopical appearance of the cells; mode of sexual reproduction; certain physiological activities and certain biochemical features. Each characteristic is recorded as having a particular response . The responses can be "positive", "negative", "positive, weak, delayed","negative, weak, delayed", "weak", "delayed" and "other". Explanation of characteristics used for the classification of yeasts and the meaning of their responses is given in Barnett et al. (2000).

When comparing characteristic values for yeast, only negative and positive responses to characteristics are considered traditionally as establishing a difference, all other responses are considered as equivocal (Barnett, 1971a;b). The differentiability function for the yeast is shown in Table 16.2. Note that this differentiability function is applicable to all characteristics recorded since all draw their values from the same domain. This *diff* function produces a "crisp" differentiation of species, i.e. minor differences are not considered as sufficient for differentiation. For example, the difference between a positive and a delayed response is not considered significant. Yet, with this differentiability function, we found that many species cannot be differentiated using conventional tests. In particular, none of the spoilage yeasts present in the database were differentiable according to this criterion. This motivated the need for "fuzzy" differentiability criterion, which would allow minor differences (for example, the difference between a positive or a delayed response to a test) to count partially toward the differentiation of two species or strains. A fuzzy classification model results in higher differentiability between species/strains. For example using the crisp *diff* function provided in Table 16.2, only 544 out of 745 yeast species are identifiable, whereas using the fuzzy *diff* function provided in Table 16.3 643 species are identifiable using a threshold parameter M of 0.5 for calculating the total differentiability of an organism, and 687 are identifiable with a threshold $M = 0.25$. The t-conorm operator used to compute the differentiability between two organisms in these experiments is $x * y = x + y - xy$, as that gives better results than the max-operator. This may be due to the fact that the combination of two values using the first operator results in a higher value of the differentiability between two organisms. An adequate value of M is obviously specific to each particular application, and can only be established by experimentation and by interpretation of the results.

Table 16.3 Fuzzy differentiability matrix for the yeast

Responses	+	$+, w, d$	$-, w, d$	w	d	others
-	1	0.5	0	0.5	0.5	0
+		0.25	0.5	0.5	0.5	0
+,w,d			0.25	0	0	0
-,w,d				0	0	0
w					0.25	0
d						0

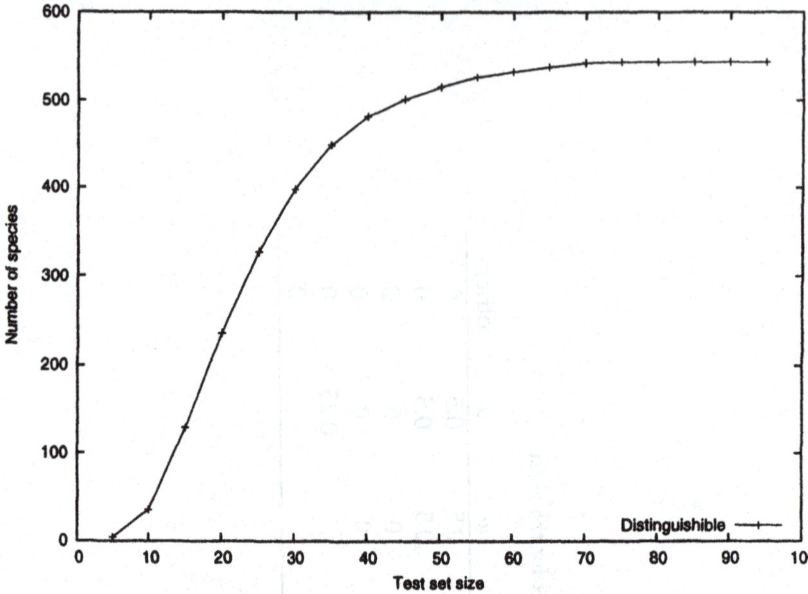

Figure 16.2 Some results for yeast data obtained with SA

Traditional classification methods for yeast use the different responses to conventional characteristics tests as a criterion for differentiability of species/ strains. At present, comparisons of genomes in terms of base sequences are increasingly used for classification. The creation of a classification system based entirely on DNA sequence data is being investigated separately, and some of the work is reported in Wesselink et al. (2002). Combination of both approaches is expected in the future.

Initial experiments with the CBS species data have highlighted sets of tests that are redundant to the classification model. This results were obtained using crisp differentiability, so they represent a lower bound. Figure 16.2 shows the number of differentiable species (as weights are assumed equal for this problem) for a single set of tests of varying sizes. The Minimum Test Set, according to this results, contains 75 tests. If minimizing cost is of essence, a test set of 45 tests would achieve differentiability of most of the species in the dataset.

16.7 CONCLUSIONS AND FURTHER RESEARCH

In this paper we have discussed how to model classification and identification of organisms as combinatorial optimization problems. The concept of fuzzy differentiability, introduced in our modeling, should result in higher differentiability of organisms.

We have shown how heuristic algorithms, and in particular Simulated Annealing, can be used to solve the formulated problem. We have also shown that trying to group organisms according to the results of a batch of tests may lead to an exponential partitioning step, and hence cannot generally be considered. We have therefore introduced a different problem formulation for batches of tests which is tractable and can be solved using heuristics.

Finally, we have shown some preliminary results of applying the new approach to yeast species data, which has strengthen the case for this research.

Work remains to be done at a practical level, to fully implement and test all the discussed approaches. Further experimental results may allow us to establish comparisons about variations in differentiability criteria and their effect on organism differentiability. Results of applying algorithms to yeast data may be interesting to the yeast expert community. We believe the approach can be extended to other domains, and so application to other areas such as medical data may play a part in future research. Further work on this problem is in preparation and is reported in Reynolds et al. (2003)

Acknowledgments

The research presented in this paper is part of a project funded by the BBSRC (Grant No: 83/BIO 12037).

Bibliography

J. A. Barnett. Identifying yeasts. *Nature*, 229(578), 1971a.

J. A. Barnett. Selection of tests for identifying yeasts. *Nature*, 232:221–223, 1971b.

J. A. Barnett, R. W. Payne, and D. Yarrow. *Yeasts: Characteristics and identification, Third Edition*. Cambridge University Press, Cambrige, UK, 2000.

B. de la Iglesia, J. C. W. Debuse, and V. J. Rayward-Smith. Discovering knowledge in commercial databases using modern heuristic techniques. In E. Simoudis, J. W. Han, and U. M. Fayyad, editors, *Proceedings of the Second Int. Conf. on Knowledge Discovery and Data Mining*. AAAI Press, 1996.

B. de la Iglesia and V. J. Rayward-Smith. The discovery of interesting nuggets using heuristic techniques. In H. A. Abbass, R. A. Sarker, and C. S. Newton, editors, *Data Mining: a Heuristic Approach*. Idea group Publishing, USA, 2001.

K. Deb, S. Agrawal, A. Pratap, and T. Meyarivan. A fast elitist non-dominated sorting genetic algorithm for multi-objective optimization: NSGA-II, 2000.

J. C. W. Debuse and V. J. Rayward-Smith. Feature subset selection within a simulated annealing data mining algorithm. *Journal of Intelligent Information Systems*, 9:57–81, 1997.

J.C.W. Debuse, B. de la Iglesia, C. M. Howard, and V. J. Rayward-Smith. Building the KDD roadmap: A methodology for knowledge discovery. In R. Roy, editor, *Industrial Knowledge Management*, pages 179–196. Springer-Verlag, London, 2000.

C.M. Fonseca and P. J. Fleming. An overview of evolutionary algorithms in multiobjective optimisation. *Evolutionary Comp.*, 3:1–16, 1995.

M. R. Garey and D. S. Johnson. *Computers and intractability: A guide to the theory of NP-completeness*. Freeman, New York, 1979.

D. E. Goldberg. *Genetic Algorithms in Search, Optimisation and Machine Learning*. Addison-Wesley, Reading, Massachusetts, 1989.

M. Hall. Correlation-based feature selection for machine learning, 1998.

J. Horn and N. Nafpliotis. Multiobjective optimisation using the niched pareto genetic algorithm. Technical Report Illigal Report 93005, Illinois Genetic Algorithms Laboratory, University of Illinois, Urbana, Champaign, 1994.

L. Hyafil and R. L. Rivest. Constructing optimal binary decision trees is np-complete. *Information Processing Letters*, 5:15–17, 1976.

W. Jakob, M. Gorges-Schleuter, and Blume C. Applications of genetic algorithms to task planning and learning. In R. Manner and B. Manderick, editors, *Parallel problem solving from Nature, 2*, pages 291–300. North-Holland, 1992.

R. M. Karp. Reducibility among combinatorial problems. In *Complexity of Computer Communications*. Plenum Press, New York, 1972.

Igor Kononenko. Estimating attributes: Analysis and extensions of RELIEF. In *European Conference on Machine Learning*, pages 171–182, 1994.

Huan Liu, Hiroshi Motoda, and Manoranjan Dash. A monotonic measure for optimal feature selection. In *European Conference on Machine Learning*, pages 101–106, 1998.

J. W. Mann. X-SAmson v1.5 developers manual. *School of Information Systems Technical Report, University of East Anglia, UK*, 1996.

A. Osyczka. Computer aided multicriterion optimisation method. *Advances in Modelling and Simulation*, 3(4):41–52, 1985.

R. J. Pankhurst, editor. *Systematics Association Special Volume No. 7, Biological Identification with Computers*. Academic Press, New York, 1975.

G. T. Parks and I. Miller. Selective breeding in a multiobjective genetic algorithm. In A. E. Eiben, editor, *Proceedings of the Fifth International Conference on Parallel Problem Solving from Nature*. Springer-Verlag, 1998.

R. W. Payne. Selection criteria for the construction of efficient diagnostic keys. *Journal of Statistical Planning and Inference*, 5:27–36, 1981.

R. W. Payne. Construction of irredundant test sets. *Applied Statistics*, 40:213–229, 1991.

R. W. Payne. The use of identification keys and diagnostic tables in statistical work. In *COMPSTAT 1992: Proceedings in Computational Statistics*, volume 2, Heidelberg, 1992. Physica-Verlag.

R. W. Payne. Genkey, a program for construction and printing identification keys and diagnostic tables. Technical Report m00/42529, Rothamsted Experimental Station, Harpenden, Hertfordshire, 1993.

R. W. Payne and T. J. Dixon. A study of selection criteria for constructing identification keys. In T. Havranek, Z. Sidak, and M. Novak, editors, *COMPSTAT 1984: Proceedings in Computational Statistics*, Vienna, 1984. Physica-Verlag.

R. W. Payne and D. A. Preece. Identification keys and diagnostic tables: a review (with discussion). *Journal of the Royal Statistical Society*, 143:253–292, 1981.

R. W. Payne and C. J. Thompson. A study of criteria for constructing identification keys containing tests with unequal costs. *Computational Statistics Quarterly*, 1:43–52, 1989.

J. I. Pitt and A. D. Hocking. *Fungi and food spoilage 2nd Edition*. Blackie Academic and Professional, London, 1997.

J. R. Quinlan. *C4.5: Programs for Machine Learning*. Morgan Kaufmann, San Mateo, CA, 1993.

J. R. Quinlan. Bagging, boosting, and C4.5. In *Proc. of the Thirteenth National Conference on A.I.* AAAI Press/MIT Press, 1996.

A. P. Reynolds, J. L. Dicks, I. N. Roberts, J. J. Wesselink, B. de la Iglesia, V. Robert, T. Boekhout, and V.J Rayward-Smith. Algorithms for identification key generation and optimization with application to yeast identification. In *Proceedings of EvoBIO-2003 LNCS, Volume 2611*. Springer, 2003. (To appear).

J. D. Schaffer. Multiple objective optimisation with vector evaluated genetic algorithms. In J. J. Grefenstette, editor, *Proceedings of the First International Conference on Genetic Algorithms*, pages 93–100, San Mateo, California, 1985. Morgan Kaufmann Publishers Inc.

N. Srinivas and K. Deb. Multiobjective optimisation using non-dominated sorting in genetic algorithms. *Evolutionary Computation*, 2(3):221–248, 1994.

J. J. Wesselink, B. de la Iglesia, S. A. James, J. L. Dicks, I. N. Roberts, and V. J. Rayward-Smity. Determining a unique defining dna sequence for yeast species using hashing techniques. *Bioinformatics*, 18(7):1004–1010, 2002.

W. R. Willcox and S. P. Lapage. Automatic construction of diagnostic tables. *Computer Journal*, 15:263–267, 1972.

H. J. Zimmermann. *Fuzzy Set Theory and its applications*. Kluwer Academic Publishers, London, 1991.

Metaheuristics: Computer Decision-Making, pp. 369-384
Mikkel T. Jensen

17 A NEW LOOK AT SOLVING MINIMAX PROBLEMS WITH COEVOLUTIONARY GENETIC ALGORITHMS

Mikkel T. Jensen[1]

[1]EVALife, Department of Computer Science
University of Aarhus
Ny Munkegade bldg. 540
DK-8000 Aarhus C, Denmark
mjensen@daimi.au.dk

Abstract: In recent work coevolutionary algorithms have been used to solve minimax problems from mechanical structure optimization and scheduling domains. The applications have been quite successful, but the algorithms used require the search-space of the minimax problem to have a certain symmetric property. The present article argues that the proposed algorithms will fail to converge if the problem does not have the symmetric property. The difficulty is demonstrated to come from the fitness evaluation of the previous algorithms, and a new kind of fitness evaluation for minimax problems is proposed. Experiments reveal that an algorithm using the new fitness evaluation clearly outperforms the previously proposed algorithms.
Keywords: Minimax problems, Coevolution, Optimization.

17.1 INTRODUCTION

Barbosa (1996; 1997; 1999), Herrmann (1999) and Jensen (2001a), in recent papers, use coevolutionary genetic algorithms to solve so-called minimax problems. A minimax problem is an optimization problem in which the task is to find the *solution* $x \in X$ with the minimum worst case cost, where some problem parameter $s \in S$ (usually called the *scenario*) is chosen by an adversary. The minimax problem is often formulated: Minimize

$$\varphi(x) = \max_{s \in S} F(x, s) \quad \text{subject to } x \in X \tag{17.1}$$

or simply: find

$$\min_{x \in X} \max_{s \in S} F(x, s). \tag{17.2}$$

The function $F(x, s)$ represents the cost of scenario s if solution x is implemented. The minimax problem was originally formulated by game theorists, and can be seen as an antagonist game in which two players have a set of options. The player trying to find the solution, x, tries to minimize the cost, while the player determining the scenario, s, tries to maximize the cost. The minimax problem can be exemplified by an engineer designing a communication network while not violating a set of constraints (e.g. a budget). The network capacity is to be as high as possible when the network is subjected to the worst of a set of possible failures. Minimax problems are found in many different areas, and are known to be relevant to research in scheduling (Herrmann, 1999; Jensen, 2001a; Kouvelis and Yu, 1997), network design (Kouvelis and Yu, 1997), mechanical engineering (Barbosa, 1997; Polak, 1987), constrained optimization (Barbosa, 1996; 1999) and function approximation (Dem'yanov and Malozemov, 1974).

A minimax problem can be treated as a standard optimization problem if the function $\varphi(x)$ can be evaluated in an exact way for all the solutions considered. However, if the set of scenarios S is large this can be a very time-consuming approach, since it may require evaluating $F(x, s)$ for all $s \in S$ for every x considered. Clearly, a more efficient approach is desirable.

Minimax problems seem well suited for coevolutionary algorithms, since the two search-spaces can be searched with two different populations. A predator-prey interaction can take place between the two populations: the solution population (P_X) needs to find solutions that evaluate to low F values with the scenario population (P_S), which needs to find scenarios s that evaluate to high values of F with the solution population. In this way, the objective function $\varphi(x)$ of a solution can be estimated by evaluating it against a small set of scenarios (P_S).

This approach has been used by Herrmann (1999) on a simple parallel machine scheduling problem, where the solution space is the assignment of tasks to machines and the scenario space is the processing time of the tasks. The objective of the algorithm is to minimize the worst case makespan of the schedule. The genetic algorithm is demonstrated to converge to the most robust schedule (the schedule with the best worst case performance) and the worst case processing time scenario, which is trivially known beforehand. Barbosa (1997) uses an equivalent approach to design mechanical structures facing external forces. A system coevolving design and the external forces is used to find structures with the minimal worst case deformation given the possible forces. The structures evolved are found to be similar to the known optimal designs. Barbosa (1996; 1999) solves constrained optimization problems using coevolution by transforming them to minimax problems using a Lagrangian formulation.

When solving minimax problems, it is often required of the problem that the solution (x^*, s^*) satisfies

$$F(x^*, s) \le F(x^*, s^*) \le F(x, s^*) \quad \forall x \in X, s \in S, \tag{17.3}$$

which can be shown (Jensen, 2001b) to be equivalent to

$$\min_{x \in X} \max_{s \in S} F(x, s) = \max_{s \in S} \min_{x \in X} F(x, s). \tag{17.4}$$

This is the case for the problems treated by Barbosa (1996; 1997; 1999) and Herrmann (1999). However, not all minimax problems satisfy this constraint. Jensen (2001a) presents a minimax problem originating from the job shop scheduling domain. This problem does not satisfy eq. (17.4). The present paper will demonstrate that the algorithms proposed by Barbosa (1996; 1997; 1999) and Herrmann (1999) are likely to fail if eq. (17.4) is not satisfied. The problem in the existing algorithms is demonstrated to be located in the fitness evaluation, and a new kind of fitness evaluation is proposed. Experiments demonstrate that the new fitness evaluation solves the problem.

The outline of the paper is as follows: The next section introduces the algorithms of Herrmann (1999) and Barbosa (1996; 1997; 1999) and discusses the difficulties expected on problems not satisfying eq. (17.4). In Section 17.3 a new fitness evaluation to solve the difficulties is introduced. Section 17.4 describes experiments on a number of simple numerical examples, demonstrating the superiority of the new fitness evaluation. Section 17.5 concludes the paper.

17.2 PREVIOUSLY PROPOSED ALGORITHMS

The algorithm proposed by Herrmann (1999) has the form

(1) Create initial populations $P_X(0)$ and $P_S(0)$. Set $t = 0$.
(2) For each $x \in P_X(t)$ evaluate $h[x] = \max_{s \in P_S(t)} F(x, s)$.
(3) For each $s \in P_S(t)$ evaluate $g[s] = \min_{x \in P_X(t)} F(x, s)$.
(4) Create new generation $P_X(t+1)$ from $P_X(t)$ using selection based on $-h[x]$.
(5) Create new generation $P_S(t+1)$ from $P_S(t)$ using selection based on $g[s]$.
(6) Set $t = t + 1$. If $t < t_{max}$ go to step 2.
(7) Do steps 2 and 3. Return $x_0 \in P_X(t)$ with lowest $h[x]$ and $s_0 \in P_s t$ with highest $g[s]$.

Step 7 was left out in the formulation of Herrmann (1999), but it has been added here since it is necessary to find a return value.

The algorithm used by Barbosa (1996; 1997; 1999) is slightly different from the above algorithm since it has sub-cycles in which the scenario population is fixed and a number of solution generations are generated and vice versa. The fitness evaluation (steps 2 and 3), which is the focus of this paper, is identical to the one used above.

In terms of the fitness evaluation, the two populations are treated symmetrically in the algorithm, since both kinds of individuals get their fitness assigned based on the extremal values found when evaluating them against the other population. In what follows, algorithms with this kind of fitness evaluation are termed *symmetric evaluation algorithms*. The symmetric evaluation is reasonable if the solution to the problem satisfies eq. (17.4). If this is not the case, the two populations are not symmetrical in the search-space, and treating them symmetrically in the algorithm can cause serious problems. To realize this consider the very simple function of x and s:

$F(x,s)$	$s=1$	$s=2$	$\max_{s\in S} F(x,s)$
$x=1$	3	2	3
$x=2$	1	4	4
$\min_{x\in X} F(x,s)$	1	2	

The function can take two values for x, represented by two rows, and two values for s, represented by two columns. In addition to the function values $F(x,s)$, the table shows $\max_{s\in S} F(x,s)$ for each x and $\min_{x\in X} F(x,s)$ for each s. The function does not satisfy eq. (17.4). Running a symmetric evaluation algorithm on this function will put an evolutionary pressure on the solution population P_X to converge to the solution $x=1$, since this minimizes $\max_{s\in S} F(x,s)$. It will also put pressure on the scenario population P_S to converge to the scenario $s=2$, since this maximizes $\min_{x\in X} F(x,s)$. However, the solution $F(1,2)=2$ does not correspond to $\min_{x\in X}\max_{s\in S} F(x,s)$, which is $F(1,1)=3$.

The difficulty stems from the fact that for a problem not satisfying eq. (17.4), the fitness of a scenario s *cannot be found only by considering the performance of* s *on* P_X. A scenario with a low $\min_{x\in X} F(x,s)$ value may deserve a high fitness if it causes a solution to get a high $\max_{s\in S} F(x,s)$ value. For this reason, when assigning fitness to the scenarios, it has to be taken into account how well the other scenarios perform against the solutions in P_X. If there is a solution in the population for which the scenario is the worst possible (or if it is close to being the worst possible), then the scenario should be given a high fitness.

More generally, consider a problem for which the solution to the problem $\min_{x\in X}\max_{s\in S} F(x,s)$ is found at x_0, s_0, while the solution to the problem $\max_{s\in S}\min_{x\in X} F(x,s)$ is located at x_1, s_1. Let us examine the behavior of a symmetric fitness evaluation algorithm on this kind of problem.

While the scenario population P_S is diverse enough, the solution population P_X will converge to $x=x_0$, attempting to minimize $\max_{s\in S} F(x,s)$.

When P_X converges to $x = x_0$, P_S will begin to converge to the corresponding worst case scenario $s = s_0$.

However, at the same time, provided the solution population P_X is diverse enough, the scenario population P_S will maximize $\min_{x \in P_X} F(x, s)$ by converging to $s = s_1$. This will cause the solution population P_X to converge to the corresponding best case solution $x = x_1$.

In other words, the coevolutionary algorithm will have at least two different attractors, x_0, s_0 and x_1, s_1. If the solutions x_0, s_0 and x_1, s_1 are stable, the coevolutionary algorithm may find either one. However, there is no guarantee that this will happen, since coevolutionary dynamics can also cause the algorithm to never converge.

17.3 THE ASYMMETRIC FITNESS EVALUATION

A fitness evaluation capable of solving the problems described above is the following. It can be inserted in the algorithm of the previous section, replacing steps 2 and 3.

```
(2)   forall x ∈ P_X(t) set h[x] = −∞.
      forall s ∈ P_S(t) set g[s] = −∞.
(3)   forall x ∈ P_X(t) do
          forall s ∈ P_S(t) do
              Set h[x] = max(h[x], F(x, s)).
              Set k[s] = F(x, s).
          od
          Sort P_S(t) on k[s] in ascending order.
          forall s ∈ P_S(t) do
              if ⌊g[s]⌋ ≠ (index of s in P_S(t)) then
                  Set g[s] = max(g[s], index of s in P_S(t)).
              else
                  Set g[s] = g[s] + 1/(|P_X(t)|+1).
              end if
          end forall
      end forall
```

The use of the $h[x]$ array is unchanged; it simply holds the worst performance (highest $F(x, s)$) found for each solution. The array $g[s]$ still holds the fitness of the scenarios, but the calculation of $g[s]$ has been changed. In step 2 the arrays $h[x]$ and $g[s]$ are initialized. In each sub-step of step 3 a solution x is tested against all scenarios in $P_S(t)$. The performance of each s against x is recorded in $k[s]$, while $h[x]$ is updated to always hold the worst performance of x found so far. After this, the scenario population $P_S(t)$ is sorted on $k[s]$ in such a way that the scenario that caused the highest $F(x, s)$ for the solution x is located at the last position of $P_S(t)$.

At the end of step 3, the fitness $g[s]$ of each scenario s is set to the highest index seen so far for s. Every time s is observed to have the highest index observed so far for s in $P_S(t)$, $g[s]$ is increased slightly (by $1/(|P_X(t)| + 1)$) to reward scenarios that several times get the same position in the sorting of $P_S(t)$.

After step 3 is complete the list $g[s]$ contains the best index of each scenario s, modified slightly for multiple occurrences of the same index. The scenario that was found to be the worst for the most solutions will have the highest possible fitness, followed by the other scenarios that were found to be the worst for at least one solution. The scenario found to be the second-worst for the highest number of solutions will be ranked immediately after these, etc.

Consider the use of this fitness evaluation on the simple problem of Section 17.2. Assume only one of each phenotype is present in the populations. The solution $x = 1$ is tested against both scenarios, and $s = 1$ is found to be the worst, setting $g[1] = 2$ and $g[2] = 1$. $h[1]$ is set to the worst performance found, $F(1, 1) = 3$. After this, the solution $x = 2$ is tested against both scenarios, and $s = 2$ is found to be the worst, updating $g[2] = 2$, while still $g[1] = 2$. $h[2]$ is set to $F(2, 2) = 4$. After this, the solution $x = 1$ will be preferred in the reproductive step of the algorithm, while neither $s = 1$ nor $s = 2$ are preferred over the other, since $g[1] = g[2] = 2$. The scenarios $s = 1$ and $s = 2$ will be assigned the same fitness as long as both $x = 1$ and $x = 2$ are present in the P_X population.

This fitness evaluation ensures that scenarios that are very bad for at least one solution will be kept in the population, while scenarios that may be "quite bad" for all solutions but not "very bad" for at least one will be removed. This is opposed to the symmetric fitness evaluation, which prefers scenarios which are "quite bad" for all solutions, but removes solutions which are "not so bad" for some solutions, even if they are "very bad" for others.

To reflect the change of the fitness evaluation, step 7 of the algorithm from the previous section is changed to

(7) Do the new version of steps 2 and 3.
(8) Return the $x_0 \in P_X(t)$ with minimal $h[x]$,
and the $s_0 \in P_S(t)$ which maximizes $F(x_0, s)$.

Because of the use of sorting in step 3, the asymmetric fitness evaluation needs more processing time than the symmetric evaluation used in Section 17.2.

17.4 EXPERIMENTS

The algorithm of Section 17.2 has been implemented both with the symmetric fitness evaluation and the asymmetric evaluation of Section 17.3. The implementation was programmed to work on functions of real variables, using a real valued encoding. A linear ranking based selection scheme was used. The crossover rate was 0.8, crossover placing the offspring at a uniformly distributed point between the parents. Offspring created by crossover were subject to mutation with a probability of 0.2. Offspring not generated by crossover were generated using the mutation operator, mutation adding a uniformly distributed value in the range $(-1, 1)$ to the genotype. A population size of 50 was used for both populations. At each generation after reproduction 10% of each population was set to completely random individuals, except in the last two generations, were this feature was disabled to help the populations converge. The random individuals were added since in some experiments it turned out to be necessary to keep the diversity of the populations at a high level. As the populations converged, coevolutionary dynamics would go rampant and drive both populations away from the optimum, see Paredis (1997) and the section on the cosine wave function below. In the experiments, the genetic algorithms were run for 100 generations.

The errors given in the following are *mean square errors* (MSE), defined as

$$MSE(x) = \frac{1}{n}\sum_{i=1}^{n}(x_i - x^*)^2, \; MSE(s) = \frac{1}{n}\sum_{i=1}^{n}(s_i - s^*)^2, \tag{17.5}$$

where x_i, s_i is the solution and scenario found in the ith experiment, and x^*, s^* are the optimal values.

In the problem with two optima x_1^*, s_1^* and x_2^*, s_2^*, the MSE is calculated as

$$MSE(x) = \frac{1}{n}\sum_{i=1}^{n}(x_i - x_{(i)}^*)^2, \; MSE(s) = \frac{1}{n}\sum_{i=1}^{n}(s_i - s_{(i)}^*)^2, \tag{17.6}$$

where $x_{(i)}^*, s_{(i)}^*$ is the optimum that minimizes $(x_i - x_{(i)}^*)^2 + (s_i - s_{(i)}^*)^2$.

For every function of the next sections the $\min_{x\in X}\max_{s\in S} F(x, s)$ problem has been solved using the symmetric evaluation algorithm and the asymmetric evaluation algorithm. For all experiments the errors were calculated based on 1000 independent runs of each algorithm.

17.4.1 A saddle-point function

The first experiment was made on a function satisfying eq. (17.4), simply to observe that both algorithms worked on that kind of problem, and to observe any kind of difference in the performance of the two algorithms. The function was

$$F(x, s) = (x - 5)^2 - (s - 5)^2, \quad x \in [0; 10], s \in [0; 10] \tag{17.7}$$

The optimal minimax solution is known to be $x = 5$, $s = 5$. As can be seen in Table 17.1, the experiments showed that both the symmetric and the

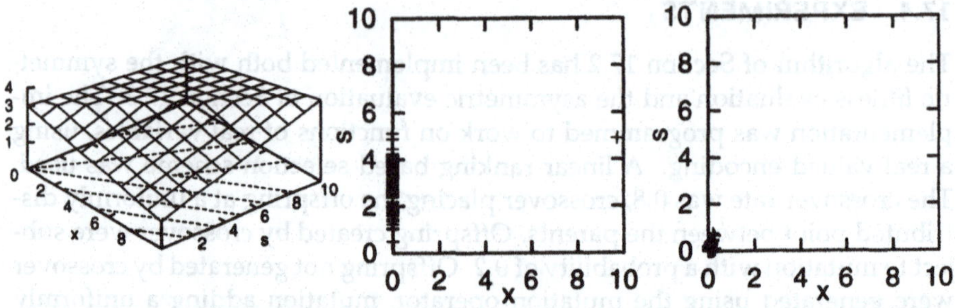

Figure 17.1 **Left:** Surface plot of the two-plane function. **Middle:** The solutions found by the symmetric evaluation algorithm. **Right:** The solutions found be the asymmetric evaluation algorithm. The optimum is located at $(0,0)$.

asymmetric algorithm solved the problem to a very high average precision. In both cases the mean square error was found to be less than 10^{-11}. There did not seem to be a performance difference between the two algorithms for this problem.

17.4.2 A two-plane function

The function

$$F(x,s) = \min(3 - \tfrac{2}{10}x + \tfrac{3}{10}s, 3 + \tfrac{2}{10}x - \tfrac{1}{10}s),\ x \in [0;10], s \in [0;10] \qquad (17.8)$$

has been constructed from two planes in such a way that the corners $(0,0)$, $(10,0)$, $(0,10)$, $(10,10)$ have the same values as the function from the table of Section 17.2. As argued in Section 17.2, the function can be expected to cause trouble for the symmetric algorithm, since there is an evolutionary pressure on the scenario population to move away from the minimax optimum at $(0,0)$ toward $s = 10$. A surface plot of the function can be seen on the left-hand-side of Figure 17.1. The first 500 solutions found by each algorithm have been plotted in the middle and right-hand parts of the figure. As can be seen, the asymmetric algorithm comes very close to the optimum every time, while the symmetric algorithm consistently finds the optimum solution but fails to find the optimum scenario. This is also evident from the mean square errors of Table 17.1.

17.4.3 A damped sinus function

The function

$$F(x,s) = \frac{\sin(x-s)}{\sqrt{x^2+s^2}}, x \in (0;10], s \in (0;10] \qquad (17.9)$$

is antisymmetric around the line $x = s$. The function has been designed in such a way that the solution to $\min_{x \in X} \max_{s \in S} F(x,s)$ is located at $x = 10, s = 2.125683$, while the solution to $\max_{s \in S} \min_{x \in X} F(x,s)$ is located at

Table 17.1 Summarized results of the experiments. For each problem the mean square error (MSE) of solutions (x) and scenarios (s) for the symmetric evaluation and asymmetric evaluation algorithm is reported.

Function	Satisfies eq. (17.4)	Symmetric algorithm MSE(x)	Symmetric algorithm MSE(s)	Asymmetric algorithm MSE(x)	Asymmetric algorithm MSE(s)
Saddle-point	yes	2.15 E−12	2.04 E−12	2.05 E−12	2.03 E−12
Two-plane	no	0.0000	10.987	0.0036576	0.017004
Sine	no	6.2345	37.724	0.35056	0.78844
Cosine	no	14.386	0.15923	0.037081	0.21164

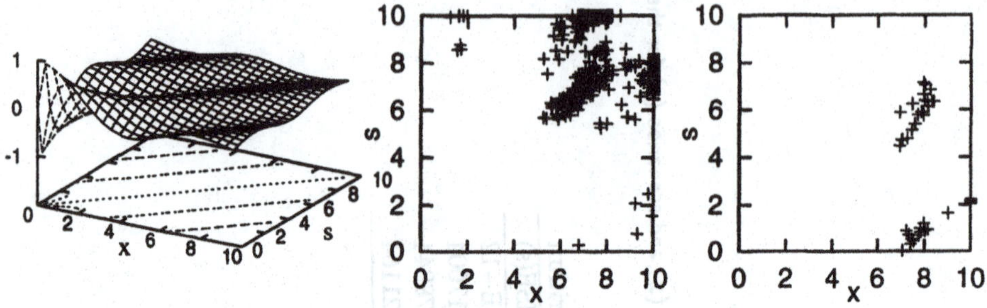

Figure 17.2 **Left:** Surface plot of the antisymmetric sine function. **Middle:** The solutions found by the symmetric evaluation algorithm. **Right:** The solutions found by the asymmetric evaluation algorithm. Most of the solutions of the asymmetric algorithm are located at the optimum point $(10, 2.125683)$.

$x = 2.125683, s = 10$. As suggested in Section 17.2 the problem causes serious trouble for the symmetric algorithm. From Table 17.1 and the middle plot of Figure 17.2 it is evident that the symmetric algorithm rarely comes close to the optimum solution. The symmetric distribution of solutions and scenarios in the plot is caused by the symmetry in the function and the symmetrical treatment of solutions and scenarios in the algorithm. The asymmetric algorithm comes close to the optimum in more than 90% of the cases (all of these runs are located at a single point on the plot at the right hand side of Figure 17.2), but also fails in some cases.

17.4.4 A damped cosine wave

An experiment was made on the function

$$F(x,s) = \frac{\cos(\sqrt{x^2+s^2})}{\sqrt{x^2+s^2}+10}, \; x \in [0;10], s \in [0;10]. \tag{17.10}$$

This function does not satisfy eq. (17.4). A surface plot can be seen on the left hand side of Figure 17.3. The function is known to have two optimal minimax solutions, one at $x = 7.04414634, s = 10$ and one at $x = 7.04414634, s = 0$.

The performance of the symmetric evaluation algorithm is displayed in Table 17.1 and the middle of Figure 17.3. The solutions found are observed to generally be far from the optimum, while usually an optimum scenario is found. The asymmetric evaluation algorithm is observed to perform somewhat better. The solutions are usually close to the optimum, while the scenarios are sometimes a bit off. The plots show two part-circle shapes, rooted in each of the two optima.

The problem is difficult to solve, since both optima are needed in the scenario population to keep coevolutionary dynamics from generating suboptimal solutions. Consider what happens if $x = 7.04$ and $s = 10$ are present in the populations, but $s = 0$ is not. It will be favorable for the solution $x = 7.04$

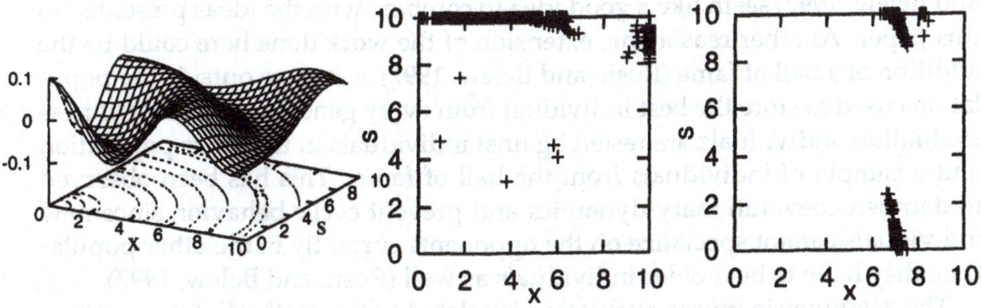

Figure 17.3 **Left:** Surface plot of the cosine-wave function. **Middle:** The solutions found be the symmetric evaluation algorithm. **Right:** The solutions found be the asymmetric evaluation algorithm.

to decrease, since this minimizes the observed worst case cost. The observed best solution x may continue to decrease for a while, until a suitable low value (e.g. $s = 1.0$) appears in the scenario population. In the same way, if $s = 10$ is not present in the population, the observed optimal value for x may start to increase, until a high value appears in the scenario population.

These problems indicate that maintaining population diversity can be crucial for the success of coevolutionary algorithms applied to minimax problems.

17.5 DISCUSSION

This paper has argued that previously published coevolutionary approaches to solving minimax problems are likely to fail on problems where the two search-spaces do not have a symmetric property. In the previous section the algorithm published by Herrmann (1999) was investigated. A similar investigation has been performed for the closely related algorithm published by Barbosa (1996; 1997; 1999). It did not perform significantly better than Herrmann's algorithm. The experiments with Barbosa's algorithm can be found in Jensen (2001b). It has been argued that the fitness evaluation used in previous approaches is fundamentally wrong, and a new procedure for fitness evaluation has been suggested. Experiments have demonstrated that the previously used approach fails for a number of simple problems, while the new approach performs much better.

The experiments also demonstrate that keeping the diversity of the populations high can be crucial for coevolution to work when solving minimax problems. If the diversity of the scenario population gets too low suboptimal solutions can be able to take over in the solution population, or coevolutionary dynamics can lead the search away from the optimum. The importance of diversity in coevolutionary optimization has been observed for other problems as well (Paredis, 1997; Rosin and Belew, 1997). For this reason diversity maintaining measures such as crowding, tagging or sharing (Bäck et al., 1997; Rosin and Belew, 1997) seem like a good idea to combine with the ideas presented in this paper. Another reasonable extension of the work done here could be the addition of a hall of fame (Rosin and Belew, 1997); a storage outside the populations used to store the best individual from every generation. During fitness evaluation, individuals are tested against individuals in the other population and a sample of individuals from the hall of fame. This has been observed to dampen coevolutionary dynamics and prevent cyclic behavior, since new individuals cannot specialize on the opponents currently in the other population; they have to beat older individuals as well (Rosin and Belew, 1997).

The asymmetric fitness evaluation is related to the method of competitive fitness sharing presented by Rosin and Belew (1997). Rosin and Belew use competitive fitness sharing as a means of preserving diversity in coevolutionary systems evolving nim and tic-tac-toe players. The idea in competitive fitness sharing is to assign to each individual i a fitness reward, which is to be shared among all players able to beat i. This means that if player j is the only one to beat i, the fitness contribution of i to j will be much higher than if j is just one of many players to beat i. While this idea is closely related to the asymmetric fitness evaluation, it is fundamentally different in the sense that both populations are treated identically in competitive fitness sharing. The two methods also differ in their areas of applicability. Rosin and Belew (1997) use competitive fitness sharing in a win-or-lose setting, while asymmetric fitness evaluation has been designed for cost-minimization problems. It is not strait-forward to generalize competitive fitness sharing to work for cost-minimization problems, or to generalize asymmetric fitness evaluation to work for win-or-lose problems. Thus, the two methods should rather be seen as complementary ideas than as competitors.

Another approach related to the asymmetric fitness evaluation is the Pareto-coevolution approach developed by Noble and Watson (2001). Noble and Watson study a simplified poker game and use a coevolutionary algorithm to develop a good poker strategy. In a game, a perfect strategy is a strategy that will always beat all other strategies. However, in poker no perfect strategy exists; it may be the case that strategy A beats strategy B, that B beats C, but that C beats A. Instead of searching for a perfect solution, the authors take a multi-objective approach to the problem: they focus on finding a set of Pareto-optimal (non-dominated) strategies, strategies that cannot be improved in their performance against one opponent without deteriorating the performance against some other opponent. The approach is novel and highly relevant for a game such as poker but not directly applicable to minimax problems since the Pareto-perspective on fitness contradicts the minimax definition of always minimizing the worst possible cost.

Apart from the work on minimax problems, coevolutionary ideas have previously been applied on numerical optimization problems by Potter and Jong (1994). They designed a cooperative coevolutionary algorithm in which there was a population for each of the free variables of the objective function. Thus, each individual represented a partial solution (a value for one of the free variables). Full solutions to the optimization problem were created by combining individuals from all of the populations. The approach was found to outperform a traditional GA on a number of standard benchmarks, and was concluded to perform well for problems in which the variables are independent.

To achieve good design of coevolutionary algorithms such as the one presented in this paper, a much better understanding of coevolutionary dynamics is needed. This issue is touched by Ficici and Pollack (2000). They present Markov-chain models of coevolutionary algorithm dynamics and use evolutionary game theory to discuss various problems with coevolutionary algorithms used for optimization. However, the results presented by Ficici and Pollack (2000) are only a beginning, and much more work is needed before a good understanding of coevolutionary dynamics can be achieved.

Jensen (2001a) uses a coevolutionary algorithm based on the asymmetric fitness evaluation to solve a minimax problem derived from job shop scheduling. The objective in (Jensen, 2001a) is to minimize the consequences of the worst possible machine disruption. This problem is much more difficult than the numerical benchmarks treated in this paper, and the experiments indicate that the approach presented here is applicable also to more difficult combinatorial problems.

Bibliography

H. J. C. Barbosa. A Genetic Algorithm for Min-Max Problems. In V. Uskov, B. Punch, and E. D. Goodman, editors, *Proceedings of the First International Conference on Evolutionary Computation and Its Applications*, pages 99–109, 1996.

H. J. C. Barbosa. A Coevolutionary Genetic Algorithm for a Game Approach to Structural Optimization. In *Proceedings of the seventh International Conference on Genetic Algorithms*, pages 545–552, 1997.

H. J. C. Barbosa. A Coevolutionary Genetic Algorithm for Constrained Optimization. In P. J. Angeline, Z. Michalewicz, M. Schoenauer, X. Yao, and A. Zalzala, editors, *Proceedings of the 1999 Congress of Evolutionary Computation*, volume 3, pages 1605–1611. IEEE Press, 1999.

T. Bäck, D. B. Fogel, and Z. Michalewicz, editors. *Handbook of Evolutionary Computation*, chapter C6 - population structures. IOP Publishing and Oxford University Press, 1997.

V. F. Dem'yanov and V. N. Malozemov. *Introduction to minimax*. John Wiley & sons, 1974.

S. G. Ficici and J. B. Pollack. A Game-Theoretic Approach to the Simple Coevolutionary Algorithm. In M. Schoenauer, K. Deb, G. Rudolph, X. Yao, E. Lutton, J. J. Merelo, and H. Schwefel, editors, *Parallel Problem Solving from Nature - PPSN VI*, volume 1917 of *Lecture Notes in Computer Science*, pages 467–476. Springer, 2000.

J. W. Herrmann. A Genetic Algorithm for Minimax Optimization Problems. In P. J. Angeline, Z. Michalewicz, M. Schoenauer, X. Yao, and A. Zalzala, editors, *Proceedings of the 1999 Congress on Evolutionary Computation*, volume 2, pages 1099–1103. IEEE Press, 1999.

M. T. Jensen. Finding Worst-Case Flexible Schedules using Coevolution. In *GECCO-2001 Proceedings of the Genetic and Evolutionary Computation Conference*, pages 1144–1151, 2001a.

M. T. Jensen. *Robust and Flexible Scheduling with Evolutionary Computation*. PhD thesis, Department of Computer Science, University of Aarhus, 2001b.

P. Kouvelis and G. Yu. *Robust Discrete Optimization and Its Applications*. Kluwer Academic Publishers, 1997.

J. Noble and R. A. Watson. Pareto coevolution: Using performance against coevolved opponents in a game as dimensions for Pareto selection. In L. Spector et al., editors, *Proceedings of GECCO'2001*, pages 493–500. Morgan Kaufmann, 2001.

J. Paredis. Coevolving Cellular Automata: Be Aware of the Red Queen! In Thomas Bäck, editor, *Proceedings of the seventh International Conference on Genetic Algorithms*, pages 393–400. Morgan Kaufmann, 1997.

E. Polak. On the mathematical foundations of nondifferentiable optimization in engineering design. *SIAM review*, 29:21–89, 1987.

M. A. Potter and K. A. De Jong. A Cooperative Coevolutionary Approach to Function Optimization. In Y. Davidor, H. Schwefel, and R. Männer, editors, *Parallel Problem Solving from Nature - PPSN III*, LNCS vol. 866, pages 249–257. Spring Verlag, 1994.

C. D. Rosin and R. K. Belew. New Methods for Competitive Coevolution. *Evolutionary Computation*, 5:1–29, 1997.

Metaheuristics: Computer Decision-Making, pp. 385-404
Joanna Józefowska, Grzegorz Wailgóra, and Jan Węglarz

18 A PERFORMANCE ANALYSIS OF TABU SEARCH FOR DISCRETE-CONTINUOUS SCHEDULING PROBLEMS

Joanna Józefowska[1], Grzegorz Waligóra[1], and Jan Węglarz[1,2]

[1]Institute of Computing Science
Poznań University of Technology
Piotrowo 3A, 60-965 Poznań, Poland
Joanna.Jozefowska@cs.put.poznan.pl, Grzegorz.Waligora@cs.put.poznan.pl
Jan.Weglarz@cs.put.poznan.pl

[2]Poznań Supercomputing and Networking Center
Noskowskiego 10, 61-704 Poznań, Poland

Abstract: Problems of scheduling jobs on parallel, identical machines under an additional continuous resource are considered. Jobs are non-preemptable and independent, and all are available at the start of the process. The total amount of the continuous resource available at a time is limited, and the resource is a renewable one. Each job simultaneously requires for its processing a machine and an amount (unknown in advance) of the continuous resource. The processing rate of a job depends on the amount of the resource allotted to this job at a time. Three scheduling criteria are considered: the makespan, the mean flow time, and the maximum lateness. The problem is to find a sequence of jobs on machines and, simultaneously, a continuous resource allocation that minimize the given criterion. A tabu search metaheuristic is presented to solve the problem. A computational analysis of the tabu search algorithm for the considered discrete-continuous scheduling problems is presented and discussed. Three different tabu list management methods are tested: the tabu navigation method, the cancellation sequence method, and the reverse elimination method.
Keywords: Discrete-continuous scheduling problems, Makespan, Mean flow time, Maximum lateness, Tabu search, Tabu navigation method, Cancellation sequence method, Reverse elimination method.

18.1 INTRODUCTION

In this paper we consider discrete-continuous scheduling problems defined in Józefowska and Węglarz (1998), where general results and methodology have been presented as well. These problems are characterized by the fact that each job simultaneously requires for its processing at a time discrete and continuous (i.e. continuously-divisible) resources. They arise, e.g., when jobs have to be assigned to multiple parallel processors driven by a common electric, hydraulic or pneumatic power source, or to CPU's of multiprocessor systems with a common primary memory treated as a continuous resource. On the other hand, in scalable (SPP) or massively parallel (MPP) systems with hundreds or even thousands of processors, the processors themselves can be considered as a continuous resource while discrete resources can be, e.g., disk drives. We deal with a class of these problems where there are: one discrete resource, which is a set of parallel, identical machines, and one continuous, renewable resource whose total amount available at a time is limited. Jobs are independent and non-preemptable, and their ready times are equal to 0. The processing rate of each job depends on the amount of the continuous resource assigned to this job at a time. The makespan, the mean flow time, and the maximum lateness are considered as the scheduling criteria. The tabu search (TS) metaheuristic to solve the considered problems is discussed. A performance analysis of tabu search for the considered discrete-continuous scheduling problems is presented. Results obtained by tabu search are compared to optimal solutions, as well as to results produced by two other metaheuristics: simulated annealing (SA) and genetic algorithms (GA). Moreover, three tabu list management methods: the tabu navigation method (TNM), the cancellation sequence method (CSM), and the reverse elimination method (REM) are compared. Several computational experiments are described and some conclusions, and final remarks are included.

18.2 DISCRETE-CONTINUOUS SCHEDULING PROBLEMS

18.2.1 General model

We consider n independent, non-preemptable jobs, each of them simultaneously requiring for its processing at time t a machine from a set of m parallel, identical machines (the discrete resource), and an amount (unknown in advance) $u_i(t) \in [0,1], i = 1, 2, \dots, n$, of a continuous, renewable resource. The job model is given in the form (Burkow, 1966):

$$\dot{x}_i(t) = \frac{dx_i(t)}{dt} = f_i[u_i(t)], x_i(0) = 0, x_i(C_i) = \tilde{x}_i \tag{18.1}$$

where $x_i(t)$ is the state of job i at time t, f_i is an increasing, continuous function, $f_i(0) = 0$, C_i is a (unknown in advance) completion time of job i, and $\tilde{x}_i$ is its processing demand (final state). We assume, without loss of generality, that $\sum_{i=1}^{n} u_i(t) = 1$ for every t.

State $x_i(t)$ of job i is an objective measure of work related to the processing of job i up to time t. It may denote, for example, the number of man-hours already spent on processing a job, the number of standard instructions in processing a computer program, and so on.

The problem is to find a sequence of jobs on machines and, simultaneously, a continuous resource allocation that minimize the given scheduling criterion.

18.2.2 Metaheuristic approach

The defined problem can be decomposed into two interrelated subproblems: (i) to find a feasible sequence of jobs on machines, and (ii) to allocate the continuous resource among jobs already sequenced. The notion of a *feasible sequence* is of crucial importance. Let us observe that a feasible schedule can be divided into $p \leq n$ intervals defined by completion times of consecutive jobs. Let Z_k denote the combination of jobs processed in parallel in the k-th interval. Thus, in general, a feasible sequence S of combinations $Z_k, k = 1, 2, \ldots, p$, can be associated with each feasible schedule. Feasibility of such a sequence requires that the number of elements in each combination does not exceed m, and that each job appears exactly in one or in consecutive combinations in S (non-preemptability).

For a given feasible sequence S of jobs on machines, we can find an optimal continuous resource allocation, i.e. an allocation that leads to a schedule minimizing the given criterion from among all feasible schedules generated by S. To this end, a non-linear mathematical programming problem has to be solved, in the general case (Józefowska and Węglarz, 1998). An optimal schedule for a given feasible sequence (i.e. a schedule resulting from an optimal continuous resource allocation for this sequence) is called a *semi-optimal schedule*. In consequence, a globally optimal schedule can be found by solving the continuous resource allocation problem optimally for all feasible sequences. Unfortunately, in general, the number of feasible sequences grows exponentially with the number of jobs. Therefore it is justified to apply local search metaheuristics, such as tabu search, operating on the set of all feasible sequences (Józefowska et al., 1998a).

18.2.3 Minimizing the makespan

In discrete-continuous scheduling problems with the makespan as the scheduling criterion, the problem is to minimize the makespan $M = \max\{C_i\}, i = 1, 2, \ldots, n$, where C_i is the completion time of job i. It has been shown in Józefowska and Węglarz (1998) that for processing rate functions f_i of the form $f_i \leq c_i u_i, c_i = f_i(1), i = 1, 2, \ldots, n$, (and thus, in particular, for convex functions) the makespan is minimized by scheduling all jobs on one machine, and allotting each job the total available amount of the continuous resource. In consequence, we consider concave processing rate functions which, in addition, have much bigger practical applications. For concave job models and the makespan minimization problem, it is sufficient to consider feasible sequences of combinations $Z_k, k = 1, 2, \ldots, n-m+1$, composed of m jobs each (Józefowska and Węglarz, 1998).

18.2.4 *Minimizing the mean flow time*

In this case the problem is to minimize the mean flow time $\bar{F} = \frac{1}{n}\sum_{i=1}^{n} C_i, i = 1, 2, \ldots, n$, where C_i is the completion time of job i. It has been shown in Józefowska and Węglarz (1996) that for processing rate functions f_i of the form $f_i \leq c_i u_i, c_i = f_i(1), i = 1, 2, \ldots, n$, (and thus, in particular, for convex functions) the mean flow time is minimized by scheduling all jobs on one machine according to non-decreasing values of $\tilde{x}_i/c_i$, and allotting each job the total available amount of the continuous resource. It has been proved in Józefowska and Węglarz (1996) that for concave job models, feasible sequences composed of n combinations of jobs, such that first $n-m+1$ combinations contain exactly m elements and next combinations $n-m+2, n-m+3, \ldots, n-1, n$ contain $m-1, m-2, \ldots, 1$ element respectively, should be considered.

18.2.5 *Minimizing the maximum lateness*

In this case the problem is to minimize the maximum lateness $L_{max} = \max\{C_i - d_i\}, i = 1, 2, \ldots, n$, where C_i is the completion time of job i and d_i is its due-date. It has been shown in Józefowska et al. (1997a) that for processing rate functions f_i of the form $f_i \leq c_i u_i, c_i = f_i(1), i = 1, 2, \ldots, n$, (and thus, in particular, for convex functions) the maximum lateness is minimized by scheduling all jobs on one machine according to non-decreasing values of d_i, and allotting each job the total available amount of the continuous resource. For concave job models, feasible sequences of the same form as for the mean flow time have to be considered.

18.3 TABU SEARCH

Tabu search is a metaheuristic based on neighborhood search with overcoming local optimality. It works in a deterministic way, trying to model human memory processes. Memory is implemented by the implicit recording of previously seen solutions, using simple but effective data structures. Tabu search was originally developed by Glover (1986); Glover (1989); Glover (1990), and a comprehensive report of the basic concepts and recent developments was given in Glover and Laguna (1997).

18.3.1 *Solution representation*

A feasible solution for tabu search is represented by a feasible sequence defined in the previous section, i.e. for the makespan minimization problem it is a sequence of $n-m+1$ m-element combinations of jobs, and for the mean flow time and maximum lateness minimization problems it is a sequence of n combinations of jobs, such that first $n-m+1$ combinations contain exactly

m elements and consecutive combinations $n-m+2, n-m+3, \ldots, n-1, n$ contain $m-1, m-2, \ldots, 1$ element, respectively.

E.g. if $n = 6$ and $m = 3$, then in the makespan minimization problem a feasible sequence may have the form:

$$S = \{1,2,3\}, \{2,3,4\}, \{2,3,5\}, \{3,5,6\}$$

which means that jobs 1, 2 and 3 are processed in parallel in the first interval, jobs 2, 3 and 4 – in the second interval, jobs 2, 3 and 5 – in the third one and jobs 3, 5 and 6 in the last interval.

In the case of the mean flow time or the maximum lateness, the form of a feasible sequence may be the following:

$$S = \{1,2,3\}, \{2,3,4\}, \{2,3,5\}, \{3,5,6\}, \{3,5\}, \{5\}$$

which means that after job 6 is finished in the fourth interval, only jobs 3 and 5 are processed in the fifth interval, and job 5 is processed alone in the last (sixth) interval.

18.3.2 *Starting solution*

A starting solution is generated in two steps. In the first step jobs in particular combinations are generated randomly. In the second step the obtained solution is transformed according to the vector of processing demands (or due dates) in order to construct a solution with a better value of the considered scheduling criterion. More precisely, the following transformation is made:

- for the makespan minimization problem, the job which occurs the largest number of times in the random solution is replaced by the job with the largest processing demand and so on; as a consequence, a job with a larger processing demand appears in a larger number of combinations which, intuitively, should lead to better schedules;
- for the mean flow time minimization problem, the job which ends in the first combination is replaced by the job with the smallest processing demand and so on; in consequence, a job with a smaller processing demand is finished earlier which should be advantageous in this case;
- for the maximum lateness minimization problem, the job which ends in the first combination is replaced by the job with the earliest duedate and so on; as a consequence, a job with an earlier duedate is finished earlier which is, obviously, profitable in the case of this scheduling criterion.

18.3.3 *Objective function*

The value of the objective function for a feasible solution is defined as the value of the considered scheduling criterion in the semi-optimal schedule for the corresponding feasible sequence. This value is calculated by solving an appropriate non-linear mathematical programming problem. More details on finding semi-optimal schedules (i.e. solving the continuous part of the problem) for the considered scheduling criteria can be found in Józefowska and Węglarz (1996; 1998) and Józefowska et al. (1997a).

18.3.4 Neighborhood generation mechanism

A neighbor of a current solution is obtained by replacing a job in a chosen combination by another job. A job may be replaced only in either the first or the last combination in the sequence of combinations it occurs (non-preemptability), provided that these combinations are different (each job must be executed). In the case of the makespan minimization problem the following rule is applied:

- if Z_k is the first (but not the only) combination in which job i occurs, then job i in Z_k is replaced by the only job j that belongs to Z_{k+1} and does not belong to Z_k;
- if Z_k is the last (but not the only) combination in which job i occurs, then job i in Z_k is replaced by the only job j that belongs to Z_{k-1} and does not belong to Z_k.

A neighborhood consists of feasible solutions obtained by performing all such replacements.

E.g. let $n = 5, m{=}3$ and $S{=}\{1,2,3\}, \{2,3,4\}, \{3,4,5\}$. The neighboring solutions are:

$$\{1,4,3\},\{2,3,4\},\{3,4,5\}$$
$$\{1,2,4\},\{2,3,4\},\{3,4,5\}$$
$$\{1,2,3\},\{1,3,4\},\{3,4,5\}$$
$$\{1,2,3\},\{2,3,5\},\{3,4,5\}$$
$$\{1,2,3\},\{2,3,4\},\{2,4,5\}$$
$$\{1,2,3\},\{2,3,4\},\{3,2,5\}.$$

Let us observe that exactly one job is completed in the first combination of a feasible solution. Thus, this job occurs in exactly one combination and so does exactly one job in the last combination. As a consequence, at most $n - 2$ jobs may occur in multiple combinations. Each of them may be replaced in exactly two positions (in its first and its last occurrence). As a result, the maximal possible number of neighbors for a given feasible solution is independent from the number of machines and is equal to $(n - 2) \cdot 2$.

In the case of mean flow time and maximum lateness minimization problems, the above mechanism has been adapted to the slightly different form of a feasible solution and the following rule is applied:

- if Z_k is the last (but not the only) combination in which job i occurs, then job i in Z_k is replaced by the only job j that belongs to Z_{k-1} and does not belong to Z_k;
- if Z_k is the first (but not the only) combination in which job i occurs and $k < n - m + 1$, then job i in Z_k is replaced by the only job j that belongs to Z_{k+1} and does not belong to Z_k.

A neighborhood consists of feasible solutions obtained by performing all such replacements.

E.g. let $n = 5, m = 3$ and S={1,2,3},{2,3,4},{3,4,5},{4,5},{5}. The neighboring solutions are:

{1,4,3},{2,3,4},{3,4,5},{4,5},{5}
{1,2,4},{2,3,4},{3,4,5},{4,5},{5}
{1,2,3},{1,3,4},{3,4,5},{4,5},{5}
{1,2,3},{2,3,5},{3,4,5},{4,5},{5}
{1,2,3},{2,3,4},{2,4,5},{4,5},{5}
{1,2,3},{2,3,4},{3,4,5},{3,5},{5}
{1,2,3},{2,3,4},{3,4,5},{4,5},{4}.

Observe that exactly one job is completed in the first combination of the feasible sequence. Thus, this job occurs in exactly one combination and must not be replaced. Moreover, exactly one job started in combination $n - m + 1$ may be replaced only in its last combination. In consequence, at most $n-2$ jobs may occur in multiple combinations. Each of them may be replaced in exactly two positions (in its first and its last occurrence). As a result, the maximum possible number of neighbors for a given feasible solution is independent from the number of machines and is equal to $2(n - 2) + 1 = 2n - 3$.

The neighbor structure described above enables searching the whole space of feasible solutions (feasible sequences). In each iteration the algorithm moves to the best solution from the neighborhood (in terms of the objective function value) that is not tabu (next move selection rule). A move leading from a solution to a neighboring one is described by 3 attributes:

(*number of a combination, job replaced, job introduced*).

E.g. if the algorithm moves from a solution {1,2,3}, {2,3,4} to a solution {1,2,4}, {2,3,4}, then the move representing this transition has the following attributes: (1,3,4).

In consequence, the reverse move to move (k,i,j) is move (k,j,i). In the above example it is the move (1,4,3) which should be tabu in the next iteration in order not to come back to the same solution again. The representation of a move is common for all the problems considered.

18.3.5 Tabu list management

In the tabu search implementations for all the three scheduling criteria, the tabu list is managed according to the reverse elimination method – REM (Glover, 1989). However, in the case of the makespan further research has been carried out, and also two other methods have been tested: the tabu navigation method – TNM (Skorin-Kapov, 1990) and the cancellation sequence method – CSM (Glover, 1989).

The tabu navigation method corresponds to a simple (static) version of tabu search. There is only one list, so-called *tabu list* (TL), which is a queue of a given length. Whenever a move is performed, its reverse is added to the TL in order to avoid going back to a solution already visited. All moves existing on the TL are tabu. However, the tabu status of a move may be canceled according to so-called *aspiration criterion*. The aspiration criterion allows the algorithm to move to a tabu solution (perform a tabu move) if this solution is better than the best found so far. The important parameter in the TNM is the length of the TL. In our implementation it has been set at 7.

The cancellation sequence method (CSM) is based on so-called *cancellation sequences* (CS) which utilize some sort of memory structure imposed on the sequence of moves already performed throughout the search process. A cancellation sequence may be defined as a sequence of moves between two solutions. In other words, if a list contains moves performed throughout the search, then any connected part of it is a cancellation sequence.

The CSM embodies the concept of combining the static and the dynamic tabu processes. Moves are set tabu in two ways. The simple (static) tabu process is performed on the tabu list (TL) which contains moves reverse to the moves performed most recently. Moves occurring on the TL are certainly tabu. The main (dynamic) tabu process is performed on so-called *active tabu list* (ATL) which also contains reverse moves, but they are set tabu depending on the cancellation sequences created on the ATL. The following three parameters have to be defined for a TS algorithm with the CSM:

- the length of the ATL,
- the number of iterations for which the tabu status of a move set by the ATL process is maintained,
- the length of the TL.

In our implementation the above parameters have been set at 7, 5 and 1, respectively. These values have been applied on a basis of the computational experiments presented in Józefowska et al. (2002).

The reverse elimination method (REM) is based on so-called *residual cancellation sequences* (RCS). A residual cancellation sequence results from a cancellation sequence (CS) by canceling (i.e. eliminating) some of its moves. The REM differs from the CSM in the way of carrying out those cancellations, but both the methods match in their fundamental principle: any move whose execution would lead to canceling an RCS is set tabu. However, whereas the CSM in one iteration computes and stores RCSs created by the present move (older RCSs are updated which may result in tabu settings), the REM in each iteration recomputes all RCSs between the current solution and any previous one. There is only one tabu list, the active tabu list (ATL), whose elements (the RCSs) are minimal sets of moves leading back from the current solution to solutions visited in the past. If the starting solution is denoted by #1, then RCS #j (the j-th element of the ATL) is the minimal set of moves by which the current solution differs from solution #j (i.e. visited in the j-th iteration). The REM takes advantage of the fact that a solution can be revisited in the next iteration only if it is a neighbor of the current solution (i.e. if the set of moves leading back to it from the current solution consists of one element only). In consequence, every move occurring on the ATL as a one-element RCS is forbidden in the next iteration. Thus, in order to define the status of a move, it is sufficient to check only those elements of the ATL which contain exactly one move. The most important parameter in the REM is the length of the ATL. In our implementation it has been set at 7.

More details concerning the tabu list management methods for discrete-continuous scheduling problems can be found in Józefowska et al. (1996; 1998b; 2002).

18.3.6 Stopping criterion

The stopping criterion is defined as an assumed number of visited solutions.

18.4 SIMULATED ANNEALING AND GENETIC ALGORITHMS

In order to evaluate the efficiency of the proposed tabu search implementations, they are compared with two other well-known metaheuristics: simulated annealing and genetic algorithms. Simulated annealing (SA) is local search metaheuristic which belongs to a class of the threshold algorithms, and can be viewed as a special case of the First Fit Strategy, where the next solution is accepted with certain probability. Genetic algorithms (GA) operate on a population of individuals, each one representing a feasible solution of the considered problem. The process of looking over new solutions is made by using recombination operators. Below we briefly describe the most important features of both the metaheuristics applied to the considered classes of problems.

18.4.1 Simulated annealing

A feasible solution for SA is represented by two n-element sequences: the first one is a permutation of jobs and defines the order in which jobs are started, the second one for each job defines the machine it will be processed on. These two sequences allow to create the first $n-m+1$ combinations of the corresponding feasible sequence. In the case of the mean flow time and the maximum lateness, the last $m-1$ combinations are created according to the assumption that the job on the first machine is completed first, the job on the second machine is completed second, and so on.

A neighbor of the current solution may be generated in two ways. The first operator consists in shifting a job randomly chosen from one position in the first sequence to another position also randomly chosen. As a result, the order in which jobs are started is changed. The second operator chooses randomly an element in the second sequence and replaces it by another element also randomly chosen. In consequence, the assignment of jobs to machines is changed. The probability of applying either of the operators is equal.

The simple cooling scheme (Aarts and van Laarhoven, 1987) has been applied in our implementation of SA. The method of generating a starting solution, as well as the way of calculating the objective function, and the stopping criterion are identical as for TS.

18.4.2 Genetic algorithms

The representation of a feasible solution for GA is identical as for SA but the corresponding sequences are here called *chromosomes.* The simplest method of creating the initial population has been chosen – the random creating. The population size has been set at 50. The fitness of an individual depends on the value of the objective function, the smaller value of the objective function, the higher fitness of the individual. The linear ranking with elitism has been chosen as the selection operator.

Four recombination operators especially suited for the assumed representation of a feasible sequence are used. Two of them operate on a single parent individual (mutation type), and the remaining two work on pairs of parent individuals (crossover type). The first mutation type operator we call *mutation I*. It is an elementary change made in the second chromosome of an individual. An element in a randomly chosen position of the second chromosome is replaced by another one.

The second mutation type operator is *mutation II*. This operator exchanges a random number of randomly selected jobs. In other words, a number of jobs which will be exchanged, as well as these jobs, are chosen randomly. Of course, this operator concerns only the first chromosome of the selected individual. As a result of mutation II we get a new order of jobs.

The first crossover operator, the *head crossover*, operates on the first chromosome only. This operator has to preserve the form of the first chromosome (a permutation of jobs) in both the offspring individuals. It uses two crossing points. First, the segments between crossing points are copied into offspring individuals. Next, starting from the second crossing point of one parent, the jobs from the other parent are copied in the same order, omitting symbols already present. Reaching the end of the string, copying is continued from the first place of the string.

The second crossover operator, the *tail crossover*, concerns the second chromosome only. The feasibility conditions of this chromosome allow to use its simplest version. The tail crossover cuts the second chromosomes of both parent individuals at the same randomly chosen cut point, and exchanges the resulting segments.

The way of calculating the objective function, and the stopping criterion for GA are identical as for TS and SA.

For extensive descriptions of our SA and GA implementations for discrete continuous-scheduling problems see Józefowska et al. (1997a;b; 1998a).

18.5 COMPUTATIONAL ANALYZES

Several computational experiments have been performed to evaluate the efficiency of the proposed tabu search algorithms for the considered classes of discrete-continuous scheduling problems. The algorithms have been implemented in C++ and have run on an SGI PowerChallenge XL computer with 12 RISC 8000 processors in the Poznań Supercomputing and Networking Center. As we have mentioned before, for each solution visited in the solution space, the corresponding value of the objective function was found by solving a non-linear mathematical programming problem. In this step specially adopted solver CFSQP (A C Code for Solving (Large Scale) Constrained Nonlinear (Minimax) Optimization Problems, Generating Iterates Satisfying all Inequality Constraints) 2.3 (Lawrence et al., 1995) was applied. The solver stopped when the absolute difference in consecutive values of the objective function was less than or equal to 10^{-3}.

The experiment was carried out for job processing rate functions of the form $f_i = u_i^{1/\alpha_i}$ with $\alpha_i \in \{1, 2\}, i = 1, 2, \ldots, n$. The values of α_i were generated randomly with equal probabilities, whereas processing demands of jobs as well as due dates were generated from the uniform distribution in the interval [1,100].

In order to ensure a comparable computational effort devoted to all the algorithms tested, they started with identical initial solutions, and visited the same number of solutions (the stop criterion) set at 1000.

In the case of the makespan, the results are compared to optimal solutions (found by the full enumeration procedure), in the cases of the mean flow time and the maximum lateness, the results are compared to the best solutions found by any of the metaheuristics tested. In the latter case a truncated experiment has been performed because of the difficulty in solving the continuous part of the problem.

Let us now discuss in detail the results obtained by the presented tabu search algorithms for the considered discrete-continuous scheduling problems, starting with the makespan minimization problem (Józefowska et al., 1998a).

In this case the experiment was divided into two parts. The first experiment was performed for three groups of instances with two machines and 10, 15, and 20 jobs. The second experiment was carried out for n=10 jobs and the number of machines m=2, 3, and 4. For each problem size, 100 instances were generated randomly, i.e. for each job its processing demand and processing rate function were generated randomly as described above. The fact of using for testing problems of rather small sizes, as well as a not very large number of configurations visited for each instance, follows from two factors. The first one is a huge number of all feasible sequences existing for each problem instance (and, of course, that number grows exponentially with the number of jobs). Since the full enumeration procedure was applied to find optimal solutions, the problem size could not have been very large. The second factor is a massive computational effort required by the non-linear solver to calculate an optimal continuous resource allocation. The time used by the solver is about 0.6 sec. for one solution of the problem with 10 jobs and 2 machines, and grows up to about 8 sec. for the problem with 20 jobs. This also results in almost identical computational times required by all the metaheuristics tested since the majority of their computational effort is devoted to the calculation of the objective function.

The results of the experiments are presented in Tables 18.1, 18.2, and 18.3. For each problem size these are shown:

- the number of instances for which the algorithm found an optimal solution,
- the average relative deviation from optimum,
- the maximal relative deviation over all the instances tested.

The relative deviation from optimum in percent is calculated according to the following formula:

$$RD = \frac{M - M^*}{M^*} \cdot 100\% \qquad (18.2)$$

where M is the value obtained by the relevant algorithm and M^* is the optimal value of the makespan.

Table 18.1 Computational results for the makespan - the number of optimal solutions found

(n,m)	*TS*	*SA*	*GA*
(10,2)	46	24	26
(10,3)	53	5	13
(10,4)	65	15	16
(15,2)	47	4	5
(20,2)	54	1	0

Table 18.2 Computational results for the makespan - the average relative deviation

(n,m)	*TS*	*SA*	*GA*
(10,2)	0.008%	0.011%	0.017%
(10,3)	0.044%	0.153%	0.126%
(10,4)	0.077%	0.219%	0.242%
(15,2)	0.030%	0.094%	0.036%
(20,2)	0.028%	0.098%	0.078%

The obtained results prove that the presented tabu search algorithm is a very efficient algorithm for the considered discrete-continuous scheduling problems. For each problem size, the algorithm finds an optimal solution in about

Table 18.3 Computational results for the makespan - the maximal relative deviation

(n,m)	*TS*	*SA*	*GA*
(10,2)	0.122%	1.018%	2.705%
(10,3)	0.685%	6.449%	11.467%
(10,4)	1.028%	11.397%	10.290%
(15,2)	0.675%	18.003%	2.985%
(20,2)	1.132%	13.544%	6.164%

50% of instances, reaching 65% for 10 jobs and 4 machines. The average relative deviation from optimum does not exceed 0.1% in any case, and the maximum relative deviation oscillates around 1%. The results obtained by tabu search are significantly better than the ones obtained by the two other metaheuristics, both in terms of the number of optimal solutions found, and in terms of the average relative deviation. It is worth noticing that the growth of the number of jobs does not result in a significant deterioration in the average relative deviation but the maximal relative deviation clearly increases. Under a fixed number of jobs, the growth of the number of machines has a greater influence on the produced results. Although the number of optimal solutions found grows, the deviations from optimum grow as well. This suggests that if the algorithm does not find an optimal solution, the best solution found can be more distant from optimum. It means that along with the growth of the number of machines, not finding an optimal feasible sequence leads to a relatively big deterioration in the schedule quality, despite an optimal allocation of the continuous resource.

Józefowska et al. (1997c; 1999) describe computational analyzes of tabu search for special cases of discrete-continuous scheduling problems to minimize the makespan.

The experiment for the mean flow time (Józefowska et al., 1997b) was performed for the same set of instances as for the makespan. The results of the experiment are presented in Tables 18.4, 18.5, and 18.6 in the same form as for the makespan, but in this case optimal solutions are not known and they are replaced by best solutions known (i.e. found by any of the metaheuristics).

Table 18.4 Computational results for the mean flow time - the number of best solutions found

(n,m)	*TS*	*SA*	*GA*
(10,2)	94	1	6
(10,3)	95	1	4
(10,4)	77	3	43
(15,2)	99	0	1
(20,2)	100	0	0

Table 18.5 Computational results for the mean flow time - the average relative deviation

(n,m)	TS	SA	GA
(10,2)	0.492%	2.536%	1.582%
(10,3)	2.207%	3.528%	1.990%
(10,4)	1.672%	3.567%	1.910%
(15,2)	0.0004%	4.801%	4.134%
(20,2)	0.000%	7.640%	7.152%

Table 18.6 Computational results for the mean flow time - the maximal relative deviation

(n,m)	TS	SA	GA
(10,2)	0.617%	5.991%	6.586%
(10,3)	4.782%	10.245%	7.336%
(10,4)	7.630%	11.781%	9.147%
(15,2)	0.041%	14.291%	10.152%
(20,2)	0.000%	21.817%	14.351%

The results show clearly that also in this case tabu search performs best, finding the greatest number of solutions better than found by the two other metaheuristics, and achieving the smallest average and relative deviations from best solutions known for all the problem sizes (the only exception is the case of 10 jobs and 3 machines where genetic algorithm obtained a slightly better average deviation). The advantage of the tabu search algorithm increases with the growth of the number of jobs, and reaches 100% for 20 jobs (i.e. for each instance of the problem, tabu search finds the best solution). The obtained results allow to state that also for the mean flow time minimization problem, the presented tabu search algorithm is very efficient and finds very good solutions. However, in order to confirm that statement, it would be necessary to compare the obtained results to optimal solutions.

For the maximum lateness minimization problem (Józefowska et al., 1997a), only a limited experiment was performed because of a huge computational effort needed for solving the continuous part of the problem. Therefore, the experiment was carried out for 10 jobs and the number of machines equal to 2, 3, and 4. The number of generated instances equals 21, 14, and 7, respectively. In the case of maximum lateness, it is not justified to calculate the relative deviations from the best solution known. Therefore we present the number of best solutions found by each of the three metaheuristics, as well as the maximal absolute deviation from the best solution found (best maximum lateness). Since for all the instances tested the values of the maximum lateness were positive, the maximal deviation is simply the maximal difference over all instances between the maximum lateness found by a metaheuristic and the best maximum lateness known. Moreover, the corresponding L_{max} is the best maximum lateness known for the instance where the maximum deviation occurred. These values are shown in Tables 18.7 and 18.8.

Table 18.7 Computational results for the maximum lateness - the number of best solutions found

(n,m)	TS	SA	GA
(10,2)	16	2	7
(10,3)	9	0	5
(10,4)	7	0	0

Table 18.8 Computational results for the maximum lateness - the maximal absolute deviation / corresponding L_{max}

(n,m)	TS	SA	GA
(10,2)	7.317/310.539	0.910/210.346	0.870/314.492
(10,3)	5.974/257.193	7.323/216.810	2.365/268.847
(10,4)	0/-	6.460/348.622	4.901/243.037

The results show that tabu search finds a solution with the best maximum lateness most often. For the case of 4 machines, no other algorithm was able to find an equally good solution in any instance. However, for 2 and 3 machines, genetic algorithm achieves a better maximal deviation. It means that in case the tabu search algorithm fails to find a solution with the best maximum lateness, the obtained solution may be relatively distant from the best one. Nevertheless, let us remind that for the two previous criteria, the advantage of tabu search increased with the growth of the number of jobs. It would be very useful to perform computational experiments for larger problem sizes, as well as to compare the results to optimal solutions. Unfortunately, this will be very difficult since the solution of the mathematical programming problem needed for finding an optimal continuous resource allocation is too time consuming in this case (about 24 hours for one instance with 10 jobs when 1000 solutions are visited).

Let us pass now to the analysis of the tabu search metaheuristic with different tabu list management methods (Józefowska et al., 2002). In this case the experiment was performed for the same set of instances as in the case of the makespan. Also in this case the makespan was the scheduling criterion, and the results are compared to optimal solutions. The results of the experiment are presented in Tables 18.9, 18.10, and 18.11 in the same form as in the case of the makespan minimization problem.

The results show that the tabu navigation method performs best for the considered problem. The number of optimal solutions found by this method is the greatest for each problem size, whereas the average relative deviation from optimum is the smallest (with the only exception of $n = 20, m = 2$ where

Table 18.9 Computational results for tabu list management methods - the number of optimal solutions found

(n,m)	*TNM*	*CSM*	*REM*
(10,2)	59	52	46
(10,3)	61	53	53
(10,4)	73	66	65
(15,2)	70	56	47
(20,2)	67	57	54

Table 18.10 Computational results for tabu list management methods - the average relative deviation

(n,m)	*TNM*	*CSM*	*REM*
(10,2)	0.002%	0.006%	0.008%
(10,3)	0.027%	0.066%	0.044%
(10,4)	0.048%	0.103%	0.077%
(15,2)	0.006%	0.025%	0.030%
(20,2)	0.030%	0.041%	0.028%

Table 18.11 Computational results for tabu list management methods - the maximal relative deviation

(n,m)	*TNM*	*CSM*	*REM*
(10,2)	0.041%	0.289%	0.122%
(10,3)	0.494%	1.706%	0.685%
(10,4)	1.028%	2.480%	1.028%
(15,2)	0.315%	0.675%	0.675%
(20,2)	1.132%	1.132%	1.132%

the reverse elimination method obtained a slightly better result but the difference is rather insignificant). The maximal relative deviation from optimum for the TNM is smaller or at most equal to the deviations obtained for the two other methods. Let us stress again that the results obtained by all the three methods are, in general, very good. About 50 – 70 optimal solutions found out of 100 for each problem size, in combination with the average relative deviation well below 0.1%, prove that our tabu search algorithms are very effective for the considered class of problems. This is even more impressive when taking into account the fact that the algorithms visited only 1000 solutions, which is not a large number comparing to the number of all feasible sequences. On the other hand, it should be stressed that many different feasible solutions give, in fact, the same value of the objective function. This is the specificity of the problem that an optimal allocation of the continuous resource may lead to schedules of the same quality for many different feasible sequences. Furthermore, let us repeat that the overwhelming majority of computational effort in the algorithms is devoted to the calculation of the value of the objective function (solving the non-linear mathematical programming problem). As a consequence, the effort put into managing the tabu list is not so significant, which makes the computational time almost identical for all the three implementations. However, if heuristic algorithms were applied for allocating the continuous resource instead of finding an optimal allocation for each feasible sequence, the tabu navigation method, as the least complicated one, might get another advantage in terms of computational time. Finally, let us notice that although the TNM performs best for our problem, the results obtained for all the three methods are, in general, of a similar quality, which seams to suggest that other aspects of the tabu search strategy (solution representation, neighborhood generation mechanism, good starting solution) have stronger impact on the effectiveness of this metaheuristic than the tabu list management method itself.

18.6 SUMMARY

In this paper discrete-continuous problems of scheduling non-preemptable, independent jobs on parallel, identical machines have been considered. The makespan, the mean flow time, and the maximum lateness have been considered as the scheduling criteria. A tabu search approach to the defined problems has been discussed. Three tabu list management methods have been tested on the makespan minimization problem. A performance analysis of the tabu search metaheuristic for the considered problems has been discussed. Several computational experiments have been described in the paper. The results obtained by tabu search have been compared to the results obtained by two other metaheuristics: simulated annealing and genetic algorithms. In the case of the makespan minimization problem the results have been also compared to optimal solutions. The computational analyzes confirm a very good efficiency of the proposed tabu search algorithms for the discrete-continuous scheduling problems. As far as the tabu list management methods are concerned, they do not seem to have a crucial impact on the efficiency of the algorithm, however, the tabu navigation method performs best. Generally, the results obtained by tabu search are superior to the results produced by both simulated annealing and genetic algorithms. On the other hand, it is still possible to extend the performance analysis of tabu search for discrete-continuous scheduling problems, e.g. in the case of mean flow time or, especially, maximum lateness.

Acknowledgments

This research has been supported by the State Committee for Scientific Research (KBN), Poland, Grant no. 4T11F 001 22.

Bibliography

E.H.L. Aarts and P.J.M. van Laarhoven. *Simulated Annealing: Theory and Applications*. Reidel, Dordrecht, 1987.

W.N. Burkow. Raspriedielenije riesursow kak zadacza optimalnogo bystrodiejstwia. *Avtomat. I Tielemieh.*, 27(7), 1966.

F. Glover. Tabu search - Part 1. *ORSA J. Computing*, 1:190–206, 1989.

F. Glover. Future path for integer programming and links to artificial intelligence. *Computers & Operations Research*, 5:533–549, 1986.

F. Glover. Tabu search - Part 2. *ORSA J. Computing*, 2:4–32, 1990.

F. Glover and M. Laguna. *Tabu Search*. Kluwer Academic Publishers, Norwell, 1997.

J. Józefowska, M. Mika, R. Różycki, G. Waligóra, and J. Węglarz. Discrete-continuous scheduling to minimize maximum lateness. In *Proc. of the Fourth International Symposium on Methods and Models in Automation and Robotics MMAR'97, Międzyzdroje 26-29.08.1997*, pages 947–952, 1997a.

J. Józefowska, M. Mika, R. Różycki, G. Waligóra, and J. Węglarz. Discrete-continuous scheduling to minimize the mean flow time - computational experiments. *Computational Methods in Science and Technology*, 3:25–37, 1997b.

J. Józefowska, M. Mika, R. Różycki, G. Waligóra, and J. Węglarz. Discrete-continuous scheduling with identical processing rates of jobs. *Foundations of Computing and Decision Sciences*, 22(4):279–295, 1997c.

J. Józefowska, M. Mika, R. Różycki, G. Waligóra, and J. Węglarz. Local search metaheuristics for some discrete-continuous scheduling problems. *European Journal of Operational Research*, 107(2):354–370, 1998a.

J. Józefowska, M. Mika, R. Różycki, G. Waligóra, and J. Węglarz. Discrete-continuous scheduling to minimize the makespan with power processing rates of jobs. *Discrete Applied Mathematics*, 94:263–285, 1999.

J. Józefowska, G. Waligóra, and J. Węglarz. Tabu search algorithm for some discrete-continuous scheduling problems. In *Modern Heuristic Search Methods*, pages 169–182. V.J. Rayward-Smith (ed.), Wiley, Chichester, 1996.

J. Józefowska, G. Waligóra, and J. Węglarz. Tabu search with the cancellation sequence method for a class of discrete-continuous scheduling problems. In *Proc. of the Fifth International Symposium on Methods and Models in Automation and Robotics MMAR'98, Międzyzdroje 25-29.08.1998*, pages 1047–1052, 1998b.

J. Józefowska, G. Waligóra, and J. Węglarz. Tabu list management methods for a discrete-continuous scheduling problem. *European Journal of Operational Research*, 137(2):288–302, 2002.

J. Józefowska and J. Węglarz. Discrete-continuous scheduling problems - mean completion time results. *European Journal of Operational Research*, 94 (2):302–309, 1996.

J. Józefowska and J. Węglarz. On a methodology for discrete-continuous scheduling. *European Journal of Operational Research*, 107(2):338–353, 1998.

C. Lawrence, J.L. Zhou, and A.L. Tits. *Users guide for CFSQP Version 2.3 (Released August 1995)*, 1995.

J. Skorin-Kapov. Tabu search applied to the quadratic assignment problem. *ORSA J. Computing*, 2:33–45, 1990.

Metaheuristics: Computer Decision-Making, pp. 405-420
Pascale Kuntz, Bruno Pinaud, and Rémi Lehn

19 ELEMENTS FOR THE DESCRIPTION OF FITNESS LANDSCAPES ASSOCIATED WITH LOCAL OPERATORS FOR LAYERED DRAWINGS OF DIRECTED GRAPHS

Pascale Kuntz[1], Bruno Pinaud[1], and Rémi Lehn[1]

[1]Ecole Polytechnique de L'Université de Nantes (Polytech'Nantes)
Institut de Recherche en Informatique de Nantes (IRIN)
Nantes, France
pascale.kuntz@irin.univ-nantes.fr, bruno.pinaud@irin.univ-nantes.fr
remi.lehn@irin.univ-nantes.fr

Abstract: Minimizing arc crossings for drawing acyclic digraphs is a well-known *NP*-complete problem for which several local-search approaches based on local transformations (switching, median, ...) have been proposed. Their adaptations have been recently included in different metaheuristics. As an attempt to better understand the dynamics of the search processes, we study the fitness landscapes associated with these transformations. We first resort to a set of multi-start descents to sample the search space for three hundred medium-sized graphs. Then, we investigate complete fitness landscapes for a set of 1875 smaller graphs, this aims at showing some instance characteristics that influence search strategies. The underlying idea is to consider a fitness landscape as a graph whose vertices are drawings and arcs representing a transformation of a drawing into another. We confirm that the properties of basins of attraction closely depend on the instances. Also, we show that the probability of being stuck on a local optimum is linked to the specific shapes of the basins of attraction of global optima which may be very different from the regular image of the continuous case generally used as a reference.

Keywords: Layered digraph drawing, Fitness landscape, Metaheuristics, Local search.

19.1 INTRODUCTION

Layered drawing of directed acyclic graphs is a classical problem of the huge graph drawing literature (Di-Battista et al., 2000; 1999). For the applications, this problem has known renewed interest in the past years in particular in the field of "information visualization," as in Web browsing and cartography (Herman et al., 2000). It consists in intelligibly arranging vertices on vertical layers and representing arcs by oriented line segments which flow in the same direction. Experimental observations have recently confirmed that one of the most important criteria for readability is arc crossing minimization (Purchase, 2000).

Minimizing arc crossing for a layer digraph could a priori seem easier than the general problem of minimizing arc crossing on a plane since the choice of geometric coordinates is here transformed into a choice of vertex ordering on each layer. Yet, it remains *NP*-complete even if there are only two layers (Garey and Johnson, 1983; Eades and Wormald, 1994). Several deterministic local transformations have been proposed. The easiest ones, which have originally been defined for 2-layered drawings (Eades and Kelly, 1986; Jünger and Mutzel, 1997), are based on simple permutations on each layer. For instance, the greedy-switch heuristic iteratively switches consecutive pairs of vertices if this decreases the crossing number, and the split heuristic assigns the vertices above or below a given pivot vertex according to the same criterion. But probably, the most popular transformations are the so-called "averaging heuristics" including the barycenter heuristic (Sugiyama et al., 1981) and the median heuristic (Eades and Wormald, 1994). The basic underlying idea is that arc crossings tend to be minimized when connected vertices are placed facing each other. Roughly speaking, these approaches compute the average positions, i.e. the barycenter or median of their neighbors, for the vertices on each layer and sort them according to these values. Several variants and implementations of these operators have been defined (a chronological review is provided in Laguna et al. (1997)).

More recently, these transformations have been combined for different metaheuristics, in particular Tabu search (Laguna et al., 1997), GRASP (Martí, 2001) and Evolutionary algorithms (Utech et al., 1998). These approaches have been essentially validated by numerical experiments on test sets constructed with the graph size and the density criteria. However, as far as we know, the limits of the influence of these criteria on the search processes have not been precisely studied. Moreover, some hypotheses on the properties of the search spaces –on the multimodality notably– have been voiced without proof. We propose a descriptive analysis of fitness landscapes associated with the classical operators introduced above. In particular, we aim at precising the distribution of the optima, and defining the shapes of their basins of attraction.

In the first part of the paper, we resort to a set of multi-start descents to explore the search space for 300 medium-sized graphs which are characteristic of the graphs we used for statistical rule visualization in data mining (Kuntz et al., 2000). Previous numerical experiments with one thousand descents on another graph family have underlined the great variability of the results (Lehn and Kuntz, 2001); the number of descents converging on an "optimal" solution, i.e. the best solution among the computed solution set, followed a distribution close to a Poisson. For validating the robustness of the first tests we here extend the experiments with a set of 5000 descents. The results confirm the dispersion previously observed with differences in the distribution, but they show that the graph density criterion can partly explain the observed features only.

In the second part, we complete these statistical results by computing the exact fitness landscapes of a graph family from the graph-based definition introduced by Jones and Forrest (Jones, 1995; Jones and Forrest, 1995b). Due to the combinatorial explosion of problems of that nature, we were obliged to restrict ourselves to small instances, but, we have preserved a variability (1875 graphs) representative of the general problem. Besides the classical fitness-distance correlation, we compute the probability of being stuck on a local optimum with a descent and show that this result is closely linked to the specific shapes of the basins of attraction which are very different from the regular image of the continuous case generally used as a reference.

19.2 PROBLEM FORMULATION

Hereafter we consider an acyclic digraph $G = (V, A)$ with a set V of n vertices and a set A of m arcs, a set $L = \{l_1, l_2, ..., l_k\}$ of K layers, and a given distribution $V_1, V_2, ..., V_K$ of V on L with respectively $n_1, n_2, ..., n_K$ vertices. The vertex ordering on l_k is defined by $\sigma_k : V_K \rightarrow \{1, 2, ..., n_K\}$: $\sigma_k(u) = i$ which means that the vertex $u \in l_k$ is on the ith position on l_k. In a layered drawing, every arc $(u, v) \in A$ flows in the same direction: if $u \in V_i$ then $v \in V_j$ where $i < j$. Moreover, we suppose that the graph is proper i.e. each arc $(u, v) \in A$ is connected to vertices on consecutive layers: $u \in V_i$ and $v \in V_{i+1}$. We reach this hypothesis by replacing an arc whose length λ is greater than one by a path of $\lambda - 1$ dummy vertices on consecutive layers.

The problem is rewritten as defining a vertex ordering σ_k on each layer l_k so that the associated drawing minimizes the arc crossing number. We denote by Ω^G the set of all layered drawings of G and $f(D_i)$ the arc crossing number for a drawing $D_i \in \Omega^G$.

A neighbor of a drawing $D_i \in \Omega^G$ is a drawing $D_j \in \Omega^G$ deduced from D_i by the application of a local operator O which acts on one layer at a time. The set of neighbors of D_i is denoted by $N_O(D_i)$. For considerations discussed in Section 19.3 we here retain two local operators: the classical switching and a variant of the median.

Switching ($O1$). $D_j = O1\,(D_i)$ means that D_j is deduced from D_i by swapping two adjacent vertices in the same layer $l_k \in L$ if $f\,(D_j) < f\,(D_i)$.

Median ($O2$). We consider a definition inspired by the barycenter definition introduced by Sugiyama et al. (1981) in a variant of their well-known method: the median of a vertex u on a layer l_k depends on the connected vertices in both layers l_{k-1} and l_{k+1}. Let us suppose that the connected vertices of u on

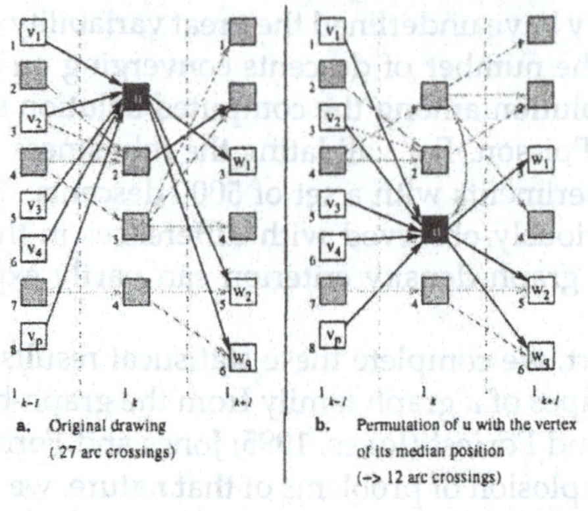

Figure 19.1 Permutation with the median operator ($O2$).

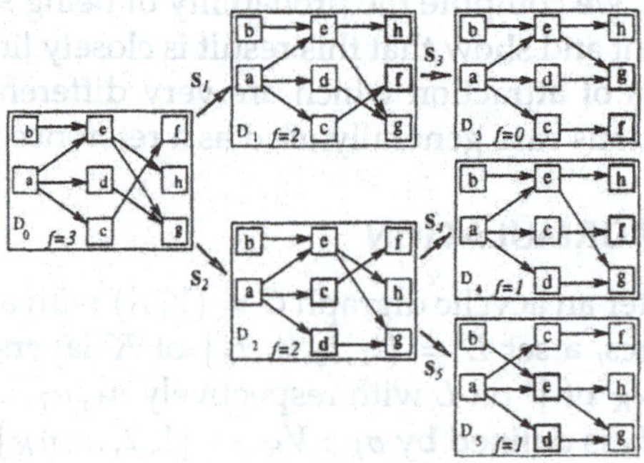

Figure 19.2 Series of transformations of a drawing D_0 on the left with the switch operator ($O1$). The drawing D_3 is a global optimum.

l_{k-1} (resp. l_{k+1}) are $v_1, ..., v_p$ (resp. $w_1, ..., w_q$). The median position $m(u)$ is the median of the following sorted set of the normalized position of each neighbor of u in l_{k-1} and l_{k+1} :

$$\left\{ \frac{\sigma_{k-1}(v_1)}{n_{k-1}}, ..., \frac{\sigma_{k-1}(v_p)}{n_{k-1}}, \frac{\sigma_{k+1}(w_1)}{n_{k+1}}, ..., \frac{\sigma_{k+1}(w_q)}{n_{k+1}} \right\}$$

The median is computed for each vertex on l_k and the transformation $O2$ is defined by the new vertex ordering σ'_k on l_k obtained after sorting the median values in increasing order: $\sigma'_k(u) > \sigma'_k(v)$ if $m(u) > m(v)$ (see Figure 19.1). This transformation is applied if it improves f.

19.3 EXPLORATION WITH A MULTI-START DESCENT

Let us first remark that the search space is intrinsically multimodal in that there may exist some local optima. For instance, the drawings D_4 and D_5 of Figure 19.2 are local optima for $O1$; different local permutations (e.g. c and d on D_2) necessary to reach the global optimum D_3 lead separately to an arc crossing increasing each of them, and are consequently not selected. Moreover, the global optimum is not necessarily unique (mainly because of global or local symmetries in the layout).

19.3.1 Experimental conditions

To precise the characteristics of this multimodality, we use an extended multi-start on a graph family. Let Δ be a set of acyclic digraphs which are not trees s.t.[1] $10^6 \leq \Pi_{k=1}^{K} n_k! \leq 10^{14}$. We have randomly generated a set $\Delta' \subset \Delta$ of 300 graphs with $4 \leq K \leq 40$ and $n_k \leq 15$. For each random graph G, a set of 5000 layouts is randomly generated, and each layout is improved by an iterative application of an operator O: O is applied on each layer taken one after the other and this loop goes on until the objective function f stabilizes.

As the objective here is not to develop a new heuristic but to better know search spaces associated with operators close to those used in the literature, we have just tested different simple variants of O easy to compute in the above multi-start descent. We first restricted ourselves to $O = O1 + O2$: for a layer l_k the ordering is first improved by the switching operator $O1$ then by the median operator $O2$. And we have experimentally observed that the obtained results can be slightly improved when a barycenter operator is applied after the median. As for the median, we have considered that the barycenter of a vertex u on a layer l_k depends on the connected vertices in both layers l_{k-1} and l_{k+1}. This definition has proved less performing in the Sygiyama's heuristic when vertices whose orders have not yet been improved are considered in the computation (Sugiyama et al., 1981). However, since the operator is not applied alone here, the situation is different. The improvement is probably a combination of two effects. It is well-known that for some subgraphs, the barycenter is better adapted than the median and vice versa (Di-Battista et al., 1999). And there may also be a side effect due to the chosen median definition which does not explicitly take the parity into account. The following results include improvements caused by the barycenter operator.

For each graph G of Δ', the minimum arc crossing reached on the 5000 runs is denoted by $\widehat{f_G}$. We cannot affirm that this value is actually a global optimum on Ω^G but the application of a genetic algorithm described in Lehn (2000) confirms that no better optimum is reached in 92% of the cases.

19.3.2 Results

The percentage of descents converging on $\widehat{f_G}$ solutions for the graph set Δ' is given in Figure 19.3. The average rate of descents converging on a $\widehat{f_G}$ solution is quite low (28.6%) but there are important differences on Δ' (the standard

[1] $\Pi_{k=1}^{K} n_k!$, where n_k is the number of vertices in layer k of a particular instance of the problem, represents the number of vertices of the fitness landscape associated with the instance i.e. the size of the associated search space.

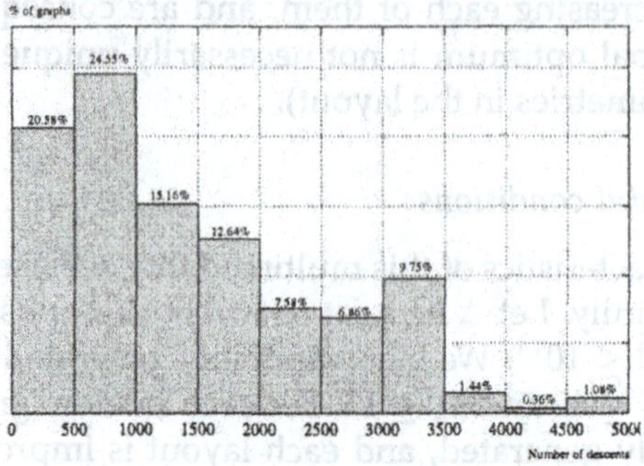

Figure 19.3 Distribution of the number of descents converging on $\widehat{f_G}$ solutions.

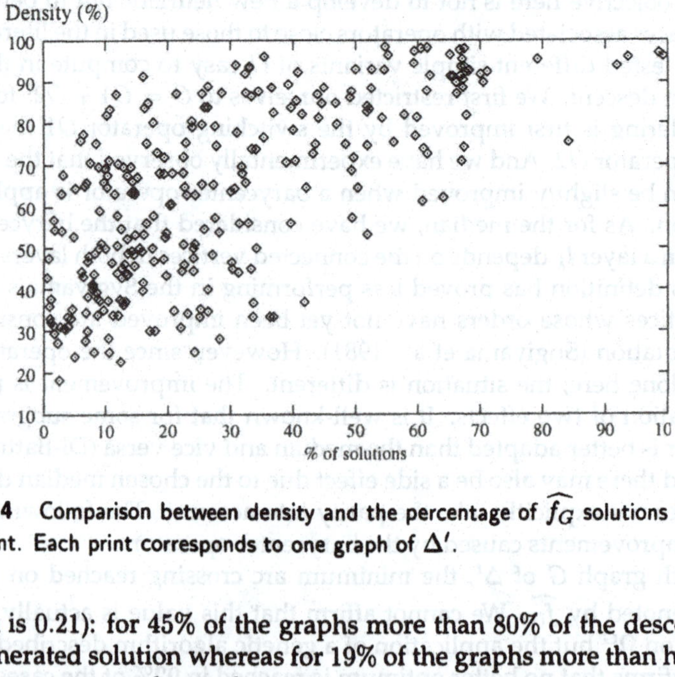

Figure 19.4 Comparison between density and the percentage of $\widehat{f_G}$ solutions in a multi-start descent. Each print corresponds to one graph of Δ'.

deviation is 0.21): for 45% of the graphs more than 80% of the descents lead to a degenerated solution whereas for 19% of the graphs more than half of the descents converge to a best solution.

For graph drawing, a density coefficient which is function of the arc and vertex cardinalities is often introduced for graph class discrimination. The usual definition of the density $d(G)$ of a graph G is given by $d(G) = m/n(n-1)$ where the denominator is the arc number in a complete graph of size n. However, this definition is not well adapted to layered digraphs where the maximal graph has $m_{max} = \sum_{k=2}^{K} n_{k-1} \times n_k$ arcs. Hence, we here consider the following density function $d'(G) = m/m_{max}$.

For graphs very close to the two extrema of $d'(G)$ (which correspond respectively to a tree and a maximally connected digraph), the drawing problem is quite simple as few permutations are possible. But, the situation is more confused for intermediate values. The relationship between the percentage of descents leading to a $\widehat{f_G}$ solution (x-axis) and the density of the graphs (y-axis) is described in Figure 19.4. The correlation between these two variables is non-null ($\rho = 0.35$). The results confirm that the density is an important factor of discrimination but, as the dispersion is high, it should be completed by other criteria to explain the results.

19.4 FITNESS LANDSCAPE STRUCTURE

The concept of fitness landscape, generally attributed to Wright (1932) for his studies on evolution, is an efficient model for understanding the dynamic of various heuristics. The definition proposed by Jones and Forrest (1995b) in the context of Genetic Algorithms is based on three components: a solution coding, a fitness function which associates each solution with a numerical value, and an operator which defines a neighborhood in the solution set. The fitness landscape can be represented by a graph LG_O, which we call *landscape-graph*, whose vertices are solutions and where arcs describe an O-transformation of a solution into another: there exists an arc between two vertices s and s' if $s' = O(s)$.

19.4.1 *Landscape-graph for the drawing problem*

In our case, vertices of LG_O are drawings of Ω^G and arcs represent transformations of a drawing into another by operators O described in Section 19.2. A *local optimum* of a landscape is a vertex $D_i \in \Omega^G$ s.t. $f(D_i) \leq f(D_j)$ for all $D_j \in N_O(D_i)$ and a *global optimum* is a vertex $\widehat{D} \in \Omega^G$ s.t. $f(\widehat{D}) \leq f(D_i)$ for all layouts $D_i \in \Omega^G$. The *basin of attraction* of a vertex D_n is the set of vertices

$$B_O(D_n) = \{D_0 \in \Omega^G; \exists D_1, ..., D_{n-1} \in \Omega^G \\ \text{with } D_{i+1} \in N_O(D_i) \text{ and } f(D_{i+1}) < f(D_i), \forall i = 0, n-1\}$$

For instance, Figure 19.5 represents the landscape-graph of the graph represented in Figure 19.7 for the operator $O1$. In this representation, vertices are arranged on layers: left vertices are drawings with numerous arc crossings whereas right vertices are drawings with fewer arc crossings. Let us remark that graph-landscapes here are not necessarily connected.

Local optima are represented by ovals and global optima by lozenges. It is very important to note that, according to the above definition adopted by many authors, the basins of attraction of the local and global optima may have here a non-null intersection. For instance, the gray vertex $[[a, b][c, e, d][h, g, f]]$ in Figure 19.5 with four arc crossings belongs to the basin of attraction of the local optimum $[[a, b][c, e, d][f, h, g]]$ and to the basin of attraction of the global optimum $[[b, a][e, d, c][h, g, f]]$ (see Figure 19.6). It is obviously possible to restrict the previous definition by considering that a drawing D is in the basin of attraction $B_O(D_n)$ if local search starting from D leads to D_n only. However, this formal change must not make us forget that the situation is here

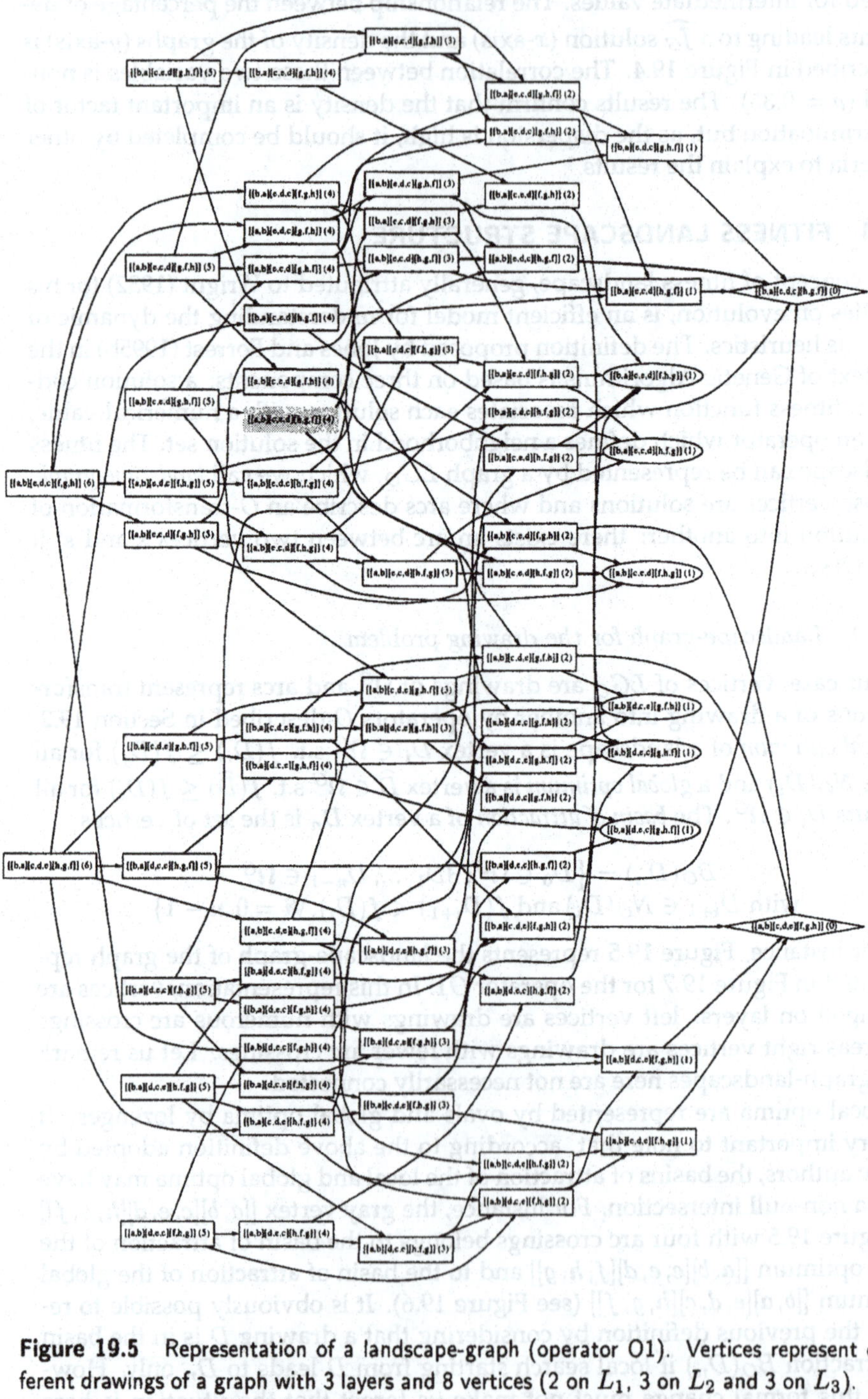

Figure 19.5 Representation of a landscape-graph (operator O1). Vertices represent different drawings of a graph with 3 layers and 8 vertices (2 on L_1, 3 on L_2 and 3 on L_3). For instance, the vertex $[[b,a]\,[e,d,c]\,[h,g,f]]$ (on the right) represents the drawing in Figure 19.7. The f value is given in parenthesis (drawing with the aid of *graphviz* ©*AT&T Bell Labs*).

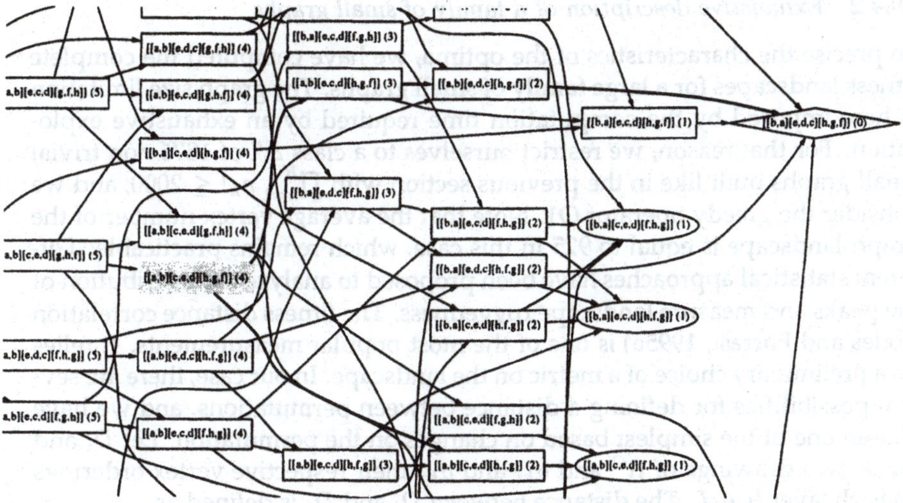

Figure 19.6 Zoom on the gray vertex of Figure 19.5

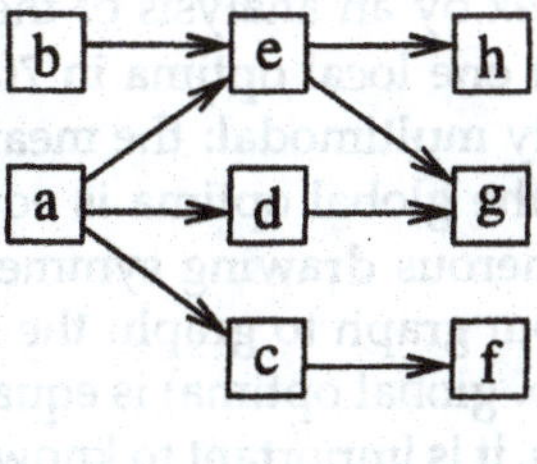

Figure 19.7 A vertex of the landscape-graph of Figure 19.5 which is a drawing of a global optimum.

completely different from the continuous case generally taken as reference. Hereafter, we study the consequence of this characteristic on the search process.

19.4.2 Exhaustive description of a family of small graphs

To precise the characteristics of the optima, we have computed the complete fitness landscapes for a large family of small graphs. The graph size limitation is here imposed by the computation time required by an exhaustive exploration. For that reason, we restrict ourselves to a class Δ'' of 1875 non trivial small graphs built like in the previous section with $\prod_{k=1}^{K} n_k! \leq 2000$, and we consider the greedy operator O1. Note that the average vertex number of the graph-landscape is equal to 925 in this case, which remains practicable. Different statistical approaches have been proposed to analyze the distribution of the peaks and measure landscape ruggedness. The fitness distance correlation (Jones and Forrest, 1995a) is one of the most popular measurements. It relies on a preliminary choice of a metric on the landscape. In our case, there are several possibilities for defining a distance between permutations, and we have chosen one of the simplest based on changes on the permutation. Let D_i and D_j be two drawings of Ω^G and σ_{ki} and σ_{kj} their respective vertex orderings on each layer $l_k \in L$. The distance between D_i and D_j is defined as

$$d(D_i, D_j) = \sum_{k \in K} |C(\sigma_{ki}, \sigma_{kj})| \text{ where } C(\sigma_{ki}, \sigma_{kj}) = \{u; \sigma_{ki}(u) \neq \sigma_{kj}(u)\}).$$

The mean fitness distance correlation is equal to 0.47. This value is quite high which is partly explained by the size of the instances. Nevertheless, even for these small instances, the variation may be important (Figure 19.8). Our exhaustive computation of the landscapes allows to complete the information measured by this global index by an analysis of the exact distribution of the optima. There exists at least one local optima in 76% of the cases, and most of the landscapes are strongly multimodal: the mean number of local optima is equal to 34.9 and that of the global optima is equal to 36.1, which can be partly explained by the numerous drawing symmetries. The distribution of the optima greatly varies from graph to graph: the standard deviation of the number of local optima (resp. global optima) is equal to 37.6 (resp 79.8).

In an optimization process, it is important to know whether the local optima values are close or not to the global one. Hence, we introduce the relative height $h(D_i)$ of a local optimum D_i : $h(D_i) = 1 - \frac{f(D_i) - \widehat{f}}{f_w - \widehat{f}}$ where $\widehat{f}$ (resp. f_w) is the optimal (resp. the worst) f value on the landscape. If $h(D_i)$ is very close to 1 then the local optimum D_i can be considered as an acceptable solution for some applications. The distribution of the relative heights on Δ'' is given in Figure 19.9.

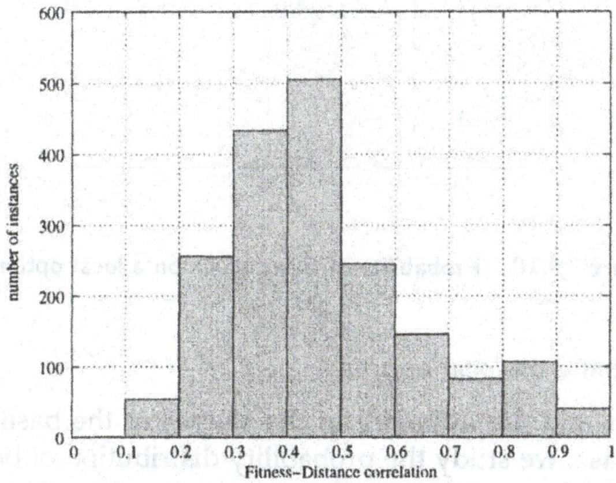

Figure 19.8 Distribution of the fitness distance correlation

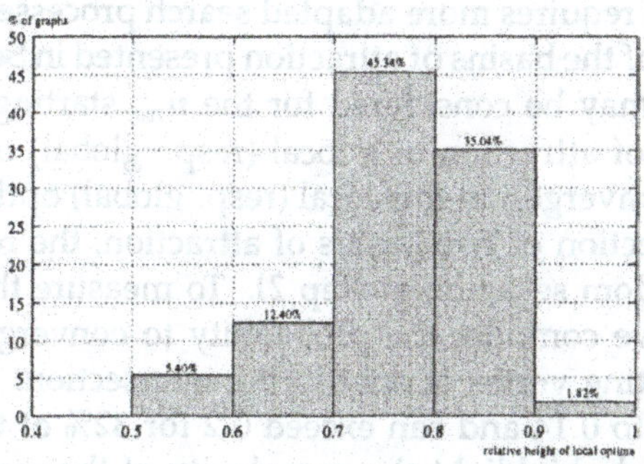

Figure 19.9 Relative height of local optima

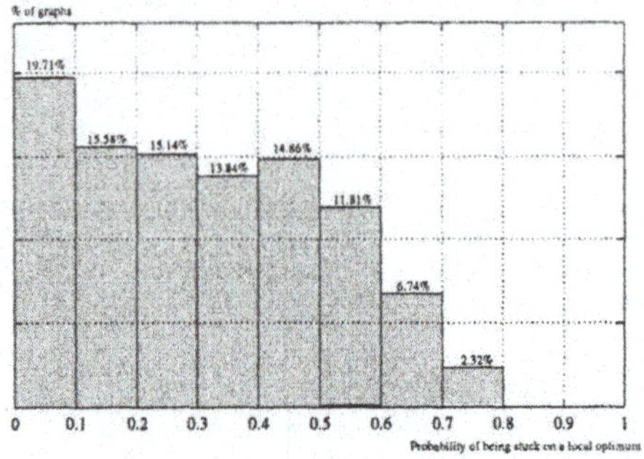

Figure 19.10 Probability of being stuck on a local optimum

19.4.3 Convergence on local optima

To better understand the influence of the shapes of the basins of attraction in the discrete case, we study the probability distribution of being stuck on a local optima. Here, we simply consider a greedy heuristic: *1)* select a vertex of LG_O different from an optimum; *2)* choose the best neighbor. If two neighbors have the same f value, select one of them randomly; *3)* iterate the process until it converges. This heuristic is applied for each vertex of LG_O different from an optimum. Let n_{no} be their number and n_l be the number of runs leading to a local optimum. The probability p_l of being stuck on a local optimum is given by n_l/n_{no}. The average probability on Δ'' is equal to 0.31. Since this calculation considers small graphs only and that all neighbors are known for each vertex, this value is far from being negligible. Again, we find important differences between graphs (Figure 19.10).

In particular, two classes are opposite: an easy one where 19.7% of the graphs have a probability p_l less than 0.1 and a difficult one (20.9% of graphs with $p_l > 0.5$) which requires more adapted search processes.

Due to the shape of the basins of attraction presented in Section 19.4.1, three different situations may be considered for the n_{no} starting vertices. If they belong to the basin of attraction of a local (resp. global) optimum only, the process obviously converges on this local (resp. global) optimum. But, if they belong to the intersection of two basins of attraction, the result can vary depending on the random selection at step *2)*. To measure the consequence of this characteristic, we compute the probability to converge on a local optimum when the starting vertex is only in the intersection. The new average probability is equal to 0.14 and can exceed 0.2 for 32% of the graphs. These non-null values clearly highlight the complexity of the search space for this kind of problem.

19.5 CONCLUSION

The main conclusions which can be drawn from the numerical experiments presented in this paper are concerned with two points: the variability of the solution quality depending on the instance characteristics, and the properties of the basins of attraction of the optima.

Our experiments with multi-start descents have confirmed that the heuristic performances closely depend on the processed instances. Due to the huge size of the search spaces, the conclusions of such a statistical study must obviously be cautiously considered. However, they allow to show that the debate recently opened on the respective performances of Evolutionary Algorithms and Stochastic Hill Climbing for graph drawing (Rosete-Suarez et al., 1999) cannot dispense with a preliminary study of the instance characteristics. In this perspective, we have studied the influence of the graph density on the descent convergence. The correlation between the density and the descent performance is not null but it is too low to discriminate different graph categories. A manual perusal of a part of the results shows that symmetries seem to play a non negligible role for this problem. However, the detection of symmetries in graphs is generally a difficult problem (Manning, 1990) and the definition of a global measure is not easy. Nevertheless, we believe that determining a typology based on a small set of discriminant quantitative criteria is a necessary preliminary work to seriously compare the different metaheuristics mentioned in the introduction.

Our exhaustive study of the graph-landscapes has highlighted the complexity of the basins of attraction, and it contributes to justify the use of a "highly" stochastic approach. We have explicitly showed that, contrary to the continuous image where all basins of attraction are clearly separated, they are here interwoven. Consequently, for some drawings, a simple descent can lead either to a global optimum or a local one depending on a random choice of the neighbors in case of equality of the fitness. For the paradigmatic continuous case, most strategies focus on the search of the global optimum attraction set, and also sometimes, on the reduction of the number of moves necessary to reach the optimum (Johnson et al., 1988). However, in addition, it is here necessary to resort to a strategy able to move away from neighbors of a local optimum "late" in the search process.

Putting this property into practice depends on the metaheuristic. For instance, for Tabu Search, the evolution of the tabu list during the search process, which has not still been applied to this problem, may be promising: the list size or the criterion for selecting forbidden directions could vary to get out basins of attraction of local optima very close to basins of attraction of global optima. In the perspective of the development of a genetic algorithm, which has initially stimulated the work presented here, it is clear that the adjustment of the mutation criterion could play a particularly important part. Nevertheless, a similar study should be carried for the crossover landscapes. Indeed, it is easy to find examples where the combination of two sub-graphs (for instance exchange right and left parts of a graph) can lead to a better drawing. Consequently, for the layered drawing problem, GA could additionally fully exploit the property of reconstruction by recombination.

G. Di-Battista, P. Eades, R. Tamassia, and I.-G. Tollis. *Graph drawing – Algorithms for the visualization of graphs*. Prentice-Hall, 1999.

G. Di-Battista, P. Eades, R. Tamassia, and I.-G. Tollis. Algorithms for drawing graphs: an annotated bibliography. *IEEE Trans. on Visualization and Computer Graphics*, 6(11):24–43, 2000.

P. Eades and D. Kelly. Heuristics for reducing crossings in 2-layered networks. *Ars Combinatorics*, 21:89–98, 1986.

P. Eades and N. Wormald. Edge crossings in drawings of bipartite graphs. *Algorithmica*, 11:379–403, 1994.

M.-R. Garey and S. Johnson. Crossing number is *NP*-complete. *SIAM J. of Algebraic and Discrete Methods*, 4(3):312–316, 1983.

I. Herman, G. Melançon, and M.S. Marshall. Graph visualization and navigation in information visualization: a survey. *IEEE Trans. on Visualization and Computer Graphics*, 6(11):24–43, 2000.

M. Jünger and P. Mutzel. 2-layer straightline crossing minimization: performance of exact and heuristic algorithms. *J. of Graph Algorithms and Applications*, 1(1):1–25, 1997.

D.S. Johnson, C.H. Papadimitriou, and M. Yannakakis. How easy is local search? *J. of Computer and System Sciences*, 17:79–100, 1988.

T. Jones and S. Forrest. Fitness distance correlation as a measure of problem difficulty for genetic algorithms. In *L. Eshelman (ed.): Proc. of the Sixth Int. Conf. on Genetic Algorithms*, pages 184–192. Morgan Kaufmann, 1995a.

T. Jones and S. Forrest. Genetic algorithms and heuristic search. In *Santa Fe Institute Tech. Report 95-02-021*. Santa Fe Institute, 1995b.

T.-C. Jones. *Evolutionary Algorithms, Fitness Landscapes and Search*. PhD thesis, University of New Mexico, Alburquerque, 1995.

Pascale Kuntz, Fabrice Guillet, Rémi Lehn, and Henri Briand. A user-driven process for mining association rules. In *Principles of data mining and knowledge discovery*, pages 483–489. LNCS 1910, 2000.

M. Laguna, R. Marti, and V. Valls. Arc crossing minimization in hierarchical design with Tabu Search. *Computers and Operations Res.*, 24(12):1175–1186, 1997.

R. Lehn. *An interactive system of visualization and rule mining for knowledge discovery in databases (in French)*. PhD thesis, University of Nantes, France, 2000.

R. Lehn and P. Kuntz. A contribution to the study of the fitness landscape for a graph drawing problem. In *Proc. of EvoCop2001 - Applications of Evolutionary Computing*, pages 172–181. LNCS 2037, 2001.

J. Manning. *Geometry symmetry in graphs*. PhD thesis, University of Purdue, New-York, 1990.

R. Martí. Arc crossing minimization in graphs with GRASP. *IIE Trans.*, 33(10): 913–919, 2001.

H. Purchase. Effective information visualisation: a study of graph drawing aesthetics and algorithms. *Interacting with computers*, 13(2):477–506, 2000.

A. Rosete-Suarez, A. Ochoa-Rodriguez, and M. Sebag. Automatic graph drawing and stochastic hill-climbing. In *Proc. of the Genetic and Evolutionary Conf.*, pages 1699–1706. Morgan Kaufmann, 1999.

K. Sugiyama, S. Tagawa, and M. Toda. Methods for visual understanding of hierarchical systems. *IEEE Trans. Sys. Man and Cybernetics*, 1981.

J. Utech, J. Branke, H. Schmeck, and P. Eades. An evolutionary algorithm for drawing directed graphs. In *Proc. of the Int. Conf. on Imaging Science, Systems and Technology*, pages 154–160. CSREA Press, 1998.

S. Wright. The roles of mutation, inbreeding, crossbreeding and selection in evolution. In *Proc. of the Sixth Int. Conf. on Genetics*, volume 1, pages 356–366, 1932.

Metaheuristics: Computer Decision-Making, pp. 421-448
Heikki Maaranen, Kaisa Miettinen and Marko M. Mäkelä

20 TRAINING MULTI LAYER PERCEPTRON NETWORK USING A GENETIC ALGORITHM AS A GLOBAL OPTIMIZER

Heikki Maaranen[1], Kaisa Miettinen[1] and Marko M. Mäkelä[1]

[1]Department of Mathematical Information Technology
University of Jyväskylä
P.O. Box 35 (Agora)
FIN-40014 University of Jyväskylä, Finland
Heikki.Maaranen@mit.jyu.fi, Kaisa.Miettinen@mit.jyu.fi
Marko.Makela@mit.jyu.fi

Abstract: In this paper, we introduce an approach for solving a regression problem. In regression problems, one tries to reconstruct the original data from a noisy data set. We solve the problem using a genetic algorithm and a neural network called Multi Layer Perceptron (MLP) network. By constructing the neural network in an appropriate way, we are able to form an objective function for the regression problem. We solve the obtained optimization problem using a hybrid genetic algorithm and compare the results to those of a simple multistart method. The hybrid genetic algorithm used is a simple hybridization of a genetic algorithm and a Nelder-Mead simplex method.
Keywords: Hybrid method, Regression problem, Neural networks, Genetic algorithms, Multi layer perceptron.

20.1 INTRODUCTION

We introduce an approach for solving a regression problem utilizing the tools of neural networks and genetic algorithms. As opposed to interpolation, in regression problems there exists noise in the observed data. Hence, the aim in regression is not to find a curve running through all the observed data points, but to recover the original data from a noisy data set. The challenge in regression is to find the right complexity for the approximation so that the approximated curve best corresponds to the original data ignoring the noise. This sort of estimation problems form one of the most important application fields of neural networks. An appropriate construction of a neural network provides an objective function for an optimization problem that describes the regression problem. This objective function is typically nonconvex.

Many real-life optimization problems are nonconvex which means that they may have several local optima. Therefore, global search methods are needed to solve them. Metaheuristics are efficient global optimizers. They include a metastrategy that guides a heuristic search. Genetic algorithms, simulated annealing, tabu search and scatter search are probably the best-known metaheuristics. In general, they work even for problems involving discontinuous functions and disconnected feasible regions. Hence, metaheuristics are applicable to a large number of problems.

Genetic algorithms and neural networks both have their roots in nature. Neural networks imitate the functioning of a human brain (Greenberg, 2001; Rojas, 1996). Roughly speaking, we can say that the human brain consists of neurons, that is, brain cells and synaptic connections between the neurons. Through the synaptic connections, the neurons receive signals from and send signals to other neurons. In neural networks, nodes correspond to neurons and arcs correspond to synaptic connections. Correspondingly, genetic algorithms imitate evolution process in a population (Goldberg, 1989). Many concepts, such as population, generation, crossover, mutation and selection, are borrowed directly from biology. Since the natural evolution process is widely known, it is easy to understand the basic idea of genetic algorithms. This is probably one of the reasons why genetic algorithms are so widely used today.

In this paper, we formulate a regression problem using Multi Layer Perceptron neural networks and obtain a related global optimization problem. Because local optimizers are widely available, global problems are often solved using a multistart method where the local optimizer is applied starting from several different points. Unfortunately, this is not a reliable way to find the global optimum because the success depends on the number of local optima and on how well the starting points are selected. As already mentioned, genetic algorithms are widely used in global optimization. They are considered to be rather reliable but may involve a lot of computation. Here our aim is to decrease the computational demand of genetic algorithms by a hybridization with a local optimizer. Somewhat similar ideas have been used, for example, in Yao and Lui (1997) and Liu (1999). The performance of our hybrid approach is compared to a multistart method. In both our cases, the well-known Nelder-Mead simplex method is used as the local optimizer.

In our computational experiments, we use a simple regression problem where we know the original data. This enables a more profound numerical comparison for the results, which is essential when developing new approaches. The promising results of the experiments indicate the applicability of the approach in a wider class of problems.

The rest of this paper is organized as follows. In the next section, we introduce the regression problem and neural networks in general. In Section 20.3, we construct a neural network along with the objective function we aim to optimize in our approach. Section 20.4 describes the genetic algorithm that will be used as a part of the hybrid method for solving the optimization problem constructed in Section 20.3. Section 20.5 is devoted to the implementation aspects related to combining neural network and the hybrid of the genetic algorithm and the Nelder-Mead simplex method. The test problem is presented in Section 20.6 and the following section presents the results obtained. After a discussion about some details of the modeling of the regression problem, we conclude the paper in Section 20.9.

20.2 REGRESSION PROBLEM AND NEURAL NETWORKS

For various reasons, people in industry and the academic world have to deal with incomplete and/or noisy data. One of the main sources of incomplete data is the future, which can never be known in detail but can only be predicted. Noisy data, on the other hand, is a common problem in experimental sciences. In many cases, there is a need to estimate the original data from the noisy data set or to approximate the unobserved data. Several tools have been developed for that purpose. Neural networks are among the most general solvers for different kinds of estimation problems. Next, we introduce regression problem, which includes estimation.

In the regression problem, given a noisy data set $\{(x^i, y^i)\}$, $x^i \in \mathbb{R}^n$, $y^i \in \mathbb{R}$, $i = 1, \ldots, N$, called the training data, the aim is to find a mapping $m : \mathbb{R}^n \mapsto \mathbb{R}$ such that m approximates the original data $\{(x^i, y^{*i})\}$ without the noise on the observed data points as well as outside the observed points. In the training data set, the vectors x^i are the points where the noisy function is evaluated and real numbers y^i are the corresponding noisy function values. The differences between original data y^{*i} and noisy data y^i are called *residues* $e^i = y^{*i} - y^i$.

The traditional way to solve regression problems is to try to fit a line (linear regression) or a polynome (polynomial regression) to the training data. Figure 20.1 illustrates polynomial fitting.

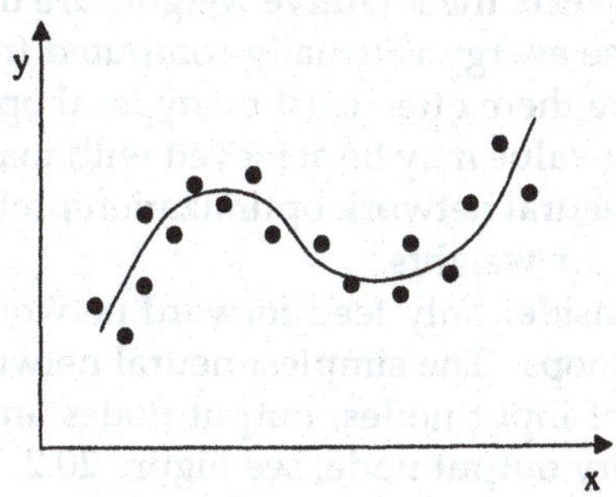

Figure 20.1 An example of the regression problem.

Regression problem can also be solved using neural networks. Neural networks are computationally intelligent methods, whose sketches were published as early as in the 1950's, see Rosenblatt (1958) and references therein. The basic idea was introduced shortly after that in the mid 1960's, but the so-called teaching algorithm did not become widely known until the late 1980's (Rumelhart et al., 1996).

Typically, neural networks are used to solve a generalized, vector valued regression problems for mappings $m : \mathbb{R}^n \mapsto \mathbb{R}^p$, where n and p are positive integers. Neural networks with certain properties are able to approximate any continuous mapping (Bishop, 1995). This is called *universal approximation*. With neural networks one could, therefore, find a mapping that goes through every point given as training data. However, this may not be desirable if the data we are dealing with includes noise. Instead of a perfect fit, one is often interested in finding a good fit with a high degree of smoothness.

As a solution we do not have an explicit equation of the curve, but a neural network with specific weights. The weights define the mapping uniquely, even though the weights may not be unique. To compute the value of the approximated mapping m at a specific point $\tilde{x}$, one gives the vector $\tilde{x}$ as the input vector for the neural network and receives the corresponding output $\tilde{o}$, that is, $m(\tilde{x}) = \tilde{o}$.

When validating a network, the residues $e^i = y^{*i} - y^i$ are computed. In practical problems, the original function values y^{*i} are not known and the residues are computed using the network output instead: $e^i = o^i - y^i$. Then, additional analysis for the residues is needed to check whether the statistical properties of the residues match with the statistical properties known for the noise.

As mentioned in the introduction, neural networks are a sort of data structures consisting of nodes and arcs (Greenberg, 2001; Rojas, 1996). Each node has a state variable, whose value depends on the weights of the arcs connected to that node. The optimization problem in neural networks is to change the arc weights in order to minimize the total energy of the network. The energy, that is, the objective function value, depends on the data vectors $\{x^i\}$. In addition, the energy depends on the arc weights. The data vectors are fixed, but the weights are adaptive, hence, the adaptive weights are the variables of the optimization problem. The energy is usually computed from the residues. Due to the network structure there often exist many local optima. In addition, the same objective function value may be achieved with many different combination of arc weights. A neural network optimization problem is often sensitive to small changes in the arc weights.

In this paper, we consider only feed forward networks, which means that there are no feedback loops. The simplest neural network is a one-layer network consisting only of input nodes, output nodes and the arcs connecting each input node to every output node, see Figure 20.2. It is a well know fact, however, that one-layer networks are not able to approximate general continuous functions (Rumelhart et al., 1996). Therefore, multiple layer networks are important for many practical implementations of neural networks. In the next section, we consider one type of multi layered neural networks.

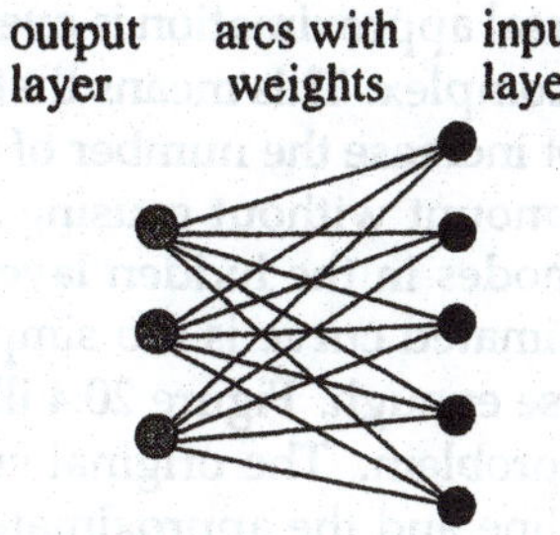

Figure 20.2 One-layer network.

20.3 MULTI LAYER PERCEPTRON NETWORK

Multi Layer Perceptron (MLP) networks are, when considered from a statistical point of view, tools for nonlinear regression analysis. As the name suggests, an MLP network has more than one layer of weights. In this paper, we restrict our consideration to a two-layer MLP network, since it is suitable for our testing purposes. The nodes that are neither input nor output nodes are called hidden nodes and the corresponding layer is called a hidden layer, see Figure 20.3. The complexity of an MLP network is highly dependent on the number of hidden nodes.

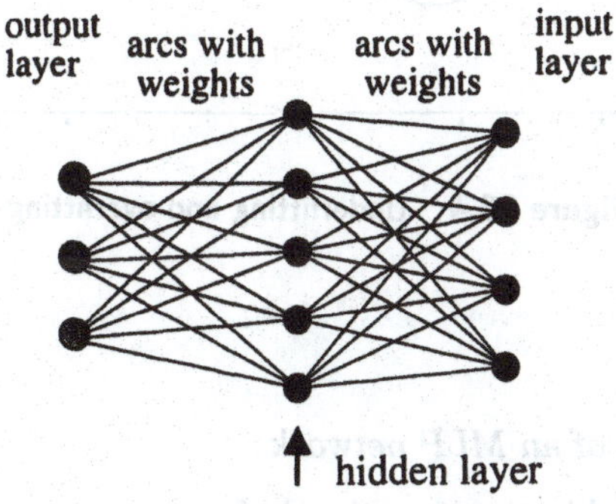

Figure 20.3 A two-layer network.

One reason why we are particularly interested in two-layer networks is that they are important for practical applications. Two-layer networks are the simplest multi layer networks with the universal approximation property, which was already mentioned in Section 20.2: a two-layer MLP network with an arbitrarily large hidden layer is able to approximate any continuous function from one finite-dimension space to another (Bishop, 1995). In practice, the number of nodes in the hidden layer is always limited and a residual error will occur.

When dealing with noisy data in the regression problem, a disadvantage that follows from the universal approximation is overfitting. In overfitting, the approximated curve is too complex. This means that the curve starts to follow the noise. Hence, we cannot increase the number of nodes in the hidden layer above a problem-specific amount without causing overfitting. On the other hand, if there are too few nodes in the hidden layer, underfitting will occur. In underfitting, the approximated curve is too simple and it does not follow the training data points close enough. Figure 20.4 illustrates underfitting and overfitting in a regression problem. The original curve is a sinus curve and it is plotted with a dashed line and the approximated curve is plotted with a continuous line.

Due to the above-mentioned facts, we must try to find a number of nodes in the hidden layer that balances between under- and overfitting. In two dimensions, it is easy to see when under- or overfitting occurs, since it is possible to illustrate the situation by plotting a curve. In a general case, this is not trivial. One way to tackle the overfitting is regularization. We will return to it at the end of this section.

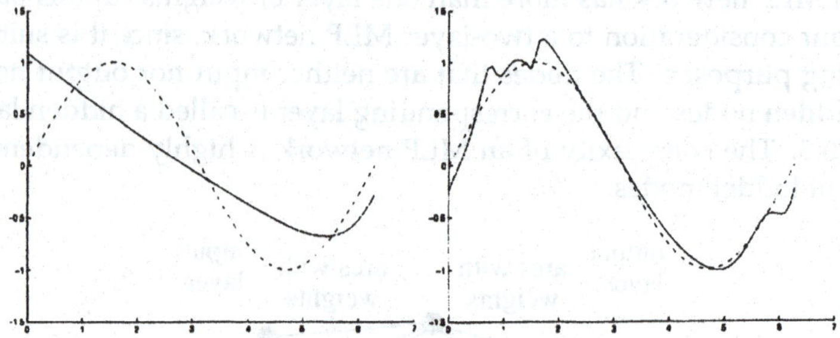

Figure 20.4 Underfitting and overfitting.

20.3.1 *Construction of an MLP network*

We need to have an objective function before we can optimize anything. In order to understand how the objective function is formed in regression problems, we briefly construct an MLP network (Hagan and Mohammad, 1994; Rojas, 1996). We follow closely the four steps described in Kärkkäinen (2002). We construct a general two-layer MLP network with input vectors of dimension n_0, n_1 nodes in the hidden layer, and output vectors of dimension n_2.

Step 1: We have the input vectors $\{x^i\}$, $i = 1, ..., N$. Let us consider one vector x^i at a time. For the sake of simplicity we note this vector as $x =$

$(x_1, ..., x_{no})^T$. We start by defining a linear transformation of the components $x_1, ..., x_{no}$

$$a(x) = w_0 + w_1 x_1 + w_2 x_2 + ... + w_{no} x_{no} = w^T \hat{x},$$

where

$$w = \begin{bmatrix} w_0 \\ w_1 \\ \vdots \\ w_{no} \end{bmatrix} \quad \text{and} \quad \hat{x} = \begin{bmatrix} 1 \\ x_1 \\ \vdots \\ x_{no} \end{bmatrix}.$$

For $no = 1$ the transformation defines a straight line, for $no = 2$ a plane and for $no = 3, 4, ...$ a hyperplane. The weight w_0 is called a *bias* node and the original vector x is extended to $\hat{x}$ to allow the short notation $w^T \hat{x}$.

Step 2: We apply m linear transformations to the same vector x to obtain:

$$\begin{aligned} a_1(x) &= w_{1,0} + w_{1,1}x_1 + w_{1,2}x_2 + ... + w_{1,no}x_{no} &= w_1^T \hat{x} \\ &\vdots & \vdots \\ a_m(x) &= w_{m,0} + w_{m,1}x_1 + w_{m,2}x_2 + ... + w_{m,no}x_{no} &= w_m^T \hat{x} \end{aligned}$$

The number of nodes in the network determines the number of linear transformations, that is, the number of weights in the network. We will discuss this in more detail later. By setting the vectors $w_1, ..., w_m$ as the rows of the matrix W

$$W = \begin{bmatrix} w_1^T \\ \vdots \\ w_m^T \end{bmatrix},$$

we can present the m linear transformations in a compact form

$$a = W\hat{x},$$

where the vector a contains the obtained real numbers $a_1, ..., a_m$. This defines a one-layer network with the input layer represented by vector x and the output layer by vector a.

Step 3: So far, we have used only linear transformations, which will cause the approximation to be linear. However, we are interested in a more general approximation. To this end, we add a nonlinear component to the linear transformation. Since it is of our interest to keep the notation compact we define a function $F = F(\cdot) = (f_i(\cdot))^T$

$$F(a) = \begin{bmatrix} f_1(a_1) \\ \vdots \\ f_m(a_m) \end{bmatrix}.$$

The components f_i are called activation functions (Bishop, 1995). Using function F, we can now write a new transformation

$$F(a) = F(W\hat{x}).$$

In the early implementations of neural networks, the activation function was a *step function*

$$f_i(a_i) = \begin{cases} 1, & \text{if } a_i \geq 0 \\ 0, & \text{if } a_i < 0 \end{cases}.$$

However, the on-off property of the step function makes it difficult to train those networks (Bishop, 1995). Later, several other activation functions have been defined that allow the outputs to be given a probabilistic interpretation. One such activation function is a *logistic sigmoid* (Bishop, 1995)

$$f_i(a_i) = \frac{1}{1+\exp(-a_i)}.$$

In our implementation, we use an activation function shown in Figure 20.5. This function is called *generalized logistic sigmoid* (Rojas, 1996), and is of the form

$$f_i(a_i) = \frac{1}{1+\exp(-ia_i)}.$$

For other activation functions, see Bishop (1995).

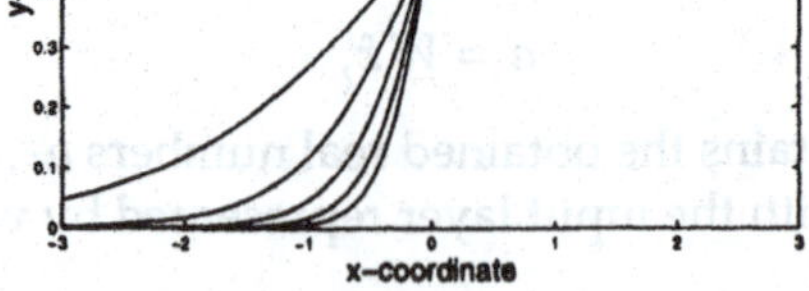

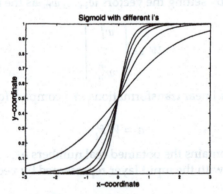

Figure 20.5 Generalized logistic sigmoid with $i = 1, ..., 5$.

The advantage of the sigmoidal functions over the step function is that they are differentiable. In addition, according to the present knowledge (Rojas, 1996), they also correspond better to the human brain structure.

Step 4: In previous three steps, we have constructed a one-layer neural network with one nonlinear transformation. As mentioned in Section 20.2, one-layer neural networks are not able to approximate general continuous functions. At the beginning of this section, however, we noted that the two-layer networks have that desired feature. Therefore, we add another layer of weights W^2 and apply another nonlinear transformation F^2 similar to F. We obtain the form

$$o = Z(x) = F^2(W^2 \hat{F}^1(W^1 \hat{x})),$$

where F^1 and W^1 correspond to what we earlier denoted with F and W, respectively, and $\hat{F}^1$ is extended from F^1 by adding an identity function as its first element.

Now, we have constructed a two-layer MLP network, that is, an MLP network with one hidden layer. This structure is illustrated in Figure 20.6, where $x = (x_1, ..., x_{n_0})$ is the input vector and $o = (o_1, ..., o_{n_2})$ is the output vector and the layer of nodes between them is the hidden layer. Hence, in Figure 20.6, n_1 is the size of the hidden layer and n_0 and n_2 are the dimensions of input vectors and output vectors, respectively. In this general structure, input and output vectors do not necessarily have to be of the same dimension. The MLP networks can directly be extended to include more hidden layers and more nonlinear transformations.

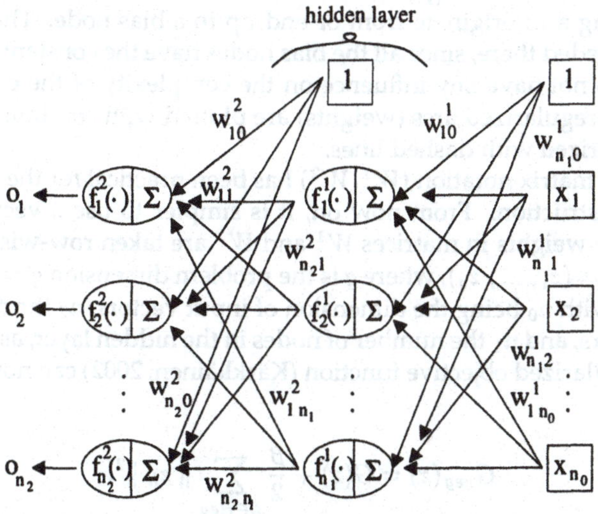

Figure 20.6 An MLP network with one hidden layer.

Now, being familiar with the MLP network structure, we are able to form an objective function for the regression problem using the above-defined notations. Our objective is to train the network using the known data and the image vectors. The most natural way is to give the data vectors x^i as input and try to minimize the average error between network output vectors o^i and the known images y^i. Denoting the objective function by G we have an optimization problem

$$\min_{(W^1,W^2)} G(W^1,W^2),$$

where

$$G(W^1,W^2) = \frac{1}{2N}\sum_{i=1}^{N} \|Z(x^i) - y^i\|^2 = \frac{1}{2N}\sum_{i=1}^{N} \|F^2(W^2\hat{F}^1(W^1\hat{x}^i)) - y^i\|^2 \quad (20.1)$$

is the sum-of-squares error function (Bishop, 1995) in its basic form, $\|\cdot\|$ is an l^2-norm, and N is the number of points. The sum-of-squares error function is particularly well suited for regression problems (Bishop, 1995).

20.3.2 Regularization

At the beginning of this section, we introduced overfitting and mentioned regularization as a possible tool to tackle it. The objective in regularization is to smooth the solution curve. In the neural network context, regularization usually refers to a penalty technique, where an extra term is added to the objective function. The added term penalizes large weights and, hence, makes the network more stable. The regularization does not concern the weights, whose corresponding arcs originate from or end up to a bias node. The regularization is not needed there, since all the bias nodes have the constant value 1 and, therefore, do not have any influence on the complexity of the curve. In Figure 20.7, the regularized arcs (weights) are plotted with continuous lines and the unregularized with dashed lines.

So far, the matrix notation (W^1, W^2) has been practical for the sake of MLP network construction. From now on, it is simpler to use a vector notation, where all the weights in matrices W^1 and W^2 are taken row-wise and put in one vector $z = (z_1, \ldots, z_q)$, where q is the problem dimension $q = (n_0+1)n_1 + (n_1+1)n_2$, with n_0 being the dimension of input vectors, n_2 the dimension of output vectors, and n_1 the number of nodes in the hidden layer, as defined earlier. The regularized objective function (Kärkkäinen, 2002) can now be written in the form

$$G_{reg}(z) = G(z) + \frac{\beta}{2}\sum_{h\in Reg} \| z_h \|^2, \quad (20.2)$$

where β is a regularization coefficient, and the set *Reg* includes all the indices of the weights that we wish to regularize. (Based on numerical experiments we used $\beta = 0.0002$ in our implementation.) The regularization term

$$\frac{\beta}{2}\sum_{h\in Reg} \| z_h \|^2$$

penalizes large weights and, therefore, favors values close to zero.

Let us next justify the use of regularization. We also need to show why regularization can be used without loosing the possibility of finding a global optimum by restricting the solutions close to the origin.

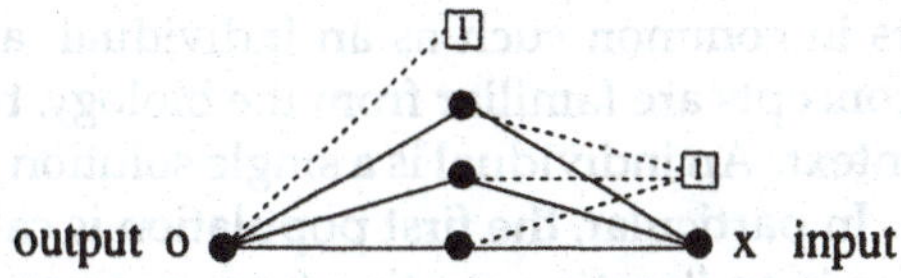

Figure 20.7 Regularized weights.

By examining Figure 20.5, we notice that the sigmoidal functions are nearly linear in the neighborhood of the origin. When restricting the weight values close to the origin we, therefore, make the approximation simpler and the approximate curve smoother. It is always advisable to favor simpler approximations, if the solution quality does not suffer. In addition, a smooth approximation levels off the peaks resulting from the noise. The combined effect is that we may increase the number of nodes in the hidden layer without causing overfitting.

Now we know that regularization favors smooth and simple approximations. This alone does not guarantee good solutions, and regularization cannot be justified without further knowledge about the problem structure. It may happen be that all the good solutions were far away from the origin and were not found due to the regularization. In MLP networks, however, the activation functions saturate far from the origin. In addition, the same objective function values can be achieved with many different combinations of weights as already mentioned in Section 20.2. Therefore, one may assume that all possible objective function values can be found in the neighborhood of the origin. The size of the area around the origin, where all the good solutions lie, varies. Hence, the regularization parameter has to be chosen with care.

20.4 GENETIC ALGORITHM

As the main optimization method we use a real-coded genetic algorithm, which we apply to the regularized objective function $G_{reg}(z)$. Genetic algorithms in their basic form require box constraint for all the variables, that is, each variable must have an upper bound and a lower bound. Therefore, we consider a minimization problem of the form

$$\begin{aligned}&\text{minimize } G_{reg}(z)\\&\text{subject to } z \in S,\end{aligned}$$

where $z = (z_1, \ldots, z_q)$ is the vector containing all the variables, q is the problem dimension and $S \subset R^q$ is the feasible region defined by box constraints with lower and upper bounds LB_i and UB_i, $i = 1, \ldots, q$.

Genetic algorithms are popular metaheuristics (Pham and Karaboga, 2000; Goldberg, 1989). Due to their popularity, there is a variety of different implementations available. The implementation used in this paper is discussed in more detail in Miettinen et al. (2000). All the genetic algorithms, however, have some concepts in common such as an individual, a population, and a generation. These concepts are familiar from the biology, but need to be redefined in this new context. An individual is a single solution and a population is a set of individuals. In particular, the first population is called the initial population. A generation is an iteration starting from one population and ending up, after the genetic operations, to another population.

A general four-step genetic algorithm can be formulated as follows. First, we set all the parameters for the genetic operations and stopping criteria. Second, we initialize the population with random individuals. Third, we perform the genetic operations in order to create a new population. Finally, if the stopping criterion is not satisfied, we return to the previous step. Otherwise, we choose the best individual found as the final solution.

All the genetic operators can be implemented in a number of different ways, see for example Pham and Karaboga (2000) and references therein. In this paper, we discuss only the implementations that we have chosen and give some reference to others in the literature. In the following five subsection, we present the genetic operations and the stopping criteria used.

20.4.1 Selection

Selection means selecting so-called parents, that is, individuals that are to be used when generating the new population. From many different types of selection we chose *tournament selection*. In tournament selection, a number of random individuals are selected from the population and the best, that is, the one with the lowest objective function value, will be chosen as the parent. The parameter *tournament size* determines how many individuals are compared when choosing the parent.

20.4.2 Crossover

There are numerous ways to perform crossover, where the selected parents are cross-bred. We used *heuristic crossover* that has provided superior results for the real-coded genetic algorithms when compared to single-point, uniform, and arithmetical crossover in Miettinen et al. (2000). In heuristic crossover, a new offspring $\hat{z}$ is produced from the parents z^1 and z^2 using the formula:

$$\hat{z} = r(z^2 - z^1) + z^2,$$

where r is a random number between 0 and 1 and the parent z^2 is not worse than the parent z^1. If the offspring is not feasible, that is, $\hat{z} \notin S$, a new random number is generated up to four times until a feasible solution is found. If after four repetitions still no feasible solution is found, new parents are selected.

20.4.3 Mutation

In mutation, the new offspring may be modified. Each element in the real-coded string are considered and the modification takes place if a randomly generated number is lower than the predefined value called a *mutation rate*.

We use non-uniform mutation (Michalewicz et al., 1994). Let us suppose that the element z_i is selected to be modified. Then the mutated element $\hat{z}_i$ is computed using the formula:

$$\hat{z}_i = z_i + \tau(UB_i - LB_i)(1 - r_i^{(t/t_{max})^b}),$$

where τ takes a boolean value, -1 or 1, each with probability 0.5, r_i is a random number between 0 and 1, b is a *mutation exponent* defined by the user and t and t_{max} are the current generation and the maximum number of generations, respectively. In our implementation, we set the mutation exponent b to 2.0 as in Michalewicz et al. (1996).

The above-defined non-uniform mutation acts like a uniform mutation during the first generations, but the distribution changes gradually to resemble the Dirac's function (Deb, 2001). Hence, in later generations, the probability to create a solution closer to the parent solution is larger than the probability of creating one away from it.

20.4.4 Elitism

Elitism guarantees that the best individuals found are not lost in the evolution process. The integer value of the parameter *elitism size* defines how many of the current generation's best solutions will automatically be included in the next generation.

20.4.5 Stopping criterion

The stopping criterion for our genetic algorithm is twofold. The search process is stopped after the maximum number of generations is reached or if the objective value of the best solution found has not changed enough in the last iterations. The minimum amount of change is defined in the parameter *tolerance* and the number of last iterations studied is defined in the parameter *step*.

20.5 IMPLEMENTATION

The primary goal of our implementation was to find out whether metaheuristics, and genetic algorithms in particular, are useful tools for training an MLP network.

20.5.1 Implementation structure

One of the main ideas in the implementation was to use modules already available, whenever possible. This resulted in carrying out the implementation using three different programming languages: C, Fortran and MatLab. The main program was written in MatLab, the genetic algorithm module was in Fortran, and the MLP network module in C.

MatLab was chosen as the main environment for the sake of visualization and for a large number of MatLab routines available. The global optimizer was attached to the main program using MatLab's Application Program Interface (API). This arrangement allowed us to use Fortran-routines for heavy computing and still enjoy the benefits of MatLab.

In practice, the communication between the external Fortran code and MatLab's main program was done with MEX-files (EXternal M-files). In MEX-files, the Fortran routines can be written in the normal way. The MEX-file includes a gateway function that passes information from the main function to the external functions and back.

The computation of the objective function value was done using an MLP network. Since the genetic algorithm module needs the objective function value information, another interface between Fortran and C was needed. This interface was provided by a Fortran module. Figure 20.8 illustrates the implementation structure.

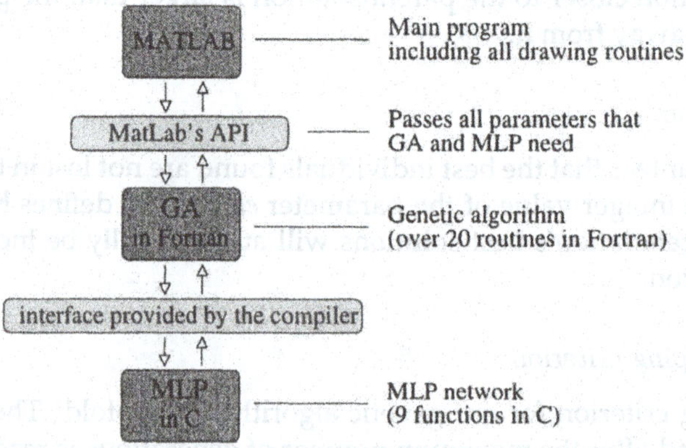

Figure 20.8 Implementation structure.

20.5.2 *Simple hybridization of a local search and GA*

A simple hybridization was carried out to improve the performance of the genetic algorithm. In the first phase of the hybrid algorithm, a local search method was run 10 times from different starting points inside a unit hypercube and the 10 local minima were added to the initial population of the genetic algorithm. The use of the unit hypercube as a starting region for the local search was justified following the similar reasoning as when justifying regularization: the activation function saturates far from the origin and, therefore, all the possible objective function values can be expected to be found close to the origin.

Adding the high quality solutions to the initial population helped to improve the quality of the results. When adding several local minima to a population there is a danger of reducing the diversity within the population. This might cause premature convergence leading to non-optimal final solution.

However, since the number of the added high quality solutions was only approximately 10% of the total population size, enough randomness remained in the population.

In the second phase, the genetic algorithm was run as a whole. In the end, the local search method was run again starting from the best solution found by the genetic algorithm. The 10 local minima that were added to the initial population also defined the box constraints for the regularized variables (weights). This was done in the following way:

$$\begin{aligned} LB_j &= \min_k\{z_j^k, -1\} \\ UB_j &= \max_k\{z_j^k, 1\}, \end{aligned}$$

where LB_j and UB_j are the lower and upper bounds for the variable j, respectively, and z^k, $k = 1, ..., 10$ are the 10 local minima. This definition guaranteed that all the 10 local minima were within the feasible region of the genetic algorithm and also that the feasible region contained the unit hypercube. The unregularized variables were, in our formulation, restricted with box constraints to the interval $[-10, 10]$.

This kind of a simple hybridization gave the genetic algorithm a jump start by increasing the solution quality in the initial population. Due to the problem structure, it can be assumed that there were not many duplicates in the 10 solutions added to the initial population. Hence, the diversity was not lost. The extra run of the local search in the end simply ensured that the final solution was at least a local minimum. The genetic algorithm with the local search will from now on be referred to as the *hybrid genetic algorithm*.

The local search method used was the well-known Nelder-Mead simplex method of Lagarias et al. (1998) (available in MatLab). It is one of the most efficient pattern search methods for relatively low dimensions. By utilizing this method not requiring any gradient information we did not restrict the applicability and generality of the genetic algorithm in solving a wide class of optimization problems. Besides, the Nelder-Mead simplex method is robust with respect to noise or inaccuracies in the objective function values, as stated in Dennis, Jr. and Schnabel (1989).

20.6 TEST PROBLEM

For testing the hybrid genetic algorithm we chose a scalar valued regression problem that was introduced in Section 20.2. In testing a new method, it is important to know as much as possible about the problem. In particular, we should know the original data in order to be able to judge the goodness of the results. Therefore, the most suitable test problem, for our purpose, is a problem, where the training data is derived by adding normally distributed noise to the original data. This setup enables us to examine the goodness of the final solutions of the implementation. In practical cases, the original data is not known and the goodness of the solution is often hard to verify.

As the original data we chose a sinus curve on the interval $[0, 2\pi]$. The discrete noisy image set $\{y^i\}$ was computed using the noisy function μ_n,

$$y^i = \mu_n(x^i) = \sin(x^i) + \delta \cdot r^i,$$

where x^i is the input, δ is the noise parameter and r^i is a normally distributed random number with a mean value zero and a variance one. During the tests the noise parameter δ was set to 0.3 and the number of training data points N was set to 50, that is, the index i runs from 1 to 50. All together we generated 20 noisy sinus curves resulting in 20 regression problems. Figure 20.9 illustrates a typical problem. The noisy curve is plotted with a continuous line and the original sinus curve with a dashed line.

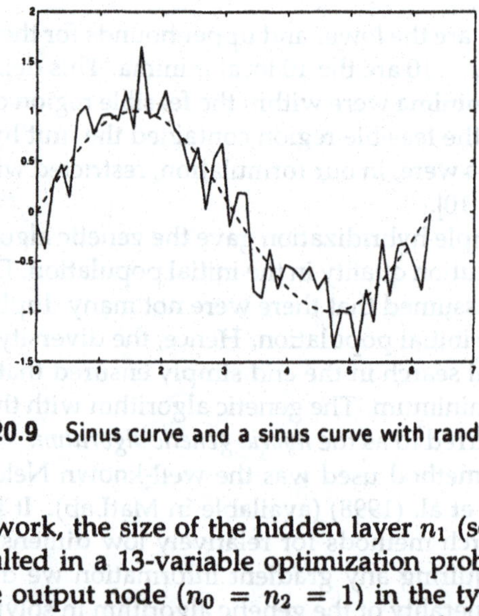

Figure 20.9 Sinus curve and a sinus curve with random noise.

In the MLP network, the size of the hidden layer n_1 (see Figure 20.6) was set to 4. This resulted in a 13-variable optimization problem, since we had one input and one output node ($n_0 = n_2 = 1$) in the type of problem that was considered. The parameters for the genetic algorithm were set according to Table 20.1. The parameter values are the same as in Miettinen et al. (2000) providing relatively good results, on the average, for a number of numerical experiments. However, we changed the mutation operation and after running more extensive empirical tests we increased the elitism size from 1 to 10 and decreased the maximum number of generations from 500 to 250.

The parameter values for the Nelder-Mead simplex method were set according to the Table 20.2. The scalars ρ, χ, ψ, and σ are the reflection, expansion, contraction, and shrinkage parameters (Lagarias et al., 1998), respectively. The parameters TolFun, TolX, and MaxFunEvals are related to the stopping criteria. The search is stopped if the diameter of the simplex is less than TolX and the function values on the corner points of the simplex differ less than TolFun, or the number of function evaluations exceeds MaxFunEvals. The scalars ρ, χ, ψ, and σ are fixed parameters in MatLab, whereas the parameters TolFun, TolX and MaxFunEvals are user defined. We use the MatLab de-

Table 20.1 Genetic algorithm parameter values.

Parameter	Value	Short description
Population	101	Number of individuals in population.
Elitism	10	Number of best solutions copied to new generation.
Generations	250	Maximum number of generations.
Tournament size	3	Number of randomly chosen individuals. The best becomes parent.
Crossover rate	0.8	Probability for crossover.
Mutation rate	0.1	Probability for mutation.
Step	100	Parameter for stopping criterion.
Tolerance	10^{-5}	Parameter for stopping criterion.

Table 20.2 MatLab parameter values for the Nelder-Mead simplex method.

Parameter	Value	Short description
ρ	1	Reflection parameter.
χ	2	Expansion parameter.
ψ	0.5	Contraction parameter.
σ	0.5	Shrinkage parameter.
TolFun	10^{-4}	Tolerance for the fcn value (stopping criterion).
TolX	10^{-4}	Tolerance for the variable value (stopping criterion).
MaxFunEvals	10000	Maximum number of fcn evaluations (stopping criterion).

fault values for TolFun and TolX, but increase the maximum number of function evaluations from the original 2600 (200 times the number of variables) to 10000.

The initial simplex is generated from the starting point by multiplying one component at a time by 1.05, and thus generating n new points near the starting point. If a component of the starting point is zero, then it is not multiplied by 1.05 but instead a value of 0.00025 is added to it.

20.7 RESULTS

We compare the results of the hybrid genetic algorithm to those of a simple multistart method. In this way, we can judge the performance of our method as opposed to a standard method when a global solver is not available. In the multistart method, a local search is applied starting from 30 random points within the unit hypercube. The local search method for the multistart method is the same as in the hybrid genetic algorithm, that is, the Nelder-Mead simplex method.

As mentioned earlier, 20 noisy sinus curves were generated for testing purposes. Each of these 20 curves corresponds to a separate problem, which was solved independently. Because both the hybrid genetic algorithm and the simple multistart method involve randomness, they were applied 20 times for each problem.

Each run of the Nelder-Mead simplex method required approximately 5.000 function evaluations. The corresponding value for the pure genetic algorithm was 10.000. Hence, the objective function was evaluated about 150.000 times when using multistart method ($30 \cdot 5.000$) and about 65.000 times when using the hybrid genetic algorithm ($10 \cdot 5.000 + 10.000 + 5.000$), which uses local search in the initialization and on the final result.

During the optimization process the objective function values were computed using the noisy data. After the network was fully trained it was validated also by replacing the noisy data in the objective function with the actual sinus curve values at the corresponding points. This enabled us to compare the results more extensively. In Table 20.3 both the average objective function values and average errors to the sinus curve are reported.

In Table 20.3, *HGA* stands for the hybrid genetic algorithm and *MS* for the multistart method. The first column *Prob #* lists the problem number. Each of the 20 problems was generated by adding normally distributed noise to the sinus function. The second and the third column, jointly denoted as *Avg. objective function value*, report the average objective function values of 20 runs for the hybrid genetic algorithm and the multistart method, respectively, for each of the 20 problems. Correspondingly, the last two columns, denoted as *Avg. error to the sinus curve* report the average error to the sinus curve for the hybrid genetic algorithm and the multistart method, respectively. We wish to emphasize that the MLP network was not trained again for the values in the last two columns, but the solutions were re-evaluated replacing the noisy data in the objective function with the actual sinus curve values.

In the last two rows at the bottom of Table 20.3, *Avg.* is the average objective function value over all the ($20 \cdot 20 = 400$) runs and *Std.* is the standard deviation[1], respectively.

As mentioned earlier, the actual data is not always available and the goodness of the solution has to be evaluated using only the noisy data. In our case, both the objective function values and the errors to the sinus curve give essentially the same information. This is because there is no overfitting. If overfitting had occurred, the objective function values would not have been good estimates for solution quality.

Large differences on the fourth and the fifth columns of Table 20.3 are mainly caused by the difference in the noise. This can be seen by studying Table 20.3 more closely. For example, on row 9, when considered column-wise, the average error to the sinus curve is very large for both the hybrid genetic algorithm

[1]The standard deviation was not computed from the mean values shown in Table 20.3, but from the original 400 objective function values.

Table 20.3 The results of the test runs

	Avg. objective function value		Avg. error to the sinus curve	
Prob #	HGA	MS	HGA	MS
1	0.0391	0.0410	0.0088	0.0119
2	0.0553	0.0595	0.0068	0.0084
3	0.0525	0.0544	0.0081	0.0106
4	0.0508	0.0493	0.0081	0.0075
5	0.0407	0.0419	0.0062	0.0074
6	0.0440	0.0469	0.0062	0.0092
7	0.0382	0.0420	0.0079	0.0091
8	0.0304	0.0327	0.0116	0.0101
9	0.0360	0.0381	0.0163	0.0136
10	0.0322	0.0377	0.0108	0.0060
11	0.0392	0.0417	0.0070	0.0079
12	0.0501	0.0510	0.0096	0.0123
13	0.0391	0.0405	0.0098	0.0107
14	0.0254	0.0268	0.0076	0.0083
15	0.0352	0.0386	0.0195	0.0151
16	0.0361	0.0383	0.0063	0.0084
17	0.0438	0.0473	0.0079	0.0104
18	0.0356	0.0383	0.0145	0.0141
19	0.0505	0.0536	0.0028	0.0042
20	0.0473	0.0476	0.0103	0.0112
Avg.	0.0411	0.0434	0.0093	0.0098
Std.	0.0078	0.0080	0.0038	0.0037

and the multistart method. The objective function values on row 9, however, are not especially large. In fact, they are even smaller than the respective values on row 19, where the average error to the sinus curve reaches its minimum. Therefore, it can be assumed that the hybrid genetic algorithm works equally well for all the problems, even though the average error to the sinus curve ranges from 0.0028 to 0.0195. These results indicate that the objective function value is a better measure for the stability of the optimization method than the error to the sinus curve (again assuming that no overfitting occurs). However, the error to sinus curve is a more reliable measure, because the possibility of overfitting cannot be easily excluded.

In Figure 20.10, 20 solutions found by the hybrid genetic algorithm and the multistart method for a typical problem are plotted. The solution curves are plotted with continuous lines, the original sinus curve is plotted with a dot-dashed line and the noisy curve with a dashed line. By looking at the curves it can be verified that the solution quality for the hybrid genetic algorithm is much better than for the multistart. The solution curves for both the methods are perfectly smooth, but the solution curves of the hybrid genetic algorithm follow the training data more closely than the ones of the multistart.

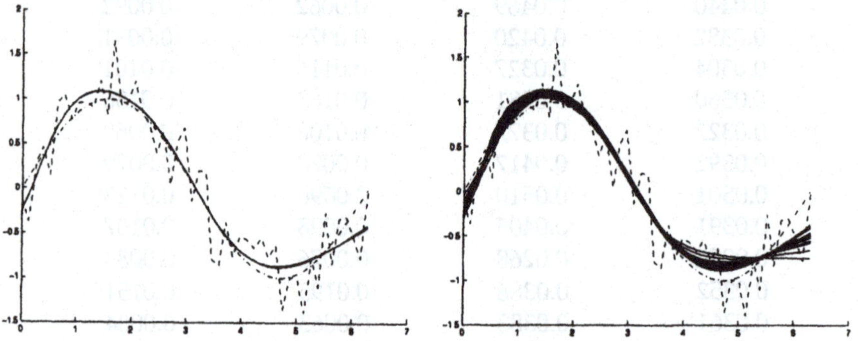

Figure 20.10 20 solution curves of the HGA (left) and the multistart method (right) for a typical problem.

So far, we have studied the global behavior of the optimization methods. Next, we study the local behavior and validate the proposed method by testing autocorrelation and by analyzing the variance and normality of the residues. For that purpose we examine closer the results of the typical problem illustrated in Figure 20.10.

The autocorrelation was tested using the Durbin-Watson test (Harvey, 1990). In the Durbin-Watson test, a *d-statistic* is computed by the formula

$$d = \sum_{i=2}^{N} (e^i - e^{i-1})^2 / \sum_{i=1}^{N} (e^i)^2.$$

Table 20.4 The Durbin-Watson test for a typical problem

	Hybrid GA	Multistart
smallest *d*	2.32	1.93
largest *d*	2.35	2.35
average *d*	2.33	2.23

The *d*-statistic values lie on the interval $[0,4]$. If $d = 2$, there is no autocorrelation. If d is larger or smaller than two, there exists negative or positive autocorrelation, respectively. In Table 20.4, there are the smallest, the largest, and the average value for the *d*-statistic for a typical problem. The values for d mainly indicate slight negative autocorrelation for both the methods. Negative autocorrelation indicates a tendency to underfitting. However, the *d*-statistic values are not too large. In practice, slight tendency for underfitting makes it safe to compare objective function values also when the original date is not known.

The range for the *d*-statistic values in Table 20.4 for the hybrid genetic algorithm is much smaller than for the multistart. This indicates that the hybrid genetic algorithm is more stable than the multistart. Therefore, the hybrid genetic algorithm is easier to adjust so that no under- or overfitting occurs. The multistart, on the other hand, is more likely to sway between under- and overfitting even when all the parameter values are kept constant. It is also worth noting that even though the maximum d-values are the same for both the methods, the solution curves in Figure 20.10 show no underfitting for hybrid genetic algorithms, whereas there is evident underfitting for at least two solutions of the multistart. Hence, the d-value alone is not a reliable measure for under- or overfitting.

In the variance analysis, we compute the variance of residues $var(e^i)$ at each point x^i and plot them in the same picture with the original sinus curve to see, whether the variance is somehow dependent on the original data. In an ideal case for a stable method, the variance of the residues should be small. Moreover, the residues should correspond to the normally distributed noise and should, therefore, not depend on the original data. The variance consideration is illustrated in Figure 20.11. In the two upper plots, the variance related to the hybrid genetic algorithm and the multistart method is multiplied with a constant 100, and in the lower plot the variance for the hybrid genetic algorithm is multiplied with a constant 5000. We can, thus, say that the variance of residues for the multistart method is approximately 50 times larger than the corresponding variance for the hybrid genetic algorithm. The variance analysis confirms the stability of the hybrid genetic algorithm. Moreover, based on this analysis we may say that for both the methods the variance is not strongly dependent on the local shape of the original data.

Finally, we analyze the normality of the residues e^i. The normality was tested using the *quantile-quantile plot* (QQ plot) (Kotz and Johnson, 1982). The

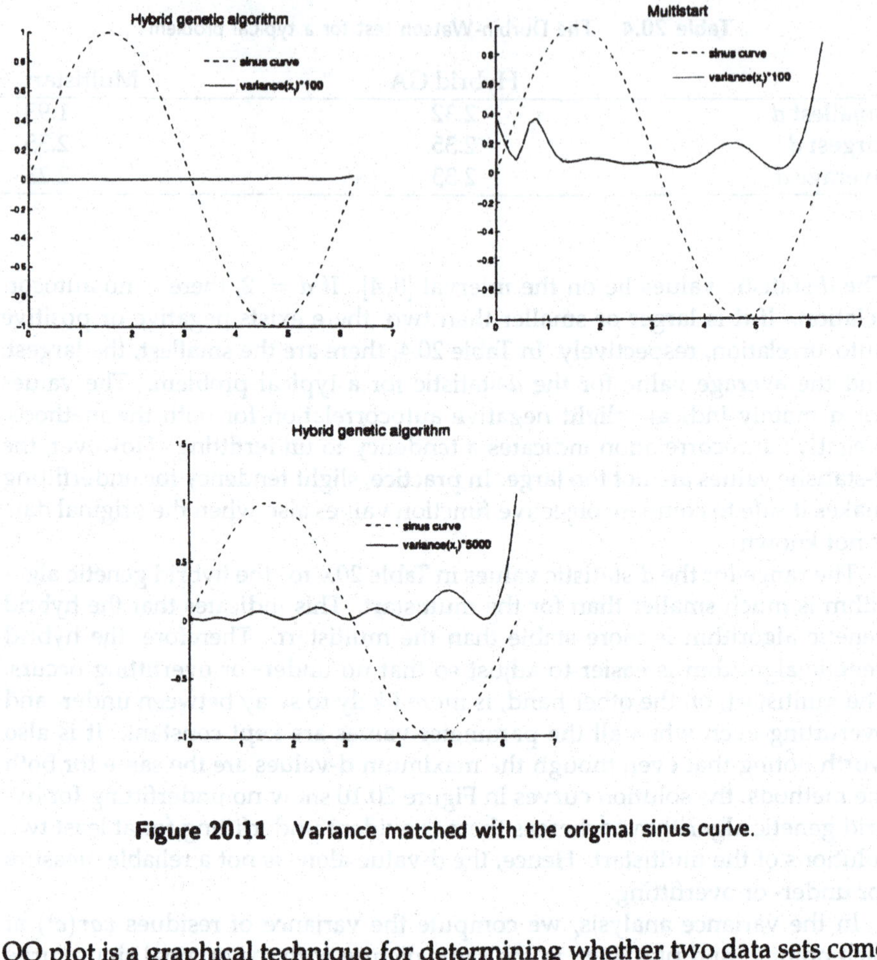

Figure 20.11 Variance matched with the original sinus curve.

QQ plot is a graphical technique for determining whether two data sets come from populations with a common distribution. We examine whether the residues are normally distributed, since the noise added to the sinus curve was from a normal distribution. Hence, we plot the quantiles of the residues versus the quantiles of a standard normal distribution. A 45 degree reference line is also plotted. If the residues follow a standard normal distribution, the points will fall approximately on the reference line. The QQ plots for the residues of the hybrid genetic algorithm and the multistart method are given in Figures 20.12 and 20.13, respectively. In Figure 20.12 only one QQ plot is given for the hybrid genetic algorithm, since the variance in the QQ plots was very small. For the multistart method, there was more variance and, therefore, four distinct QQ plots are given in Figure 20.13. The QQ plots show that the residues for both the hybrid genetic algorithm and the multistart method follow standard normal distribution.

Now, we have validated the proposed method by analyzing the global and the local behavior of the solutions obtained. The global behavior was vali-

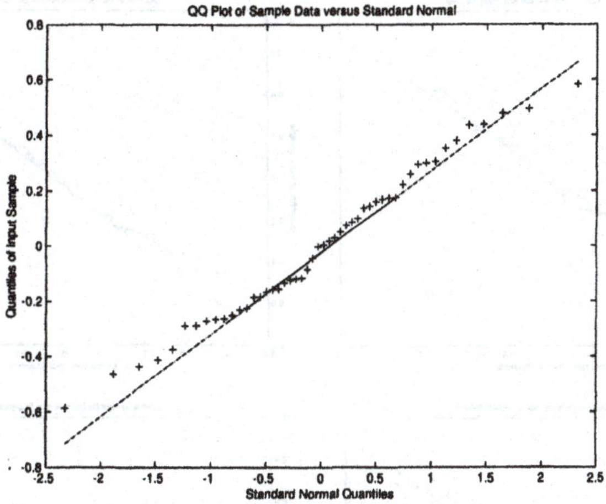

Figure 20.12 An example of QQ plot for the residues of the hybrid genetic algorithm.

dated by solving each of the 20 problems 20 times, and by analyzing the objective function values and the average errors to sinus curve. In addition, the global results were compared to those of the multistart. In the comparison, hybrid genetic algorithm was considerably faster than the multistart. Moreover, the objective function values and the average errors to sinus curves were better for the hybrid genetic algorithm than for the multistart.

The local behavior of the solutions obtained by the methods were validated by analyzing the autocorrelation and the variance and normality of the residues. The autocorrelation test and the variance analysis indicated better stability for the hybrid genetic algorithm than to the multistart. This was even more evident, when the solution curves were plotted for both the methods for a typical problem. The normality analysis showed that the residues of both the methods were normally distributed and, hence, corresponded well to the noise that was added to the sinus curve when constructing the problem.

We had the original data and, therefore, were able to validate the proposed hybrid method extensively. The performed analyzes of autocorrelation and of the variance and normality of the residues reveal the difficulty of analyzing results in a case when no original data is available and when there is no possibility to plot the solution curves. With the help of the detailed analysis similar to what we here performed to one typical problem it is possible to validate the local and global behavior of the solutions to some extend, but one is not able to state preference over any two solutions obtained unless the difference in solution quality is large. Therefore, when presenting a new method it is important to first test it with academic problems, whose original data is known.

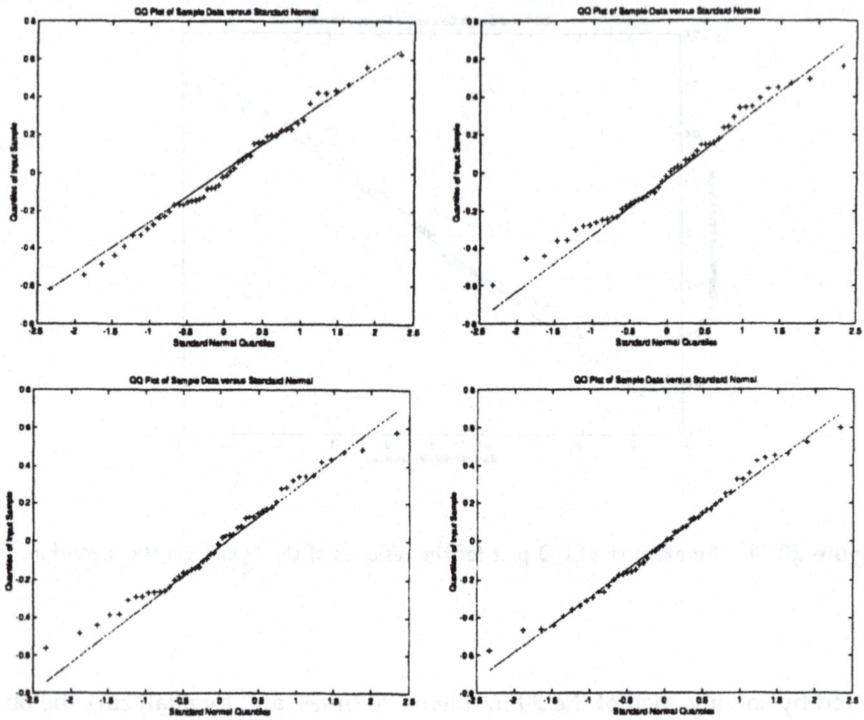

Figure 20.13 Some examples of QQ plot for the residues of the multistart.

20.8 DISCUSSION

In the future research, the use of the hybrid genetic algorithm will be expanded for larger problems. Also different types of problems will be tested to verify whether the promising results received with the above-defined test problem can be generalized for a larger group of problems.

It is also be interesting to compare different strategies that have been designed to guide the training process to avoid overfitting. We used regularization and were able to improve the solution quality considerably compared to our earlier experiments without regularization. However, regularization is not the only popular tool for that purpose. A common method is to try to predict when overfitting occurs and stop the training process and maybe go back a few iterations. In that approach, one assumes that the MLP network learns gradually. Consequently, by going back a few iteration after overfitting has occurred, one will find a solution with a relatively low objective function value and no overfitting.

To predict overfitting and to determine where to return, one uses an independent validation set, which is a set of different noisy data for the same problem. In practice, this means, for example, that one makes several measurements resulting in several noisy data sets. Then one solution is chosen as the current solution and the others form the validation set. During the training process the objective function value is, in an appropriate interval, also computed using the validation set. Until a certain point, the objective value for the validation set decreases along with the current solution objective function value. However, when the overfitting starts to be prominent, the objective function value for the validation set starts increasing although the objective function value of the current solution continues to decrease. The solution that gives the minimum objective function value for the validation set is usually chosen as the final solution.

The predicting process is sequential with two alternating phases. First, one fixes the training data as the current solution and lets the weights change freely in the training process. Second, after a few iterations of training the network, one fixes the weights and computes the objective function value using the different noisy data in the validation set. This sequence repeats until the objective function value for the validation set starts to increase. The advantage for the predicting process over the regularization is that one may use a network that is capable of overfitting. In this way, one avoids the difficult balancing between over- and underfitting. The downside is that the predicting process is computationally more expensive than the use of regularization.

20.9 SUMMARY

We have introduced an approach for solving a regression problem. This approach leads to a nonconvex optimization problem involving multi layer perceptron neural networks. It is still a very common practice to use a simple multistart method for training neural networks. The multistart method is easy to implement, since there are a number of well-tested local search methods available. Therefore, the multistart method is well suited as a first level implementation. However, there are faster and more accurate methods for training neural networks as shown in this paper.

We have trained the neural network using a hybrid genetic algorithm and compared its performance to the multistart method based on the Nelder-Mead simplex method. In the comparison, a simple regression problem has been studied, where the original data was known. This has enabled an extensive comparison of the results and the performance of the methods. To conclude, we can say that the hybrid genetic algorithm was considerably faster than the multistart method. In addition, the objective function values for the hybrid genetic algorithm were better than for the multistart, on the average. The hybrid genetic algorithm was also more robust and, therefore, easier to adjust so that no under- or overfitting occurs. Based on these encouraging results we can next apply our approach to more complicated real-life problems, and the presented study provides a good basis for further research.

Acknowledgments

The authors are indebted to Dr. Jari Toivanen for providing the genetic algorithm code and Dr. Erkki Heikkola and Prof. Tommi Kärkkäinen for the MLP network code. In addition, we wish to express our gratitude to Prof. Tommi Kärkkäinen for valuable pieces of advice concerning the nature of MLP network and to Ph.Lic. Salme Kärkkäinen and Prof. Antti Penttinen for the help with the statistical analysis. This research was supported by the Academy of Finland grants #71822, #65760 and #8583 and the Finnish Cultural Foundation.

Bibliography

C. M. Bishop. *Neural Networks for Pattern Recognition*. Clarendon Press, Oxford, 1995.

K. Deb. *Multi-Objective Optimization using Evolutionary Algorithms*. John Wiley & Sons, 2001.

J. E. Dennis, Jr. and R. B. Schnabel. A view of unconstrained optimization. In *Optimization*, pages 1–72. Elsevier North-Holland, New York, 1989.

D. E. Goldberg. *Genetic Algorithms in Search, Optimization, and Machine Learning*. Addison-Wesley, New York, 1989.

H. Greenberg. Mathematical programming glossary. http://carbon.cudenver.edu/~hgreenberg/glossary/glossary.html, 2001.

M. Hagan and M. Mohammad. Training feedforward networks with the Marquardt algorithm. *IEEE Transactions on Neural Networks*, 5(6):989–993, 1994.

A. C. Harvey. *The Econometric Analysis of Time Series*. MIT Press, 2nd edition, 1990.

T. Kärkkäinen. MLP-network in a layer-wise form with applications to weight decay. *Neural Computation*, 14(6):1451–1480, 2002.

S. Kotz and N. L. Johnson, editors. *Encyclopedia of Statistical Sciences*, volume 3 and 7. John Wiley & Sons, New York, 1982.

J. C. Lagarias, J. A. Reeds, M. H. Wright, and P. E. Wright. Convergence properties of the Nelder–Mead simplex method in low dimensions. *SIAM Journal on Optimization*, 9(1):112–147, 1998.

H.-L. Liu. A hybrid AI optimization method applied to industrial processes. *Chemometrics and Intelligent Laboratory Systems*, 45(1–2):101–104, 1999.

Z. Michalewicz, T. D. Logan, and S. Swaminathan. Evolutionary operators for continuous convex parameter spaces. In A. V. Sebald and L. J. Fogel, editors,

Proceedings of the 3rd Annual Conference on Evolutionary Programming, pages 84–97. World Scientific Publishing, River Edge, NJ, 1994.

Z. Michalewicz, G. Nazhiyath, and M. Michalewicz. A note on usefulness of geometrical crossover for numerical optimization problems. In L. J. L.J. Fogel, P. J. Angeline, and T. Back, editors, *Proceedings of the 5th Annual Conference on Evolutionary Programming*, pages 305–312. MIT Press, Cambridge, MA, 1996.

K. Miettinen, M. M. Mäkelä, and J. Toivanen. Comparison of four penalty function-based methods in handling constraints with genetic algorithms. Reports of the Department of Mathematical Information Technology, Series B. Scientific Computing No. B 17/2000, University of Jyväskylä, Department of Mathematical Information Technology, 2000.

D. T. Pham and D. Karaboga. *Intelligent Optimization Techniques: Genetic Algorithms, Tabu Search, Simulated Annealing and Neural Networks*. Springer-Verlag, London, 2000.

R. Rojas. *Neural Networks: A Systematic Introduction*. Springer-Verlag, Berlin Heidelberg, 1996.

F. Rosenblatt. The perceptron: A probabilistic model for information storage and organization in the brain. *Psychological Review*, 65(6):386–408, 1958.

D. E. Rumelhart, G. E. Hinton, and R. J. Williams. Learning internal representation by error propagation. In J. L. McClelland, D. E. Rumelhart, and the PDP Research Group, editors, *Parallel Distributed Processing Explorations in the Microstructure of Cognition*, volume 1: Foundations, pages 318–362. MIT Press, Cambridge, MA, 1996.

X. Yao and Y. Lui. New evolutionary systems for evolving artificial neural networks. *IEEE Transactions on Neural Networks*, 8(3):694–713, 1997.

Metaheuristics: Computer Decision-Making, pp. 449-464
Manuel A Matos, M. Teresa Ponce de Leão, J. Tomé Saraiva, J. Nuno Fidalgo,
Vladimiro Miranda, J. Peças Lopes, J. Rui Ferreira, Jorge M. C. Pereira,
L. Miguel Proença, and J. Luís Pinto

21 METAHEURISTICS APPLIED TO POWER SYSTEMS

Manuel A. Matos[1,3], M. Teresa Ponce de Leão[1,3],
J. Tomé Saraiva[1,3], J. Nuno Fidalgo[1,3], Vladimiro Miranda[1,3],
J. Peças Lopes[1,3], J. Rui Ferreira[1,3], Jorge M. C. Pereira[2,3],
L. Miguel Proença[3], and J. Luís Pinto[3]

[1]FEUP – Faculty of Engineering
University of Porto
Porto, Portugal
mmatos@inescporto.pt

[2]FEP – Faculty of Economics
University of Porto
Porto, Portugal
jpereira@inescporto.pt

[3]INESC Porto
Power Systems Unit
Porto, Portugal
lproenca@inescporto.pt

Abstract: Most optimization and decision problems in power systems include integer or binary variables, leading to combinatorial problems. In this paper, several approaches using metaheuristics and genetic algorithms are presented that deal with real problems of the power industry. Most of these methodologies are now implemented in distribution management systems (DMS) used by several utilities.
Keywords: Metaheuristics, Power systems, Simulated annealing, Genetic algorithms.

21.1 INTRODUCTION

This paper describes a number of real applications of metaheuristics (in this case, simulated annealing) and genetic algorithms to power system problems. The research work was developed in the framework of European projects and industrial contracts and addresses areas like planning and operation of electrical distribution systems, wind park layout, unit commitment of isolated systems with renewable energy sources and voltage collapse in systems. The combinatorial nature comes naturally in power systems, since most of the decision variables are binary or integer due to technical reasons. On the other hand, a common characteristic of these problems is the presence of technical constraints, which poses difficulties to the application of metaheuristics, leading to the need of penalty factors in the evaluation functions. The paper also includes feature selection for security analysis using artificial neural networks, a related topic, although not really an application of metaheuristics. The paper is organized as follows. Regarding each topic, the corresponding problem is briefly described, followed by the presentation of the approach and, in some cases, a summary of the results. Global conclusions and references complete the paper.

21.2 DISTRIBUTION SYSTEMS PLANNING

21.2.1 Problem description

A main issue in planning electric distribution systems is to decide what new branches and injection points (substations) to construct. The decisions must be taken to meet future demand at minimum cost, without violating technical constraints (maximum current flow and maximum voltage drop). These networks are characterized by a great number of nodes and possible branches. The nodes are the injection points of the network, consumer points (loads), substations and generation from independent producers. The problem involves many integer-valued (0-1) variables related to the decisions to build or not build facilities. In addition, the overall problem is dynamic, since the decisions about investments must be made over a multi-period horizon. Besides investment cost, other criteria are generally considered (operating cost, reliability) and uncertainty is present due to forecasting.

21.2.2 Approach

In the simulated annealing strategy, the analogy between the electric distribution planning problem and the Metropolis algorithm can be stated as follows. The *alternative solutions* or *configurations* of the combinatorial electric distribution planning problem are equivalent to the physical system *states*. The network configurations (alternative solutions) (attributes) are equivalent to the different *energy* states. The *control parameter*, which is such that about half the new configurations found are accepted at the start of the process, is equivalent to the *temperature parameter*. The SA cooling schedule integrates several aspects that are detailed in the next paragraphs:

- The choice of the initial temperature was such that about half the configurations were accepted;
- The number of iterations during which the temperature is kept constant was approximately 10 times the average number of possible neighbors. This value can be obtained from initial temperature calculations where the possible neighbors are explored;
- A stepwise temperature reduction scheme with $T(i) = T(i-1)/\beta$ was used;
- From our experience an average value for β was derived between 0.85 and 0.92. Larger values yield high computation times, and smaller values, below 0.85, lead to poorer solutions;
- The freezing point is determined where the number of acceptances is very small. This point is highly dependent on the problem size and must be determined for each problem.

The initial solution is generated by an auxiliary algorithm that finds a shortest spanning tree, the weights used for the edges usually being the edges (branches) costs although other criteria could be used. A structure of neighborhood generation mechanism is then created. To eliminate a great number of trials only feasible solutions S' from the solution space S are considered. This procedure consists of choosing all the possible combinations of connected trees that result from removing a branch to the loop created by each edge of the co-tree entering the initial configuration. The alternatives are evaluated according to the criterion to be optimized.

21.2.3 *Results*

A simulated annealing procedure was combined with the ϵ-constraint search to form a methodology for solving the multi-objective planning problem. In practice, the initial multi-objective problem was split into several single objective problems, successively solved to generate non-dominated solutions. In Ponce de Leão and Matos (1999), the detailed procedure to generate the set of efficient solutions is explained in detail.

21.3 DISTRIBUTION SYSTEMS OPERATION - RECONFIGURATION

21.3.1 *Problem description*

Distribution networks are usually operated radially, but they have a meshed structure (with open branches) that allows several configurations. The reconfiguration of these networks by changing the switching devices statuses aims at optimizing the network configuration for a given load profile, by minimizing the power losses. Figure 21.1 shows a typical instance of this problem, where we can see the selected trees (solid) and the potential branches, not used in the final solution. A related problem is the service restoration problem, in the sequence of a faulty situation, when one tries to re-supply at least part of the loads. Besides the main objective of each problem (minimizing losses or minimizing load not supplied), the minimization of the total number of switching operations is also included as a criterion. In either case, the network reconfiguration is a multi-objective optimization problem. An algorithm to deal with this problem is detailed in Matos and Melo (1999).

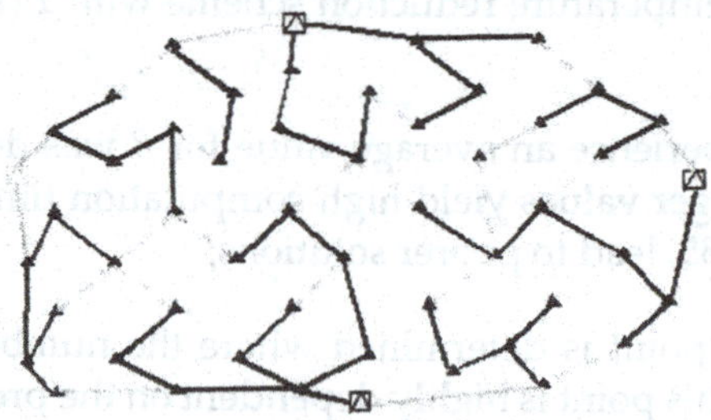

Figure 21.1 Distribution network reconfiguration problem.

21.3.2 Approach

The basic simulated annealing algorithm was used for this problem, with the following characteristics:

- When restoring service, the initial configuration represents the distribution network with the fault isolated and some unsupplied consumers. When minimizing losses, the initial configuration is the present status of the network;

- The neighborhood structure implemented depends on the problem. For the loss minimization problem, when a branch is Opened another one is Closed, in such a way that the network remains radial and all the consumers remain served. For the service restoration problem, the rule is as follows: Close a branch that close no loops *or* Open any branch;

- The evaluation includes a cost attribute (losses or power not supplied, depending on the problem) and penalties related to overload limits and maximum voltage drop. For the multi-objective case, a penalty relative to the maximum number of allowed switching operations is also included. Variation of this limit allows the generation of all the efficient solutions.

21.3.3 *Results*

Figure 21.2 shows the efficient solutions for a restoration situation. Note that similar solutions in the attribute space may be very different in the decision space. As can be observed, the power not supplied decreases as the number of switching operations increases. The operator may now choose among the given set for the most satisfying solution, according to his preferences and temporary constraints not included in the model.

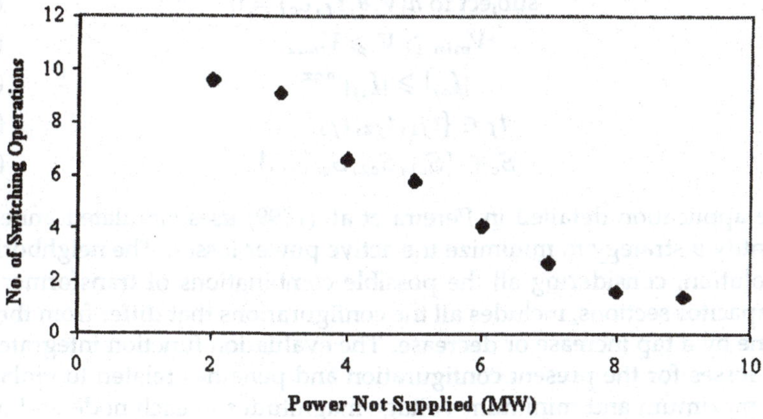

Figure 21.2 Efficient solutions for the restoration problem

21.4 DISTRIBUTION SYSTEMS OPERATION - REACTIVE POWER CONTROL

21.4.1 *Problem description*

In this section, we present a model to identify optimal operation strategies of distribution networks, considering that one wants to minimize active losses. This objective can be achieved by adequately selecting transformer taps and sections of capacitor banks that are in operation if one assumes that the topology in operation is fixed. To be more realistic, the model allows the specification of voltage ranges for nodes and overload limits for branches.

21.4.2 *Approach*

The problem of minimizing active power losses can be formulated by (21.1–21.2). In this formulation, $h()$ represents the AC power flow equations, t_f stands for the available values of transformer taps, e_c represents the available reactive power steps of capacitor banks, V are the voltage magnitudes, I are the branch currents, and gij corresponds to the conductance of branch ij. Traditionally, this problem has been solved assuming that transformer taps and sections of capacitor banks are represented by continuous variables. Such formulation consists of a nonlinear optimization problem which can be solved using, for instance, gradient methods. The solution obtained this way must be rounded to the closest discrete solution, which may, or may not, be the optimal solution.

$$\text{Minimize } z = \sum_{nr} g_{ij} \cdot (V_i^2 + V_j^2 - 2 \cdot V_i \cdot V_j \cdot \cos\theta_{ij}) \tag{21.1}$$

$$\text{subject to } h(V, \theta, t_f, e_c) = 0 \tag{21.2}$$

$$V_{min} \geqslant V \geqslant V_{max} \tag{21.3}$$

$$|\underline{I}_{ij}| \geqslant |\underline{I}_{ij}|^{max} \tag{21.4}$$

$$t_f \in \{t_{f1}, t_{f2}, t_{f3}, \ldots\} \tag{21.5}$$

$$S_c \in \{S_{c1}, S_{c2}, S_{c3}, \ldots\} \tag{21.6}$$

The application detailed in Pereira et al. (1999) uses simulated annealing to identify a strategy to minimize the active power losses. The neighborhood of a solution, considering all the possible combinations of transformer taps and capacitor sections, includes all the configurations that differ from the current one by a tap increase or decrease. The evaluation function integrates the active losses for the present configuration and penalties related to violations of the maximum and minimum voltage magnitudes in each node and of the maximum current flow in each branch.

21.4.3 Results

Tests with the IEEE24 system (24 nodes, 31 lines, 5 transformers) were conducted considering 11 taps in each transformer. Each tap corresponds to a 0.01 increment within the interval [0.95,1.05]. Initially, the transformer taps were considered at the nominal values (position 6). The selected simulated annealing parameters were:

- the number of iterations for the same temperature level is 25;
- the initial temperature level was set to 1.0. The lowering step determining the cooling scheme (parameter β) is 95% of previous temperature;
- the maximum number of iterations without improvement of the evaluation function is set to 75;

With these parameters, the algorithm converged in 312 iterations and the temperature was reduced till 0.54. Figure 21.3 shows the evolution, along the search process, of the present evaluation function value (continuous line) and for the best solution found till the current iteration (dotted line).

To evaluate the efficiency of the search process, we simulated all the possible combinations for the 5 transformer taps. This enumeration process required 161051 power flow studies, to obtain the true optimal solution, that

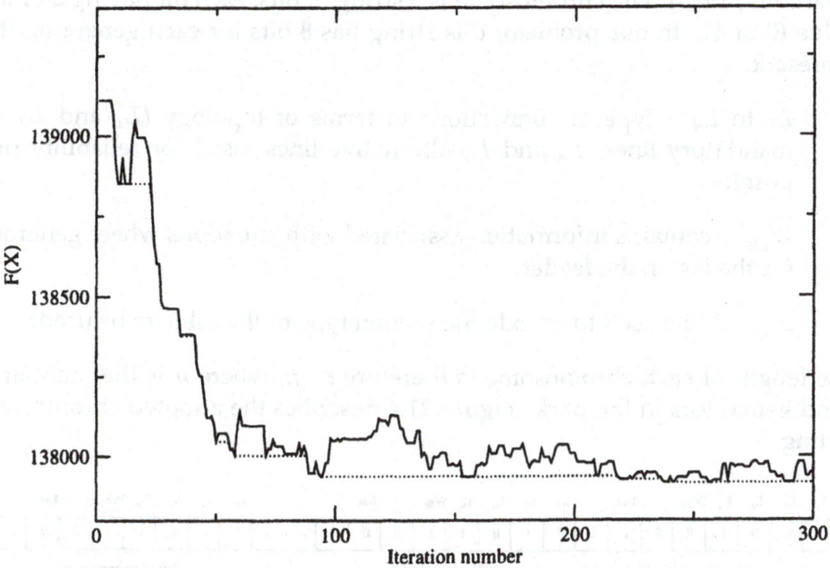

Figure 21.3 Evolution of the evaluation function for the current solution (continuous line) and for the best-identified solution (dotted line) for the 24 node network.

happened to be the same that was obtained with our approach, with only 312 power flow studies. Additional tests with a larger distribution system (645 nodes, 59 million solutions) showed also a good performance, with convergence obtained after only 256 iterations.

21.5 WIND PARK LAYOUT

21.5.1 Problem description

In a wind park the location of wind generators is defined having in mind the wind and terrain characteristics. The internal electrical network of the wind park is a problem that involves the network topology determination and the definition of the cable sections to be used. The solution of this problem can be obtained by solving a global optimization problem that involves minimization of investments and minimization of operation costs. If reliability is taken into account in the design procedure, an evaluation of the possible benefits of supplementary connections that provide alternative paths for the delivery of the energy is performed.

This optimization problem has combinatorial characteristics, due to the possibilities of connecting the several wind generators in the field and the type of discrete cable sections available (Resende, 1999). The identification of solutions for the problem described was performed using a genetic algorithm that implements the concept of geographic isolation, based on a *niching* scheme (Mahfoud, 1995). The chromosome is a string of bits, each bit having a binary value (0 or 1). In our problem, this string has 8 bits for each generator, that represent:

- L_1 to L_4 – type of connections in terms of topology (L_1 and L_3 are mandatory lines, L_2 and L_4 alternative lines, used for reliability purpose);
- W_{gui} – contains information associated with situations where generator i is the last in the feeder;
- S_{eci} – 3 bits used to encode the section type of the cable to be used;

The length of each chromosome is therefore $8 \cdot n$, where n is the number of wind generators in the park. Figure 21.4 describes the adopted chromosome coding.

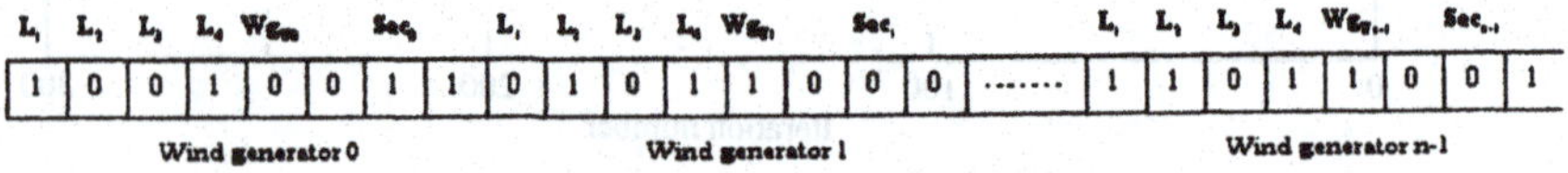

Figure 21.4 Chromosome coding

The fitness function corresponds to the total cost, but it includes penalty terms related to the technical constraints, as shown in (21.7), where nf is the number of feeders, ngu in the number of generators in the end of the feeder, L_{NC} is the number of lines not connected correctly, V_c and V_T are the number of violations of the voltage profile and of the thermal limits, C_C is the cost of cables (including protections), C_{JL} is the cost of energy losses, C_{ND} is the cost of energy not delivered to the grid, due to internal faults on the wind park, and $k1 \ldots k4$ are penalty factors.

$$\begin{aligned} FIT \quad = \quad & k_1 \cdot nf + k_2 \cdot L_{NC} + k_3 |ngu - nf| + \\ & k_4 (V_T + V_C) + C_C + C_{JL} + C_{ND}. \end{aligned} \tag{21.7}$$

This approach was tested in the electrical design of some wind parks, namely in the case of a 10 MW wind farm with twenty 500kW wind generators. This corresponds to a real wind park presently in operation in Portugal. In the genetic algorithm approach, the values adopted for the crossover probability and the mutation probability were respectively 0.8 and 0.04, and the process was found to converge after 1500 generations. The optimal solutions obtained are, for the tests performed, systematically better than the ones provided by the traditional engineering judgment and rules. Figure 21.5 shows two of the best solutions for this problem.

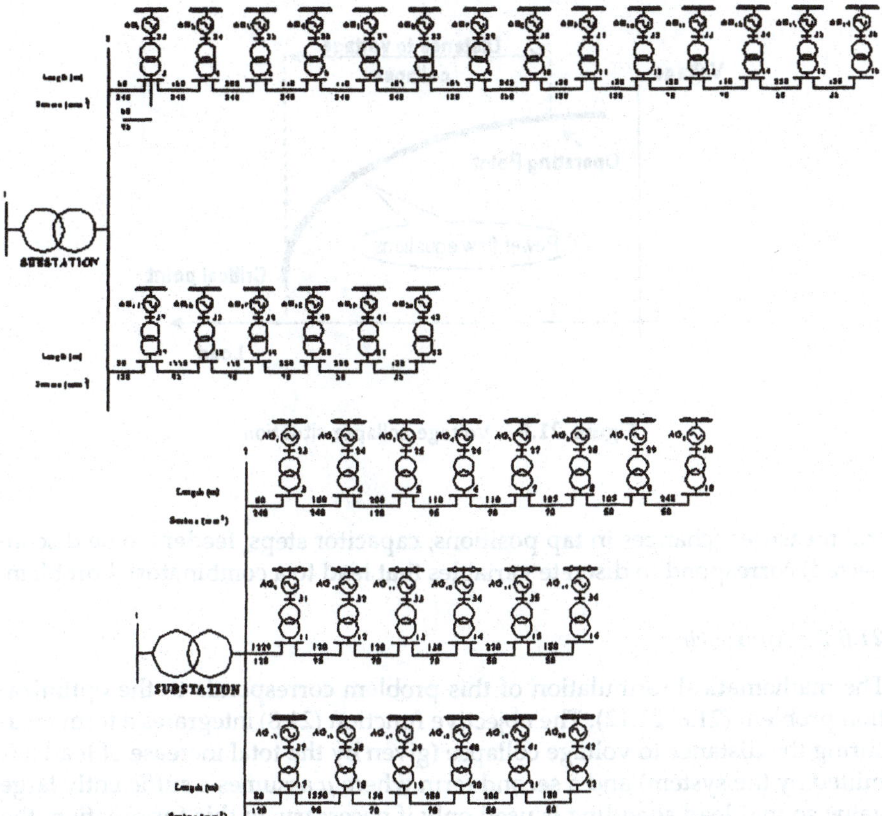

Figure 21.5 Two of the best possible layouts for the wind park

21.6 PREVENTIVE CONTROL PROCEDURE TO AVOID VOLTAGE COLLAPSE

21.6.1 Problem description

Most power systems are presently operated close to their operating limits, due to liberalization and increase in consumption, while economic and environmental constraints have limited the construction of new generation facilities and lines. Therefore, it is important to evaluate the distance to voltage collapse (see Figure 21.6) and to identify preventive measures that avoid this problem (Lemâitre et al., 1989).

The model includes the possibility of modifying the current positions of transformer taps as well as the steps of capacitor banks or performing load shedding. It is obvious that load shedding is understood as an end of line resource, in the sense that it will only be used if there is no other way of supplying loads even changing transformer taps or capacitor steps. Most of the con-

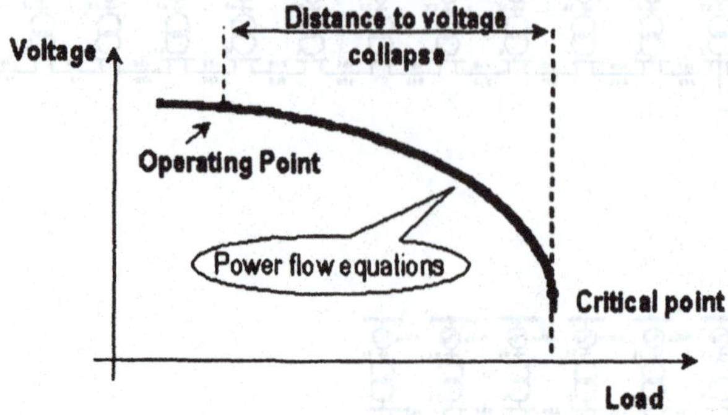

Figure 21.6 Voltage collapse situation

trol measures (changes in tap positions, capacitor steps, feeders to be disconnected) correspond to discrete variables that lead to a combinatorial problem.

21.6.2 Approach

The mathematical formulation of this problem corresponds to the optimization problem (21.8–21.12). The objective function (21.8) integrates a term measuring the distance to voltage collapse (given by the total increase of load admitted by the system) and a second term where μ assumes a sufficiently large value so that load shedding is used only if necessary. In this formulation, the equality constraints $f(X,u)$ represent the power flow equations, X is the state vector and u is the vector of control variables. Constraint (21.10) puts a hard limit on the distance to the collapse, and (21.11) and (21.12) constrain transformer taps and capacitor steps to their feasible (discrete) values.

$$\text{Maximize } C = Dist\_colp - \mu \cdot \sum P_{shed} \tag{21.8}$$

$$\text{subject to } f(X,u) = 0, \tag{21.9}$$

$$Dist\_colp \geqslant \epsilon \tag{21.10}$$

$$t_{min} \geqslant t_i \geqslant t_{max} \tag{21.11}$$

$$Y_{min} \geqslant Y_i \geqslant Y_{max} \tag{21.12}$$

To avoid local minimum solutions, a deterministic crowding genetic algorithm was used to solve the optimization problem described. The fitness function corresponds to the distance to collapse (in MVA) plus a penalty factor proportional to the summation of the load shed in each bus (MVA). Details of the methodology can be seen in Ferreira (1999).

21.6.3 Results

To evaluate the performance of the previously described approach, we conducted several numerical tests considering the Reduced Mato Grosso system (Granville et al., 1996). The results obtained were excellent in quality, matching with the ones obtained when the optimization problem in a continuous space was solved using the interior point method. When curtailment of loads was needed, the approach adopted was able to provide better results than the traditional methods, due to the fact that it was dealing with the real specificity and characteristics of the amount of load supplied by each feeder in a substation.

21.7 UNIT COMMITMENT

21.7.1 Problem description

The unit commitment problem is related to the hourly scheduling of the generators that will be running to satisfy the entire load, in a horizon of several hours (at least 48 hours). The problem is not separable in the intervals, since there are transition costs (due to start-up and shut-down costs) and constraints, and has been addressed traditionally by dynamic programming (small problems) and Lagrangian relaxation.

21.7.2 Approach

The genetic algorithm used to solve the unit commitment problem is based on the general scheme of genetic algorithm, but includes several enhancements (Hang, 1999). It replaces the standard selection scheme with a dedicated niching selection scheme – deterministic crowding selection. In addition, a dynamic mutation rate technique, chromosome repair and neighborhood digging schemes are also included, in order to accelerate convergence and enforce robustness. Encoding the chromosome plays an important role in the genetic algorithm. In the UC module, an efficient way is proposed to compress the length of the chromosome. The main idea comes from the observation that the generators will not switch on/off frequently and thus some of the variables can be reduced. With this method, the length of the chromosome can be reduced a lot. For example, in a problem with 17 generators and an hourly scheduling for 48 hours, the direct encoding will lead to 816 bits. On the other hand, if divide the 48 hours into 16 sections, the length of the chromosome will only be 368 bits.

The fitness function includes the generation and start-up and the shutdown costs of generators. To identify an infeasible chromosome a penalty cost is added when the schedule violates the constraints. To avoid trapping in local optima, dynamic mutation rate adjustment is used. Deterministic crowding (DC) selection scheme is a niching method, which is used to preserve the diversity of the genetic algorithm. Chromosome repair is a corrective algorithm. The algorithm will correct slightly an infeasible chromosome, which is near

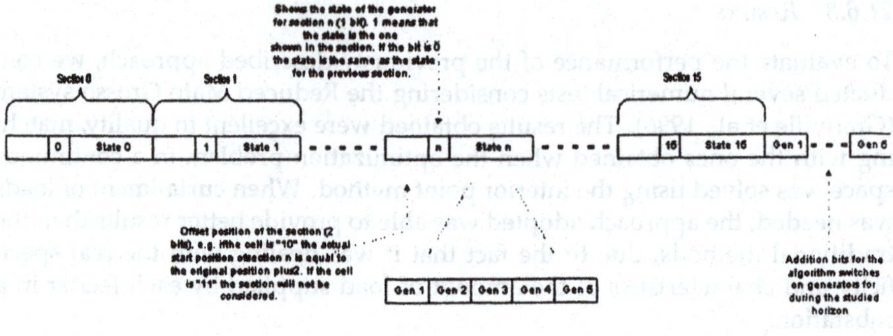

Figure 21.7 Compression scheme for the Unit Commitment problem

the feasible region. Afterwards, the corrected chromosome will place it back to the population and enrich the genetic pool. Neighborhood digging concentrates mainly in feasible chromosomes. By searching the neighbor of the chromosome, the algorithm will determine a better chromosome to replace the original. By experience, we find that this genetic operator not only accelerates the convergence of the process, but also gives better results regarding both the economic optimization and the execution time, which was an important project constraint.

21.7.3 Results

The implementation of this method lead to very good results regarding both the economic optimization and the execution time, which was an important limitation in the project where it was developed. Figure 21.8 shows a typical solution for a 48 hour problem.

21.8 FEATURE SELECTION USING ANN

21.8.1 Problem description

Feature subset selection (FSS) is a central issue in a vast diversity of problems including classification, function approximation, machine learning, and adaptive control. In this study, the selection of adequate variable subset is accomplished by sensitivity analysis of trained artificial neural networks (ANN). Data sets concern to power systems of islands of Crete (Greece) and Madeira (Portugal). The problem under analysis refers to the dynamic security classification. This section describes how FSS and ANN tools may be used for enhancing classification performance. Results attained so far support the validity of the developed approach.

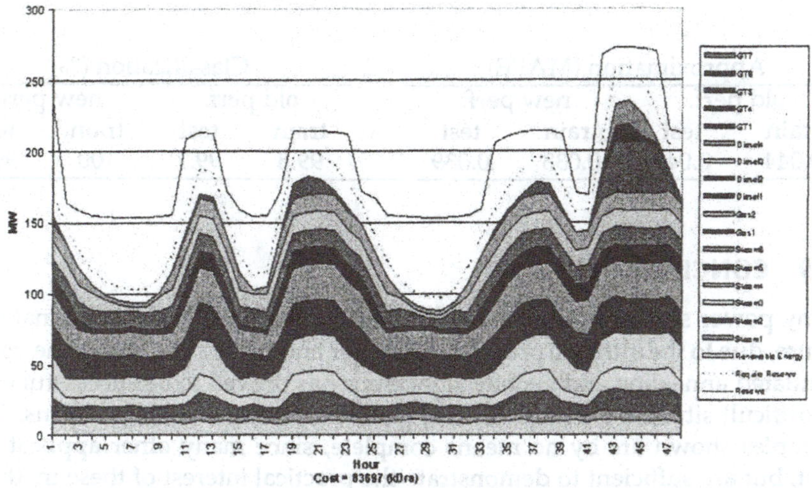

Figure 21.8 Unit commitment for a 48 hour horizon

21.8.2 Approach

The FSS methodology applied in Fidalgo (2001) consists basically of three steps:

1. Use of correlation analysis to discard linearly correlated features;

2. Train an ANN with the remaining ones, compute sensitivities of its output (dynamic security index) with respect to all inputs and discard inputs with lowest sensitivity s_i indexes;

3. Train a second ANN using only the remaining variables.

21.8.3 Results

Tests with a data set of the power system of Crete, initially with $N1 = 60$ attributes, lead to $N2 = 20$ after correlation analysis. An ANN was then trained and s_i calculus were performed, leading to further elimination of 14 variables. A new ANN was finally trained with the remaining $N3 = 6$ features. Performance was computed and results were compared to the ones reported in Peças Lopes et al. (1999), were 22 attributes were used, as shown in the Table 21.8.3. Results show that some real data applications may be amazingly simplified without loss of performance.

Table 21.1 Error comparison.

Approximation (MAPE)				Classification (%)			
old perf.		new perf.		old perf.		new perf.	
train	test	train	test	train	test	train	test
0.044	0.043	0.033	0.039	99.8	99.7	100	99.8

21.9 CONCLUSIONS

Many power system optimization or decision problems are of combinatorial nature, due to the intrinsic presence of integer and binary variables. The use of simulated annealing and genetic algorithms has proved to be successful even on difficult situations, related to real planning and operation problems. The examples shown are by no means complete, since many other applications exist, but are sufficient to demonstrate the practical interest of these methodologies to a demanding engineering field.

Bibliography

J. R. Ferreira. *Voltage stability evaluation considering fuzzy injection scenarios and identification of control procedures (in Portuguese).* PhD thesis, University of Porto, 1999.

J. N. Fidalgo. Feature selection based on ANN sensitivity analysis – A practical study. In *International Conference on Neural Networks and Applications WSES2001*, February 2001.

S. Granville, J. C. O. Mello, and A. C. G. Melo. Application of interior point methods to power flow unsolvability. *IEEE Transactions on PWRS*, 11(1), February 1996.

P. S. Hang. Evolutionary computing and fuzzy logic applied to generation scheduling in systems with renewables. Master's thesis, University of Porto, December 1999.

C. Lemâitre, J. P. Paul, J. M. Tesseron, Y. Harmand, and Y. S. Zhao. An indicator of the risk of voltage profile instability for real-time control applications. In *IEEE/PES Summer Meeting*. IEEE/PES, 1989.

S. Mahfoud. *Niching methods for genetic algorithms*. PhD thesis, University of Illinois at Urbana-Champaign, 1995.

M. Matos and P. Melo. Multi-objective reconfiguration for loss reduction and service restoration using simulated annealing. In *Proceedings of IEEE Budapest Power Tech'99*. IEEE, August 1999.

J. A. Peças Lopes et al. On-line dynamic security assessment of isolated networks integrating large wind power production. *Wind Engineering Review*, 23(2), 1999.

Jorge Pereira, J. Tomé Saraiva, and M. T. Ponce de Leão. Identification of operation strategies of distribution networks using a simulated annealing approach. In *Proceedings of the IEEE Budapest Power Tech'99*. IEEE, August 1999.

M. T. Ponce de Leão and M. Matos. Multicriteria distribution network planning using simulated annealing. *International Transactions on Operational Research*, 6:377–391, 1999.

F. Resende. Optimization of the electrical internal network of wind farms. Master's thesis, University of Porto, December 1999. In Portuguese.

Metaheuristics: Computer Decision-Making, pp. 465-480
Daniel Merkle and Martin Middendorf

22 ON THE BEHAVIOR OF ACO ALGORITHMS: STUDIES ON SIMPLE PROBLEMS

Daniel Merkle[1] and Martin Middendorf[1]

[1] Department of Computer Science
University of Leipzig
D-04109 Leipzig, Germany
merkle@informatik.uni-leipzig.de, middendorf@informatik.uni-leipzig.de

Abstract: The behavior of Ant Colony Optimization (ACO) algorithms is studied on simple problems which allow us to identify characteristic properties of these algorithms. In particular, ACO algorithms using different pheromone evaluation methods are investigated. A new method for the use of pheromone information by artificial ants is proposed. Experimentally it is shown that an ACO algorithm using the new method performs better than ACO algorithms using other known methods for certain types of problems.
Keywords: Ant colony optimization, Pheromone information, Permutation problems.

22.1 INTRODUCTION

The Ant Colony Optimization (ACO) metaheuristic has been applied successfully to several combinatorial optimization problems (for an overview see Dorigo and Di Caro (1999)). Usually ACO algorithms have been tested on benchmark problems or real world problems. In this paper, we start studying the behavior of ACO algorithms on simple problems. The aim of this study is to show that several interesting properties of ACO algorithms can be observed much more clearly when applying them to quite simple problems instead of large irregular problems. Therefore, our observations help to understand the behavior of ACO algorithms also on more complex problems.

ACO algorithms are based on the principle of communication by pheromones that has been observed by real ants. In ACO algorithms, artificial ants that found a good solution mark their paths through the decision space by putting some amount of pheromone along the path. Alternatively, all ants are allowed to mark their paths but the better the solution of an ant is the more pheromone it uses. The following ants of the next generation are attracted by the pheromone so that they will search in the solution space near good solutions. A characteristic feature of the original ACO metaheuristic is that the ants make their decisions based on local pheromone information that correspond directly to the actual decision (Dorigo et al., 1999). It was shown in Merkle and Middendorf (2000) that ACO algorithms with a global use of the pheromone information can improve the optimization behavior for some types of problems. In this paper, we propose a new global pheromone evaluation method, called relative pheromone evaluation. It is shown experimentally that relative pheromone evaluation performs better than the standard local evaluation method and the global summation evaluation method on some of our test problems. We argue that relative pheromone evaluation will be interesting also for solving more complex problems.

The paper is organized as follows. All test problems are permutation problems which are introduced in the next Section 22.2. In Section 22.3 we describe the ACO algorithm with the different pheromone evaluation methods. The choice of the parameter values of the algorithms that are used in the test runs and the test instances are described in Section 22.4. Results are reported in Section 22.5 and a conclusion is given in Section 22.6.

22.2 PROBLEMS

We consider the following type of permutation problems which is known as the Linear Assignment problem. Given are n items $1, 2, \ldots, n$ and an $n \times n$ cost matrix $C = [c(i,j)]$ with $c(i,j) \geq 0$. Let $\mathcal{P}_n$ be the set of permutations of $(1, 2, \ldots, n)$. For a permutation $\pi \in \mathcal{P}_n$ let $c(\pi) = \sum_{i=1}^{n} c(i, \pi(i))$ be the cost of the permutation. The problem is to find a permutation $\pi \in \mathcal{P}_n$ of the items that has minimal costs, i.e., $c(\pi) = \min\{c(\pi') \mid \pi' \in \mathcal{P}_n\}$. It should be noted, that the Linear Assignment problem is not an NP-hard problem and can be solved in polynomial time (Papadimitriou and Steiglitz, 1982).

22.3 THE ACO ALGORITHMS

The ACO algorithm for permutation problems consists of several iterations where in every iteration each of m ants constructs one permutation. Every ant selects the items for the places of the permutation in some order. For the selection of an item the ant uses pheromone information which stems from former ants that have found good solutions. In addition an ant may also use heuristic information for its decisions. The pheromone information, denoted by τ_{ij}, and the heuristic information, denoted by η_{ij}, are indicators of how good it seems to have item j at place i of the permutation.

The next item is chosen from the set S of items that have not been placed so far according to a probability distribution. In this paper, we use three different methods to determine the probabilities. The first method is the standard *local evaluation* (L-Eval) of the pheromone values that is used in nearly all ACO algorithms in the literature (e.g. Dorigo et al. (1996)). According to this method, the probability p_{ij} to select item $j \in S$ on place i depends only on the local pheromone values and heuristic values in the same row, i.e., τ_{il}, η_{il} $l \in [1:n]$, and is determined by

$$p_{ij} = \frac{\tau_{ij} \cdot \eta_{ij}}{\sum_{h \in S} \tau_{ih} \cdot \eta_{ih}}. \tag{22.1}$$

The second method is a form of global pheromone evaluation where p_{ij} can depend on the pheromone τ_{kl} or heuristic values η_{kl} with $k \neq i$. This method is called *summation evaluation* (S-Eval). The use of global pheromone evaluation and the summation evaluation have been proposed by Merkle and Middendorf (2000). For summation evaluation an ant considers for deciding which item to put on place i the sums of all pheromone values in the columns up to the row i. Summation evaluation is attractive for problems like tardiness problems in machine scheduling where there is only a small interval of places in the schedule which are good for a job. The probability p_{ij} to select item $j \in S$ on place i is determined by

$$p_{ij} = \frac{\left(\sum_{k=1}^{i} \tau_{kj}\right) \cdot \eta_{ij}}{\sum_{h \in S}\left(\sum_{k=1}^{i} \tau_{kh}\right) \cdot \eta_{ih}} \tag{22.2}$$

According to this method, an item j that has high pheromone values τ_{kj} for $k < i$, but has yet not been scheduled, is likely to be scheduled soon. This property is desirable for tardiness problems since it is important that jobs are scheduled not too late (before their deadline). Summation evaluation has been shown to perform well for the Single Machine Weighted Total Tardiness problem (Merkle and Middendorf, 2000), the Job Shop problem (Teich et al., 2001), the Flow-Shop problem (Rajendran and Ziegler, 2003), and the Assembly Line problem (Bautista and Pereira, 2002). A combination of local evaluation and summation evaluation performed best for the Resource-Constrained Project Scheduling problem (Merkle et al., 2002).

The third rule is a newly proposed pheromone evaluation rule that we call *relative (pheromone) evaluation* (R-Eval). Relative evaluation is motivated by the following observation: Assume that an item j has not been placed by an ant on a place $< i$ although a large part of the pheromone in column j lies on the first $i-1$ elements $\tau_{1j}, \ldots, \tau_{(i-1)j}$ in the column. In this case, all values τ_{kj} for $k \geq i$ are small. Hence, it is unlikely that the ant will place item j on place i when some other item has a high pheromone value for place i. This is true even when τ_{ij} is the highest of all pheromone values $\tau_{ij}, \ldots, \tau_{nj}$ on the remaining places. However, for many optimization problems it might still be important on which place $l \geq i$ item j is, even when these places are not the most favorite ones. For such problems (i.e. when there are several good places for an item) we propose not to use directly the pheromone values in each row but to normalize them with the relative amount of pheromone in the rest of their columns. Hence, the probability p_{ij} to select item $j \in S$ on place i is determined by

```
procedure ANTALGORITHM(X,y)
parameter X ∈ {L,S,R}, y ∈ {f,b,p});
repeat
   foreach ant k ∈ {1,...,m}
      if y = p
         Permute the rows of the pheromone and cost matrix
         according to the given permutation π_k;
      if y = b
         Reverse the order of the rows of the pheromone and
         cost matrix;
      S = {1,2,...,n};                         set of available items
      for i = 1 to n
         Choose item j ∈ S with probability p_ij;
         S := S − {j};
      endfor
      Undo the permutation of the pheromone and cost ma-
      trix;
   endfor
   foreach (i,j)
      τ_ij ← (1 − ρ) · τ_ij;                   evaporate pheromone
   endfor
   foreach (i,j) ∈ best solution
      τ_ij ← τ_ij + 1/(cost of best solution+1);   update pheromone
   endfor
until stopping criterion is met
```

Algorithm 22.1 Ant algorithm.

$$p_{ij} = \frac{\tau^*_{ij} \cdot \eta_{ij}}{\sum_{h \in S} \tau^*_{ih} \cdot \eta_{ih}} \quad \text{with} \quad \tau^*_{ij} := \left(\frac{\sum_{k=1}^{n} \tau_{kj}}{\sum_{k=i}^{n} \tau_{kj}} \right)^{\gamma} \cdot \tau_{ij} \tag{22.3}$$

where $0 \leq \gamma \leq 1$ is a parameter that allows to determine the strength of normalization.

In those cases where a heuristic was used, the values η_{ij} were computed according the following formula $\eta_{ij} = 1/c(i,j)$.

The best solution found by an ant in the current iteration is then used to update the pheromone matrix by adding some amount of pheromone for every element corresponding to the solution. But before the update is done a percentage of the old pheromone is evaporated according to the standard formula $\tau_{ij} = (1 - \rho) \cdot \tau_{ij}$. Parameter ρ allows to determine how strongly old pheromone influences the future. Then, for every item j of the best permutation found so far some amount of pheromone Δ is added to element τ_{ij} of the pheromone matrix where i is the place of item j in the solution. The algorithm stops when some stopping criterion is met, e.g. a certain number of generations has been done.

We extend the ACO algorithm described above by possibly changing the sequence in which the ants assign items to the places. For each ant k in a generation a permutation π_k is given that is used by the ant to permute the rows of the pheromone matrix and the cost matrix before the ant starts constructing a solution. We consider the following cases i) no permutation of the rows (forward ants), ii) reversing the sequence of rows in the matrices (backward ants), and iii) performing a random permutation before an ant starts (permutation ants). The advantage of using forward and backward ants for the Shortest Common Supersequence problem was shown by Michels and Middendorf (1999). Merkle and Middendorf (2001) showed that using different random permutations for the ants can lead to an improved optimization behavior for ACO algorithms solving permutation scheduling problems because then the probabilities used by the ants to select the items usually better reflect the relative size of the corresponding products of pheromone value with heuristic value. Note, that when using summation evaluation it is not reasonable to use random permutations for the ants.

The different combinations of matrix permutation and pheromone evaluation method are denoted by X-y-Eval with $X \in \{L, S, R\}$ and $y \in \{f, b, p\}$ where f=forward, b=backward, and p=permutation.

22.4 TESTS

All results in the following section are averages over 25 runs, every run was stopped when the quality of the best found solution has not changed for 500 iterations or after 10000 generations. All test problems have $n = 50$ items. Evaporation rate was set to $\rho = 0.01$ and in some cases we also tested $\rho = 0.005$ in order to make sure that our conclusions do not depend too much on the actual value of ρ. Pheromone update was done with $\tau_{ij} = \tau_{ij} + 1/(\text{cost of solution}+1)$ and the pheromone matrix was initialized with $\tau_{ij} = 1/(\rho \cdot c(\pi_{opt}))$ (this is also the maximal possible value of an element in the pheromone matrix) where $c(\pi_{opt})$ is the quality of an optimal solution, $i, j \in [1, 50]$. The tests have been done with and without using the heuristic, i.e., $\eta_{ij} = 1$ for $i, j \in [1, 50]$. The results with heuristic are similar to the results without heuristic only the convergence is faster. Therefore we describe only the results obtained without using the heuristic in the next section. Note also, that X-b-Eval with $X \in \{L, S, R\}$ is tested only for asymmetric instances were an interesting difference between the results for X-b-Eval and X-f-Eval might occur.

Table 22.1 Cost matrix and results for an instance of type1: brighter gray pixel indicate smaller costs; given are average quality of best solution (Best), average iteration number when best solution was found (Gen.).

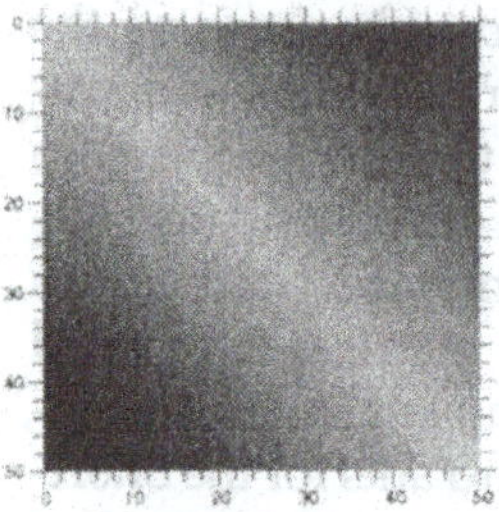

Evaluation		Type 1 Best	Gen.
L-Eval	f	287.4	3456
	p	50.4	3547
R-Eval	f	70.1	2421
	p	50.1	3415
S-Eval	f	51.9	2932

22.5 RESULTS

Problem instances of type 1 reflect the situation that there exists only a single optimal solution. Without loss of generality we assume that the optimal solution is to place item i at place i for $i \in [1 : n]$. We consider the case where the costs for placing an item grow the more its actual place differs from its optimal place. The cost matrix is rather simple (see Table 22.1). For each element (i, j) of the cost matrix the cost $c(i, j)$ equals the distance from the main diagonal, i.e., $c(i, j) = |i - j| + 1$. Clearly, to place item i, $i = 1, \ldots, n$ on place i is the only optimal solution with costs n.

Table 22.1 shows the results of different X-y-Eval ACO algorithms on the problem instance of size 50. L-p-Eval, R-p-Eval, and S-f-Eval are the best methods while L-f-Eval is worst. This confirms that summation evaluation is a good method for problems where there exists (more or less) a single good place for every item. Both other methods perform worse when using forward ants. Figure 22.1 shows that L-f-Eval tends to choose items too early (in the upper 2/3rds of the pheromone matrix the values above the diagonal are higher than below). Note that this effect has been observed before by Merkle and Middendorf (2001). It can also be seen that L-f-Eval converges earlier in the uppermost rows of the matrix than in the more lower rows. It is interesting that the new proposed relative evaluation method performs much better when using forward ants than L-Eval. Compared to L-f-Eval the results of all other methods depicted in Figure 22.1 show a very symmetric behavior and do not have the undesired bias. Local evaluation and relative evaluation profit clearly from using permutation ants for this type of problem.

Problem instances of type 2 are similar to problem instances of type 1 but the places of some of the items are not relevant now. Hence there exist several optimal solutions which differ only for some of the items. For the test instances the elements in the first $k < n$ rows of the cost matrix are one and the other rows of the cost matrix are the same as for the cost matrix of in-

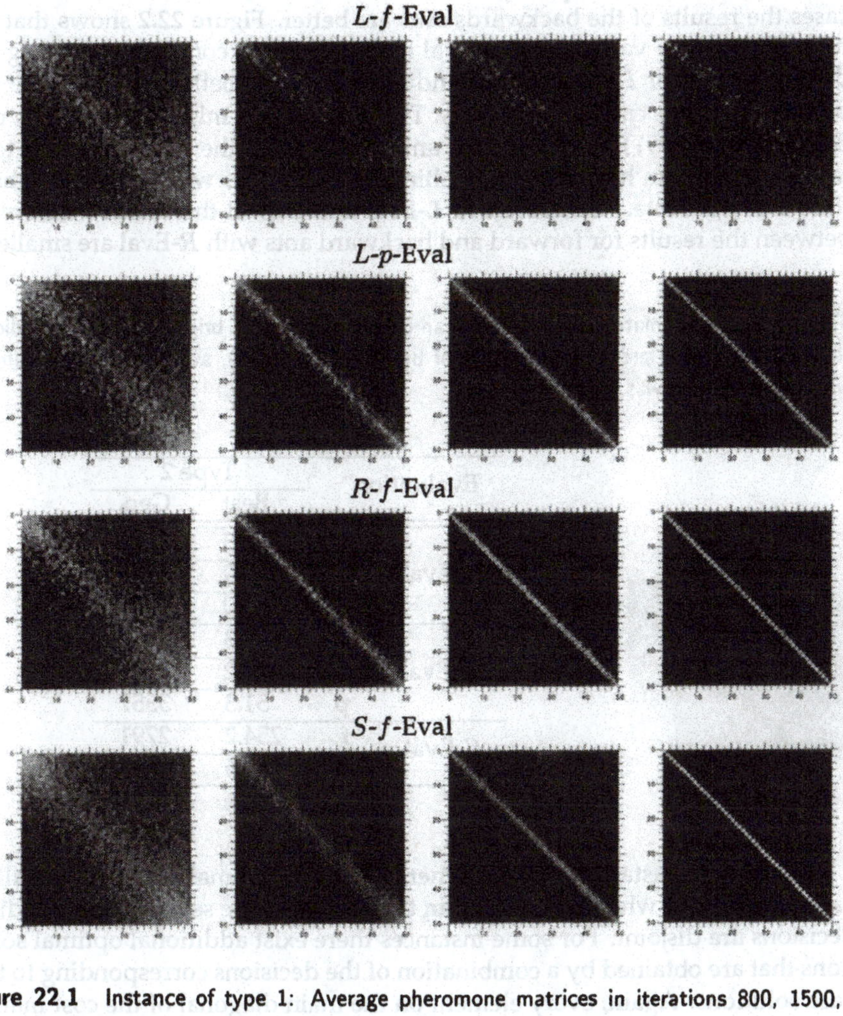

Figure 22.1 Instance of type 1: Average pheromone matrices in iterations 800, 1500, 2200, and 2900: brighter gray colors indicate higher pheromone values.

stance of type 1 (see Table 22.2). Hence, every optimal solution has item i on place i for $i > k$ and the places for items $1, \ldots, k$ are not relevant. The results for $n = 50$ and $k = 10$ that are presented in Table 22.2 show that similar as for problem instance 1 very good results are found by L-p-Eval and R-p-Eval. Also S-b-Eval found very good results. It can be seen that there is a significant difference between the quality obtained by backward or forward ants. In all cases the results of the backwards ants are better. Figure 22.2 shows that the high pheromone values for L-b-Eval are much more concentrated along the diagonal than for L-f-Eval. This indicates that it is better for when the important decisions can be made early. This observation might explain why den Besten et al. (1999) found that their ant algorithm for the single machine total tardiness problem had more difficulties with instances where most jobs have their due date late. Compared to L-Eval and S-Eval the relative difference between the results for forward and backward ants with R-Eval are smaller.

Table 22.2 Cost matrix and results for an instance of type 2: brighter gray pixel indicate smaller costs; given are average quality of best solution (Best), average iteration number when best solution was found (Gen.).

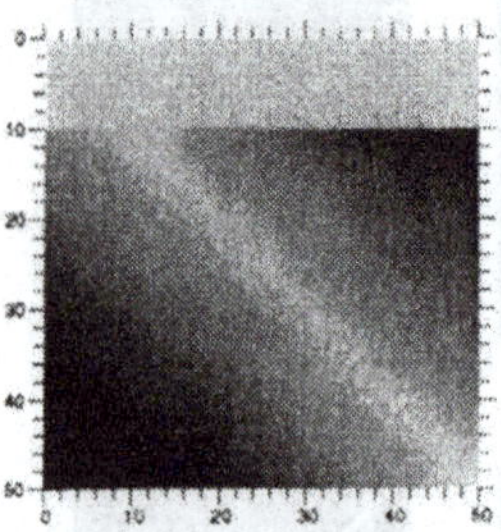

Evaluation		Type 2 Best	Type 2 Gen.
L-Eval	f	270.5	3490
	b	207.9	3183
	p	50.0	3308
R-Eval	f	62.3	2256
	b	70.2	2345
	p	51.3	3557
S-Eval	f	254.2	2791
	b	50.6	2604

For problem instances of type 3 there exist two optimal solutions — called base solutions — which are disjoint in the sense that the sets of corresponding decisions are disjoint. For some instances there exist additional optimal solutions that are obtained by a combination of the decisions corresponding to the base solutions. Again, every element on the main diagonal of the cost matrix is one. Moreover, all elements on another line (called 1-line) that is parallel to the main diagonal are one, i.e., all elements (i, j) with $j - i = k \bmod n$ for a fixed $k \in [1, n]$ are one. All other elements of the cost matrix are two. The instances of type 3 differ only in the distance k of the 1-line from the diagonal. For $n = 50$ we tested the algorithms on matrices with values $k \in \{3, 23, 25\}$.

L-Eval and R-Eval perform quite well on instances of type 3 and mostly the algorithm found optimal solutions (see Table 22.3). S-Eval does not perform very well since for each item there are two good places which are not

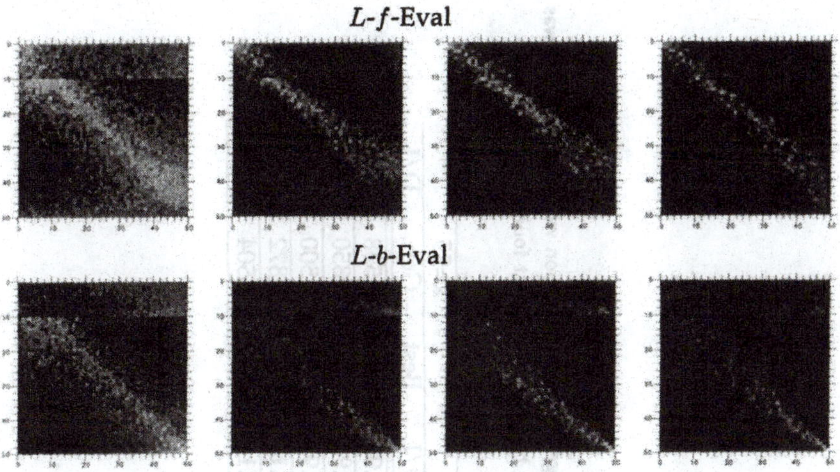

Figure 22.2 Instance of type 2: Average pheromone matrices in iterations 800, 1500, 2200, and 2900: brighter gray colors indicate higher pheromone values.

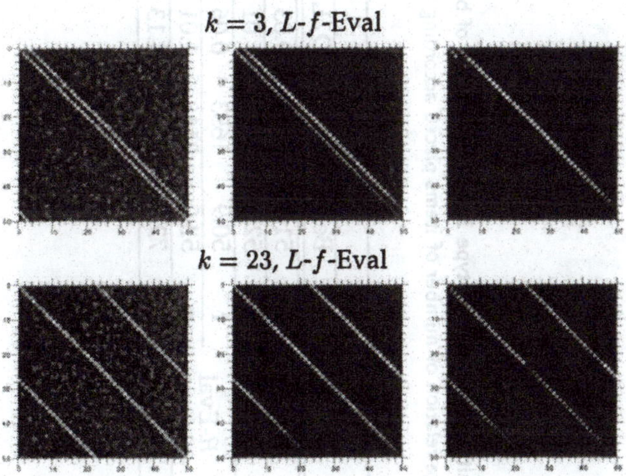

Figure 22.3 Instances of type 3 with $k = 3$ and $k = 23$: Average pheromone matrices in iterations 100, 800, 1500: brighter grey colors indicate higher pheromone values.

Table 22.3 Results on instances of type 3: average quality of best solution (Best), average number of iteration when best solution was found (Gen.), and average relation of number of items placed according to diagonal and according to 1-line in the best found solution (D/L).

Evaluation		$k=3$			k=23			k=25		
		Best	Gen.	D/L	Best	Gen.	D/L	Best	Gen.	D/L
L-Eval	f	51.2	1770	0.03	53.3	1919	0.93	50	1538	0.89
	p	52.0	3011	1.01	52.1	3012	0.93	50	1850	0.95
R-Eval	f	50.9	1653	0.06	52.2	2009	0.95	50	1500	0.93
	p	52.0	3118	1.01	51.8	3094	1.11	50	1872	0.98
S-Eval	f	72.5	6193	4.13	72.4	6502	1.15	72.6	6504	1.01

Table 22.4 Cost matrix and results for an instance of type 4: brighter gray pixel indicate smaller costs; given are average quality of best solution (Best), average iteration number when best solution was found (Gen.), results with $\rho = 0.005$ are in brackets.

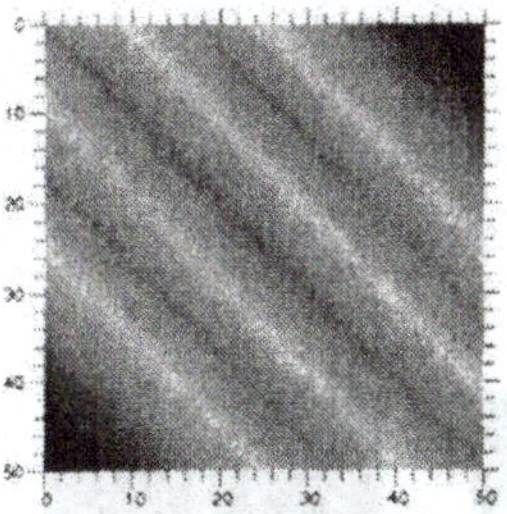

Evaluation		Type 4 Best	Gen.
L-Eval	f	109.0 (105.4)	3200 (8977)
	p	50.0	4493
R-Eval	f	55.4 (50.8)	2657 (5164)
	p	50.2	3941
S-Eval	f	194.4 (192.7)	1993 (2970)

neighbored. It is interesting that the instance with $k = 25$ could always be solved optimal when using *L*-Eval and *R*-Eval. The reason is that instances with $k = 3$ and $k = 23$ have only two different optimal solutions: i) the diagonal with item $i \in [1, 50]$ on place i ii) the 1-line with item $i \in [1, 50]$ on place $i - k \bmod 50$ where k is the distance between the diagonal and the 1-line. For the instance with $k = 25$ there exist 2^{25} optimal solutions: for $j \in [1, 25]$ items j and $j + 25$ are placed either on places j and $j + 25$ or on places $j - k \bmod 50$ and $j + 25 - k \bmod 50$. Another interesting observation is that *L*-*f*-Eval and *R*-*f*-Eval prefer to place elements according to the 1-line for $k = 3$. But this is not the case for $k = 23$ (see Figure 22.3). The reason is that choosing item 4 for place 1 will not allow to place item 4 on place 4. Hence it is likely that item 7 is placed on place 4. In other words choosing an item according to the 1-line in the first generations enforces that another item is placed according to the 1-line k decision later. So for small k this effect is of importance when using forward ants.

Similar as for instances of type 3 there exist several optimal solutions for instances of type 4. But for instances of type 4 it is more difficult to find these optimal solutions. The matrix has four diagonal lines with elements of cost one. For all other elements in the matrix the cost is equal to the minimal distance from the next diagonal line plus one (see Table 22.4). *L*-*p*-Eval and *R*-Eval performed best and find always or nearly always an optimal solution while — as expected for this type of problems — *S*-Eval performs much worse even for $\rho = 0.005$ (see Table 22.4). It is interesting that *L*-*f*-Eval performed much worse than *L*-*p*-Eval which performed best (even for $\rho = 0.005$ the results for *L*-*f*-Eval do not improve much). One reason is that *L*-*f*-Eval tends to converge early in the upper part of the matrix which is not the case for *L*-*p*-Eval (see Figure 22.4). For this type of problem starting to converge in the upper rows early is dangerous because there exist many solutions that have small costs in the upper rows but which can not be extended to an optimal or nearly optimal solution. *R*-Eval performs very well, no matter whether forward or

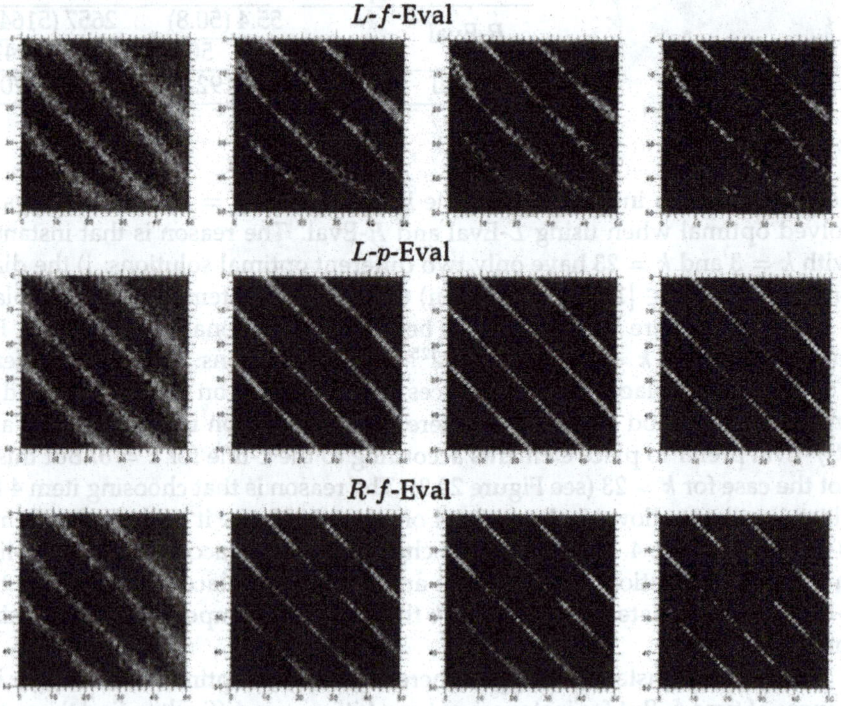

Figure 22.4 Instance of type 4: Average pheromone matrices in iterations 800, 1500, 2200, and 2900: brighter gray colors indicate higher pheromone values.

permutation ants are used. R-f-Eval does not show early convergence in the upper rows (see Figure 22.4) since it has the property that pheromone differences in lower rows have an influence even for items which have some high pheromone values in some of the upper rows. Thus when in early phases of a run an item has been placed mostly onto one of the first places this decision has good chances to be revised when later during a run a few ants found that it is better to place this item late.

22.6 CONCLUSION

We have shown that simple permutation problems are well suited to identify and explain some characteristic properties of Ant Colony Optimization (ACO) algorithms. In particular, we identified characteristic properties of different pheromone evaluation methods. A new method to use the pheromone information (relative evaluation) was proposed and compared to known standard methods. Experiments on the simple test problems show that the new method performs best on a certain type of problems for which it was designed. It is interesting to study whether relative evaluation is also a good method for harder and real world problems.

22.6.1 A final note

Independently Stützle and Dorigo (2001) studied ACO algorithms on simple problems. In particular, they investigate the influence of pheromone information and evaporation on the convergence on simple graph problems.

Bibliography

J. Bautista and J. Pereira. Ant algorithms for assembly line balancing. In M. Dorigo, G. Di Caro, and M. Sampels, editors, *Ant algorithms*, number 2463 in Lecture Notes in Computer Science (LNCS), pages 65–75, Berlin, 2002. Springer.

M. den Besten, T. Stützle, and M. Dorigo. Scheduling single machines by ants. Technical Report IRIDIA/99-16, IRIDIA, Université Libre de Bruxelles, Belgien, 1999.

M. Dorigo, G. Di Caro, and L. M. Gambardella. Ant algorithms for discrete optimization. *Artificial Life*, 5(2):137–172, 1999.

M. Dorigo and G. Di Caro. The ant colony optimization meta-heuristic. In D. Corne, M. Dorigo, and F. Glover, editors, *New ideas in optimization*, pages 11–32. McGraw-Hill, London, 1999.

M. Dorigo, V. Maniezzo, and A. Colorni. The ant system: optimization by a colony of cooperating agents. *IEEE Trans. Systems, Man, and Cybernetics – Part B*, 26:29–41, 1996.

D. Merkle and M. Middendorf. An ant algorithm with a new pheromone evaluation rule for total tardiness problems. In S. Cagnoni, R. Poli, Y. Li, G. Smith, D. Corne, M.J. Oates, E. Hart, P.L. Lanzi, E.J.W. Boers, B. Paechter, and T.C. Fogarty, editors, *Real-World Applications of Evolutionary Computing, Proceedings of the EvoWorkshops 2000*, number 1803 in Lecture Notes in Computer Science (LNCS), pages 287–296, Berlin, 2000. Springer.

D. Merkle and M. Middendorf. A new approach to solve permutation scheduling problems with ant colony optimization. In E.J.W. Boers, J. Gottlieb, P.L. Lanzi, R.E. Smith, S. Cagnoni, E.Hart, G.R. Raidl, and H. Tijink, editors, *Applications of Evolutionary Computing, Proceedings of the EvoWorkshops 2001*, number 2037 in Lecture Notes in Computer Science (LNCS), pages 484–493. Springer Verlag, 2001.

D. Merkle, M. Middendorf, and H. Schmeck. Ant colony optimization for resource-constrained project scheduling. *IEEE Transactions on Evolutionary Computation*, 6(4):333–346, 2002.

R. Michels and M. Middendorf. An ant system for the shortest common supersequence problem. In D. Corne, M. Dorigo, and F. Glover, editors, *New Ideas in Optimization*, pages 51–61. McGraw-Hill, London, 1999.

C.H. Papadimitriou and K. Steiglitz. *Combinatorial Optimization: Algorithms and Complexity*. Prentice-Hall, 1982.

C. Rajendran and H. Ziegler. Ant-colony algorithms for permutation flowshop scheduling to minimize makespan / total flowtime of jobs. *European Journal of Operational Research*, 2003. To appear.

T. Stützle and M. Dorigo. An experimental study of the simple ant colony optimization algorithm. In N. Mastorakis, editor, *Proceedings of the 2001 WSES International Conference on Evolutionary Computation (EC'01)*, pages 253–258. WSES-Press International, 2001.

T. Teich, M. Fischer, A. Vogel, and J. Fischer. A new ant colony algorithm for the job shop scheduling problem. In L. Spector, D. Whitley, D. Goldberg, E. Cantu-Paz, I. Parmee, and H.-G. Beyer, editors, *Proceedings of the Genetic and Evolutionary Computation Conference (GECCO-2001)*, page 803, San Francisco, CA, USA, 2001. Morgan Kaufmann.

Metaheuristics: Computer Decision-Making, pp. 481-500
Nenad Mladenović and Dragan Urošević

23 VARIABLE NEIGHBORHOOD SEARCH FOR THE K-CARDINALITY TREE

Nenad Mladenović[1,2] and Dragan Urošević[1]

[1]Mathematical Institute of Serbian Academy of Sciences and Arts
Knez Mihailova 35
11000 Belgrade, Serbia and Montenegro
nenad@mi.sanu.ac.yu, draganu@mi.sanu.ac.yu

[2]GERAD, H.E.C
3000, chemin de la Côte–Sainte–Catherine
Montreal H3T 2A7, Canada
nenad@crt0.crt.umontreal.ca

Abstract: The minimum k-cardinality tree problem on graph G consists in finding a subtree of G with exactly k edges whose sum of weights is minimum. In this paper we propose variable neighborhood search heuristic for solving it. We also analyze different shaking strategies and their influence on the final solution. New methods are compared favorably with other heuristics from the literature.
Keywords: Graphs, k-cardinality tree, Variable neighborhood search, Optimization.

23.1 INTRODUCTION

An undirected weighted graph $G = (V, E)$ is given with the vertex set V, edge set E and weights $w_i \in \mathbb{R}$ associated to V or to E. The *minimum weighted k-cardinality tree problem* (k-CARD problem for short) consists in finding a subtree of G with exactly k edges (and exactly $k+1$ vertices) whose sum of weights is minimum (Hamacher et al., 1991). There are two versions of this problem: *vertex-weighted* and *edge-weighted*, if the weights are associated with V or with E, respectively.

In this paper, we consider the edge-weighted k-CARD problem, i.e., for each $e \in E$, a weight $w_e \in \mathbb{R}$ is given. By $w(E')$ we denote the *weight* of E', such that

$$w(E') = \sum_{e \in E'} w_e \quad for\ each\ E' \subset E.$$

A tree in G is a subgraph $T = (V_T, E_T)$ of G so that T is connected and contains no cycles. For a given k $(1 \le k \le |V| - 1)$, a *k-cardinality* tree is a tree T such that $|E_T| = k$ (if $|E_T| = |V| - 1$, then T is a spanning tree of G). The k-CARD problem is as follows:

$$\min\{w(E_T) \mid T = (V_T, E_T) \text{ is a tree, } |E_T| = k\}.$$

We also use $w(T)$ instead $w(E_T)$. The k-CARD problem is strongly NP-hard (Fischetti et al., 1994). However, if G is a tree then there exists a polynomial algorithm which is based on dynamic programming (Fischetti et al., 1994). The 0-1 integer programming formulation can be found in (Fischetti et al., 1994). Let us define the 0-1 variables x and y in the following way:

$$x_e = \begin{cases} 1 & \text{if } e \in E_T \\ 0 & \text{if } e \notin E_T \end{cases} \quad \text{and} \quad y_u = \begin{cases} 1 & \text{if } u \in V_T \\ 0 & \text{if } u \notin V_T \end{cases}$$

The k-CARD problem can be formulated as:

$$\min \sum_{e \in E} w_e x_e \tag{23.1}$$

subject to

$$\sum_{e \in E} x_e = k \tag{23.2}$$

$$\sum_{u \in V} y_u = k + 1 \tag{23.3}$$

$$\sum_{e \in E(U)} x_e \le \sum_{u \in U} y_u - y_w \quad \forall U \subset V, |U| \ge 2, w \in U \tag{23.4}$$

$$y_u, x_e \in \{0, 1\} \quad \forall u \in V, \forall e \in E \tag{23.5}$$

Note that the number of constraints (23.4) that assures connectivity and acyclity of the solution is exponential. There are several applications of the k-CARD problem. The first one was in the oil field leasing in the Norwegian part of the North Sea (Fischetti et al., 1994). Another application comes from the layout of office buildings (Foulds and Hamacher, 1992).

Simple *lower bounds* can be obtained by using a greedy procedure (Kruskal, 1956): (i) rank the edges of G in a nondecreasing order of their weights; (ii) add an edge (one at a time with respect to that order) to the already selected edges so that they do not create a cycle. (iii) stop as soon as the cardinality k is reached. This procedure produces a forest with k edges and hence a lower bound of the problem. An *upper bound* can be obtained, of course, by any heuristic method. Ehrgott *et al.* (1997) propose several constructive heuristics: (i) modified Prim's algorithm (1957) for the minimum spanning tree (MST) which terminates when the number of edges is equal to k, i.e., not all n outer loop steps are used; (ii) modified Dijkstra's algorithm for the shortest paths so that the shortest path with m edges ($1 \leq m < k$) is extended to the tree with k edges; (iii) the heuristic that suggests finding a minimum k-cardinality subtree of the MST by the exact (polynomial) method proposed in Fischetti et al. (1994) (which is based on Dynamic programming); (iv) the combinations of previous constructive heuristics such as the Dijkstra-Prim, Dynamic-Prim, Dynamic-Dijkstra-Path and Dynamic-Dijkstra-Tree. It appeared that Dynamic-Dijkstra-Path (DDP for short) had the best performance on average. Jörnsten and Løkketangen (1997) propose a Tabu search for the k-CARD. They use two tabu lists and a few parameters that control the intensification and diversification of the search.

Two variable neighborhood search (VNS) heuristics are proposed in the next section and compared with the existing heuristics in Section 23.3. Several different strategies for the *shaking step* are analyzed in Section 23.3 as well. We also develop a tabu search (TS) (Glover, 1989) based on the chain-interchange idea proposed in Mladenović et al. (1996). Our TS is very similar to Jörnsten's and Lokketangen's TS (both methods use two tabu lists) and we shall not give its details in the text. Its advantage is the use of no parameters in the search, i.e. the lengths of two tabu lists are changed in each iteration at random.

23.2 THE VNS FOR THE MINIMUM K-CARDINALITY TREE

The basic idea of the VNS is a change of the neighborhoods in the search for a better solution (see e.g. Hansen and Mladenović (1999; 2001b; 2002)). To construct different neighborhood structures and to perform a systematic search, one needs to have a way of finding the distance between any two solutions, i.e. one needs to supply the solution space with some metrics (or quasi-metrics) and then induce the neighborhoods from them.

Now we shall give the rules for two VNS variants developed for the k-CARD problem. They both start with an initial solution T obtained by a modified Prim's algorithm (Ehrgott et al., 1997), where the first vertex of a tree is chosen at random. Moreover, both variants use the maximum CPU time allowed in the search (t_{max}) as a stopping condition.

The differences between two proposed VNS heuristics are in the *shaking* and *local search* steps.

23.2.1 The basic VNS (VNS-1)

The solution space $\mathcal{S}_k$ for the k-CARD problem can be represented as a set of all subtrees of G with exactly k edges (or $k+1$ nodes). It holds $|\mathcal{S}_k| = n/(k+1) \cdot O(k^k)$, since there are $n/(k+1)$ different sets with $k+1$ vertices and $O(k^k)$ spanning trees in the subgraph with $k+1$ vertices. Let us introduce a distance $\rho(T_1, T_2)$ between any two solutions (the trees with cardinality k) T_1

procedure VNS-1( )
Set $p \leftarrow p_{min}$;
repeat
(a) *Shaking*: Generate a tree T' at random from the pth neighborhood of T ($T' \in \mathcal{N}_p(T)$);
(b) *Local search*: Apply any local search method with T' as an initial solution; denote so obtained local optimum with T'';
(c) *Move or not*: If this local optimum is better than the incumbent, move there ($T \leftarrow T''$), and continue the search with $\mathcal{N}_{p_{min}}$ ($p \leftarrow p_{min}$); otherwise, set $p \leftarrow p + p_{step}$;
until $p = p_{max}$

Algorithm 23.1 The main step of the VNS, for the k-CARD problem.

and T_2 as a cardinality of difference between their edge sets, i.e.

$$\rho(T_1, T_2) = |E_{T_1} \setminus E_{T_2}| = |E_{T_2} \setminus E_{T_1}|.$$

The neighborhood $\mathcal{N}_p(T_1)$ consists of all solutions (subtrees) having distance p from T_1: $T_2 \in \mathcal{N}_p(T_1) \iff \rho(T_1, T_2) = p$. It is clear that $\rho(\cdot, \cdot)$ is a metric function and thus $\mathcal{S}_k$ is the metric space $(\mathcal{S}_k, \rho)$. The maximum distance p_{max} (a parameter) used in this step is set to $k/10$.

23.2.1.1 Shaking. The distance function ρ is used in our *Shaking* step.

Shaking-1. Repeat both a random deletion of an edge that belongs to the tree so that the resulting set is also a tree (i.e. is connected), followed by a random addition of an edge whose exactly one endpoint is in the current tree p times. In that way a tree T' that belongs to $\mathcal{N}_p(T)$ is obtained.

23.2.1.2 Local search. Note that $\mathcal{N}_1(T)$ corresponds to the neighborhood usually used in the local and tabu search methods for solving the k-CARD, and we use it in our *local search* step within the VNS-1 as well (Step (b) in Algorithm 23.1). In other words, if a tree T is given, we generate a new one $T' \in \mathcal{N}_1(T)$ by replacing one edge in the tree with one out of it while maintaining the tree condition. If $w(T') < w(T)$ a better solution is found ($T \leftarrow T'$) and the search continues. Otherwise, it stops and the local minimum T has been reached. We develop both the *first* and *best improvement* strategies. However, in Section 23.3, the results obtained by the *first improvement* are reported since it is faster and thus more useful for the VNS.

23.2.2 The VNS with the VND as a local search (VNS-2)

A deterministic change of neighborhoods within a local search leads to the so-called Variable Neighborhood Descent (VND) heuristic (see e.g. Hansen and

Mladenović (1999; 2001b; 2002)). It has been shown that the solution obtained by the basic VNS may be improved if the VND is used in its local search step (see e.g. Brimberg et al. (2000) and Hansen and Mladenović (2001b)). Here we propose such a heuristic for solving the k-CARD problem. In addition, we analyze several shaking strategies that all use the distance $\rho(T_1, T_2)$ introduced into the solution space S_k.

23.2.2.1 Shaking. In the shaking step of the VNS-1, a cardinality of the set of candidate solutions T' from $\mathcal{N}_p(T)$ is relatively large since both deletion and addition of edges are obtained at random. Such a solution T' has two disadvantages: $w(T')$ is large and local search lasts longer. We now suggest several variants where the leaves are removed at random, but the edges to be added are chosen by a partly deterministic strategy. In this way, a smaller subset of better quality solutions from $\mathcal{N}_p(T)$ is generated.

Let $T' = (V', E')$ be a tree so that $|V'| = k$ and $|E'| = k - 1$. A new edge (v', v''), $v' \in V'$, $v'' \in V \setminus V'$ may be added in one of the following ways (i.e. $v' \leftarrow V' \cup \{v'\}$, $E' \leftarrow E' \cup \{(v', v'')\}$).

Shaking-2. Choose a vertex v' from V' at random and compute the shortest edge (v', v'') where $v'' \in V \setminus V'$.

Shaking-3. Choose a vertex $v'' \in V \setminus V'$ and compute the shortest edge $e = (v'', v')$ so that $v' \in V'$.

Shaking-4. Choose a vertex $v' \in V'$ at random and find the set

$$E'' = \{(v', v'') | v'' \in V \setminus V'\}.$$

Rank edges of the set E'' in the nondecreasing order of their weights and choose one at random among $q\%$ shortest (the detailed description of this variant is given in Algorithm 23.2). For parameter q, we found in testing that $q = 50\%$ is a good choice for graphs with different densities, and we always use this value. Note that ranking of the edges (Step (d) in Algorithm 23.2) can be performed in the preprocessing.

In these three strategies new edges are selected among those with small weights. Therefore, we except that the resulting tree does not have very large weight.

In Figure 23.1, we illustrate the performance of four shaking procedures on a random graph with 1000 nodes and with $k = 300$ (in the next section we give more details on how the test problems are generated). It appears that for $\rho = 20$ the solutions obtained by *Shaking 1* and *Shaking 3* could be 25% and 5% above the incumbent, respectively. The influence of these different shaking strategies on the final solution (applied within the VNS) will be shown in the next section.

procedure VNS-2( )
for $t = 1$ **to** p
(*a*): Find the leaf vertex $v' \in V'$ and the vertex $v'' \in V'$ at random so that $(v', v'') \in E'$, then remove v' from V' and (v', v'') from E';
(*b*): Generate a vertex v' from V' at random;
(*c*): Find a set of edges E'' whose one endpoint is v' and another does not belong to V';
(*d*): Rank the edges of E'' in the nondecreasing order of their weights;
(*e*): Find a vertex v'', where (v', v'') is a random edge among q % (a parameter) smallest from E'';
(*f*): Set $V' \leftarrow V' \cup \{v''\}$ and $E' \leftarrow E' \cup \{(v', v'')\}$;
endfor

Algorithm 23.2 A low-neighborhood shaking.

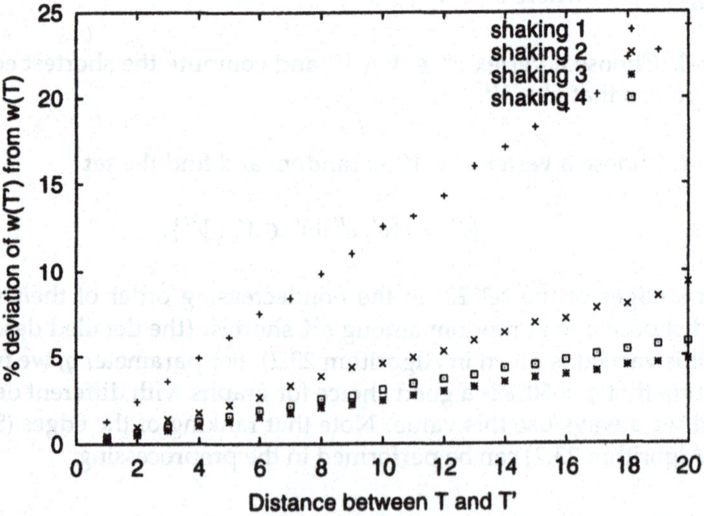

Figure 23.1 The average % deviations as a function of the distance to the incumbent.

23.2.2.2 Local search (VND). The steps of the basic VND are given in Algorithm 23.3. The idea is to also use of several neighborhoods in a local search phase.

To solve the *k*-CARD, we use four neighborhood structures, N_1, N_2, N_3 and N_4, i.e., the parameter p'_{max} is equal to 4. We denote by LS–1, LS–2, LS–3, and

procedure VND( )
Initialization: Select the set of neighborhood structures $N_p, p = 1, \ldots, p'_{max}$, that will be used in the descent; Find an initial solution T;
repeat
(1): Set $p \leftarrow 1$;
repeat
(2a) Exploration of neighborhood: Find the best neighbor T' of T ($T' \in N_p(T)$);
(2b) Move or not: If the solution T' thus obtained is better than T, set $T \leftarrow T'$ and $p \leftarrow 1$; Otherwise, set $p \leftarrow p + 1$;
until $p = p'_{max}$
until no further improvement is obtained

Algorithm 23.3 The steps of the basic VND for the k-CARD problem.

LS–4 the local search procedures that use N_1, N_2, N_3, and N_4 neighborhood structures, respectively.

Neighborhoods $N_1(T)$ and $N_2(T)$: Neighborhood $\mathcal{N}_1(T)$ may be split into two disjoint subsets

$$\mathcal{N}_1(T) = N_1(T) \cup N_2(T), \quad N_1(T) \cap N_2(T) = \emptyset,$$

such that $N_1(T)$ and $N_2(T)$ consist of trees whose new edge e has one or both endpoints in T, respectively. This splitting is based on the fact that the exploration of $N_2(T)$ takes longer than that of $N_1(T)$. If both endpoints of the added edge e belong to T, then the new tree is obtained after detecting a cycle and removing an edge from the cycle with the largest weight. This *elementary tree transformation* should be performed for each such edge e. On the other hand, adding and removing of an edge e that has one common vertex with T is much easier. Add an edge with the smallest (or very small) weight to T and remove the leaf with the largest weight from T. Note that the ranking of edges (from each vertex) is done in the preprocessing, which makes a local search within $N_1(T)$ very fast.

The neighborhood $N_2(T)$ contains trees T' whose vertex set is the same as the vertex set of tree T ($V_{T'} = V_T$). In this way, by a local search within $N_2(T)$ we may only move to the minimum spanning tree on a subgraph of G induced by V_T.

Neighborhoods $N_3(T)$ and $N_4(T)$: Since the minimal weighted k-cardinality tree is a spanning tree on any subgraph of G with $k+1$ vertices, the solution space S_k may be reduced to the set of all minimal spanning subtrees $\mathcal{T}_k \subseteq S_k$. Its cardinality is obviously $|\mathcal{T}_k| = n/(k+1)$. Moreover, the set of minimal spanning trees T_k may be supplied with another metric η as a cardinality of the difference of their vertex sets:

$$\eta(T_1, T_2) = |V_{T_1} \setminus V_{T_2}|.$$

This allows us to develop a new local search for the k-CARD problem based on *vertex interchange*. The new neighborhood $\mathcal{N}'(T)$ consists of all trees obtained by adding to and removing from one vertex of T.

Note that updating a minimal spanning tree after inserting new vertex has complexity $O(k)$. However, updating a minimal spanning tree after removing any vertex (if the removed vertex is not a leaf) has a higher complexity. Therefore, we can split the neighborhood $\mathcal{N}'(T)$ into two neighborhoods: $N_3(T)$ and $N_4(T)$. Neighborhood $N_3(T)$ consists of spanning trees T' which can be obtained from tree T by inserting a new vertex v_{in} ($T'' = T \cup \{v_{in}\}$) and removing a leaf vertex v_{out} from tree T'' ($T' = T'' \setminus \{v_{out}\}$). Neighborhood $N_4(T)$ consists of spanning trees T' which can be obtained from tree T by inserting a new vertex v_{in} and removing a vertex v_{out} that is not a leaf vertex in T''.

A local search with respect to (w.r.t.) neighborhood $N_3(T)$ (LS–3 for short) is given in Algorithm 23.4, where we assume that the initial tree $T = (V_T, E_T)$ is known.

> **procedure** LS–3( )
> **repeat**
>   **foreach** vertex v_{in} that is not in T but connected with T
>   (1): Add v_{in} to T and find a subgraph $G'' = (V_T \cup \{v_{in}\}, E_T \cup \{(v, v_{in}) | v \in T\})$;
>   (2): Find (update) the minimum spanning tree of G'' to get T'' ($|E_{T''}| = k + 1$);
>   (3): Remove a leaf e_{out} from T'' with the maximum weight and remove the corresponding vertex v_{out}, to get a new tree T': ($E_{T'} \leftarrow E_{T''} \setminus \{e_{out}\}$; $V_{T'} \leftarrow V_{T''} \setminus \{v_{out}\}$);
>   (4): If $w(T') < w(T)$, then $T \leftarrow T'$;
>   **endfor**
> **until** there is an improvement in $w(T)$

Algorithm 23.4 A local search w.r.t. $N_3(T)$ (LS–3).

A local search w.r.t. $N_4(T)$ neighborhood structure (LS–4 for short) differs from the local search w.r.t. $N_3(T)$ in step *(3)* of Algorithm 23.4. Instead of removing a leaf in T'', we try to drop each possible vertex v_{out} from T''.

Assume that the degree of $v_{out} \in T$ is equal to m. Then, by removing v_{out}, we get a forest with m trees $T_1, T_2, ..., T_m$. Let us consider T_i, $i = 1, ..., m$, as a super–node of a graph G_s, i.e. let (v_i, v_j) be an edge from G_s such that $v_i \in T_i$ and $v_j \in T_j$ ($V_s = \{T_1, T_2, ..., T_m\}$, $E_s = \{(v_i, v_j) | v_i \in T_i, v_j \in T_j\}$). The weights on $G_s(V_s, E_s)$ are defined as

$$w_{ij}^{(s)} = \min_q \min_r \{w(v_q, v_r) | v_q \in T_i, v_r \in T_j\}.$$

To get the tree T' in step *(3)*, we can apply Kruskal's algorithm to graph G_s.

However, the calculation of weights for all spanning trees that belong to $N_4(T)$ can be performed in a more efficient way (Das and Loui, 2001). Let e be an edge that does not belong to the tree, but whose endpoints v_1 and v_l are in the tree and let $(v_1, v_2, ..., v_l)$ be a cycle such obtained. For each vertex $v \in \{v_2, v_3, ..., v_{l-1}\}$, the edge e connects two super–nodes T_1^v and T_2^v, produced after removing vertex v from the tree. T_1^v and T_2^v may already be connected by any previously analyzed edge and in this case we simply ignore edge e. If T_1^v and T_2^v are not connected, then we connect them with edge e. In this way we form (update) the minimal spanning tree over the set $V_{T'} \setminus \{v\}$ for all $v \in V_{T'}$.

Some advantages can be reached by reducing all neighborhood solutions from $N_4(T)$ to some subset of it. To speed up the local search, we consider the subset of $N_4(T)$ that consists of the edges whose weights are smaller than $2w_L$, where w_L denotes the largest edge in the current tree, i.e. we expect that the new edge with weight larger than $2w_L$ cannot improve the current solution.

23.3 THE COMPUTATIONAL RESULTS

In this section, we compare our new heuristics with two tabu search methods (TS-1 and TS-2, for short) and with the Dynamic-Dijkstra-Path constructive heuristic (DDP for short). TS-1 represents our implementation of a chain interchange of TS rules suggested in Mladenović et al. (1996) and the second TS (denoted by TS-2) is taken from Jörnsten and Løkketangen (1997).

All methods, except TS-2 (whose executable version was kindly provided to us by the authors) have been coded in C++ and run on a Pentium II (450MHz, 64Mb) computer. The test instances include random graphs with 500, 1000, 1500, and 2000 vertices, where the degree of each node is set to 20 (as suggested in Jörnsten and Løkketangen (1997)). The integer weights of edges are generated with a uniform distribution from the interval $[50, 500]$ for $n =$ 500, 1000, and 1500, while for $n = 2000$, the interval is $[100, 1000]$.

23.3.1 Local searches

We first investigate the differences between the solutions obtained by the local searches LS-1, LS-3, LS-4, and the VND on the same random instance with 1000 vertices and $k = 300$. Each local search procedure is restarted 1000 times from the same randomly generated initial solution: (i) choose a vertex at random; (ii) take the vertex connected with the set of previously selected vertices at random and find the shortest corresponding edge; (iii) stop as soon as the cardinality condition is reached.

In Figure 23.2, each local minimum is represented as a point (x, y), where x is its distance from the best known solution (previously found by our VNS)

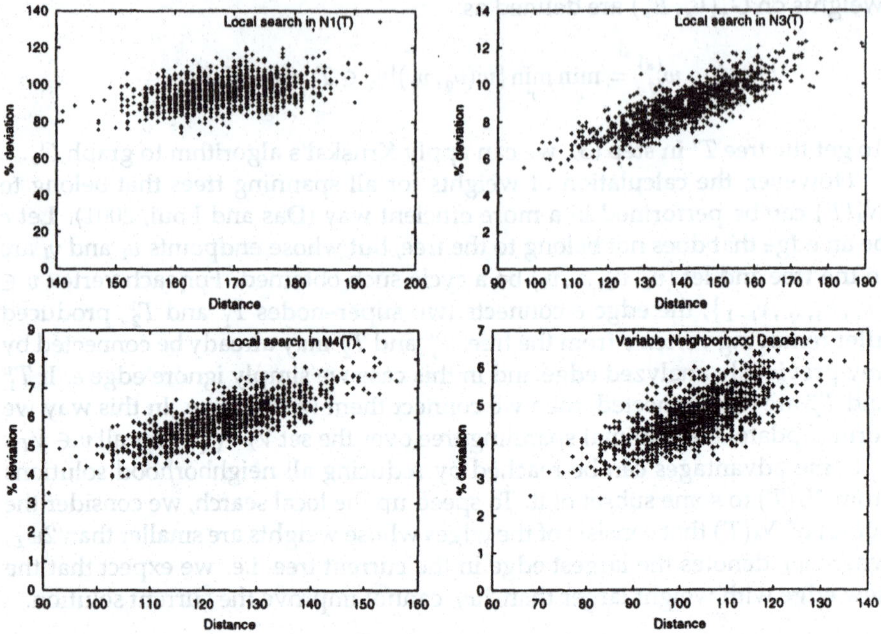

Figure 23.2 Multistart local searches LS–1, LS–3, LS–4 and VND

and y is the corresponding % error. Therefore the origin corresponds to the best known solution. Some results of this experiment are also given in Table 23.1, where the values obtained by using the modified Prim's algorithm for an initial solution are also included.

The results obtained with both initialization procedures have similar characteristics:

- Multi–start local searches are not able to reach the best known solution;
- The VND local search performs best in terms of solution quality; its average distance to the best known solution is the smallest;
- The average CPU time that the VND spends in the descent is less than that of LS–4 and even less than that of LS–3 for a random initial solution, despite the fact that the VND includes LS–3 and LS–4 in its list. This property of the VND has already been observed in solving other combinatorial optimization problems (see e.g. Hansen and Mladenović (2001a) and Ribeiro and Souza (2002)). Indeed, most of the descent steps within the VND, are performed by a very fast LS–1 procedure.

The major differences between two different initial solutions are:

- % errors for LS-1 are between 96% and 130% and between 4% and 6% for a random and Prim's initialization respectively;

- The number of different local minima for all neighborhoods is significantly reduced when Prim's initial solution is used;
- The standard deviation of objective values obtained by LS-3, LS-4 and the VND are smaller than the corresponding deviations obtained by a random initial solution, but the best solutions found are of similar qualities.

23.3.2 Shaking strategies

In Table 23.2, the average results on 10 random instances for different shaking strategies (proposed in Section 23.2) are reported.

Columns 1 and 2 of Table 23.2 contains problem parameters (n – the size of graph, and k – the cardinality of the tree). In column 3 the average results for the Dynamic Dijkstra Path heuristic are given. Columns 4–7 contain results for the VNS obtained by the corresponding shaking strategy. Only the average values on 10 different random instances (with the same n and k) are reported. It appears that the best performance (within VNS-2) uses the Shaking-3 strategy, and we use it in the computational analysis below.

23.3.3 Parameter settings

As indicated before, the maximum time allowed in the search (t_{max}) is chosen to be a stopping rule for all methods compared (except for the DDP, of course). When solving the instances with $n = 500$ and $n = 1000$, t_{max} was set to be $2 \cdot time(DDP)$. For $n = 1500$ and $n = 2000$, where DDP was not able to solve the k-CARD in a reasonable period of time, t_{max} values are indicated in the last column of Table 23.5.

Besides the maximum time allowed in the search (t_{max}), the VNS may have three more parameters: p_{min}, p_{max}, and p_{step}. To increase the user-friendliness property of our VNS, we try to reduce the number of parameters, i.e. we try to express them as functions of n and k.

We first set the parameter p_{max} to $\lfloor k/10 \rfloor$ (i.e. 10% of a tree size). We also set the parameters p_{min} and p_{step} to $\lfloor p_{max}/10 \rfloor$. For these parameter values and relatively small k (e.g. $k = \lfloor n/10 \rfloor$ or $\lfloor n/5 \rfloor$) we did not get good results. Indeed, for small k, the optimal subtree T^* and our initial solution T may be in two disjoint parts of graph G and no perturbation of T can be performed to reach the neighborhood of T^*, or the probability that local search solution T'' (applied to any perturbed solution) is close to the global optimum is very small. If the parameter k is large, then the distance between the initial solution and the global optimum is not necessary large and hence, the VNS has a chance to find T^*.

Then we try to increase the parameter p_{max}, i.e., we set $p_{max} = \lfloor k/2 \rfloor$ and $p_{max} = \lfloor 8k/10 \rfloor$. As expected, better results for small k are obtained, but not for large k. Based on the experimental observations above (see also Table 23.3) in all experiments bellow, we use $p_{max} = \min\{20, \lfloor 8k/10 \rfloor)\}$.

Table 23.1 The comparison of multistart local searches with two different initial solutions (1000 trials).

	Random initial solution				Prim's initial solution			
	LS-1	LS-3	LS-4	VND	LS1	LS-3	LS-4	VND
# Loc. minima	907	897	589	462	109	78	77	72
Av. CPU time	1.3	1.8	17.6	1.6	0.10	0.15	0.95	0.85
Av. % error	96.00	8.81	6.00	5.00	4.71	4.69	3.98	3.96
Best % error	73.00	5.37	3.00	2.00	3.79	4.04	3.09	2.17
Worst % error	130.00	12.74	9.00	7.00	5.70	5.57	4.54	4.61

Table 23.2 The average values on 10 random instances with different shaking strategies. The best values are boldfaced.

n	k	DDP	VNS–2			
			Shaking-1	*Shaking-2*	*Shaking-3*	*Shaking-4*
500	50	1535.90	1562.60	1546.40	**1525.40**	1528.30
	100	3397.50	3380.00	3379.50	**3367.50**	3372.90
	150	5541.50	5522.30	5527.00	**5514.10**	5515.60
	200	7980.70	7946.20	7944.20	7943.10	**7942.60**
	250	10829.40	10797.70	10795.50	10793.30	**10792.20**
1000	100	5847.20	5833.30	5831.70	**5830.30**	5830.60
	200	11922.00	11898.90	11896.50	**11891.40**	11892.50
	300	18222.60	18230.10	18219.80	**18215.40**	18216.80
	400	24763.90	24758.00	24751.70	**24748.30**	24749.10
	500	31591.90	31584.80	31583.10	31567.40	**31566.60**

Table 23.3 The average results on 10 random instances for different values of p_{max} ($p_{max} \in \{\lfloor k/10 \rfloor, \lfloor k/2 \rfloor, \lfloor 8k/10 \rfloor\}$). The best values are boldfaced.

		Objective			Time (sec.)		
n	k	$\lfloor k/10 \rfloor$	$\lfloor k/2 \rfloor$	$\lfloor 8k/10 \rfloor$	$\lfloor k/10 \rfloor$	$\lfloor k/2 \rfloor$	$\lfloor 8k/10 \rfloor$
500	50	1562.40	1547.10	**1540.70**	21.276	31.056	31.557
	100	3370.30	3363.90	**3339.80**	77.885	105.925	185.078
	150	5522.00	**5518.50**	5526.90	115.453	121.323	157.823
	200	7946.60	**7945.50**	7952.60	74.367	181.423	188.716

23.3.4 Comparison

In Tables 23.4 and 23.5 the average results on 10 random instances for each n and k are reported. Table 23.5 reports on the average computing time in seconds spent to get the best solution by each method, within a given maximum time limit t_{max}. To solve the instances with $n = 500$ and $n = 1000$ with VNS-1, VNS-2, and TS-1, t_{max} was set to be as twice as that of the DDP. The computing times of TS-2 are not reported since that code was not made to be efficient, but rather to visualize the various aspects of the search. For the same reasons on the larger test instances, we did not consider that code.

It appears that

- VNS-2 shows the best performance on average;
- The results obtained by VNS-1 are of a slightly better quality than the results obtained by the DDP. However, large problem instances could not be solved by the DDP in a reasonable period of time and therefore we did not report any in Tables 23.4 and 23.5.

23.4 CONCLUSION

In this paper, we propose two Variable Neighborhood Search (VNS) heuristics for solving the k-cardinality tree problem. In the first one (denoted by VNS-1), the basic VNS rules are applied. In the second (VNS-2), several neighborhood structures are also explored in the descent phase. The results obtained by our methods outperform those obtained by previous heuristics.

The future work could include the following:

- Develop new neighborhoods within the VND based on $(\mathcal{T}, \eta)$;
- Develop a variable neighborhood decomposition search (VNDS) and a parallel VNS for larger instances;
- Adapt the methods suggested here for solving the vertex k-CARD, as well as k-subgraph problems.

Acknowledgments

This paper is partly supported by the Serbian Ministry of Science, Project No. 1583. We thank Matthias Ehrgott and Arne Løkketangen for providing us with the codes for the k-CARD, written with their colleages.

Table 23.4 Average objective value results on 10 random instances. Best values are boldfaced.

n	*k*	VNS-1	VNS-2	TS-1	TS-2	DDP
500	50	1535.90	**1525.40**	1600.40	1713.70	1535.90
	100	3380.00	**3367.50**	3604.80	3736.40	3397.50
	150	5522.30	**5514.10**	5826.10	5915.80	5541.50
	200	7946.20	**7943.10**	8231.50	8626.80	7980.70
	250	10797.70	**10793.30**	11203.30	11349.10	10829.40
1000	100	5833.30	**5830.30**	5912.50		5847.20
	200	11898.90	**11891.40**	12079.60		11922.00
	300	18230.10	**18215.40**	18393.60		18222.60
	400	24758.00	**24748.30**	25002.30		24763.90
	500	31584.80	**31567.40**	31886.10		31591.90
1500	150	8197.40	**8183.20**	8260.90		
	300	16542.20	**16537.60**	16686.40		
	450	25081.10	**25039.80**	25293.04		
	600	33817.10	**33802.10**	34099.20		
	750	42820.10	**42785.40**	43118.30		
2000	200	23722.60	**23586.30**	23924.90		
	400	48188.40	**48077.80**	48867.90		
	600	73486.50	**73344.10**	74461.30		
	800	99670.40	**99535.80**	100950.40		
	1000	126938.20	**126804.70**	128311.60		

Table 23.5 Average running time (secs.) on 10 random instances.

n	k	VNS-1	VNS-2	TS-1	TS-2	DDP	t_{max}
500	50	14.38	29.79	39.40		38.97	78.
	100	38.88	47.53	48.24		46.66	93.
	150	29.23	50.93	75.39		58.11	117.
	200	59.21	75.91	113.11		73.53	148.
	250	67.79	54.23	72.39		84.51	170.
1000	100	192.85	255.12	426.61		287.86	585.
	200	348.40	275.86	647.56		327.23	655.
	300	622.77	425.71	656.78		387.62	775.
	400	661.82	183.14	895.43		466.85	930.
	500	744.84	447.71	920.34		568.98	1138.
1500	150	1419.05	1448.92	567.36			3000.
	300	1542.80	1936.50	1892.70			3500.
	450	2960.40	2884.51	2254.73			4000.
	600	4309.23	4096.55	3482.97			5000.
	750	4238.21	4143.40	2623.85			6000.
2000	200	1244.30	1920.67	1381.18			3000.
	400	3134.59	2615.00	1776.35			3500.
	600	3357.24	3589.81	1544.88			4000.
	800	3929.33	3868.16	2954.05			5000.
	1000	4130.51	4427.17	3088.95			6000.

Bibliography

J. Brimberg, P. Hansen, N. Mladenović, and É. Taillard. Improvements and comparison of heuristics for solving multisource weber problem. *Oper. Res.*, 48(3):444–460, 2000.

B. Das and M. C. Loui. Reconstructing a minimum spanning tree after deletion of any node. *Algorithmica*, 31(4):530–547, 2001.

M. Ehrgott, J. Freitag, H. Hamacher, and F. Maffioli. Heuristics for the k-cardinality tree and subgraph problems. *Asia-Pacific J. of Op. Res.*, 14:87–114, 1997.

M. Fischetti, H. Hamacher, K. Jornsten, and F. Maffioli. Weighted k-cardinality trees: Complexity and polyhedral structure. *Networks*, 24:11–21, 1994.

L.R. Foulds and H.W. Hamacher. A new integer programming approach to (restricted) facilities layout problems allowing flexible facility shapes. Technical Report 1992–3, University of Waikato, 1992.

F. Glover. Tabu search – Part I. *ORSA, J. Comput.*, 1:190–206, 1989.

H. Hamacher, K. Jornsten, and F. Maffioli. Weighted k-cardinality trees. Technical Report 91.023, Politecnico di Milano, 1991.

P. Hansen and N. Mladenović. An introduction to variable neighborhood search. In S. Voss et al., editor, *Meta-heuristics, Advances and Trends in Local Search Paradigms for Optimization*, pages 433–458, Dordrecht, 1999. Kluwer Academic Publishers.

P. Hansen and N. Mladenović. J-means: A new local search heuristic for minimum sum–of-squares clustering. *Pattern Recognition*, 34:405–413, 2001a.

P. Hansen and N. Mladenović. Variable neighborhood search: Principles and applications. *European J. Oper. Res.*, 130(3):449–467, 2001b.

P. Hansen and N. Mladenović. Developments of variable neighborhood search. In C. Ribeiro and P. Hansen, editors, *Essays and Surveys in Metaheuristics*, pages 415–440, Dordrecht, 2002. Kluwer Academic Publishers.

K. Jörnsten and A. Løkketangen. Tabu search for weighted *k*-cardinality trees. *Asia-Pacific J. of Op. Res.*, 14(2):9–26, 1997.

J. B. Kruskal. On the shortest spanning subtree of a graph and the traveling salesman problem. *Proc. Am. Math. Soc.*, 7:48–50, 1956.

N. Mladenović, J.P. Moreno, and J. Moreno-Vega. A chain-interchange heuristic method. *Yugoslav J. Oper. Res.*, 6(1):45–54, 1996.

C. Ribeiro and C. Souza. Variable neighborhood descent for the degree-constrained minimum spanning tree problem. *Discrete Applied Mathematics*, 118:43–54, 2002.

Metaheuristics: Computer Decision-Making, pp. 501-522
Christine L. Mumford-Valenzuela, Janis Vick and Pearl Y. Wang

24 HEURISTICS FOR LARGE STRIP PACKING PROBLEMS WITH GUILLOTINE PATTERNS: AN EMPIRICAL STUDY

Christine L. Mumford-Valenzuela[1], Janis Vick[2] and Pearl Y. Wang[2]

[1]Department of Computer Science
Cardiff University CF10 3XQ, Wales, UK
christine@cs.cf.ac.uk

[2]Department of Computer Science
George Mason University
Fairfax, VA 22030-4444 USA
jvick@gmu.edu, pwang@cs.gmu.edu

Abstract: In this paper, we undertake an empirical study which examines the effectiveness of eight simple strip packing heuristics on data sets of different sizes with various characteristics and known optima. We restrict this initial study to techniques that produce guillotine patterns (also known as slicing floor plans) which are important industrially. Our chosen heuristics are simple to code, have very fast execution times, and provide a good starting point for our research. In particular, we examine the performance of the eight heuristics as the problems become larger, and demonstrate the effectiveness of a preprocessing routine that rotates some of the rectangles by 90 degrees before the heuristics are applied. We compare the heuristic results to those obtained by using a good genetic algorithm (GA) that also produces guillotine patterns. Our findings suggest that the GA is better on problems of up to about 200 rectangles, but thereafter certain of the heuristics become increasingly effective as the problem size becomes larger, producing better results much more quickly than the GA.
Keywords: Strip packing, Heuristics, Genetic algorithm.

24.1 INTRODUCTION

We undertake an empirical study in which we compare the performance of some well-known strip packing heuristics and a good genetic algorithm on a range of two-dimensional rectangular packing problems. We restrict our study to problems involving *guillotine patterns* which are produced using a series of vertical and horizontal edge–to–edge cuts. Many applications of two-dimensional cutting and packing in the glass, wood, and paper industries are restricted to guillotine patterns. Packing heuristics that produce these patterns are generally simple to code and fast to execute.

The packing problem under consideration involves the orthogonal placement of a set of rectangles into a rectangular strip (or bin) of given width and infinite height so that no rectangles are overlapping. The goal of this strip packing problem is to minimize the height of the packing. We compare the performance of our GA, adapted from Valenzuela and Wang (2000; 2002), with the performance of eight strip packing heuristics (Coffman Jr. et al., 1984; Baker and Schwarz, 1983; Sleator, 1980; Golan, 1981) on data sets of various sizes with a variety of characteristics.

These data sets were generated using a suite of data generation programs developed in Valenzuela and Wang (2001) that produce data sets with optimal guillotine packings of zero waste. The software allows a set of basic rectangles to be cut from a large enclosing rectangle of given dimensions, and options are available which allow the user to control various characteristics: the number of rectangles in the data set, the maximum and minimum height–to–width ratios of the rectangles, and the ratio of the area of the largest rectangle to that of the smallest rectangle.

Several recent comparative studies on strip packing (Hopper and Turton, 2001; Kröger, 1995; Hwang et al., 1994) have reported superior performances for metaheuristic algorithms over simple heuristic approaches. We note, however, that in most of these cases, the problem sizes are restricted to 100 rectangles or less. Our interests lie in examining a larger range of problem sizes and types.

24.2 SIMPLE HEURISTIC ALGORITHMS FOR GUILLOTINE PACKING

In this section, we review the heuristic algorithms used for our comparative study. The sets of rectangles to be packed may first be pre-sequenced by arranging them in order of non-increasing height or width. They are then placed in the bin, one at a time, in a deterministic fashion. Pre-sequencing is employed by four of the heuristic algorithms we use: those based on the *level oriented* approach introduced by Coffman Jr. et al. (1980). To avoid pre-sequencing, two similar approaches utilize shelves for packing the rectangles and were proposed by Baker and Schwarz (1983). A seventh algorithm was formulated by Sleator (1980) and splits the rectangular bin vertically into two sub-bins after packing some initial rectangles. This idea is further extended by Golan (1981) who repeatedly splits the bin into smaller sub-bins and packs the rectangles, sorted by decreasing width, into ever narrower bins as the algorithm progresses. All eight heuristic algorithms produce guillotine patterns. Other algorithms which produce non-guillotine patterns are reviewed in Coffman Jr. et al. (1984).

24.2.1 The level oriented algorithms

To implement a level oriented algorithm, the items are first pre-sequenced by non–increasing height, and then the packing is constructed as a series of *levels*, each rectangle being placed so that its bottom rests on one of these levels. The first level is simply the bottom of the bin. Each subsequent level is defined by a horizontal line drawn through the top of the tallest rectangle on the previous level.

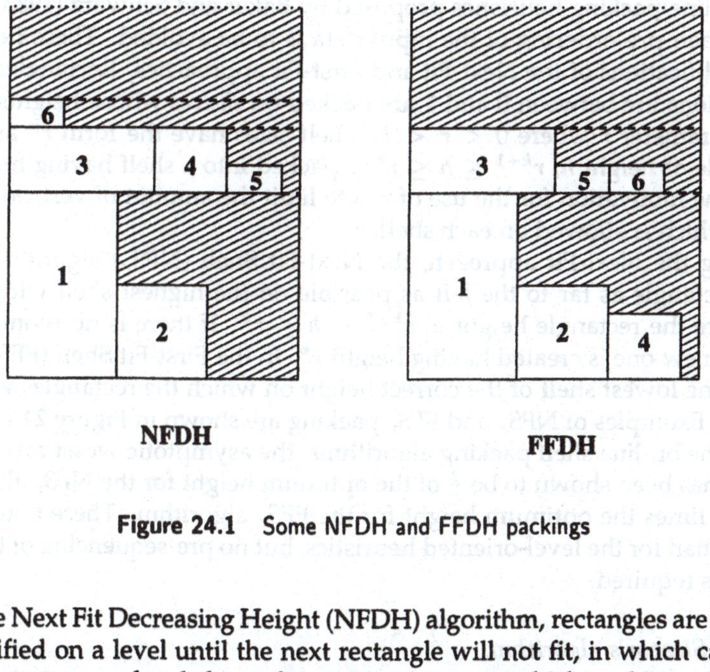

Figure 24.1 Some NFDH and FFDH packings

In the Next Fit Decreasing Height (NFDH) algorithm, rectangles are packed left justified on a level until the next rectangle will not fit, in which case it is used to start a new level above the previous one, on which packing proceeds. The run time complexity of NFDH (excluding the sort) is linear, just placing one rectangle after another in sequence. The First Fit Decreasing Height (FFDH) algorithm places each rectangle left justified on the first (i.e. lowest) level in which it will fit. If none of the current levels has room, then a new level is started.

The FFDH heuristic can be easily modified into a Best Fit Decreasing Height (BFDH) and a Worst Fit Decreasing Height (WFDH) heuristic. BFDH packs each rectangle onto the best level (i.e. the one with least remaining horizontal space) on which it will fit. WFDH packs the rectangle into the largest remaining horizontal space where it will fit. These four approaches are two-dimensional analogues of classical one-dimensional packing heuristics. The run time complexity for FFDH, BFDH and WFDH is $O(n \lg n)$ where n is the number of rectangles being packed. Figure 24.1 illustrates the NFDH and FFDH packings for the same six rectangles.

For the level packing algorithms, the asymptotic worst case performance has been shown to be twice the optimum height for the NFDH algorithm and 1.7 times the optimum height for the FFDH algorithm. The asymptotic worst case performance bounds for BFDH and WFDH are, respectively, at least twice and 1.7 times the optimum height. These are the asymptotic bounds for their one-dimensional analogues and easily serve as lower bounds on their worst case performance in two dimensions as level-oriented algorithms.

24.2.2 The shelf algorithms

The on-line packing heuristics proposed by Baker and Schwarz (1983) can be used when pre-ordering of the input data is to be avoided. These heuristics are modifications of the Next-Fit and First-Fit approaches. Rectangles whose dimensions are between 0 and 1 are packed on *shelves* whose heights are set by a parameter r where $0 < r < 1$. Shelf sizes have the form r^k and each rectangle of height h, $r^{k+1} \leq h \leq r^k$ is packed into a shelf having height r^k. Thus, the motivation for the use of r is to limit the amount of vertical wasted space which is allowed on each shelf.

Using the Next-Fit approach, the Next-Fit Shelf (NFS_r) algorithm packs each rectangle as far to the left as possible on the highest shelf with height r^k where the rectangle height h, $r^{k+1} \leq h \leq r^k$. If there is no room on this shelf, a new one is created having height r^k. In the First-Fit Shelf (FFS_r) algorithm, the lowest shelf of the correct height on which the rectangles will fit is chosen. Examples of NFS_r and FFS_r packing are shown in Figure 24.2.

For the on-line shelf packing algorithms, the asymptotic worst case performance has been shown to be $\frac{2}{r}$ of the optimum height for the NFS_r algorithm and $\frac{1.7}{r}$ times the optimum height for the FFS_r algorithm. These bounds are higher than for the level-oriented heuristics, but no pre-sequencing of the rectangles is required.

24.2.3 Sleator's algorithm

Unlike the shelf heuristics, the heuristic proposed by Sleator (1980) is not an on-line algorithm. It also packs rectangles having width at most one. Initially, the rectangles with width greater than $\frac{1}{2}$ are stacked vertically beginning at the bottom left corner of the bin. A horizontal line (h_0) is then drawn through the top of the last packed rectangle. The area of the rectangles below this line is denoted by A_0.

Next, the remaining rectangles of the list are sorted by decreasing height. They are placed along the h_0 line until a rectangle is encountered which won't fit, or until no more rectangles remain in the list. In the first case, the bin is next split in half vertically into two open-ended bins with width exactly $\frac{1}{2}$. We let

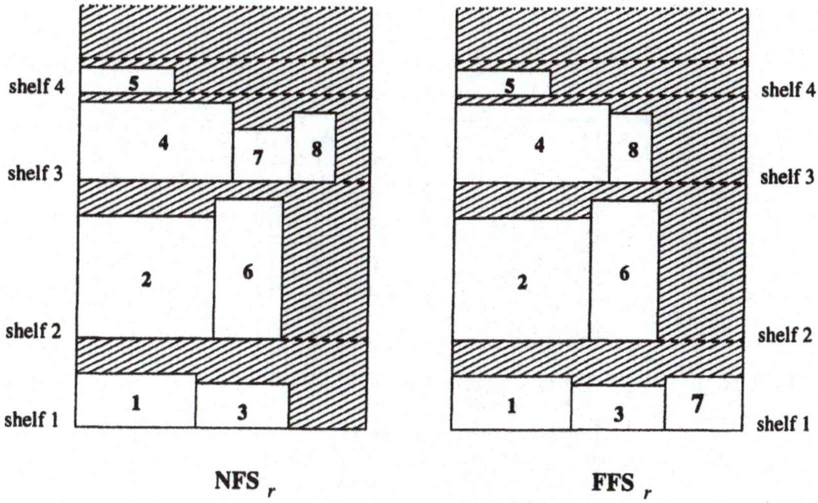

Figure 24.2 Two shelf packing algorithms

A_1 be the area of the rectangles in the left half-bin. All remaining packed and unpacked rectangles have area designated as A_2. Notice that one rectangle may be "split" in two, with part of it residing in the left half bin and the other part in the right half bin.

The remaining unpacked rectangles are packed into the two half-bins using a level approach: the half-bin with the current lowest level (defined by the tallest rectangle) is packed from left to right by placing rectangles along the level until a half-bin edge is reached. The first rectangle packed this way defines a new level in this half-bin. If it is lower than the current level in the other half-bin, then packing resumes on this level in this half-bin. Otherwise, packing proceeds using the current level of the other half-bin as the site where the remaining rectangles should be placed. An illustration of this process is shown in Figure 24.3.

Sleator (1980) was able to prove that the absolute worst case error bound for the packing height of this heuristic was 2.5 times the optimal height and that this is a tight bound.

24.2.4 *The split algorithm*

The Split algorithm is an extension of Sleator's approach. It packs the rectangles in order of non–increasing width. We can imagine that for each rectangle that is packed the original bin is split into two, and then into two again when the next rectangle is packed, and so on. As the rectangles to be packed are sequenced according to width, after packing some rectangles, those left to be packed are narrower and thus easy to fit into one of newly created bins. If possible, we pack rectangles side by side with previously packed rectangles.

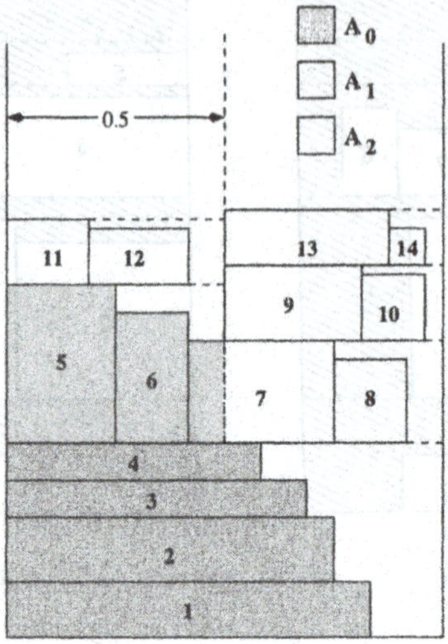

Figure 24.3 The Sleator packing algorithm

When this is not possible, we pack rectangles on top of previously packed rectangles. When bins are split, closed bins (M) are created below in which no further rectangles can be packed.

A brief example is shown above where the seven rectangles are to be packed. Rectangle 1 is packed first. Then rectangle 2 is placed to its right, splitting the open-ended bin into two sub-bins. (There is no M in this case.) Rectangle 3 is placed in the right sub-bin (it can't fit next to rectangle 1 in the left sub-bin), causing another split of the right sub-bin. When the fourth rectangle is packed, it will not fit to the right of the last rectangle packed in any sub-bin. Hence it is stacked on top of rectangle 2 in the middle sub-bin because that is the lowest position where a rectangle can be stacked in the three sub-bins. Similarly rectangles 5 and 6 will be stacked in the first and third sub-bins.

Subsequent to this, the heuristic checks the three sub-bins to see if rectangle 7 can be packed next to the last rectangle packed into each sub-bin. Rectangle 7 can be placed next to rectangle 6 in the third sub-bin but is too wide to be placed next to either rectangle 5 or 4 in the first and second sub-bins. Thus, the placement of rectangle 7 splits the third sub-bin into three parts: two narrower sub-bins and a lower rectangle. Finally, rectangle 8 could be placed to the right of either rectangle 5, 4, or 7. However, the second choice reflects the sub-bin with the lowest current level and so rectangle 8 is packed there which causes another split.

Figure 24.4 illustrates the splitting process that is encountered when packing a bin using this approach. The worst–case performance of the Split algo-

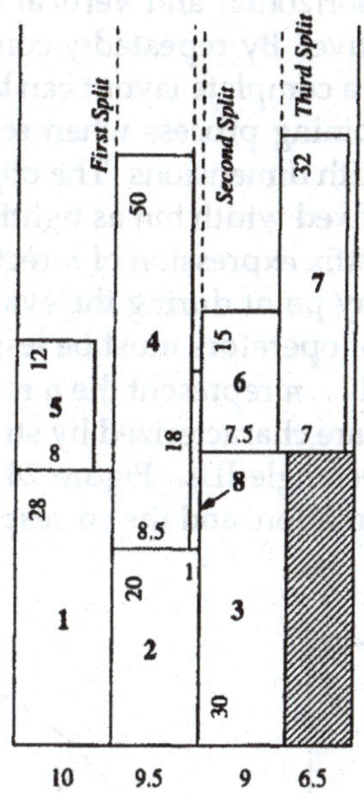

Figure 24.4 The Split algorithm packing process

rithm is three times the optimum height. Full details of the Split algorithm are given in Golan (1981).

24.3 OUR GENETIC ALGORITHM

Our genetic approach to solving packing and placement problems is based on a normalized postfix representation which offers a unique encoding for each guillotine pattern and covers the search space efficiently. Our postfix representation provides a blueprint for a recursive bottom-up construction of a packing or placement by combining rectangles together in pairs. The general technique that we use has proved effective for the VLSI floor planning problem (Valenzuela and Wang, 2000; 2002). However, that problem is more complex than the strip packing problem of the present study: although the individual rectangular components for VLSI placement have fixed areas, it is possible to vary their height and width dimensions to obtain closer packings. The GA that we developed for VLSI floor planning incorporates a sophisticated area optimization routine that is not needed for the present strip packing study.

24.3.1 The representation and decoder

The postfix expressions that we use to encode placements for our GA utilize the binary operators '+' and '*' which signal that one rectangle should be placed on top of the other (+), or by the side of the other (*). Alternatively, + and * represent horizontal and vertical cuts, when viewed from a top-down, cutting perspective. By repeatedly combining pairs of rectangles or sub-assemblies together, a complete layout can be generated. Wasted space will be created in the combining process when rectangles or sub-assemblies have different height or width dimensions. The objective of the GA is to pack the rectangles into a given fixed width bin as tightly as possible.

Note that a complete postfix expression of n rectangles will contain exactly $n - 1$ operators. Also, at any point during the evaluation of a postfix expression, the cumulative total of operators must be less than the cumulative total of rectangles. The integers $1 \ldots n$ represent the n rectangles in a problem. *Normalized* postfix expressions are characterized by strings of alternating * and + operators separating the rectangle IDs. Figure 24.5 illustrates a slicing floor plan, its slicing tree representation, and the corresponding normalized postfix expression.

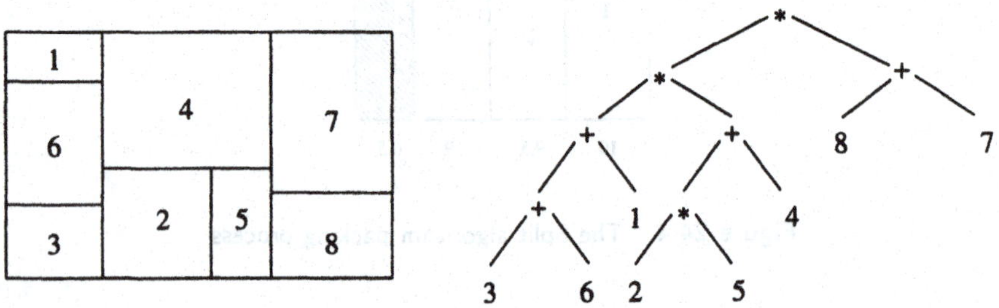

3 6 + 1 + 2 5 * 4 + * 8 7 + *

Figure 24.5 Slicing Trees and Normalized Postfix Strings

The representation used for our GA is order based and consists of an array of records, with one record for each of the basic rectangles of the data set. Each record contains three fields:

- *a rectangle ID field*: this identifies one of the basic rectangles from the set {1, 2, 3,..., n}
- *an op-type flag*: this boolean flag distinguishes two types of normalized postfix chains, + = + * + * + * + ... and * = * + * + * + * ...
- *a chain length field*: this field specifies the maximum length of the operator chain consecutive with the rectangle identified in the first field.

Starting with a given array of records produced by the GA, our decoder will construct a legal normalized postfix expression. It does this by transcribing the rectangular IDs given in the first field of each record in the sequence specified, while at the same time inserting chains of alternating operators following each rectangle ID, as specified in the second field of each record (i.e., either + * + * +... or * + * + *...). The length of each individual chain of alternating operators is determined by examining the third field of each record which specifies chain length.

The chain lengths specified in the records are adopted whenever possible. If such an adoption produces an illegal postfix expression or partial expression, however, the decoder steps in to correct it, either by adding operators to, or subtracting operators from, an offending chain. Below is an example showing an encoded string and its normalized postfix interpretation:

rectangle 5	rectangle 2	rectangle 4	rectangle 1	rectangle 3
op-type *	op-type +	op-type *	op-type *	op-type +
length 2	length 1	length 0	length 2	length 0

Postfix expression generated: 5 2 + 4 1 * + 3 +

24.3.2 *The genetic algorithm*

```
procedure GA( )
Generate N random permutations {N is the population size};
Evaluate the objective function (i.e. bin height)for each struc-
ture and store it;
Store best-so-far;
while stopping condition not satisfied
   foreach member of the population
      This individual becomes the first parent
      Select a second parent at random
      Apply crossover to produce offspring
      Apply a single mutation to the offspring
      Evaluate the objective function produced by offspring
      if offspring is a duplicate
         Delete the offspring;
      else
         if offspring better than weaker parent
            Offspring replaces parent in population;
         if offspring better than best-so-far
            Offspring replaces best-so-far;
      endif
   endfor
endwhile
print best-so-far;
```

Algorithm 24.1 A simple steady state genetic algorithm

Algorithm 24.1 gives the simple genetic algorithm (GA) which we use. It is an example of a steady state GA, and it uses the weaker parent replacement strategy first described in Cavicchio (1970). In this scheme, a newly generated offspring will replace the weaker of its two parents if it has a better fitness value than that parent. The current GA is similar to the GAs used in Valenzuela and Wang (2000; 2002), but differs in the following important ways:

- The area optimization routine designed for rectangles with fixed area but flexible height and width dimensions is not needed here because all rectangles have fixed dimensions (as mentioned earlier).
- Duplicates in the population are deleted.
- Selection based on fitness values has been abandoned in favor of simple uniform selection. The weaker parent replacement strategy seems to be sufficient to advance the genetic search, thus reducing the computational requirement of the GA.
- A rotation heuristic has been added which will rotate rectangles through 90 degrees when this is locally effective.

In the current GA population, duplicates are deleted as soon as they arise. To save computation time, we delete *phenotypic* duplicates i.e., new offspring are deleted if they duplicate a current population member's height (within sampling error). Ideally we should delete *genotypic* duplicates, but this would involve performing time consuming pairwise comparisons on the genetic array of records encoding the postfix expressions.

Each time the GA iterates through its inner loop, it selects two parents and applies *cycle crossover*, CX (Oliver et al., 1987) to the permutations of records to produce a single offspring. A single mutation is then applied which is selected from three alternatives M1, M2 or M3: M1 swaps the position of two rectangles, M2 switches the op-type flag from + to $*$ or vice versa, and M3 increments or decrements (with equal probability) the length field. The pairs of parents are selected in the following way: the first parent is selected deterministically in sequence, but the second parent is selected uniformly, at random.

For the present study, we have incorporated a rotation heuristic into our GA. This performs 90 degree rotations when it is locally effective during the construction of the layout from the postfix expression. This means that individual rectangles and sub-assemblies can be rotated when they are combined in pairs, if this results (locally) in a reduction of wasted space. For each pairwise combination of rectangles A and B, there are four alternative configurations: A combined with B, 'A rotated' combined with B, A combined with 'B rotated', and 'A rotated' combined with 'B rotated'. All four alternatives are tried each time a pair or rectangles or sub-assemblies are combined, and the best configuration selected.

The GA handles the fixed width constraint in the strip packing problem (i.e. we are packing rectangles into a bin of fixed width and infinite height) by rejecting packings that are too wide, and repeatedly generating new ones until the width constraint is satisfied.

Our current goal is to compare the performance of a good genetic algorithm with the classical strip packing heuristics on a range of data sets of different

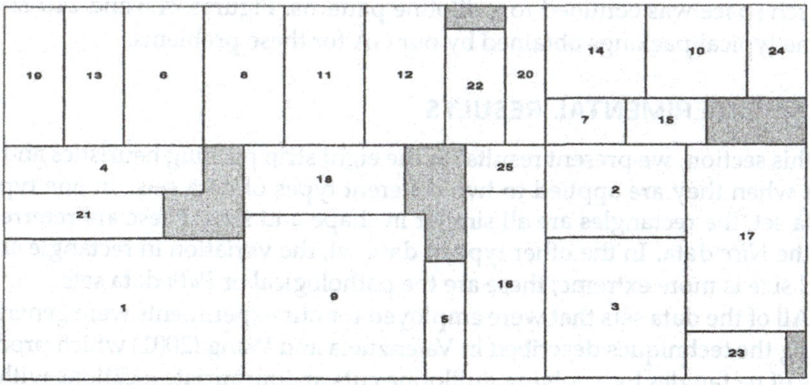

Figure 24.6 GA packing for the 25 rectangle problem from (Jakobs, 1996): width = 40, height = 16

sizes with different characteristics. We do not claim that our GA is the best in existence; we do not know whether it is or not. Evidence that our genetic packing techniques are good, however, was provided first by the results obtained for VLSI floor planning in Valenzuela and Wang (2000; 2002), and more directly for some small strip packing results: we matched the best solutions obtained by Liu and Teng in a recent paper (Liu and Teng, 1999).

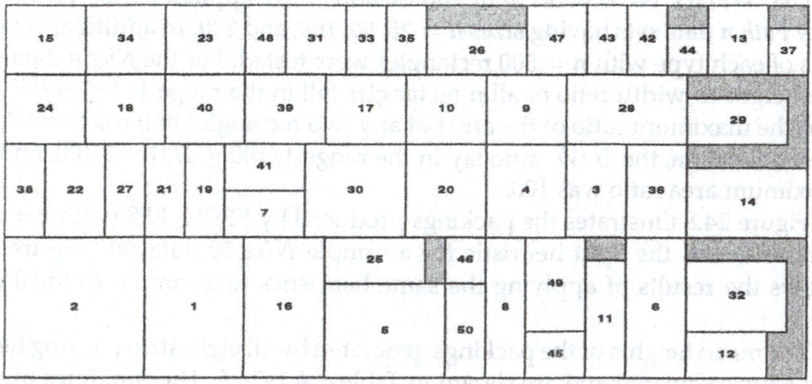

Figure 24.7 GA packings for the 50 rectangle problem from (Jakobs, 1996): width = 40, height = 16

For these initial packing experiments, the 25 and 50 rectangle problems from Jakobs (1996) were used. We were able to match the results of Liu and Teng, despite our use of a much more restrictive search engine: in Liu and Teng (1999), a non-guillotine Bottom Left placement heuristic is used; our search space was confined to guillotine patterns. Figures 24.6 and 24.7 shows some typical packings obtained by our GA for these problems.

24.4 EXPERIMENTAL RESULTS

In this section, we present results for the eight strip packing heuristics and our GA when they are applied to two different types of data sets. In one type of data set, the rectangles are all similar in shape and size. These are referred to as the *Nice* data. In the other type of data set, the variation in rectangle shape and size is more extreme; these are the pathological or *Path* data sets.

All of the data sets that were employed for our experiments were generated using the techniques described in Valenzuela and Wang (2001) which produce sets of rectangles by applying guillotine cuts at appropriate locations within a 100×100 rectangle. Data sets of size n are created by making $n - 1$ guillotine cuts, and each resulting data set can be packed into a strip of width 100 with an optimum height of 100 (and zero waste). Simple constraints may be applied during the cutting process to guarantee that the data sets have the different shape and size characteristics that were mentioned above. For the majority of our experiments, we used data sets with sizes $n = 25, 50, 100, 200$ and 500 for both the *Nice* and *Path* types, but in a final set of experiments, we applied only the most successful of the heuristics to some larger problems.

24.4.1 *Comparing the eight heuristics*

In the first set of experiments, the eight packing heuristics (NFDH, FFDH, BFDH, WFDH, NFS_r,FFS_r, Split and Sleator) were applied to fifty *Nice.n* and fifty *Path.n* data sets having sizes $n = 25, 50, 100,$ and 200. In addition, ten data sets of each type with $n = 500$ rectangles were tested. For the *Nice.n* data sets, the height-to-width ratio of all n rectangles fell in the range $1/4 \leq H/W \leq 4$, and the maximum ratio of the areas of any two rectangles in the set was 7. For the *Path.n* data, the H/W ratio lay in the range $1/100 \leq H/W \leq 100$, and the maximum area ratio was 100.

Figure 24.8 illustrates the packings produced by FFDH, FFS (with $r = 0.5$), Sleator's, and the Split heuristic for a sample *Nice*.50 data set. Figure 24.9 shows the results of applying the same heuristics to a sample *Path*.50 data set.

The mean heights of the packings generated by all eight strip packing heuristics were calculated and are shown in Tables 24.1–24.6. The run times of each of the eight algorithms took less than one second when applied to each data set on a Toshiba Tecra computer with a 1.2 GHz Pentium III processor and 256 MBytes RAM.

Tables 24.1 and 24.2 give mean packing heights for both *Nice* and *Path* data sets. Recall that the rectangles in a data set are reordered by decreasing height in the level algorithms and by decreasing width in the Split algorithm. Tables 24.3 and 24.4 show the mean packing heights when a preprocessing of the sets of rectangles is applied so that tall rectangles are rotated through

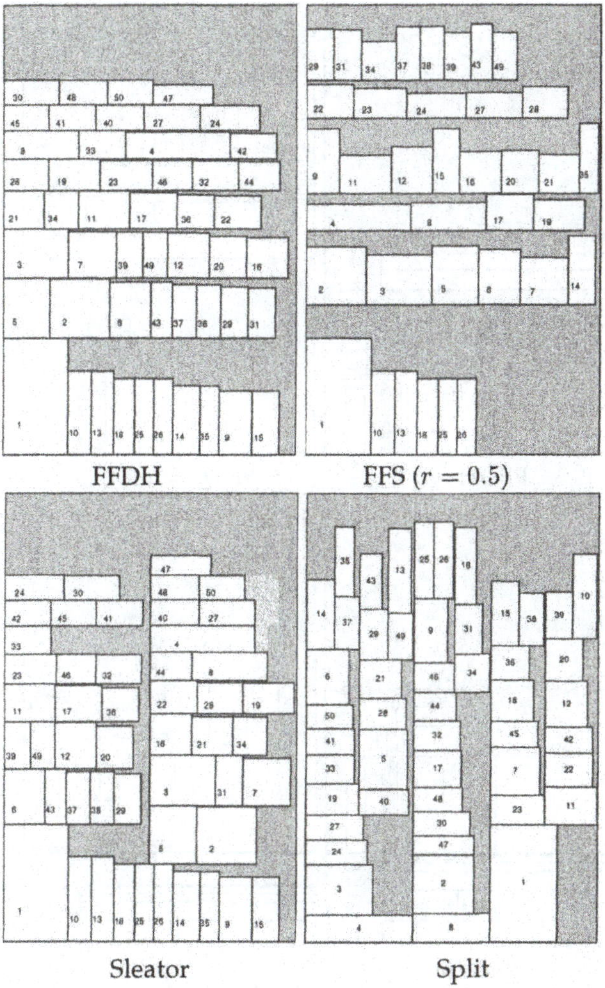

Figure 24.8 Packings of a $Nice.50$ data set

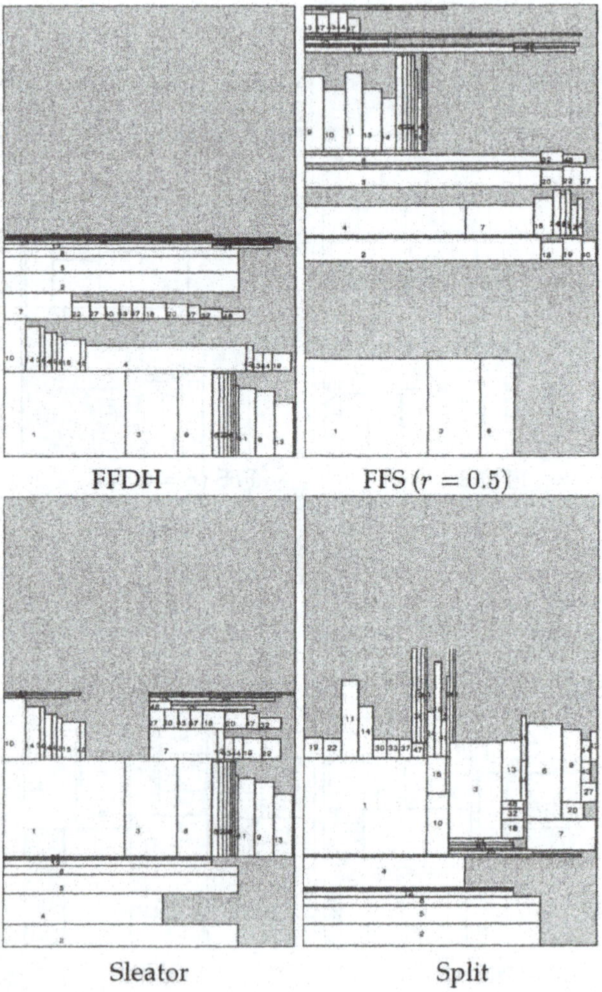

Figure 24.9 Packings of a $Path.50$ data set

Table 24.1 Mean packing heights for eight heuristics applied to *Nice.n* data sets. Optimum height is 100.

Heuristic	*Nice.25* (50 sets)	*Nice.50* (50 sets)	*Nice.100* (50 sets)	*Nice.200* (50 sets)	*Nice.500* (10 sets)
NFDH	134.2	125.9	120.4	115.0	109.0
FFDH	130.4	121.9	117.7	113.1	108.2
BFDH	130.4	121.9	117.7	113.1	108.2
WFDH	132.6	123.6	119.0	114.0	109.0
NFS 0.5	210.8	195.7	193.3	173.4	168.2
NFS 0.6	210.2	189.9	181.3	165.3	154.7
NFS 0.7	211.5	192.3	176.5	165.3	151.2
FFS 0.5	206.5	189.1	189.4	169.8	167.4
FFS 0.6	207.8	185.5	176.9	162.8	151.7
FFS 0.7	209.4	190.1	174.9	163.2	150.2
Split	141.8	140.4	138.1	138.6	138.0
Sleator	133.8	125.5	119.2	113.9	108.5

Table 24.2 Mean packing heights for eight heuristics applied to *Path.n* data sets. Optimum height is 100.

Heuristic	*Path.25* (50 sets)	*Path.50* (50 sets)	*Path.100* (50 sets)	*Path.200* (50 sets)	*Path.500* (10 sets)
NFDH	152.9	157.0	154.9	152.4	144.7
FFDH	149.4	149.9	149.6	147.8	142.0
BFDH	149.4	149.9	149.6	147.8	142.0
WFDH	150.4	152.3	151.6	148.8	142.8
NFS 0.5	223.1	241.3	250.7	255.8	246.7
NFS 0.6	241.9	266.8	268.4	269.1	280.0
NFS 0.7	290.5	310.6	316.3	308.8	294.0
FFS 0.5	219.4	238.7	245.8	252.6	242.0
FFS 0.6	240.7	265.9	266.1	266.1	276.5
FFS 0.7	289.7	309.4	314.9	307.8	293.2
Split	169.4	170.2	168.3	165.1	158.6
Sleator	145.7	139.4	137.9	134.3	125.4

Table 24.3 Mean packing heights for eight heuristics applied to preprocessed ($W \geq H$) *Nice.n* data sets. Optimum height is 100.

Heuristic	*Nice.25* (50 sets)	*Nice.50* (50 sets)	*Nice.100* (50 sets)	*Nice.200* (50 sets)	*Nice.500* (10 sets)
NFDH	126.1	120.5	114.9	110.9	105.4
FFDH	118.9	115.5	110.7	108.2	105.4
BFDH	118.9	115.5	110.7	108.3	105.4
WFDH	121.4	117.3	111.7	109.2	105.7
NFS 0.5	193.2	175.2	172.4	161.8	163.3
NFS 0.6	177.2	170.7	160.8	150.0	144.0
NFS 0.7	167.4	160.2	150.9	144.5	134.0
FFS 0.5	184.8	168.6	167.7	158.4	160.2
FFS 0.6	168.6	164.0	157.3	146.5	141.5
FFS 0.7	163.4	155.9	148.0	141.6	132.2
Split	138.3	137.1	137.4	138.2	138.9
Sleator	138.0	127.7	119.3	113.6	108.6

Table 24.4 Mean packing heights for eight heuristics applied to preprocessed ($W \geq H$) *Path.n* data sets. Optimum height is 100.

Heuristic	*Path.25* (50 sets)	*Path.50* (50 sets)	*Path.100* (50 sets)	*Path.200* (50 sets)	*Path.500* (10 sets)
NFDH	132.2	131.2	130.7	125.8	118.0
FFDH	121.4	120.6	118.1	115.3	110.6
BFDH	121.1	120.3	118.0	115.2	110.6
WFDH	123.0	117.3	120.8	117.5	112.6
NFS 0.5	193.7	197.5	199.9	191.3	174.0
NFS 0.6	180.9	183.8	184.0	179.2	159.2
NFS 0.7	174.2	183.7	179.0	172.6	166.7
FFS 0.5	187.2	188.5	187.6	180.0	166.0
FFS 0.6	177.5	177.7	174.0	170.5	153.0
FFS 0.7	172.5	181.0	173.8	167.7	159.9
Split	129.6	134.3	136.4	137.4	137.9
Sleator	131.3	134.8	133.7	130.0	122.8

90°, i.e. the rectangles are reoriented so that their width W is greater than or equal to their height H for each rectangle. Tables 24.5 and 24.6 show the mean packing heights when rectangles are first preprocessed by rotating wide rectangles through 90°. Following the preprocessing in the two later cases, the level-oriented, Sleator, and Split algorithms will again sort the rectangles by decreasing height or width as before.

Table 24.5 Mean packing heights for eight heuristics applied to preprocessed ($H \geq W$) *Nice.n* data sets. Optimum height is 100.

Heuristic	*Nice.25* (50 sets)	*Nice.50* (50 sets)	*Nice.100* (50 sets)	*Nice.200* (50 sets)	*Nice.500* (10 sets)
NFDH	131.5	124.7	118.9	113.5	109.2
FFDH	125.9	123.8	117.3	112.5	108.5
BFDH	125.9	123.8	117.3	112.5	108.5
WFDH	127.7	124.3	117.8	112.8	108.7
NFS 0.5	212.4	188.9	196.9	174.6	168.5
NFS 0.6	198.6	192.0	178.2	163.6	158.8
NFS 0.7	202.4	191.4	181.8	161.0	149.9
FFS 0.5	208.5	187.0	194.8	171.8	167.2
FFS 0.6	197.2	190.2	176.2	161.8	157.3
FFS 0.7	202.4	191.0	180.5	160.2	148.1
Split	137.4	136.6	136.5	137.0	138.6
Sleator	127.8	121.7	116.5	112.1	108.1

Table 24.6 Mean packing heights for eight heuristics applied to preprocessed ($H \geq W$) *Path.n* data sets. Optimum height is 100.

Heuristic	*Path.25* (50 sets)	*Path.50* (50 sets)	*Path.100* (50 sets)	*Path.200* (50 sets)	*Path.500* (10 sets)
NFDH	151.6	151.9	153.6	150.6	147.3
FFDH	149.1	149.3	150.3	148.9	146.5
BFDH	149.1	149.3	150.3	148.9	146.5
WFDH	149.5	149.7	151.0	149.3	146.8
NFS 0.5	209.5	226.6	241.1	261.6	251.0
NFS 0.6	240.1	259.2	252.8	264.5	272.0
NFS 0.7	300.9	313.8	304.1	301.8	301.8
FFS 0.5	209.5	224.6	239.6	260.9	251.1
FFS 0.6	240.1	259.2	252.2	264.4	270.1
FFS 0.7	300.9	313.8	304.1	301.8	301.8
Split	163.4	162.2	160.2	155.6	153.7
Sleator	143.1	135.0	131.7	126.8	124.1

From the tables, we can see that for the unrotated data sets, the NFDH, FFDH, BFDH and WFDH level heuristics and Sleator's heuristic tend to perform the best. Since the shelf heuristics are on-line algorithms, they are expected to do worse than the others. The shelf heuristics perform less effectively when there are larger variations of rectangle shapes in the pathological data sets as compared to the *Nice.n* sets.

It is also clear from examining Tables 24.3 and 24.4, that preprocessing the data by rotating the rectangles to ensure that $W \geq H$ will improve the results significantly for the level-oriented and shelf heuristics. Sleator's algorithm produces consistent solutions despite the changes in orientation for both types of data sets.

In addition, one can observe that the level-oriented, shelf, and Sleator's heuristics become more effective as the problem size increases, but the Split algorithm performs consistently throughout the range of problem instances tested. Overall, the best solutions to the problems were obtained when the rectangles were rotated so that $W \geq H$ and the FFDH or BFDH heuristic is used. They produced very similar packings.

24.4.2 Comparing the best heuristics with our GA

Table 24.7 compares the results of our GA to the best packings obtained by the simple heuristic algorithms for the test data sets. For each category of data, single runs of the GA were applied to each problem, and the means and standard deviations calculated. In each case, the GA was allowed to run until no improvement to the best solution had been observed for 100 generations. Using a population size of 1,000, the run times for the GA took only a few minutes for the smaller problems of up to 100 rectangles, but for the larger problems, run times were unpredictable because of our flexible stopping condition and could take several hours.

Parameters such as the population size and stopping criterion were arrived at after some initial experimentation with our GA, with the aim of producing good quality solutions in a reasonable amount of time. These settings could be viewed as somewhat arbitrary and better results can certainly be obtained if longer run times are used. However, in our view it was difficult to justify the vast computing resources that would be required to lift the solution by a small percentage when considering that the simple heuristics ran in a fraction of a second.

Each row of Table 24.7 compares the GA with the best performing heuristic for a particular category of problem. A Student's t-test reveals highly significant (0.1 %) differences between the mean packing heights of the GA and the best performing heuristic in all cases except for the *Path.200* data sets. The GA outperforms the heuristics for the smaller problems up to and including *Nice.100* and *Path.100* (although its runtime is very much longer), while the heuristics do better on the larger problems. The best performing heuristics are usually FFDH and BFDH, as mentioned previously.

Table 24.7 Comparison of GA with results from the eight heuristics on rotated ($W \geq H$) data sets. Optimum height is 100.

Data set type	*No. of sets*	*GA mean (std)*	*Heuristic mean (std)*	*Heuristic used*	*sig at* 0.1%
Nice.25	50	107.3 (1.6248)	118.9 (4.085)	F/BFDH	yes
Nice.50	50	107.8 (2.0806)	115.5 (2.9384)	F/BFDH	yes
Nice.100	50	108.6 (1.6111)	110.7 (2.1615)	F/BFDH	yes
Nice.200	50	111.3 (2.2485)	108.2 (1.1170)	FFDH	yes
Nice.500	10	120.8 (2.7234)	105.4 (0.5788)	N/F/BFDH	yes
Path.25	50	104.4 (1.6950)	121.1 (7.8641)	BFDH	yes
Path.50	50	108.5 (3.9577)	120.3 (6.6813)	BFDH	yes
Path.100	50	112.6 (7.4351)	118.0 (5.4290)	BFDH	yes
Path.200	50	116.7 (7.6316)	115.2 (4.0875)	BFDH	no
Path.500	10	120.8 (4.3667)	110.6 (1.6088)	F/BFDH	yes

24.4.3 Applying the heuristics to some large problems

To complete our study, the best performing heuristics (FFDH and BFDH) were applied to some very large data sets containing 1,000–5,000 rectangles. Ten data sets were generated for each size and type category. The resulting packing heights are presented in Table 24.8. The heuristics were each applied to the unrotated and rotated data sets ($W \geq H$ and $H \geq W$) as before.

Table 24.8 Results for heuristics on large data sets. Means are tabulated for 10 data sets in each category. Optimum height is 100.

Data set	*Unrotated*		$W \geq H$		$H \geq W$	
type	*FFDH*	*BFDH*	*FFDH*	*BFDH*	*FFDH*	*BFDH*
Nice.1000	106.1	106.1	104.2	104.2	105.8	105.8
Nice.2000	104.5	104.5	103.0	103.0	104.2	104.2
Nice.5000	103.5	103.5	101.9	101.9	103.4	103.4
Path.1000	141.9	141.9	109.4	109.4	141.3	141.3
Path.2000	134.4	134.4	107.2	107.2	135.2	135.2
Path.5000	130.5	130.5	105.1	105.1	131.9	131.9

It is clear from Table 24.8 that the results for the FFDH and BFDH heuristics continue to improve (i.e. get closer to the optimum of 100) as the data sets get larger. Furthermore, better packing results are observed when the data sets are first preprocessed so that $W \geq H$ for each rectangle, although the preprocessing appears to have much more effect on the *Path.n* data sets than it does on the *Nice.n* data sets. Preprocessing the data set to make $H \geq W$ for each rectangle produces results very similar to those obtained when the heuristics are applied directly to the unrotated data.

24.5 SUMMARY AND FUTURE WORK

In this paper, we have tested a number of well-known classical on-line and off-line strip packing heuristics on a range of data sets for which the optimum is known, and we have compared these results with those produced by a good genetic algorithm (GA). Although the GA found better solutions to these problems for the data set sizes up to about $n = 100$ rectangles, some of the classical strip packing heuristics, particularly Best Fit Decreasing Height and First Fit Decreasing Height, performed much better than the GA on these types of data sets as the size of the set increased. Furthermore, the simple heuristics run in a fraction of the time that it takes the GA. We also discovered that the performance of many of the simple heuristics could be enhanced if the data sets undergo a simple preprocessing routine involving the rotation of some of the rectangles by 90°.

Probabilistic studies have been performed, e.g. (Karp et al., 1984) which analyze the expected wasted space of some of these heuristic algorithms when n gets large. As an example, it is known (Coffman Jr. and Shor, 1993) that the asymptotic packing efficiency of BFDH approaches 100% for data sets whose rectangles have uniformly distributed heights and widths. Our experiments appear to confirm this predicted asymptotic behavior. A larger study is currently underway to determine if the individual characteristics of these *Nice.n* and *Path.n* data sets have any impact upon the observed performance of the strip heuristics in the average and worst cases. We also plan to experiment with techniques for seeding an initial population with packings produced using simple heuristics, and then applying our GA with a view to improving upon the heuristic output. In order to do this effectively, however, we may have to remove the locally applied rotation heuristic used by our current GA, as this will probably change the orientations of some of the rectangles packed by the simple strip packing heuristics and produce different results. We also need to investigate suitable methods for injecting the small number of seeded individuals that can be produced using the simple strip packing heuristics into a larger and varied population.

Acknowledgments

The research of Pearl Y. Wang was partially supported by the NASA Goddard Space Flight Center (NAG-5-9781).

Bibliography

B. S. Baker and J. S. Schwarz. Shelf algorithms for two-dimensional packing problems. *SIAM Journal of Computing*, 12(3):508–525, August 1983.

D. J. Cavicchio. *Adaptive Search Using Simulated Evolution*. PhD thesis, University of Michigan, Ann Arbor, 1970.

E. G. Coffman Jr., M. R. Garey, and D. S. Johnson. Approximation algorithms for bin packing – an updated survey. In G. Ausiello, N. Lucertini, and P. Serafini, editors, *Algorithm Design for Computer Systems Design*, pages 49–106. Springer-Verlag, Vienna, 1984.

E. G. Coffman Jr., M. R. Garey, D. S. Johnson, and R. E. Tarjan. Performance bounds for level-oriented two-dimensional packing algorithms. *SIAM Journal of Computing*, 9:808–826, 1980.

E. G. Coffman Jr. and P. W. Shor. Packings in two dimensions: Asymptotic average-case analysis of algorithms. *Algorithmica*, 9:253–277, 1993.

I. Golan. Performance bounds for orthogonal oriented two-dimensional packing algorithms. *SIAM Journal of Computing*, 10(3):571–581, August 1981.

E. Hopper and B. C. H. Turton. An empirical investigation of meta-heuristic and heuristic algorithms for a 2d packing problem. *European Journal of Operational Research*, 128:34–57, 2001.

S. M. Hwang, C. Y. Kao, and J. T. Horng. On solving rectangle bin packing problems using genetic algorithms. In *Proceedings of the 1994 IEEE International Conference on Systems, Man and Cybernetics*, pages 1583–1590, 1994.

S. Jakobs. On genetic algorithms for the packing of polygons. *European Journal of Operational Research*, 88:165–181, 1996.

R. M. Karp, M. Luby, and A. Marchetti-Spaccamela. Probabilistic analysis of multi-dimensional bin-packing problems. In *Proceedings of the 16th ACM Symposium on the Theory of Computing*, pages 289–298, 1984.

B. Kröger. Guillotineable bin packing: A genetic approach. *European Journal of Operational Research*, 84:645–661, 1995.

D. Liu and H. Teng. An improved BL-algorithm for genetic algorithm of the orthogonal packing of rectangles. *European Journal of Operational Research*, 112:413–420, 1999.

I. M. Oliver, D. J. Smith, and J. R. C. Holland. A study of permutation crossover operators on the travelling salesman problem. In *Genetic Algorithms and their Applications: Proceedings of the Second International Conference on Genetic Algorithms*, pages 224–230, 1987.

D.D.K. Sleator. A 2.5 times optimal algorithm for packing in two dimensions. *Information Procesing Letters*, 10(1):37–40, February 1980.

C. L. Valenzuela and P. Y. Wang. A Genetic Algorithm for VLSI Floorplanning. In *Parallel Problem Solving from Nature – PPSN VI*, Lecture Notes in Computer Science 1917, pages 671–680, 2000.

C. L. Valenzuela and P. Y. Wang. Data set generation for rectangular placement problems. *European Journal of Operational Research*, 134(2):150–163, 2001.

C. L. Valenzuela and P. Y. Wang. VLSI Placement and Area Optimization Using a Genetic Algorithm to Breed Normalized Postfix Expressions. *IEEE Transactions on Evolutionary Computation*, 6(4):390–401, 2002.

Metaheuristics: Computer Decision-Making, pp. 523-544
Alexander Nareyek

25 CHOOSING SEARCH HEURISTICS BY NON-STATIONARY REINFORCEMENT LEARNING

Alexander Nareyek[1]

[1]Computer Science Department
Carnegie Mellon University
5000 Forbes Avenue
Pittsburgh, PA 15213-3891, USA
alex@ai-center.com

Abstract: Search decisions are often made using heuristic methods because real-world applications can rarely be tackled without any heuristics. In many cases, multiple heuristics can potentially be chosen, and it is not clear a priori which would perform best. In this article, we propose a procedure that learns, during the search process, how to select promising heuristics. The learning is based on weight adaptation and can even switch between different heuristics during search. Different variants of the approach are evaluated within a constraint-programming environment.
Keywords: Non-stationary reinforcement learning, Optimization, Local search, Constraint programming.

25.1 INTRODUCTION

All kinds of search techniques include choice points at which decisions must be made between various alternatives. For example, in refinement search, an extension step from a variables' partial assignment toward a complete assignment must be chosen. In local search methods, it must be decided how a complete but suboptimal/infeasible assignment is to be changed toward an optimal/feasible assignment.

However, for large and complex real-world problems, decisions can rarely be made in an optimal way. Especially for local search techniques, this is a very critical issue because they do not normally incorporate backtracking mechanisms. Many different metaheuristic techniques have therefore been developed to handle the complications involved when choosing a decision alternative.

Figure 25.1 shows a choice point, representing the current state (i.e., the current variable assignment) of local search, and multiple decision alternatives, representing the so-called *neighbor states* that can be reached within an iteration.

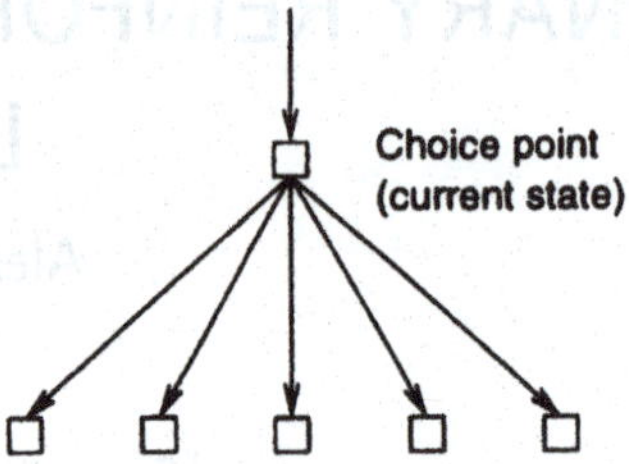

Figure 25.1 A decision point in local search.

Nearly all local search methods evaluate all neighbor states in a kind of look-ahead step to choose the most beneficial alternative. However, complex real-world problems – such as action planning including time, resources and optimization – often have utility functions whose computation requires a great deal of computing power. Analyzing large neighborhoods is mostly out of the question, and even smaller neighborhoods are difficult to check. Techniques like simulated annealing (Kirkpatrick et al., 1983) are highly suitable for these tasks because *only one neighbor is analyzed* for a choice decision (though more neighbors may be analyzed if the current neighbor appears to be unsuitable).

For the purposes of this paper, we go one step further, not analyzing any neighbor, but choosing a neighbor according to learned utility values. In addition, we do not choose a specific neighbor state but a *transformation heuristic* that will be applied to create the new state. Unlike other reinforcement learning approaches for learning which heuristics perform well, our approach allows the search to switch between different heuristics during search to adapt to specific regions of the search space.

Section 25.2 introduces the constraint-programming environment that is applied in our experiments, and details the use of heuristics. Weights and their adaptation are presented in Section 25.3. The scheme is evaluated in Section 25.4. Conclusions and related work are discussed in Section 25.5.

25.2 SEARCH DECISIONS

As an example of local search, we give a brief description of the search method applied in the **DragonBreath** engine. The underlying paradigm is presented in detail in Nareyek (2001b).

The problem is specified as a so-called constraint satisfaction problem (CSP). A CSP consists of

- a set of variables $x = \{x_1, \ldots, x_n\}$
- where each variable is associated with a domain $d_1, ..., d_n$
- and a set of constraints $c = \{c_1, ..., c_m\}$ over these variables.

The domains can be symbols as well as numbers, continuous or discrete (e.g., "door", "13", "6.5"). Constraints are relations between variables (e.g., "x_a is a friend of x_b", "$x_a < x_b \times x_c$") that restrict the possible value assignments. Constraint satisfaction is the search for a variable assignment that satisfies the given constraints. Constraint optimization requires an additional function that assigns a quality value to a solution and tries to find a solution that maximizes this value.

In our local search approach, a specific cost function is specified for every constraint (so-called *global constraints*), which returns a value that represents the constraint's current inconsistency/optimality with respect to the connected variables. For example, a simple Sum constraint with two variables a and b to be added and an s variable for the sum could specify its costs as $\text{Sum}_{costs} = |a + b - s|$. The total costs o_{now} of a current variable assignment (which is often also called *objective function value*) is a function of all constraints' costs, e.g., a simple addition.

In addition, a constraint has a number of heuristics to improve its cost function. For example, a heuristic for the Sum constraint could randomly choose one of the related variables and change it such that there are no more costs. Another heuristic might resolve the inconsistency by distributing the necessary change such that all variables are changed by the same (minimal) amount. The constraint must make the choice as to which heuristic to apply on its own.

On top of all constraints is a *global search control* which selects, in each iteration of local search, one of the constraints which is to perform a change, i.e., the transition to a neighbor state. Figure 25.2 shows the control flow.

The global search control possesses qualitative and quantitative information from the constraints' cost functions to decide which constraint to choose (e.g., the constraint with the maximal costs), but a constraint itself has little guidance as to which of its heuristics to choose. This choice point — for choosing one of the constraint's heuristics — is investigated below.

25.3 UTILITY WEIGHT

For a choice point, a *utility value* $\omega_a \geq 1$ is computed/maintained for every alternative a (an *alternative* stands for a *heuristic* here) that expresses the expected benefit of choosing this alternative.

The utility values are subject to learning schemes, which change the values based on past experiences with choosing this alternative. In many cases, an appropriate balance of the utility values will depend on the area of the search space that the search is currently in. We will therefore focus on schemes that dynamically adapt the weights *during* search and not only after a complete run.

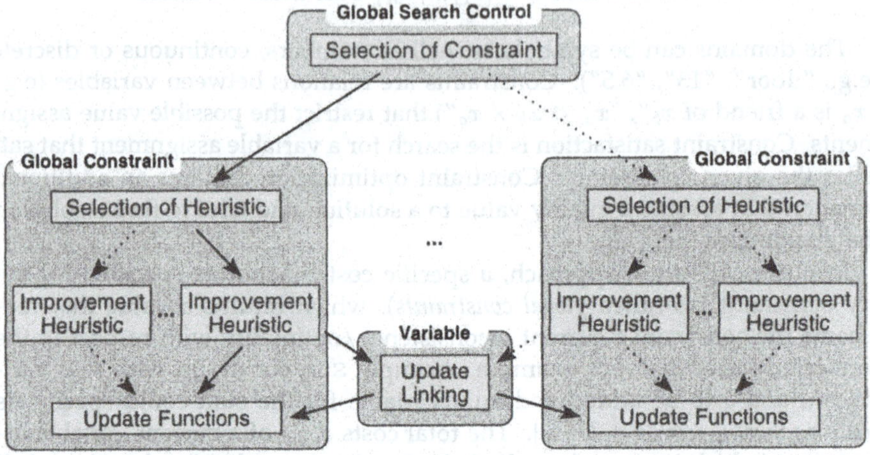

Figure 25.2 Using global constraints for local search.

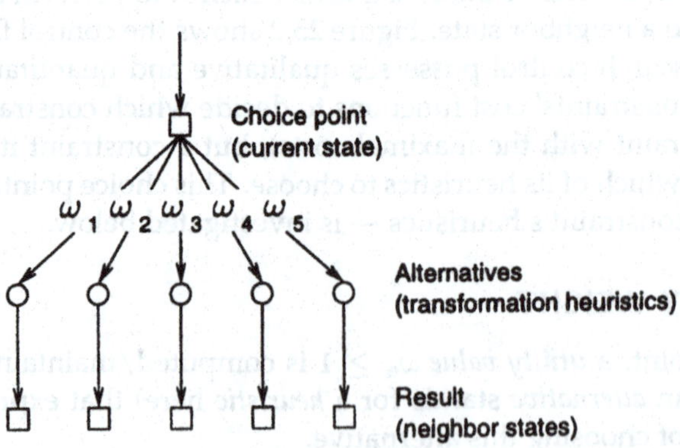

Figure 25.3 A decision point in our approach.

25.3.1 Selection function

Selection between possible alternatives is done based on the alternative's utility values. Here, we look at two simple ones. The first is a *fair random choice* (often called a *softmax* kind of selection rule), referred to below as `M:0`, which selects an alternative a from the choice point's alternatives $\mathcal{A}$ with a choice probability p_a in proportion to the alternative's utility value ω_a:

$$\texttt{M:0}: \quad p_a = \frac{\omega_a}{\sum_{i \in \mathcal{A}} \omega_i}$$

Another possibility is to make a random choice between the alternatives with *maximal* utility values:

$$\texttt{M:1}: \quad p_a = \frac{\begin{cases} 0 & : \exists i \in \mathcal{A} : \omega_a < \omega_i \\ 1 & : \forall i \in \mathcal{A} : \omega_a \geq \omega_i \end{cases}}{\sum_{i \in \mathcal{A} \mid \forall j \in \mathcal{A} : \omega_i \geq \omega_j} 1}$$

25.3.2 Weight adaptation

All utility weights have integer domains and are initially set to 1. If the choice point is selected, the utility weight of the alternative is changed that was chosen by the choice point when it was called last time. The kind of change depends on the relation of the current cost function value o_{now} to the cost function value when the choice point was called last time o_{before} (i.e., if there is a positive or negative reinforcement). The update schemes below can be combined to give many different strategies, e.g., a simple `P:1-N:1` strategy.

o_{now} `better-than` o_{before}:
(positive reinforcement)

`P:1` (Additive adaptation): $\omega_a \leftarrow \omega_a + 1$

`P:2` (Escalating additive adaptation): $\omega_a \leftarrow \omega_a + m_{promotion}$

`P:3` (Multiplicative adaptation): $\omega_a \leftarrow \omega_a \times 2$

`P:4` (Escalating multiplicative adaptation): $\omega_a \leftarrow \omega_a \times m_{promotion}$

`P:5` (Power adaptation): $\omega_a \leftarrow \begin{cases} \omega_a \times \omega_a & : \omega_a > 1 \\ 2 & : \omega_a = 1 \end{cases}$

o_{now} `worse-than-or-equal-to` o_{before}:
(negative reinforcement)

`N:1` (Subtractive adaptation): $\omega_a \leftarrow \omega_a - 1$

`N:2` (Escalating subtractive adaptation): $\omega_a \leftarrow \omega_a - m_{demotion}$

`N:3` (Divisional adaptation): $\omega_a \leftarrow \frac{\omega_a}{2}$

N:4 (Escalating divisional adaptation): $\omega_a \leftarrow \frac{\omega_a}{m_{demotion}}$

N:5 (Root adaptation): $\omega_a \leftarrow \sqrt{\omega_a}$

If a utility value falls below 1, it is reset to 1; if a utility value exceeds a certain max_ω, it is reset to max_ω; if a utility value is assigned a non-integer value, it is rounded down. In the case of an escalating adaptation, each time there is a consecutive improvement/deterioration, the $m_{promotion}/m_{demotion}$ value is doubled. Otherwise, it is reset to 1 (for P:2 and N:2) or 2 (for P:4 and N:4).

25.3.3 A weight-adaptation example

For the illustration of the weight-adaptation mechanism, we follow the development of the weights of a choice point's six heuristics below. The heuristics A to F change the values of the problem definition's variables in different ways, e.g., by additions and subtractions. The chosen weight-adaptation strategy is P:1-N:2 with a fair random choice M:0. The current search situation is to be

ω_A	ω_B	ω_C	ω_D	ω_E	ω_F
1	2	1	3	3	1

last choice: E $\quad o_{before} = 20$

$m_{demotion} = 1 \quad o_{now} = 12$

and one of the heuristics is to be selected for execution.

Entering this choice point, the heuristics' weights will be updated at first. The cost function value is now better (assuming that we want to minimize here) than the last time this decision had to be made, and the last choice was heuristic E. According to strategy P:1, the weight of heuristic E is rewarded by increasing it by one.

Next, a heuristic is to be selected for execution. The choice probability for the E option is the highest (weight value divided by the sum of all weight values; 4/12 = 33%), and we assume that this alternative is chosen by strategy M:0. Heuristic E is executed.

Other changes may follow, and after some iterations, our choice point might be called again. The situation is now:

ω_A	ω_B	ω_C	ω_D	ω_E	ω_F
1	2	1	3	4	1

last choice: E $\quad o_{before} = 12$

$m_{demotion} = 1 \quad o_{now} = 17$

The cost function value has deteriorated since the last time the choice point was entered, and — according to strategy N:2 — the last decision's weight value ω_E is decreased by the $m_{demotion}$ value of 1. The choice probability for heuristic E is now about 27% (3/11) and we assume that this alternative is chosen again.

After some time, our choice point is called once more:

ω_A	ω_B	ω_C	ω_D	ω_E	ω_F	last choice: E	$o_{before} = 17$
1	2	1	3	3	1	$m_{demotion} = 1$	$o_{now} = 17$

The cost function value is the same as at the last call. Stagnation is considered to be as bad as deterioration, and because this is a consecutive deterioration, the $m_{demotion}$ value is doubled. Heuristic E's weight value is therefore decreased by 2. We assume that heuristic D is chosen this time (probability of 3/9 = 33%) for execution.

At the next call of the choice point, the situation is:

ω_A	ω_B	ω_C	ω_D	ω_E	ω_F	last choice: D	$o_{before} = 17$
1	2	1	3	1	1	$m_{demotion} = 2$	$o_{now} = 16$

The cost function value has improved, and so, the weight value of heuristic D is increased by one, and the $m_{demotion}$ value is set back to 1.

25.3.4 *Invalid alternatives*

For some choice points, more than one alternative must be tested. For example, an alternative may turn out to be infeasible (i.e., the corresponding transformation is not applicable). An *applicability flag* f with a value of 0 or 1 is introduced for every alternative, indicating whether the alternative is still a valid option:

$$\omega_a \leftarrow f_a \times \omega_a$$

By the option of setting an applicability flag to 0, alternatives can often be ruled out a priori by simple feasibility tests.

However, in some cases, the infeasibility of an alternative will only become apparent during the state-transformation process of the chosen heuristic, i.e., after the choice has been made. In such a case, all changes in the current state that were made after the choice point are reversed, the corresponding applicability flag is set to 0 and the choice process is repeated. If no alternative remains applicable, the choice point's improvement fails.

If the choice point's selection is subject to the learning scheme, applicability flags are not set to 0 if an alternative fails. The failure may be caused by a bad random decision during the alternative's computations and the alternative may not be *fully* inapplicable. The learning process can handle this situation more appropriately than in a non-learning case, skipping the usual update of the utility weights and *temporarily* dividing the failed alternative's utility weight by two (though no weight may fall below one). If one of the alternatives has been successfully applied, all adaptations of the utility weights that were done for the restarts are undone.

25.4 EMPIRICAL EVALUATION

Two optimization problems — the Orc Quest problem and a modification of the Logistics Domain — are evaluated with different learning/selection schemes. The problems are only roughly described here because the actual problems are not important with respect to our following analysis. A detailed presentation of the problems can be found in Nareyek (2001a).

The Orc Quest problem's solving process involves only three constraints with six heuristics each. Each of these heuristics applies a specific set of additions and subtractions to the problem variables. There is a hierarchical cost function, demanding the minimization/maximization of specific problem variables. The learning scheme is applied to the choice points of all three constraints.

The Logistics Domain is a classical benchmark in the action-planning community. The problem investigated here is enriched so that actions have durations (more specifically, the duration minimization of Problem 6-1a is analyzed). The problem involves a large number of constraints, which are even created and deleted during the solving process. The learning scheme is applied to all constraints of the STATE RESOURCE type. This constraint type is responsible for projecting a specific state of the environment and has to ensure that all related preconditions of actions are fulfilled. Such a constraint can for example be responsible for the location of a truck, and must ensure that the truck is at the right location when a loading action is to take place. The constraint type includes five alternative heuristics, e.g., to create a new action, to temporally shift an action, and to delete an action.

25.4.1 *Results*

A strategy is denoted by `P-N-M`, $P \in \{1..5\}$ indicating the adaptation scheme that is applied in the case of an improvement, $N \in \{1..5\}$ the adaptation scheme for non-improvement, and $M \in \{0, 1\}$ if the fair random choice is applied or a maximal value is chosen.

The results for some strategies for the Orc Quest problem are shown in Figure 25.4 as the percentage of test runs (100 % = 100,000 test runs) that found the optimal solution after a specific number of iterations[1]. For example, in case of strategy `1-5-1`, about 98 % of the test runs found the optimal solution after 2,000 iterations while 2 % were still running. The strategies' curves are overlapping in some cases, which means that for these strategies, the strategy dominance is dependent on available computation time.

The problem from the Logistics Domain is much harder, so only the best solution (minimal duration) found after 100,000 iterations is shown in Figure 25.5 (100 % test runs = 1,000 test runs).

[1]While presenting results as so-called *run-time distributions* is not widespread in the Operations Research community, it addresses a number of serious issues related to result presentation and analysis (see Hoos and Stützle (1998)).

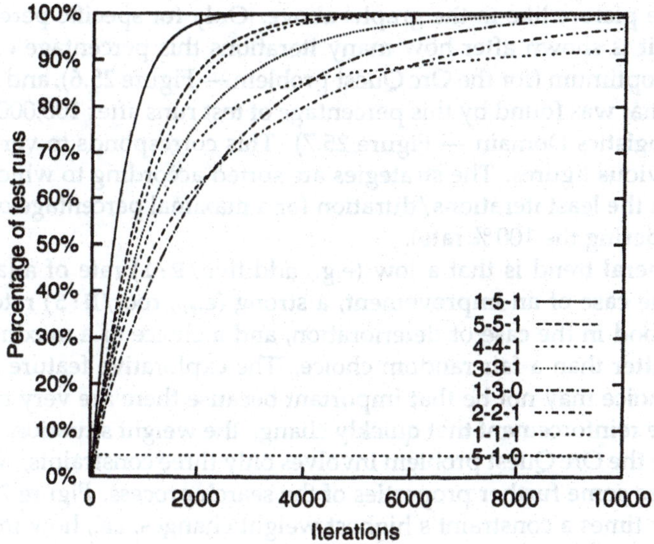

Figure 25.4 Sample strategies for the Orc Quest problem.

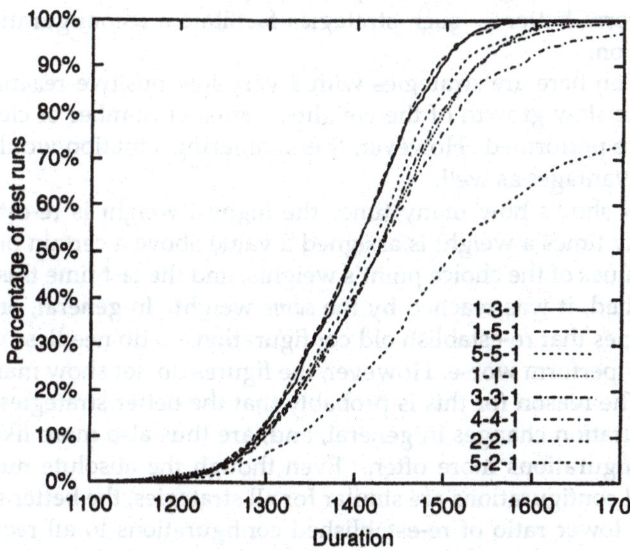

Figure 25.5 Sample strategies for the Logistics Domain.

The following figures will display more detailed results, i.e., for all possible strategy mixes. The amount of data to be shown does however not allow for a complete picture like in the graphs above. Only for specific percentages of test runs, it is shown after how many iterations this percentage of test runs found the optimum (for the Orc Quest problem — Figure 25.6), and the lowest duration that was found by this percentage of test runs after 100,000 iterations (for the Logistics Domain — Figure 25.7). This corresponds to vertical slices of the previous figures. The strategies are sorted according to which strategy resulted in the least iterations/duration for a maximal percentage of test runs (not considering the 100 % rate).

The general trend is that a low (e.g., additive/`P:1`) rate of adaptation is good in the case of an improvement, a strong (e.g., root/`N:5`) rate of adaptation is good in the case of deterioration, and a choice of a maximal weight is often better than a fair random choice. The explorative feature of the fair random choice may not be that important because there are very often cases of negative reinforcement that quickly change the weight situation.

Because the Orc Quest problem involves only three constraints, we can easily visualize some further properties of the search process. Figure 25.8 shows how many times a constraint's highest weight changes, i.e., how many times a weight is assigned a value above a certain percentage of the total values of the choice point's weights, and the last time this percentage threshold was reached, it was reached by another weight. Strategies that perform many changes in the configuration appear to perform better. This might be an indication of why strategies with a low rate of adaptation in the case of an improvement and a strong rate of adaptation in the case of a deterioration are likely to perform better — such strategies facilitate a reconfiguration of the weight situation.

An exception here are strategies with a very low positive reaction (`P:1`). Because of the slow growth of the weights, a smaller number of clear reconfigurations are performed. However, this simmering situation would appear to have its advantages as well.

Figure 25.9 shows how many times the highest weight is re-established, i.e., how many times a weight is assigned a value above a certain percentage of the total values of the choice point's weights, and the last time this percentage was reached, it was reached by the *same* weight. In general, one would expect strategies that re-establish old configurations to do needless work and thus, possibly perform worse. However, the figures do not show many differences here. The reason for this is probably that the better strategies perform many configuration changes in general, and, are thus also more likely to re-establish configurations more often. Even though the absolute numbers of re-established configurations are similar for all strategies, the better strategies have a much lower ratio of re-established configurations to all reconfigurations.

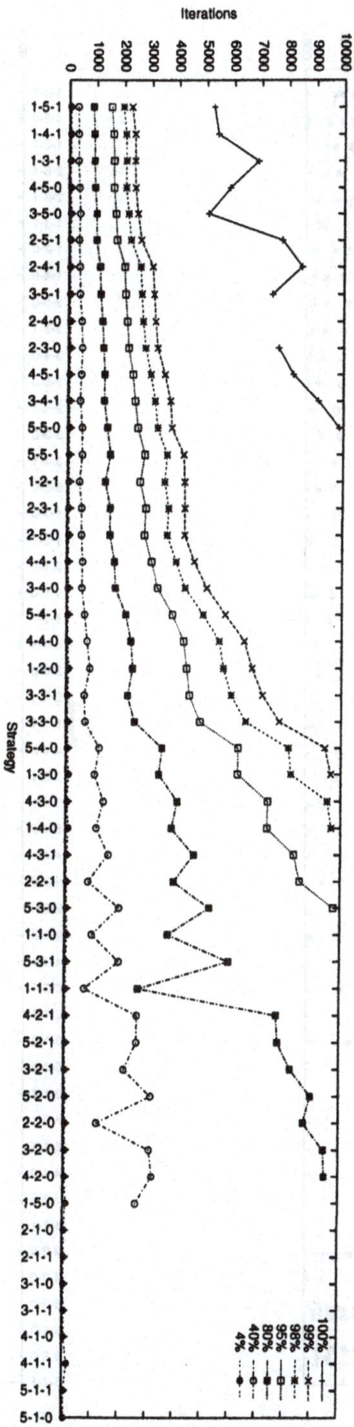

Figure 25.6 Weight-adaptation results for the Orc Quest problem.

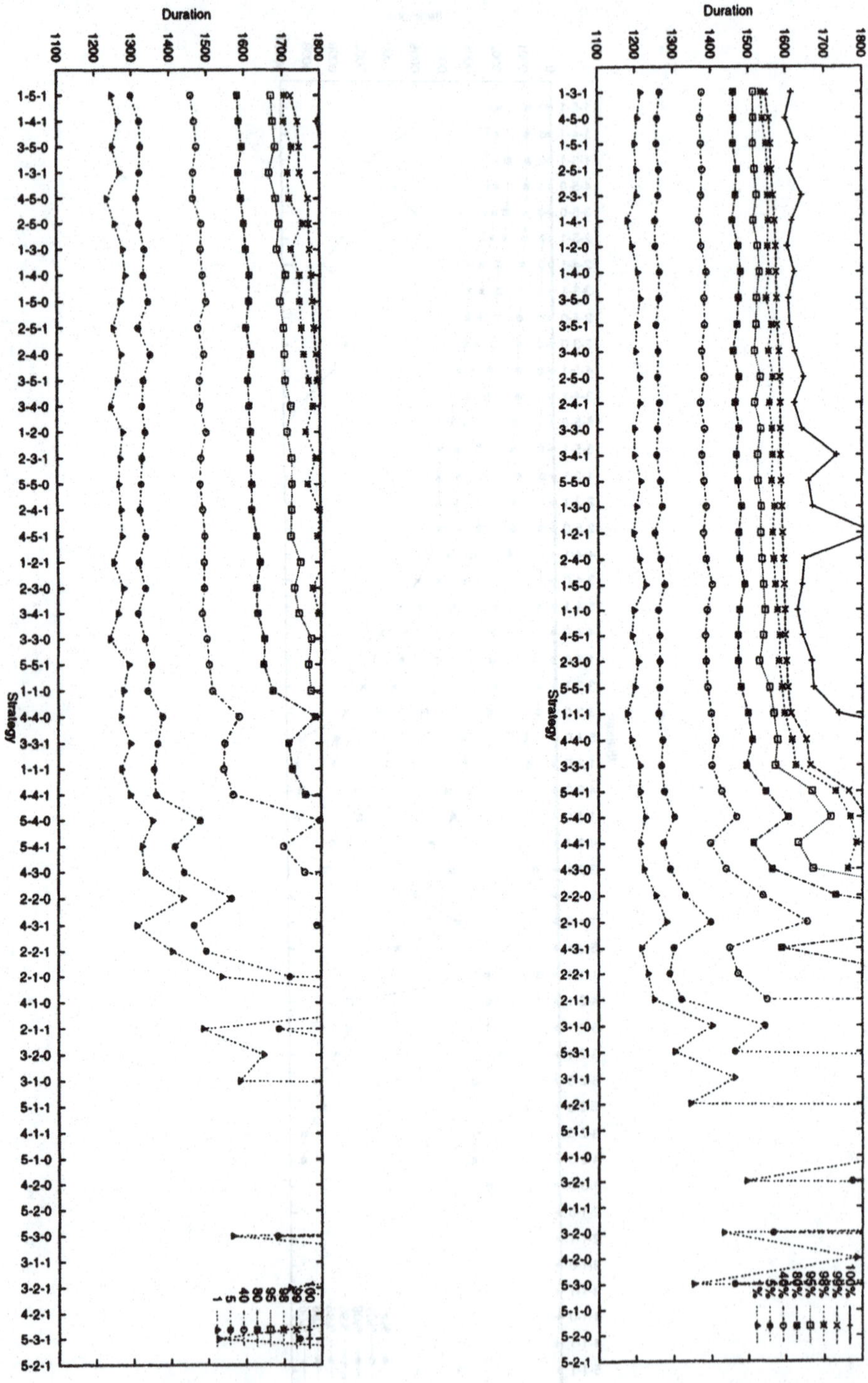

Figure 25.7 Weight-adaptation results for the Logistics Domain after 25,000 (left) and 100,000 (right) iterations.

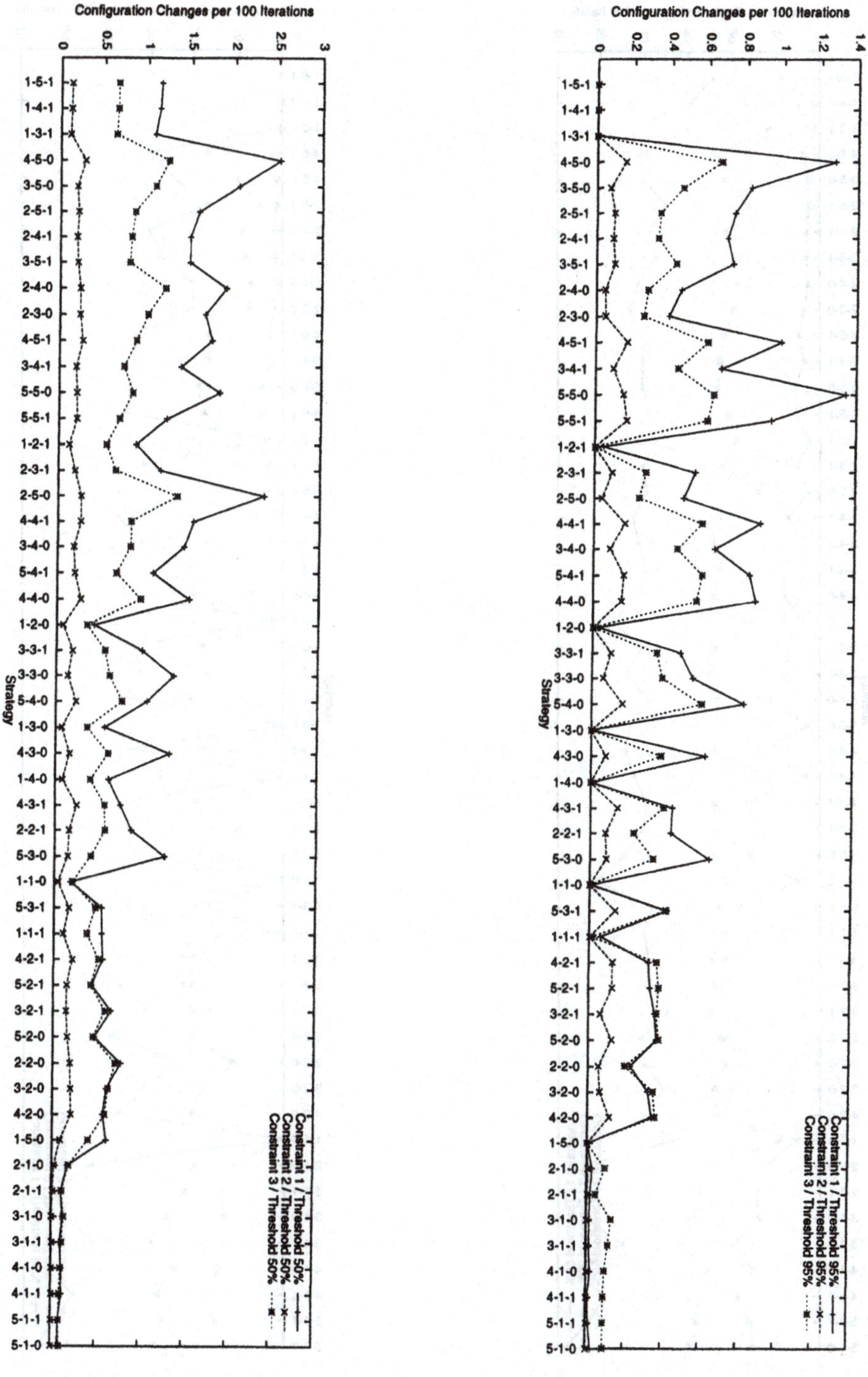

Figure 25.8 Number of configuration changes for the Orc Quest problem.

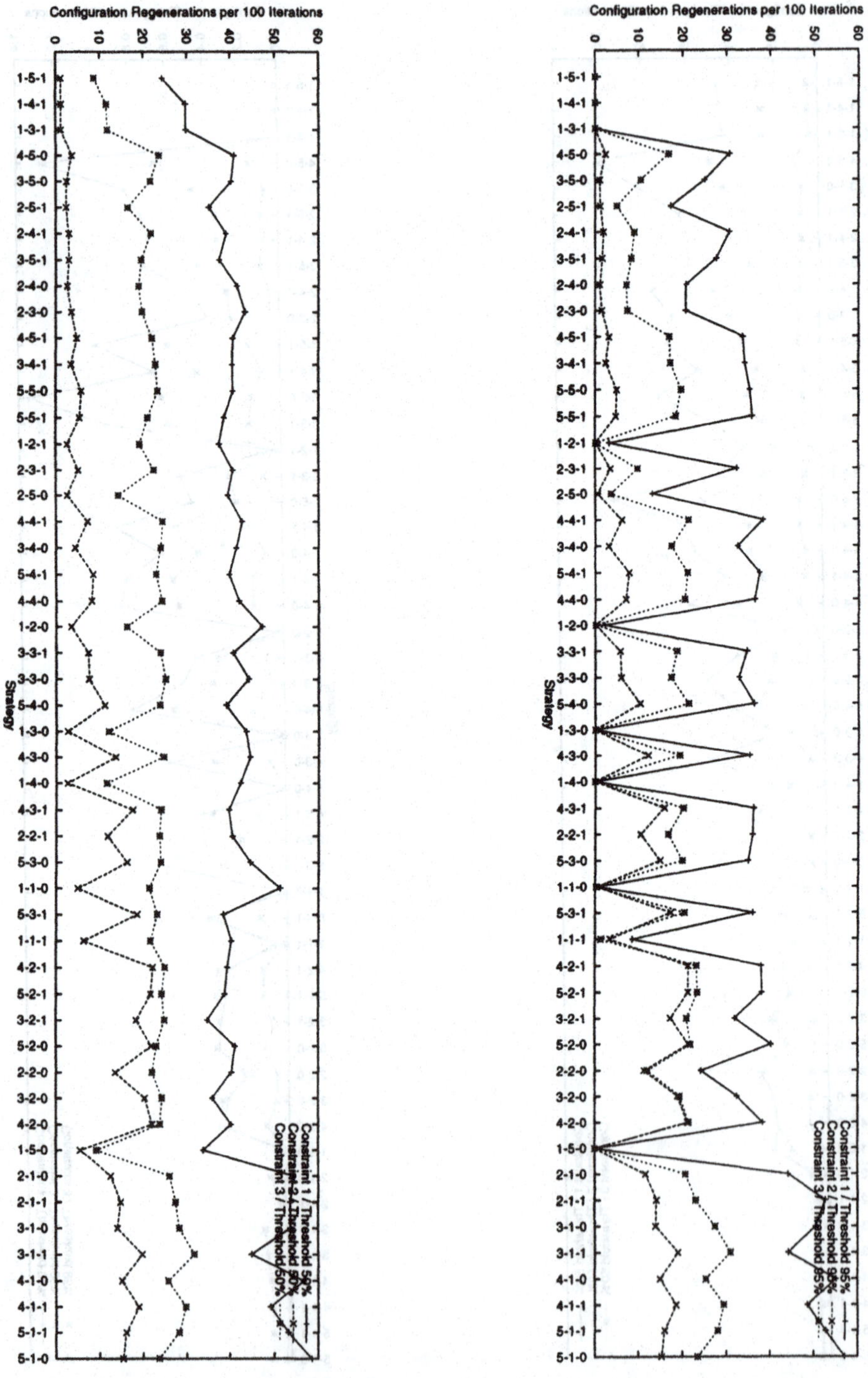

Figure 25.9 Number of configuration regenerations for the Orc Quest problem.

25.4.2 *Extended experiments*

Following the observed trend, we can extend our experiments by more extreme options in this direction:

`P:0` (No adaptation): $\omega_a \leftarrow \omega_a$
To enable weight increases for this option at all, in the case of a negative change, the decrease of ω_a is distributed as an increase to all $\omega_{i \neq a}$ (starting with high initial weights).

`N:6` (Total loss adaptation): $\omega_a \leftarrow 1$

Figure 25.10 shows that these options do not improve performance for the Orc Quest problem. However, as shown in Figure 25.11, strategies with an `N:6` option appear to work well for the early phase of search, i.e., for less constrained problems.

25.4.3 *Stationary reinforcement learning*

So far, we have looked at different methods to adapt the weights during search, assuming that different areas of the search space can be handled more efficiently using different search strategies. Although this assumption seems to be intuitively correct, it remains to be shown to be true. This section, then, compares adaptive non-stationary learning with stationary approaches.

Previous approaches adapted learning parameters *after a complete run* or *when a local minimum was reached*. Of these two options, only an adaptation after a complete run (with an upper bound of a specific number of iterations) is applicable here because we do not evaluate the whole neighborhood and cannot therefore tell if we are in a local minimum. The time taken by the learning process to find an optimal stationary weight distribution is not measured because the results may be very different for different learning techniques. Thus, the adaptive learning strategies are compared with the optimal stationary distribution.

For the Orc Quest problem, we can actually find an optimal static weight distribution such that all test runs find the optimum in about **35** iterations. With the most simple, non-stationary `1-1-0` strategy, 50 % of the test runs found the optimum after **1,305** iterations, and after **405** iterations for strategy `1-5-1`. Thus, an adaptive strategy would seem to perform very poorly for the simple Orc Quest problem. This is not completely true, however, given the time that would be required to learn the optimal stationary distribution. For example, using a simple static distribution such that every heuristic is chosen equally often, **none** of the 100,000 test runs found the optimum within 100,000 iterations. We conclude that, if the problem (or very "similar" problems) is solved very often, a stationary reinforcement learning approach will ultimately perform much better; but for a short time-frame, the non-stationary approach is probably much superior.

For the more complex Logistics Domain, our findings are different. The performance of even the most simple, non-stationary `1-1-0` strategy is sim-

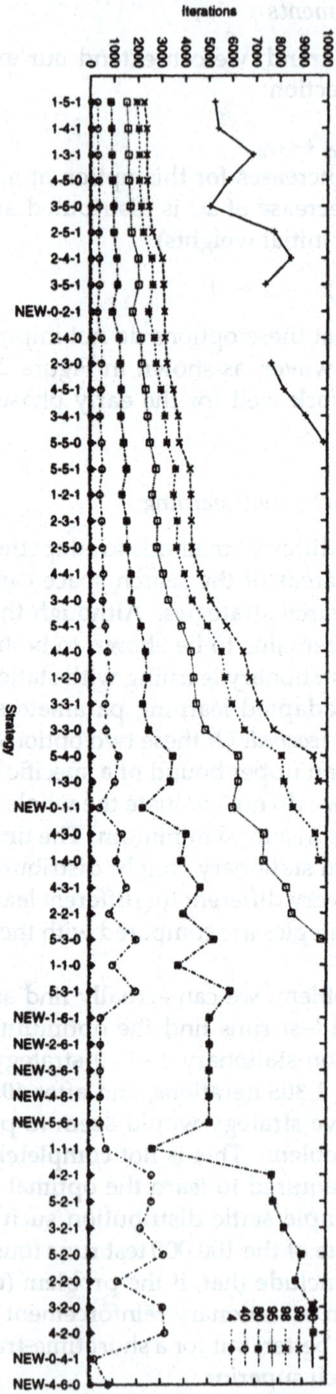

Figure 25.10 Extended weight-adaptation results for the Orc Quest problem; showing only the 50 best strategies.

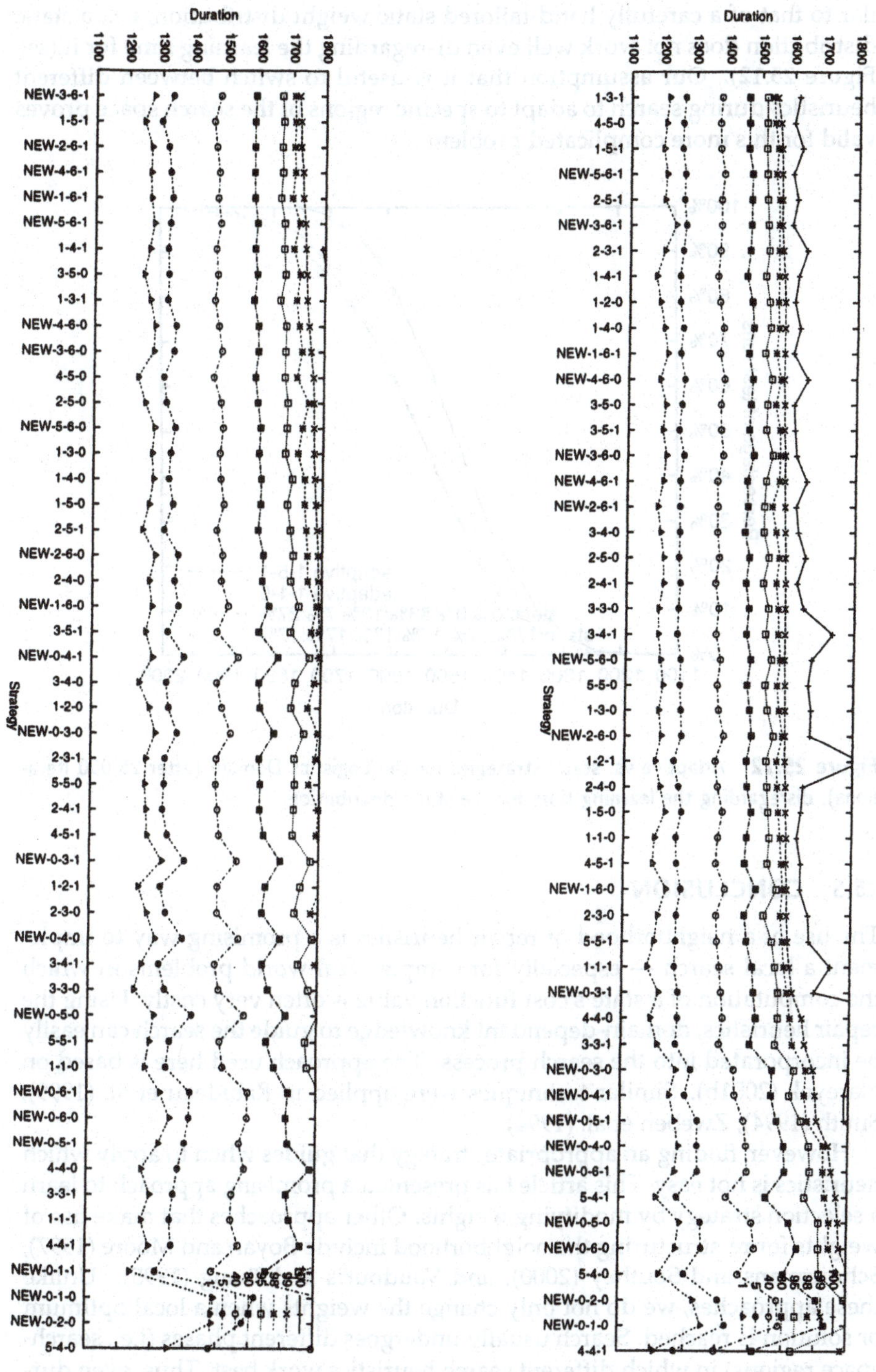

Figure 25.11 Extended weight-adaptation results for the Logistics Domain after 25,000 (left) and 100,000 (right) iterations; showing only the 50 best strategies.

ilar to that of a carefully hand-tailored static weight distribution, i.e., a static distribution does not work well even disregarding the learning time for it (see Figure 25.12). Our assumption that it is useful to switch between different heuristics during search to adapt to specific regions of the search space proves valid for this more complicated problem.

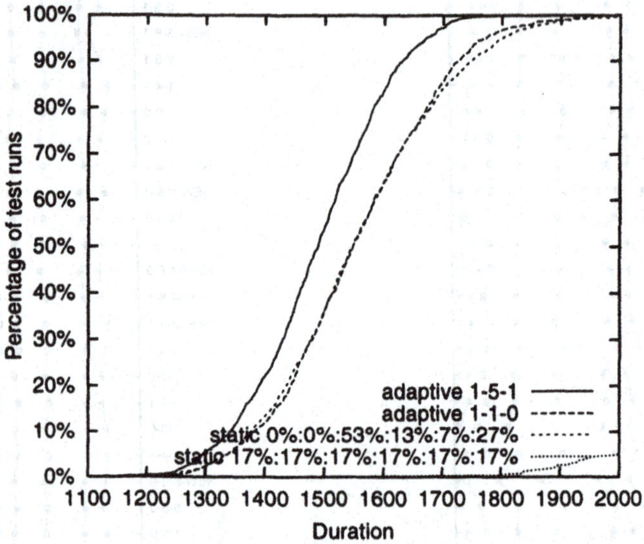

Figure 25.12 Adaptive vs. static strategies for the Logistics Domain (after 25,000 iterations), disregarding the learning time for the static distribution.

25.5 CONCLUSION

The use of a neighborhood of repair heuristics is a promising way to implement a local search — especially for complex real-world problems in which the computation of a state's cost function value is often very costly. Using the repair heuristics, domain-dependent knowledge to guide the search can easily be incorporated into the search process. The approach used here is based on Nareyek (2001b). Similar techniques were applied in Rabideau et al. (1999); Smith (1994); Zweben et al. (1994).

However, finding an appropriate strategy that guides when to apply which heuristics is not easy. This article has presented a promising approach to learn a selection strategy by modifying weights. Other approaches that make use of weights for re-structuring the neighborhood include Boyan and Moore (1997), Schuurmans and Southey (2000), and Voudouris and Tsang (1995). Unlike these approaches, we do not only change the weights when a local optimum or solution is reached. Search usually undergoes different phases (i.e., search-space regions) in which different search heuristics work best. Thus, even during search, the heuristics' configuration is constantly updated. For this purpose, a less-carrot-more-stick strategy seems to be most adequate, allowing for quick configuration changes and preventing the old configuration from being re-established too quickly.

WGSAT (Frank, 1997) follows a similar strategy, even though weights are connected to problem features (i.e., SAT clauses) instead of improvement heuristics. The approach is not really comparable, but it is close to a `P:0-N:1` strategy with an additional decay factor.

In reinforcement learning, non-stationary environments (such as the search-space region) are only rarely considered. Examples include approaches based on supervised techniques (Schmidhuber, 1990), evolutionary learning (Littman and Ackley, 1991) and model-based learning (Michaud and Matarić, 1998). Unlike these approaches, we have used a modification of standard action-value methods (Sutton and Barto, 1998), applying functional updates instead of cumulative value additions in order to influence the impact of the already learned reinforcements. This simple method enables the search to compute weight updates very quickly – which is very important for a local search environment because a single iteration should consume only very little computing power.

Adaptive weighs are not restricted to local search; they can also be used for (esp. restart-based) refinement search. Examples include the pheromone trails in ant colony optimization (Dorigo et al., 1999), the use of domain-specific prioritizers (Joslin and Clements, 1999) and action costs in adaptive probing (Ruml, 2001). The results obtained in this study may be transferred to these areas, and techniques like pheromone evaporation are worth studying for neighborhoods of heuristics as well.

So far, we have not considered quantitative cost-function effects of decisions. Improvement or non-improvement was the only criterion for learning. However, "good" heuristics may not be equally good on the quantitative level and incorporating mechanisms to exploit the quantitative differences is a promising idea for future work. An interesting approach in this direction was presented at this conference by Cowling et al. (2001).

The presented techniques are integrated into the **DragonBreath** [2] engine, which is a free optimization engine based on constraint programming and local search. It can be obtained via:

Acknowledgments

Thanks to Michael Littman for his feedback.

[2]Can be obtained at `http://www.ai-center.com/projects/dragonbreath/`.

Bibliography

J. A. Boyan and A. W. Moore. Using prediction to improve combinatorial optimization search. In *Proceedings of the Sixth International Workshop on Artificial Intelligence and Statistics (AISTATS-97)*, 1997.

P. Cowling, G. Kendall, and E. Soubeiga. A parameter-free hyperheuristic for scheduling a sales summit. In *Proceedings of the Fourth Metaheuristics International Conference (MIC'2001)*, pages 127–131, 2001.

M. Dorigo, G. Di Caro, and L. M. Gambardella. Ant algorithms for discrete optimization. *Artificial Life*, 5(3):137–172, 1999.

J. Frank. Learning short-term weights for GSAT. In *Proceedings of the Fifteenth International Joint Conference on Artificial Intelligence (IJCAI-97)*, pages 384–391, 1997.

H. H. Hoos and T. Stützle. Evaluating las vegas algorithms — pitfalls and remedies. In *Proceedings of the Fourteenth Annual Conference on Uncertainty in Artificial Intelligence (UAI-98)*, pages 238–245, 1998.

D. E. Joslin and D. P. Clements. Squeaky wheel optimization. *Journal of Artificial Intelligence Research*, 10:353–373, 1999.

S. Kirkpatrick, C. D. Gelatt, and M. P. Vecchi. Optimization by simulated annealing. *Science*, 220(4598):671–680, 1983.

M. L. Littman and D. H. Ackley. Adaptation in constant utility non-stationary environments. In *Proceedings of the Fourth International Conference on Genetic Algorithms*, pages 136–142, 1991.

F. Michaud and M. J. Matarić. Learning from history for behavior-based mobile robots in non-stationary environments. *Machine Learning (Joint Special Issue on Learning in Autonomous Robots)*, 31:141–167, 1998.

A. Nareyek. *Constraint-Based Agents – An Architecture for Constraint-Based Modeling and Local-Search-Based Reasoning for Planning and Scheduling in Open and Dynamic Worlds*, volume 2062 of *Lecture Notes in Artificial Intelligence*. Springer, 2001a.

A. Nareyek. Using global constraints for local search. In E. C. Freuder and R. J. Wallace, editors, *Constraint Programming and Large Scale Discrete Optimization*, volume 57 of *DIMACS Series on Discrete Mathematics and Theoretical Computer Science*, pages 9–28. American Mathematical Society Publications, 2001b.

G. Rabideau, R. Knight, S. Chien, A. Fukunaga, and A. Govindjee. Iterative repair planning for spacecraft operations in the aspen system. In *Proceedings of the International Symposium on Artificial Intelligence Robotics and Automation in Space (iSAIRAS 99)*, 1999.

W. Ruml. Incomplete tree search using adaptive probing. In *Proceedings of the Seventeenth International Joint Conference on Artificial Intelligence (IJCAI-01)*, pages 235–241, 2001.

J. Schmidhuber. Making the world differentiable: On using self-supervised fully recurrent neural networks for dynamic reinforcement learning and planning in non-stationary environments. Technical Report TR FKI-126-90, Department of Computer Science, Technical University of Munich, Germany, 1990.

D. Schuurmans and F. Southey. Local search characteristics of incomplete SAT procedures. In *Proceedings of the Seventeenth National Conference on Artificial Intelligence (AAAI-2000)*, pages 297–302, 2000.

S. F. Smith. OPIS: A methodology and architecture for reactive scheduling. In M. Zweben and M. S. Fox, editors, *Intelligent Scheduling*, pages 29–66. Morgan Kaufmann, 1994.

R. S. Sutton and A. G. Barto. *Reinforcement Learning: An Introduction*. MIT Press, 1998.

C. Voudouris and E. Tsang. Guided local search. Technical Report CSM-247, University of Essex, Department of Computer Science, Colchester, United Kingdom, 1995.

M. Zweben, B. Daun, E. Davis, and M. Deale. Scheduling and rescheduling with iterative repair. In M. Zweben and M. S. Fox, editors, *Intelligent Scheduling*, pages 241–255. Morgan Kaufmann, 1994.

Metaheuristics: Computer Decision-Making, pp. 545-573
Teresa Neto and João Pedro Pedroso

26 GRASP FOR LINEAR INTEGER PROGRAMMING

Teresa Neto[1] and João Pedro Pedroso[1]

[1]Departamento de Ciência de Computadores
Faculdade de Ciências da Universidade do Porto
Rua do Campo Alegre, 823
4150-180 Porto, Portugal
tneto@mat.estv.ipv.pt, jpp@ncc.up.pt

Abstract: In this paper, we introduce a GRASP for the solution of general linear integer problems. The strategy is based on the separation of the set of variables into the integer subset and the continuous subset. The integer variables are fixed by GRASP and replaced in the original linear problem. If the original problem had continuous variables, it becomes a pure continuous problem, which can be solved by a linear program solver to determine the objective value corresponding to the fixed variables. If the original problem was a pure integer problem, simple algebraic manipulations can be used to determine the objective value that corresponds to the fixed variables. When we assign values to integer variables that lead to an impossible linear problem, the evaluation of the corresponding solution is given by the sum of infeasibilities, together with an infeasibility flag. We report results obtained for some standard benchmark problems, and compare them to those obtained by branch-and-bound and to those obtained by an evolutionary solver.
Keywords: GRASP, Linear integer programming.

26.1 INTRODUCTION

A wide variety of practical problems can be solved using integer linear programming. Typical problems of this type include lot sizing, scheduling, facility location, vehicle routing, and more; see for example Nemhauser and Wolsey (1988) and Wolsey (1998).

In this paper, we introduce a GRASP (greedy randomized adaptive search procedure) for the solution of general linear integer programs. The strategy is based on the separation of the set of variables into the integer subset and the continuous subset (if some). The procedure starts by solving the linear programming (LP) relaxation of the problem. Values for the integer variables are then chosen, through a semi-greedy construction heuristic based on rounding around the LP relaxation, and fixed. The continuous variables (if some) can then be determined in function of them, by solving a linear program where all the integer variables have been fixed by GRASP. Afterward, local search improvements are made on this solution; these improvements still correspond to changes made exclusively on integer variables, after which the continuous variables are recomputed through the solution of an LP. When the linear program leads to a feasible solution, the evaluation of the choice of the variables is determined directly by the objective function. If the choice of the variables induces an infeasible problem, its evaluation is measured by the sum of infeasibilities.

26.2 BACKGROUND

In this paper we focus on the problem of optimizing a linear function subject to a set of linear constraints, in the presence of integer and, possibly, continuous variables. The more general case, where there are integer and continuous variables, is usually called *mixed integer* (MIP).

The general formulation of a mixed integer linear program is

$$\max_{x,y}\{cx + hy : Ax + Gy \leq b, x \in \mathbb{Z}_+^n, y \in \mathbb{R}_+^p\}, \tag{26.1}$$

where $\mathbb{Z}_+^n$ is the set of nonnegative, n-dimensional integral vectors, and $\mathbb{R}_+^p$ is the set of nonnegative, p-dimensional real vectors. A and G are $m \times n$ and $m \times p$ matrices, respectively, where m is the number of constraints. There are n integer variables (x), and p continuous variables (y).

If the subset of continuous variables is empty, the problem is called *pure integer* (IP); its formulation is

$$\max_{x}\{cx : Ax \leq b, x \in \mathbb{Z}_+^n\}. \tag{26.2}$$

In general, there are additional bound restrictions on the integer variables, stating that $l_i \leq x_i \leq u_i$, for $i = 1, \ldots, n$.

The main idea for the conception of the algorithm described in this paper is provided in Pedroso (1998; 2002). It consists of fixing the integer variables of a linear integer program by a metaheuristic—in this case GRASP. For MIP, by replacing these variables on the original formulation, we obtain a pure, continuous LP, whose solution provides an evaluation of the fixed variables. On the case of pure IP, we can compute directly the corresponding objective. We can also directly check feasibility, and compute the constraints' violation.

Notice that this algorithm, as opposed to branch-and-bound, does not work with the solution of continuous relaxations of the initial problem. The solution of LPs is only required for determining the value of the continuous variables, and of the objective that corresponds to a particular instantiation of the integer variables.

Instances of integer linear problems correspond to specifications of the data: the matrix A and the vectors b and c in Equations 26.1 and 26.2 for IPs, and also the matrix G and the vector h for MIPs. The most commonly used representation of instances of these problems is through *MPS* files, which is the format used on this GRASP implementation. We have tested GRASP with a subset of the benchmark problems that are available in this format in the *MIPLIB* (Bixby et al., 1998).

26.3 GRASP

GRASP (Feo and Resende, 1989; 1995; Pitsoulis and Resende, 2002; Resende and Ribeiro, 2002) is a metaheuristic based on a multi-start procedure where each iteration has two phases: construction and local search. In the construction phase, a solution is built, one element at a time. At each step of this phase, the candidate elements that can be incorporated to the partial solution are ranked according to a greedy function. The evaluation of these elements leads to the creation of a restricted candidate list (RCL), where a selection of good variables, according to the corresponding value of the greedy function, are inserted. At each step, one element of the RCL is randomly selected and incorporated into the partial solution (this is the probabilistic component of the heuristic). The candidate list and the advantages associated with every element are updated (adaptive component of the heuristic). The number of elements of the RCL is very important for GRASP: if the RCL is restricted to a single element, then only one, purely greedy solution will be produced; if the size of the RCL is not restricted, GRASP produces random solutions. The mean and the variance of the objective value of the solutions built are directly affected by the cardinality of the RCL: if the RCL has more elements, then more different solutions will be produced, implying a larger variance.

The solutions generated in the construction phase generally are not local optima, with respect to some neighborhood. Hence, they can often be improved by means of a local search procedure. The local search phase starts with the constructed solution and applies iterative improvements until a locally optimal solution is found.

The construction and improvement phases are repeated a specified number of times. The best solution over all these GRASP iterations is returned as the result.

In Algorithm 26.1 we present the structure of a general GRASP algorithm.

GRASP has been applied successfully to numerous combinatorial optimization problems in different areas, including routing (Kontoravdis and Bard, 1995; Carreto and Baker, 2002), scheduling (Feo et al., 1991; Binato et al., 2002), logic (Resende and Feo, 1996; Resende et al., 1997), assignment (Li et al., 1994; Robertson, 2001). An annotated GRASP bibliography is available in Festa and Resende (2002).

```
procedure GRASP()
while stopping criterion is not satisfied
    x ← SEMIGREEDY()
    x ← LOCALSEARCH(x)
    if x* is not initialized or x is better than x*
        x* ← x
endwhile
return x*
```

Algorithm 26.1 A general GRASP algorithm.

26.4 GRASP IMPLEMENTATION

In this section we specialize GRASP for the solution of general integer linear problems. We describe the fundamental aspects taken into account in the GRASP implementation, which are presented in Algorithm 26.2. The parameters of this procedure are the number of iterations, N (used as a stopping criterion), the largest type of neighborhood, k_{max} (see Section 26.4.4), the seed for initializing the random number generator, and the name of the *MPS* file containing the instance's data.

```
procedure GRASP(N, k_max, seed, MPSfile)
read data A, G, b, c, and h from MPSfile
initialize random number generator with seed
for k = 1 to N
    x̄ ← SEMIGREEDY(x̄^LP)
    x̄ ← LOCALSEARCH(x̄, k_max)
    if x̄* is not initialized or x̄ is better than x̄*
        x̄* ← x̄
endfor
return x̄*
```

Algorithm 26.2 A GRASP for integer programming.

26.4.1 Representation of the solutions

The part of the solution that is determined by GRASP is the subset of integer variables x in Equations 26.1 or 26.2. The data structure representing a solution used by GRASP is therefore an n-dimensional vector of integers, $\bar{x} = (\bar{x}_1 \dots \bar{x}_n)$.

26.4.2 *Evaluation of solutions*

The solutions on which the algorithm works may be feasible or not. For the algorithm to function appropriately it has to be able to deal with both feasible and infeasible solutions in the same framework. We describe next the strategies used for tackling this issue on MIPs and IPs.

26.4.2.1 Mixed integer programs (MIP). In the process of evaluation of a solution, we first formulate an LP by fixing all the variables of the MIP at the values determined by GRASP:

$$z = \max_{y}\{c\bar{x} + hy : Gy \leq b - A\bar{x}, y \in \mathbb{R}^p_+\}. \qquad (26.3)$$

We are now able to solve this (purely continuous) linear problem using a standard algorithm, like the simplex.

Feasible solutions. If problem 26.3 is feasible, the evaluation given to the corresponding solution is the objective value z, and the solution is labeled feasible.

Infeasible solutions. If problem 26.3 is infeasible, we formulate another LP for the minimization of the infeasibilities. This is accomplished by setting up artificial variables and minimizing their sum (a procedure that is identical to the phase I of the simplex algorithm):

$$\zeta = \min_{s}\{\sum_{k=1}^{m} s_k : Gy \leq b - A\bar{x} + s, y \in \mathbb{R}^p_+ \ , \ s \in \mathbb{R}^m_+\}, \qquad (26.4)$$

where m is the number of constraints.

The evaluation attributed to such a solution $\bar{x}$ is the value ζ of the optimal objective of the LP of Equation 26.4, and the solution is labeled infeasible.

26.4.2.2 Pure integer programs (IP). Fixing all the integer variables in Equation 26.2 leads to no free variables. Feasibility and the objective value can be inspected directly.

Feasible solutions. If the solution $\bar{x}$ fixed by GRASP does not violate any constraint, the evaluation attributed to the corresponding solution is the objective value $z = c\bar{x}$, and the solution is labeled feasible.

Infeasible solutions. If $\bar{x}$ is infeasible, its evaluation is given by the sum of constraint violations. For problems stated in the canonic form of Equation 26.2, this is done by determining:

$$\zeta = \sum_{k=1}^{m} s_k, \text{ with } s_k = \max\{A_k\bar{x} - b_k, 0\} \text{ for } k = 1, \dots m. \qquad (26.5)$$

The evaluation given to the solution is the value ζ and the solution is labeled infeasible.

26.4.2.3 Evaluation data structure. For a feasible solution of an integer linear program, the evaluation is denoted by $\bar{z}$, a data structure consisting of the objective value z and a flag stating that the solution is feasible. For an infeasible solution, $\bar{z}$ consists of the value ζ and an infeasibility flag.

26.4.2.4 Comparison of solutions. We have to provide a way of comparing solutions, either they are feasible or not. What we propose is to classify solutions in such a way that:

- feasible solutions are always better than infeasible ones;
- feasible solutions are ranked among them according to the objective of the integer linear problem;
- infeasible solutions are ranked among them according to the sum of infeasibilities (i.e., according to a measure of their distance from the feasible set).

We say that a solution structure i *is better* than another structure j if and only if:

- $\zeta^i < \zeta^j$ (i is closer to the feasible region than j);
- $\zeta^i = \zeta^j$, and (for maximization) $z^i > z^j$ (i has a better objective).

26.4.3 Construction phase

We propose two construction methods, differing in the greedy function and in the number of elements of the restricted candidate list (RCL). Both are based on the solution of the LP relaxation, which we denote by $x^{LP} = (x_1^{LP}, \ldots, x_n^{LP})$. We use the term RCL with a meaning slightly different of the currently used in the GRASP literature. For each index $k \in \{1, \ldots, n\}$ of the variables, we set up a list of the values that we can potentially assign to it. In a purely greedy construction, we always assign the integer closest to the value of the LP relaxation, x_k^{LP}. Hence, the RCL for a variable x_k would have a single element, the closest integer to x_k^{LP}. The RCL for semi-greedy construction has more elements, as explained below.

In Algorithm 26.3, we present the construction algorithm. Since there are n variables in the solution, each construction phase consists of n steps. The two constructions differ on steps 2 and 3.

26.4.3.1 Probabilistic rounding construction. This semi-greedy construction is inspired in an algorithm provided in Lengauer (1990). It consists of rounding each variable i to the integer closest to its value on the LP relaxation, x_i^{LP}, in a randomized way, according to some rules. The probabilities

```
procedure SemiGreedy(x̄^LP)
    for k = 1 to n
        RCL ← {values allowed to x̄_k }
        Select an r ∈ RCL with some probability
        x̄_k ← r
    endfor
    return x̄
```

Algorithm 26.3 Semi-greedy solution construction.

for rounding up or down each of the variables are given by the distance from the fractional solution x_k^{LP} to its closest integer.

For all the indices $k \in \{1, \ldots, n\}$, the variable $\bar{x}_k$ is equal to the corresponding LP relaxation value rounded down with probability

$$P(\bar{x}_k = \lfloor x_k^{LP} \rfloor) = \lceil x_k^{LP} \rceil - x_k^{LP},$$

or rounded up with probability $1 - P(\bar{x}_k = \lfloor x_k^{LP} \rfloor)$.

The RCL is built with the two values that each variable $\bar{x}_k$ can take.

26.4.3.2 Bi-triangular construction. The goal of this strategy is to increase the diversification of the solutions. In the previous construction, we only have two possibilities when we round the value of the LP relaxation of a variable. We here extend this for having the possibility of assigning each variable to any integer between its lower-bound, l_k, and its upper-bound, u_k. The RCL is built with these values, and, again, we give a probability of assignment to each of them based on the solution of the LP relaxation.

The probability density function that we considered is composed by two triangles (*bi-triangular distribution*) and is defined by three parameters: the minimum $a = l_k - 0.5$, the maximum $b = u_k + 0.5$, and the mean $c = x_k^{LP}$. The values a and b were considered in order to have a non-zero probability of rounding to l_k and u_k. The bi-triangular density function is represented in Figure 26.1.

The area of the left triangle is proportional to the distance between c and b and the area of the right triangle is proportional to the distance between a and c. The combined areas must be one, since it is a density function. The probability density function is given by:

$$f(x) = \begin{cases} \frac{2(b-c)(x-a)}{(b-a)(c-a)^2} & \text{if } a \leq x \leq c, \\ \frac{2(c-a)(b-x)}{(b-a)(b-c)^2} & \text{if } c < x \leq b, \\ 0 & \text{otherwise,} \end{cases}$$

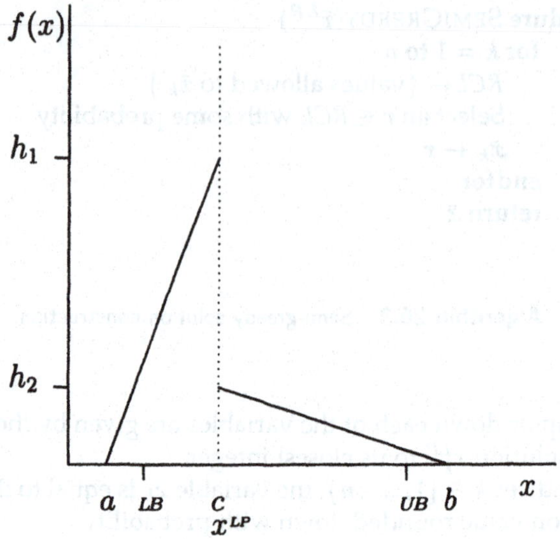

Figure 26.1 Bi-triangular density function.

and the distribution function is:

$$F(x) = \begin{cases} 0 & \text{if } x < a, \\ \frac{(b-c)(x-a)^2}{(b-a)(c-a)^2} & \text{if } a \le x \le c, \\ 1 - \frac{(c-a)(b-x)^2}{(b-a)(b-c)^2} & \text{if } c < x \le b, \\ 1 & \text{if } x > b. \end{cases}$$

The mean for this distribution is c, which corresponds to the value of the LP relaxation.

With this construction, the value for each variable x_k is obtained by drawing a random number with this distribution with $a = l_k - 0.5$, $b = u_k + 0.5$, and $c = x_k^{LP}$, and then rounding it to the closest integer.

26.4.4 Local search

Local search tries to improve the quality of a solution by hill climbing on its neighborhood, according to one of the improvement methods that are described next. For this purpose we propose neighborhoods that consist of incrementing or decrementing variables, one at a time or simultaneously. The main idea behind the definition of these neighborhoods is the extension to the case of integer values of the idea presented in Resende and Feo (1996) for the case of binary variables. The local search procedure iterates up to the point where the improvement procedure does not lead to a better solution.

26.4.4.1 Increment neighborhood. The *increment* neighborhood of a solution x, $N^1(x)$, is composed of solutions which differ from x on one element x_j, whose value is one unit above or below x_j. Hence y is a neighbor solution of x if for one index i, $y_i = x_i + 1$, or $y_i = x_i - 1$, with $y_j = x_j$ for all indices $j \neq i$:

$$N^1(x) = \{\, y \in \mathbb{Z} : y \text{ can be obtained from } x \text{ by adding or subtracting one unit to an element of } x\}.$$

The idea used on this neighborhood can be extended to the case where we change more than one variable at the same time. For example, the *2-increment* neighborhood of a solution x, $N^2(x)$, is composed of solutions which differ from x on two elements x_i and x_j, whose values are one unit above or below the original ones. Hence y is a neighbor solution of x if for two indices i, j we have $y_i = x_i + 1$ or $y_i = x_i - 1$, and $y_j = x_j + 1$ or $y_j = x_j - 1$, with $y_l = x_l$ for all indices $l \neq i, l \neq j$.

More generally, we can define the *k-increment* neighborhood (for problems with $n \geq k$ integer variables) as:

$$N^k(x) = \{\, y \in \mathbb{Z} : y \text{ can be obtained from } x \text{ by adding or subtracting one unit to } k \text{ of its elements}\}.$$

When k increases, the number of neighbors of a solution increases exponentially. In order to reduce the size of the set of neighbors that is more frequently explored, we devised the following strategy. Let O be the set of indices of variables which have coefficients that are different of zero in the objective function:

$$O = \{i : c_i \neq 0, \text{ for } i = 1, \ldots, n\}.$$

Let $V(x)$ be the set of indices of variables (if some) which have coefficients different of zero on constraints violated by x:

$$V(x) = \{j : a_{ij} \neq 0, \text{ for } i \in \{\text{constraints violated by } x\}\}.$$

For a solution x, the subset of neighbors with at least one index in the sets O (for feasible x) or $V(x)$ (for infeasible x) compose the neighborhood $N^{k^*}(x) \subseteq N^k(x)$ which is explored first. The subset $N^{k'}(x) = N^k(x) \setminus N^{k^*}(x)$ is explored when a local optimum of N^{k^*} has been found. Neighborhoods N^k are explored in increasing order on k.

This strategy for exploring the neighborhoods is called *variable neighborhood search* (Hansen and Mladenovic, 2001). It consists in searching first restricted neighborhoods which are more likely to contain improving solutions; when there are no better solutions in a restricted neighborhood, this is enlarged, until having explored the whole, unrestricted neighborhood.

26.4.4.2 Improvements. There are two methods for updating the best solution when searching a particular neighborhood.

The first one, called *breadth-first*, consists of searching the best solution y in the entire neighborhood of a solution x. If it is better than the current solution

x, then x is replaced by y; otherwise, x is a local optimum. This method will hence search the whole neighborhood of the current solution, even if improving solutions can be found on the early exploration of the neighborhood.

The second method, called *depth-first*, consists of replacing x whenever the neighbor generated, y, is better than x. In this case, the subsequent neighbor z is generated from y.

Empirically, the computational time required to obtain a local optimum is longer with the first method, and premature convergence is more likely to occur. Therefore, in this implementation the search is made depth-first.

26.4.4.3 Local search algorithm. The local search method is presented in Algorithm 26.4. The improvement of a solution is made in the routine IMPROVE(x, k_{max}). This procedure, presented in Algorithm 26.5, searches first the neighborhood $N^{1^*}(x)$ and returns the first neighbor better than x found. If no such a neighbor is found, it switches to $N^{1'}(x)$. When no improving solution is found also in this neighborhood, the method explores $N^2(x)$ (first checking in $N^{2^*}(x)$), and so on, until having explored $N^{k_{max}}$.

```
procedure LOCALSEARCH(x, k_max)
s ← IMPROVE(x, k_max)
while s ≠ x
    x ← s
    s ← IMPROVE(x, k_max)
endwhile
return s
```

Algorithm 26.4 Local search main cycle.

Algorithms 26.6 and 26.7 present a strategy called *hunt search*. It was originally conceived for locating values in an ordered table, and is used here for quickly exploring large ranges, when lower and upper bounds of some variable are far apart. The step added to a variable, Δ, is initially $+1$ or -1, and is doubled until no improvements are obtained, or until reaching a bound of the variable.

26.5 BENCHMARK PROBLEMS

Languages for mathematical programming are the tools more commonly used for specifying a model, and generally allow transforming the mathematical model into an *MPS* file. As the heuristic that we describe can be used for any MIP model that can be specified as a mathematical program, we have decided to provide the input to the heuristic through *MPS* files. GRASP starts by reading an *MPS* file, and stores the information contained there into an internal representation. The number of variables and constraints, their type and bounds, and all the matrix information is, hence, determined at runtime.

```
procedure IMPROVE(x, k_max)
k ← 1
while k ≤ k_max
    S ← N^{k*}(x)
    while S ≠ ∅
        s ← RandomChoice(S)
        if s is better than x
            return s
        S ← S \ {s}
    endwhile
    S ← N^{k'}(x)
    while S ≠ ∅
        s ← RandomChoice(S)
        if s is better than x
            return s
        S ← S \ {s}
    endwhile
    k ← k + 1
endwhile
return x
```

Algorithm 26.5 Improvements without hunt search.

We report results obtained for some standard benchmark problems. The instances of MIP and IP problems used as benchmarks are defined in the *MIPLIB* (Bixby et al., 1998) and are presented in Table 26.1. They were chosen to provide an assortment of MIP structures, with instances coming from different applications.

Notice that the MIPLIB problems are minimizations[1].

26.6 COMPUTATIONAL RESULTS

We compare the results obtained with GRASP to those obtained by branch-and-bound (B&B), the classical algorithm for solving general linear integer programs, and to those obtained by another metaheuristic based on evolutionary computation (Pedroso, 1998). B&B starts with a continuous relaxation of the integer linear program and finds the optimal solution by a systematic division of the domain of the relaxed problem. The evolutionary solver is based on ideas similar to those employed here for solution representation and improvement, but uses population-based methods.

[1]The GRASP implementation works for minimization and maximization, by adapting the meaning of *is better* (see Section 26.4.2.4).

Table 26.1 Set of benchmark problems used: application, number of constraints, number of variables and optimal solutions as reported in MIPLIB.

Problem name	Application	Number of variables total	integer	binary	Number of constraints	Optimal solution
bell3a	fiber optic net. design	133	71	39	123	878430.32
bell5	fiber optic net. design	104	58	30	91	8966406.49
egout	drainage syst. design	141	55	55	98	568.101
enigma	unknown	100	100	100	21	0
flugpl	airline model	18	11	0	18	1201500
gt2	truck routing	188	188	24	29	21166
lseu	unknown	89	89	89	28	1120
mod008	machine load	319	319	319	6	307
modglob	heating syst. design	422	98	98	291	20740508
noswot	unknown	128	100	75	182	-43
p0033	unknown	33	33	33	16	3089
pk1	unknown	86	55	55	45	11
pp08a	unknown	240	64	64	136	7350
pp08acut	unknown	240	64	64	246	7350
rgn	unknown	180	100	100	24	82.1999
stein27	unknown	27	27	27	118	18
stein45	unknown	45	45	45	331	30
vpm1	unknown	378	168	168	234	20

```
procedure IMPROVE(x, k_max)
k ← 1
while k ≤ k_max
    S ← N^{k*}(x)
    while S ≠ ∅
        s ← RandomChoice(S)
        if s is better than x
            i ← any index such that s_i ≠ x_i
            Δ ← s_i − x_i
            l_i ← lower bound of the variable s_i
            u_i ← upper bound of the variable s_i
            s ← HUNTSEARCH(s, i, Δ, l_i, u_i)
            return s
        endif
        S ← S \ {s}
    endwhile
    S ← N^{k'}(x)
    while S ≠ ∅
        s ← RandomChoice(S)
        if s is better than x
            i ← any index such that s_i ≠ x_i
            Δ ← s_i − x_i
            l_i ← lower bound of the variable s_i
            u_i ← upper bound of the variable s_i
            s ← HUNTSEARCH(s, i, Δ, l_i, u_i)
            return s
        endif
        S ← S \ {s}
    endwhile
    k ← k + 1
endwhile
return x
```

Algorithm 26.6 Improvements with hunt search.

The software implementing the branch-and-bound algorithm used in this experiment is called *lp_solve* (Berkelaar and Dirks, 1994). It also comprises a solver for linear programs based on the simplex method, which was used both on this GRASP implementation and in Pedroso (1998) for solving Equa-

```
procedure HUNTSEARCH(x, i, Δ, l_i, u_i)
while true
    s ← x
    Δ ← Δ × 2
    s_i ← s_i + Δ
    if s_i + Δ ≥ u_i
        s_i ← u_i
    else if s_i + Δ ≤ l_i
        s_i ← l_i
    if s is better than x
        x = s
    else
        return x
endwhile
```

Algorithm 26.7 Hunt search on a given index.

tions 26.3 and 26.4[2]. Notice that the LPs solved by GRASP at the time of solution evaluation (if some) are often much simpler than those solved by B&B; as all the integer variables are fixed, the size of the LPs may be much smaller. Hence, numerical problems that the LP solver may show up in B&B, do not arise for LPs formulated by GRASP. Therefore, the comparison made in terms of objective evaluations/LP solutions required favors B&B.

This section begins presenting the results obtained using B&B. Statistical measures used in order to assess the empirical efficiency of GRASP are defined next. Follow the results obtained using GRASP, and a comparison of results of GRASP, B&B, and an evolutionary algorithm.

26.6.1 Results obtained using branch-and-bound

The results obtained using B&B on the series of benchmark problems selected are provided in the Table 26.2. The maximum number of LPs solved in B&B was limited to 100 million; in cases where this was exceeded, the best solution found within that limit is reported.

26.6.2 Statistical measures

In order to assess the empirical efficiency of GRASP, we provide measures of the expectation of the number of LP solutions and CPU time required for finding a feasible solution, the best solution found, and the optimal solution,

[2]This software has the advantage of being free; on the other hand, it does not have some important components, like the dual simplex method, which would allow to quickly reoptimize Equation 26.3 from a dual solution after a change in the right hand side.

Table 26.2 Results obtained using branch-and-bound: best solution found, number of LPs solved and CPU time.

Problem name	Best solution found	Number of LPs solved	CPU time	Remarks
bell3a	878430.32	438587	170.55	Optimal solution
bell5	8966406.49	420499	81.61	Optimal solution
egout	562.273	55057	6.06	Incorrect solution
enigma	0	8947	1.8	Optimal solution
flugpl	1201500	1588	0.06	Optimal solution
gt2	-	-	-	Failed (unknown error)
lseu	1120	236207	23.97	Optimal solution
mod008	307	2848585	844.09	Optimal solution
modglob	26308600	> 1.00E+08	47678	Stopped
noswot	-25	3753	3.72	Incorrect solution
p0033	3089	7393	0.25	Optimal solution
pk1	11	3710343	2467.23	Optimal solution
pp08a	9770	> 1.00E+08	29924	Stopped
pp08acut	8110	> 1.00E+08	161880	Stopped
rgn	82.1999	4963	1.2	Optimal solution
stein27	18	11985	2.62	Optimal solution
stein45	30	236453	218.57	Optimal solution
vpm1	22	18354	24.95	Incorrect solution

for MIP problems. These measures are similar for IP problems, but instead of being developed in terms of the number of LP solutions, they are made in terms of the number of calls to the objective function (Equations 26.2 or 26.5).

The number of GRASP independent iterations (or *runs*) for each benchmark problem in a given experiment is denoted by N.

26.6.2.1 Measures in terms of the number of LP solutions. Let r^f, r^b and r^o be the number of runs in which a feasible, the best, and the optimal solution were found, respectively. Let n_k^f be the number of objective evaluations required for obtaining a feasible solution in iteration k, or the total number of evaluations in that run if no feasible solution was found. Identical measures for reaching optimality and the best solution found by GRASP are denoted by n_k^o and n_k^b, respectively. Then, the expected number of evaluations for reaching feasibility, based on these N iterations, is:

$$E[n^f] = \sum_{k=1}^{N} \frac{n_k^f}{r^f}.$$

Equivalently, the expected number of evaluations for reaching the best GRASP solution is

$$E[n^b] = \sum_{k=1}^{N} \frac{n_k^b}{r^b},$$

and the expected number of evaluations for reaching optimality is

$$E[n^o] = \sum_{k=1}^{N} \frac{n_k^o}{r^o}.$$

In case $r^o = 0$, the sum of the evaluations of the total experiment (N iterations) provides a lower bound on the expectations for optimality. The same for feasibility, when $r^f = 0$.

26.6.2.2 Measures in terms of CPU time. Let t_k^f be the CPU time required for obtaining a feasible solution in iteration k, or the total CPU time in that iteration if no feasible solution was found. Let t_k^o and t_k^b be identical measures for reaching optimality, and the best solution found by GRASP, respectively. Then, the expected CPU time required for reaching feasibility, based on these N iterations, is:

$$E[t^f] = \sum_{k=1}^{N} \frac{t_k^f}{r^f},$$

while

$$E[t^b] = \sum_{k=1}^{N} \frac{t_k^b}{r^b}$$

is the expected CPU time for finding the best GRASP solution, and the expected CPU time required for reaching optimality is

$$E[t^o] = \sum_{k=1}^{N} \frac{t_k^o}{r^o}.$$

For $r^f = 0$ and $r^o = 0$, the sums provide respectively a lower bound on the expectations of CPU time required for feasibility and optimality.

26.6.3 Results obtained using GRASP

In this section, we provide a series of results comparing the several strategies that were implemented: the two construction methods, the k-increment neighborhood (N^k) for $k = 1$ and $k = 2$, explored with and without hunt search.

The computer environment used on this experiment is the following: a Linux Debian operating system running on a machine with an AMD Athlon processor at 1.4 GHz, with 512 Gb of RAM. GRASP was implemented on the C++ programming language.

26.6.3.1 Hunt search. We started making an experiment for assessing the validity of hunt search (HS), with 1000 GRASP iterations with the 1-increment neighborhood (N^1), and 100 iterations with the 2-increment neighborhood (N^2) (the total time spent in each of the two cases is roughly the same).

We used both probabilistic rounding construction (PRC) and bi-triangular construction (BTC).

We report results obtained for instances with non-binary integer variables—*bell3a, bell5, flugpl, gt2* and *noswot* (hunt search does not apply when all the integer variables are binary). Table 26.3 reports the percent distance from the best solution found to the optimal solution[3], for N^1 and N^2, respectively ("-" means that the best solution found is not feasible). Table 26.4 reports the expected number of LP solutions/evaluations for obtaining feasibility. The same results for obtaining optimality are reported in Table 26.5.

Table 26.3 Percent distance above optimum observed with and without hunt search, for N^1 (on 1000 iterations) and N^2 (on 100 iterations), and for both construction methods: probabilistic rounding and bi-triangular.

Problem	Probabilistic rounding without HS	Probabilistic rounding with HS	Bi-triangular without HS	Bi-triangular with HS	Neighborhood
bell3a	0.4676	0.2762	1.171	1.325	
bell5	1.365	-	-	-	
flugpl	-	-	-	-	N^1
gt2	-	198.8	222.9	151.8	
noswot	4.651	4.651	4.651	4.651	
bell3a	0	0	0	0	
bell5	0.4012	-	-	-	
flugpl	-	-	-	-	N^2
gt2	141.9	94.06	154.4	102.5	
noswot	4.651	4.651	4.651	4.651	

Table 26.4 Expected number of objective evaluations for finding a feasible solution with and without hunt search, for N^1 and N^2, and for both construction methods: probabilistic rounding and bi-triangular.

Problem	Probabilistic rounding without HS	Probabilistic rounding with HS	Bi-triangular without HS	Bi-triangular with HS	Neighborhood
bell3a	60.11	27.45	178.01	43.92	
bell5	730448	>174782	>3957140	>187457	
flugpl	>44720	>47649	>112475	>78854	N^1
gt2	>521465	14450	189306	9538	
noswot	68.29	79.26	40.11	37.14	
bell3a	60.54	27.77	181.63	39.47	
bell5	331254	>226731	>624409	>213884	
flugpl	>24328	>24857	>35634	>33071	N^2
gt2	57324	8990	39384	6948	
noswot	190.46	163.01	90.56	36.32	

[3]Let f^b be the objective value for the best feasible solution, and f^o for the optimal solution. The percent distance above the optimum is given by $|100 \times (f^b - f^o)/f^o|$.

Table 26.5 Expected number of objective evaluations for finding an optimal solution with and without hunt search, for N^1 and N^2, and for both construction methods: probabilistic rounding and bi-triangular.

Problem	Probabilistic rounding without HS	with HS	Bi-triangular without HS	with HS	Neighbor-hood
bell3a	>205578	>351353	>1964395	>612769	
bell5	>730591	>174782	>3957140	>187457	
flugpl	>44720	>47649	>112475	>78854	N^1
gt2	>521465	>506547	>568958	>539824	
noswot	>294254	>439764	>643568	>342003	
bell3a	6513	10630	17147	33637	
bell5	>352317	>226731	>624409	>213884	
flugpl	>24328	>24857	>35634	>33071	N^2
gt2	>4752588	>5594115	>4824420	>5563503	
noswot	>765014	>876859	>871802	>778050	

Comparing these two strategies (GRASP with and without hunt search), we conclude that, in general, hunt search slightly improves the results. This improvement is more significant for the bi-triangular construction: as the constructed solutions are more likely to be far away from local optima, hunt search has more potential for operating.

In the experiments that follow, GRASP was implemented with hunt search.

26.6.3.2 Construction methods. The next experiment was conceived in order to assess the influence of the construction method in the GRASP performance, and to choose one of the methods for later comparison with other methods. The results presented are based on a sample obtained with 1000 GRASP iterations for N^1, and 100 iterations for N^2.

In Tables 26.6 and 26.7 we compare probabilistic rounding to bi-triangular construction. The comparison is made in terms of the percent distance above optimum, and on the expected number of LP solutions for reaching feasibility and optimality.

The results show that probabilistic rounding is in general preferable to the bi-triangular construction, if we take into account the computational burden. Hence, probabilistic rounding is the construction method used for comparing GRASP to other approaches.

26.6.3.3 Neighborhoods. We now present results of an experiment conceived for assessing the influence of the neighborhoods used, and to choose one of N^1 and N^2 for comparing GRASP to other methods. As the distinction between these results is less clear than the preceding, they are now based on a larger sample of 10000 GRASP iterations for N^1, and 1000 iterations for N^2. The results are reported in the Tables 26.8 and 26.9.

The results show a superiority of the N^2 neighborhood for most of the instances, both in terms of solution quality and expected evaluations, or CPU time, required to obtain them. Therefore, for comparison with other methods,

Table 26.6 Comparison between probabilistic rounding and bi-triangular construction: percent distance above optimum observed for neighborhoods N^1 (on 1000 iterations) and N^2 (on 100 iterations).

	Neighborhood N^1		Neighborhood N^2	
Problem	PRC	BTC	PRC	BTC
bell3a	0.2762	1.325	0	0
bell5	-	-	-	-
egout	10.12	11.22	0	1.387
enigma	-	-	-	-
flugpl	-	-	-	-
gt2	198.8	151.8	94.06	102.5
lseu	2.589	4.821	0	1.429
mod008	0	0.3257	0	0
modglob	0.09060	0.08530	0	0
noswot	4.651	4.651	4.651	4.651
p0033	-	-	0.1942	0
pk1	100	100	63.64	45.45
pp08a	1.769	0.1361	0.1361	0.6803
pp08acut	0	0.5442	0.1361	0
rgn	0	0	0	0
stein27	0	0	0	0
stein45	3.33	3.33	3.33	3.33
vpm1	0	0	0	0

we decided to use the results obtained with a GRASP implementation using probabilistic rounding construction, and the N^2 neighborhood.

26.6.3.4 Comparison of GRASP with other methods. For comparing GRASP to other methods, the criteria used are the best solution found and its distance to the optimum, the actual or expected number of LP solutions required, and the actual or expected CPU time used. A comparison between GRASP and B&B is presented in Table 26.10. Table 26.11 reports a comparison between GRASP and an evolutionary solver.

The comparison with B&B indicates that each algorithm works well on rather different instances: GRASP determines a good feasible solution in all the cases where B&B failed, and B&B quickly determines the optimal solution for the two instances where GRASP could not find any feasible solution (*enigma* and *flugpl*). The expected number of LP solutions and the expected CPU time is many times smaller for GRASP than the number of LPs and CPU time required by B&B. For larger problems, like *modglob* and *vpm1* GRASP seems to have some advantage.

The results obtained with the evolutionary solver (ES) were reported in Pedroso (1998). The comparison between GRASP and this metaheuristic is made in terms of the percent distance to optimum, and the expected number of LP solutions for reaching feasibility and optimality. The results are reported in Table 26.11. Notice that the termination criteria for GRASP and for the ES are very different, and hence the comparison in terms of distance to optimality is

Table 26.7 Comparison between probabilistic rounding and bi-triangular construction: expected number of objective evaluations for obtaining feasibility and optimality, for neighborhoods N^1 and N^2.

Problem	Feasibility ($E[n^f]$) PRC	Feasibility ($E[n^f]$) BTC	Optimality ($E[n^o]$) PRC	Optimality ($E[n^o]$) BTC	Neighborhood
bell3a	27.45	43.92	>351353	>612769	
bell5	>174782	>187457	≫174782	≫187457	
egout	142.35	148.12	>240030	>244815	
enigma	>296625	>297595	≫296625	≫297595	
flugpl	>47649	>78854	≫47649	≫78854	
gt2	14450	9538	>506547	>539824	
lseu	4587.68	3410.4	>208294	>213632	
mod008	3.33	3.24	692264	>681598	
modglob	1	1	>442000	>444842	N^1
noswot	79.26	37.14	>439764	>342003	
p0033	>76324	>73073	≫76324	≫73073	
pk1	1	1	>115125	>112925	
pp08a	30.01	37.02	>275048	>287272	
pp08acut	21.87	27.57	260959	>270440	
rgn	12.02	4.94	2774	3337	
stein27	12.89	18.85	182.34	201.43	
stein45	24.61	35.39	>121756	>134237	
vpm1	9.61	9.29	1292.02	931.99	
bell3a	27.77	39.47	10630	33637	
bell5	>226731	>213884	≫226731	≫213884	
egout	143.17	150.76	490209	>499191	
enigma	>729342	>698885	≫729342	≫698885	
flugpl	>24857	>33071	≫24857	≫33071	
gt2	8989	6948	>5594115	>5563503	
lseu	669.22	681	439230	>850069	
mod008	3.6	2.91	922149	1008139	
modglob	1	1	12724	12564	N^2
noswot	163.01	36.32	>876859	>778050	
p0033	3439.42	3250.79	>105066	100819	
pk1	1	1	>346686	>334838	
pp08a	30.39	37.02	>424686	>431121	
pp08acut	20.23	29.25	>394332	219656	
rgn	12.95	3.82	2275	2415	
stein27	13.21	19.22	1109	1027	
stein45	24.86	34.79	>111285	>112306	
vpm1	9.51	8.18	57224	53631	

Table 26.8 Best solution found, percent distance above optimum, and expected number of LP solutions for reaching the best solution, feasibility and optimality. Results obtained with 10000 iterations for N^1 and 1000 iterations for N^2.

Problem name	Best sol. found	% dist to opt	$E[n^b]$	$E[n^f]$	$E[n^o]$	Neigh-borhood
bell3a	880857	0.2763	116312	26.52	>3376050	
bell5	0.31887 (inf.)	-	165.10	>1744874	>1744874	
egout	615.719	8.382	2391640	141.85	>2391720	
enigma	4 (inf.)	-	74585.23	>2988675	>2988675	
flugpl	0.6 (inf.)	-	43411.73	>477819	>477819	
gt2	49556	134.1	5068183	14034	>5068412	
lseu	1149	2.589	2110868	5119	>2110972	
mod008	307	0	2346518	3.42	2346518	
modglob	20755100	0.07036	4436169	1.00	>4436268	N^1
noswot	-41	4.651	3249.46	81.61	>4372469	
p0033	3095	0.1942	764869	764869	>764916	
pk1	19	72.73	381708	1.00	>1145420	
pp08a	7380	0.4082	925905	30.04	>277800	
pp08acut	7350	0	435974	21.46	435974	
rgn	82.1999	0	2784.93	12.33	2784.93	
stein27	18	0	184.03	12.93	184.03	
stein45	30	0	93568	24.31	93568	
vpm1	20	0	1203	9.90	1202.59	
bell3a	878430.32	0	11216	27.14	11216	
bell5	9030230	0.7118	2601738	2256575	>2604076	
egout	568.101	0	977203	141.48	977203	
enigma	4 (inf.)	-	671243	>7439411	>7439411	
flugpl	0.7 (inf.)	-	6861	>248127	>248127	
gt2	36131	70.70	55164302	9194	>55183799	
lseu	1120	0	1245115	693.27	1245115	
mod008	307	0	871031	3.42	871031	
modglob	20740508	0	12381	1.00	12381	N^2
noswot	-41	4.651	24533	140.36	>8714283	
p0033	3089	0	1066553	3322	1066553	
pk1	16	45.45	435039	1.00	>3492631	
pp08a	7350	0	2161662	30.16	2161662	
pp08acut	7350	0	4173685	21.18	4173685	
rgn	82.1999	0	2469	12.09	2469	
stein27	18	0	1130	12.91	1130	
stein45	30	0	1110292	23.97	1110292	
vpm1	20	0	64719	9.97	64719	

Table 26.9 Comparison of expected CPU time (in seconds) required using neighborhoods N^1 and N^2, in order to obtain the best solution, feasibility and optimality, using probabilistic rounding construction.

Problem	Neighborhood N^1			Neighborhood N^2		
name	$E[t^b]$	$E[t^f]$	$E[t^o]$	$E[t^b]$	$E[t^f]$	$E[t^o]$
bell3a	299.3	0.08	>8687	34.52	0.08	34.52
bell5	0.3	>3197	>3197	4880	4289	>4884
egout	5547	0.37	>5547	2394	0.38	2394
enigma	7.23	>289.7	>289.7	85	>935.3	>935.3
flugpl	12.82	>141.1	>141.1	2.12	>76.69	>76.69
gt2	392.2	1.09	>392.2	4297	0.5517	>4298
lseu	95.99	0.23	>96	90.72	0.0328	90.72
mod008	552.4	0.001	552.4	298.9	0.0013	298.9
modglob	57039	0.02	>57040	146.7	0.0133	146.7
noswot	3.13	0.10	>4198	28.83	0.1788	>10244
p0033	17.41	17.41	>17.41	25.66	0.074	25.66
pk1	152.9	0.0004	>458.8	181.9	0.0005	>1460
pp08a	5209	0.31	>15627	12571	0.3244	12571
pp08acut	3762	0.4	3762	37617	0.41	37617
rgn	2.93	0.01	2.93	2.77	0.0134	2.77
stein27	0.01	0.0008	0.01	0.08	0.0008	0.077
stein45	19.61	0.01	19.61	248.9	0.0053	248.9
vpm1	6.9	0.1	6.88	415.9	0.1007	415.9

not very meaningful. Still, it shows that the two metaheuristics have difficulties on roughly the same instances. This is not surprising, as the ES has an improvement routine based on a neighborhood similar to N^1. On the other hand, the expected times required for obtaining feasibility and optimality can be dramatically different, indicating that population-based routines and recombination are a good complement to moves within the neighborhoods N^1 and N^2. Instances where the ES is much slower than GRASP are probably due to the absence of improvements based on N^2 on that solver, or to the lack of diversity generated by construction on GRASP. (Comparisons based on CPU time were not made, as the ES results were obtained on very different machines.)

Comparing GRASP to the state-of-the-art commercial solver *Xpress-MP Optimizer, Release 13.02* indicated a clear advantage of that solver, which in most cases could find an optimal solution one to four orders of magnitude faster. Still, this solver had problems on some instances: *bell5* and *noswot* could not be solved in 24 hours of CPU time. For some other instances (*bell3a, pk1, stein27*), Xpress-MP required more LP solutions than GRASP.

26.7 CONCLUSION

In this paper we present a GRASP for the solution of integer linear problems. The algorithm starts by reading an *MPS* file with the instance data. When the problem is a MIP, the integer variables are fixed by GRASP and replaced in the original problem, leading to a pure continuous problem. This problem

Table 26.10 Comparison between GRASP and B&B: percent distance above optimum, expected number of LP solutions and CPU time for GRASP to obtain its best solution, and number of LPs and CPU required by B&B.

Problem name	GRASP % to opt.	B&B % to opt.	GRASP $E[n^b]$	B&B # LPs	GRASP $E[t^b]$	B&B CPU time
bell3a	0	0	11216	438587	34.52	170.56
bell5	0.7118	0	2601738	420499	4880	81.61
egout	0	1.03	977203	55057	2394	6.06
enigma	-	0	671243	8947	85	1.8
flugpl	-	0	6861	1588	2.12	0.06
gt2	70.70	-	55164302	-	4297	-
lseu	0	0	1245115	236207	90.72	23.97
mod008	0	0	871031	2848585	298.9	844.1
modglob	0	26.85	12381	1.00E+08	146.7	47678
noswot	4.651	40.48	24533	3753	28.83	3.72
p0033	0	0	1066553	7393	25.66	0.25
pk1	45.45	0	435039	3710343	181.9	2467
pp08a	0	32.93	2161662	1.00E+08	12571	29924
pp08acut	0	10.34	4173685	1.00E+08	37617	161880
rgn	0	0	2469	4963	2.77	1.2
stein27	0	0	1130	11985	0.08	2.62
stein45	0	0	1110292	236453	248.9	218.6
vpm1	0	10	64719	18354	415.9	24.95

Table 26.11 Comparison between GRASP and an evolutionary solver: percent distance to optimum, and expected number of LP solutions for reaching feasibility and optimality.

Problem	% dist. to optimum		expected number $E[n^f]$		expected number $E[n^o]$	
name	GRASP	ES	GRASP	ES	GRASP	ES
bell3a	0	0.3990	27.14	2053	11216	>18246645
bell5	0.712	0.7143	2256575	33738	>2604076	>18024642
egout	0	0	141.48	423	977203	133764
enigma	-	-	>7439411	>11876637	≫7439411	≫11876637
flugpl	-	0	>248127	29048	≫248127	91004
gt2	70.70	5.556	9194	6383	>55183799	>37665907
lseu	0	0	693.27	1985	1245115	10269416
mod008	0	0	3.4	17	871031	2557585
modglob	0	0	1	3	12381	99478
noswot	4.651	4.651	140.36	33627	>8714283	>34335094
p0033	0	0	332	8350	1066553	93571
pk1	45.46	72.72	1	3	>3492631	>6259152
pp08a	0	0	30.16	49	2161662	177969
pp08acut	0	0	21.18	33	4173685	45582
rgn	0	0	12.09	21	2469	8050
stein27	0	0	12.91	41	1130	286
stein45	0	0	23.97	61	1110292	54791
vpm1	0	0	9.97	123	64719	7397

can be solved by a liner program solver, to evaluate the corresponding fixed variables. When the original problem is an IP, simple algebraic manipulations can be used to evaluate the fixed variables.

The algorithm works with feasible and infeasible solutions. If the solution is feasible, its evaluation is determined directly by the objective function. If the solution is infeasible, the evaluation is given by the sum of constraint violations, which is determined by solving an LP (for MIP problems) or by simple algebraic manipulations (for IP problems).

The results obtained with GRASP for some benchmark problems were compared to those obtained by B&B and to those obtained by an evolutionary solver. The comparison with B&B shows that GRASP has a very interesting behavior, as it determines good feasible solutions in the cases where B&B fails. In the comparison with the evolutionary solver, we could verify that the population-based methods used there could lead many times to substantial reductions on the CPU time required to obtain a given solution. On other cases, substantial CPU time reductions are on the side of GRASP; therefore, no clear conclusion about which of the metaheuristics is better could be drawn. GRASP being simpler, it might be the appropriate choice if implementation burden is to be avoided.

Michel R. Berkelaar and Jeroen Dirks. lp_solve - a solver for linear programming problems with a callable subroutine library. Internet repository, version 2.2, 1994. ftp://ftp.es.ele.tue.nl/pub/lp_solve: Last visited on March 25, 2003.

S. Binato, W. J. Henry, D. Loewenstern, and M. G. C. Resende. A greedy randomized adaptive search procedure for job scheduling. In C. C. Ribeiro and P. Hansen, editors, *Essays and surveys on metaheuristics*, pages 58–79. Kluwer Academic Publishers, 2002.

Robert E. Bixby, Sebastiàn Ceria, Cassandra M. McZeal, and Martin W. P. Savelsbergh. An updated mixed integer programming library. Technical report, Rice University, 1998. TR98-03.

C. Carreto and B. Baker. A GRASP interactive approach to the vehicle routing problem with backhauls. In C. C. Ribeiro and P. Hansen, editors, *Essays and Surveys on Metaheuristics*, pages 185–199. Kluwer Academic Publishers, 2002.

T. A. Feo and M. G. C. Resende. A probabilistic heuristic for a computacionally difficult set covering problem. *Operations Research Letters*, 8:67–71, 1989.

T. A. Feo and M. G. C. Resende. Greedy randomized adaptive search procedures. *J. of Global Optimization*, 6:109–133, 1995.

T. A. Feo, K. Venkatraman, and J. F. Bard. A GRASP for a difficult single machine sheduling problem. *Computers & Operations Research*, 18:635–643, 1991.

P. Festa and M. G. C. Resende. GRASP: an annotated bibliography. In C. C. Ribeiro and P. Hansen, editors, *Essays and Surveys on Metaheuristics*, pages 325–367. Kluwer Academic Publishers, 2002.

Pierre Hansen and Nenad Mladenovic. Variable neighborhood search: Principles and applications. *European Journal of Operational Research*, 130:449–467, 2001.

G. Kontoravdis and J. F. Bard. A GRASP for the vehicle routing problem with time windows. *ORSA J. on Computing*, 7:10–23, 1995.

Thomas Lengauer. *Combinatorial Algorithms for Integrated Circuit Layout*, chapter 8, pages 427–446. Applicable Theory in Computer Science. John Wiley and Sons, 1990.

Y. Li, P. M. Pardalos, and M. G. C. Resende. A greedy randomized adaptive search procedure for the quadratic assignment problem. In P.M. Pardalos and H. Wolkowicz, editors, *Quadratic assignment and related problems*, volume 16 of *DIMACS Series on Discrete Mathematics and Theoretical Computer Science*, pages 237–261. American Mathematical Society, 1994.

George L. Nemhauser and Laurence A. Wolsey. *Integer and Combinatorial Optimization*. Wiley-Interscience in Discrete Mathematics and Optimization, 1988.

João P. Pedroso. An evolutionary solver for linear integer programming. BSIS Technical Report 98-7, Riken Brain Science Institute, Wako-shi, Saitama, Japan, 1998.

João P. Pedroso. An evolutionary solver for pure integer linear programming. *International Transactions in Operational Research*, 9(3):337–352, May 2002.

L. S. Pitsoulis and M. G. C. Resende. Greedy randomized adaptive search procedures. In P. M. Pardalos and M. G. C. Resende, editors, *Handbook of Applied Optimization*, pages 168–183. Oxford University Press, 2002.

M. G. C. Resende and T. A. Feo. A GRASP for satisfiability. In D. S. Johnson and M. A. Trick, editors, *Cliques, Coloring and Satisfiability: The second DIMACS Implementation Challenge*, volume 26 of *DIMACS Series on Discrete Mathematics and Theoretical Computer Science*, pages 499–520. American Mathematical Society, 1996.

M. G. C. Resende, L. S. Pitsoulis, and P. M. Pardalos. Approximate solution of weighted MAX-SAT problems using GRASP. In *Satisfiability problems*, volume 35 of *DIMACS Series on Discrete Mathematics and Theoretical Computer Science*, pages 393–405. American Mathematical Society, 1997.

M. G. C. Resende and C. C. Ribeiro. Greedy randomized adaptive search procedure. In F. Glover and G. Kochenberger, editors, *Handbook of Metaheuristics*, pages 219–249. Kluwer Academic Publishers, 2002.

A.J. Robertson. A set of greedy randomized adaptive local search procedure (GRASP) implementations for the multidimensional assignment problem. *Computational Optimization and Applications*, 19:145–164, 2001.

Laurence A. Wolsey. *Integer Programming*. Wiley-Interscience in Discrete Mathematics and Optimization, 1998.

Metaheuristics: Computer Decision-Making, pp. 575-600
Ana Periera, Filipe Carvalho, Miguel constantino, and João Pedro Pedroso

27 RANDOM START LOCAL SEARCH AND TABU SEARCH FOR A DISCRETE LOT-SIZING AND SCHEDULING PROBLEM

Ana Pereira[1], Filipe Carvalho[1], Miguel Constantino[1], and João Pedro Pedroso[2]

[1]Centro de Investigação Operacional
Faculdade de Ciências da Universidade de Lisboa
Campo Grande, 1749-016 Lisboa, Portugal
anapereira@fc.ul.pt, filipe.carvalho@fc.ul.pt
miguel.constantino@fc.ul.pt

[2]LIACC and DCC
Faculdade de Ciências da Universidade do Porto
Rua do Campo Alegre, 823
4150-180 Porto, Portugal
jpp@ncc.up.pt

Abstract: In this paper we describe random start local search and tabu search for solving a multi-item, multi-machine discrete lot sizing and scheduling problem with sequence dependent changeover costs. We present two construction heuristics with a random component; one of them is purely random and another is based on the linear programming relaxation of the mixed integer programming model. They are used to generate initial solutions for random start local search and tabu search. We also propose two ways of exploring the neighborhoods, one based on a random subset of the neighborhood, and another based on exploring the whole neighborhood. Construction and improvement methods were combined on random start local search and tabu search, leading to a total of eight different methods. We present results of extensive computer experiments for analyzing the performance of all methods and their comparison with branch-

and-bound, and conclude with some remarks on the different approaches to the problem.
Keywords: Lot sizing, Scheduling, Metaheuristics, Tabu search, Random start local search.

27.1 INTRODUCTION

In this paper we consider a discrete lot sizing and scheduling problem (DLSP) with some features that may appear in real life, namely multiple items, multiple machines, sequence dependent changeover costs and convex piece-wise linear backlogging costs, among other. The model is based on one of the instances of the LotSizeLib (Belvaux and Wolsey, 1999). Our purpose is to develop several variants of local search metaheuristics, and compare their performance in several instances of this benchmark.

Many models and variants of lot sizing problems have been proposed by several authors. Kuik et al. (1994) present a classification of models and provide a list of references. The problem considered here falls in the category of discrete lot sizing and scheduling problems. For a survey on models for this class of problems, see for instance (Drexl and Kimms, 1997).

Multi-item, multi-machine discrete lot sizing and scheduling problems consist of finding a minimal cost production schedule over a finite discrete horizon, where in each period a machine may either be producing one item at full capacity or be idle. The demand for each item in each period is known, and can be satisfied either by production in the period or from stock carried from the previous periods.

Here we allow the demand to be satisfied from backlogging from subsequent periods as well. Machine capacity (or the quantity produced) varies with the items and the periods. We consider costs for production, storage, backlogging and changeover. Production costs are fixed for each item. Storage costs are proportional to the quantities left in stock at the end of each period. Backlogging costs are defined by a two step piecewise linear convex function, corresponding to the situation where a large amount of backlogging is highly penalized. Changeover costs are incurred when a machine produces some item (or is idle) in some period and is in a different situation in the subsequent period. They are sequence dependent i.e., for a particular machine and period, they depend on the order in which the items are produced. There is a sale revenue for products in stock at the end of the last period.

This problem is, in general, NP-hard. Even the single machine version of the problem, without backlogging, zero production costs and constant inventory and changeover costs is NP-hard (van Eijl, 1996). Many heuristics have been applied for this kind of problems; see for instance (Barnes and Laguna, 1993).

In this paper, we describe two metaheuristics for this problem: random start local search and a special version of tabu search. The main idea behind random start local search is to repeatedly obtain a local optimum, using different solutions as the starting point for the search, and select the best found local optimum on all the iterations. On simple versions of tabu search, one uses in general a single starting solution, and local optima are avoided by allowing non-improving moves. The inverse of these moves are kept in a tabu list, and become forbidden (tabu) for some iterations (Glover, 1989; 1990). In our paper we use a different version of tabu search, which has been proposed in the same papers but, to the best of our knowledge, was implemented here for the first time for solving practical problems. In this version, we keep track of all the accepted solutions, by means of a hash table. What is forbidden (tabu) is moving back to a solution in that table. We keep track of all the previously accepted solutions, using what is called *rigid memory*. In a recent paper, Hanafi (2000) proved that this version has properties that allow it to converge to the optimal solution, though this is not achievable in realistic problems.

Next we introduce a MIP formulation for the *DLSP* variant considered. We then describe some heuristics for this problem. The heuristics consist in obtaining an initial solution, which may be feasible or not, and then use local search on some neighborhoods for improving the solution. We describe several methods for obtaining initial solutions (construction methods), as well as some improvement methods which use neighborhoods and different ways of exploring these neighborhoods. We present the results of extensive computer experiments for analyzing the performance of the heuristics, and conclude with some remarks and directions for further research.

27.2 MIP PROBLEM FORMULATION

To write a mixed integer programming (MIP) formulation for the problem it is necessary to define some parameters and decision variables. Item $j = 0$ is used with the meaning of *no-production*, i.e., a machine producing item 0 in a given period is idle in that period. All the machines are at no-production in period 0, as well as after the last period of the model. We refer to period 0 as the period before the beginning of the planning horizon. Similarly, we refer to period $T + 1$ as the period after the end of the planning horizon.

Parameters:

- P - number of items/products
- K - number of machines
- T - number of periods
- C_{ikt} - quantity produced in a batch of item i on machine k in period t ($i = 1, \ldots, P$; $k = 1, \ldots, K$; $t = 1, \ldots, T$)
- d_{it} - demand for item i in period t ($i = 1, \ldots, P$; $t = 1, \ldots, T$)
- e_i^0 - initial inventory of item i ($i = 1, \ldots, P$)
- f_i - production cost of item i per batch ($i = 1, \ldots, P$)
- h_i - unit holding cost of item i ($i = 1, \ldots, P$)
- R_i - backlog breakpoint of item i ($i = 1, \ldots, P$)

b_i^0 - unit backlog lower level cost of item i ($i = 1, \dots, P$)

b_i^1 - unit backlog upper level cost of item i ($i = 1, \dots, P$). We assume that $b_i^1 \geq b_i^0$

v_i - unit sale revenue of item i in the last period T ($i = 1, \dots, P$)

u_{ij} - changeover cost for producing item j after item i ($i = 0, \dots, P$; $j = 0, \dots, P$). This may include the *start-up* cost u_{0j} incurred when the machine changes from no-production ($i = 0$) to production (if $j > 0$), and the *switch-off* cost u_{j0} incurred when the machine changes from production (if $j > 0$) to no-production ($j = 0$)

Decision variables:

y_{ikt} - binary variable indicating whether a batch of item i is produced in a machine k in period t ($y_{ikt} = 1$) or not ($y_{ikt} = 0$) ($i = 0, \dots, P$; $k = 0, \dots, K$; $t = 1, \dots, T$)

w_{ijkt} is 1 if on machine k item i is produced on period $t - 1$ and item j is produced on period t, 0 otherwise ($i = 0, \dots, P$; $j = 0, \dots, P$; $k = 1, \dots, K$; $t = 1, \dots, T + 1$)

s_{it} - amount of stock for item i at the end of period t ($i = 1, \dots, P$; $t = 1, \dots, T$)

r_{it}^0 - backlog quantity at lower cost for item i at the end of period t ($i = 1, \dots, P$; $t = 1, \dots, T$)

r_{it}^1 - total backlog quantity for item i at the end of period t ($i = 1, \dots, P$; $t = 1, \dots, T$)

Objective function:

The objective is to minimize the total costs of production, holding, backlogging and changeover, minus the total sale revenues.

For each item i, the backlogging cost is a piece-wise linear convex function, with one breakpoint R_i. In a given period, if r_i is the backlog for product i, the backlog cost is given by

$$g(r_i) = \begin{cases} b_i^0 r_i & \text{if } r_i \leq R_i \\ b_i^0 R_i + b_i^1 (r_i - R_i) & \text{if } r_i > R_i \end{cases}$$

27.2.1 *Mixed integer programming formulation*

The problem is to:

$$\begin{aligned}\min z \;=\; & \sum_{i=1}^{P}\sum_{k=1}^{K}\sum_{t=1}^{T} f_i y_{ikt} + \sum_{i=1}^{P}\sum_{t=1}^{T}\left(h_i s_{it} + b_i^1 r_{it}^1 - (b_i^1 - b_i^0)\, r_{it}^0\right) + \\ & + \sum_{i=0}^{P}\sum_{k=1}^{K}\left(u_{0i} w_{0ik1} + \sum_{j=0}^{P}\sum_{t=2}^{T} u_{ij} w_{ijkt} + u_{i0} w_{i,0,k,T+1}\right) - \\ & - \sum_{i=1}^{P} v_i s_{iT}\end{aligned}$$

subject to:

$$s_{i,t-1} - r_{i,t-1}^1 + \sum_{k=1}^{K} y_{ikt} C_{ikt} = d_{it} + s_{it} - r_{it}^1, i = 1, \ldots, P, t = 1, \ldots, T \quad (27.1)$$

$$s_{i0} = e_i^0,\ i = 1, \ldots, P \quad (27.2)$$

$$r_{i0}^1 = 0,\ i = 1, \ldots, P \quad (27.3)$$

$$\sum_{i=0}^{P} y_{ikt} = 1,\ k = 1, \ldots, K,\ t = 1, \ldots, T \quad (27.4)$$

$$\sum_{i=1}^{P} r_{iT}^1 = 0 \quad (27.5)$$

$$r_{it}^1 \geq r_{it}^0,\ i = 1, \ldots, P,\ t = 1, \ldots, T \quad (27.6)$$

$$r_{it}^0 \leq R_i,\ i = 1, \ldots, P,\ t = 1, \ldots, T \quad (27.7)$$

$$\sum_{i=0}^{P} w_{ijkt} = y_{jkt},\ j = 0, \ldots, P,\ k = 1, \ldots, K,\ t = 1, \ldots, T \quad (27.8)$$

$$\sum_{i=0}^{P} w_{j,i,k,t+1} = y_{jkt},\ j = 0, \ldots, P, k = 1, \ldots, K, t = 1, \ldots, T-1 \quad (27.9)$$

$$w_{0jk1} = y_{jk1},\ j = 0, \ldots, P, k = 1, \ldots, K \quad (27.10)$$

$$w_{j,0,k,T+1} = y_{jkT},\ j = 0, \ldots, P, k = 1, \ldots, K \quad (27.11)$$

$$r_{it}^0 \geq 0,\ i = 1, \ldots, P,\ t = 1, \ldots, T \quad (27.12)$$

$$r_{it}^1 \geq 0,\ i = 1, \ldots, P,\ t = 1, \ldots, T \quad (27.13)$$

$$s_{it} \geq 0,\ i = 1, \ldots, P,\ t = 1, \ldots, T \quad (27.14)$$

$$y_{ikt} \in \{0, 1\},\ i = 0, \ldots, P, k = 1, \ldots, K,\ t = 1, \ldots, T \quad (27.15)$$

$$w_{ijkt} \in \{0,1\}, i,j = 0,\ldots,P,\ k = 1,\ldots,K, t = 1,\ldots,T \tag{27.16}$$

Constraints (27.1) represent the inventory balance for periods $t = 1,\ldots,T$, and constraints (27.2) and (27.3) the initialization of the inventory and backlog variables. They assure that stock and backlog are correctly determined, and that the amount of each item is conserved, for all periods.

Constraints (27.4) guarantee that exactly one item must be produced on a machine in each period (although it may be the no-production item). With constraint (27.5) we assure that at the end all demand must be satisfied, i.e., we cannot backlog at the last period.

Constraints (27.6) and (27.7) are used for correctly calculating the backlog costs as a piecewise linear function of the corresponding quantities, as mentioned before. Constraints (27.7) guarantee that the backlog quantity at the less expensive rate is below the break point, and (27.6) guarantee that the backlog upper level for each item is greater than or equal to its lower backlog level.

Constraints (27.8) and (27.9) define variables used to calculate changeover costs. This way to model the changeovers was proposed in Belvaux and Wolsey (2000). They assure that if an item j is produced on a machine k in period t there must be two changeovers implied: a changeover cost for changing the item being produced in the previous period to item j (27.8), and another cost for changing the production of item j to another item in the following period (27.9). The items produced before or after product j can be the no-production item; in that case, we incur the start-up and switch-off costs, respectively. In the first period it is known that the item preceding item j is the no-production item (27.10), which is also the item following product j in the last period (27.11).

27.3 SOLUTION REPRESENTATION AND COST CALCULATION

For the heuristics presented in this work, the production plan is represented by a matrix x of order $T \times K$, where rows correspond to the periods, columns to the machines, and each element of the matrix holds an integer corresponding to the item being produced; hence $x_{tk} = i$ means that product i is being produced on the machine k at period t. As in Section 27.2.1, $x_{tk} = 0$ means that machine k is stopped at period t.

Let $x = [x_{tk}]$ be a solution; the search space $\mathcal{X}$ is the set of all the solutions (feasible and infeasible) considered for this problem,

$$\mathcal{X} = \left\{ x \in \{0,1,\ldots,P\}^{T\times K} \right\}$$

To each solution there is associated a stock matrix and a backlog matrix. Let the stock for a solution x be a $T \times P$ matrix $s = [s_{ti}]$, where each element $s_{ti} \in \mathbb{R}_0^+$ represents the stock of item i in period t. Similarly, the backlog is a $T \times P$ matrix $r = [r_{ti}]$, with $r_{ti} \in \mathbb{R}_0^+$.

For a given solution x_{tk} we can calculate stock and backlog at all periods $t = 1, \ldots, T$, determining a value

$$\Delta_{it} = s_{t-1,i} - r_{t-1,i} - d_{it} + \sum_{\substack{k=1 \\ x_{tk}=i}}^{K} C_{ikt}$$

which is the inventory of product i if positive, or its backlog (in absolute terms) if negative. Therefore

$$s_{ti} = \begin{cases} \Delta_{it} & \text{if } \Delta_{it} > 0 \\ 0 & \text{otherwise} \end{cases}$$

and

$$r_{ti} = \begin{cases} 0 & \text{if } \Delta_{it} > 0 \\ -\Delta_{it} & \text{otherwise} \end{cases}$$

The stock cost is $\sum_{i=1}^{P} \sum_{t=1}^{T} s_{ti} h_i$ and the sale revenue is $\sum_{i=1}^{P} s_{Ti} v_i$.

The backlog cost is given by $\sum_{i=1}^{P} \sum_{t=1}^{T} g_i(r_{ti})$ where $g_i(r_{it})$ is determined by the following expression:

$$g(r_{it}) = \begin{cases} b_i^0 r_{it} & \text{if } r_{it} \leq R_i \\ b_i^0 r_{it} + b_i^1 (r_{it} - R_i) & \text{if } r_{it} > R_i \end{cases} \tag{27.17}$$

In this model there are changeover costs. These correspond to start-up and switch-off costs when, respectively, the previous or next items to be produced in a given machine is 0 (i.e., the no-production item). The initial and final state of all the machines is no-production; taking this into consideration, we can calculate the changeover costs as:

$$\sum_{k=1}^{K} \left(u_{0,x_{1k}} + \sum_{t=2}^{T} u_{x_{t-1,k},x_{tk}} + u_{x_{TK},0} \right)$$

Finally, there are fixed production costs associated to each batch. Whenever a machine produces an item, there is a fixed production cost f_i ($i = 1, \ldots, P$), which varies according to the item. Therefore, the batches production cost is given by

$$\sum_{t=1}^{T} \sum_{k=1}^{K} f_{x_{tk}}$$

The total cost of a solution x can be calculated as:

$$\begin{aligned} c(x) &= \sum_{i=1}^{P}\sum_{t=1}^{T}(s_{ti}h_i + g_i(r_{ti})) - \sum_{i=1}^{P} s_{Ti}v_i \\ &+ \sum_{k=1}^{K}\left(u_{0,x_{1k}} + \sum_{t=2}^{T} u_{x_{t-1,k},x_{tk}} + u_{x_{TK},0}\right) \\ &+ \sum_{t=1}^{T}\sum_{k=1}^{K} f_{x_{tk}} \end{aligned}$$

A solution is feasible if and only if the demand for all items is completely satisfied within the planning horizon, i.e, there is no backlog at the last period. Solution infeasibility is measured as the sum of the backlog variables at the last period:

$$\mathrm{Inf}(x) = \sum_{i=1}^{P} r_{Ti}.$$

The set of feasible solutions $\mathcal{F}$ is defined as

$$\mathcal{F} = \{x \in \mathcal{X} : \mathrm{Inf}(x) = 0\}.$$

When comparing two solutions, we need to decide which one is preferable based on their objective value and on their deviation from feasibility. In the context of this paper, for two solutions $x, y \in \mathcal{X}$ we say that x *is better* than y iff:

- $\mathrm{Inf}(x) < \mathrm{Inf}(y)$ or
- $\mathrm{Inf}(x) = \mathrm{Inf}(y)$ and $c(x) < c(y)$.

This means that the less infeasible solution is always preferable. If both solutions have the same infeasibility then the best solution is the one with the smallest objective value.

27.4 CONSTRUCTION HEURISTICS

We present two heuristics that randomly construct a solution. They require a very short computational time, but in general fail to produce feasible solutions for difficult instances. We have tried several other, more traditional, construction procedures, but have not obtained significantly different results. Construction heuristics are used to generate initial solutions for local search (metaheuristics); actually, as the quality of their solutions can be rather poor, they simply are expected to provide good starting solutions for the local search procedures.

27.4.1 Random construction

This procedure constructs a solution in a random way. It randomly chooses an item to be produced on every machine, for all time periods. The choice is made taking into account the possibility of leaving the machine switched-off (i.e., product zero can be chosen with the same probability of any other product).

27.4.2 Construction based on the linear programming relaxation

The information provided by the solution of the linear relaxation of a mixed integer programming model can be used in a construction heuristic to obtain an integer solution. An initial application of the idea that we are using was published in Lengauer (1990).

This heuristic takes the solution of the linear programming (*LP*) relaxation of the MIP problem and constructs an integer solution with the information obtained on the variables y_{ikt}. In the MIP formulation, y_{ikt} is a binary variable which indicates whether an item i is produced in a machine k in period t ($y_{ikt} = 1$) or not ($y_{ikt} = 0$). Furthermore, the sum of the variables concerning a machine in a time period is constrained to 1. Let us call y_{ikt}^{LP} the optimal value of these variables on the LP relaxation of the MIP. To obtain an integer solution, we create a solution where item i is produced on machine k in time t with probability y_{ikt}^{LP}:

$$\mathrm{Prob}(x_{tk} = i) = y_{ikt}^{\mathrm{LP}}.$$

With this heuristic we build solutions that, although still infeasible in general, lead to better solutions than the previous methods when used as initial solutions in local search metaheuristics.

27.5 NEIGHBORHOODS

The metaheuristics considered here use three types of neighborhoods: *n-change*, *swap* and *insert*. The *n-change* neighborhood takes n elements of the solution matrix, and for each of them (independently) change the item being produced into another item. The changes induced by the other two neighborhoods imply swapping elements of the matrix or its combination with shifting some elements in the matrix.

27.5.1 n-change neighborhood

Let x be a solution on the search space $\mathcal{X}$. We say that a solution $y \in N_{change}^{n}(x)$ iff y and x differ in at most n elements:

$$N_{change}^{n}(x) = \{y \in \mathcal{X} : y \text{ can be obtained from x by changing at most n elements } x_{tk} \text{ from their current value into an element of the set } \{0, 1, \ldots, P\}\}.$$

The *n-change* neighborhood is only computationally affordable for small values of n, otherwise they are too large to explore. In this paper we are using n=1. This neighborhood is particularly useful when there is an extreme excess or lack of certain products, as usually occurs on randomly constructed solutions, because it helps in balancing the products accordingly to their demand. However, it is not effective when the solution needs to be improved on machine changeover costs.

27.5.2 Swap neighborhood

The *swap* neighborhood of a solution x, $N_{swap}(x)$, is composed of solutions with all the elements identical to those of x, except for two elements which are swapped.

$$N_{swap}(x) = \{y \in \mathcal{X} : y \text{ can be obtained from } x \text{ by interchanging the position of two of its elements}\}.$$

Exploring this neighborhood of a solution is useful when the solution contains all the products in similar quantities, and needs to improve locally on some product's backlog and stock. It is also very effective on minimizing machine changeovers.

27.5.3 Insert neighborhood

The *insert* neighborhood of a solution x consists on solutions obtained by removing an element of a column of x and inserting it on another position of the same column, by shifting all the lines of that column that were between the two positions. Hence, the production plan is modified only on one machine, between the two periods concerned.

$$N_{insert}(x) = \{y \in \mathcal{X} : y \text{ can be obtained from } x \text{ by removing an element } x_{t_1,k}, \text{ shifting all the elements next to } t_1 \text{ up to another position on the same column } t_2, \text{ and inserting } x_{t_1,k} \text{ on } t_2\}.$$

The movement can be made upwards or downwards, which means that an element can be moved to a preceding or succeeding period.

This neighborhood was defined to allow the search for solutions that introduce small improvements on a product's backlog and stock costs. But it is especially useful to minimize machine changeover costs.

27.5.4 How to make computationally efficient changes

A solution is represented by a matrix holding the production plan, but more data structures are kept updated during the search. This is done to efficiently perform rather complicated functions of the solution, such as its cost, or stock and backlog for all periods. Updating all this data can be very time consuming. However, when we want to make a change on a solution, using one of the mentioned procedures, *3-change*, *swap* or *insert*, part of the original solution is preserved. Ignoring this fact when updating a solution can be critical to the computational performance of the algorithm. After a move, all that should be updated in a solution is what has effectively changed. For instance, in the *insert* change the production of a single machine is affected and only between two time periods. In terms of costs it is not necessary to recalculate the setup costs of other machines, as they are not affected by the change. On the other hand, only stock or demand variables related to products being changed need to be updated.

27.6 RANDOM START LOCAL SEARCH AND TABU SEARCH

Four versions of random start local search (RSLS) have been implemented. Each of them repeatedly constructs a solution with one of the construction methods, and improves it using one of the methods described below, until reaching a local optimum. They hence differ on the construction procedures and on the way improvements are made. A high level description of the required procedures is presented in Algorithms 27.1, 27.2, and 27.3. The method stops at the end of a predetermined CPU time T.

procedure IMPROVE(x)
$S \leftarrow N(x)$
while $S \neq \{\}$
  $s \leftarrow$ RANDOMCHOICE(S)
  if s is better than x
    return s
  $S \leftarrow S \setminus \{s\}$
endwhile
return x

Algorithm 27.1 Solution improvement for random start local search.

On the first RSLS, improvements are made as described in Section 27.6.1, and on the second one as in Section 27.6.2.

We also propose four versions of tabu search (TS) implementations. These are based on keeping track of tabu solutions (as opposed to tabu moves) as proposed in Hanafi (2000); Glover and Hanafi (2002). In this case, whenever a solution that improves the current best solution is found, we are sure that it cannot be on the tabu list; therefore, there will be no need for aspiration criteria. In our implementation, there is not an intensification/diversification strategy; the search space is so large, that for a reasonable amount of CPU time we do not observe stagnation. The tabu search methods differ on the

```
procedure LOCALSEARCH(x)
s ← IMPROVE(x)
while s ≠ x
    x ← s
    s ← IMPROVE(x)
endwhile
return s
```

Algorithm 27.2 Local search routines for random start local search.

```
procedure RANDOMSTARTLOCALSEARCH(T)
t ← TIME()
while TIME() − t <= T
    x ← CONSTRUCT()
    x ← LOCALSEARCH(x)
    if x* not initialized or x is better than x*
        x* ← x
endwhile
return x*
```

Algorithm 27.3 Random start local search.

construction procedures and on the way a solution is tentatively improved on each iteration.

To escape from local optima, on tabu search, we have sometimes to make non-improving moves. To prevent cycling (falling on the same local optimum after a non-improving move), we keep a list of all the previously accepted solutions, by means of a hash table. Solutions on this table are tabu. The only exception is a 'blockage' situation. If this situation happens then all neighbors are tabu. To escape this situation we accept the oldest neighbor in the tabu list. However, in all our computational experiences a 'blockage' situation never happened. This may be due to the combined use of several neighborhoods, thus enlarging the search space very considerably, making the possibility of having visited all solutions in a neighborhood very unlikely.

A TS procedure takes an initial solution, constructed using one of the methods described in Section 27.4, and iteratively tries to improve that solution according to one of the methods that are described below. If it fails to improve a solution, then it escapes to a random non-tabu solution in one of the neighborhoods, and tries to improve again from that point. The method stops at the end of a predetermined CPU time T.

A high level description of the procedures required for tabu search is presented in Algorithms 27.4, 27.5, and 27.5.

```
procedure IMPROVE(x, L)
S ← N(x) \ L
while S ≠ ∅
    s ← RandomChoice(S)
    if s is better than x
        L ← L ∪ {s}
        return s
    endif
    S ← S \ {s}
endwhile
return x
```

Algorithm 27.4 Solution improvement for tabu search.

```
procedure LOCALSEARCH(x, L)
s ← IMPROVE(x, L)
while s ≠ x
    x ← s
    s ← IMPROVE(x, L)
endwhile
return s
```

Algorithm 27.5 Local search routines for tabu search.

One difference between RSLS and TS is that on the TS the set of solutions that can be used in each movement excludes the tabu solutions (which correspond to the list of all the previously accepted solutions). Another difference is the way of escaping local optima. On RSLS we move to a newly constructed solution, while on TS we move to a random, non-tabu neighbor solution.

27.6.1 Triple improvements

Note that there are two possibilities for updating the best solution when searching a particular neighborhood.

The first one, called *best-updating* or *breadth-first*, consists of searching the best solution y^* in the entire neighborhood of the current solution. If it is better than our current solution x, then replace x by y^*. This method will hence search the whole neighborhood of the current solution, even if improving solutions can be found on the early exploration of the neighborhood.

The second method, called *first-updating* or *depth-first*, consists of replacing x during the local search whenever the currently generated neighbor, y^i, is better than x. In our implementation we used first-updating, because it generally provides superior results in terms of efficiency and quality of solutions.

```
procedure TABUSEARCH(T)
t ← TIME()
x ← CONSTRUCT()
L ← ∅
while TIME() − t <= T
    x ← LOCALSEARCH(x, L)
    if x* not initialized or x is better than x*
        x* ← x
        if RandomChoice(N(x) \ L) = ∅
            x ← first element of L
            move first element of L to the end
        else
            x ← RandomChoice(N(x) \ L)
        endif
    endif
endwhile
return x*
```

Algorithm 27.6 Tabu search.

The explanation might be related to the number of solutions "tried" in each local search, which is much larger in first-updating.

Local search using *triple improvement* takes a solution and runs sequentially through three kinds of neighborhoods – *insert*, *swap* and *1-change*. For each neighborhood, it searches an improved solution in a first-updating way.

On tabu search, solutions in the tabu list are not accepted (except in a blockage situation), and all accepted solutions are inserted in the tabu list.

If on a triple improvement we were not able to improve the current solution, then we are on a local optimum of the three combined neighborhoods. In this situation, on RSLS we start a new iteration, and on TS we make an escape, non-improving move.

27.6.2 Probabilistic improvements

In this method there is a parameter R which determines the number of moves that is tried on the current solution. For each of the R attempts, a neighborhood is drawn from one of the *insert*, *swap* and *1-change* neighborhoods, according to the empirical probabilities of success given by the following expression:

$$P(N_i) = \frac{S_{N_i}/T_{N_i}}{\sum_j S_{N_j}/T_{N_j}}$$

In this expression S_{N_i} is the number of successful attempts with neighborhood N_i, and T_{N_i} is the total number of attempts with this neighborhood, with i identifying neighborhoods *insert*, *swap*, and *1-change*. All the variables S_{N_i}, T_{N_i} are initialized with the value one.

After a neighborhood is chosen, a solution is randomly drawn from this neighborhood. If during these attempts an improving move is found, the method (immediately) returns it. Otherwise, we assume that we have reached a local optimum, and we apply an escaping procedure.

27.6.3 Method combination

All the heuristics presented in this paper are combinations of improvement methods applied to solutions resulting from construction heuristics. We described two construction heuristics: random (RND) in 27.4.1, and based on the linear programming relaxation (LP) in 27.4.2. We also described methods for searching the different neighborhoods: probabilistic (PROB) in 27.6.2, and triple (TRPL) in 27.6.1. Random start local search (RSLS) and tabu search (TS) variants were implemented for all the combinations of construction with improvement methods. This means that there are 8 combined methods: RSLS/RND/PROB, RSLS/RND/TRPL, RSLS/LP/PROB, RSLS/LP/TRPL, TS/RND/PROB, TS/RND/TRPL, TS/LP/PROB and TS/LP/TRPL.

27.7 COMPUTATIONAL RESULTS

We generated 18 instances of 3 different sizes. All of the 8 methods were tested on these instances 20 times. We also tried to obtain optimal values (or good lower bounds) running branch and bound (B&B) on these instances. Optimal values were obtained for smaller instances; for mid-size and larger instances with 5 hours of computation we obtained feasible integer solutions, but could not prove optimality. Actually, we checked empirically that for a mid-size instance, running B&B for more than a week led to a solution that was hardly better than that computed in 5 hours.

27.7.1 Test problems

The size of an instance is measured by the amount of products, machines and time periods. The six smaller instances have 4 products, 4 machines and 8 time periods, and were solved to optimality by B&B. However, a small increase in the problem size is enough for the B&B to be unable to solve those problems. The increase considered is to 5 products, 5 machines and 12 time periods; we have also generated six of these mid-size instances. The six larger instances have 10 products, 10 machines and 25 time periods.

Each of these 18 instances was generated using several statistical distributions for each product's demand along the time periods, and they all are available in Pereira et al. (2001).

27.7.2 *Quality of solutions*

For the analysis of the solutions, we have considered the three different sizes separately, as the behavior of the algorithm for small and mid-size instances differs considerably when compared to that of large instances.

The lack of good lower bounds does not allow us to properly evaluate the quality of the solutions obtained for mid and large-size instances.

On small instances the performance of the different metaheuristic methods do not differ much. Some runs achieve optimality but most of them stay at a near-optimal solution.

Branch and bound did not achieve optimal solutions for mid-size instances, but analysis of the B&B tree showed that the range between upper and lower bounds varied from 5% to 20%. These solutions are very close to the optimum, and it was difficult for the metaheuristic methods to find better solutions on these instances. Notice that even though B&B solutions are better, its computational time is considerably larger.

On the other hand, on large instances B&B does not obtain good solutions (here, lower to upper bound ranges were from 32% to 45%). In these cases, it is easier for the metaheuristics to get better results, even within a much smaller CPU time. Figure 27.1 graphically compares the different metaheuristic methods with B&B. For a given CPU time (x-axis) we plot the percentage of runs of a given metaheuristic where the best solution found was better than the final solution of B&B. Ties are broken with computational time, but there remains some ambiguity because at some time instants the solutions obtained by both methods were identical. Graphical results for small instances are not shown as they were very similar to those of mid-size instances.

The graphic for mid-size instances shows that for all metaheuristics only a very small percentage of runs performed better than B&B. However, for large instances the corresponding graphic shows that most of the metaheuristics have more than 50% of runs finding better solutions than B&B after some computational time.

27.7.3 *Computational effort*

Each of the metaheuristics is allocated a run time of 1 minute for small instances, 5 minutes for mid-size instances and 30 minutes for large instances, after which they are stopped. The time to reach the best solution found for each of the runs averages 30 seconds for small instances, 3 minutes for mid-size instances and 16 minutes for large instances. This represents about 50% of the total running time.

Feasibility is obtained on the first seconds of computation and it was always achieved by all methods on all instances.

27.7.4 *Comparisons between metaheuristics*

27.7.4.1 Initialization methods. The construction methods that provide initial solutions have a crucial relevance on the performance of the metaheuris-

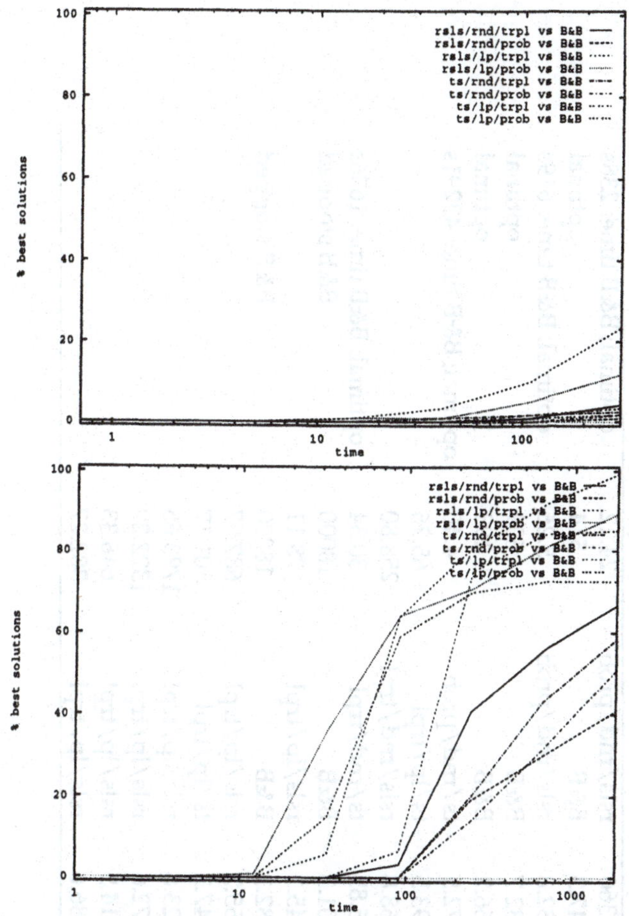

Figure 27.1 Percentage of runs where metaheuristics were better than B&B for mid-size instances (top) and for large instances (bottom).

Table 27.1 Best known results for all benchmark instances.

instance	*best solution*	*best method*	time (s)	*remarks*
p4x4x8-1	14366	rsls/rnd/prob	24.12	optimal; B&B time: 238s
p4x4x8-2	15347.3	B&B	514	optimal
p4x4x8-3	4732.8	rsls/rnd/prob	5.26	optimal; B&B time: 619s
p4x4x8-4	8682.7	B&B	609	optimal
p4x4x8-5	48166.2	B&B	193	optimal
p4x4x8-6	24572.5	ts/rnd/prob	4.40	optimal; B&B time: 43291s
p5x5x12-1	27893.5	ts/lp/trpl	65.26	
p5x5x12-2	21168.8	rsls/rnd/trpl	258.80	
p5x5x12-3	12935.85	ts/rnd/trpl	30.34	optimal; B&B time: 1636s
p5x5x12-4	14401.3	B&B	18000	B&B stopped
p5x5x12-5	70545.1	rsls/lp/trpl	78.11	
p5x5x12-6	40082.1	B&B	18000	B&B stopped
p10x10x25-1	63055.7	rsls/lp/trpl	687.97	
p10x10x25-2	53347.2	ts/lp/trpl	501.17	
p10x10x25-3	22323.6	rsls/lp/trpl	1793.65	
p10x10x25-4	35972.6	rsls/lp/trpl	1322.29	
p10x10x25-5	204014.5	rsls/lp/trpl	646.33	
p10x10x25-6	113786.6	rsls/lp/trpl	382.27	

tics. The probabilistic construction heuristic (in Section 27.4.2) based on the linear programming solution is better than the random heuristic presented (in Section 27.4.1), and this superiority is especially noticed as the size of the instances increases. Figure 27.2 shows graphical comparisons between the runs with RND initialization and LP initialization. It states the percentage of runs with RND initialization better than the runs with LP initialization. In these graphics we compare the results of both metaheuristics through time, for small and large instances. The results for mid-size instances are not shown as they are very similar to those presented for small instances.

The graphic for small instances shows that only a small percentage of runs of metaheuristics with RND initialization are better than runs of metaheuristics with LP initialization in the first seconds of computation, but as the methods evolve through time the initialization influence tends to vanish. In large instances the difference between RND and LP initializations is emphasized as methods with LP initialization obtain better solutions through the entire computational experience.

27.7.4.2 Improvement methods. In terms of empirical performance, it is difficult to choose the best improvement method. The probabilistic search method (in Section 27.6.2) gets good solutions faster, but the solutions obtained by the triple search method (in Section 27.6.1) become better after of a certain amount of time. However, if we can say that for small and mid-size instances these methods tend to perform similarly, we remark a considerable difference for larger instances. Here we notice that the triple search is better than the probabilistic search after an initial phase. Figure 27.3 graphically compares runs with TRPL and PROB search methods for small and large instances. It was built according to the same parameters as the graphics in the previous figure.

27.7.4.3 RSLS and TS metaheuristics. Solutions obtained by methods on the RSLS implementation are better, in average, than those obtained by TS; still, it is not possible to decide for one of the methods empirically. Figure 27.4 graphically compares the runs of RSLS and TS for mid-size and large instance. The results for small instances are not shown as they were very similar to those presented for mid-size instances.

In both graphics of Figure 27.4 we can observe that there are two pairs of lines with similar appearance. One pair of lines for methods with RND construction and another for methods with LP construction. Their curves show that, when using LP construction, RSLS is usually better than TS but when they use RND construction then TS is better than RSLS.

We have to notice that RSLS is always restarting from new solutions (sometimes very good solutions, when given by the LP initialization), and TS only applies an escaping move, therefore remaining closer to the previous solution.

RSLS has a faster implementation, as it does not have to check for solutions in the tabu table; this might allow RSLS to explore more solutions than TS,

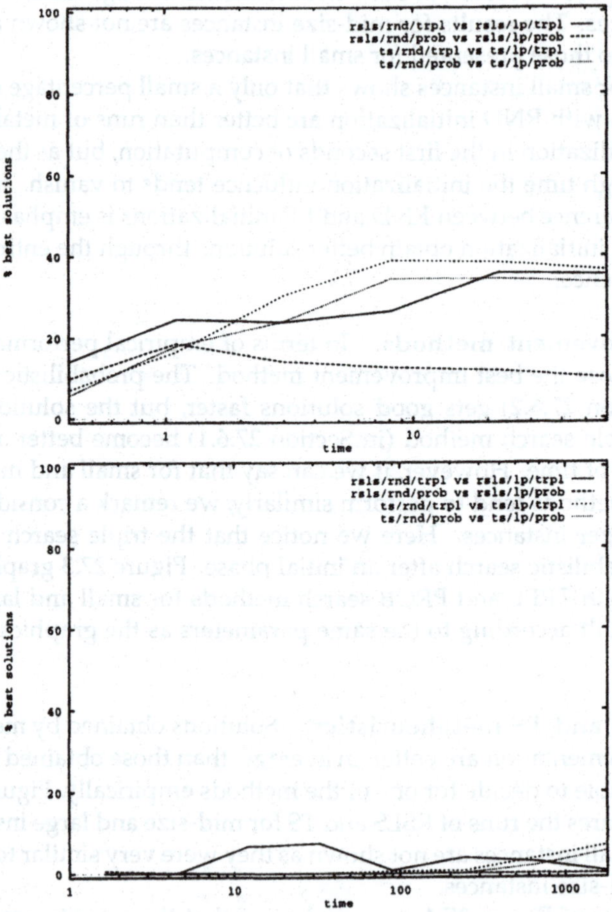

Figure 27.2 Percentage of runs with RND initialization better than the runs with LP initialization for small instances (top) and for large instances (bottom).

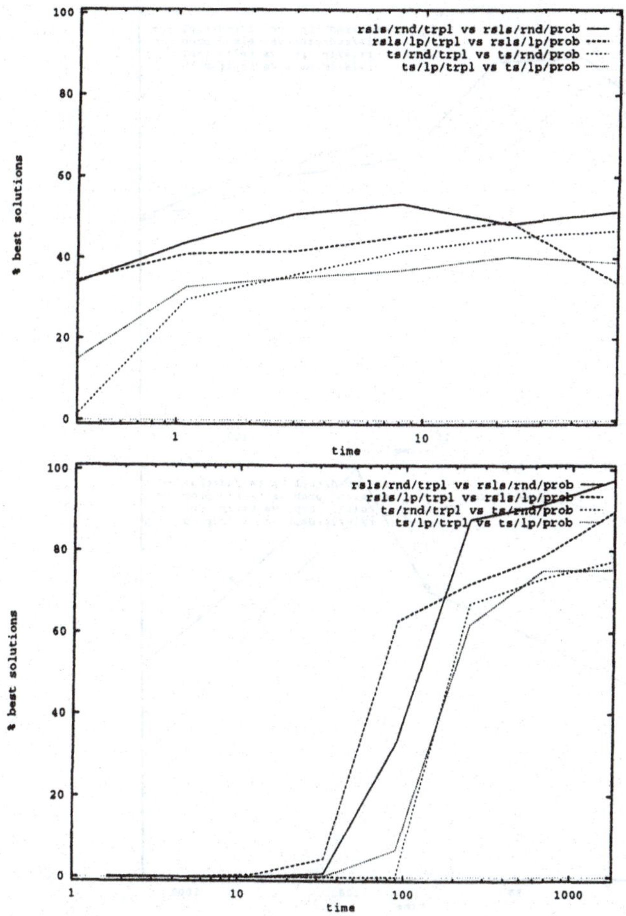

Figure 27.3 Percentage of runs with TRPL improvements better than the runs with PROB improvements for small instances (top) and for large instances (bottom).

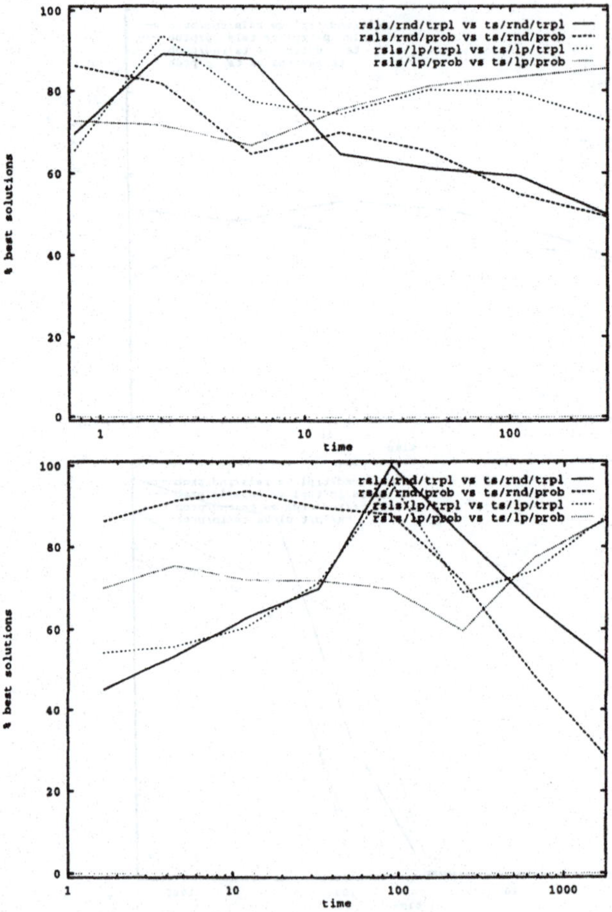

Figure 27.4 Percentage of RSLS runs that were better than TS runs for mid-size instances (top) and for large instances (bottom).

and in some cases be more competitive. For the CPU times used, and due to the complexity of the instances studied in this paper, neither TS nor RSLS have reached a stagnation point, where no improvements were obtained for a large number of iterations. It would be interesting to study the behavior of the methods for much larger CPU times.

It would also be interesting to apply this TS approach to other problems as it is easy to use and implement, and the results obtained with this experience are very promising.

27.7.5 Implementation Environment

All the metaheuristics were implemented in the interpreted language "Python" version 2.0. The MIP solver used is the DASH XPRESS system, which consists of C and Fortran routines. All the tests were performed on a dedicated Pentium 800 MHz PC with 1Gb of RAM and the Win2000 operating system.

27.8 CONCLUSION

In this paper, two construction methods and two improvement methods have been proposed for a Discrete Lot-sizing and Scheduling Problem. These methods were combined on Random Start Local Search and Tabu Search, leading to eight different metaheuristics variants.

The variants of Tabu Search implemented are based on a version proposed by Hanafi (2000), and are, to the best of our knowledge, implemented for the first time by the work presented in this paper.

The results obtained allowed us to conclude that for small instances the B&B method is faster and better than the metaheuristics, but when the size of the instances becomes larger the metaheuristics become comparatively better. A method using a construction based on the Linear Programming relaxation usually achieves better solutions than a method using a random construction. Selecting a random subset of neighborhoods does improve the quality of the solutions initially, but this method is overtaken by the exploration of the whole neighborhood when time increases.

We also proposed a graphical representation to compare non-deterministic algorithms, even when some solutions may be infeasible, based on the number of times that a given method obtains a better solution than another, for a fixed CPU time. This allowed the comparison of the very complex behavior of the algorithms, which would be extremely difficult to analyze with traditional methods.

Table 27.2 Expected run time to reach feasibility for all RSLS and TS methods, after 20 runs. Note: (1) RND/TRPL; (2) RND/PROB; (3) LP/TRPL; (4) LP/PROB.

	RSLS				*TS*			
instances	$rt_{(1)}$	$rp_{(2)}$	$lt_{(3)}$	$lp_{(4)}$	$rt_{(1)}$	$rp_{(2)}$	$lt_{(3)}$	$lp_{(4)}$
p4x4x8-1	0.67	0.60	0.21	0.22	3.36	1.15	1.31	0.33
p4x4x8-2	2.56	0.96	0.56	0.34	4.15	1.48	2.46	0.77
p4x4x8-3	1.54	1.33	0.83	0.12	3.48	2.94	2.87	0.54
p4x4x8-4	1.13	0.74	0.47	0.39	5.13	3.51	6.05	1.50
p4x4x8-5	1.13	0.89	0.39	0.34	1.52	1.32	0.83	0.68
p4x4x8-6	28.63	41.5	16.19	4.42	31.61	40.46	32.46	21.75
p5x5x12-1	1.82	0.47	0.40	0.14	6.27	1.14	1.24	0.42
p5x5x12-2	2.81	0.54	1.64	0.20	3.67	1.07	3.37	0.36
p5x5x12-3	0.98	0.47	0.32	0.08	3.82	1.17	0.59	0.17
p5x5x12-4	1.35	0.41	1.11	0.20	3.78	1.06	2.21	0.33
p5x5x12-5	2.58	0.67	1.82	0.22	3.59	0.93	3.50	0.30
p5x5x12-6	3.57	0.94	2.23	0.37	7.69	1.43	4.28	0.52
p10x10x25-1	27.30	3.23	21.77	0.92	112.08	7.21	75.01	1.76
p10x10x25-2	73.11	4.06	41.21	1.70	113.67	5.21	82.65	2.23
p10x10x25-3	34.96	2.22	25.87	0.57	98.74	4.67	64.66	0.97
p10x10x25-4	36.79	2.69	24.06	0.98	109.26	5.23	79.75	1.73
p10x10x25-5	70.53	3.88	40.91	1.71	112.04	4.84	80.09	2.10
p10x10x25-6	70.29	4.56	38.03	1.81	113.27	5.97	82.50	1.97

Bibliography

J. Wesley Barnes and Manuel Laguna. A tabu search experience in production scheduling. In F. Glover, M. Laguna, E. Taillard, and D. de Werra, editors, *Annals of Operations Research*, volume 41, pages 141–156. J. C. Baltzer AG, 1993.

G. Belvaux and L. A. Wolsey. Lotsizelib: A library of models and matrices for lot-sizing problems. Internal report, Center for Operations Research and Econometrics, Universite Catholique de Louvain, 1999.

Gaetan Belvaux and Laurence A. Wolsey. Modelling practical lot-sizing problems as mixed integer programs. Discussion paper, Center for Operations Research and Econometrics, Belgium, 2000.

A. Drexl and A. Kimms. Lot sizing and scheduling – Survey and extensions. *European Journal of Operational Research*, 99:221–235, 1997.

Fred Glover. Tabu search – Part I. *ORSA Journal on Computing*, 1:190–206, 1989.

Fred Glover. Tabu search – Part II. *ORSA Journal on Computing*, 2:4–32, 1990.

Fred Glover and Saïd Hanafi. Tabu search and finite convergence. *Discrete Applied Mathematics*, 119(1–2):3–36, 2002.

Saïd Hanafi. On the convergence of tabu search. *Journal of Heuristics*, 7:47–58, 2000.

R. Kuik, M. Salomon, and L.N. van Wassenhove. Batching decisions: structure and models. *European Journal of Operational Research*, 75:243–263, 1994.

Thomas Lengauer. *Combinatorial Algorithms for Integrated Circuit Layout*, chapter 8, pages 427–446. John Wiley and Sons, 1990.

A. Pereira, F. Carvalho, J. Pedroso, and M. Constantino. Benchmark instances for dynamic lot sizing and scheduling problems. Internet repository, 2001. http://www.ncc.up.pt/ jpp/dlsp.

C. van Eijl. *A Polyhedral Approach to the Discrete Lot-Sizing and Scheduling Problem*. PhD thesis, Technische Universiteit Eindhoven, 1996.

Metaheuristics: Computer Decision-Making, pp. 601-614
Isabel Rosseti, Marcus Poggi de Aragão, Celso C. Ribeiro, Eduardo Uchoa,
and Renato F. Werneck

28 NEW BENCHMARK INSTANCES FOR THE STEINER PROBLEM IN GRAPHS

Isabel Rosseti[1], Marcus Poggi de Aragão[1], Celso C. Ribeiro[1],
Eduardo Uchoa[1], and Renato F. Werneck[1]

[1]Department of Computer Science
Catholic University of Rio de Janeiro
Rua Marquês de São Vicente, 225
Rio de Janeiro, 22453-900, Brazil
rosseti@inf.puc-rio.br, poggi@inf.puc-rio.br, celso@inf.puc-rio.br
uchoa@inf.puc-rio.br, rwerneck@inf.puc-rio.br

Abstract: We propose in this work 50 new test instances for the Steiner problem in graphs. These instances are characterized by large integrality gaps (between the optimal integer solution and that of the linear programming relaxation) and symmetry aspects which make them harder to both exact methods and heuristics than the test instances currently in use for the evaluation and comparison of existing and newly developed algorithms. Our computational results indicate that these new instances are not amenable to reductions by current preprocessing techniques and that not only do the linear programming lower bounds show large gaps, but they are also hard to be computed. State-of-the-art heuristics, which found optimal solutions for almost all test instances currently in use, faced much more difficulties for the new instances. Fewer optimal solutions were found and the numerical results are more discriminant, allowing a better assessment of the effectiveness and the relative behavior of different heuristics.
Keywords: Steiner problem in graphs, Benchmark instances, Test instances, Algorithms.

28.1 INTRODUCTION

Let $G = (V, E)$ be a connected undirected graph, where V is the set of nodes and E denotes the set of edges. Given a non-negative weight function $w : E \rightarrow$

R_+ associated with its edges and a subset $X \subseteq V$ of terminal nodes, the Steiner problem in graphs (SPG) consists of finding a minimum weighted subtree of G spanning all nodes in X. The solution of SPG is a Steiner minimum tree. This is one of the most widely studied NP-hard problems, with many applications.

Three sets of benchmark instances are commonly used in the literature to assess the performance of algorithms for the SPG: instances available from the online repository OR-Library (Beasley, 1990), the "incidence" instances of Duin (1994) (see also (Duin and Voss, 1997)), and the VLSI instances of Koch and Martin (1998). All these instances are available from the SteinLib repository (Martin and Voss, 2000). However, these instances are rapidly becoming inadequate for benchmarking, for the following reasons:

- They have already been solved to optimality, with the exception of a few of the larger incidence instances (Duin, 1994; Koch and Martin, 1998; Lucena and Beasley, 1998; Uchoa et al., 2002; Polzin and Vahdati, 2001; de Aragão et al., 2001b). In fact, optimality can be proved within seconds in most cases. These instances are not challenging enough to stimulate further development of exact algorithms.

- Metaheuristics can easily find optimal solutions to a large portion of these instances (Duin and Voss, 1997; 1999; Gendreau et al., 1999; Martins et al., 2000; Ribeiro and Souza, 2000; Bastos and Ribeiro, 2002; Ribeiro et al., 2002). It is becoming increasingly hard to compare different metaheuristics properly. Deciding among different variants, parameter values, and implementation choices is also difficult, since they often lead to quite similar solutions, which are frequently optimal.

This paper proposes three new series of benchmark instances that will hopefully lead to a better assessment of exact and approximate algorithms for the SPG. Even though some of them are somewhat artificial, we believe they can play an important role in the development of algorithms whose ultimate goal is solving real-world instances. Algorithms that can cope with a variety of artificially hard instances tend to be very robust.

Our first goal was to design instances hard to be solved exactly. Current state-of-the-art exact algorithms are based on linear programming formulations that yield very tight bounds for most existing benchmark instances. They can solve several of these instances to optimality without branching. Even when branching is necessary, it is often possible to reduce the problem size significantly using reduction tests and fixation by reduced costs. Therefore, we tried to create instances with large duality gaps. To make them even harder, we introduced various degrees of symmetry, both in the structure of the underlying graphs and in terminal placement. In practice, symmetry increases the number of nodes in the search trees of branch-and-bound algorithms, since LP-based lower bounds tend to improve very slowly as branchings are performed.

The other design goal was to make the instances challenging also for metaheuristics. For each instance originally created, we generated another one with the same structure, but with perturbed edge weights. Since this results in instances with a much smaller number of optimal solutions, finding one of them by usual heuristic search methods tends to be a harder task.

Three classes of instances with a total of 50 reasonably small test problems are proposed and described in Section 28.2. The number of nodes in each graph ranges from 64 to 4096, while the number of edges ranges from 128 to 28512. This means that, although much harder, our instances are not bigger than those currently in use. Section 28.3 presents some computational results and discusses the effectiveness of the new instances in terms of achieving their goal. Concluding remarks are made in Section 28.4.

28.2 INSTANCES

28.2.1 Hypercube (hc)

Graphs in this series are d-dimensional hypercubes, with $d \in \{6, \ldots, 12\}$. For each value of d, the corresponding graph has 2^d nodes and $d \cdot 2^{d-1}$ edges. These graphs are bipartite (because all cycles are even, see e.g. (West, 2000)) and both partitions have the same number of nodes. The vertices in one of such subsets become terminals ($|X| = 2^{d-1}$). Edge weights in the originally created instances are unitary. The perturbed instances have integer edge weights randomly chosen from a uniform distribution in the interval [100, 110]. These instances seem to be extremely difficult for existing algorithms. For example, using the branch-and-ascent algorithm proposed by de Aragão et al. (2001b), we could not solve to optimality the instances with more than 128 nodes. Duality gaps are large and symmetry makes traditional branching schemes much less effective.

The naming convention is hc$d$[u|p], where u stands for "unperturbed" and p for "perturbed". For example, hc8u corresponds to an 8-dimensional hypercube with unperturbed weights. The pseudocode in Figure 28.1 describes the algorithm applied for the construction of these instances. We denote by $a \otimes b$ the integer number obtained by the exclusive-or operation between the binary representations of the integers a and b. To make the description of the algorithm simpler, the nodes of the d-dimensional hypercube are indexed by $i = 0, 1, \ldots, 2^{d-1} - 1$, instead of by $i = 1, 2, \ldots, 2^{d-1}$. Then, $i \otimes 2^k$ gives the index of each neighbor of node i, for $k = 0, \ldots, d-1$. Figure 28.1 shows what hc3u (a three-dimensional hypercubic instance) would look like.

28.2.2 Code covering (cc)

Let V_q^n be the set of all n-dimensional vectors whose components are integers in the interval $[0, q-1]$. The *Code Covering Problem* (CCP) is defined as follows: given V_q^n and a positive integer r, find a minimum cardinality subset C of V_q^n such that there exists a vector $x \in C$ with $d(x, y) \leq r$ for all $y \in V_q^n$ (where d denotes the Hamming distance). This NP-hard problem (van Lint, 1975) is equivalent to finding a minimum dominating set in a graph in which there is

```
E ← ∅;
V ← ∅;
X ← {0};
for i = 0 to 2^d − 1
    V ← V ∪ {i};
    for k = 0 to d − 1
        j ← i ⊗ 2^k;
        if i ∉ X
            X ← X ∪ {j};
        w_{i,j} ← 1;
        if (i, j) ∉ E
            E ← E ∪ {(i, j)};
    endfor
endfor
```

Algorithm 28.1 Algorithm for generation of hypercube instances.

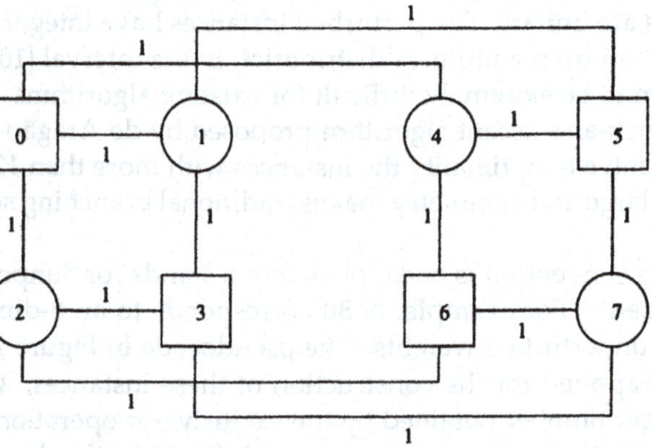

Figure 28.1 Graph associated with the hypercube instance hc3u.

a vertex associated to each element of V_q^n and an edge connecting the nodes associated to every pair $x, y \in V_q^n$ such that $d(x, y) \leq r$.

We created instances for the SPG using 13 of such graphs and taking as the set of terminals a solution (i.e., a dominating set) obtained by the tabu search algorithm of de Aragão and de Souza (1999). Edges have incidence weights so as to make preprocessing ineffective, following Duin (1994) and Duin and Voss (1997). In the unperturbed case, this means that an edge has weight equal to 1 if it connects non-terminals, 2 if it is incident to a single terminal, and 3 if it connects two terminals. The perturbed case follows the same princi-

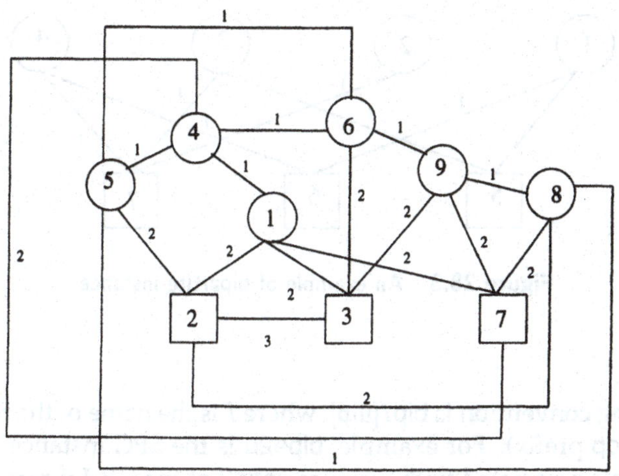

Figure 28.2 Code covering instance cc3-2u.

ple, but with integral weights uniformly distributed in the intervals [100, 110], [200, 210], or [300, 310], depending on the number of terminals incident to an edge.

Although these instances do challenge current exact algorithms, they are not as hard as the hc instances. The graphs in this case are still very symmetric, but terminal placement is not.

The naming convention is cc$n$-q[u|p] (the value of r is omitted because we have used only $r = 1$). For example, cc3-4u corresponds to an unperturbed instance derived from a CCP with $n = 3$ and $q = 4$ (and $r = 1$). Figure 28.2 shows what the code covering instance cc3-2u with $q = 2$ and $n = 3$ would look like.

28.2.3 Bipartite (bip)

This series contains ten instances defined on irregular bipartite graphs, with one of the bipartition-defining vertex subsets acting as the set of terminals. These graphs are derived from random instances of the set covering problem (SCP) from the OR-Library (Beasley, 1990). Given an instance I of SCP with m rows and n columns, we build an instance I' for the SPG as follows. There is a terminal in I' associated with each row i in I. Similarly, each non-terminal node in I' is associated with a column j in I. Whenever row i covers column j, there is an edge linking terminal i to the non-terminal node j in I'. The resulting graph has $m+n$ vertices, m terminals, and as many edges as nonzero entries in the coefficient matrix of the SCP instance I. Edge weights are unitary in the unperturbed case and uniformly distributed in the interval [100, 110] for the instances with perturbations.

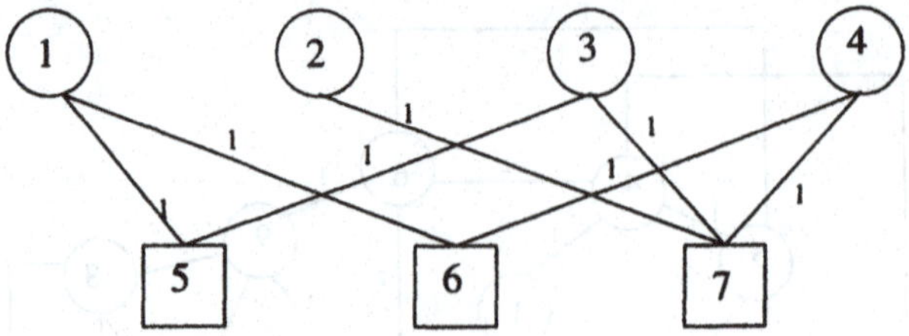

Figure 28.3 An example of bipartite instance.

The naming convention is bip$I$[u|p], where I is the name of the SCP instance (minus the scp prefix). For example, bipe2u is the SPG instance with unperturbed weights associated with instance scpe2 of the OR-Library. Figure 28.3 shows what a bipartite instance with $m = 3$ and $n = 4$ would look like.

Our motivation in creating this series was to have instances with a certain structure (bipartite), but without the artificial symmetry found in the previous two classes. Using SCP instances is no better or worse than generating random bipartite graphs from scratch, but the OR-Library instances have the advantage of being already publicly available.

28.3 EXPERIMENTAL RESULTS

This section presents some preliminary results on the new instances proposed. We have conducted some experiments to determine if they are indeed hard to be solved.

Table 28.1 Dimensions of the bipartite instances before and after preprocessing.

Instance	before preprocessing			after preprocessing		
	$\|V\|$	$\|E\|$	$\|X\|$	$\|V\|$	$\|E\|$	$\|X\|$
bipe2u	550	5013	50	550	5013	50
bip42u	1200	3982	200	990	3610	200
bip62u	1200	10002	200	1199	10000	200
bip52u	2200	7997	200	1819	7326	200
bipa2u	3300	18073	300	3140	17795	300
bipe2p	550	5013	50	550	5013	50
bip42p	1200	3982	200	990	3619	200
bip62p	1200	10002	200	1199	10000	200
bip52p	2200	7997	200	1819	7336	200
bipa2p	3300	18073	300	3140	17797	300

One of the most important practical techniques for the solution of SPG instances is the preprocessing phase, which consists of applying tests that try to eliminate some edges or vertices from the instance before any optimization algorithm is invoked. The hypercube and code covering instances could not be reduced, either by the traditional reduction tests of Duin and Volgenant (1989) or by the recent tests proposed by Uchoa et al. (2002). The sparse bipartite instances can be slightly reduced by some very simple tests. For instance, in the case of instance depicted in Figure 28.3, vertex 2 and its incident edge can be removed by the NTD1 Non-Terminal Degree 1 test. In the same figure, vertex 3 could be removed and its incident edges replaced by a single edge with cost equal to 2 from vertex 5 to 7 by the NTD2 test. Table 28.1 shows the dimensions (number of nodes, number of edges, and number of terminals) of the bipartite instances before and after preprocessing. Although all tests were applied, only the TD1, NTD1, and NTD2 tests described by Duin and Volgenant (1989) lead to some reductions.

Tables 28.2 to 28.4 present the results obtained for each of the new classes of instances. For each instance, we first report its dimensions (number of nodes, number of edges, and number of terminals). Next, we give the weights of the solutions found by six different heuristics: (SPH) shortest-path heuristic (Takahashi and Matsuyama, 1980), (HGP) hybrid GRASP with perturbations (Ribeiro et al., 2002), (HGP-PR) hybrid GRASP with perturbations and adaptive path-relinking (Ribeiro et al., 2002), (RTS) reactive tabu search (Bastos and Ribeiro, 2002), (RTS-PR) reactive tabu search with path-relinking (Bastos and Ribeiro, 2002), and (GHLS3) a variant (de Aragão et al., 2001a) of HGP-PR which uses a more powerful local search based on key-nodes and a larger pool of elite solutions. These implementations were chosen only because they were readily available to us. The shortest-path heuristic is included in this study only to indicate to the reader how a simple and fast approximate algorithm performs for these instances. Finally, we give the linear programming lower bound (LP) obtained by solving the linear relaxation of the directed cut formulation (Wong, 1984) and an estimate (gap) of the duality gap, given by the percentage difference between the LP bound and the best (GHLS3) upper bound. Results for the bipartite instances in Table 28.4 were obtained after preprocessing. Results in bold face indicate optimal values obtained by GHLS3 proved by the algorithm described by de Aragão et al. (2001b).

We now comment on the relative effectiveness of the heuristics when applied to the new classes of instances. We stress that our goal here is not to try to establish the superiority of one heuristic over the others. Instead, we want to show that the new instances provide a test bed on which insightful conclusions can be easily drawn, which is not always the case for the instances currently available in the literature.

We first compare the five first heuristics: SPH, HGP, HGP-PR, RTS, and RTS-PR. HGP-PR is clearly the literature heuristic with the best performance among them, finding the best solution for all but three instances. Comparing the implementations of metaheuristics which do not make use of path-relinking, HGP is never less effective than RTS for the unperturbed instances, which indicates that it is more suited to instances with many global optima. On the other hand, RTS performs better than HGP for most perturbed instances. Finally, GHLS3 outperforms all other heuristics and finds solutions which are always at least as good as those produced by the best among the first five heuristics.

28.4 CONCLUDING REMARKS

The results in Tables 28.1 to 28.4 clearly indicate that the new instances proposed in this work are not amenable to reductions by current preprocessing techniques and that the linear programming upper bounds have large gaps and are hard to be computed. State-of-the-art heuristics, which found optimal solutions for almost all instances currently in use, faced much more difficulties for the new instances. Fewer optimal solutions were found and the numerical results are more discriminant, allowing a better assessment of the effectiveness and the relative behavior of different heuristics. The new instances have been available at the SteinLib repository (Martin and Voss, 2000) since July 2001.

Table 28.2 Hypercube instances

Instance	$\|V\|$	$\|E\|$	$\|X\|$	SPH	HGP	HGP-PR	RTS	RTS-PR	GHLS3	LP	gap (%)
hc6u	64	192	32	41	39	39	39	39	39	37.1	4.87
hc7u	128	448	64	80	77	77	77	77	77	73.4	4.68
hc8u	256	1024	128	157	149	149	151	151	148	145.1	$\leq$ 1.96
hc9u	512	2304	256	319	296	296	304	304	292	286.8	$\leq$ 1.78
hc10u	1024	5120	512	627	588	588	606	606	582	567.7	$\leq$ 2.46
hc11u	2048	11264	1024	1219	1173	1173	1200	1200	1162	$\geq$1125.2	$\leq$ 3.17
hc12u	4096	24576	2048	2427	2336	2336	2396	2396	2303	$\geq$2201.1	$\leq$ 4.42
hc6p	64	192	32	4302	4017	4003	4003	4003	4003	3867.6	3.38
hc7p	128	448	64	8384	7932	7905	7909	7909	7905	7646.8	3.27
hc8p	256	1024	128	16746	15637	15376	15573	15526	15322	15115.7	$\leq$ 1.35
hc9p	512	2304	256	32509	31108	30572	30996	30920	30258	29877.6	$\leq$ 1.26
hc10p	1024	5120	512	64472	61905	61030	61633	61605	60494	59213.4	$\leq$ 2.12
hc11p	2048	11264	1024	128204	123129	120804	122741	122741	120096	117388.7	$\leq$ 2.25
hc12p	4096	24576	2048	253825	244674	243390	244477	244477	238673	$\geq$232709.0	$\leq$ 2.50

The time limit for the computation of the LP bound was set at 6 hours on a 400 MHz Pentium II machine.

Table 28.3 Code cover instances

Instance	$\|V\|$	$\|E\|$	$\|X\|$	SPH	HGP	HGP-PR	RTS	RTS-PR	GHLS3	LP	gap (%)
cc6-2u	64	192	12	32	32	32	32	32	**32**	29.8	6.88
cc3-4u	64	288	8	23	23	23	23	23	**23**	21.0	8.70
cc3-5u	125	750	13	36	36	36	36	36	**36**	≥32.8	≤ 8.89
cc5-3u	243	1215	27	76	72	71	74	73	**71**	69.5	2.11
cc9-2u	512	2304	64	187	171	171	178	171	167	≥162.7	≤ 2.57
cc6-3u	729	4368	76	217	201	198	209	206	197	≥194.1	≤ 1.47
cc3-10u	1000	13500	50	132	126	126	130	129	**125**	123.8	0.96
cc10-2u	1024	5120	135	381	349	346	362	359	342	≥332.4	≤ 2.81
cc3-11u	1331	19965	61	163	154	154	158	157	153	≥151.0	≤ 1.31
cc3-12u	1728	28512	74	191	186	186	187	187	186	≥182.0	≤ 2.15
cc11-2u	2048	11263	244	687	624	619	669	651	614	≥600.5	≤ 2.20
cc7-3u	2187	15308	222	612	562	554	598	588	552	≥534.1	≤ 3.24
cc12-2u	4096	24574	473	1315	1201	1184	1287	1255	1179	≥1141.0	≤ 3.22
cc6-2p	64	192	12	3388	3271	3271	3271	3271	**3271**	3078.3	5.89
cc3-4p	64	288	8	2349	2338	2338	2338	2338	**2338**	2194.0	6.16
cc3-5p	125	750	13	3673	3667	3661	3664	3664	**3661**	≥3384.6	≤ 7.55
cc5-3p	243	1215	27	8266	7491	7404	7484	7484	7299	≥7117.8	≤ 2.48
cc9-2p	512	2304	64	18704	17836	17376	17946	17904	17296	≥16766.0	≤ 3.06
cc6-3p	729	4368	76	22680	20850	20554	21060	20657	20458	≥20064.1	≤ 1.93
cc3-10p	1000	13500	50	14149	13084	13061	13118	13003	12860	≥12663.6	≤ 1.53
cc10-2p	1024	5120	135	38608	36552	35867	37234	36545	35466	≥34297.2	≤ 3.30
cc3-11p	1331	19965	61	17111	15924	15728	15940	15917	15609	≥15436.8	≤ 1.10
cc3-12p	1728	28512	74	20626	19285	19167	19201	19159	18838	≥18634.9	≤ 1.08
cc11-2p	2048	11263	244	70666	66073	64334	68486	67328	63841	≥61905.0	≤ 3.03
cc7-3p	2187	15308	222	63339	59005	57601	61190	60303	57339	≥55071.9	≤ 3.95
cc12-2p	4096	24574	473	135953	126541	122928	131778	128015	121772	≥117884.5	≤ 3.19

The time limit for the computation of the LP bound was set at 6 hours on a 400 MHz Pentium II machine.

Table 28.4 Bipartite instances

Instance	$\|V\|$	$\|E\|$	$\|X\|$	SPH	HGP	HGP-PR	RTS	RTS-PR	GHSL3	LP	gap (%)
bipe2u	550	5013	50	60	55	55	55	55	54	52.4	$\leq$ 2.96
bip42u	1200	3982	200	270	248	239	259	258	237	232.0	$\leq$ 2.11
bip62u	1200	10002	200	247	229	227	233	231	221	213.3	$\leq$ 3.48
bip52u	2200	7997	200	275	248	247	261	261	235	229.1	$\leq$ 2.51
bipa2u	3300	18073	300	385	358	356	371	367	342	$\geq$329.3	$\leq$ 3.71
bipe2p	550	5013	50	6334	5721	5684	5687	5666	5660	5515.4	$\leq$ 2.55
cdcd bip42p	1200	3982	200	28758	25737	25175	25833	25285	24818	24373.8	$\leq$ 1.79
bip62p	1200	10002	200	25267	23877	23291	23779	23500	22944	22445.2	$\leq$ 2.17
bip52p	2200	7997	200	29616	26137	25415	26079	25884	24936	24186.3	$\leq$ 3.01
bipa2p	3300	18073	300	40457	37368	36439	37415	37345	35774	34685.7	$\leq$ 3.04

The time limit for the computation of the LP bound was set at 6 hours on a 400 MHz Pentium II machine.

Bibliography

M.P. Bastos and C.C. Ribeiro. Reactive tabu search with path-relinking for the Steiner problem in graphs. In C.C. Ribeiro and P. Hansen, editors, *Essays and Surveys in Metaheuristics*, pages 39–58. Kluwer, 2002.

J.E. Beasley. OR-Library: Distributing test problems by electronic mail. *Journal of the Operational Research Society*, 41:1069–1072, 1990.

M. Poggi de Aragão and C.C. de Souza. Upper bounds for minimum covering codes via tabu search. In *Extended Abstracts of the 3rd Metaheuristics International Conference*, pages 359–364. Angra dos Reis, 1999.

M. Poggi de Aragão, C.C. Ribeiro, E. Uchoa, and R.F. Werneck. Hybrid local search for the Steiner problem in graphs. In *Extended Abstracts of the 4th Metaheuristics International Conference*, pages 429–433. Porto, 2001a.

M. Poggi de Aragão, E. Uchoa, and R.F. Werneck. Dual heuristics on the exact solution of large Steiner problems. *Electronic Notes in Discrete Mathematics*, 7, 2001b.

C.W. Duin. Steiner's problem in graphs: Approximation, reduction, variation. Doctorate Dissertation, Institute of Actuarial Science and Economics, University of Amsterdam, 1994.

C.W. Duin and T. Volgenant. Reduction tests for the Steiner problem in graphs. *Networks*, 19:549–567, 1989.

C.W. Duin and S. Voss. Efficient path and vertex exchange in Steiner tree algorithms. *Networks*, 29:89–105, 1997.

C.W. Duin and S. Voss. The Pilot method: A strategy for heuristic repetition with application to the Steiner problem in graphs. *Networks*, 34:181–191, 1999.

M. Gendreau, J.-F. Larochelle, and B. Sansó. A tabu search heuristic for the Steiner tree problem. *Networks*, 34:163–172, 1999.

T. Koch and A. Martin. Solving Steiner tree problems in graphs to optimality. *Networks*, 32:207–232, 1998.

A.P. Lucena and J.E. Beasley. A branch and cut algorithm for the Steiner problem in graphs. *Networks*, 31:39–59, 1998.

T. Koch A. Martin and S. Voss. Steinlib: An updated library on Steiner tree problems in graphs, 2000. Online document at http://elib.zib.de/steinlib-/steinlib.php, last visited on March 20, 2003.

S.L. Martins, P. Pardalos, M.G. Resende, and C.C. Ribeiro. A parallel GRASP for the Steiner tree problem in graphs using a hybrid local search strategy. *Journal of Global Optimization*, 17:267–283, 2000.

T. Polzin and S. Vahdati. Improved algorithms for the Steiner problem in networks. *Discrete Applied Mathematics*, 112:263–300, 2001.

C.C. Ribeiro and M.C. Souza. Tabu search for the Steiner problem in graphs. *Networks*, 36:138–146, 2000.

C.C. Ribeiro, E. Uchoa, and R.F. Werneck. A hybrid GRASP with perturbations for the Steiner problem in graphs. *INFORMS Journal on Computing*, 14:228–246, 2002.

H. Takahashi and A. Matsuyama. An approximate solution for the Steiner problem in graphs. *Math. Japonica*, 24:573–577, 1980.

E. Uchoa, M. Poggi de Aragão, and C.C. Ribeiro. Preprocessing Steiner problems from VLSI layout. *Networks*, 40:38–50, 2002.

J.H. van Lint. Recent results on perfect codes and related topics. In M. Hall Jr. and eds. J.H. van Lint, editors, *Combinatorics*, pages 163–183. Kluwer, 1975.

D.B. West. *Introduction to graph theory*. Prentice-Hall, 2000.

R. Wong. A dual ascent approach for Steiner tree problems on a directed graph. *Mathematical Programming*, 28:271–287, 1984.

Metaheuristics: Computer Decision-Making, pp. 615-626
Suwan Runggeratigul

29 A MEMETIC ALGORITHM FOR COMMUNICATION NETWORK DESIGN TAKING INTO CONSIDERATION AN EXISTING NETWORK

Suwan Runggeratigul[1]

[1] Telecommunications Program
Sirindhorn International Institute of Technology
Thammasat University
Pathumthani 12121, Thailand
suwan@siit.tu.ac.th

Abstract: This paper applies a memetic algorithm (MA) to solve a communication network design problem taking into consideration existing network facilities. The link capacity assignment problem in packet-switched networks (CA problem) is studied as an example of the network design problem. In the CA problem, we focus on the case in which link cost functions are piecewise linear concave, where the unit cost of the newly-installed link capacity is smaller than that of the existing link capacity. The MA in this paper is constructed by combining a genetic algorithm (GA) with a local search operator, which is a heuristic design algorithm previously developed for the CA problem. Experimental results show that MA solves the CA problem very efficiently, and the solutions obtained by MA are better than those by GA.

Keywords: Communication network design, Packet-switched networks, Link capacity assignment, Genetic algorithms, Memetic algorithms.

29.1 INTRODUCTION

It is widely known that utilizing existing network facilities in the construction of a new network is very important in the case of short-term network design (Kawashima and Kobayashi, 1993). The reason is that existing network facilities are ready to be used and cannot be changed easily in a short period of time. This leads to the possibility of modeling the cost functions of network facilities as convex functions. For the short-term network design, the circuit dimensioning in circuit-switched networks (Shinohara, 1990), and the link capacity assignment in packet-switched networks (Runggeratigul et al., 1995) have been studied with convex piecewise linear cost functions, where the unit cost of using an existing facility is smaller than that of installing a new facility. However, in other cases of network design such as the long-term design, it is proper to consider the cost functions as concave functions (Kershenbaum, 1993) since economy of scale is often present in communication resources (Kleinrock, 1976). Some good examples are concave cost functions resulting from the envelopes of several cost functions according to the development in communication transmission/switching technologies (Gersht and Weihmayer, 1990; Verma, 1989; Yaged, 1971).

The link capacity assignment problem (CA problem) in packet-switched networks taking into consideration an existing network where the link cost functions are piecewise linear concave has also been studied, and a heuristic design algorithm (we refer to it as HDA in this paper) has been developed to solve the problem (Runggeratigul and Tantaratana, 1999). Genetic algorithms (GAs) have also been applied to the CA problem, and it has been shown that GAs solve the problem more efficiently than the multi-start version of HDA (Runggeratigul, 2000). Although GAs efficiently solve several optimization problems, it has been reported that the search ability of GAs to find optimal solutions in some problems is a bit inferior to local search techniques (Ishibuchi et al., 1997). Consequently, there is an enormous effort to consider hybrid version of GAs by combining the evolutionary operators with local search operators. This new class of methods is known as memetic algorithms (MAs) (Merz and Freisleben, 1999; Moscato, 1999). Since MAs have been successfully applied to a wide range of problems, this paper investigates the application of MA to the CA problem. To design an MA, we consider the HDA as a local search operator since it can work as an improvement heuristic for the CA problem, where HDA is implemented after every evolutionary operator in the MA. Experimental results show that MA solves the CA problem very efficiently, and it outperforms GA, especially in the case of large networks.

29.2 PROBLEM DESCRIPTION

The packet-switched network model in this paper is the same as the one used in the early work of packet-switched network design (Gerla and Kleinrock, 1977; Kleinrock, 1976; Runggeratigul et al., 1995). In this model, the average packet delay throughout the network, T, can be given by Eq.(29.1).

$$T = \frac{1}{\gamma} \sum_{L} \frac{f_i}{C_i - f_i}, \tag{29.1}$$

where L is the set of links in the network, f_i is the traffic flow (in bits/second) on link i, C_i is the capacity (in bits/second) of link i, and γ is the overall traffic (in packets/second) in the network. The total network cost D is defined as the sum of all link costs,

$$D = \sum_{L} D_i. \tag{29.2}$$

For link i, link cost D_i is a piecewise linear function of link capacity as follows.

$$D_i = \begin{cases} d_{0i} C_i & , \quad C_i \le C_{0i}, \\ d_{1i} C_i - (d_{1i} - d_{0i}) C_{0i} & , \quad C_i > C_{0i}, \end{cases} \tag{29.3}$$

where d_{0i} and d_{1i} are respectively the unit cost of existing and newly-installed capacity, and C_{0i} is the value of the existing capacity of link i. In this paper, we focus on the case of concave cost functions, where $d_{0i} > d_{1i}$, $d_{0i} > 0$, $d_{1i} > 0$, $\forall i \in L$. This means the unit cost of the existing link capacity is greater than that of the newly-installed link capacity. The concave link cost function is depicted in Figure 29.1.

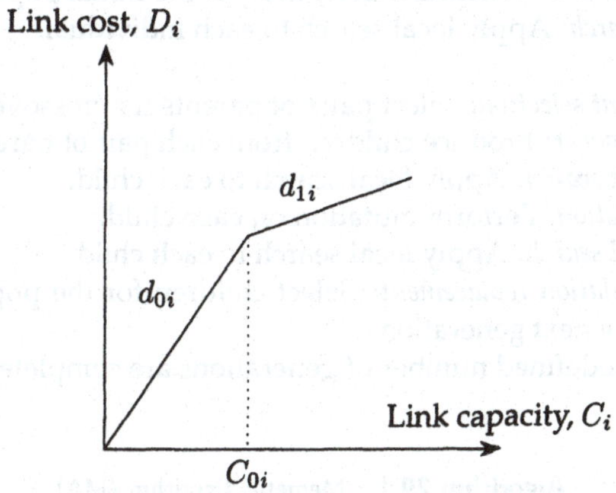

Figure 29.1 A piecewise linear concave link cost function.

This type of cost function arises in many communication network design problems. One example is the facility selection problem (Gersht and Weihmayer, 1990) where economy of scale is present in communication resources. Given f_i, C_{0i}, d_{0i}, and d_{1i}, we then formulate the CA problem as follows.

$$\min D = \sum_{L} D_i$$

subject to

$$T = \frac{1}{\gamma} \sum_{L} \frac{f_i}{C_i - f_i} \le T_{\max}$$

$$C_i - f_i > 0, \quad C_i > 0, \quad \forall i \in L,$$

where $T_{\max}$ is the maximum allowable packet delay in the network.

In the CA problem, $\{f_i\}$ is a new traffic pattern applied to an existing network. Link capacity $\{C_i\}$ is determined to cope with the new traffic pattern such that the total network cost D is minimized. Design constraints are the upper limit of packet delay, and the relationship between link flow and link capacity. Obviously, link capacity must be greater than link flow so that packet delay will not grow indefinitely.

29.3 MEMETIC ALGORITHM

Memetic algorithms (MAs) are metaheuristics that take advantage of evolutionary operators in determining interesting regions of the search space, as well as the local search operator in rapidly finding good solutions in small regions of the search space (Merz and Freisleben, 1999). In this paper, we propose the following memetic algorithm for the CA problem.

procedure MA( )
Initialization: Generate individuals for the initial population.
Local search: Apply local search to each individual.
repeat
    Parent selection: Select pairs of parents for crossover.
    Crossover: Produce children from each pair of parents.
    Local search: Apply local search to each child.
    Mutation: Perform mutation on each child.
    Local search: Apply local search to each child.
    Population replacement: Select children for the population in the next generation.
until predefined number of generations are completed

Algorithm 29.1 Memetic algorithm (MA).

The implementation of MA is described below in two parts: genetic algorithm, and local search operator.

29.3.1 Genetic Algorithm

The part of the genetic algorithm (GA) in the MA is the same as the one used in Runggeratigul (2000) for the CA problem. Each candidate solution of the CA problem in the search space is encoded as a binary string of length n, where n is the total number of links in the network. The fitness value of an individual string is defined as the network cost D corresponding to the bits in the string. To calculate the fitness of an individual, we firstly determine the link capacity C_i by Eq.(29.4).

$$C_i = f_i + \frac{\sum_{L} \sqrt{f_j d_j}}{\gamma T_{\max}} \sqrt{\frac{f_i}{d_i}}, \qquad \forall i \in L, \tag{29.4}$$

where

$$d_i = \begin{cases} d_{0i}, & \text{when bit } i \text{ is } 0, \\ d_{1i}, & \text{when bit } i \text{ is } 1. \end{cases} \tag{29.5}$$

Link cost D_i can then be calculated by Eq.(29.3), and the sum of all link costs is the network cost D, which is used as the fitness value of the individual.

A population size of 100 is used, and 100 binary strings are randomly generated for the initial population, where each bit in the string is set to be 0 or 1 with equal probability, i.e., 0.5.

Since the CA problem is a minimization problem, we avoid the problem of fitness value scaling by using tournament selection to select 100 pairs of parents for the crossover process. Each individual in the population has an equal probability of being chosen to enter the tournament comparison, where the tournament size is set to be 2. We use uniform crossover to create two children from each pair of parents. The crossover mask is constructed as a binary string of length n, where each bit is randomly set to 0 or 1 with the same probability. The bit mutation operator is applied to all bits in all children with a mutation probability of 0.01.

To determine the population of the next generation, we compare the fitness value of two children produced from the same pair of parents, and the one with better fitness value (smaller D) is survived to the next generation. Elitism is also adopted by randomly discarding one individual in the new population, and replacing it with the best individual in the previous generation.

29.3.2 Local Search Operator

To solve the CA problem with concave cost functions, a heuristic design algorithm (HDA) derived from the Lagrange multiplier method has been proposed (Runggeratigul and Tantaratana, 1999). The algorithm starts with an initial binary string, where each bit i gives the value of d_i as in Eq.(29.5). The Lagrange multiplier β for the CA problem is defined as

$$\beta = \left(\frac{\sum_L \sqrt{f_i d_i / \gamma}}{T_{\max}} \right)^2 . \tag{29.6}$$

The link capacity C_i is then determined by

$$C_i = f_i + \sqrt{\frac{\beta f_i}{\gamma d_i}}, \qquad \forall i \in L. \tag{29.7}$$

If the value of C_i calculated above is less than C_{0i}, we set $d_i = d_{0i}$, and bit i to be 0; otherwise, we set $d_i = d_{1i}$, and bit i to be 1. The algorithm iteratively determines an improving solution string, and terminates when the solution converges.

Table 29.1 An example of the CA problem with $n = 3$ ($\gamma = 382$, $T_{\max} = 20$ ms.).

i	f_i	C_{0i}	d_{0i}	d_{1i}
0	78400	27135.680	1.2	0.9
1	70000	91716.312	0.7	0.1
2	4400	28285.469	2.0	0.2

Let us examine an example of the application of HDA. We consider a CA problem with $n = 3$ and network parameters given in Table 29.1.

For this CA problem, we let the input binary string to the HDA be 000. Consequently, the initial values of d_i are set to be d_{0i} for all i. This initial solution string has $\beta = 2531087.500$ by Eq.(29.6). Using Eq.(29.7), the link capacity values are obtained as follows: $C_0 = 99206.039$, $C_1 = 95740.797$, and $C_2 = 8217.977$. This gives the network cost $D = 178465.953$. For each link i, we compare C_i with C_{0i}, and obtain a new solution string 110. By this procedure, HDA finally converges at the binary string 110 with $D = 175229.781$ as shown in Table 29.2.

Table 29.2 The application of HDA to the CA problem in Table 29.1.

iteration	*current string*	β	D	*new string*
1	000	2531087.500	178465.953	110
2	110	1284964.875	175229.781	110

Since HDA can work as an improvement method as illustrated above, we then consider this algorithm as the local search operator in the proposed MA by applying it to every string newly generated by GA. In the MA, the output of HDA to each input string is an improving solution of the string.

29.4 NUMERICAL RESULTS

In this section, we compare the performance of MA and GA for the CA problem, where GA can be considered as an MA without local search. For a fair comparison, we let the two methods solve a given problem instance in almost the same running time (see Table 29.3). This makes the number of generations of applying MA and GA to a problem be 1000 and 2500, respectively.

For each of the CA problem instances, a network with fully connected topology is considered. The number of nodes, N, is from 10 to 100, which gives the number of links, n, to be from 45 to 4950. For each network size, we examine two types of traffic models as follows. The first type is the uniform random traffic model, where the rate of traffic flow between each pair of nodes is a uniformly distributed random number over the range 0 and 100 kbits/second.

The second type is a non-uniform random traffic model based on the traffic model in client-server systems. In this model, the number of server nodes is set to be 0.1 N, and each server node provides services to a disjoint set of 9 client nodes, i.e., there are 10 nodes in each client-server group. The rate

Table 29.3 Running time of MA and GA when solving a CA problem (the machine used in the experiment is a personal computer with 450 MHz Pentium III processor).

n	*MA running time (sec.)*	*GA running time (sec.)*
45	52.270	61.098
190	225.282	257.288
435	534.394	594.974
780	1006.092	1080.442
1225	1627.152	1705.980
1770	2409.410	2475.956
2415	3320.456	3375.578
3160	4364.488	4415.630
4005	5555.108	5603.156
4950	6868.022	6931.902

of traffic flow between each pair of nodes is set at random following a uniform distribution over the range 0 and 100 kbits/second, except that between each pair of client and server nodes in the same group where the range of the uniform random number is 0 and 1 Mbits/second.

For each type of traffic model, we generate 100 traffic patterns each of which is treated as a problem instance and solved by both GA and MA. The performance comparison of the two methods are given in Tables 29.4 and 29.5 for the case of uniform random traffic model and the traffic model based on client-server systems, respectively. The results in the first part of each table are the number of cases that MA gives better, the same, or worse solutions when compared with GA. The second part is the results on the ratio between the network cost D of solutions by MA and GA (average, minimum, maximum). Note that GA is the current best known method for the CA problem (Runggeratigul, 2000).

It can be seen from Table 29.4 that the solutions by MA are never worse than those by GA. Furthermore, MA always gives better solutions for large networks, and the rate of improvement in the solution quality increases as the network size increases. In Table 29.5, although the improvement in the performance of GA is realized, the method is still outperformed by MA.

In the above results, it can be seen that the quality of solutions by GA and MA are quite close to each other. This can be explained by the flat property of global optima in large network design problems (Gersht and Weihmayer, 1990), which is similar to the asymptotic behavior of large-sized combinatorial optimization problems (Burkard, 1984). For this reason, it implies that GA is an efficient method for solving the CA problem. However, as can be seen clearly from the experimental results, the method is inferior to its hybrid version, i.e., MA.

Table 29.4 Performance comparison between MA and GA: the case of uniform random traffic model.

	MA is			ratio		
n	*better*	*same*	*worse*	*avg*	*min*	*max*
45	0	100	0	1.00000	1.00000	1.00000
190	30	70	0	0.99999	0.99982	1.00000
435	95	5	0	0.99997	0.99985	1.00000
780	100	0	0	0.99990	0.99972	0.99998
1225	100	0	0	0.99979	0.99944	0.99994
1770	100	0	0	0.99967	0.99929	0.99985
2415	100	0	0	0.99954	0.99917	0.99976
3160	100	0	0	0.99937	0.99907	0.99963
4005	100	0	0	0.99919	0.99892	0.99940
4950	100	0	0	0.99902	0.99870	0.99926

Table 29.5 Performance comparison between MA and GA: the case of traffic model based on client-server systems.

	MA is			ratio		
n	*better*	*same*	*worse*	*avg*	*min*	*max*
45	0	100	0	1.00000	1.00000	1.00000
190	12	88	0	1.00000	0.99997	1.00000
435	82	18	0	0.99998	0.99991	1.00000
780	100	0	0	0.99996	0.99979	1.00000
1225	100	0	0	0.99991	0.99974	0.99999
1770	100	0	0	0.99985	0.99967	0.99996
2415	100	0	0	0.99974	0.99955	0.99984
3160	100	0	0	0.99963	0.99948	0.99977
4005	100	0	0	0.99951	0.99926	0.99964
4950	100	0	0	0.99940	0.99905	0.99959

29.5 CONCLUSIONS

In this paper, we have proposed the application of a memetic algorithm (MA) to the link capacity assignment problem (CA problem) in packet-switched networks with the consideration of existing network facilities. The link cost functions are considered to be piecewise linear concave. The MA proposed in this paper is designed by combining a local search operator to evolutionary operators of genetic algorithm (GA), where the local search is a heuristic algorithm previously developed for solving the CA problem. Experimental results show that MA is superior to GA in all test problems.

For further studies, the applications of other types of local search operators in the MA can be examined. Some good examples are basic techniques such as bit-flipping methods, local search metaheuristics such as simulated annealing and tabu search, etc.

R. E. Burkard. Quadratic assignment problems. *European Journal of Operational Research*, 15:283–289, 1984.

M. Gerla and L. Kleinrock. On the topological design of distributed computer networks. *IEEE Trans Commun*, COM-25:48–60, 1977.

A. Gersht and R. Weihmayer. Joint optimization of data network design and facility selection. *IEEE J Select Area Commun*, 8:1667–1681, 1990.

H. Ishibuchi, T. Murata, and S. Tomioka. Effectiveness of genetic local search algorithms. In *Proceedings of the Seventh International Conference on Genetic Algorithms*, pages 505–512, 1997.

K. Kawashima and H. Kobayashi. Network planning tools. *Journal of IEICE*, 76:141–148, 1993.

A. Kershenbaum. *Telecommunications Network Design Algorithms*. McGraw-Hill, New York, 1993.

L. Kleinrock. *Queueing Systems, Volume II*. John Wiley & Sons, New York, 1976.

P. Merz and B. Freisleben. A comparison of memetic algorithms, tabu search, and ant colonies for the quadratic assignment problem. In *Proceedings of the 1999 Congress on Evolutionary Computation*, pages 2063–2070, 1999.

P. Moscato. Memetic algorithms: A short introduction. In D. Corne, M. Dorigo, and F. Glover, editors, *New Ideas in Optimization*, pages 219–234. McGraw-Hill, London, 1999.

S. Runggeratigul. Application of genetic algorithms to communication network design taking into consideration an existing network. In *Proceedings of the Second Asia-Pacific Conference on Genetic Algorithms and Applications*, pages 323–331, 2000.

S. Runggeratigul and S. Tantaratana. Link capacity assignment in packet-switched networks: the case of piecewise linear concave cost function. *IEICE Trans Commun*, E82-B:1566–1576, 1999.

S. Runggeratigul, W. Zhao, Y. Ji, A. Aizawa, and S. Asano. Link capacity assignment in packet-switched network with existing network consideration. *IEICE Trans Commun*, E78-B:709–719, 1995.

M. Shinohara. Generalization of Pratt's formula for dimensioning circuit-switched network. *Computer Networks and ISDN Systems*, 20:115–126, 1990.

P. K. Verma. *Performance Estimation of Computer Communication Networks*. Computer Science Press, Rockville, 1989.

B. Yaged. Minimum cost routing for static network models. *Networks*, 1:139–172, 1971.

Metaheuristics: Computer Decision-Making, pp. 627-657
Mauricio C. de Souza, Christophe Duhamel, and Celso C. Ribeiro

30 A GRASP HEURISTIC FOR THE CAPACITATED MINIMUM SPANNING TREE PROBLEM USING A MEMORY-BASED LOCAL SEARCH STRATEGY

Mauricio C. de Souza[1], Christophe Duhamel[1], and Celso C. Ribeiro[2]

[1]LIMOS, Université Blaise Pascal
BP 125, 63173 Aubière Cedex, France
mauricio@isima.fr, christophe.duhamel@isima.fr

[2]Department of Computer Science
Catholic University of Rio de Janeiro
Rio de Janeiro, 22453-900, Brazil
celso@inf.puc-rio.br

Abstract: We describe a new neighborhood structure for the capacitated minimum spanning tree problem. This neighborhood structure is used by a local search strategy, leading to good trade-offs between solution quality and computation time. We also propose a GRASP with path-relinking heuristic. It uses a randomized version of a savings heuristic in the construction phase and an extension of the above local search strategy, incorporating some short term memory elements of tabu search. Computational results on benchmark problems illustrate the effectiveness of this approach, which is competitive with the best heuristics in the literature in terms of solution quality. The GRASP heuristic using a memory-based local search strategy improved the best known solution for some of the largest benchmark problem.
Keywords: Capacitated minimum spanning tree, Metaheuristics, GRASP, Local search, Neighborhood reduction, Short term memory, Path-relinking.

30.1 INTRODUCTION

Let $G = (V, E)$ be a connected undirected graph, where $V = \{0, 1, \ldots, n\}$ denotes the set of nodes and E is the set of edges. Non-negative integers c_e and b_i are associated respectively with each edge $e \in E$ and with each node $i \in V$. Given an integer Q and a special *central node* $r \in V$, the Capacitated Minimum Spanning Tree (CMST) problem consists of finding a minimum spanning tree T of G in terms of the edge costs, such that the sum of the node weights in each connected component of the graph induced in T by $V \setminus \{r\}$ is less than or equal to Q.

The CMST problem is NP-hard (Papadimitriou, 1978) for $3 \leq Q \leq \lfloor |V|/2 \rfloor$ and has applications in the design of communication networks, see e.g. Amberg et al. (1996); Gavish (1982); Gouveia and Martins (2000). Gouveia and Martins (1999) proposed a hop-indexed flow model which is a generalization of a single-commodity flow model proposed by Gavish (1983) and reviewed exact and lower bounding schemes, including earlier works of Gavish (1982; 1983), the branch-and-bound algorithm of Malik and Yu (1993), the Lagrangean relaxation approach of Gouveia (1995), and the cutting plane method of Hall (1996). Gouveia and Martins (2000) proposed an iterative method for computing lower bounds for the CMST problem, based on a hierarchy of hop-indexed linear programming models. Amberg et al. (1996) reviewed exact and approximate algorithms. Among the main heuristics, we find the savings procedure EW of Esau and Williams (1966) and the tabu search algorithms of Amberg et al. (1996) and Sharaiha et al. (1997). The neighborhood structure used in Amberg et al. (1996) is based on exchanging single nodes between subtrees of the current solution. The neighborhood used in Sharaiha et al. (1997) is an extension of the latter, in which parts of a subtree are moved from one subtree to another or to the central node. More recently, Ahuja et al. (2001) proposed new neighborhoods based on the cyclic exchange neighborhood described in Thompson and Orlin (1989); Thompson and Psaraftis (1993) and developed GRASP and tabu search heuristics based on the concept of improvement graphs.

In this paper, we propose a new GRASP with path-relinking heuristic for the capacitated minimum spanning tree problem. This heuristic uses a new local search strategy based on a different neighborhood structure defined by path exchanges, as described in the next section. Numerical results on benchmark problems are reported in Section 30.3, showing that the proposed local search strategy leads to good trade-offs between solution quality and computation time. The GRASP with path-relinking heuristic using a memory-based local search strategy is described in Section 30.4. Further computational results are reported in Section 30.5, illustrating the effectiveness of the heuristic. Concluding remarks and extensions are discussed in the last section.

30.2 LOCAL SEARCH WITH A PATH-BASED NEIGHBORHOOD

Feasible initial solutions are constructed by the savings heuristic EW of Esau and Williams (1966). Starting from the r-centered star, this procedure joins the two components which yield maximum savings with respect to the edge costs and capacity constraints. The process iterates until no further savings can be achieved. It runs in $O(|V|^2 \log |V|)$ time.

Let T be a spanning tree of G, satisfying the capacity constraints. We define the components $T_i, i = 1, \ldots, q$ of T as the subtrees induced in T by $V \setminus \{r\}$. Solutions in the neighborhood of T are obtained as follows. Every edge $e \notin T$ with extremities in two different components T_i and T_j is considered as a candidate for insertion in the current tree. To speed up the search, we consider only a subset of candidate edges and we investigate the insertion of each of them. For each edge $e \in E$ considered for insertion, different forests can be obtained by removing each of at most Q^2 candidate paths with extremities in T_i and T_j. A candidate path may be removed if the tree containing edge e in the resulting forest does not violate the capacity constraint. The example in Figure 30.1 (a) shows a feasible solution for a CMST instance with $Q = 3$ and a candidate edge e for insertion. Figures 30.1 (b) to (d) show the only three (out of nine) paths that may be removed without violating the capacity constraints in the resulting forest. Instead of evaluating all paths satisfying the above condition, we use an estimation of the reconnection costs to select a unique path for each candidate edge. Each component of the resulting forest is reconnected to the central node by the least cost edge and the savings procedure EW is applied from that solution. The structure of each move is summarized in Figure 30.2. The best improving neighbor generated by the above scheme is chosen and the local search resumes from the new solution.

Each solution has $O(|E|Q^2)$ neighbors. For each edge $e \in E$ considered for insertion and for each path considered for removal, we evaluate in time $O(Q)$ whether this move is acceptable or not, by examining the nodes within the tree containing edge e. Since the reconnection of each acceptable move is computed by the savings procedure in $O(|V|^2 \log |V|)$ time, the complexity of the investigation of the neighborhood of each solution is $O(|E||V|^2 Q^2 \log |V|)$, with $Q \leq \lfloor |V|/2 \rfloor$.

30.2.1 Selection of candidate edges

The neighborhood size has a strong influence on the computational performance of local search procedures. We use some heuristic selection rules to speed up the search by restricting the neighbor solutions investigated.

Each local search iteration starts by a depth-first search traversal of the current solution. For each node $i \in V \setminus \{r\}$, we obtain its predecessor j in the incoming path from the central node and the capacity of the subtree rooted at i (i.e., the sum of the weights of all nodes in this subtree) if edge (i, j) is eliminated from the current tree.

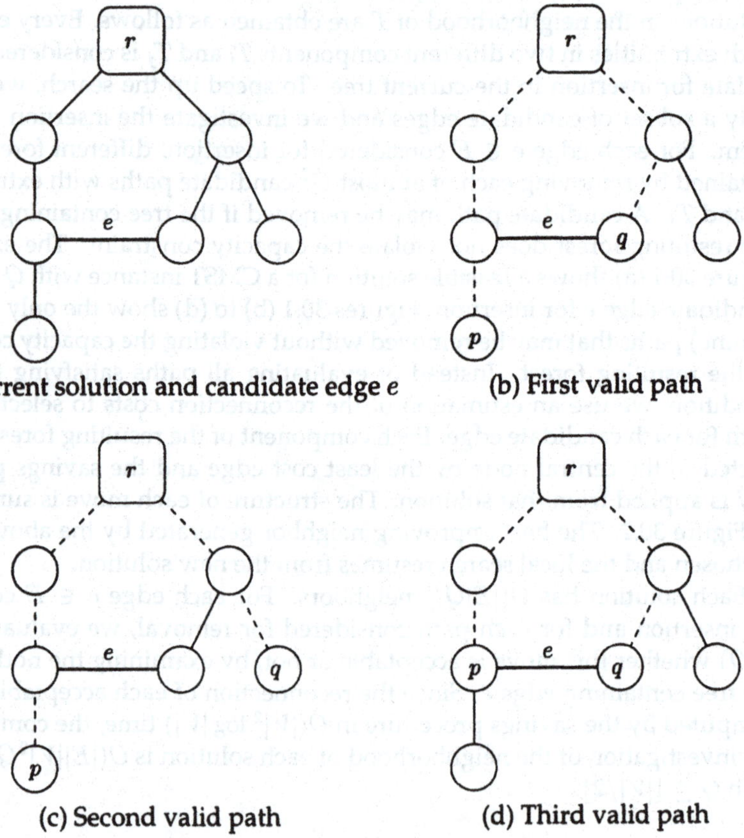

Figure 30.1 The three valid paths for removal ($Q = 3$).

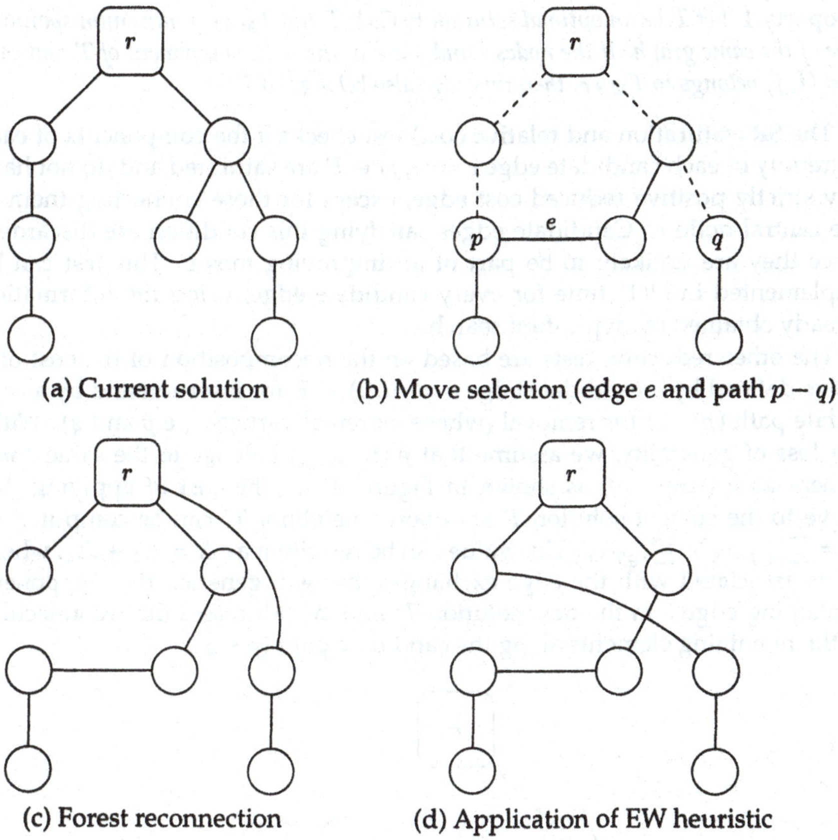

Figure 30.2 The four steps of a move.

Each move is based on the selection of an edge $e \in E$, followed by that of a path $(p - q)$ between two different components of the current solution. We propose three tests to reduce the number of moves evaluated. The first of them makes use of the relative costs associated with the edges:

Definition 1 (Relative cost) *The relative cost r_e of an edge $e \in E$ is the difference between the weight of a minimum spanning tree of G and the weight of a constrained minimum spanning tree of G in which the presence of edge e is enforced.*

The relative costs can be efficiently computed once for all. We start by the application of Kruskal's algorithm (Kruskal, 1956) to build a minimum spanning tree T_{MST} of G in time $O(|E||V|)$. The relative cost of every edge $e \in T_{MST}$ is $r_e = 0$. Otherwise, $r_e = c_e - c_{e'}$, where e' is the edge with maximum weight among those in the cycle generated by inserting e into T_{MST}. The rationale for the use of the relative cost comes from a property defined in Chandy and Lo (1973); Chandy and Russel (1972):

Property 1 *Let T be an optimal solution to CMST and T_{MST} a minimum spanning tree of the same graph. If the nodes i and j are in the same component of T and edge $e = (i, j)$ belongs to T_{MST}, then this edge also belongs to T.*

The SR (saturation and relative cost) test checks if the components of each extremity of each candidate edge $e = (i, j) \in E$ are saturated and do not have any strictly positive reduced cost edge, except for those connecting them to the central node r. Candidate edges satisfying this condition are discarded, since they are unlikely to be part of an improving move. This test can be implemented in $O(1)$ time for every candidate edge, using the information already obtained by depth-first search.

The other reduction tests are based on the decomposition of the cost of a move defined by a candidate edge $e = (u, v) \in E$ for insertion and by a candidate path $(p - q)$ for removal (whose extremal vertices are p and q). Without loss of generality, we assume that p (resp. q) belongs to the same component as u (resp. v), as shown in Figure 30.3. The cost of applying this move to the current solution T to obtain a neighbor T' can be computed as $\Delta = \sum_{e \in T'} c_e - \sum_{e \in T} c_e$. This value can be rewritten as $\Delta = \Delta_1 + \Delta_2$, where Δ_1 is associated with the edge exchanges that will generate the component containing edge e in the new solution T' and Δ_2 addresses the reconnection of the remaining elements along the candidate path $(p - q)$.

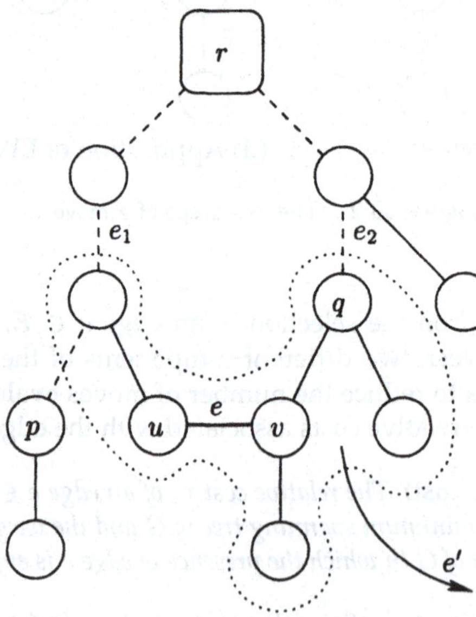

Figure 30.3 Selection of a candidate edge.

The computation of Δ_1 is based on four particular edges, as illustrated in Figure 30.3. We first define $\hat{T}$ as the tree containing edge e in the forest resulting from the removal of path $(p - q)$. Let e_1 (resp. e_2) be the first edge in the component shared by the paths going from u and p (resp. v and q) to the central node r. Among those in the path $(p-q)$, edges e_1 and e_2 are the most appropriate be involved in the computation of Δ_1. The removal of edges between e_1 and e_2 (e.g. edges incident to r in Figure 30.3) would disconnect nodes of T which do not belong to $\hat{T}$. The removal of other edges (e.g. edges incident to p in Figure 30.3) do not disconnect $\hat{T}$ from T. We now consider the forest obtained once path $(p-q)$ has been removed. After reconnecting the forest and applying the savings heuristic EW, at least one edge e' will be incident to $\hat{T}$ in the new solution T'. Then, $\Delta_1 = c_e + c_{e'} - c_{e_1} - c_{e_2}$. The other component Δ_2 addresses the cost of first disconnecting, then reconnecting all other subtrees affected by the deletion of path $(p-q)$.

We now introduce the reduction tests LB1 and LB2. Both of them discard the cost Δ_2 and make use of a lower bound $\underline{\Delta}_1 \leqslant \Delta_1$. Indeed, a candidate edge e can be eliminated if $\underline{\Delta}_1 \geqslant 0$ for every possible candidate path $(p-q)$ that can be used in conjunction with e. Test LB1 is based on a weak estimation of the weights of the three actual edges e', s_1, and s_2 to compute $\underline{\Delta}_1$ in time $O(1)$. The reconnection cost $c_{e'}$ can be underestimated by the minimum edge weight $\underline{c}$ over all edges of the graph. The disconnection cost c_{s_1} (resp. c_{s_2}) can be overestimated by the largest edge weight $\underline{c}_u$ (resp. $\underline{c}_v$) among those in the path from vertex u (resp. v) to the central vertex. The bound $\underline{\Delta}_1 = c_e + \underline{c} - \underline{c}_u - \underline{c}_v$ can be computed in time $O(1)$, since the computation of the value $\underline{c}_u$ for every $u \in V$ can be performed along the application of depth-first search.

Test LB2 is based on a stronger estimation of the weights of the three actual edges e', e_1, and e_2. The disconnection cost c_{e_1} (resp. c_{e_2}) can be exactly computed in time $O(Q)$. The reconnection cost $c_{e'}$ can be underestimated by the minimum edge weight among all those incident to the new component containing e. This underestimation can be computed in time $O(Q)$ if the incident edges to each node are ordered by increasing weights. Test LB1 is clearly dominated by LB2. However, the computation of LB1 is still of interest to perform a first round of quick reductions. Test LB2 is applied afterwards, to the candidates that passed test LB1.

30.2.2 Selection of path extremities

Given a candidate edge e considered for insertion, there are $O(Q^2)$ possible choices for the associated path $(p-q)$ involved in the move. Each move can be evaluated in time $O(|V|^2 \log |V|)$ using the EW procedure. Substantial savings in computation time can be achieved by evaluating only a small fraction of these paths. We propose heuristic rules to restrict the exact move evaluation to only two paths for each candidate edge e.

We first consider the selection of the first path, as illustrated in Figure 30.4. Given a candidate edge $e = (u, v) \in E$, we identify in time $O(Q)$ the nodes belonging to the paths from u and v to the central node. We start by applying a depth-first search to the component containing u. For each node visited along the depth-first search, another depth-first search is applied to the component containing v. Let p (resp. q) be the node being visited by depth-first search in the component containing u (resp. v). The path connecting p and q may be removed if the tree containing $e = (u, v)$ in the resulting forest does not violate the capacity constraint. If p is in the path from u to the central node, this move will carry to the component containing e in the new solution all nodes currently in the subtree rooted at p. Otherwise, let p' be the node in the path from p to the central node whose predecessor p'' belongs to the path from u to the central node. In this case, the move will carry all nodes in the subtree rooted at p'', except those also belonging to the subtree rooted at p'. A $(p - q)$ path is said to be feasible if the total weight of the nodes carried from the components containing u and v is less than or equal to Q. Since the capacities of each subtree are systematically computed along the application of depth-first search at the beginning of each local search iteration, the feasibility of each path can be evaluated in time $O(1)$.

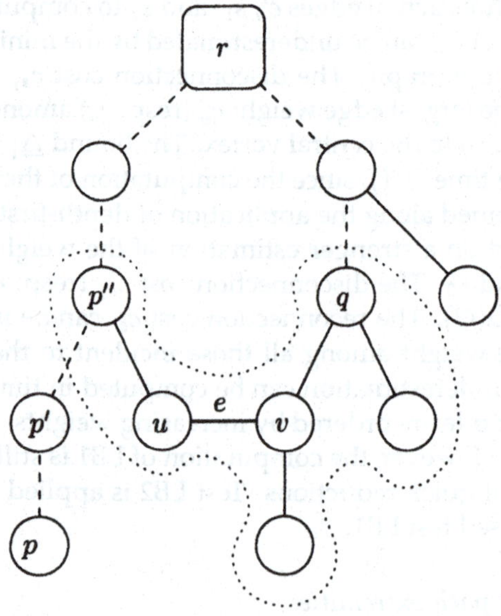

Figure 30.4 Selection of path extremities.

Once a feasible $(p - q)$ path has been identified, we can compute an estimation of the reconnection cost. Let c_p (resp. c_q) be the weight of the minimum weight edge incident to p (resp. q) among all those that can be used to reconnect the subtree rooted at p (resp. q) either to the central node or to another component not involved in the move, without violating the capacity constraints. If nodes p and q are both in the tree containing e in the forest obtained by the removal of path $(p - q)$, the estimation of the reconnection costs is given by $\min\{c_p, c_q\} + c_e$. Otherwise, it is given by $c_p + c_q$. These computations can be performed in time $O(|V|)$, since all edges incident to p (resp. q) may have to be inspected.

The first selected path is that with the smallest estimation of the reconnection costs (ties are randomly broken). To avoid the repeated selection of the same path when different candidate edges connect the same components, a second path for each candidate edge is always randomly chosen.

30.2.3 *Local search procedure*

In this section, we summarize the main steps of the local search procedure for CMST. Algorithm 30.1 gives the algorithmic description of this procedure, which receives an initial solution T_0 as a parameter. We denote by $c(T)$ the cost of any feasible solution T. The current solution T is initialized with the starting solution T_0. The external loop is performed as long as improvements in the current solution are possible. The best neighbor $\bar{T}$ of T is initialized with T itself. Every edge $e \in E$ with extremities in two different components of T is considered as a candidate for insertion into the current tree. An edge e is discarded if it fails to pass any of the tests SR, LB1, and LB2. Tests SR and LB1 are applied first, since their complexity is smaller.

In case a candidate edge e is not eliminated by the reduction tests, two possible paths $(p_1 - q_1)$ and $(p_2 - q_2)$ are considered for removal, as described in Section 30.2.2. $\bar{T}$ is eventually substituted by the best among the two solutions obtained by inserting e and removing either path $(p_1 - q_1)$ or path $(p_2 - q_2)$. Finally, the current solution T is replaced by $\bar{T}$ in case an improvement was detected, and a new local search iteration resumes.

30.3 EFFECTIVENESS OF THE LOCAL SEARCH STRATEGY

The local search algorithm described in the previous section was implemented in C, using version 2.8.1 of the gcc compiler for AIX, with the optimization flag set to -O. All computational experiments were performed on a 200 MHz RISC 6000 workstation model 43P-140 with 64 Mbytes of RAM memory.

We report numerical results obtained by our heuristic on a subset of the test problems described by Gouveia and Martins (2000). The instances have 40 and 80 nodes, besides the central node. They are divided into two classes, in which the central node is placed either in the middle (tc) or at one of the corners (te) of an 100 × 100 grid. The nodes are randomly generated within this grid and edge costs are equal to the integer part of the Euclidean distance between their extremities. All node weights are equal to one. For each class there are ten graphs with 40 nodes and five with 80 nodes. For each graph with 40 nodes, there are two different instances with $Q = 5$ and $Q = 10$. For each of the larger graphs with 80 nodes, there are three instances with $Q = 5$, $Q = 10$, and $Q = 20$. Optimal solutions are known for all problems with 40 nodes and for 14 out of the 30 problems with 80 nodes. For the 16 remaining problems for which an optimal solution is not known, the best known upper

```
procedure LS_CMST(T0)
T ← T0;
improvement ← .TRUE.;
while improvement
    improvement ← .FALSE.;
    T̄ ← T;
    foreach edge e ∈ E with extremities in different compo-
    nents
        if edge e satisfies tests SR, LB1, and LB2
            Select the candidate paths (p1 − q1) and (p2 − q2);
            Obtain T' by performing the move defined by e and
            (p1 − q1);
            if c(T') < c(T̄)
                T̄ ← T';
            Obtain T' by performing the move defined by e and
            (p2 − q2);
            if c(T') < c(T̄)
                T̄ ← T';
            if c(T̄) < c(T)
                T ← T̄;
                improvement ← .TRUE.;
            endif
        endif
    endfor
endwhile
return T
```

Algorithm 30.1 Local search procedure.

bounds are within less than 2% of the lower bound described in Gouveia and Martins (2000).

Tables 30.1 to 30.3 show the computational results obtained over ten runs of each instance. For each instance, we first give the minimum and the average deviations in percent of the solution obtained by the savings heuristic EW, with respect to the best known values given in Gouveia and Martins (2000), together with the average computation time in seconds. The same information is given for solutions obtained by the local search procedure described in the previous section. To illustrate the effectiveness of the neighborhood reduction strategy, we give results for two variants: without neighborhood reduction (complete LS) and with neighborhood reduction (reduced LS).

The results in these three tables show that the local search with neighborhood reduction significantly improves the solutions found by the savings procedure. The average improvements are summarized in Table 30.4. For each class and for each value of the instance size (40 nodes and 80 nodes), we give

Table 30.1 Computational results: Local search on problems with 40 nodes (central node in the middle.)

Instance	Q	Esau-Williams			Complete LS			Reduced LS		
		best %	avg. %	s	best %	avg. %	s	best %	avg. %	s
tc40-1	5	0.68	1.89	0.0	0.00	0.77	2.1	0.00	1.59	0.0
tc40-2	5	1.73	3.39	0.0	0.00	0.67	2.2	0.00	1.57	0.0
tc40-3	5	2.25	2.63	0.0	0.00	0.26	2.1	0.00	0.45	0.0
tc40-4	5	1.62	2.92	0.0	0.00	0.10	2.5	0.00	0.42	0.1
tc40-5	5	0.83	3.00	0.0	0.50	0.93	1.6	0.83	1.97	0.0
tc40-6	5	1.53	2.78	0.0	0.00	0.00	2.3	0.85	1.39	0.0
tc40-7	5	0.00	2.78	0.0	0.00	0.10	2.1	0.00	0.10	0.1
tc40-8	5	0.90	1.79	0.0	0.36	0.36	1.9	0.36	0.94	0.0
tc40-9	5	5.18	6.78	0.0	0.00	1.79	1.8	0.00	2.42	0.1
tc40-10	5	0.00	3.15	0.0	0.00	0.17	2.9	0.00	1.52	0.0
tc40-1	10	0.00	1.53	0.0	0.00	0.00	22.0	0.00	0.16	0.0
tc40-2	10	0.00	1.18	0.0	0.00	0.00	12.2	0.00	0.49	0.0
tc40-3	10	1.60	3.40	0.0	1.60	2.00	18.4	1.60	2.08	0.0
tc40-4	10	0.00	1.45	0.0	0.00	0.00	27.4	0.00	0.00	0.1
tc40-5	10	0.00	0.00	0.0	0.00	0.00	13.4	0.00	0.00	0.0
tc40-6	10	0.80	0.80	0.0	0.00	0.00	26.0	0.00	0.20	0.0
tc40-7	10	0.00	1.04	0.0	0.00	0.00	19.2	0.00	0.00	0.0
tc40-8	10	0.00	0.00	0.0	0.00	0.00	10.7	0.00	0.00	0.0
tc40-9	10	0.00	0.99	0.0	0.00	0.00	16.6	0.00	0.08	0.0
tc40-10	10	1.93	3.13	0.0	0.00	0.12	32.2	0.00	1.41	0.1

Table 30.2 Computational results: Local search on problems with 40 nodes (central node at one of the corners.)

		Esau-Williams			Complete LS			Reduced LS		
Instance	Q	best %	avg. %	s	best %	avg. %	s	best %	avg. %	s
te40-1	5	2.77	4.59	0.0	0.24	1.18	2.3	1.33	2.90	0.1
te40-2	5	1.77	3.96	0.0	0.00	1.64	2.1	0.51	2.90	0.1
te40-3	5	2.89	4.28	0.0	0.00	1.23	2.6	0.00	1.63	0.2
te40-4	5	3.07	4.78	0.0	0.00	1.44	2.4	0.12	2.67	0.2
te40-5	5	1.40	2.91	0.0	0.26	0.31	2.1	0.51	0.82	0.1
te40-6	5	0.61	4.34	0.0	0.00	1.67	2.6	0.00	1.80	0.2
te40-7	5	2.07	4.63	0.0	0.73	1.28	1.9	0.85	1.66	0.2
te40-8	5	4.72	5.11	0.0	0.12	0.27	2.5	1.09	1.78	0.1
te40-9	5	3.08	3.18	0.0	0.13	0.13	2.1	0.13	0.27	0.1
te40-10	5	1.68	2.42	0.0	0.52	0.52	2.3	0.52	0.76	0.2
te40-1	10	2.85	7.57	0.0	0.34	2.10	41.2	1.01	4.75	0.3
te40-2	10	5.58	7.96	0.0	0.00	1.26	49.9	0.00	2.67	0.6
te40-3	10	2.11	5.16	0.0	0.00	0.97	39.2	0.35	1.90	0.4
te40-4	10	0.67	0.67	0.0	0.00	0.00	19.5	0.34	0.34	0.2
te40-5	10	3.67	3.67	0.0	0.00	0.00	50.0	0.00	0.93	0.3
te40-6	10	2.43	7.29	0.0	0.00	0.62	38.4	0.00	5.12	0.3
te40-7	10	3.38	7.53	0.0	2.88	2.88	20.1	2.88	3.67	0.1
te40-8	10	4.92	5.25	0.0	0.66	0.66	43.9	0.98	1.90	0.2
te40-9	10	4.27	8.75	0.0	0.18	0.18	35.8	0.18	2.85	0.4
te40-10	10	3.83	5.66	0.0	0.00	1.88	44.4	1.28	2.76	0.5

Table 30.3 Computational results: Local search on problems with 80 nodes (continues on next page).

Instance	Q	Esau-Williams			Complete LS			Reduced LS		
		best %	avg. %	s	best %	avg. %	s	best %	avg. %	s
tc80-1	5	5.91	7.36	0.0	0.45	1.90	32.2	2.46	4.15	0.3
tc80-2	5	5.36	6.36	0.0	1.64	1.96	31.9	2.00	4.27	0.2
tc80-3	5	4.29	6.00	0.0	0.00	0.84	35.1	1.68	3.13	0.3
tc80-4	5	5.46	6.66	0.0	0.46	1.64	38.2	3.15	4.55	0.2
tc80-5	5	3.03	6.04	0.0	0.08	0.95	31.4	2.10	3.43	0.4
tc80-1	10	4.84	6.27	0.0	0.00	0.82	593.2	0.90	2.38	0.7
tc80-2	10	2.05	4.50	0.0	0.91	1.46	665.2	1.60	2.20	0.7
tc80-3	10	2.73	5.26	0.0	0.00	0.64	466.0	0.46	2.74	0.5
tc80-4	10	5.41	6.19	0.0	0.92	1.61	676.5	3.23	4.49	0.3
tc80-5	10	5.39	6.97	0.0	0.20	1.77	477.1	0.30	2.66	0.8
tc80-1	20	0.96	2.58	0.0	0.00	0.24	4739.2	0.48	0.70	0.4
tc80-2	20	1.95	2.84	0.0	0.00	0.80	3289.0	0.73	1.27	0.6
tc80-3	20	0.48	1.30	0.0	0.00	0.19	2811.2	0.48	0.92	0.2
tc80-4	20	1.22	2.34	0.0	0.00	0.61	3219.4	0.00	1.34	0.4
tc80-5	20	3.49	3.49	0.0	0.00	0.61	5088.4	0.87	2.54	0.8

Table 30.3 Computational results: Local search on problems with 80 nodes (continued from previous page).

Instance	Q	Esau-Williams			Complete LS			Reduced LS		
		best %	avg. %	s	best %	avg. %	s	best %	avg. %	s
te80-1	5	2.91	3.20	0.0	1.06	1.30	21.0	0.94	1.70	1.5
te80-2	5	1.49	2.69	0.0	0.20	1.09	14.1	0.55	1.52	1.0
te80-3	5	2.60	3.12	0.0	0.57	1.18	21.4	0.88	1.63	1.8
te80-4	5	2.35	2.97	0.0	0.59	1.04	15.3	1.13	1.54	1.2
te80-5	5	2.07	3.09	0.0	0.00	0.88	17.2	0.73	2.07	0.5
te80-1	10	3.56	3.74	0.0	0.06	1.36	321.4	1.39	2.01	4.6
te80-2	10	2.81	4.08	0.0	1.65	2.25	189.5	2.56	3.56	0.9
te80-3	10	3.68	5.05	0.0	0.77	1.13	511.5	0.95	2.03	5.4
te80-4	10	7.55	9.15	0.0	1.72	2.56	662.4	2.39	6.17	4.7
te80-5	10	6.55	6.87	0.0	0.44	0.51	570.3	4.80	5.44	1.2
te80-1	20	0.78	3.14	0.0	0.16	0.75	4678.4	0.78	1.30	6.7
te80-2	20	5.23	7.71	0.0	0.16	1.58	5132.1	0.74	4.12	10.7
te80-3	20	5.13	6.57	0.0	0.00	0.40	5745.7	0.08	0.85	11.5
te80-4	20	7.59	8.23	0.0	0.16	0.61	4614.8	0.40	1.52	15.8
te80-5	20	0.89	2.15	0.0	0.00	0.02	3641.0	0.00	0.40	4.8

the average percentage reduction in the deviation from the best known solution obtained by the complete and the reduced local search procedures, with respect to the savings heuristic. Over the 70 instances, the latter found nine best known solutions, while the complete and reduced local search procedures obtained respectively 38 and 23. In particular, the savings procedure did not find any best known solution for the harder problems with the central node at one of the corners of the grid, while the complete and reduced local search procedures found respectively 13 and six best known solutions for the 30 problems in this class.

Table 30.4 Average reductions with respect to the savings heuristic.

Instances	Complete LS (%)	Reduced LS (%)
tc40	1.87	1.39
te40	3.99	2.80
tc80	3.93	2.28
te80	3.89	2.61

The computation times observed with the neighborhood reduction strategies are significantly smaller, without too much deterioration in solution quality. For the problems with $n = 80$, $Q = 10$, and the central node at one of the corners of the grid, Figure 30.5 illustrates the effect of tests SR, LB1, and LB2 described in Section 30.2.1. For each instance, we give the fraction in percent of the number of candidate edges (i.e., those edges with both extremities in different components) which are not eliminated by each combination of tests (SR alone, LB1 alone, SR together with LB1, and LB2 after SR and LB1). None of the $O(1)$ tests SR and LB1 dominate the other. They eliminate at least 37% of the candidate edges of any of the five instances in this figure. The fraction of remaining edges range from 10 to 30% after the application of the three tests.

For the same problems, Figure 30.6 shows the fraction of the iterations in which the best movement was that obtained with the path chosen by the first criterion, by the random choice, or by both of them. It illustrates that none of the criteria is dominated by the other and that both should be used.

The results in this section show that the reduced local search procedure compares favorably with the savings heuristic of Esau and Williams. Local search obtains much better solutions in marginal time, due to the effectiveness of the reduction procedures. Therefore, the reduced local search procedure represents a good trade-off between solution quality and computation time. In the next section, we describe an extension of this approach, embedding the reduced local search strategy into a GRASP heuristic.

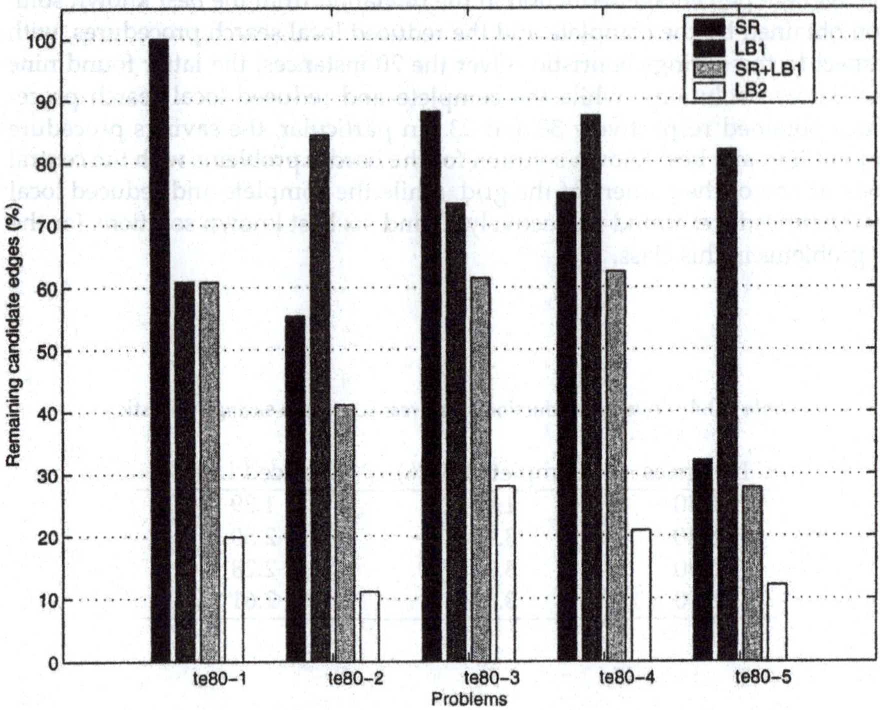

Figure 30.5 Effectiveness of the reductions tests SR, LB1, and LB2.

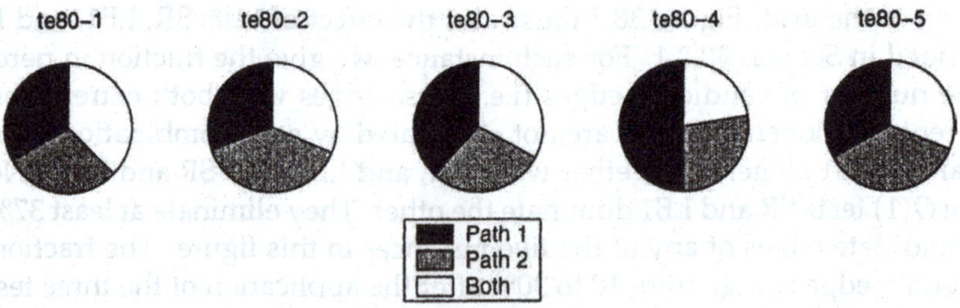

Figure 30.6 Path selection criteria.

30.4 GRASP WITH A MEMORY-BASED LOCAL SEARCH STRATEGY

A GRASP is a multistart process, in which each iteration consists of two phases: construction and local search (Feo and Resende, 1995). The best solution found after `Max_Iterations` iterations is returned. The reader is referred to Festa and Resende (2002) for a recent survey of applications and to Resende and Ribeiro (2003b) for implementations strategies, variants, and extensions.

A feasible solution is iteratively constructed in the first phase, one element at a time. At each construction iteration, the choice of the next element to be added is determined by ordering all candidate elements (i.e. those that can be added to the solution) in a candidate list, with respect to its contribution to the objective function. The list of best candidates is called the *restricted candidate list* (RCL). The random selection of an element from the RCL allows for different solutions to be obtained at each GRASP iteration. Since solutions generated at the construction phase are not guaranteed to be locally optimal, it is almost always beneficial to apply a local search as an attempt to improve each constructed solution.

In the remainder of this section, we customize a GRASP for the capacitated minimum spanning tree problem. We describe the construction and local search procedures, as well as a path-relinking intensification strategy.

30.4.1 Construction phase

The construction phase is a randomized version of the savings heuristic of Esau and Williams. It starts with an r-centered star, in which all nodes are directly connected to the central node. Each iteration is characterized by the merging of two components into a new one. Two components may be merged if the sum of the weights of their nodes is less than or equal to Q. This operation is performed in two steps. First, we select an edge $(i, j) \in E$ connecting nodes i and j from two different components T_i and T_j. Let i' (resp. j') be the node connecting T_i (resp. T_j) to the central node r. Next, we eliminate the maximum weight edge among the two that connect T_i and T_j to the central node. This operation leads to saving $s_{ij} = \max\{c_{i'r}, c_{j'r}\} - c_{ij}$ cost units with respect to the current solution. Let C be the subset formed by all edges $(i, j) \in E$ whose associated components T_i and T_j can be merged. For a given parameter value $\alpha \in [0, 1]$ (we used $\alpha = 0.9$ in our implementation), we insert in the RCL all edges from C with savings $s_{ij} \geq s^{min} + \alpha \cdot (s^{max} - s^{min})$, where $s^{min} = \min\{s_{ij} > 0 : (i, j) \in C\}$ and $s^{max} = \max\{s_{ij} > 0 : (i, j) \in C\}$. An edge is randomly selected from the RCL, the components associated to its extremities are merged, and a new iteration resumes, until no further savings are possible.

30.4.2 Local search phase

The local search phase seeks to improve each solution built in the construction phase. Basically, the same strategy described in Section 30.2 is used.

As a memoryless approach, the basic GRASP procedure does not make use of the information collected along the search. Also, the local search phase of a GRASP iteration very often repeatedly spends too much time to stop in the first same local optimum, eventually missing better, but hidden solutions which might be quite close to this local optimum. As an attempt to improve the effectiveness of the local search phase, we propose to add to the latter some elements of very short term memory taken from tabu search (Glover and Laguna, 1997).

However, instead of maintaining a list of forbidden attributes, we keep the last `Tabu_Tenure` visited solutions in the memory. We always move to the best neighbor which is not stored in this memory, regardless of whether it is an improving solution or not. The search stops if the incumbent is not improved after a sequence of `Tabu_Tenure` local search iterations. A small value is used for this parameter, to keep the computation times limited.

30.4.3 Path-relinking

Path-relinking was originally proposed by Glover (1996), as an intensification strategy exploring trajectories connecting elite solutions obtained by tabu search or scatter search (Glover and Laguna, 1997; Glover, 2000; Glover et al., 2000). The use of path-relinking within a GRASP procedure was first proposed by Laguna and Martí (1999), followed by several extensions, improvements, and successful applications (Aiex et al., 2000; Canuto et al., 2001; Resende and Ribeiro, 2003a; Ribeiro et al., 2002). Implementation strategies are investigated in detail by Resende and Ribeiro (2003b).

In this context, path-relinking is applied to pairs of solutions, one obtained by local search and the other randomly chosen from a pool with a limited number `Max_Elite` of elite solutions found along the search. The pool is originally empty. Each locally optimal solution is considered as a candidate to be inserted into the pool if it is different from every other solution currently in the pool. If the pool already has `Max_Elite` solutions and the candidate is better than the worst of them, then the former replaces the latter. If the pool is not full, the candidate is simply inserted.

Let z (resp. y) be the an elite solution in the pool (resp. a locally optimal solution). Resende and Ribeiro (2003b) observed that better solutions are usually found in smaller computation times if z is used as the initial solution, and y as the guiding one. The algorithm starts by computing the edges that belong to either y or z, but not to the other. These computations amount to determining a set of moves $\Delta(z, y)$ which should be applied to the initial solution to reach the guiding one. Each move is characterized by a pair of edges, one to be inserted and the other to be eliminated from the current solution. The best solution $\bar{y}$ along the new path to be constructed is initialized with the initial solution. Each path-relinking iteration has three steps: the best remaining move leading to a feasible tree is applied to the current solution, the incumbent $\bar{y}$ is updated, and the selected move is removed from $\Delta(z, y)$. The algorithm stops when the guiding solution is reached or in case all remaining moves lead to infeasible solutions. The incumbent $\bar{y}$ is returned as the best solution found by path-relinking and inserted into the pool if it satisfies the membership conditions.

We recall that we denote by $c(x)$ the weight of a solution x. The pseudo-code with the complete description of the procedure `GRASP+PR_CMST` for the capacitated minimum spanning tree problem is given in Algorithm 30.2. This description incorporates the following phases: construction, local search, and path-relinking.

```
procedure GRASP+PR_CMST( )
c* ← ∞;
Pool ← ∅;
for k = 1 to Max_Iterations
    Construct a greedy randomized solution x;
    Find y by applying local search to x;
    if y satisfies the membership conditions
        Insert y into Pool;
    Randomly select a solution z ∈ Pool (z ≠ y) with uniform
    probability;
    if c(z) > c(y)
        y ↔ z;
    Compute Δ(z, y);
    Let ȳ be the best solution found by applying path-
    relinking to (y, z);
    if ȳ satisfies the membership conditions
        Insert ȳ into Pool;
    if c(ȳ) < c*
        x* ← ȳ;
        c* ← c(ȳ);
    endif
endfor
return x*;
```

Algorithm 30.2 Pseudo-code of the GRASP with path-relinking procedure for the capacitated minimum spanning tree problem.

30.5 COMPUTATIONAL RESULTS

We report in this section the numerical results obtained with the algorithm GRASP+PR_CMST described in the previous section. This heuristic was also implemented in C, using version 2.8.1 of the gcc compiler for AIX, with the optimization flag set to -O. Again, all experiments were performed on a 200 MHz RISC 6000 workstation model 43P-140 with 64 Mbytes of RAM memory.

In addition to the test problems already described in Section 30.3, we considered some larger instances with 120 and 160 nodes, besides the central node, also described by Gouveia and Martins (2000). These instances are generated as the previous ones. They are also divided into two classes, in which the central node is placed either in the middle (tc) or at one of the corners (te) of an 100×100 grid. There is only one graph with 120 nodes and another one

Table 30.5 Computational results: Memoryless GRASP vs. GRASP with path-relinking and short term memory on problems with 40 nodes (central node in the middle.)

		Memoryless GRASP			GRASP+PR_CMST		
Instance	Q	deviation %	iteration	s	deviation %	iteration	s
tc40-1	5	0.00	181	9.9	0.00	1	51.4
tc40-2	5	0.00	6	13.5	0.00	4	42.2
tc40-3	5	0.00	7	8.1	0.00	1	30.4
tc40-4	5	0.00	13	16.4	0.00	2	48.8
tc40-5	5	0.33	90	15.0	0.00	15	50.0
tc40-6	5	0.00	1	17.1	0.00	1	64.9
tc40-7	5	0.00	2	23.7	0.00	2	103.1
tc40-8	5	0.00	43	9.0	0.00	1	45.4
tc40-9	5	0.00	1	14.0	0.00	1	53.3
tc40-10	5	0.00	3	21.1	0.00	1	72.6
tc40-1	10	0.00	13	8.4	0.00	4	46.8
tc40-2	10	0.00	1	10.1	0.00	1	25.8
tc40-3	10	0.00	53	11.1	0.00	3	53.4
tc40-4	10	0.00	4	27.8	0.00	2	110.0
tc40-5	10	0.00	30	14.2	0.00	1	60.2
tc40-6	10	0.00	14	17.5	0.00	3	68.5
tc40-7	10	0.00	19	17.3	0.00	4	65.3
tc40-8	10	0.00	3	14.1	0.00	1	69.7
tc40-9	10	0.00	8	6.9	0.00	1	41.0
tc40-10	10	0.00	6	22.8	0.00	2	112.8

with 160 nodes for each class: tc-120, te-120, tc-160, and te-160. There are three different instances in each case, with $Q = 5$, $Q = 10$, and $Q = 20$.

The computational experiments reported in Amberg et al. (1996) and Ahuja et al. (2001) used instances with 40 and 80 nodes. Amberg et al. (1996) considered 60 instances: those indexed from 1 to 5 with 40 nodes and all instances with 80 nodes. The results presented by Ahuja et al. (2001) comprise 65 instances: all those with 40 nodes and the central node in the middle, those indexed from 1 to 5 with 40 nodes and the central node at one of the corners, and all instances with 80 nodes. Results for the instances with 120 nodes are reported by Gouveia and Martins (2000), while those for the instances with 160 nodes appeared exclusively in Martins (1999).

The best known solutions values (i.e., the best known upper bounds) for all problems with 40, 80, and 120 nodes are reported in Gouveia and Martins (2000). The best known solution values for instance te120 with $Q = 20$ and for all instances with 160 nodes were informed by Martins (2001). Amberg et al. (1996) found the best known upper bounds for 67 out of the 70 instances with 40 and 80 nodes, using three variants of tabu search (not all best upper bounds are found by the same variant). Ahuja et al. (2001) improved the previous results for three instances with 80 nodes, finding the best known upper bounds for all these instances using the same tabu search algorithm.

We report computational results for all test instances with 40 to 160 nodes. One possible shortcoming of the standard memoryless GRASP metaheuristic is the independence of its iterations, i.e., the fact that it does not learn from the history of solutions found in previous iterations. This is so because it discards information about any solution that does not improve the incumbent. However, the use of a very short term memory in an extended local search procedure can improve the solutions found, by adding some elements of tabu search, as proposed in Section 30.4.2. Also, long term memory information gathered from good solutions can be used to implement extensions based on path-relinking, as described in Section 30.4.3. With these extensions, we were able to build a GRASP heuristic using a memory-based local search incorporating short- and long-term memory features. To illustrate the effectiveness of these two extensions of the basic algorithm, we report in Tables 30.5 to 30.8 the results obtained for both the memoryless algorithm and the full procedure `GRASP+PR_CMST` with path-relinking and short term memory. For each instance and for each approach, we give the deviation in percent of the best solution value found with respect to the best known upper bound, the GRASP iteration in which this solution was found, and the total computation time in seconds. Parameters `Max_Iterations`, `Max_Elite`, and `Tabu_Tenure` were set once for all and without tuning, respectively at 200, 20, and 6.

Procedure `GRASP+PR_CMST` improved many solutions found by the memoryless approach, finding the optimal values for all instances with 40 nodes. Although the overall computation times increased, we notice that the best solutions were found much earlier. The best solution is found at the first iteration for 13 instances and in at most 20 iterations for 37 instances, among the

Table 30.6 Computational results: memoryless GRASP vs. GRASP with path-relinking and short term memory on problems with 40 nodes (central node at one of the corners.)

		Memoryless GRASP			GRASP+PR_CMST		
Instance	Q	deviation %	iteration	s	deviation %	iteration	s
te40-1	5	0.00	35	78.3	0.00	13	208.2
te40-2	5	0.00	8	53.8	0.00	6	176.7
te40-3	5	0.00	61	69.4	0.00	2	215.0
te40-4	5	0.00	84	58.7	0.00	5	211.7
te40-5	5	0.00	6	58.7	0.00	5	189.2
te40-6	5	0.00	30	66.2	0.00	9	178.6
te40-7	5	0.24	20	67.1	0.00	19	215.4
te40-8	5	0.00	25	62.2	0.00	8	179.8
te40-9	5	0.13	3	67.4	0.00	24	195.3
te40-10	5	0.13	125	60.5	0.00	50	192.5
te40-1	10	0.34	18	177.7	0.00	13	549.8
te40-2	10	0.00	1	132.7	0.00	1	489.3
te40-3	10	0.35	40	135.8	0.00	28	410.0
te40-4	10	0.00	18	105.2	0.00	2	382.4
te40-5	10	0.00	13	104.0	0.00	1	272.1
te40-6	10	0.00	32	125.9	0.00	2	340.1
te40-7	10	0.34	3	108.2	0.00	1	423.0
te40-8	10	0.33	12	160.6	0.00	17	446.5
te40-9	10	0.00	23	107.8	0.00	2	334.1
te40-10	10	0.00	177	124.7	0.00	3	389.5

Table 30.7 Computational results: Memoryless GRASP vs. GRASP with path-relinking and short term memory on problems with 80 nodes (continues on next page).

Instance	Q	Memoryless GRASP			GRASP+PR_CMST		
		deviation %	iteration	s	deviation %	iteration	s
tc80-1	5	0.36	105	154.6	0.00	7	325.4
tc80-2	5	0.73	33	165.8	0.09	159	401.5
tc80-3	5	0.00	161	126.0	0.00	8	294.2
tc80-4	5	0.93	191	134.9	0.56	56	309.0
tc80-5	5	0.08	107	195.1	0.00	122	432.8
tc80-1	10	0.45	14	214.7	0.00	14	448.3
tc80-2	10	1.37	49	346.4	0.00	28	785.0
tc80-3	10	0.69	8	188.0	0.23	3	390.1
tc80-4	10	0.92	5	222.9	0.00	49	517.5
tc80-5	10	1.00	45	368.3	0.00	52	820.7
tc80-1	20	0.00	47	136.1	0.00	2	239.7
tc80-2	20	0.73	57	227.5	0.00	65	571.5
tc80-3	20	0.00	5	87.3	0.00	2	187.4
tc80-4	20	0.49	46	151.2	0.00	2	262.3
tc80-5	20	0.22	169	421.8	0.00	2	1023.5

Table 30.7 Computational results: Memoryless GRASP vs. GRASP with path-relinking and short term memory on problems with 80 nodes (continued from previous page).

Instance	Q	Memoryless GRASP			GRASP+PR_CMST		
		deviation %	iteration	s	deviation %	iteration	s
te80-1	5	0.08	44	802.5	0.00	83	1609.2
te80-2	5	0.39	98	940.6	0.00	81	1938.4
te80-3	5	0.31	41	1039.9	0.00	88	2228.2
te80-4	5	0.51	192	987.3	0.08	55	2128.9
te80-5	5	0.00	111	741.2	0.00	19	1658.0
te80-1	10	0.06	183	2480.9	0.00	5	5387.8
te80-2	10	0.37	36	2528.5	0.00	21	6172.2
te80-3	10	0.36	117	2825.2	0.00	4	6088.5
te80-4	10	0.43	12	2677.6	0.00	5	6311.5
te80-5	10	0.00	106	2434.5	0.00	39	5465.1
te80-1	20	0.15	191	3582.3	0.00	15	6423.5
te80-2	20	0.00	65	5040.1	0.00	11	11034.1
te80-3	20	0.24	80	5139.1	0.00	79	10608.2
te80-4	20	0.00	79	4632.7	0.00	2	10038.2
te80-5	20	0.00	15	3039.2	0.00	2	5676.7

Table 30.8 Computational results: Memoryless GRASP vs. GRASP with path-relinking and short term memory on problems with 120 and 160 nodes.

Instance	Q	Memoryless GRASP dev. %	iter.	s	GRASP+PR_CMST dev. %	iter.	s	Best value previous	new
tc120	5	0.23	185	342.3	0.00	64	665.4	1291	same
tc120	10	0.22	55	1111.1	0.00	3	2368.0	904	same
tc120	20	0.52	23	1314.4	0.00	4	3075.5	768	same
te120	5	0.59	7	1323.3	0.22	38	2118.5	2197	same
te120	10	0.45	103	3672.0	0.08	111	6974.8	1329	same
te120	20	0.65	87	8297.1	0.00	82	15509.9	921	same
tc160	5	0.58	173	2834.5	(-0.10)	43	4662.6	2083	2081
tc160	10	0.45	195	8159.7	(-0.30)	20	13957.6	1323	1319
tc160	20	1.03	25	19999.4	(-0.62)	197	37139.9	966	960
te160	5	0.46	137	4139.6	0.07	129	5954.8	2790	same
te160	10	0.18	108	10784.2	(-0.24)	83	17257.9	1650	1646
te160	20	0.18	166	27678.6	(-0.27)	57	45129.8	1101	1098

40 problems in Tables 30.5 and 30.6. We recall that the computation times are reported for a total of 200 iterations. Similar results are observed in Table 30.7 for the test problems with 80 nodes. The improved GRASP procedure missed only four of the best known upper bounds (we recall that Amberg et al. (1996) missed three upper bounds, but using three different algorithms). In these few cases, the deviation from the best upper bound was always less than 0.60%. For more than 50% of the problems with results reported in this table, the best solution is found in at most 20 iterations.

The instances with 120 and 160 nodes were run on a Celeron 800 MHz machine with 128 Mbytes of RAM memory. The program was compiled with version 2.96 of the `gcc` compiler for Linux, once again with the optimization flag set to `-O`. The computational results for these larger instances are given in Table 30.8. Besides the statistics given in the previous tables, we also inform the currently best known solution value according to Martins (2001) and the new best solution value for the instances in which the latter was improved by the GRASP heuristic with the memory-based local search. Once again, we notice that algorithm GRASP+PR_CMST improved the solutions found by the basic GRASP procedure for all 12 instances in this table. In particular, the GRASP heuristic with the memory-based local search improved the best known solution for five out of these 12 instances (tc160 with $Q = 5$, $Q = 10$, and $Q = 20$; te160 with $Q = 10$ and $Q = 10$). For the other problems, the deviation from the best upper bound was always less than 0.25%.

30.6 CONCLUDING REMARKS

We first presented a local search heuristic for the capacitated minimum spanning tree problem, using a new path-based neighborhood structure and reduction techniques to speed up the search. Computational results on a set of standard benchmark problems showed that the proposed local search considerably improved the solutions found by a well known savings heuristic, leading to good trade-offs between solution quality and computation time.

We also proposed a GRASP heuristic using this neighborhood structure within the local search phase. One possible shortcoming of the standard memoryless GRASP metaheuristic being the independence of its iterations, we improved the local search strategy by incorporating short- and long-term memory features to it. A very short term memory is used to implement an extended local search going beyond the first local optimum, by the incorporation of some elements of tabu search. Long term memory information gathered from good solutions found along the search is used to implement a path-relinking procedure for intensification. Computational results obtained for benchmark problems illustrated the effectiveness of the proposed heuristic, which is competitive with the best heuristic in the literature in terms of solution quality. The GRASP heuristic using a memory-based local search strategy improved the best known solution for five out of the six largest benchmark problems.

Acknowledgments

The authors are grateful to L. Gouveia and P. Martins for kindly making available their test problems. Work of the first author was sponsored by CNPq in the framework of the doctorate scholarship grant number 200133/98-05. Work of the third author was sponsored by FAPERJ grant 150966/99 and by CNPq grants 302281/85-1, 202005/89-5, and 910062/99-4.

Bibliography

R.K. Ahuja, J.B. Orlin, and D. Sharma. Multi-exchange neighborhood structures for the capacitated minimum spanning tree problem. *Mathematical Programming*, 91:71–97, 2001.

R.M. Aiex, M.G.C. Resende, P.M. Pardalos, and G. Toraldo. GRASP with path-relinking for the three-index assignment problem. Technical report, AT&T Labs-Research, Florham Park, NJ, 2000. To appear in *INFORMS J. on Computing*.

A. Amberg, W. Domschke, and S. Voss. Capacitated minimum spanning trees: Algorithms using intelligent search. *Combinatorial Optimization: Theory and Practice*, 1:9–40, 1996.

S.A. Canuto, M.G.C. Resende, and C.C. Ribeiro. Local search with perturbations for the prize-collecting steiner tree problem in graphs. *Networks*, 38: 50–58, 2001.

K.M. Chandy and T. Lo. The capacitated minimum spanning tree. *Networks*, 3:173–181, 1973.

K.M. Chandy and R.A. Russel. The design of multipoint linkages in a teleprocessing tree network. *IEEE Transactions on Computers*, 21:1062–1066, 1972.

L.R. Esau and K.C. Williams. On teleprocessing system design. *IBM Systems Journal*, 5:142–147, 1966.

T.A. Feo and M.G.C. Resende. Greedy randomized adaptive search procedures. *Journal of Global Optimization*, 6:109–133, 1995.

P. Festa and M.G.C. Resende. GRASP: An annotated bibliography. In C.C. Ribeiro and P. Hansen, editors, *Essays and Surveys in Metaheuristics*, pages 325–367. Kluwer, 2002.

B. Gavish. Topological design of centralized computer networks: Formulations and algorithms. *Networks*, 12:355–377, 1982.

B. Gavish. Formulations and algorithms for the capacitated minimal directed tree problem. *Journal of the ACM*, 30:118–132, 1983.

F. Glover. Tabu search and adaptive memory programing – Advances, applications and challenges. In R.S. Barr, R.V. Helgason, and J.L. Kennington, editors, *Interfaces in Computer Science and Operations Research*, pages 1–75. Kluwer, 1996.

F. Glover. Multi-start and strategic oscillation methods – Principles to exploit adaptive memory. In M. Laguna and J.L. Gonzáles-Velarde, editors, *Computing Tools for Modeling, Optimization and Simulation: Interfaces in Computer Science and Operations Research*, pages 1–24. Kluwer, 2000.

F. Glover and M. Laguna. *Tabu Search*. Kluwer, 1997.

F. Glover, M. Laguna, and R. Martí. Fundamentals of scatter search and path relinking. *Control and Cybernetics*, 39:653–684, 2000.

L. Gouveia. A 2*n* formulation for the capacitated minimal spanning tree problem. *Operations Research*, 4:130–141, 1995.

L. Gouveia and P. Martins. The capacitated minimum spanning tree problem: An experiment with a hop-indexed model. *Annals of Operations Research*, 86: 271–294, 1999.

L. Gouveia and P. Martins. A hierarchy of hop-indexed models for the capacitated minimum spanning tree problem. *Networks*, 35:1–16, 2000.

L. Hall. Experience with a cutting plane approach for the capacitated spanning tree problem. *INFORMS Journal on Computing*, 8:219–234, 1996.

J.B. Kruskal. On the shortest spanning tree of a graph and the traveling salesman problem. In *Proceedings of the American Mathematical Society*, volume 7, pages 48–50, 1956.

M. Laguna and R. Martí. GRASP and path relinking for 2-layer straight line crossing minimization. *INFORMS Journal on Computing*, 11:44–52, 1999.

K. Malik and G. Yu. A branch and bound algorithm for the capacitated minimum spanning tree problem. *Networks*, 23:525–532, 1993.

P. Martins. *Problema da árvore de suporte de custo mínimo com restrição de capacidade: Formulações com índice de nível*. PhD thesis, Departamento de Estatística e Investigação Operacional, Universidade de Lisboa, 1999.

P. Martins, 2001. Personal communication.

C. Papadimitriou. The complexity of the capacitated tree problem. *Networks*, 8:217–230, 1978.

M.G.C. Resende and C.C. Ribeiro. A GRASP with path-relinking for permanent virtual circuit routing. *Networks*, 41:104–114, 2003a.

M.G.C. Resende and C.C. Ribeiro. Greedy Randomized Adaptive Search Procedures. In F. Glover and G. Kochenberger, editors, *Handbook of Metaheuristics*, pages 219–249. Kluwer, 2003b.

C.C. Ribeiro, E. Uchoa, and R.F. Werneck. A hybrid GRASP with perturbations for the steiner problem in graphs. *INFORMS Journal on Computing*, 14:228–246, 2002.

Y. Sharaiha, M. Gendreau, G. Laporte, and I. Osman. A tabu search algorithm for the capacitated shortest spanning tree problem. *Networks*, 29:161–171, 1997.

P.M. Thompson and J.B. Orlin. The theory of cyclic transfer. Technical Report OR200-89, MIT, Operations Research Center, 1989.

P.M. Thompson and H.N. Psaraftis. Cyclic transfer algorithms for multi-vehicle routing and scheduling problems. *Operations Research*, 41:935–946, 1993.

Metaheuristics: Computer Decision-Making, pp. 659-672
Marcone Jamilson Freitas Souza, Nelson Maculan, and Luis Satoru Ochi

31 A GRASP-TABU SEARCH ALGORITHM FOR SOLVING SCHOOL TIMETABLING PROBLEMS

Marcone Jamilson Freitas Souza[1], Nelson Maculan[2], and Luis Satoru Ochi[3]

[1]Department of Computer Science
Federal University of Ouro Preto,
Ouro Preto, MG Brazil
marcone@iceb.ufop.br

[2]Systems Engineering and Computer Science Program
Federal University of Rio de Janeiro
Rio de Janeiro, RJ Brazil
maculan@cos.ufrj.br

[3]Department of Computer Science
Fluminense Federal University
Niterói, RJ Brazil
satoru@dcc.ic.uff.br

Abstract: This paper proposes a hybrid approach to solve school timetabling problems. This approach is a GRASP that uses a partially greedy procedure to construct an initial solution and attempts to improve the constructed solution using Tabu Search. When an infeasible solution without overlapping classes is generated, a procedure called Intraclasses-Interclasses is activated, trying to retrieve feasibility. If successful, it is reactivated, in an attempt to improve the timetable's compactness as well as other requirements. Computational results show that the Intraclasses-Interclasses procedure speeds up the process of obtaining better quality solutions.
Keywords: Metaheuristics, GRASP, Tabu search, School timetabling.

31.1 INTRODUCTION

The school timetabling problem (STP) regards the weekly class schedule. The problem consists in coordinating lessons with periods, satisfying a set of requirements. The difficulty of solving a real case is well known, since the problem has a huge search space and is highly constrained. As it is NP-hard (Even et al., 1976), a heuristic approach to solve it is justified.

Among recent techniques that have been successfully used to solve the school timetabling problem, the following metaheuristics techniques stand out: Tabu Search (Schaerf, 1996; Alvarez-Valdes et al., 1996; Costa, 1994), Simulated Annealing (Abramson, 1991), and Genetic Algorithms (Carrasco and Pato, 2001; Colorni et al., 1998).

In this paper, a local search technique, called Intraclasses-Interclasses (II), is developed. It is introduced in the course of the local search phase of a Greedy Randomized Adaptive Search Procedure, or GRASP (Feo and Resende, 1995), to try speed up the process of obtaining better quality solutions. Procedure II attempts to improve a given timetable without overlapping classes in 2 steps. First, it tries to retrieve feasibility and if successful, it is reactivated, seeking to improve timetabling quality requirements. This procedure can be applied to solve school timetabling problems, as described in Section 31.2, in which the number of lessons and the periods set aside for classes are the same.

This paper is organized as follows. In Section 31.2, the problem is described; the following sections regard the used representation, the neighborhood structure, and the objective function. Section 31.5 describes, in detail, the proposed algorithm, and Section 31.6 presents some experimental results. Concluding remarks are made in the last section.

31.2 PROBLEM DESCRIPTION

The school timetabling problem (STP) in question consists of a set of m teachers, n classes, s subjects and p weekly periods which are set aside for classes. Periods are distributed in d week days and h daily periods, which occur during the same shift, i.e., $p = d * h$. Classes, which are always available, are disjoint sets of students who take the same subjects. Each subject of a given class is associated with only one teacher, who is previously determined. Furthermore, the number of weekly lessons for each class is precisely p. The following requirements must be satisfied:

(a) a teacher cannot teach more than one class at the same time, i.e, a timetable cannot have an overlapping of teachers;

(b) a class cannot have a lesson with more than one teacher at the same period, i.e, a timetable cannot have an overlapping of classes;

(c) each teacher must fulfill his/her weekly number of lessons;

(d) a teacher cannot be scheduled for a period in which he/she is not available;

(e) a class cannot have more than two lessons a day with the same teacher;

(f) teachers' request for double lessons (lessons conducted in two consecutive periods) should be granted as often as possible;

(g) teacher's schedule should be as compact as possible.

We make the following definitions.

- A timetable Q that does not satisfy at least one of the requirements (a), ..., (e), is considered *infeasible* and cannot be implemented by the school.
- A timetable Q is considered to be *type-1 infeasible* if one of the following conditions is satisfied in at least one period: (i) There is a class having a lesson with more than one teacher (constraint (b) is not verified); (ii) There is a class having no lesson.
- A timetable Q is considered to be *type-2 infeasible* if constraint (e) is not satisfied by a teacher.

31.3 PROBLEM REPRESENTATION AND NEIGHBORHOOD STRUCTURE

The timetable is represented as a matrix $Q_{m \times p}$ of integer values. Each row i in Q represents the weekly schedule for teacher i. Each element $q_{ik} \in \{-1, 0, 1, 2, \ldots, n\}$ indicates the activity of teacher i in period k. Positive values represent the classes of the teachers in the period. Negative values indicate the teacher is unavailable, whereas null values indicate inactivity in the period.

A neighbor of a timetable Q is a timetable Q' that can be reached from Q through a movement consisting of a mere change of two distinct and non-negative values of a given row of Q. This movement is identified by the triplet $< i, k, \bar{k} >$, where k and $\bar{k}$ represent the periods in which the activities q_{ik} and $q_{i\bar{k}}$ of teacher i will be interchanged.

One can observe that this kind of movement may produce type-1 or type-2 infeasibility. Nevertheless, the possibility of having a teacher that teaches more than one class at the same time (violation of constraint (a)) is automatically rejected by this representation. In Section 31.5.1 we note that constraint (c) is always satisfied by the representation. Constraint (d) is also always satisfied because movements involving non-negative values are not permitted.

31.4 THE OBJECTIVE FUNCTION

A timetable Q is evaluated according to the following objective function, based on penalties, and should be minimized:

$$f(Q) = \omega * f_1(Q) + \delta * f_2(Q) + \rho * f_3(Q) \tag{31.1}$$

The first two components respectively measure, type-1 and type-2 infeasibility levels, and the third component measures the level of satisfaction regarding the granting of teachers' requests. The weights ω, δ, and ρ are chosen to satisfy the condition: $\omega > \delta \gg \rho$ and according to the definition on page 661, Q is feasible if $f_1(Q) = f_2(Q) = 0$.

Type-1 infeasibility level of Q, $f_1(Q)$, is measured by adding to each period k: (a) the number l_k of times a class has no activity in k; and (b) the number s_k of times more than one teacher teaches the same class in period k.

With respect to type-2 infeasibility, timetable Q is evaluated by adding the number e_i of times constraint (e) is not satisfied by each teacher i.

The satisfaction of the teachers' requests is measured with respect to the compactness of the timetable (constraint (g)) as well as the compliance with the number of double lessons required (constraint (f)). More precisely,

$$f_3(Q) = \sum_{i=1}^{m} (\alpha_i * b_i + \beta_i * v_i + \gamma_i * c_i) \tag{31.2}$$

where α_i, β_i, and γ_i are weights that reflect, respectively, the relative importance of the number b_i of "holes" (periods of no activity between two lesson periods during the same day), the number of week days v_i each teacher is involved in any teaching activity during the same week, and the non-negative difference c_i between the minimum required number of double lessons and the effective number of double lessons in the current schedule of each teacher i. We note that the definition of b_i implies that no difference is made between a hole of two periods and two holes of one period.

31.5 THE ALGORITHM

The proposed algorithm, called GTS-II, is a GRASP in which an initial solution is generated by a partially greedy constructive procedure (see Section 31.5.1). Refinement is obtained by means of a Tabu Search method (Section 31.5.2). When a solution without type-1 infeasibility is generated, the TS method activates the Intraclasses-Interclasses procedure (see Section 31.5.3). This construction and refinement sequence is repeated for $GTSmax$ iterations. The best overall solution is kept as the result. The pseudo-code of GTS-II algorithm for minimization is presented in Algorithm 31.1.

31.5.1 *Generating an initial solution*

An initial solution is generated in a constructive way by means of a partially greedy procedure, according to the description that follows.

At first, the periods are sorted according to the number of available teachers, that is, the periods with the smallest number of available teachers are on the top. Next, the unscheduled lessons are sorted according to the activity degree of each teacher, that is, the lessons, whose teachers have a larger number of lessons and unavailable periods, have priority. Then, well-ranked lessons are placed in a restricted candidate list (RCL) and a lesson is selected randomly from the RCL. Next, using the critical period order, the selected lesson is scheduled, so that there is no type of infeasibility (in this case, violation to constraints (b) and (e)). In the event this is impossible, only violation to con-

```
procedure GTS-II(())
    Let Q* be the best timetable and f* its value;
    f* ← ∞;
    GTSmax ← Maximum number of GTS-II iterations;
    for i = 1 to GTSmax
        Q⁰ ← GraspConstruction();
        Q ← TS − II(Q⁰);
        if (f(Q) < f*)
            Q* ← Q;
            f* ← f(Q);
        endif
    endfor
    return Q*;
```

Algorithm 31.1 Pseudo-code of GTS-II algorithm.

straint (e) is admitted at this time. If the impossibility still persists, the lesson will be scheduled by admitting violation to constraint (b) as well. Every time a lesson is scheduled, critical periods are updated as well as the list of remaining unscheduled lessons.

Since all lessons will be scheduled, even though some sort of infeasibility may occur, constraint (c) is automatically satisfied.

31.5.2 Tabu Search

Starting from an initial solution generated by the constructive procedure, Tabu Search (TS) iteratively explors the whole neighborhood $N(Q)$ of the current solution Q by means of movements defined in Section 31.3 and guided by the objective function described in (31.1). Then, the search moves to the neighbor Q' which produces the smallest $f(Q')$ value, which may be worse than the current $f(Q)$ value.

To prevent the occurrence of cycling, each time a movement is done, it is stored in a tabu list T. This list contains the $|T|$ most recent movements and reduces the risk of revisiting one of the $|T|-1$ last solutions previously visited. Since the tabu list can be very restrictive (Glover and Laguna, 1997), the TS algorithm also uses an aspiration criteria by objective. This way, a movement loses its tabu status if it produces a solution Q' whose value is smaller than the best solution value, Q^*, that had been obtained so far.

Whenever a timetable without overlapping classes is generated (i.e., $f_1(Q) = 0$), the Intraclasses-Interclasses procedure is activated (lines 12 to 15 in Algorithm 31.2) if no solution with the same value has yet been submitted to II. Reverse movements to those made by procedure II are introduced into the tabu list. The TS procedure continues the search from a solution produced by

II and is performed for $TSmax$ iterations without improving the value of the best solution.

The pseudo-code of TS algorithm using the procedure II is presented in Algorithm 31.2.

```
procedure TS-II(Q)
    Let Q be a initial timetable;
    Q* ← Q;                                   /* Best timetable reached so far */
    Iter ← 0;                                 /* Iteration counter */
    BestIter ← 0;    /* Iteration at which Q* has been found */
    T ← ∅;                                    /* Tabu List */
    TSmax ← Maximum number of consecutive itera-
    tions without improving f(Q*);
    while (Iter − BestIter < TSmax)
        Iter ← Iter + 1;
        Let Q' ← Q ⊕ m the best neighbor in N(Q) such
        that either the move m does not be tabu (m ∉
        T) or Q' satisfies the aspiration criteria (f(Q') <
        f(Q*));
        Update the tabu list T;
        Q ← Q';
        if (f1(Q')==0)
            Q ← Intraclasses-Interclasses(Q');
            Update the tabu list T;
        endif
        if (f(Q) < f(Q*))
            Q* ← Q;
            BestIter ← Iter;
        endif
    endwhile
    return Q*;
```

Algorithm 31.2 Pseudo-code of Tabu Search procedure using II.

31.5.3 Intraclasses-Interclasses Procedure

The Intraclasses-Interclasses procedure is based on shortest paths. It is activated when a solution without type-1 infeasibility is available. First, II attempts to make the f_2 component equal to zero. Should it succeed, improvement of the f_3 component, respecting constraints (a) through (e), is attempted, i.e., considering only feasible solutions. Alvarez-Valdes et al. (1996) use a similar procedure, but only to improve a feasible timetable. The type of movement under consideration in their study is also more restricted than II's is.

Since II acts in a similar manner to either retrieve feasibility or to improve a timetable, only the working principles for the latter case will be presented. Let us assume that a solution with no infeasibility is available. Thus, given a timetable Q under these conditions ($f_1(Q) = f_2(Q) = 0$), the graph of class j is defined by $G_j = (V_j, A_j)$, where V_j is the set of periods reserved for class j. A_j is a set of oriented arcs, and is defined as follows: $A_j = \{(k, \bar{k})$: the teacher who will teach class j in period k is available in period $\bar{k}$ and requirement (e) in Section 31.2 is respected in period $\bar{k}\}$.

To each arc $(k, \bar{k}) \in G_j$, a cost $\Delta f_i(k, \bar{k})$ is associated. It represents the cost variation of transferring teacher i from period k to period $\bar{k}$, taking only the f_3 component of the objective function into consideration. Thus, the cost is obtained by calculating the difference between the values of the objective function, regarding the teacher, in the old and new configurations, i.e.

$$\Delta f_i(k, \bar{k}) = f_i(\bar{k}) - f_i(k), \tag{31.3}$$

where $f(\cdot) = (\rho * f_3)(\cdot)$.

Table 31.1 shows a fragment of a timetable. Each row i represents a teacher ($i = T_1, T_2, T_3, T_4$) and each column k represents a period ($k = P_1, P_2, P_3, P_4, P_5$) of the same day. Each element (i, k) in this table represents the activity of teacher i in period k. A, B, C and D are classes. A dash (–) means that the teacher is unavailable, whereas an empty cell indicates there is no activity in the period. Column f_i indicates the value of the objective function of each teacher, as defined in (31.1) and (31.2) taking $\rho = 1$, $\alpha_i = 1$ and $\beta_i = \gamma_i = 0\ \forall i$, i.e. only holes in the schedule are relevant. The cost of this timetable is $f(Q_1) = f_{T_1} + f_{T_2} + f_{T_3} + f_{T_4} = 1 + 1 + 0 + 0 = 2$.

Table 31.1 Timetable Q_1.

	P_1	P_2	P_3	P_4	P_5	f_i
T_1	A		B	B		1
T_2	B	C		A	A	1
T_3		B	A	C	B	0
T_4	C	A	C	D	–	0

Figure 31.1 represents class A graph, G_A. Each period is represented by a vertex to which a teacher is associated. The -1 cost arc (P_1, P_5) indicates that in a case where teacher T_1 changes his/her lesson from period P_1 to period P_5, the value of the objective function will be reduced by 1 unit ($\Delta f_{T_1}(P_1, P_5) = f_{T_1}(P_5) - f_{T_1}(P_1) = 0 - 1 = -1$).

To find a timetable that presents a smaller value for the objective function, one should only search for a negative cost cycle in G_j. In the example under consideration, the arc sequence $\{(P_1, P_5), (P_5, P_3), (P_3, P_1)\}$ forms a cycle that has a total cost of -1 ($= -1 + (-1) + 1$). This sequence defines a set of *intraclasses* movements.

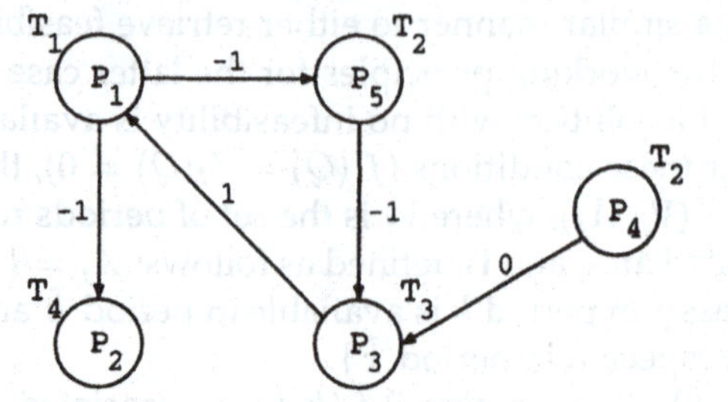

Figure 31.1 G_A: class A graph.

After updating graph G_A and timetable Q_1 with these movements, it is necessary to check once again whether there are negative cost cycles in the class A graph. In the event that there are no such cycles, the idea is, then, to repeat the procedure with another class and so forth, until it is no longer possible to improve the teachers' timetable by means of intraclasses movements.

Nevertheless, the existence of a negative cost cycle cannot guarantee a better value of the objective function. Moreover, it can generate type-2 infeasible solutions, as demonstrated by Souza (2000). However, such situations only happen when the same teacher is associated with more than one vertex in a cycle. Thus, it is necessary to check the feasibility and the value of the objective function after candidate movements. To find other negative cost cycles while using a similar type graph, one should proceed as follows: Choose any arc from the cycle, $(k, \bar{k}) \in G_j$, and insert it in a forbidden movement list L. Then, update the graph of the class under evaluation, excluding the arcs belonging to L from G_j, and search for another negative cost cycle. When it is no longer possible to find negative cost cycles, one should move to another class and eliminate list L.

At the end of the Intraclasses procedure, negative cost arcs may still remain in the graphs of the classes. Figure 31.2, which considers the graphs of classes j and $\bar{\jmath}$, illustrates this situation.

In Figure 31.2, there is a negative cost arc in class j from k to $\bar{k}$, i.e, $\Delta f_i^j(k, \bar{k})$. This indicates that the value of the objective function is likely to be improved, if the lesson assigned to teacher i in period k is transferred to period $\bar{k}$. However, such change cannot occur, because teacher $\bar{\imath}$, who teaches in period $\bar{k}$, is unavailable in period k (he/she is teaching class $\bar{\jmath}$). The idea is to switch the teacher's $\bar{\imath}$ lesson periods, so as to enable the search for a negative cost cycle, involving the graphs of both classes that are connected to periods k and $\bar{k}$. Let $c^{\bar{\jmath}}(\bar{k}, k)$ be the cost of the shortest path from $\bar{k}$ to k in $G_{\bar{\jmath}}$. The existence of a negative cost cycle involving both classes may be verified by checking whether the condition $\Delta f_i^j(k, \bar{k}) + c^{\bar{\jmath}}(\bar{k}, k) < 0$ is satisfied while transferring teacher $\bar{\imath}$ from class j to class $\bar{\jmath}$ in period $\bar{k}$ and from class $\bar{\jmath}$ to class j in period k. It is noted that there are no costs involved in moving one teacher from one class to another in the same period (null cost arcs in Figure 31.2).

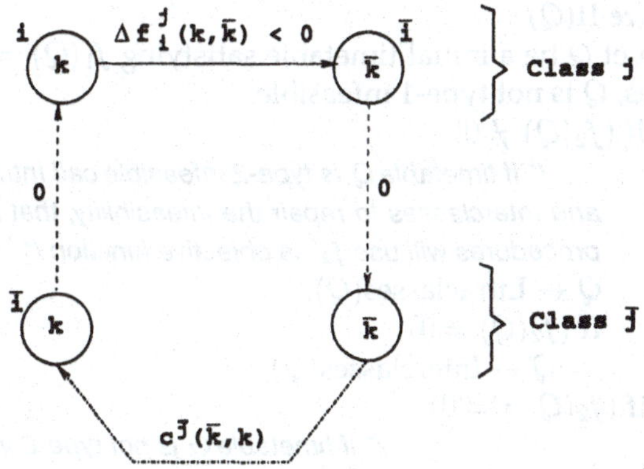

Figure 31.2 Graphs of j and $\bar{j}$ classes after Intraclasses procedure.

Thus, to each negative cost arc $(k, \bar{k}) \in G_j$, one should investigate whether there is a negative cost cycle involving this arc of G_j, the graph $G_{\bar{j}}$ and the null cost arcs that connect them. This arc sequence defines the so-called *interclasses* movements. Like the intraclasses movements, one must check the feasibility and the value of the objective function after candidate movements.

The Intraclasses-Interclasses procedure is therefore composed of two steps. First, the Intraclasses procedure is applied, resulting in n graphs G_j, $\forall j = 1, \ldots, n$ and, possibly, negative cost arcs. When the first step is finished, Interclasses procedure is applied.

When II procedure is activated, in an attempt to retrieve the feasibility of a solution which has only type-2 infeasibility, the cost of an arc (equation 31.3) is evaluated by assuming that $f(.) = f_2(.)$ Once this attempt is finished, the arc cost is evaluated again based on the cost variation of function (31.1).

Algorithm 31.3 illustrates how Intraclasses-Interclasses procedure works.

31.6 EXPERIMENTAL RESULTS

We have evaluated the performance of the GTS-II Algorithm using data from a Brazilian public high school, Escola Dom Silvério, located in Mariana, Minas Gerais State. The requirements for a timetable in this school are displayed in Section 31.2.

The GTS-II algorithm was implemented in C language and tested on a Pentium II PC (350 MHz, 64 MB RAM) running Linux operating system. To determine negative cost cycles, Floyd's algorithm (see Ahuja et al. (1993)) was implemented.

To test the efficiency of the II procedure, the GTS-II algorithm was compared to the GTS algorithm, which does not include such procedure (in GTS lines 12 to 15 in Algorithm 31.2 are not present). The following parameters were set, taking into consideration that *#lessons* represent the number of remaining lessons to be scheduled: $|RCL| = \max\{1, (\#lessons)/10\}, TSmax = 500,$

```
procedure II(Q)
    Let Q be a initial timetable satisfying f1(Q) = 0, that
    is, Q is not type-1 infeasible;
    if (f2(Q) ≠ 0)
        /* If timetable Q is type-2 infeasible call Intraclasses
        and Interclasses to repair the infeasibility, that is, these
        procedures will use f2 as objective function /*
        Q ← Intraclasses(Q);
        if (f2(Q) ≠ 0)
            Q ← Interclasses(Q);
    if (f2(Q) == 0)
        /* If timetable Q is not type-2 infeasible
        call Intraclasses and Interclasses to improve teachers'
        personal requests, that is, these procedures will use f3
        as objective function /*
        Q ← Intraclasses(Q);
        if (f2(Q) ≠ 0)
            Q ← Interclasses(Q);
    return Q;
```

Algorithm 31.3 Pseudo-code of Intraclasses-Interclasses procedure.

$GTSmax = 1, \omega = 100, \delta = 30, \rho = 1, \alpha_i = 3, \beta_i = 9$ and $\gamma_i = 1$ $\forall i = 1, ..., m$.

Table 31.2 presents a few characteristics of the data sets and summarizes the results. Columns m, n, and *#lessons* represent, respectively, the number of teachers, classes, and lessons to be scheduled. Column *Sparseness ratio* indicates the sparseness ratios of the data sets, i.e. *Sparseness* $= (m \times p - (\#lessons + u))/(m \times p)$, where u is the number of unavailable periods and $p = 25$ weekly periods are set aside for classes. Lower values of *Sparseness* indicate a harder problem. The results of Table 31.2 are based on 25 runs, each one initialized by a different random number generator seed. However, each run had the same seed for both algorithms. For each 25 run set, the average best solution $\lceil f^* \rceil$ and the average CPU time (minutes:seconds) are shown.

Figure 31.3 illustrates a typical evolution of the best solution in the early seconds of a GRASP iteration in GTS and GTS-II algorithms. As it can be observed, better quality solutions are obtained faster when II is used.

31.7 CONCLUSIONS

A hybrid metaheuristic is developed for solving the school timetabling problem. While the partially greedy constructive procedure generates good initial solutions and diversifies the search, the Tabu Search procedure refines the search.

Table 31.2 Computational results.

Data sets	m	n	*#lessons*	*Sparseness ratio*	GTS $\lceil f^* \rceil$	GTS CPU time	GTS-II $\lceil f^* \rceil$	GTS-II CPU time
DS00A	14	6	150	0.50	368	01:53	356	01:41
BR89M	16	8	200	0.30	491	02:17	481	01:24
DS00M	23	12	300	0.13	749	06:01	741	06:02
DS98M	31	13	325	0.58	831	15:36	826	08:09
DS00N	30	14	350	0.52	847	16:10	710	09:40
DS00D	33	20	500	0.39	1164	38:45	1101	29:00

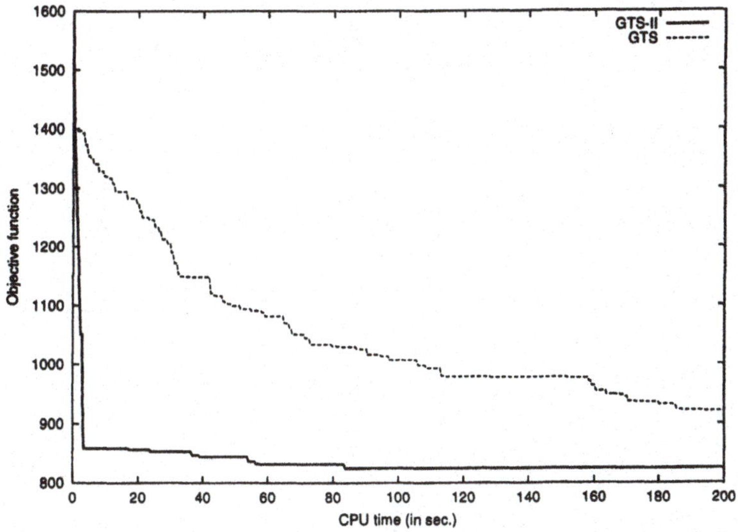

Figure 31.3 Typical performance of the GTS and GTS-II algorithms.

One of the main contributions of our research was the development of the Intraclasses-Interclasses procedure. It is applied to intensify the search when TS generates a timetable without overlapping classes. In this situation, if the solution has some other type of infeasibility, it will try to retrieve feasibility. If successful, it is applied again, to try to improve the compactness of the timetable as well as the other requirements.

Finally, besides the requirements mentioned in Section 31.2, the GTS-II algorithm can also be applied to treat other requirements such as a daily limit of lessons per teacher, a teachers' period preference, and intervals between lessons of the same subject to the same class. In such cases, one should simply add the components which measure the difference between the current and the desired solution to the objective function.

Acknowledgments

This research was partially supported by CAPES, Brazil.

Bibliography

D. Abramson. Constructing school timetables using simulated annealing: Sequential and parallel algorithms. *Management Science*, 37:98–113, 1991.

R.K. Ahuja, T.L. Magnanti, and J.B. Orlin. *Network Flows: Theory, Algorithms and Applications*. Prentice-Hall, New Jersey, 1993.

R. Alvarez-Valdes, G. Martin, and J.M. Tamarit. Constructing good solutions for the Spanish school timetabling problem. *Journal of the Operational Research Society*, 47:1203–1215, 1996.

M.P. Carrasco and M.V. Pato. A multiobjective genetic algorithm for the class/teacher timetabling problem. In E.K. Burke and W. Erben, editors, *Practice and Theory of Automated Timetabling III*, volume 2079 of *Lecture Notes in Computer Science*, pages 3–17. Springer-Verlag, Konstanz, Germany, 2001.

A. Colorni, M. Dorigo, and V. Maniezzo. Metaheuristics for high school timetabling. *Computational Optimization and Applications*, 9:275–298, 1998.

D. Costa. A tabu search algorithm for computing an operational timetable. *European Journal of Operational Research*, 76:98–110, 1994.

S. Even, A. Itai, and A. Shamir. On the complexity of timetabling and multicommodity flow problems. *SIAM Journal of Computation*, 5:691–703, 1976.

T.A. Feo and M.G.C. Resende. Greedy randomized adaptive search procedures. *Journal of Global Optimization*, 6:109–133, 1995.

F. Glover and M. Laguna. *Tabu Search*. Kluwer Academic Publishers, Boston, 1997.

A. Schaerf. Tabu search techniques for large high-school timetabling problems. In *Proceedings of the 30th National Conference on Artificial Intelligence*, pages 363–368, 1996.

M.J.F. Souza. *School timetabling: An approximation by metaheuristics*. Phd thesis, Systems Engineering and Computer Science Program, Federal University of Rio de Janeiro, Rio de Janeiro, Brazil, December 2000. In Portuguese.

Metaheuristics: Computer Decision-Making, pp. 673-698
Shunji Umetani, Mutsunori Yagiura, and Toshihide Ibaraki

32 A LOCAL SEARCH APPROACH FOR THE PATTERN RESTRICTED ONE DIMENSIONAL CUTTING STOCK PROBLEM

Shunji Umetani[1], Mutsunori Yagiura[1], and Toshihide Ibaraki[1]

[1]Department of Applied Mathematics and Physics
Graduate School of Informatics
Kyoto University
Kyoto, Japan
umetani@amp.i.kyoto-u.ac.jp, yagiura@i.kyoto-u.ac.jp
ibaraki@i.kyoto-u.ac.jp

Abstract: As the setup cost of cutting patterns becomes more important in modern cutting industry, we consider the pattern restricted one dimensional cutting stock problem (1D-PRP), in which the number of stock rolls is minimized while the number of different cutting patterns is constrained within a bound given by a program parameter. For this problem, we propose a new heuristic algorithm based on local search, and incorporate a heuristic algorithm that provides a small subset of neighborhood which tends to contain good solutions in the original neighborhood. According to our computational experiments, the proposed algorithm attains a wide variety of good solutions which are comparable to the existing heuristic approaches for the one dimensional cutting stock problem (without pattern restriction).
Keywords: Cutting stock problem, Local search, Setup cost of cutting patterns, Pattern generation.

32.1 INTRODUCTION

The cutting stock problem (CSP) is one of the representative combinatorial optimization problems, which arises in many industries such as steel, paper, wood, glass and fiber. In the one dimensional cutting stock problem (1D-CSP), we are given a sufficient number of stock rolls of the same length, and several types of products with given lengths and demands. A standard formulation of 1D-CSP is to describe it in terms of the variables associated with cutting patterns, where a cutting pattern (or pattern) is a feasible combination of products cut from one stock roll. 1D-CSP asks to specify a set of cutting patterns and their frequencies (i.e., the number of times the cutting pattern is used) satisfying the demands of all products. A typical cost function in this problem is the number of required stock rolls, and the standard 1D-CSP minimizes the number of required stock rolls while satisfying the demands of all products. As this standard 1D-CSP contains the bin packing problem (BPP), which is known to be strongly NP-hard, as a special case, 1D-CSP is clearly a hard problem.

A classical approach to 1D-CSP is to formulate it as a integer linear programming problem (IP), and solve it by a heuristic method based on its linear programming problem (LP) relaxation. As it is impractical to consider all cutting patterns which correspond to these columns in the LP relaxation, Gilmore and Gomory (1961; 1963) proposed a column generation technique that generates only the columns necessary to improve the lower bound of IP by solving the associated knapsack problems. The LP relaxation often has a property that the round up of the LP lower bound is equal to the optimal value (Marcotte, 1985; Scheithauer and Terno, 1997). Based on these ideas, several branch-and-bound algorithms with column generation techniques have been developed (Vance, 1998; Vanderbeck, 1999) with certain computational success.

In recent years, however, it is argued that the setup costs for changing cutting patterns have become more dominant. Although these branch-and-bound algorithms based on the column generation techniques solve the standard 1D-CSP, their solutions are not desirable from this point of view since these algorithms tend to use many different cutting patterns. Hence, several types of algorithms have then been developed to reduce the number of different cutting patterns.

Walker (1976) introduced the fixed cost for the setup of a cutting pattern in addition to the cost for the stock rolls, and this formulation was called the fixed charge problem. He proposed a heuristic algorithm called SWIFT which is a variant of the simplex algorithm. It utilizes a solution found by the conventional simplex algorithm, and then selects a new cutting pattern to enter the basis only if the additional setup cost is less than the compensation achieved by the reduction of the stock roll cost. SWIFT repeatedly introduces new cutting patterns until no new cutting pattern satisfying the above condition is found. Farley and Richardson (1984) proposed an improvement of SWIFT to reduce the setup cost of cutting patterns in 1D-CSP. Their algorithm uses additional pivoting rules not to increase the number of basic variables corresponding to cutting patterns. However, their computational results showed that the number of stock rolls increased rapidly as the number of different cutting patterns was reduced.

Haessler (1971; 1975) proposed a pattern generating heuristic algorithm called the sequential heuristic procedure (SHP). SHP starts from empty set of cutting patterns, and sequentially adds new cutting patterns to the current solution until all demands are satisfied. In each step, it first generates the candidate list of cutting patterns which satisfy some portion of remaining demands, and then heuristically selects a cutting pattern from the candidate list, whose trim loss (i.e., the unused portion of the stock rolls) is small and frequency is high.

Johnston (1986) and Goulimis (1990) proposed a pattern combination heuristic algorithm. The algorithm starts from a solution obtained by another algorithm designed to minimize the number of stock rolls, and reduces the number of different cutting patterns by combining two patterns together; i.e., it selects a number of cutting patterns in the solution and replaces them with a smaller number of new cutting patterns such that the amount of products covered by the new patterns is equivalent to that covered by the removed cutting patterns. It repeats the pattern combination until no further reduction can be made. Recently, Foerster and Wäscher (2000) proposed KOMBI which uses many types of combinations. For example, three patterns are replaced with two new patterns, four patterns are replaced with three new patterns, etc.

Vanderbeck (2000) considered a formulation of 1D-CSP which minimizes the number of different cutting patterns while using a given number of stock rolls or less, and it was called the pattern minimization problem (1D-PMP). He first described 1D-PMP as a integer quadratic programming problem (IQP), which is then decomposed into integer linear programming problems (IP) with strong LP relaxations. Then a branch-and-bound algorithm is applied while utilizing the column generation technique. According to the computational experiment, this algorithm could solve many small problems exactly, but failed to obtain optimal solutions for several instances of moderate sizes in two hours.

In this paper, we take a new approach by considering the number of different cutting patterns n as a program parameter. We call this variant of 1D-CSP, as the pattern restricted version of 1D-CSP (1D-PRP), which minimizes the number of stock rolls while using n different cutting patterns or less. In general, it becomes easier to find a solution using a smaller number of stock rolls as the number of different cutting patterns becomes larger. In this sense, there is a trade-off between the number of stock rolls and the number of different cutting patterns. By solving the 1D-PRP for different parameter values n, we can obtain a trade-off curve as illustrated in Figure 32.1. Using this we can make a more careful analysis of the trade-off between the two objective functions: the number of different cutting patterns and the number of required stock rolls. It is also possible to minimize the number of different cutting patterns by searching the minimum feasible n by employing binary search, for example, over the space of n.

Our algorithm for 1D-PRP is based on local search. It starts from an initial solution obtained by a modified first fit heuristic (FF) known for the bin packing problem (BPP). It repeats replacing the current set of cutting patterns with a better set in its neighborhood until no better set is found in the neighbor-

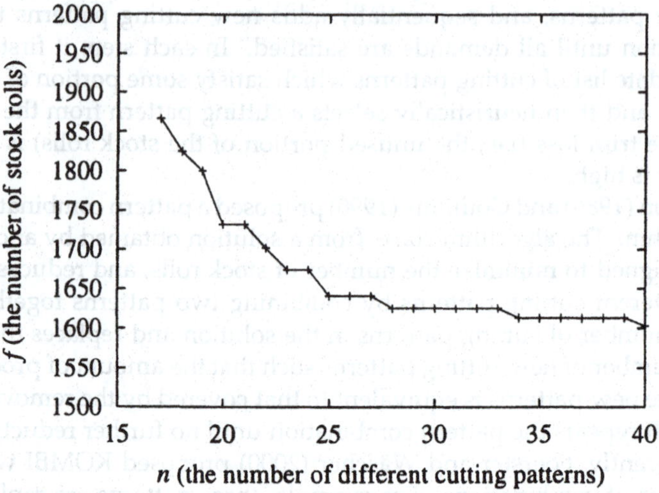

Figure 32.1 Trade-off curve between the number of different cutting patterns and the number of stock rolls.

hood. Solutions in the neighborhood are constructed from the current solution by removing two cutting patterns and adding two new cutting patterns. Since the size of the resulting neighborhood may increase exponentially in the number of product types, it is impractical to enumerate all feasible cutting patterns. To overcome this, we develop a heuristic algorithm that tries to generate only promising sets of cutting patterns. Finally, we conduct computational experiment for randomly generated problem instances, and observe that the proposed algorithm attains a wide variety of good solutions comparable to SHP and KOMBI, and our algorithm provides reasonable trade-off curves between the number of different cutting patterns and the number of required stock rolls for most instances.

32.2 FORMULATION

The one dimensional cutting stock problem (1D-CSP) is defined as follows. Given a sufficient number of stock rolls of length L, and m products $M = \{1, 2, \ldots, m\}$ which have lengths $(l_1, l_2, \ldots, l_m)$ and demands $(d_1, d_2, \ldots, d_m)$. A cutting pattern is described as $p_j = (a_{1j}, a_{2j}, \ldots, a_{mj})$, where $a_{ij} \in Z_+$ (the set of nonnegative integers) is the number of products i cut from one stock roll. We say a cutting pattern p_j satisfying

$$\sum_{i \in M} a_{ij} l_i \leq L \tag{32.1}$$

feasible, and let S denote the set of all feasible cutting patterns.

A solution of 1D-CSP consists of a set of cutting patterns $\Pi = \{p_1, p_2, \ldots, p_{|\Pi|}\} \subseteq S$, and the set of their frequencies $X = \{x_1, x_2, \ldots, x_{|\Pi|}\} \subseteq Z_+$, where frequency x_j denotes the number of times cutting pattern p_j is used. The standard 1D-CSP is formulated as follows:

$$\begin{aligned} \text{(1D-CSP)} \quad \text{minimize} \quad & f(\Pi, X) = \sum_{p_j \in \Pi} x_j \qquad (32.2) \\ \text{subject to} \quad & \sum_{p_j \in \Pi} a_{ij} x_j \geq d_i, \text{ for } i \in M \\ & \Pi \subseteq S \\ & x_j \in \boldsymbol{Z}_+ \text{ for } p_j \in \Pi. \end{aligned}$$

Now we consider the pattern minimization version of 1D-CSP (1D-PMP), which minimizes the number of different cutting patterns while using f_{UB} stock rolls or less, where f_{UB} is an input parameter.

$$\begin{aligned} \text{(1D-PMP)} \quad \text{minimize} \quad & |\Pi| \qquad (32.3) \\ \text{subject to} \quad & \sum_{p_j \in \Pi} a_{ij} x_j \geq d_i, \text{ for } i \in M \\ & \sum_{p_j \in \Pi} x_j \leq f_{UB} \\ & \Pi \subseteq S \\ & x_j \in \boldsymbol{Z}_+ \text{ for } p_j \in \Pi. \end{aligned}$$

As already mentioned in Section 32.1, we consider in this paper the pattern restricted version of 1D-CSP (1D-PRP) which minimizes the number of required stock rolls while using n different cutting patterns or less, where n is an input parameter.

$$\begin{aligned} \text{(1D-PRP)} \quad \text{minimize} \quad & f(\Pi, X) = \sum_{p_j \in \Pi} x_j \qquad (32.4) \\ \text{subject to} \quad & \sum_{p_j \in \Pi} a_{ij} x_j \geq d_i, \text{ for } i \in M \\ & \Pi \subseteq S \\ & |\Pi| \leq n \\ & x_j \in \boldsymbol{Z}_+, \text{ for } p_j \in \Pi. \end{aligned}$$

The hardness of the standard 1D-CSP arises from the fact that the number of all feasible cutting patterns may be exponential in m. Furthermore, it has to solve an integer linear programming problem (IP) which is already NP-hard (Garey and Johnson, 1979).

In the above formulation, the shortage of demands d_i is not allowed, while overproduction is permitted. In some applications, such as chemical fiber industry, however, the shortage of demands may be allowed because the additional cost due to the shortage is relatively small. In our previous paper (Umetani et al., 2003), we considered a variant of 1D-PRP that minimizes the squared deviation from the demands of all products while using a given number of different cutting patterns, where the shortage of demands is allowed as well as the overproduction.

32.3 LOCAL SEARCH

We develop a local search (LS) algorithm for 1D-PRP. LS starts from an initial solution and repeats replacing it with a better solution in its neighborhood until no better solution is found in the neighborhood. A solution of 1D-PRP is given by a set of patterns $\Pi = \{p_1, p_2, \ldots, p_n\} \subseteq S$ and the corresponding set of frequencies $X = \{x_1, x_2, \ldots, x_n\}$. The set of cutting patterns is found by repeatedly changing two patterns in the current set of patterns Π and their frequencies X is obtained by solving the associated integer programming problem. In this section, we explain the detail of three ingredients of LS for 1D-PRP: (i) how to construct an initial feasible solution, (ii) how to compute the frequencies of a given set of patterns, and (iii) how to construct the neighborhood of the current solution.

32.3.1 Construction of an initial solution

If there is no restriction on the number of different cutting patterns, as in the standard 1D-CSP, it is easy to construct a feasible solution. However, just finding a feasible solution is not trivial in 1D-PRP. The problem of finding a feasible solution under a given number of cutting patterns is equivalent to the bin packing problem (BPP) known to be NP-complete (Garey and Johnson, 1979). Here, we suppose that the number of cutting patterns n is less than that of product types m, because a feasible solution is easy to obtain in the case of $n \geq m$.

Problem BPP

> **Input:** Lengths l_i of all products $i \in M$, the number of different cutting patterns n, and the length of stock rolls L.
>
> **Output:** A partition of M into n disjoint subsets $M_1, M_2, \ldots, M_n \subseteq M$ subject to $\sum_{i \in M_j} l_i \leq L$ for all $M_j \subseteq M$.

From a solution of BPP, we can construct a pattern p_j and its frequency x_j by setting $a_{ij} \leftarrow 1$ for $i \in M_j$ (and $a_{ij} \leftarrow 0$ for $i \notin M_j$) and $x_j \leftarrow \max_{i \in M_j} d_i$. The resulting set of patterns $\Pi = \{p_1, p_2, \ldots, p_n\}$ and the set of frequencies $X = \{x_1, x_2, \ldots, x_n\}$ is obviously a feasible solution to 1D-PRP. We first consider how to construct a feasible solution by solving BPP, and then modify it to reduce the number of stock rolls.

To construct a feasible solution, we prepare a heuristic algorithm based on the first fit principle (FF), where FF is one of the representative approximation algorithms known for BPP. After preparing n stock rolls of length L, FF sequentially assigns each product i into the stock roll with the lowest index among those having the residual capacity of at least l_i. However, to guarantee that every stock roll is assigned at least one product, and to improve the quality as an initial solution of 1D-PRP, we modify FF as follows and call this algorithm as SPREAD-PACKING, shown in Algorithm 32.1.

First, we sort all product $i \in M$ in the descending order of demands d_i, where $\sigma(k)$ denotes the k-th product in the resulting order. We assign all products to stock rolls in this order. We also define an aspiration length

$$L' = \alpha L, \tag{32.5}$$

where α is a parameter satisfying $0 < \alpha \leq 1$. If the processed length of the current stock roll exceeds L' after product $\sigma(k)$ will be assigned, the next product $\sigma(k+1)$ is assigned to the next stock roll. This algorithm assigns at least one product to every stock roll if $\alpha \leq \sum_{i \in M} l_i / nL$ is used, and it is equivalent to the original FF if α is set to 1. If SPREAD-PACKING terminates before m products are assigned into n stock rolls, we conclude the failure of constructing a feasible solution.

algorithm SPREAD-PACKING( )
Input: Lengths l_i and demands d_i of products $i \in M = \{1, 2, \ldots, m\}$, the number of cutting patterns n, the length of stock rolls L and a given parameter α.
Output: n disjoint subsets $M_1, M_2, \ldots, M_n$ of M or "failure".
(1) Set $M_j \leftarrow \emptyset$ for $j \in \{1, 2, \ldots, n\}$, and $L' \leftarrow \alpha L$.
(2) Sort all products $i \in M$ in the descending order of d_i, where $\sigma(k)$ denotes the k-th product in this order. Set $k := 1$ and $j \leftarrow 1$.
(3) If $l_{\sigma(k)} \leq L - \sum_{i \in M_j} l_i$, set $M_j \leftarrow M_j \cup \{\sigma(k)\}$, $k \leftarrow k+1$ and $j \leftarrow 1$; otherwise set $j \leftarrow j+1$. If $k \leq m$ and $j \leq n$, return to Step 3.
(4) If $k > m$, output $M_1, M_2, \ldots, M_n$ and halt; otherwise output "failure" and halt.

Algorithm 32.1 Outline of SPREAD-PACKING algorithm.

In our local search algorithm, we first try SPREAD-PACKING with $\alpha = \sum_{i \in M} l_i / nL$. If SPREAD-PACKING fails to obtain a feasible solution, we switch to the first fit decreasing heuristic (FFD); i.e., it sorts all products in the descending order of l_i (not d_i) in Step 2, and uses $\alpha = 1$ in Step 1. The other part of the algorithm does not change. If this attempt also fails, we conclude the failure of our algorithm.

Now assume that SPREAD-PACKING outputs n disjoint subsets $M_1, M_2, \ldots, M_n \subseteq M$. For each M_j, we then consider the construction of pattern p_j (i.e., a_{ij} for $i \in M_j$) and the frequency x_j so that all the demands d_i ($i \in M_j$)

are met with the smallest x_j (nonnegative real number).

$$(\mathrm{P}_1(j)) \quad \text{minimize} \quad x_j \tag{32.6}$$
$$\text{subject to} \quad a_{ij}x_j \geq d_i, \text{ for } i \in M_j$$
$$\sum_{i \in M_j} a_{ij}l_i \leq L$$
$$a_{ij} \in Z_+, \text{ for } i \in M_j$$
$$x_j \geq 0.$$

For this problem, we propose an exact algorithm PAT(j), shown in Algorithm 32.2. This starts from $a_{ij} \leftarrow 1$ for all $i \in M_j$ and $x_j \leftarrow \max_{i \in M_j} d_i$, and sequentially increases a_{ij} to reduce the frequency x_j. Here we say that product i is bottleneck if $a_{ij}x_j = d_i$ holds for the current a_{ij} and x_j. Let

$$L^{res} = L - \sum_{i \in M_j} a_{ij}l_i \tag{32.7}$$

denote the current residual length of the stock roll. If $l_i \leq L^{res}$ holds for a bottleneck products i, we have to increase a_{ij} (by one) to reduce x_j. We repeat this process as long as it is possible.

algorithm PAT(j)
Input: A set $M_j \subseteq M$, demands d_i and lengths l_i of products $i \in M_j$, and the length of the stock roll L (M_j satisfies $\sum_{i \in M_j} l_i \leq L$).
Output: A feasible pattern $p_j = (a_{ij} \mid i \in M_j)$ and its frequency x_j.
(1) Set $a_{ij} \leftarrow 1$ for $i \in M_j$, $x_j \leftarrow \max_{i \in M_j} d_i$ and $L^{res} \leftarrow L - \sum_{i \in M_j} l_i$.
(2) If there is no bottleneck product $i \in M_j$ such that $l_i \leq L^{res}$, output $p_j = (a_{ij} \mid i \in M_j)$ and x_j, and halt. Otherwise take a bottleneck product $i \in M_j$ (i.e., $a_{ij}x_j = d_i$ holds) such that $l_i \leq L^{res}$, and set $a_{ij} \leftarrow a_{ij} + 1$, $x_j \leftarrow \max_{i \in M_j} \frac{d_i}{a_{ij}}$, $L^{res} \leftarrow L^{res} - l_i$ and return to Step 2.

Algorithm 32.2 Outline of PAT algorithm.

Now we prove the following theorem about PAT(j).

Theorem 1 *PAT(j) outputs an optimal solution of problem* $\mathrm{P}_1(j)$.

Proof: Let x_j and $p_j = (a_{ij} \mid i \in M_j)$ be the output of PAT(j). Since a_{ij} increases only when $a_{ij}x_j = d_i$ holds, and variable x_j is non increasing during the execution of PAT(j), we have

$$\frac{d_i}{a_{ij}-1} > x_j \text{ for all } a_{ij} \geq 2. \tag{32.8}$$

Now let x_j^* and $p_j^* = (a_{ij}^* \mid i \in M_j)$ be an optimal solution of $P_1(j)$. If there is an a_{ij}^* such that $a_{ij}^* < a_{ij}$, it follows that

$$x_j^* \geq \frac{d_i}{a_{ij}^*} \geq \frac{d_i}{a_{ij}-1} > x_j \tag{32.9}$$

which contradicts the optimality of x_j^*. Therefore, $a_{ij}^* \geq a_{ij}$ holds for all $i \in M_j$. Now if $x_j^* < x_j$ holds (i.e., x_j is not optimum), there is a bottleneck product i' for which $a_{i'j}^* > a_{i'j}$ holds. But this implies

$$0 \leq L - \sum_{i \in M_j} a_{ij} l_i - l_{i'} = L^{res} - l_{i'}, \tag{32.10}$$

and $a_{i'j}$ would have been increased in Step 2 of PAT(j). This is a contradiction, and shows that $x_j^* = x_j$ holds.

Finally, we summarize in Algorithm 32.3 the algorithm INIT that constructs an initial feasible solution.

algorithm INIT( )
Input: Lengths l_i and demands d_i of products $i \in M = \{1, 2, \ldots, m\}$, the number of cutting patterns n, and the length of stock rolls L.
Output: A set of n feasible patterns $\Pi = \{p_1, p_2, \ldots, p_n\}$ and the set of their frequencies $X = \{x_1, x_2, \ldots, x_n\}$, or "failure".
(1) Apply SPREAD-PACKING to obtain $M_1, M_2, \ldots, M_n$. (If this attempt fails, apply the FFD version of SPREAD-PACKING. If FFD still outputs "failure", we conclude the failure of INIT.)
(2) For each $j \in \{1, 2, \ldots, n\}$, compute pattern $p_j = (a_{ij} \mid i \in M_j)$ and its frequency x_j by applying PAT(j).

Algorithm 32.3 Outline of INIT algorithm.

Note that the frequencies x_j computed in INIT are real numbers, and it is necessary to round them into integers before starting LS. This rounding process will be explained in the next subsection for any set of patterns Π (not restricted to the initial set obtained by INIT).

32.3.2 Computing frequencies of cutting patterns

For a given set of patterns $\Pi = \{p_1, p_2, \ldots, p_n\}$, the problem of computing their frequencies $X = \{x_1, x_2, \ldots, x_n\}$ to minimize the cost function $f(\Pi, X)$, can be described as the following integer programming problem:

$$
\begin{aligned}
(\mathrm{IP}(\Pi)) \quad & \text{minimize} \quad f(\Pi, X) = \sum_{p_j \in \Pi} x_j \\
& \text{subject to} \quad \sum_{p_j \in \Pi} a_{ij} x_j \geq d_i, \; i = 1, 2, \ldots, m \\
& \qquad\qquad\quad x_j \in \mathbf{Z}_+, \; j = 1, 2, \ldots, n.
\end{aligned} \tag{32.11}
$$

As it contains the set covering problem (SCP) known to be NP-hard (Garey and Johnson, 1979) as a special case, we consider to find an approximate solution $\widehat{X}$ and its cost $f(\Pi, \widehat{X})$.

Our heuristic algorithm FREQ, shown in Algorithm 32.4, first solves the LP relaxation LP(Π) of IP(Π), in which the integer constraints $x_j \in Z_+$ are replaced with $x_j \geq 0$. Let $\overline{X} = \{\bar{x}_1, \bar{x}_2, \ldots, \bar{x}_n\}$ denote the LP optimal solution. To obtain an integer solution, it sorts variables $\bar{x}_j$ in descending order of the fractions $\bar{x}_j - \lfloor \bar{x}_j \rfloor$, and then rounds them up in this order until all demands are satisfied.

algorithm FREQ( )
Input: Demands d_i of products $i \in M$, and a set of patterns $\Pi = \{p_1, p_2, \ldots, p_n\}$.
Output: A set of frequencies $\widehat{X} = \{\hat{x}_1, \hat{x}_2, \ldots, \hat{x}_n\}$ or "failure".

(1) Compute an optimal solution $\overline{X} = \{\bar{x}_1, \bar{x}_2, \ldots, \bar{x}_n\}$ of the LP relaxation LP(Π) of IP(Π) (by applying the simplex method for example). If LP(Π) is infeasible, output "failure" and halt.

(2) Set $k \leftarrow 1$ and $\hat{x}_j \leftarrow \lfloor \bar{x}_j \rfloor$ for all $j \in \{1, 2, \ldots, n\}$.

(3) Sort all variables $\bar{x}_j$ in the descending order of fractions $\bar{x}_j - \lfloor \bar{x}_j \rfloor$, and let $\sigma(k)$ denotes the k-th variable in this order.

(4) If all demands are satisfied (i.e., $\sum_{j=1}^{n} a_{ij} \hat{x}_j \geq d_i$, for all $i \in M$), then output $\widehat{X} = \{\hat{x}_1, \hat{x}_2, \ldots, \hat{x}_n\}$ and halt. Otherwise, if there is at least one $i \in M$ such that $a_{i\sigma(k)} > 0$ and $\sum_{p_j \in \Pi} a_{ij} \hat{x}_j < d_i$ hold, then set $\hat{x}_{\sigma(k)} \leftarrow \lceil \bar{x}_{\sigma(k)} \rceil$. Set $k \leftarrow k + 1$ and return to Step 4.

Algorithm 32.4 Outline of FREQ algorithm.

32.3.3 Construction of neighborhood

A natural neighborhood of the current set of patterns Π may be

$$N_1(\Pi) = \bigcup_{j \in \{1,2,\ldots,n\}} N_1(\Pi, j), \tag{32.12}$$

where

$$N_1(\Pi, j) = \{\Pi \cup \{p'_j\} \setminus \{p_j\} \mid p'_j \in S\},\ j \in \{1, 2, \ldots, n\}, \tag{32.13}$$

where S is the set of all feasible patterns. That is, $\Pi' \in N_1(\Pi)$ is obtained from Π by switching one p_j with another p'_j. However, according to our preliminary computational experiment, such Π' does not appear powerful enough to generate new solutions, which are substantially different from the current solution. Therefore, our local search algorithm uses a larger neighborhood $N_2(\Pi)$ defined as follows:

$$N_2(\Pi) = \bigcup_{j_1, j_2 \in \{1,2,\ldots,n\}} N_2(\Pi, j_1, j_2), \tag{32.14}$$

where

$$N_2(\Pi, j_1, j_2) = \{\Pi \cup \{p'_{j_1}, p'_{j_2}\} \setminus \{p_{j_1}, p_{j_2}\} \mid p'_{j_1}, p'_{j_2} \in S\},\ j_1, j_2 \in \{1, 2, \ldots, n\}. \tag{32.15}$$

That is, $N_2(\Pi)$ is the set of all solutions obtainable by removing two patterns $p_{j_1}, p_{j_2} \in \Pi$ and adding two new patterns $p'_{j_1}, p'_{j_2} \in S$. Here, of course, we assume $|\Pi| \geq 2$.

As the size of $N_2(\Pi, j_1, j_2)$ is very large, however, we propose a heuristic algorithm to generate only a small subset $N'_2(\Pi, j_1, j_2)$ of $N_2(\Pi, j_1, j_2)$, containing good solutions. The resulting neighborhood is denoted

$$N'_2(\Pi) = \bigcup_{j_1, j_2 \in \{1,2,\ldots,n\}} N'_2(\Pi, j_1, j_2). \tag{32.16}$$

To explain $N'_2(\Pi, j_1, j_2)$, we define the residual demands when patterns p_{j_1} and p_{j_2} are removed from the current solution (Π, X):

$$r_i(j_1, j_2) = \max\left\{0, d_i - \sum_{p_j \in \Pi \setminus \{p_{j_1}, p_{j_2}\}} a_{ij} x_j\right\},\ i \in M. \tag{32.17}$$

Then we introduce the following problem of constructing new patterns $p'_{j_1} = (a_{ij_1} \mid i \in M(j_1, j_2))$ and $p'_{j_2} = (a_{ij_2} \mid i \in M(j_1, j_2))$, which replace the removed p_{j_1} and p_{j_2}, where $M(j_1, j_2) \subseteq M$ denotes the set of products i satisfying $r_i(j_1, j_2) > 0$.

$$
\begin{aligned}
(\mathrm{P}_2(j_1,j_2)) \quad & \text{minimize} \quad x_{j_1} + x_{j_2} && (32.18)\\
& \text{subject to} \quad a_{ij_1}x_{j_1} + a_{ij_2}x_{j_2} \geq r_i(j_1,j_2), \text{ for } i \in M(j_1,j_2)\\
& \sum_{i \in M(j_1,j_2)} a_{ij_1} l_i \leq L\\
& \sum_{i \in M(j_1,j_2)} a_{ij_2} l_i \leq L\\
& a_{ij_1}, a_{ij_2} \in \mathbf{Z}_+, \text{ for } i \in M(j_1,j_2)\\
& x_{j_1}, x_{j_2} \geq 0.
\end{aligned}
$$

Now we propose a heuristic algorithm 2-PATTERNS to obtain a good solution of this problem. It first generates a feasible pair of cutting patterns p'_{j_1}, p'_{j_2} such that every product $i \in M(j_1, j_2)$ is included in at least one of them. This part of 2-PATTERNS is done by the algorithm BALANCE-PACKING, shown in Algorithm 32.5. After sorting all products $i \in M(j_1, j_2)$ in the descending order of l_i, it sequentially assigns each product $i \in M(j_1, j_2)$ into the pattern which has longer residual length (between the two patterns p'_{j_1} and p'_{j_2} being constructed). Here L_1^{res} and L_2^{res} represent the current residual lengths of the new patterns p'_{j_1} and p'_{j_2}, respectively.

algorithm BALANCE-PACKING( )
Input: A set $M(j_1, j_2) \subseteq M$, lengths l_i of products $i \in M(j_1, j_2)$, and the length of stock rolls L.
Output: A pair of new patterns $p'_{j_1} = (a_{ij_1} \mid i \in M(j_1, j_2))$ and $p'_{j_2} = (a_{ij_2} \mid i \in M(j_1, j_2))$.
(1) Sort all $i \in M(j_1, j_2)$ in the descending order of l_i, and let $\sigma(k)$ denote the k-th product in this order. Set $L_1^{res} \leftarrow L$, $L_2^{res} \leftarrow L$ and $k \leftarrow 1$.
(2) If $L_1^{res} \geq L_2^{res}$ (resp. $L_2^{res} > L_1^{res}$), let $h \leftarrow 1$ (resp. $h \leftarrow 2$). If $L_h^{res} \geq l_{\sigma(k)}$, set $a_{\sigma(k)j_h} \leftarrow 1$ and $L_h^{res} \leftarrow L_h^{res} - l_{\sigma(k)}$; else output "failure" and halt.
(3) Set $k \leftarrow k+1$. If $k \leq |M(j_1, j_2)|$, return to Step 2. Otherwise output two patterns $p'_{j_1} = (a_{ij_1} \mid i \in M(j_1, j_2))$ and $p'_{j_2} = (a_{ij_2} \mid i \in M(j_1, j_2))$, and halt.

Algorithm 32.5 Outline of BALANCE-PACKING algorithm.

For the resulting patterns p'_{j_1} and p'_{j_2}, it holds that $a_{ij_1} + a_{ij_2} = 1$ for all $i \in M(j_1, j_2)$. Thus, the corresponding frequencies x_{j_1} and x_{j_2} can be easily calculated by

$$x_j \leftarrow \max\{r_i(j_1, j_2) \mid a_{ij} > 0, i \in M(j_1, j_2)\}, \; j \in \{j_1, j_2\}. \qquad (32.19)$$

algorithm 2-PATTERNS( )

Input: Lengths l_i and demands d_i of products $i \in M = \{1, 2, \ldots, m\}$, a set $M(j_1, j_2) \subseteq M$, residual demands $r_i(j_1, j_2)$ for all $i \in M(j_1, j_2)$, and the length of stock rolls L.

Output: A pair of patterns $p'_{j_1} = (a_{ij_1} \mid i \in M(j_1, j_2))$ and $p'_{j_2} = (a_{ij_2} \mid i \in M(j_1, j_2))$.

(1) (Initial pair of feasible patterns) Apply BALANCE-PACKING to obtain a feasible pair of patterns $p'_{j_1} = (a_{ij_1} \mid i \in M(j_1, j_2))$ and $p'_{j_2} = (a_{ij_2} \mid i \in M(j_1, j_2))$. If BALANCE-PACKING outputs "failure", then output "failure" and halt. Otherwise calculate

$$L_1^{res} \leftarrow L - \sum_{i \in M(j_1, j_2)} a_{ij_1} l_i,$$

$$L_2^{res} \leftarrow L - \sum_{i \in M(j_1, j_2)} a_{ij_2} l_i,$$

$$x_{j_1} \leftarrow \max\{r_i(j_1, j_2) \mid a_{ij_1} > 0, i \in M(j_1, j_2)\},$$

$$x_{j_2} \leftarrow \max\{r_i(j_1, j_2) \mid a_{ij_2} > 0, i \in M(j_1, j_2)\},$$

$$\delta_i \leftarrow a_{ij_1} x_{j_1} + a_{ij_2} x_{j_2} - r_i(j_1, j_2), \ i \in M(j_1, j_2),$$

$$\Delta \leftarrow \sum_{i \in M(j_1, j_2)} \delta_i.$$

(2) (Elimination of bottleneck products) Let $B \leftarrow \{i \mid i \in M(j_1, j_2)$ and $a_{ij_1} x_{j_1} + a_{ij_2} x_{j_2} = r_i(j_1, j_2)\}$ (B is the set of bottleneck products). For each $i \in B$, if $x_{j_2} < x_{j_1} - \min\left(x_{j_2}, \left\lfloor \frac{x_{j1}}{x_{j_2}} \right\rfloor\right)$ and $l_i \le L_2^{res}$, apply rule (i) and let $L_2^{res} \leftarrow L_2^{res} - l_i$. Otherwise if $l_i \le L_1^{res}$, apply rule (ii), and let $L_1^{res} \leftarrow L_1^{res} - l_i$ and $L_2^{res} \leftarrow L_2^{res} + \min\left(a_{ij_2}, \left\lfloor \frac{x_{j1}}{x_{j_2}} \right\rfloor\right) l_i$.

(3) (Frequencies x_{j_1}, x_{j_2}) Calculate the continuous frequencies x_{j_1} and x_{j_2} by solving the LP problem $\bar{P}_2(j_1, j_2)$.

(4) (Readjustment of patterns) For all $i \in M(j_1, j_2)$, apply rules (iii), (iv), (v), (vi) in this order if applicable without violating the length constraint of patterns and all the demand constraints of products.

(5) (Termination) Calculate the continuous frequencies x_{j_1} and x_{j_2} by solving the LP problem $\bar{P}_2(j_1, j_2)$. Calculate the amount of overproduction Δ. If either the number of stock rolls $x_{j_1} + x_{j_2}$ or the amount of overproduction Δ decreases in the current loop of Steps 2–5, return to Step 2. Otherwise output the current pair of patterns p'_{j_1} and p'_{j_2} together with their frequencies x_{j_1} and x_{j_2}, and halt.

Algorithm 32.6 Outline of 2-PATTERNS algorithm.

2-PATTERNS starts from this pair of feasible patterns, and repeats adding products to these patterns to reduce the number of stock rolls (i.e., $x_{j_1} + x_{j_2}$) to cover the residual demands $r_i(j_1, j_2)$ for all $i \in M(j_1, j_2)$.

Call a product i bottleneck if $a_{ij_1}x_{j_1} + a_{ij_2}x_{j_2} = r_i(j_1, j_2)$ holds for the current patterns p'_{j_1}, p'_{j_2} and frequencies x_{j_1}, x_{j_2}. In each step, 2-PATTERNS increases either a_{ij_1} or a_{ij_2} of a bottleneck product i, and correspondingly updates their frequencies x_{j_1} and x_{j_2}. This is repeated until the number of stock rolls and the amount of overproduction become unable to improve. This process is similar to that of PAT(j) because both algorithms are based on the residual demand. However, as there are two patterns involved, 2-PATTERNS has to consider the balance between them.

For the current pair of patterns p'_{j_1} and p'_{j_2}, their frequencies x_{j_1} and x_{j_2} can be computed by solving the following LP problem.

$$\begin{aligned} (\bar{\mathrm{P}}_2(j_1, j_2)) \quad & \text{minimize} && x_{j_1} + x_{j_2} \\ & \text{subject to} && a_{ij_1}x_{j_1} + a_{ij_2}x_{j_2} \geq r_i(j_1, j_2), \text{ for } i \in M(j_1, j_2) \\ & && x_{j_1}, x_{j_2} \geq 0. \end{aligned} \tag{32.20}$$

This solution yields overproduction δ_i of product i as defined as

$$\delta_i = a_{ij_1}x_{j_1} + a_{ij_2}x_{j_2} - r_i(j_1, j_2), \; i \in M(j_1, j_2). \tag{32.21}$$

As positive overproduction is not desirable in order to minimize the number of stock rolls, we try to reduce the amount of overproduction. If a_{ij} is increased to $a_{ij} + 1$ for a bottleneck product i (i.e., $\delta_i = 0$ holds before modification), the resulting overproduction becomes $\delta_i = x_j$. From this observation, a simple rule may be to select the pattern $p_j \in \{p_{j_1}, p_{j_2}\}$ with a smaller x_j and increase the a_{ij} of a bottleneck product i. But, this rule may keep selecting one particular pattern in early stage of the algorithm, because the frequency x_j decreases further when a_{ij} is increased. To alleviate this drawback, we propose the following rule of increasing a_{ij_1} and a_{ij_2}. Let i be the index of a bottleneck product, and assume $x_{j_1} \geq x_{j_2}$ without loss of generality. Consider the following two modifications.

(i) $\quad a_{ij_2} \leftarrow a_{ij_2} + 1$,

(ii) $\quad a_{ij_1} \leftarrow a_{ij_1} + 1$ and $a_{ij_2} \leftarrow \max\left(0, a_{ij_2} - \left\lfloor \frac{x_{j_1}}{x_{j_2}} \right\rfloor\right)$.

The rule (i) or (ii) can be applied if patterns have enough residual lengths to permit the increase $a_{ij_2} \leftarrow a_{ij_2} + 1$ or $a_{ij_1} \leftarrow a_{ij_1} + 1$, respectively. If rule (i) is applied, the resulting overproduction becomes $\delta_i = x_{j_2}$, while if rule (ii) is applied, we have $\delta_i \leftarrow x_{j_1} - \min(a_{ij_2}x_{j_2}, \lfloor x_{j_1} \rfloor)$. If both rules are applicable, 2-PATTERNS selects the one having smaller overproduction.

After modifying the patterns by the above rule (i) or (ii), and calculating the frequencies x_{j_1}, x_{j_2}, 2-PATTERNS tries to readjust the patterns in order either to reduce overproduction or to increase the residual lengths in the two patterns, while satisfying all the demand constraints. Increasing residual lengths may be useful to eliminate other bottlenecks in subsequent steps. This readjustment is carried out in the following manner. Assume $x_{j_1} \geq x_{j_2}$ without loss of generality. Apply the following rules in the stated order if they are applicable (i.e., if patterns have enough residual lengths in rules (v) and (vi) below) and do not violate any demand constraint.

$$\begin{array}{ll} \text{(iii)} & a_{ij_1} \leftarrow a_{ij_1} - 1, \\ \text{(iv)} & a_{ij_2} \leftarrow a_{ij_2} - 1, \\ \text{(v)} & a_{ij_1} \leftarrow a_{ij_1} - 1 \text{ and } a_{ij_2} \leftarrow a_{ij_2} + 1, \\ \text{(vi)} & a_{ij_1} \leftarrow a_{ij_1} + 1 \text{ and } a_{ij_2} \leftarrow a_{ij_2} - \left\lceil \dfrac{x_{j_1}}{x_{j_2}} \right\rceil. \end{array}$$

The whole procedure is repeated as long as either the number of stock rolls $x_{j_1} + x_{j_2}$ decreases or the amount of overproduction $\sum_{i \in M(j_1,j_2)} \delta_i$ decreases, as a result of modification. When none of the two criteria is achieved, 2-PATTERNS halts.

Now 2-PATTERNS is described in Algorithm 32.6 as follows, where L_1^{res} and L_2^{res} represent the residual lengths of p'_{j_1} and p'_{j_2}, respectively. The amount of all overproduction is denoted $\Delta = \sum_{i \in M(j_1,j_2)} \delta_i$.

After completing 2-PATTERNS, the patterns may still have residual length enough to accommodate some products $i \notin M(j_1, j_2)$. In this case, after calculating the continuous frequencies $\overline{X}' = \{\overline{x}'_1, \overline{x}'_2, \ldots, \overline{x}'_n\}$ of the new set of patterns $\Pi' = \Pi \cup \{p'_{j_1}, p'_{j_2}\} \setminus \{p_{j_1}, p_{j_2}\}$ by solving LP relaxation LP(Π'), we try to reduce the number of stock rolls further by adding bottleneck products (i.e., products i such that $\sum_{p_j \in \Pi'} a_{ij}\overline{x}'_j = d_i$ holds) to such residual lengths (see Step 3 of LS in the next subsection).

32.3.4 *Entire algorithm of local search*

We now explain the framework of local search algorithm LS using INIT, FREQ and 2-PATTERNS. Let (Π, X) be the current solution. LS uses the first admissible move strategy by searching neighborhood $N'_2(\Pi)$ as follows. It first selects an index j_1, and then tries $N'_2(\Pi, j_1, j_2)$ for all $j_2 = j_1+1, j_1+2, \ldots, n, 1, \ldots, j_1-1$ (in this order). This is repeated for all j_1. If LS finds a better solution (Π', X') in this process, it immediately moves to (Π', X'), and repeats the neighborhood search for this solution. To measure the improvement of solutions in LS, we employ the main criterion of $f(\Pi, X)$ as well as the secondary criterion of $tloss(\Pi, X)$, defined by

$$tloss(\Pi, X) = \sum_{p_j \in \Pi} \left(L - \sum_{i \in M} a_{ij} l_i \right) x_j. \tag{32.22}$$

That is, LS moves from (Π, X) to (Π', X') if either $f(\Pi', X') < f(\Pi, X)$ holds, or $f(\Pi', X') = f(\Pi, X')$ and $tloss(\Pi', X') < tloss(\Pi, X)$ hold. LS is shown in Algorithm 32.7.

algorithm LS( )
Input: Lengths l_i and demands d_i of products $i \in M = \{1, 2, \ldots, m\}$, the number of different cutting patterns n, and the length of stock rolls L.
Output: A set of cutting patterns $\Pi = \{p_1, p_2, \ldots, p_n\}$ and the corresponding set of frequencies $X = \{x_1, x_2, \ldots, x_n\}$.

(1) (Initialization) Apply INIT to obtain an initial set of patterns $\Pi = \{p_1, p_2, \ldots, p_n\}$, and then apply FREQ to calculate the corresponding set of frequencies $X = \{x_1, x_2, \ldots, x_n\}$. Set $j^* \leftarrow 1$ and $j_1 \leftarrow j^*$.

(2) (Construction of a Π' in the neighborhood of Π) Set $j_2 \leftarrow j^* + 1 \pmod n$. Calculate $r_i(j_1, j_2)$ of (32.17) for all $i \in M$, and let $M(j_1, j_2) \leftarrow \{i \in M \mid r_i(j_1, j_2) > 0\}$. Apply 2-PATTERNS to obtain a pair of new patterns p'_{j_1} and p'_{j_2}, and let $\Pi' \leftarrow \Pi \cup \{p'_{j_1}, p'_{j_2}\} \setminus \{p_{j_1}, p_{j_2}\}$.

(3) (Elimination of bottleneck products) Let $\overline{X}'$ be the continuous set of frequencies obtained from the LP relaxation LP(Π') defined for Π'. Set $B \leftarrow \{i \in M \mid \sum_{p_j \in \Pi'} a_{ij}\bar{x}'_j = d_i\}$ and let $a_{kj} \leftarrow a_{kj} + 1$ for all $k \in B$ for which there is a pattern $p_j \in \Pi'$ such that $l_k \leq L - \sum_{i \in M} a_{ij} l_i$. Let Π' be the resulting set of patterns.

(4) (Move) Calculate X' by applying FREQ to Π'. If $f(\Pi', X') < f(\Pi, X)$ holds, or $f(\Pi', X') = f(\Pi, X)$ and $tloss(\Pi', X') < tloss(\Pi, X)$ hold, set $\Pi \leftarrow \Pi'$, $X \leftarrow X'$, $j^* \leftarrow j^* + 1 \pmod n$, $j_1 \leftarrow j^*$, and return to Step 2.

(5) (Termination) Set $j_2 \leftarrow j_2 + 1 \pmod n$. If $j_2 = j_1$ (i.e., all j_2 have been checked), set $j_1 \leftarrow j_1 + 1 \pmod n$. If $j_1 = j^*$ (i.e., all j_1 and j_2 have been checked), output Π and X, and halt. Otherwise return to Step 2.

Algorithm 32.7 Outline of LS algorithm.

32.4 COMPUTATIONAL EXPERIMENT

We conducted computational experiment for random instances generated by CUTGEN (Gau and Wäscher, 1995), to compare our LS with the existing two algorithms SHP (Haessler, 1971; 1975) and KOMBI (Foerster and Wäscher, 2000). LS and SHP were coded in C language and executed on an IBM-compatible personal computer (Pentium III 1GHz, 1GB memory). The results of KOMBI were taken from (Foerster and Wäscher, 2000), as we could not get the source code of KOMBI. KOMBI was run on an IBM-compatible 486/66 personal computer using MODULA-2 as the programming language under MS-DOS 6.0.

We generated 18 classes of random instances by CUTGEN, which are defined by combining different values of parameters $L, m, \nu_1, \nu_2, \bar{d}$. The lengths l_i are treated as random variables taken from interval $[\nu_1 L, \nu_2 L]$. $\bar{d}$ is the average of demands $(d_1, d_2, \ldots, d_m)$ (the rule of generating d_i is described in (Gau and Wäscher, 1995)). In our experiments, L was set to 1000, m was set to 10, 20 and 40, and $\bar{d}$ was set to 10 and 100. Furthermore, (ν_1, ν_2) was set to (0.01, 0.2) for classes 1–6, (0.01, 0.8) for classes 7–12, and (0.2, 0.8) for classes 13–18. The parameter *seed* for generating random numbers was set to 1994. For each class, 100 problem instances were generated and solved. These classes of problem instances were also solved by KOMBI, where 100 instances are tested for each class.

As mentioned in Section 32.1, our local search algorithm LS can obtain a trade-off curve between the number of different cutting patterns n and the number of required stock rolls f. For this, we conducted preliminary computational experiment. We took a random instance of class 12 generated by CUTGEN, and applied LS for all n between n_{LB} and m, where

$$n_{LB} = \left\lceil \frac{\sum_{i \in M} l_i}{L} \right\rceil \tag{32.23}$$

is a lower bound of different cutting patterns and m is the number of products. Figure 32.2 shows the number of stock rolls f with respect to the number of different cutting patterns n. As SHP has an input parameter MAXTL to control the maximum trim loss of cutting patterns, we also illustrated the solutions of SHP for different values of MAXTL. For this instance, we observe that LS attains comparable number of stock rolls to SHP with smaller number of different cutting patterns. Figure 32.3 shows the CPU time of LS for different n. We observe that the CPU time of LS tends to increase as n increases.

Tables 32.1 and 32.2 show computational results, where $\bar{n}_{LB}$ denotes the average of the lower bound n_{LB} on different cutting patterns, $\bar{n}$ denotes the average of different cutting pattern n, and $\bar{f}$ denotes the average of required stock rolls for each class. Table 32.1 contains the results of SHP and KOMBI, where SHP was run for three different values of MAXTL= 0.05, 0.03 and 0.01. The results of LS are shown in Table 32.2. To capture the general behavior of the trade-off curve of LS, we consider several different upper bounds of required stock rolls f_{UB}, and obtain minimum numbers of different cutting

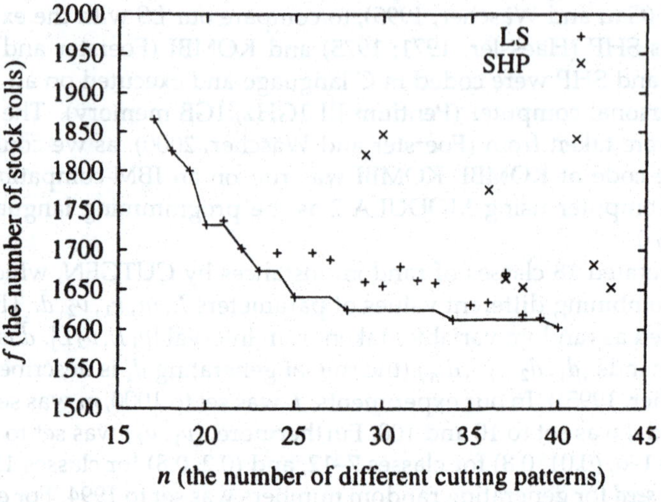

Figure 32.2 The number of stock rolls versus the number of different cutting patterns ($n_{LB} = 17$).

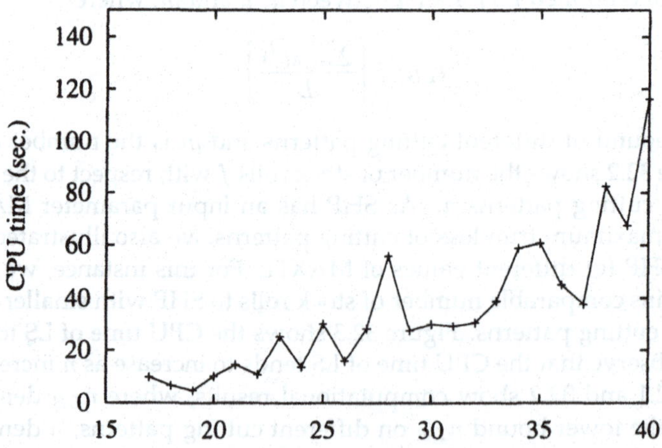

Figure 32.3 The CPU time in seconds versus the number of different cutting patterns.

patterns n that requires f_{UB} stock rolls or less. In this experiment, the upper bounds were set to as follows:

$$f_{UB} = f^* + \beta f_{LB}, \tag{32.24}$$

where f^* is the number of required stock rolls obtained by LS for $n = m$, and f_{LB} is the trivial lower bound of required stock rolls:

$$f_{LB} = \left\lceil \frac{\sum_{i \in M} d_i l_i}{L} \right\rceil. \tag{32.25}$$

We tested LS for $f_{UB} = \infty$ and $f_{UB} = f^* + \beta f_{LB}$ with $\beta = 0.05, 0.03, 0.01$.

From Tables 32.1 and 32.2, we first observe that the numbers of stock rolls attained by KOMBI are smaller than those of SHP and LS for all classes, since KOMBI is based on Stadtler's algorithm (Stadtler, 1990) which gives near optimal solutions for the standard 1D-CSP. However, SHP and LS can obtain a wide variety of solutions by controlling their input parameters, i.e., they can realize the trade-off curve between the number of different cutting patterns and the number of stock rolls by their input parameters. SHP attains smaller number of different cutting patterns than KOMBI for classes 2–6 using only a slightly larger number of stock rolls, while the solutions of SHP are much worse than those of KOMBI for classes 7–18. It shows that SHP does not provide good trade-off curves for the instances in which the ratio of product lengths l_i to the length of stock rolls L is relatively large. On the other hand, LS attains smaller number of different cutting patterns than KOMBI for all classes, without much increasing the number of additional stock rolls. Note that LS obtains feasible solutions even for very small n, close to n_{LB}, while SHP and KOMBI could not produce feasible solutions for such n. From these observation, we may conclude that LS is useful, as it can provide reasonable trade-off curves for a very wide range of n.

Table 32.3 gives the average CPU time of SHP (MAXTL=0.03), KOMBI and LS ($f_{UB} = f^* + 0.03 f_{LB}$), respectively, for all classes. For these problem instances, SHP is faster than LS except for classes 5 and 6, and KOMBI may be faster than LS, taking into consideration the power of computers in use. However, the average CPU time of LS is within 20 seconds for all classes, and it may be sufficiently short even if LS repeatedly applies for all n to obtain the trade-off curve.

32.5 CONCLUSION

The setup cost for changing cutting patterns becomes more important in recent cutting stock industry, and several algorithms have been proposed for this purpose. In this paper, instead of directly minimizing the number of different cutting patterns, we considered the pattern restricted 1D-CSP (1D-PRP) in which the number of different cutting patterns n is treated as a parameter, and the number of stock rolls is minimized under the constraint of n. We proposed a local search algorithm (LS) based on larger neighborhood than that

Table 32.1 Computational results of SHP and KOMBI for the random instances generated by CUTGEN.

class	m	$\bar{n}_{LB}$	SHP MAXTL=0.05		SHP 0.03		SHP 0.01		KOMBI	
			$\bar{n}$	$\bar{f}$	$\bar{n}$	$\bar{f}$	$\bar{n}$	$\bar{f}$	$\bar{n}$	$\bar{f}$
1	10	1.67	4.08	11.68	4.25	11.62	4.49	11.57	3.40	11.49
2	10	1.67	6.33	112.80	6.33	111.81	6.75	110.85	7.81	110.25
3	20	2.56	5.77	22.55	5.89	22.37	5.98	22.17	5.89	22.13
4	20	2.56	9.06	220.63	8.98	218.94	9.25	217.00	14.26	215.93
5	40	4.26	9.07	43.89	9.03	43.60	9.01	43.17	10.75	42.96
6	40	4.26	13.90	434.59	13.45	430.79	13.77	426.81	25.44	424.71
7	10	4.62	10.14	52.21	10.33	52.19	10.82	52.77	7.90	50.21
8	10	4.62	11.30	519.88	11.46	520.36	11.97	526.81	9.96	499.52
9	20	8.65	18.58	97.42	19.22	97.96	19.91	98.82	15.03	93.67
10	20	8.65	20.96	970.43	21.74	973.63	21.99	984.32	19.28	932.32
11	40	16.27	35.06	186.45	36.29	187.37	37.78	189.91	28.74	176.97
12	40	16.27	39.90	1854.79	40.53	1865.41	41.86	1891.90	37.31	1766.20
13	10	5.54	10.55	65.46	10.55	65.41	10.67	65.52	8.97	63.27
14	10	5.54	10.92	652.95	10.95	654.80	11.12	654.95	10.32	632.12
15	20	10.52	20.11	123.36	20.30	124.36	20.86	124.60	16.88	119.93
16	20	10.52	21.09	1232.42	21.09	1240.39	21.32	1243.43	19.91	1191.80
17	40	19.85	38.02	235.64	38.52	238.06	39.10	239.77	31.46	224.68
18	40	19.85	40.16	2351.38	40.38	2366.95	40.67	2389.49	38.28	2242.40

Table 32.2 Computational results of LS with different f_{UB} for the random instances generated by CUTGEN.

class	m	$\bar{n}_{LB}$	∞		$f^* + 0.05f_{LB}$		$f^* + 0.03f_{LB}$		$f^* + 0.01f_{LB}$	
			$\bar{n}$	$\bar{f}$	$\bar{n}$	$\bar{f}$	$\bar{n}$	$\bar{f}$	$\bar{n}$	$\bar{f}$
1	10	1.67	2.00	14.47	2.90	12.54	2.90	12.54	2.90	12.54
2	10	1.67	2.00	141.28	4.09	115.32	4.74	113.80	5.73	112.73
3	20	2.56	2.57	30.76	4.83	23.66	4.83	23.66	4.83	23.66
4	20	2.56	2.57	305.78	6.07	226.04	6.96	223.49	9.28	220.31
5	40	4.26	4.28	61.63	8.65	45.72	9.41	45.25	9.41	45.25
6	40	4.26	4.28	609.73	9.49	447.94	11.23	441.73	15.23	434.79
7	10	4.62	5.01	55.77	6.14	51.45	6.26	51.28	6.26	51.28
8	10	4.62	5.01	558.69	6.42	510.60	6.73	507.78	7.38	504.77
9	20	8.65	9.27	105.07	10.81	97.39	11.41	96.43	11.81	96.14
10	20	8.65	9.27	1053.08	11.44	962.27	12.22	953.92	13.91	944.45
11	40	16.27	16.95	201.46	19.60	185.68	20.77	183.79	23.98	181.34
12	40	16.27	16.95	2010.26	20.71	1837.90	22.44	1817.07	26.03	1787.39
13	10	5.54	6.26	68.73	7.06	64.45	7.30	64.36	7.30	64.36
14	10	5.54	6.26	68.73	7.19	644.87	7.45	641.05	7.77	638.53
15	20	10.52	11.76	129.10	12.76	123.43	13.33	122.65	14.18	121.81
16	20	10.52	11.76	1292.09	13.38	1226.05	14.01	1216.66	15.13	1205.10
17	40	19.85	21.50	246.34	22.89	235.85	23.49	233.55	25.58	230.94
18	40	19.85	21.50	2471.32	23.99	2334.88	25.35	2305.35	28.35	2274.79

of previous paper (Umetani et al., 2003), which is based on the idea of changing two patterns in the current set of patterns. To facilitate the search in the neighborhood, we introduced a heuristic algorithm to generate a small subset of the neighborhood containing good solutions. We conducted computational experiment for random instances, and observed that our LS attains a wide variety of good solutions comparable to SHP and KOMBI, and LS provides reasonable trade-off curves between the number of different cutting patterns and the number of required stock rolls for a very wide range.

It may be interesting to consider a generation of our problem, in which we are given s_i stock rolls of length L_i, $i = 1, 2, \ldots, k$. It is not difficult to see that our algorithm can be modified to handle such problem. Its implementation is, however, one of our future projects.

Table 32.3 The average CPU time in seconds for the random instances generated by CUTGEN.

class	m	d	SHP	KOMBI	LS
1	10	10	0.04	0.14	0.01
2	10	100	0.08	1.14	0.07
3	20	10	1.56	1.74	0.19
4	20	100	1.57	16.00	0.76
5	40	10	631.74	38.03	5.32
6	40	100	107.11	379.17	11.41
7	10	10	0.00	0.07	0.03
8	10	100	0.00	0.20	0.04
9	20	10	0.01	1.34	0.48
10	20	100	0.02	3.25	0.71
11	40	10	0.09	36.27	13.07
12	40	100	0.14	76.31	19.53
13	10	10	0.00	0.08	0.03
14	10	100	0.00	0.13	0.03
15	20	10	0.01	1.81	0.44
16	20	100	0.01	2.60	0.60
17	40	10	0.06	50.93	10.34
18	40	100	0.10	70.94	14.32

A.A. Farley and K.V. Richardson. Fixed charge problems with identical fixed charges. *European Journal of Operational Research*, 18:245–249, 1984.

H. Foerster and G. Wäscher. Pattern reduction in one-dimensional cutting stock problems. *International Journal of Production Research*, 38:1657–1676, 2000.

M.R. Garey and D.S. Johnson. *Computers and Intractability – A Guide to the Theory of NP-Completeness*. W.H. Freeman and Company, 1979.

T. Gau and G. Wäscher. Cutgen1: A problem generator for the standard one-dimensional cutting stock problem. *European Journal of Operational Research*, 84:572–579, 1995.

P.C. Gilmore and R.E. Gomory. A linear programming approach to the cutting-stock problem. *Operations Research*, 9(6):849–859, 1961.

P.C. Gilmore and R.E. Gomory. A linear programming approach to the cutting-stock problem — Part II. *Operations Research*, 11(6):863–888, 1963.

C. Goulimis. Optimal solutions for the cutting stock problem. *European Journal of Operational Research*, 44:197–208, 1990.

R.E. Haessler. A heuristic programming solution to a nonlinear cutting stock problem. *Management Science*, 17(12):793–802, 1971.

R.E. Haessler. Controlling cutting pattern changes in one-dimensional trim problems. *Operations Research*, 23(3):483–493, 1975.

R.E. Johnston. Rounding algorithm for cutting stock problems. In *Journal of Asian-Pacific Operations Research Societies*, volume 3, pages 166–171, 1986.

O. Marcotte. The cutting stock problem and integer rounding. *Mathematical Programming*, 33:82–92, 1985.

G. Scheithauer and J. Terno. Theoretical investigations on the modified integer round-up property for the one-dimensional cutting stock problem. *Mathematical Methods of Operations Research*, 48:105–115, 1997.

H. Stadtler. A one-dimensional cutting stock problem in the aluminium industry and its solution. *European Journal of Operational Research*, 44:209–223, 1990.

S. Umetani, M. Yagiura, and T. Ibaraki. One dimensional cutting stock problem to minimize the number of different patterns. *European Journal of Operational Research*, 146(2):388–402, 2003.

P.H. Vance. Branch-and-price algorithm for the one-dimensional cutting stock problem. *Computational Optimization and Applications*, 9:211–228, 1998.

F. Vanderbeck. Computational study of a column generation algorithm for bin packing and cutting stock problems. *Mathematical Programming*, 86:565–594, 1999.

F. Vanderbeck. Exact algorithm for minimising the number of setups in the one-dimensional cutting stock problem. *Operations Research*, 48(6):915–926, 2000.

W.E. Walker. A heuristic adjacent extreme point algorithm for the fixed charge problem. *Managemant Science*, 22(5):587–696, 1976.

Metaheuristics: Computer Decision-Making, pp. 699-719
Anne Wade and Said Salhi

33 AN ANT SYSTEM ALGORITHM FOR THE MIXED VEHICLE ROUTING PROBLEM WITH BACKHAULS

Anne Wade[1] and Said Salhi[1]

[1]Management Mathematics Group
School of Mathematics and Statistics
The University of Birmingham, U.K.
wadeac@for.mat.bham.ac.uk, S.Salhi@bham.ac.uk

Abstract: Ant system algorithms have been used successfully to solve many hard combinatorial problems. In this paper we introduce an ant system method to solve the mixed vehicle routing problem with backhauls. Some enhancements to the general characteristics of ant system algorithms are proposed. We concentrate on the way the candidate list is constructed, a look ahead scheme to incorporate into the visibility function and efficient rules to deal with local as well as global updating. Computational results on test problems are reported and compared to known results.
Keywords: Metaheuristic, Ant system, Vehicle routing, Backhauls.

33.1 INTRODUCTION

The ant system is a metaheuristic developed for the solution of hard combinatorial optimization problems. It was first proposed by Colorni et al. (1991) and Dorigo (1992) and used to solve the traveling salesman problem (TSP). The method was inspired by the observation of the behavior of real life ant colonies, in particular the way in which real ants find the shortest path between food sources and their nest. While walking ants deposit a substance called pheromone on to the ground which forms a pheromone trail. Ants can detect the pheromone and choose their way according to the level of the pheromone trail. The greater the concentration of pheromone on the ground the higher the probability that an ant will choose that path. Where there is a shorter path from a food source to the nest, ants will reach the end of the path in a quicker time compared to ants on a longer path. This means that the trail will build up at a faster rate on the shorter path which in turn causes more ants to choose the shorter path which also causes a greater level of pheromone. In time all ants will have the tendency to choose the shorter path.

This real life behavior of ants has been adapted to solve combinatorial optimization problems using simulation. A number of artificial ants build solutions in parallel using a form of indirect communication. The artificial ants cooperate via the artificial pheromone level deposited on arcs which is calculated as a function of the quality of the solution found. The amount of pheromone an ant deposits is proportional to the quality of the solution generated by that ant helping direct the search toward good solutions. The artificial ants construct solutions iteratively by adding a new node to a partial solution using information gained from both past performance and a greedy heuristic. The greedy heuristic, known as the visibility, is introduced in an attempt to guide the search.

Ant systems have been used successfully to solve many hard combinatorial problems such as the traveling salesman problem (see Dorigo et al. (1996), Dorigo and Gambardella (1997), Stützle and Hoos (1997a), and Cordon et al. (2000)), the quadratic assignment problem (see Maniezzo and Colorni (1998) and Maniezzo (1999)), the sequential ordering problem (see Gambardella and Dorigo (1997)) and the vehicle routing problem (see Bullnheimer et al. (1997), Bullnheimer et al. (1998), and Gambardella et al. (1999)).

The vehicle routing problem (VRP) is known to be NP-hard in the strong sense (see Garey and Johnson (1979)). This means that finding a solution to a problem which is an extension of the VRP, such as the VRP with backhauls (VRPB), is NP-complete. Therefore exact methods are only suitable to solve relatively small instances of vehicle routing problems. In these circumstances intelligent heuristic search methods are the best way forward to address such complex combinatorial problems. In this paper, we propose some enhancements to the ant system metaheuristic to solve one variant of the VRPB.

In the remainder of this section, we introduce the VRPB. In Section 33.2, we present the fundamentals of ant system algorithms and introduce successful ant system methods used to solve simpler versions of the VRPB namely the traveling salesman problem and the vehicle routing problem. Section 33.3 describes our basic ant system to solve the VRPB and also our modifications to the general components of ant system methods. Computational results are reported in Section 33.4 and finally conclusions and ideas for future research are presented in Section 33.5.

33.1.1 The vehicle routing problem with backhauls

The vehicle routing problem (VRP) consists of a number of homogeneous vehicles based at a single depot and a number of customers to be served. Each customer has a demand quantity to be delivered from the depot and a service time allocated. The objective is to find a set of routes with the least cost where all customer demands are met, every customer is visited exactly once, the total demand on each vehicle does not exceed the maximum vehicle capacity, the total length of each route does not exceed the maximum bound and every vehicle starts from and ends at the depot.

The vehicle routing problem with backhauls is an extension to the vehicle routing problem (VRP) where two different types of customer are served. The first type are linehaul customers, also known as delivery customers, who require a given quantity of product to be delivered. The second type are backhaul customers, also known as pickup customers, who require a quantity of goods to be picked up and sent back to the depot. The goods are transported to or from a single depot by a fleet of homogeneous vehicles. Methods in the literature to solve the VRPB either restrict all backhauls to be visited once all linehauls have been served, the classical VRPB, or allow mixed linehaul and backhaul customers along the routes, the mixed VRPB. In this paper we propose an ant system to solve the mixed VRPB. The objective is to find the set of routes with the least cost such that (i) all customer demands are satisfied, (ii) each customer is visited exactly once, (iii) all routes begin and end at the depot, and finally (iv) the load on the vehicle at any point along the route must not exceed the vehicle capacity.

The mixed problem is more complicated than the classical VRPB due to the fluctuating load on the vehicle. In the classical VRPB it is only necessary to check that the total linehaul load and the total backhaul load do not separately exceed the total vehicle capacity. This is also known as the VRP with pickups and deliveries (VRPPD) where all deliveries are performed first then pickups second. Gendreau et al. (1999) investigate the TSP with pickups and deliveries, and Osman and Wassan (2002) put forward a useful implementation of tabu search to solve the VRP with backhauling. For more details and references on the classical VRPB or VRPPD, see Toth and Vigo (1996), Toth and Vigo (1997), Goetschalckx and Jacobs-Blecha (1989) and Mingozzi et al. (1999). In the mixed VRPB, which is also referred to as the mixed VRPPD, it is necessary to check that the vehicle capacity is not exceeded at any point along the route. In the mixed VRPB the vehicle load can either decrease or increase at each customer depending on whether the customer has a linehaul or a backhaul demand, respectively. The mixed VRPB has not received as much attention in the literature as its counterpart the classical VRPB. Deif and Bodin (1984) use a savings based heuristic based on the Clarke and Wright algorithm to solve the mixed VRPB with a penalty function to delay the insertion of a backhaul in a route. Casco et al. (1988) propose an insertion based heuristic method which uses a penalty function based on delivery load after pickup. The aim of the penalty function is to encourage backhaul customers to be serviced toward the end of the routes. Salhi and Nagy (1999) develop an insertion based heuristic which considers a cluster of backhauls for insertion on to linehaul routes at each iteration. Halse (1992) adopts a cluster-first route-second heuristic using a relaxed assignment problem and modified route improvement procedures to solve the mixed VRPB. In this study we do not address the simultaneous VRPPD where customers may simultaneously receive and send goods.

The main reason for imposing the restriction that backhauls can only be served after all linehauls have been visited on a route, apart from the fact that the mixed problem is more complicated to solve, is that many of the routes that are created if mixed customers are allowed may not be practical to implement. This is due to the fact that it may be necessary to remove backhaul load from the vehicle at subsequent visits to linehaul customers. However, in many real life applications it may not be necessary to restrict all backhauls to the end of the routes. In such cases significant savings can be achieved. The improved design of vehicles, in particular vehicles that permit side loading, means that mixed routes are now a much more practical option and therefore worth investigating.

33.2 ANT COLONY OPTIMIZATION (ACO)

33.2.1 *A brief overview of ACO*

The Ant system algorithm (AS), first proposed by Dorigo (1992) and Colorni et al. (1991) to solve the traveling salesman problem uses artificial pheromone trail values, τ_{ij}, associated with each arc (i, j) where i and j denote the ith and the jth node (or customer site) of the network. Initially m ants are placed on randomly selected nodes and each ant starts constructing a tour which passes through all the nodes. The tour is built by each ant successively choosing the next customer to visit probabilistically. The probability that an ant building its tth tour currently situated at customer i will visit customer j next is given by the state transition rule:

$$p_{ij}(t) = \frac{[\tau_{ij}(t)]^{\alpha} \cdot [\eta_{ij}]^{\beta}}{\sum_{l \in F_i^k} [\tau_{il}(t)]^{\alpha} \cdot [\eta_{il}]^{\beta}} \quad \text{if} \quad j \in F_i^k, \quad 0 \quad \text{otherwise,} \tag{33.1}$$

where η_{ij} is a local heuristic function, known as the visibility, usually given as $\eta_{ij} = 1/d_{ij}$. d_{ij} is the distance from customer i to customer j, and F_i^k is the list of feasible customers not yet visited by ant k. The parameters α and β determine the relative influence of the trail value and the heuristic information. To avoid the use of two parameters, one can set say $\alpha = 1$ and choose the value of β accordingly.

The ants keep track of the customers that have been visited in the order they were served to form a partial tour. Once all ants have constructed a complete tour, i.e. all customers have been visited, then the pheromone trail levels are updated by each ant according to the global update rule given in the equation

$$\tau_{ij}(t) = (1 - \rho) \cdot \tau_{ij}(t) + \sum_{k=1}^{m} \Delta\tau_{ij}^{k}(t) \tag{33.2}$$

where

$$\Delta\tau_{ij}^k(t) = \begin{cases} Q/L_k(t) & \text{if} \quad (i,j) \in T^k(t) \\ 0 & \text{otherwise} \end{cases}$$

and $0 < \rho \leq 1$ simulates the pheromone trail evaporation. The parameter ρ is used to avoid stagnation in the solution which would occur if the trail values were unrestricted, Q is a correction factor used to normalize the data set, $L^k(t)$ is the length of the tour produced by ant k at iteration t and $T^k(t)$ is the tour produced by ant k at iteration t.

A number of modifications have been proposed to the AS algorithm in order to improve the solution quality. The elitist strategy which was proposed by Dorigo et al. (1996) is based on a modified global updating rule where the arcs belonging to the global best tour received an extra amount of pheromone. This idea is then generalized by Bullnheimer et al. (1997) to include not only the global best tour but also a number of the best tours to update trail values. The ants are ranked according to their tour length and the amount of pheromone an ant adds is proportional to its rank. The ant colony system (ACS) proposed by Dorigo and Gambardella (1997) includes three main modifications: a different transition rule based on the pseudo-random-proportional rule, a pheromone global updating rule based only on the global best solution, and a local updating rule that adjusts the pheromone level on the selected arcs. The ACS method makes use of a candidate list to solve large combinatorial problems. The candidate list is used to reduce the number of choices available to each ant when the ant is considering its next move, hence reducing the computation time. The candidate list is one of the key factors which we will investigate in this study. Stützle and Hoos (1997a;b) introduced the Max-Min ant system, known as MMAS. Their method is basically similar to AS except that pheromone trail values are restricted to the interval $[\tau_{min}, \tau_{max}]$ to avoid stagnation. The introduction of a lower limit on pheromone trail values was found to have the greatest effect on the solution quality.

For an overview on ant systems, the book of Bonabaeu et al. (1999), the review paper by Dorigo et al. (1999) and the technical report by Dorigo and Stützle (2000) are useful references.

33.2.2 Ant systems for vehicle routing problems

Bullnheimer et al. (1997; 1998) propose an ant system to solve the VRP. Their method, named AS-VRP, is based upon their AS_{Rank} method. Bullnheimer et al. (1997) extend the visibility function to be problem specific by incorporating a savings function and capacity utilization. The revised visibility function produced better results but was found to be costly in respect of computation time. In Bullnheimer et al. (1998) a parametric savings function is adopted for the visibility producing an improvement in solution quality and a reduction in computation time. A 2-opt local search procedure is implemented to improve solutions. Bullnheimer et al. (1997; 1998) adapted the list of feasible customers for each ant so that the capacity constraint and the maximum route length are taken into account. If no feasible customers are available to an ant building its tour then the ant returns to the depot so that a new route can be started.

Gambardella et al. (1999) also solve the VRP using an ant based algorithm. The problem is reformulated by adding $(M - 1)$ depots, where M is the number of vehicles. The problem therefore becomes a traveling salesman problem with added constraints. The algorithm Gambardella et al. develop, called HAS-VRP, is based upon the ACS method. Each ant builds its tour so that the vehicle capacity is not exceeded. Each tour that is generated contains M subtours each corresponding to the route of a vehicle. A local search heuristic comprising edge exchanges is applied to the solutions. Gambardella et al. also solve the vehicle routing problem with time windows (VRPTW) using an ant algorithm. The VRPTW is an extension to the VRP where each customer must be serviced within a given time interval. Two ant colonies are utilized by their method. The first ant colony is used to minimize the number of vehicles and the second colony is used to minimize the total travel time given the number of vehicles determined by the first ant colony. The pheromone trail levels are independent for each colony but the best ants from one colony are allowed to update trail values associated with the other colony.

The results reported by the methods detailed in this subsection have been shown to be competitive when compared with known methods in the literature to solve the vehicle routing problem and the vehicle routing problem with time windows.

33.3 AN ANT SYSTEM HEURISTIC FOR THE VRPB

We propose an algorithm based on the ant colony system, used successfully to solve the TSP, for the VRPB. A number of modifications to the ACS method are proposed but the basic ant algorithm can be described as follows. Initially m ants are positioned at starting customers. Each ant selects the next customer to visit according to the state transition rule which will be looked at in this section. Once an arc (i, j) has been selected by an ant then a local updating rule is applied to reduce the trail value of that arc. Each ant constructs its tour selecting one customer at a time. If no feasible customers are available to be visited due to the vehicle capacity constraint then the depot is selected by the ant. The ant then continues to build a new route repeating the process until all customers have been selected. Two local search procedures are adopted accordingly for the VRPB. A restricted 2/3-opt procedure is used to improve each solution after each ant has completed its tour (see Salhi and Rand (1987)). A shift procedure that exchanges customers between routes is also introduced. The structure of our ant algorithm is given in Algorithm 33.1.

The modifications we propose to the ant colony system are detailed in the remainder of this section.

```
procedure ASO( )
for i = 1 to number of iterations
    for m = 1 to number of ants
        for n = 1 to number of customers
            Select the next customer to visit using state transi-
            tion rule;
            Update the pheromone trail for the selected arc us-
            ing local updating rule;
        endfor
        Apply fast local search heuristics;
    endfor
    Update pheromone trail values using global updating
    rule;
endfor
```

Algorithm 33.1 Ant System Algorithm for VRPB

33.3.1 A site dependent candidate list

Many ant system procedures use a data structure known as a candidate list. A candidate list contains for each customer, i, a given number of potential customers to be visited. These favored customers can be chosen according to their distance from customer i. A candidate list is a static data structure containing a predetermined number of customers. An ant first chooses the next customer to be visited from the candidate list corresponding to the current customer. Only if it is not feasible to visit any of the customers belonging to the list then the ant selects the next customer to visit as the closest out of all feasible customers and the depot. We adopt this procedure in our algorithm unless there is only one remaining customer to be visited. In our method if only one customer remains to be visited in the problem and the customer is contained in the feasible list but not the candidate list, then that customer is forced to be visited next. This should avoid one remaining feasible customer being forced into a new route if the depot had been the closest option.

The purpose of the candidate list is to reduce the possible number of moves that an ant must evaluate. This is performed by recording for each customer only those who are not considered far away customers to it. In other words, the path between the current customer and one of those far away customers is unlikely to be chosen as such a link is unlikely to produce a good quality solution. Looking at Figure 33.1, if we consider customer A it may not be sensible for an ant currently at customer A to consider customer B as the next move in view of solution quality.

It can be observed that using a static candidate list does not take into consideration the number of customers situated in the surrounding neighborhood of customer i. For instance in Figure 33.1, customer A has only a few customers

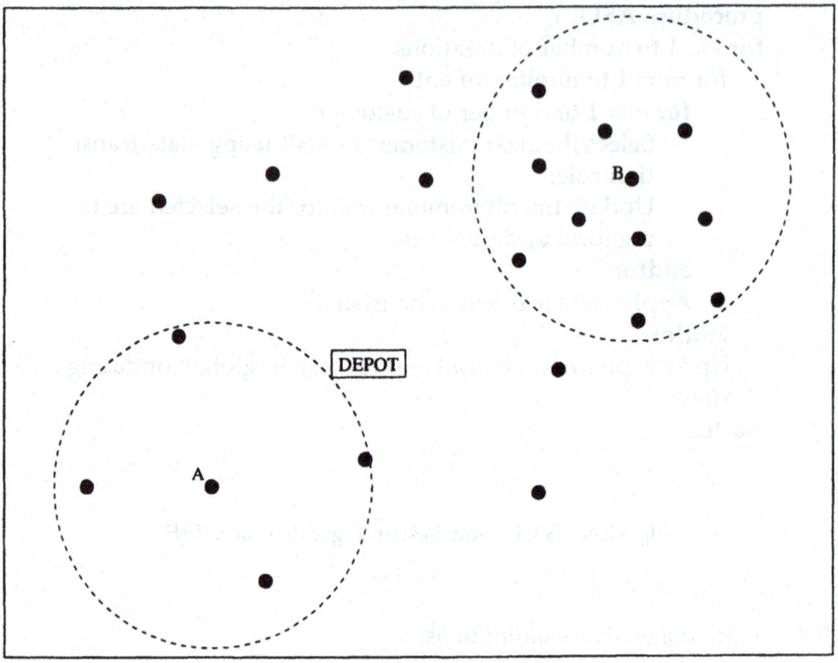

Figure 33.1 Example Problem

in the near vicinity. Therefore, if the size of the candidate list is set too high then far away customers may be considered which would not be a sensible choice. If we now consider customer B, which has many customers in the surrounding neighborhood, then if the number in the candidate list is set too low customers which may lead to a good solution could be ruled out. We put forward an approach that is site dependent, i.e. that takes into account the sparsity of customers in the surrounding neighborhood of each customer.

The minimum number, M, contained in the candidate list for each customer is given by

$$M = \min\{N/4, 10\},$$

where N is the total number of customers. The distance from customer i to all other customers, d_{ij} is calculated and the average of these distances is

$$\bar{d}_i = \frac{\sum_{j \neq i} d_{ij}}{N-1}.$$

This process is repeated for all customers and the average of these results is given by

$$D = \frac{\sum_{k=1}^{N} \bar{d}_k}{N}.$$

For each customer the distance to all other customers that fall within the range D is calculated and the average of all these distances

$$C_i = \frac{\sum_{\{j:d_{ij}<D\}} d_{ij}}{|\{j : d_{ij} < D\}|}.$$

In addition, if $|\{j : d_{ij} < D\}| = 0$, then set $C_i = 0$. The overall average is

$$R = \frac{\sum_{k=1}^{N} C_i}{N}.$$

The candidate list is then constructed for each customer i as

$$E_i = \{j = 1, \ldots, m \quad s.t. \quad d_{ij} \leq R\}.$$

In addition, if $|E_i| < M$, then the nearest $M - |E_i|$ customers not already contained in E_i are selected, say O_i, then we set $E_i = E_i \cup O_i$. In particular, if E_i is empty, we set $E_i = O_i$.

33.3.2 Strategic placement of ants

In some TSP and VRP applications, it is common to place one ant on each node. In this paper, we relax such a restriction and we adopt a more strategic approach to the placement of ants.

The whole region originated from the depot is divided into a given number of sectors which is based on the number of customers. The sectors are constructed so that each contains an equal number of customers where possible. If this is not possible then the remaining customers are allocated to the final sector. Within each sector an ant is placed at the closest and farthest customer in relation to the depot. A given number of ants are then placed randomly on nodes remaining in each sector. For each iteration of the ant algorithm the starting position in the calculation of the sectors is rotated. This procedure should result in an even spread of ants being selected for each iteration. The aim of this procedure is that the sectors should roughly correspond to the routes that would be created in a good solution. The restriction on the number of ants can be altered within the algorithm so that all customer sites can still be selected if desired. In our algorithm we use $N/2$ ants, where N is the total number of linehauls and backhauls. The restricted number of ants has the effect that fewer solutions are created which could have a detrimental effect on solution quality. However, computation time decreases which means that more iterations can be performed than would be the case if the number of ants is set equal to the number of customers as used in most applications. Because more iterations are performed to generate a given number of solutions, the ants can take advantage of a greater number of pheromone trail updates hence learning from past experience at a faster rate.

33.3.3 A look ahead based visibility

As discussed in Section 33.2, the visibility is a greedy heuristic used within the transition rule in an attempt to guide the search. For the traveling salesman problem the greedy heuristic is generally calculated as the reciprocal of the distance between the current customer where the ant is placed and the customer to which the ant is considering to move. Due to the fact that the visibility function is used to guide the search, it can have a significant influence on the quality of results. In this paper we propose a visibility that takes into account the load on the vehicle in the selection process. Two different visibility functions are used, the choice of which is dependent on the capacity remaining on the vehicle.

- If the vehicle is nearly full so that the vehicle is near to the end of its route then it would not be a sensible choice to visit a customer which is further away from the depot than the current customer. It would be more efficient to visit a customer that is between the current customer and the depot if possible. This can be seen in Figure 33.2 where the vehicle is currently positioned at customer B. The load on the vehicle at customer B is near the maximum capacity constraint and can only accommodate the load of A or C but not both. Therefore it would not be sensible to visit customer C next as there may not be enough room remaining on the vehicle to service another customer but the vehicle has traveled further away from the depot. It is more likely to be an efficient choice to visit customer A after customer B due to the fact that the cost of inserting customer A between B and the depot is relatively smaller.

  The visibility we propose uses the insertion cost between the current customer and the depot in order to encourage the vehicle to travel closer to the depot toward the end of each route. If the unused capacity on the vehicle is less than a given parameter based on the average linehaul demand then the visibility is

  $$\eta_{ij} = \frac{1}{d_{ij} + d_{j0}},$$

  where j is the customer to which the ant is considering to move and customer 0 is the depot.

- At the beginning of routes, before the above cut off point is reached, we propose a visibility that takes into account a chain of two customers. When considering the next customer j to be visited then the nearest customer k in relation to j is incorporated into the visibility detailed in Equation (33.3). Note that in previous applications κ_{il} is not included so if $\phi = 0$ our transition rule reduces to the one given in ACS by Dorigo and Gambardella (1997).

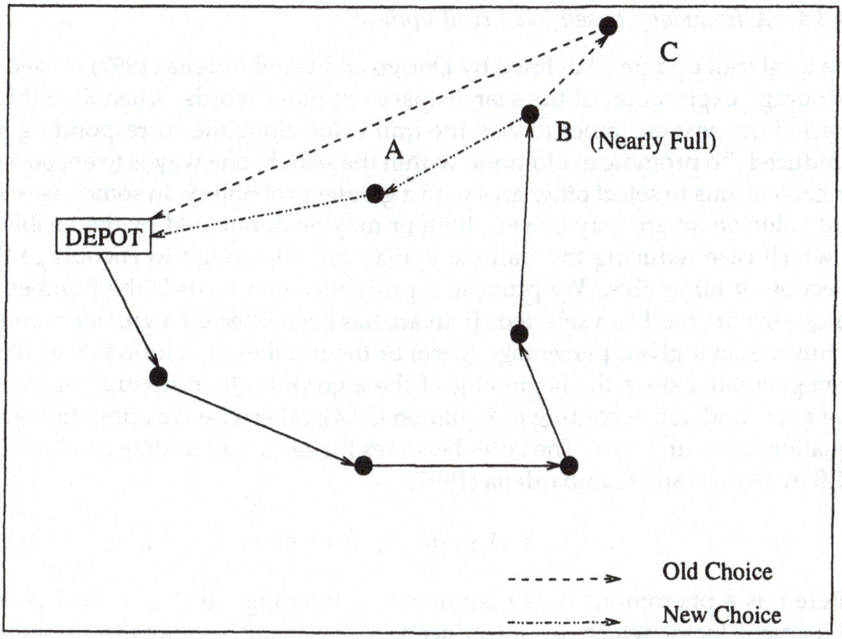

Figure 33.2 Insertion Based Visibility

$$j = \begin{cases} \text{argmax}_{u \in F_i^k} \{[\tau_{iu}(t)]^\alpha \cdot [\eta_{iu}]^\beta \cdot [\kappa_{iu}]^\phi\} & \text{if } q \le q_0 \\ S & \text{if } q > q_0 \end{cases} \tag{33.3}$$

where q is a random variable uniformly distributed in the interval [0,1], q_0 is a parameter ($0 \le q_0 \le 1$) and S is a random variable selected according to the probability distribution

$$P_{ij}^k(t) = \frac{[\tau_{ij}(t)]^\alpha \cdot [\eta_{ij}]^\beta \cdot [\kappa_{ij}]^\phi}{\sum_{l \in F_i^k} [\tau_{il}(t)]^\alpha \cdot [\eta_{il}]^\beta \cdot [\kappa_{il}]^\phi},$$

where

$$\kappa_{ij} = \frac{1}{d_{ij} + d_{jk}} \text{ and } \eta_{ij} = \frac{1}{d_{ij}}.$$

The idea is to incorporate such a look ahead strategy that takes into consideration the next possible move following the selection of customer j. In our implementation we do not check for feasibility when considering customer k in the chain due to the significant increase in computation time which would be required. Therefore for each customer, i, the value of κ_{ij} is fixed for all iterations irrespective of whether it would be feasible to visit customer k later along the route. It should also be noted that η_{ij} is also incorporated in to the visibility so that the probability function is not solely influenced by the chain of customers i, j, k.

33.3.4 A frequency based local trail update

The local trail update as defined by Dorigo and Gambardella (1997) is used to encourage exploration of the search space. In other words, when an ant has selected the next customer to visit the trail value along the corresponding arc is reduced. To promote exploration within the search, one way is to encourage successive ants to select other arcs with a greater probability. In some cases the trail value on an arc may be very high or may be dominated by the visibility in which case reducing the trail value may not be enough to encourage the selection of other arcs. We propose a procedure that records the number of times each arc has been selected. If an arc has been selected a greater number of times than a given percentage (*perc*) of the number of solutions that have been generated since the beginning of the algorithm, then the trail value on that arc is updated according to Equation (33.4), otherwise we apply the same equation but with $\gamma = 1$. The latter becomes the local trail update rule used in ACS by Dorigo and Gambardella (1997):

$$\tau_{ij} = (1 - \gamma\upsilon) \cdot \tau_{ij} + \upsilon \cdot \tau_0, \tag{33.4}$$

where υ is a pheromone decay parameter in the range ($0 < \upsilon < 1$), γ is our adjustment factor which we introduced ($\gamma > 1$), and τ_0 refers to the initial pheromone trail level. The value of τ_0 is usually set to $N/Cost(H)$ where $Cost(H)$ refers to the cost found by a given heuristic. In this study we approximate such a value based on our preliminary results on cost.

The idea is to reduce the pheromone trail value by a greater amount on arcs that have been selected many times compared to arcs that have only been picked by a few ants. This procedure is designed to reduce the possibility that arcs are selected repeatedly and to encourage exploration of the search space. It is also intended that customers which are chosen toward the beginning of a route by one ant will be selected later along a route by other ants so that mixing occurs on routes. This procedure is only implemented after a limited number of complete iterations of the algorithm have been performed. For instance, in our experiments we opted for three complete iterations based on our limited computational experience. The aim is to allow the ants to generate routes freely according to the trail values before any influence is applied.

33.3.5 Combining strategies for global trail update

In our method we use a global trail update based on the rank based version with elitist strategies, AS_{rank}, proposed by Bullnheimer et al. (1997). In their method the global best solution found so far is used to update the trail values together with a given number, λ, of iteration best solutions. The ranking system means that the trail value update is weighted according to the rank of the solution. However, this method does not take into account the quality of the λ iteration best solutions created when compared to the global best solution. This means that if λ is set too small then good quality solutions may not be used to update trail values if many good solutions have been generated. Also if λ is set too large then bad solutions may be used to update trail values if only a small number of good quality solutions have been built in the current iteration.

We propose a method that does consider the quality of the solutions generated. The aim is to reduce the risk that the number of ants used to update trail values has been set too high if we want to make sure that many good solutions are used in the updating process where appropriate. Our procedure uses the global best solution to update pheromone trail values together with a maximum number, λ, of iteration best solutions. Considering each of the λ iteration best solutions in turn, only if the solution is within a given percentage, θ, of the global best solution is the route used to update trail values. The trail values are updated according to Equation (33.6) as given by Bullnheimer et al. (1997) in their AS_{rank} except that λ may be adjusted as follows:

Let the set of the best λ solutions whose cost lie within θ% of the global best solution be defined by

$$G = \{s = 1, \ldots, \lambda \mid \frac{(cost(s) - cost(best))}{cost(best)} \cdot 100 < \theta\}. \tag{33.5}$$

If $|G| < \lambda$, then we set $\lambda = |G|$.

The constraint on cost is imposed to avoid poor quality solutions being used to update the trail values,

$$\tau_{ij}(t) = (1 - \rho) \cdot \tau_{ij}(t) + \lambda \Delta \tau_{ij}^{gb} + \Delta \tau_{ij}(t), \tag{33.6}$$

where $\Delta\tau_{ij}^{gb}(t) = Q/L^{gb}(t)$, L^{gb} being the global best solution and Q a correction factor to normalize the data, where

$$\Delta \tau_{ij}(t) = \sum_{\mu=1}^{\lambda-1} \Delta \tau_{ij}^{\mu}(t)$$

and

$$\Delta \tau_{ij}^{\mu}(t) = \begin{cases} (\lambda - \mu) Q / L^{\mu}(t) & \text{if } (i,j) \in \text{tour of ant with rank } \mu \\ 0 & \text{otherwise,} \end{cases}$$

where $L^{\mu}(t)$ is the length of the μth best tour at iteration t.

To control the quality of the solutions obtained we adopted into our ant heuristic two simple and well known local search procedures namely the 2/3-opt approach and the shift heuristic. In the 2/3-opt procedure only customers which belong to the same route are considered and hence each route is improved independently. A local search algorithm that works on the entire set of routes is obviously more powerful but can be expensive. In this work we adopted a route improvement heuristic which exchanges one or two customers between routes. These refinement procedures, which are used by Salhi and Rand (1987) for the vehicle routing problem, are modified for the vehicle routing problem with backhauls.

Note that it is not possible to apply standard post-optimize's because of the fluctuating load on the vehicle caused by the two different types of customer. In other words, exchanging a linehaul or backhaul customer in the VRPB can affect the load on the vehicle and could mean that the vehicle capacity is exceeded at other nodes along the route. It should also be noted that due to the great number of solutions that are generated it would be time prohibitive to employ a more sophisticated local search heuristic for each solution. In our implementation we used the 2/3-opt after each iteration and for all the routes generated whereas the shift procedure is used at particular times such as when the cost of a new solution is found to be very different from the previous ones.

33.4 COMPUTATIONAL RESULTS

The test problems corresponds to the VRPB instances proposed by Goetschalckx and Jacobs-Blecha (1989). The problems range in size from 25 to 150 customers. The cost of traveling from customer i to customer j is defined as the Euclidean distance between nodes i and j. The ant algorithm was implemented in Fortran90 and the experiments executed on an Ultra Enterprise 450 (300 MHz dual processor). For each test problem the ant system was run for 200 iterations. The parameters were set to the following values:

- Transition rule (Equation (33.3)): $\alpha = 1, \beta = 2, \phi = 1, q_0 = 0.5$;
- Local trail update rule (Equation (33.4)): $\upsilon = 0.4, \gamma = 2, perc = 25\%, \tau_0 = 5.10^{-5}$;
- Cost constraint (Equation (33.5)): $\theta = 10\%$;
- Global update rule (Equation (33.6)): $\rho = 0.1, Q = 1, \lambda = 10$ at the first iteration, then $\lambda = \max(10, NR)$ in subsequent iterations. NR is the number of routes found in the global best solution so far.

These values were chosen experimentally based on our limited preliminary testing and also on data from the literature whenever appropriate. To our knowledge this set of problems has only been solved for the mixed VRPB by Halse (1992). As the placement of the ants is randomly chosen within the sectors we conducted experiments using 20 runs for each test problem. For each problem we report the best solution, the average of the 20 solutions and the average computing time per run. Our summary results are given in Table 33.1. Bold refers to a new best solution. In Table 33.1 we also provide the name of the problem, together with the total number of customers N, the vehicle capacity CAP and the number of vehicles found K by both heuristics. The percentage deviation with respect to results reported by Halse is given in the two final columns (A minus sign shows that the present method outperforms Halse and vice versa). Underline shows both the best and the worst deviation. The deviation (in %) is computed as $Deviation = 100 \cdot (Cost_A - Cost_H)/Cost_H$, where $Cost_A$ and $Cost_H$ refer to the cost of the best solution found by our ant heuristic and Halse respectively.

Table 33.1 Computational Results for the Mixed VRPB (continues on next page).

			Ant results				Halse results			% Deviation w.r.t. Halse	
Prob	*N*	*CAP*	*Best*	*K*	*Avg*	*cpu*	*Halse*	*K*	*cpu*	Best	Avg
a1	25	1550	**223088**	8	223374	0.35	227725	8	0.18	-2.04	-1.91
a2	25	2550	169500	5	169797	0.31	169497	5	0.09	0.002	0.18
a3	25	4050	142034	3	142126	0.34	142032	3	0.05	0.001	0.06
b1	30	1600	**233001**	7	234514	0.58	233950	7	0.13	-0.41	0.24
b2	30	2600	**179258**	4	179760	0.76	182326	4	0.07	-1.68	-1.41
b3	30	4000	145702	3	145702	0.71	145699	3	0.30	0.002	0.002
c1	40	1800	**239192**	7	240126	1.65	242931	7	0.69	-1.54	-1.15
c2	40	2600	**196883**	5	197287	2.08	197276	5	0.74	-0.20	0.006
c3	40	4150	**164891**	3	165710	1.87	167663	4	0.06	-1.65	-1.16
d1	38	1700	**307110**	11	307383	1.56	307875	11	0.47	-0.25	-0.16
d3	38	2750	224196	7	224598	1.28	222195	7	0.18	0.90	1.08
e1	45	2650	**223774**	7	225927	2.30	222518	7	0.43	0.56	1.53
e2	45	4300	190559	4	191342	3.26	190048	4	0.07	0.27	0.68
e3	45	5225	**182804**	4	184621	2.42	187793	4	0.16	-2.66	-1.69
f1	60	3000	**248333**	7	251364	5.55	254977	6	0.53	-2.61	-1.42
f3	60	4400	217317	5	218818	6.61	215575	5	0.44	0.81	1.50
f4	60	5500	**200964**	4	202280	7.58	203448	5	0.41	-1.22	-0.57
g1	57	2700	**301235**	10	311057	7.64	304106	10	0.41	-0.94	2.28
g2	57	4300	235920	6	236623	4.26	235220	6	0.17	0.30	0.60
g3	57	5300	214534	5	215133	9.89	213757	5	0.22	0.36	0.64
g5	57	6400	203233	4	203777	5.32	202610	4	0.33	0.31	0.58
g6	57	8000	**189922**	3	190572	9.88	201875	4	0.33	-5.92	-5.59

Table 33.1 Computational Results for the Mixed VRPB (continued from previous page).

			Ant results				Halse results			% Deviation w.r.t. Halse	
Prob	N	CAP	$Best$	K	Avg	cpu	$Halse$	K	cpu	Best	Avg
h1	68	4000	241619	6	241850	9.94	235269	6	0.65	2.69	2.79
h2	68	5100	220305	5	220532	8.57	215649	5	0.18	2.15	2.26
h3	68	6100	208412	4	208693	8.80	202971	4	0.12	2.68	2.82
h5	68	7100	203193	4	203647	9.21	201896	4	0.24	0.64	0.86
i1	90	3000	**327168**	10	336501	22.78	329237	10	1.06	-0.63	2.21
i2	90	4000	**278727**	7	281864	27.78	289501	7	1.76	-3.72	-2.64
i3	90	5700	**238626**	5	241102	24.95	244782	5	3.83	-2.51	-1.50
j1	94	4400	**332471**	10	341272	28.02	337800	10	1.08	-1.58	1.03
j2	94	5600	**292698**	8	301779	32.36	298432	8	1.32	-1.92	1.12
j3	94	6600	**259243**	6	261264	32.15	280070	7	0.38	-7.44	-6.71
j4	94	8200	261066	7	278866	50.55	257895	6	0.53	1.23	8.13
k1	113	4100	**360954**	10	370134	57.71	361287	10	1.40	-0.09	2.45
k2	113	5200	323979	8	327324	57.99	320012	8	1.57	1.24	2.28
k4	113	6200	298518	7	308065	58.79	296766	7	0.66	0.59	3.81
l1	150	4000	416167	11	418644	159.97	412278	10	3.81	0.94	1.54
l2	150	5000	**360018**	8	377962	164.42	362399	8	4.01	-0.66	4.29
l4	150	6000	**337620**	7	339020	160.74	341304	7	2.09	-1.08	-0.67
m1	125	5200	**370920**	10	381006	70.79	372840	11	2.41	-0.51	2.19
m3	125	6200	**335486**	9	354797	80.21	336011	9	2.71	-0.16	5.59
m4	125	8000	310567	7	311410	83.48	305118	7	0.82	1.78	2.06
n1	150	5700	**370690**	10	387926	129.97	385978	10	2.60	-3.96	0.50
n3	150	6600	**349516**	9	349555	151.62	352992	9	2.82	-0.98	-0.97
n5	150	8500	323698	7	332754	166.68	319811	7	3.11	1.21	4.05
									Average:	**-0.62**	0.71
									Std. Dev.:	2.01	2.56
Number of Best:			25		14		20				

It can be seen from the table that the results using our ant algorithm range from −7.44% to 2.69% when compared to the solutions generated by Halse. In other terms, our worst solution is just below 3% whereas our new best is over 7%. In addition, 25 new best solutions out of 45 are obtained. The average deviation for all sets of problems is −0.62% for our best results and our average solutions were found to be on average 0.71% worst. Our ant heuristic seems to produce robust solutions as demonstrated by the low values of the standard deviations.

It is worth reporting that the enhancement to the candidate list gave an improvement of approximately 8% for the overall set of test problems when compared to the basic ant algorithm applied to the VRPB without any modifications. Also, the amendment to the visibility, local and global updating each showed an improvement of around 2% in the overall results for a given number of iterations. The restriction to the number of ants did not show an improvement when comparing results over the same number of iterations. However, by restricting the number of ants fewer routes are created, and therefore more iterations can be performed for the same computational effort. A small improvement of approximately 1% was achieved when comparing results over an equal number of solutions generated. Applying all modifications to the ant system algorithm produced an improvement of 14% taken as an average over the complete set of test problems. The detailed results are not reported here but can be found in the thesis of Wade (2002).

33.5 CONCLUSIONS

In this paper, we have presented an ant system algorithm to solve the mixed vehicle routing problem with backhauls. We have analyzed and modified a number of the basic attributes of ant system methods in order to solve the problem. It can be noted that our ant system produces encouraging results as it obtains 25 out of 45 new best results. The method performs well in most problems including the larger ones and proves to be robust.

The proposed study could make a strong basis for further enhancement. There are a number of ideas which can be incorporated into the proposed heuristic.

- One would be to adapt the greedy heuristic or visibility to be more problem specific. The visibility is used in an attempt to guide the search and hence can have a significant influence on solution quality. A revised heuristic could be based on a savings function and incorporate information concerning linehaul and backhaul load at each customer.

- As larger problems require more solutions to be generated because of the relatively larger number of ants to be used, solutions may become dominated by large trail values on certain arcs. One way that this could be addressed would be to introduce maximum and minimum values for the pheromone trail as proposed by Stützle and Hoos (1997a;b). These values would only be enforced during given stages of the iteration. Though such modifications can increase computation time a further investigation would be worthwhile.

- As the number of parameters introduced in this heuristic is increased, contrarily to what one would envisage, it would be interesting to investigate ways of eliminating some parameters especially those that could be considered as correlated. A sensitivity analysis on some of those relevant parameters as well as a statistical investigation which explores the statistical significance of each of the enhancements may be useful to conduct as part of assessing further the performance of such a population based heuristic.

Acknowledgments

The authors are grateful to both referees for their constructive comments that improved the presentation as well as the content of the paper. The authors would like also to thank the EPSRC for the sponsorship.

E. Bonabaeu, M. Dorigo, and G. Theraulaz. *Swarm Intelligence: From Natural to Artificial Systems*. Oxford University Press, 1999.

B. Bullnheimer, R. Hartl, and C. Strauss. An Improved Ant System Algorithm for the Vehicle Routing Problem. Technical Report POM-10/97, Institute of Management Science, University of Vienna, 1997.

B. Bullnheimer, R. Hartl, and C. Strauss. Applying the Ant System to the Vehicle Routing Problem. In I. Osman, S. Voss, S. Martello, and C. Roucairol, editors, *Metaheuristics: Advances and Trends in Local Search Paradigms for Optimization*, pages 109–120. Kluwer Academics, 1998.

O. Casco, B. Golden, and E. Wasil. Vehicle Routing with Backhauls: Models, Algorithms and Case Studies. In B. Golden and A. Assad, editors, *Vehicle Routing: Methods and Studies*, pages 127–147. North Holland, Amsterdam, 1988.

A. Colorni, M. Dorigo, and V. Maniezzo. Distributed Optimization by Ant Colonies. In F. Varela and P. Bourgine, editors, *Proceedings of the European Conference on Artificial Life*, pages 134–142. Elsevier Publishing, Amsterdam, 1991.

O. Cordon, I. Fernandez de Viana, F. Herrera, and L. Moreno. A New ACO Model Integrating Evolutionary Computation Concepts: The Best-Worst Ant System. In M. Dorigo, M. Middendorf, and T. Stützle, editors, *Abstract Proceedings of ANTS2000- From Ant Colonies to Artificial Ants: A Series of International Workshops on Ant Algorithms*, pages 22–29. Université Libre de Bruxelles, Belgium, 2000.

I. Deif and L. Bodin. Extension of the Clarke and Wright Algorithm for Solving the Vehicle Routing Problem with Backhauls. In A. E. Kidder, editor, *Proceedings of the Babson Conference on Software Uses in Transportation and Logistics Managemant*, pages 75–96, 1984.

M. Dorigo. *Optimization, Learning and Natural Algorithms*. PhD thesis, Politecnico di Milano, Italy, 1992.

M. Dorigo, G. Caro, and L. Gambardella. Ant Algorithms for Discrete Optimization. *Artificial Life*, 5(3):137–172, 1999.

M. Dorigo and L. Gambardella. Ant Colony System: A Cooperative Learning Approach to the Travelling Salesman Problem. *IEEE Transactions on Evolutionary Computation*, 1(1):53–66, 1997.

M. Dorigo, V. Maniezzo, and A. Colorni. The Ant System: Optimization by a Colony of Cooperating Agents. *IEEE Transactions Systems, Man, and Cybernetics- Part B*, 26(1):29–41, 1996.

M. Dorigo and T. Stützle. The Ant Colony Optimization Metaheuristic: Algorithms, Applications, and Advances. Technical Report IRIDIA-2000-32, Université Libre de Bruxelles, Belgium, 2000.

L. Gambardella and M. Dorigo. HAS-SOP: An Hybrid Ant System for the Sequential Ordering Problem. Technical Report 11-97,IDSIA, Lugano, Switzerland, 1997.

L. Gambardella, E. Taillard, and G. Agazzi. MACS-VRPTW: A Multiple Ant Colony System for Vehicle Routing Problems with Time Windows. In D. Corne, M. Dorigo, and F. Glover, editors, *New Ideas in Optimization*, pages 63–76. McGraw Hill, London, 1999.

M. Garey and D. Johnson. *Computers and Intractability: A Guide to the Theory of NP Completeness*. Freeman, San Francisco, 1979.

M. Gendreau, G. Laporte, and D. Vigo. Heuristics for the Travelling Salesman Problem with Pickup and Delivery. *Computers and Operations Research*, 26: 699–714, 1999.

M. Goetschalckx and C. Jacobs-Blecha. The Vehicle Routing Problem with Backhauls. *European Journal of Operational Reasearch*, 42:39–51, 1989.

K. Halse. *Modelling and Solving Complex Vehicle Routing Problems*. PhD thesis, The Technical University of Denmark, 1992.

V. Maniezzo. Exact and Approximate Nondeterministic Tree-Search Procedures for the Quadratic Assignment Problem. *Informs Journal of Computing*, 11(4):358–369, 1999.

V. Maniezzo and A. Colorni. The Ant System applied to the Quadratic Assignment Problem. *IEEE Trans. Knowledge and Data Engineering*, 1998.

A. Mingozzi, S. Giorgi, and R. Baldacci. An Exact Method for the Vehicle Routing Problem with Backhauls. *Transportation Science*, 33(3):315–329, 1999.

I.H. Osman and N.A. Wassan. A reactive tabu search meta-heuristic for the vehicle routing problem with backhauling. *Computers and Operations Research (to appear)*, 5, 2002.

S. Salhi and G. Nagy. A Cluster Insertion Heuristic for the Single and Multiple Depot Vehicle Routing Problems with Backhauls. *Journal of the Operational Research Society*, 50(10):1034–1042, 1999.

S. Salhi and G. Rand. Improvements to Vehicle Routing Heuristics. *Journal of the Operational Research Society*, 38:293–295, 1987.

T. Stützle and H. Hoos. Improvements on the Ant System: Introducing MAX-MIN Ant System. In *Proceedings of the International Conference on Artificial Neural Networks and Genetic Algorithms*, pages 245–249. Springer Verlag, Wien, 1997a.

T. Stützle and H. Hoos. The MAX-MIN Ant System and Local Search for the Travelling Salesman Problem. In T. Baeck, Z. Michalewicz, and X. Yao, editors, *Proceedings of IEEE-ICEC-EPS'97, IEEE International Conference on Evolutionary Computation and Evolutionary Programming Conference*, pages 309–314. IEEE Press, 1997b.

P. Toth and D. Vigo. A Heuristic Algorithm for the Vehicle Routing Problem with Backhauls. In *Advanced Methods in Transportation Analysis*, pages 585–608. Springer Verlag, Berlin, 1996.

P. Toth and D. Vigo. An Exact Algorithm for the Vehicle Routing Problem with Backhauls. *Transportation Science*, 31(4):372–385, 1997.

A.C. Wade. *Constructive and Ant System heuristics for a class of VRP with Backhauls.* PhD thesis, University of Birmingham, 2002.